Interdisciplinary Biotechnological Advances

Series Editors

Jayanta Kumar Patra, Research Institute of Integrative Life Sciences, Dongguk University, Ilsandong, Goyang, Kyonggi-do, Korea (Republic of)

Gitishree Das, Research Institute of Integrative Life Sciences, Dongguk University, Goyang, Korea (Republic of)

AF799257

This series is an authoritative resource that explains recent advances and emerging directions in biotechnology, reflecting the forefront of research clearly and reliably, without excessive hype. Each volume is written by authors with excellent reputations and acknowledged expertise in the topic under discussion. The volumes span the entire field from an interdisciplinary perspective, covering everything from biotechnology principles and methods to applications in areas including genetic engineering, transgenic plants and animals, environmental problems, genomics, proteomics, diagnosis of disease, gene therapy, and biomedicine. The significance of these applications for the achievement of UN Sustainable Development Goals is highlighted. The series will be highly relevant for Master's and PhD students in Biotechnology, Nanochemistry, Biochemical Engineering, and Microbiology, medical students, academic and industrial researchers, agricultural scientists, farmers, clinicians, industry personnel, and entrepreneurs.

Anupam Das Talukdar •
Jayanta Kumar Patra • Gitishree Das •
Deepa Nath
Editors

Traditional Resources and Tools for Modern Drug Discovery

Ethnomedicine and Pharmacology

Springer

Editors
Anupam Das Talukdar
Dept of Life Science and Bioinformatics
Assam University
Silchar, Assam, India

Gitishree Das
Dongguk University
Goyang-si, Korea (Republic of)

Jayanta Kumar Patra
Dongguk University
Goyang-si, Korea (Republic of)

Deepa Nath
Department of Botany
Gurucharan College
Silchar, Assam, India

ISSN 2730-7069 ISSN 2730-7077 (electronic)
Interdisciplinary Biotechnological Advances
ISBN 978-981-97-4602-6 ISBN 978-981-97-4600-2 (eBook)
https://doi.org/10.1007/978-981-97-4600-2

© The Editor(s) (if applicable) and The Author(s), under exclusive license to Springer Nature Singapore Pte Ltd. 2024
This work is subject to copyright. All rights are solely and exclusively licensed by the Publisher, whether the whole or part of the material is concerned, specifically the rights of translation, reprinting, reuse of illustrations, recitation, broadcasting, reproduction on microfilms or in any other physical way, and transmission or information storage and retrieval, electronic adaptation, computer software, or by similar or dissimilar methodology now known or hereafter developed.
The use of general descriptive names, registered names, trademarks, service marks, etc. in this publication does not imply, even in the absence of a specific statement, that such names are exempt from the relevant protective laws and regulations and therefore free for general use.
The publisher, the authors and the editors are safe to assume that the advice and information in this book are believed to be true and accurate at the date of publication. Neither the publisher nor the authors or the editors give a warranty, expressed or implied, with respect to the material contained herein or for any errors or omissions that may have been made. The publisher remains neutral with regard to jurisdictional claims in published maps and institutional affiliations.

This Springer imprint is published by the registered company Springer Nature Singapore Pte Ltd.
The registered company address is: 152 Beach Road, #21-01/04 Gateway East, Singapore 189721, Singapore

If disposing of this product, please recycle the paper.

Preface

In the constant developing era of drug discovery, the potential of novel therapeutics to combat diseases remains a major problem. As scientists gain a deeper understanding of biological intricacies and disease causes, they are becoming more and more adept at understanding the richness of traditional resources, which have great potential for traversing the complex paths involved in drug discovery. As implied by the title, this book *Traditional Resources and Tools for Modern Drug Discovery* aims to explore the knowledge, application, and developments of traditional resources in the context of drug delivery systems. Natural resources possess huge collection of treasures in the form of biologically active compounds since the beginning of mother Earth. Over time, civilizations have come to recognize and cherish the immense knowledge and potential that they possess. They have spread this knowledge through generations, spanning a wide range of traditions and cultures around the world. From indigenous medicinal plants and herbal remedies, age-old traditions such as Ayurveda have studied the healing capacity of natural resources and utilized in curing number of diseases.

The use of traditional tools has immensely helped in understanding the molecular mechanisms of underlying diseases along with pharmacological properties of the natural components present in them. Several techniques are developed such as ethnopharmacology which explores the traditional uses of ethnomedicinal plants and its advantages, along with other branches like phytochemistry which emphasizes on chemical constitution of plant resources and provides valuable guidance in the endeavour to identify new drugs. These tools serve as a connecting bridge between the past and future of medicine, a holistic approach which holds different aspects of a disease when viewed from social, cultural, environmental, or biochemical contexts. Limitations appeared during the merging of traditional resources into modern-day drug discovery, and significant hurdles appeared in maintaining quality control, standardization, intellectual property rights, etc. Cumulative effort from both traditional and scientific approaches will be able to aqueduct the gap and channel the potential therapeutic properties of traditional medicine before the world.

In this preface, we embark on the journey to inquire and study about the significance and impact of traditional resources in modern drug development. We seek to comprehend the evolutionary aspect, role of metabolites, markers, etc. and present in ethnic and medicinal plants through interdisciplinary collaboration, creative methodologies, and a deep respect for traditional knowledge systems. This understanding paves the way for the development of safe and effective therapeutics for the benefit of mankind and society.

This book displays an introduction on traditional resources in relation to ethnobotanical sources and their role on drug discovery. It focuses on the impacts and significance of natural products in the present time. This book has firmly been laid out into 25 well-drafted chapters, contributed by researchers/scholars from around the world.

Traditional Resources and Tools for Modern Drug Discovery is a vital reference book for those who are intrigued to know various traditional resources present in nature, their impact on management of diseases, drug discovery system, and their benefits and prospects.

We would like to extend sincere appreciation to all the esteemed authors who have actively participated and contributed their invaluable wisdom in this book. We are much obliged to the Springer Nature team, especially Dr. Emmy Lee, Associate Editor, Springer Nature Korea Limited, for her generous assistance and persistence in finalizing the edited volume.

Silchar, India — Anupam Das Talukdar
Seoul, South Korea — Jayanta Kumar Patra
Seoul, South Korea — Gitishree Das
Silchar, India — Deepa Nath

Contents

About the Editors

Anupam Das Talukdar, M.Sc., Ph.D., is currently working as a Professor in the Department of Life Science and Bioinformatics at Assam University, Silchar, Assam, India. With 21 years of teaching and research experience in phytochemistry, structure-based drug design, computer-aided drug discovery, and natural product study, he has excelled in this field. In his current research area, he has successfully completed six research projects funded by various national funding agencies of India as a Principal Investigator. He has supervised 13 Ph.D. and 5 M.Phil. scholars, and currently, 6 Ph.D. scholars are working under his guidance. He has contributed significantly to academia, with more than 99 international research articles, 44 book chapters, and 5 books to his credit. His expertise is recognized internationally, as he serves on various editorial boards and as a reviewer for many esteemed journals. He was awarded gold medals for the first-class first position in his graduation and post-graduation degrees. He did his Ph.D. degree from Assam University, Silchar, Assam, India, in collaboration with Bose Institute, Kolkata, India. He was honoured with the prestigious DBT—Overseas Associateship for his post-doctoral studies in the UK.

Jayanta Kumar Patra, M.Sc., Ph.D., PDF, is currently working as an Associate Professor at Dongguk University, Republic of Korea. He has about 14 years of research and teaching experience in the field of food, pharmacological, and nano-biotechnology. To his credit, he has published more than 180 papers in various national and international peer-reviewed journals and around 40 book chapters in different edited books. He has also authored 20 books in various academic publications. He is an editorial board member of several reputed academic journals.

Gitishree Das, M.Sc., Ph.D., PDF, is currently working as an Associate Professor at Dongguk University, Republic of Korea. She has 12 years of research experience in the field of rice molecular biology, plant breeding, endophytic bacteria, and green nanotechnology. To her credit, she has published around 100 research articles in international and national reputed journals and 25 book chapters. She has also authored eight books in various academic publications. She is an editorial board member of some national and international journals.

Deepa Nath, M.Phil., Ph.D., is currently serving as an Assistant Professor in the Department of Botany at Gurucharan College, Silchar, Assam, India. With a robust academic background and post-Ph.D. research experience in phytochemistry and plant tissue culture, she has made significant contributions to her field. She completed her MPhil and Ph.D. from the Department of Life Science and Bioinformatics, Assam University, Silchar. Throughout her academic career, she has received recognition for her excellence. She obtained a Gold Medal for securing the first-class first position in M.Sc. (botany). She has published extensively in reputed international journals and contributed to several book chapters, emphasizing her expertise in phytochemistry and natural product research. With a strong academic foundation and a passion for research, she continues to make significant strides in the field of phytochemistry and plant biology.

Introduction to Ethnobotany and Traditional Medicine

Sabyasachi Banerjee

Contents

Abstract The field of ethnobotany, as a scientific research discipline, has been extensively employed for the purpose of documenting indigenous knowledge pertaining to plant utilization, as well as for compiling a comprehensive inventory of plants that hold significance in traditional herbal medicine across various nations. These studies play a crucial role in the advancement of this field. Nevertheless, it is worth noting that in certain nations during recent times, there has been a notable utilization of ethnobotanical investigations in the realm of pharmaceutical research, specifically pertaining to the identification and development of novel therapeutic

S. Banerjee (✉)
Department of Pharmaceutical Chemistry, Gupta College of Technological Sciences, Ashram More, Asansol, West Bengal, India

© The Author(s), under exclusive license to Springer Nature Singapore Pte Ltd. 2024

A. Das Talukdar et al. (eds.), *Traditional Resources and Tools for Modern Drug Discovery*, Interdisciplinary Biotechnological Advances,
https://doi.org/10.1007/978-981-97-4600-2_1

agents. The current chapter provides a comprehensive overview of the historical progression and evolutionary trajectory of ethnobotany, elucidating its significant role in the exploration and advancement of traditional medicine within clinical contexts. This field of study has been crucial in uncovering and developing the therapeutic potential of various plant species that have long been employed in indigenous cultures for the treatment of ailments. Ethnobotany has made noteworthy contributions to the preservation of human health and well-being, particularly in underdeveloped regions of the globe. In these areas, traditional medicines derived from plants serve as the primary way of treating diseases. This chapter also encompasses an extensive examination of the current methodologies employed in ethnobotanical field research, as well as an analysis of the obstacles encountered in the ongoing pursuit of identifying novel bioactive natural compounds via ethnobotanically guided drug discovery and development initiatives.

Keywords Ethnobotany · Traditional medicine · Natural compounds · Plants · Clinical · Alternative medicine

1 Introduction

Throughout the course of human history, individuals have used a diverse range of natural materials to alleviate ailments and enhance their overall well-being. Substances were obtained from botanical sources found both in close proximity to individuals and in distant regions (Lev and Amar 2000). For centuries, nature has served as a rich reservoir of therapeutic compounds, with several contemporary pharmaceuticals being derived from natural origins, often inspired by their historical use in traditional healing practices. Traditional medical systems that are based on plants continue to have significant importance in the field of health care. Approximately 80% of the global population primarily relies on traditional medicines for their primary healthcare needs (Ghorbani et al. 2006). It is worth noting that medicinal plants constitute the core component of many traditional medical practices. In recent times, there has been a growing recognition of the need to conduct comparative analyses of plant utilization across different cultures. Additionally, it has come to light that there is a dearth of knowledge on the relative importance of medicinal plants, as well as other useful plant species, within certain cultural contexts. The ethnobotanical investigations presented herein emphasize the need of investigating the phytochemical composition of 110 specific species. These species have been identified as having a higher likelihood of containing bioactive components. Consequently, these findings have significant implications for research in the field of natural products (Pieroni 2000).

Ethnobotany is a combined name derived from the union of "ethano," denoting the study of plant culture, and "botany," referring to the study of plant species.

Ethnobotany is a field of study that investigates the interrelationship between plants and human societies, particularly focusing on traditional knowledge and practices (Birhanu et al. 2015). In 1895, at a lecture in a metropolitan area, Dr. J. W. Harshberger introduced the word "Ethnobotany" to delineate his area of research, which he defined as the investigation of "Plants utilized by primitive and aboriginal populations." In 1896, Dr. J.W. Harshberger proposed the term "ethnobotany" as an area of study aimed at understanding the cultural significance of plants in relation to their utilization by various cultures for food and clothing (Kokni et al. 2016). Due to the significant contributions of plants across various domains of human endeavor, ethnobotany encompasses a wide range of disciplines, such as biochemistry, botany, toxicology, pharmacognosy, medicine, agriculture, nutrition, ecology, cognitive studies, anthropology, archaeology, and history. The interdisciplinary aspect of ethnobotany facilitates diverse methodologies and practical applications, hence enabling several scientists to investigate plant use from various perspectives (Alexiades and Sheldon 1996). Medicinal plants have always been the primary focus of research on the subject of ethnobotany, and the examination of these resources has also yielded substantial contributions to the theoretical advancement of the discipline (Heinrich et al. 2017).

The genesis of this scientific discipline may be traced back to the period of European colonization in regions outside of Europe, during which Europeans embarked on expeditions to discover novel remedies for tropical fevers [BOU 16] (Manderson and Jolly 1997). During the first stages of colonization in the Americas, the Spanish monarchs initiated inquiries with the aim of acquiring botanical knowledge from native peoples. The field of ethnobotany saw further development with the establishment of Jesuit missions and the compilation of the first floras documenting American plant species and their traditional medicinal use (SarDesai 2018). During the time of French colonialism, French botanists conducted a study on the taxonomy of medicinal flora in Vietnam, which was subsequently documented in the publication titled "General Flora of Indo-China" (Lecomte 1912–1937) (Gagnepain 1950). Similar to the French, the Spaniards also disparaged the native inhabitants' medical expertise. Despite the existence of traditional medicine since at least the second-century B.C., with a diverse range of over 3800 traditional remedies (Le Tran Duc [LE 95], as cited in Ngoc Van [NGO 08]), the French colonial regime didn't actively promote the utilization of traditional medicine or its integration into the colonial health system (Thompson 2003).

The use of inter- and multidisciplinary methodologies has been shown to provide more productive, comprehensive, and systematic methods in the examination of interactions between plants and humans. Ethnobotany, despite its many hurdles, continues to be an intriguing and opportune field of research (Alexiades and Sheldon 1996).

(a) Quantitative ethnobotany: The loss of habitat and species-specific overexploitation are major threats to certain wild plant resources. Furthermore, there is a concerning trend of the fast erosion of indigenous knowledge pertaining to the usage of wild plant resources within traditional societies. Quantitative ethnobotany has the potential to make valuable contributions to the scientific foundation of management choices in the realm of conserving and equitable and sustainable utilization of wild plant resources. In recent years, efforts have been made to strengthen the reliability and validity of ethnobotanical studies by integrating quantitative research techniques into the conventional compilation-style approach. These approaches are used at various stages of ethnobotanical data collecting, processing, and interpretation. Quantitative methodologies are used with the objective of providing quantitative descriptions of variables and analyzing observable patterns within research while also conducting statistical tests to evaluate hypotheses.

The term "quantitative ethnobotany" was introduced by Prance and colleagues in 1987. The term "plant ethnobotany" refers to the use of quantitative methodologies in the direct examination of present-day data on plant usage. The use of quantification and the subsequent testing of related hypotheses contribute to the production of high-quality information. Various approaches are employed in this context, such as regression analysis, cluster and principal analysis, log-linear modeling, and analysis of variance (Hoft et al. 1999; Peters 1996; Phillips and Gentry 1993).

(b) Applied ethnobotany: The field of ethnobotany has broadened its purview to include many facets of the natural world, with the development of quantitative methodologies. An increasing number of professionals in the field of ethnobotany are now adopting the titles of ethnobiologists or ethnoecologists. This shift is motivated by the greater potential offered by these disciplines to investigate the intricate relationship between human societies and their natural environments. Moreover, these arenas facilitate the analysis of the impact imposed by external variables on regional economies and sociocultural patterns, such as the implementation of innovative trading frameworks.

The concept of "applied ethnobotany" has been employed by the people and plants initiative since 1992 to denote the investigation and methodologies that facilitate cooperative engagement with knowledge holders, knowledge evaluation, and the development of enhanced management systems that draw upon traditional practices and social dynamics. The field of applied ethnobotany seeks to establish a connection between scientific and traditional knowledge. This is achieved by the examination of local practices and knowledge systems in the context of international and national norms, regulations, and economic patterns (Aumeeruddy-Thomas et al. 2003).

(c) Ethnoecology: The word "ethnoecology" has emerged in the last several years. According to Martin (2004), ethnoecology is characterized as an interdisciplinary topic that incorporates a wide range of academic disciplines. The word ethnoecology is employed to denote a range of studies that examine the ways in which local communities interact with their natural environment. This includes many subdisciplines such as ethnobotany, ethnobiology, ethnozoology, and ethnoentomology (Martin 2004). Ethnoecology, as a discipline, pertains to the scientific study of the perceptions and interpretations that individuals possess about the intricate connections existing between plants, humans, and the many physical components within a certain geographical area (Davidson-Hunt 2000) (Fig. 1).

Traditional medicine refers to a combination of information, expertise, and customs that are rooted in concepts, opinions, and observations specific to many cultures. It is employed for the purpose of preserving health and addressing bodily and mental ailments through diagnosis, prevention, amelioration, or treatment (Fokunang et al. 2011). There exists substantial evidence, ranging from ancient literary works to contemporary scientific records, indicating that plants serve as the primary source of therapeutic remedies for healthcare practices in developing nations. In China, the current compilation of medicinal plants utilized in traditional Chinese medicine (TCM) encompasses a total of 11,146 species. Among them, 492 species are cultivated, while the remaining 10,654 species are classified as wild plants (Jahiel 1998). Based on data from the World Health Organization (WHO), it has been reported that a significant proportion, namely, over 80%, of the global population depends on traditional herbal medicine as their primary source of health care (Inglis 1994). The invaluable herbal traditions present in emerging nations have consistently been regarded as a significant element of the global cultural legacy.

In the past 25 years, many environmental and cultural trends, as well as an evolution from subsistence to market-based economies, have significantly influenced traditional medical systems. These changes have had a profound impact on all elements of traditional medicine, particularly by impacting its resource base and the surrounding environment. The excessive extraction of animal species and medicinal plants has contributed to the destruction of resources, the decline in biodiversity, and the erosion of indigenous medical knowledge and customs. This phenomenon has resulted in the erosion of conventional medical systems. Furthermore, the

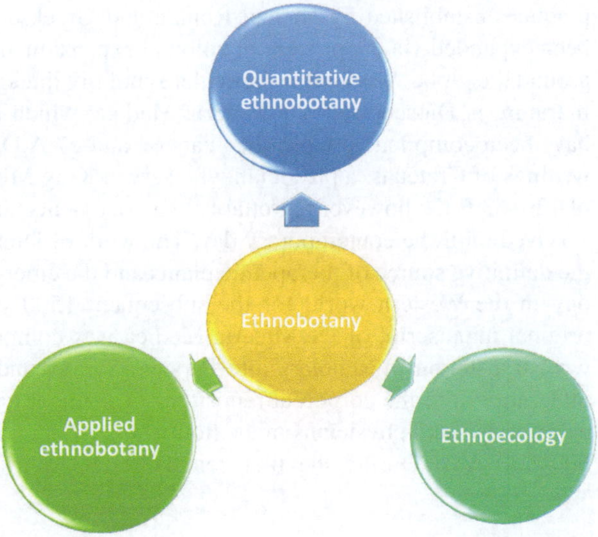

Fig. 1 Three approaches of assessing ethnobotanical knowledge

utilization of traditional herbal medicine in the formulation and advancement of novel pharmaceuticals serving to contemporary markets with allopathic remedies would ultimately result in the displacement of traditional medicine (Adnan et al. 2023) (Table 1).

2 History of Ethnobotany

2.1 Sources from Western Countries

The origins of ethnobotany may be traced back to a time before its official establishment as a separate discipline within the subject of botany. The medical capabilities of plants have captivated human curiosity for years if not millennia. The earliest documented instances of plants being utilized for medicinal purposes within the Western tradition can be traced back to ancient Egypt. Notable records include the Ebers Papyrus, with extant copies dating back to 1550 B.C., as well as various texts found in funerary monuments, sarcophagi, and wall paintings (Rahman et al. 2019). Additionally, ancient Sumer provides evidence of medicinal plant usage through a tablet containing a compilation of physician's prescriptions, estimated to be from around 3000 B.C. The aforementioned early artifacts serve as a testament to the expertise possessed by medical practitioners in ancient civilizations of the Old World. It is widely believed that these written records pertaining to the utilization of medicinal plants are indicative of a medical practice that predates the compilation of this knowledge in written form by a span of five to twenty-five centuries (Kumbhar 2022).

The examination of plants for their therapeutic properties has been a subject of extensive scholarly research in Western societies, drawing upon the knowledge and practices established by ancient Roman and Greek civilizations. This pursuit has been expanded via the process of colonial expansion and the study of many regions around the globe. Within this particular scholarly lineage, the first seminal source of literature is Dioscorides' De Materia Medica, which historians have estimated to have been compiled between the years 64 and 77 A.D. Dioscorides drew upon the writings of Crateuas, a physician who served King Mithrades VI during the period of 120–63 B.C.; however regrettably, the documents authored by Crateuas have not survived until the contemporary day. The work of Dioscorides established itself as the definitive source of therapeutic plants and the emerging field of folk pharmacology in the Western world for the subsequent 1500 years (Petrovska 2012). The original manuscript of De Materia Medica was composed in Greek, but it underwent translation by scholars into Persian, Syriac, and Arabic. This facilitated its integration into the corporeal repository of medical knowledge possessed by educated priests and physicians in the flourishing Arab civilization during the period of 600–900 A.D. During the first century A.D., Latin translations of De Materia

Table 1 Various ethnomedicinal plants recorded from diverse regions around the globe

Family	Vernacular name	Scientific name	Parts used	Medicinal use	Reference
Asteraceae	Xinyetuerfeng	*Ainsliaea bonatti*	Entire plant	Throat itching, asthma, cough	Hong et al. (2015)
	Lengana	*Artemisia afra*	Leaves	Prostatitis	Van Wyk and Wink (2018)
	Tsikitlane	*Gazania krebsiana*	Roots	Female infertility	Kose et al. (2015)
Pinaceae	Talish patra, Gobre salla	*Abies spectabilis*	Leaf needle	Bronchitis, rheumatism, cold, and asthma	Malla et al. (2015)
Annonaceae	Mufa, Gwanda	*Annona senegalensis*	Leaf, root, bark	Constipation, diarrhea, stomachache	Adjanohoun (1980)
	Seetapalam	*Annona squamosal*	Leaves	Tumor	Naini and Mamidala (2013)
Fabaceae	Dundu	*Dichrostachys cinerea*	Leaves	Headache	(Kankara et al. 2015)
	Bangahat	*Desmodium heterocarpon*	Leaves, roots	Diarrhea, skin diseases, cough	Malla et al. (2015)
Myrsinaceae	Zoumatai	*Ardisia gigantifolia*	Whole plant, rhizome	Traumatic injury, rheumatic arthritis, paralysis, leg, and waist pain	Hong et al. (2015)
Asparagaceae	Yu zhu	*Polygonatum odoratum*	Rhizomes	Skin cancer	Wujisguleng et al. (2012)
	Lehonyeli	*Asparagus microraphis*	Roots	Pain in period	Shale et al. (1999)
Poaceae	Mohloa	*Cynodon incompletus*	Root and leaves	Labor pain	Watt and Brandwijk (1927)
Combretaceae	Sabara	*Guiera senegalensis*	Leaves	Diarrhea, vomiting, nausea	Kankara et al. (2015)
Equisetaceae	Mohlaka-photoane	*Equisetum ramosissimum*	Rhizomes	Female infertility	Hutchings (1996)
Bignoniaceae	Lapacho, Pau d' arco	*Handroanthus impetiginosus*	Bark	Leukemia, tumor	Ochwang'i et al. (2014)
Ebenaceae	Tsilabelo	*Searsia erosa*	Leaves	Endometrial carcinoma	Mugomeri et al. (2016)
Lamiaceae	Mosisili	*Salvia runcinata*	Entire plant	Infertility, breast cancer	Kose et al. (2015)
Brassicaceae	Zamantarori	*Lepidium sativum*	Leaf	Jaundice	Kankara et al. (2015)
Acanthaceae	Aaraeuri	*Lepidagathis incurva*	Root, leaves, barks	Skin cancers	Esha et al. (2012)

Medica were diligently replicated and maintained inside monastic institutions across Europe (Keller 1991).

Dioscorides undertook the first endeavor to organize and consolidate the whole of botanical knowledge available to the ancient Greek society. Despite the absence of a zoological, botanical, or mineralogical classification system, the author's approach to organizing medicinal remedies based on their form, origin, and associated illness was comprehensive and highly esteemed by contemporary scholars and subsequent generations (Touwaide 2004). The surviving works of other classicists, including Pliny the Elder's Naturalis Historia, were transmitted and preserved within the scholarly and medical communities. During this period, individuals who possessed knowledge in these fields often held multiple roles as scholars, physicians, and religious figures, such as priests or nuns. These works exerted a significant influence on medical practices and the accumulation of medicinal knowledge in both the Western and Arab regions, spanning from classical Greece to the medieval era (Francia and Stobart 2014).

The writings of Dioscorides were referenced by several herbalists in the subsequent decades, such as in Hortus sanitatis (according to Jacob Meydenbach and first printed in 1491) and Herbarius latinus, which served as the pinnacle of herbal knowledge throughout the medieval period in Europe (Francia and Stobart 2014). During the medieval era and subsequent periods, there was an upsurge of "herbals," which were comprehensive collections of information pertaining to herbal treatment. These herbals had been developed and disseminated across many regions in Western Europe, with particular prominence seen in southern Italy, Naples, England, and Germany. The impact of the classical historians extended to all botanical studies until the post-Renaissance period, including notable contributions such as the herbals authored by Nicholas Culpepper and John Gerard in England, as well as Pietro Matthiolus in Italy (Tobyn 2017).

2.2 *Sources from Non-Western Countries*

The comprehensive examination of plants is not exclusively limited to Western societies. In ancient China, it is believed that Emperor Shen Nung wrote the Pen T'sao Ching, which is considered to be the first known treatise on medicinal herbs. This compilation is said to have taken place in about 2700 B.C. (Liao 2011). The Rig Vedas and Ayurvedic medicine, which were collected in ancient India, include valuable insights into several medicinal plants used for therapeutic purposes. Ayurvedic medicine is believed to have originated around 5000–10,000 years ago, with references to its principles found in the Rig Veda, a text that predates 2500 B.C. or maybe much earlier (Jaiswal and Williams 2017). Various medical traditions that use medicinal plants, which have their origins in Buddhism, have been documented in palm-leaf manuscripts discovered in regions such as Laos, Burma, Cambodia, Vietnam, Thailand, and other parts of Indonesia, including Java and Bali. Tibetan medicine, which shares its origins with Ayurvedic medicine, developed based on the foundational texts of Caraka Samhita (about 1500 B.C.), the Susruta Samhita (300–400 A.D.), and the Ashtang Hrdyam (around 500 A.D.). The existing Mayan codices, such as the Book of Chilam Balam of Kaua, include documented information pertaining to the therapeutic use of several plants. The Incas of South America and the Aztecs of Mesoamerica established botanical gardens that housed a variety of commercially and medicinally valuable plant species (Jain 1986).

2.3 Verbal History

The correlation between the discovery and exploitation of plants for therapeutic reasons and the development of sophisticated civilizations and geopolitical institutions is not always evident. The universality of medical knowledge and the pursuit of treating ailments throughout all civilizations, ranging from primitive foraging tribes to sophisticated city-state polities, are evident. However, the existence of written books exclusively dedicated to medicinal plant knowledge and therapeutic practices highlights the increasing specialization of this information (Rahman et al. 2019). Literacy, specialized training, and education have long been and are essential prerequisites for engaging in the practice of medicine within intricate societal contexts. On the other hand, nonliterate communities have historically and persistently relied on oral tradition to transmit their medicinal knowledge. This knowledge is typically passed down through apprenticeship to individuals who hold the role of traditional practitioners, healers, village shamans, or "household herbalists." These individuals are responsible for gathering and preparing remedies specifically intended for the treatment of sick children or other members of the household (Soejarto et al. 2009).

Research on the ethnobotany of cultural communities that depend on oral transmission for the preservation of traditional knowledge regarding medicinal plants reveals that these groups possess not only a substantial reservoir of information on plants with medicinal properties but also possess a comprehensive understanding of plants that have economic value, as well as the traditional methods employed to sustainably manage, harvest, and preserve these species. The aforementioned investigations have shown that several indigenous communities have a profound understanding of their surroundings and exhibit a remarkable ability to perceive nuanced alterations within the natural environment. This includes their ability to discern variations in weather patterns and identify the specific locations and seasonal availability of medicinal plant resources (Cheikhyoussef et al. 2011). The comprehensive oral literature encompasses a wealth of knowledge pertaining to illnesses, treatments, remedies, and healing techniques. This knowledge is acquired through a period of apprenticeship under the guidance of an experienced healer. Additionally, it involves developing expertise in symptom evaluation and disease diagnosis, as well as the collection, preparation, and administration of medicinal plant remedies. Furthermore, this tradition entails providing ongoing patient care and assessing the progress of the individual receiving treatment. Moreover, current research in the field of medical ethnobotany has demonstrated that the traditional medicinal systems of knowledge of nonliterate cultural groups are as comprehensive and comparable to those of civilizations with literacy (Sivasankari et al. 2014).

2.4 Modern Ethnobotany

Contemporary investigations in ethnobotany exhibit a discernible departure from previous examinations of utilitarian flora, since contemporary ethnobotanical inquiries tend to include a broader range of facts pertaining to the cultural communities that use these plants for medicinal or other reasons. Contemporary medical ethnobotanical research encompasses various aspects, such as the examination of cultural perspectives on illness, treatment, and healing, including the process of seeking medical assistance. Additionally, it explores cultural beliefs regarding the interconnections between plants and humans within their ecological context, as well as the societal significance of medicinal plants. Furthermore, it investigates the ceremonial, ritualistic, and other applications of medicinal plants. Lastly, it investigates the role of traditional healers, shamans, or other specialized individuals who employ medicinal plants to treat patients, regardless of whether the illness is perceived to have a natural or spiritual origin. Furthermore, contemporary medical ethnobotanical research often aims to provide substantiation of their effectiveness in addressing ailments and the therapeutic impacts of these plants on human physiology (Chun-Lin 2013).

The documentation of medicinally and economically valuable plants utilized by indigenous populations has been prevalent since the early exploration of the New World. However, the systematic investigation of plants for their potential in contemporary medicine has primarily emerged alongside advancements in modern medicine, which have led to a more comprehensive understanding of the effects of medicinal substances on the human body (Van Wyk 2008). Early ethnobotanical studies that concentrated on the therapeutic properties of certain plants included investigations conducted by William Withering in 1785, which centered on Digitalis purpurea, often known as foxglove. Additionally, John Harshberger, attributed with introducing the word "ethnobotany," conducted studies in 1896 that explored the objectives of this field of research (Idu 2009).

The emergence of cultural anthropology as a nascent social science in the late nineteenth century expanded the development of a contemporary focus on the many societies and civilizations that utilized medicinal plants. The impetus for this development may be attributed, in part, to a quest for enhanced comprehension of the scientific principles behind the use of traditional medicinal plants. Additionally, it was driven by the belief that traditional cultures, including the knowledge of medicinal plant usage, were rapidly diminishing due to cultural transformations (Kayser 2018). Consequently, several ethnographers and ethnobotanists throughout the early modern period meticulously recorded extensive inventories of plants used by indigenous populations for therapeutic intentions. These records not only included botanical knowledge but also incorporated cultural perspectives on the origins and remedies of certain ailments. The American Southwest has seen the emergence of notable instances of contemporary ethnography and ethnobotany, as shown by the contributions of Matilda Cox Stevens (1915) in her study on the Ethnobotany of the Zuni Indians (Soejarto et al. 2009) and Ralph Roys (1931) in his exploration of the Ethno-Botany of the Maya (Marcus 1982).

3 History of Traditional Medicine

There has been a renewed interest in herbal medicines, particularly in North America and Europe, due to the aspiration of preserving the knowledge embedded in traditional healing systems (Tyler 2000). Consequently, herbal products have been integrated into medical systems commonly referred to as "complementary," "alternative," "integrative," or "holistic."

The utilization of medicinal plants for therapeutic purposes has been documented since ancient times (Schulz et al. 2001). Herbal utilization is a prevalent practice in various global folk healing methods. Below, a brief description is provided for some of these traditions, offering examples of the diverse range of significant healing practices from throughout the globe that incorporate the utilization of herbs for therapeutic purposes.

In the latter half of the twentieth century, there was a notable surge in the popularity of traditional therapeutic practices, such as the utilization of herbal treatments, primarily observed in the United States. This trend can be attributed to the growing fascination with self-care among individuals. Consumers have expressed favorable perceptions regarding these products primarily due to their perceived natural origin, contrasting them with synthetic alternatives. This belief originates from the belief that such products are perceived as safer than pharmaceutical drugs. Additionally, these products are viewed as integral to a healthy lifestyle and are seen as a means to minimize dependency on conventional Western medicine (Adhikari and Paul 2018).

The extensive historical utilization of herbal components in traditional contexts can serve as evidence supporting their efficacy and safety. However, it is crucial to acknowledge and resolve certain challenges that arise when integrating these ingredients into contemporary practices. One issue arises from the fact that ingredients previously employed for the purpose of managing symptoms in traditional healing practices are now being utilized in developed nations as a means of promoting health or preventing diseases. Consequently, the focus has shifted from providing immediate relief to prolonged and continuous exposure, as exemplified by the use of herbal products for weight loss (Allison et al. 2001). This implies that a claim regarding the safety of a product based on historical evidence spanning thousands of years may not hold true in the context of its current usage. This does not explicitly imply that a component is hazardous; rather, it indicates that safety cannot be presumed in the contemporary setting. One additional concern pertains to the limited demonstration of usefulness and effectiveness through contemporary scientific inquiries. The use of an evidence-based approach to this topic has been a new development, and the findings indicate that there are significant information gaps that must be addressed before one can confidently ascertain the efficacy of most herbal products (Mukherjee et al. 2010).

One of the most challenging aspects of integrating traditional herbal practices into contemporary Western medicine is the customization of prescriptions that involve a combination of several herbal and non-herbal components. The lack of motivation to standardize items for a large-scale market arises from the primary objective of catering to personalized prescriptions (Qiu 2007). For small-scale growers and conventionally trained herbalists, the concept of standardization entails comprehending the optimal growth conditions, appropriate timing of harvesting, and the specific methods employed for extracting or preparing materials. This ensures the provision of a dependable albeit limited quantity of active ingredients to individuals. Standardization in the context of manufacturers or distributors dealing with large quantities intended for sale in supermarkets or health food stores pertains to the implementation of industrial production processes that adhere to established conditions. These conditions are defined by the adoption of good manufacturing practices (GMP), which are akin to the practices employed in the production of pharmaceutical drugs (Lincoln 2012).

In the United States, there exists a dual presence of both small-scale and large-scale production of herbal goods, resulting in a significant degree of heterogeneity in terms of their composition and quality within the commercial sphere. In the United States, current regulations do not mandate that producers of dietary supplements strictly adhere to standardized manufacturing procedures, resulting in a lack of assurance regarding product quality. The general public experiences disillusionment due to the dissemination of reports indicating that items procured from retail establishments lack consistency in terms of constituent composition or the specified quantities as stated on the product labels. The efficacy of commonly used herbal products can be supported by various forms of evidence, including traditional use, testimonials, clinical research encompassing both controlled and uncontrolled designs, as well as randomized, double-blind, placebo-controlled trials. In general, it is worth noting that there exists a dearth of comprehensive clinical investigations to substantiate the assertions made (Mukherjee et al. 2010).

3.1 Indian Traditional Medicine

Ayurveda, a medical practice predominantly observed in India, has garnered recognition for approximately five millennia. The approach encompasses dietary interventions and the use of herbal treatments, with a particular focus on addressing the interconnectedness of the body, mind, and spirit in the context of preventing and treating diseases (Mukherjee 2001).

3.2 Japanese Traditional Medicine

Numerous herbal medicines have been introduced from China into the ancient medical practices of Japan. The classification of herbs indigenous to Japan was documented in the initial pharmacopoeia of Japanese traditional medicine during the ninth century (Saito 2000).

3.3 Traditional Chinese Medicine (TCM)

TCM has been utilized by the Chinese populace since antiquity. While animal and mineral elements have been utilized to a certain extent, botanical resources have emerged as the predominant source of treatments. According to Li (2000), out of the extensive repertoire of over 12,000 items employed by traditional healers, approximately 500 are commonly utilized. Botanical items undergo various processing methods, such as stir-frying or immersion in vinegar or wine, prior to their utilization. In the context of clinical practice, the conventional diagnostic process is typically accompanied by the administration of a multifaceted and frequently personalized treatment (Li 2000).

The utilization of TCM remains prevalent in contemporary China. A significant proportion of the population, over 50%, exhibits regular utilization of traditional treatments, with the most pronounced prevalence observed in rural regions. China possesses a vast array of over 5000 traditional treatments, constituting roughly 20% of the whole Chinese pharmaceutical market (Li 2000).

4 Ethnobotany and Its Practical Implications

Despite the derivation and identification of numerous new chemical components from medicinal plants utilized by different cultural groups, there has been a notable absence of traditional drugs produced from such plants through ethnobotanical knowledge and subsequently monitored to be pharmaceuticals in the US over the past four decades (McClatchey 2005). This may appear remarkable, given the substantial allocation of resources and materials dedicated to the study of ethnobotanical knowledge over the previous four decades. In contrast, scientists and researchers continue to rely on indigenous knowledge regarding medicinal plants to identify novel chemical constituents and structures (Do and Bernard 2006). These findings serve as basic foundations for the production of new bioactive compounds. In contemporary times, the professional endeavors of scholars and researchers have become increasingly challenging compared to previous periods. The emphasis was not placed on the exploration of novel cultures; rather, attention was directed toward the examination of over 4000 preexisting cultures and associated information. One of the primary challenges faced by researchers in the field of medicinal plants pertains to the potential limitations in achieving a comprehensive understanding of the mechanisms of action. Instead, their studies often yield results that primarily demonstrate "in vitro pharmacological activity" or "in vivo pharmacological activity on various organs" or focus on the identification of chemical constituents and propose potential activities based on the existence of specific functional groups. The establishment of a strong collaboration between ethnobotanists engaged in research on the evaluation of medicinal plant utilization practices by indigenous communities and experimental pharmacologists investigating the pharmacological properties of traditional medicinal plants, as asserted by traditional healers, holds significant importance. This collaboration serves to enhance the current healthcare system by facilitating the discovery of innovative drugs derived from natural products and plants (McClatchey et al. 2009) (Fig. 2).

The correlation between the contemporary utilization of medicinal plants and their historical significance within ethnic communities has been extensively explored by scholars. This investigation involves examining the local names and applications of medicinal plants found in archival material and literature from ancient civilizations. Additionally, recent studies have employed linguistic analysis of common names to further understand this connection (de Santayana et al. 2005; Nebel et al. 2006). The examination of the historical utilization patterns of Rhododendron tomentosum Harmaja among the indigenous population in Norway reveals a consistent presence of vernacular names and the medicinal application of this plant from the eighteenth to the twentieth centuries. However, there is a notable

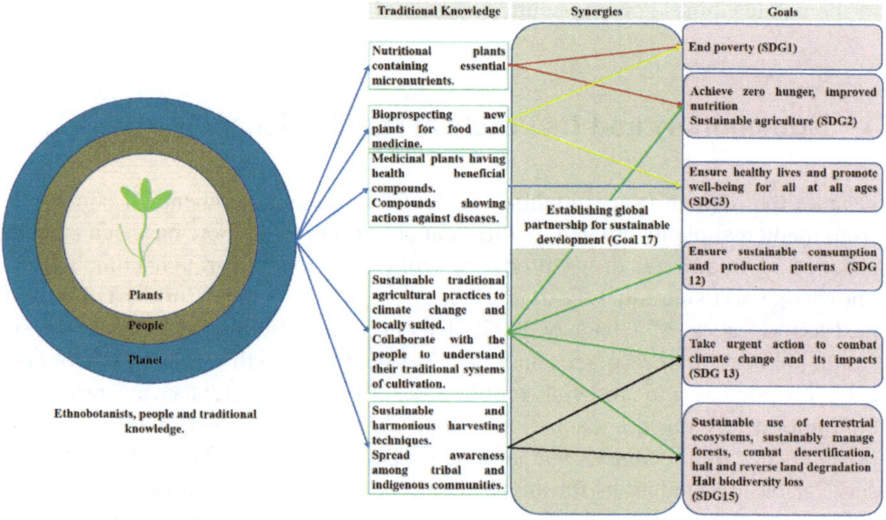

Fig. 2 A diagram illustrating the possible functions of ethnobotanists. (Obtained from Kumar et al. (2021), distributed under Creative Commons Attribution 4.0 International License, which permits unrestricted use, distribution, and reproduction in any medium. http://creativecommons.org/licenses/by/4.0/)

decline in its usage as a substitute for salt. The examination of cognates pertaining to indigenous medicinal plant names provides evidence of historical interactions between medicinal plants and indigenous societies. The specific nomenclature of these plants reveals notable attributes such as color, traditional uses, and perspectives held by indigenous communities (De Santayana 2008). Sabine Nebel conducted a study on the assessment of names for edible greens among Grecanico speakers in Italy, and the results from Greek findings reveal a remarkable persistence of language and cultural practices. As an example, the plant species *Portulaca oleracea* L., well known as purslane, is referred to as "andracla" in Gallicianò and "andrakla" in Greece. The utilization of several medicinal plants represents the enduring practice of traditional medical methods within ancient Greek society (Judd et al. 1999). Spanish ethnobotanists and researchers endeavor to establish a correlation between historical-linguistic methodology and the utilization patterns of medicinal plants known as manzanilla (chamomile) in the context of Spanish tea consumption. This practice can be traced back to the era of Moorish traditions during the twelfth and thirteenth centuries, and potentially even further to the ancient Greek physician Dioscorides (De Santayana 2008). Additionally, other scholars provide limited insight into the classification of Lithuanian botanical diversity by examining and effectively distinguishing the etymological origins of cognate indigenous names for sycamore maple (*Acer pseudoplatanus* L.), black poplar (*Populus nigra* L.), plane tree (*Platanus* spp.), guelder rose (*Viburnum opulus* L.), and sacred wreaths crafted from harvested rye. This study highlights the significance and potential pitfalls associated with the utilization of conventional texts and ethnographic data, including songs, mysteries, and children's poetry, as sources for reconstructing the etymology and figurative evolution of ethnobotanical nomenclature (Šeškauskaitė and Gliwa 2006).

Numerous academics place significant emphasis on the contemporary utilization of higher plants obtained from forests or the wild. This emphasis stems not only from the enduring historical traditions of indigenous communities and their continued presence in emerging markets but also from the notable nutritional value these plants possess. Wild plants are commonly regarded as having a high nutritional value due to their significant concentration of vitamins and minerals (Ansari et al. 2005; Pardo-de-Santayana et al. 2007). Research indicates that the utilization of the common golden thistle (*Scolymus hispanicus* L.) within the impoverished farming community of Portugal possesses a lengthy historical background, dating back to ancient Roman civilization. Furthermore, it has gained significant popularity among visitors in recent years. The utilization of this specific thistle is predicated on its ability to sustain its nutritious content irrespective of the soil conditions in which it thrives. The unique characteristics of these thistles have been recognized by indigenous agriculturalists, leading them to disregard any other thistle species that grow in the region (Pardo-de-Santayana et al. 2007). Additionally, factors such as gender are also of significant importance in the gathering and processing of indigenous plants for medicinal and culinary purposes throughout Europe. While it is widely acknowledged that women in many nations possess extensive expertise regarding collecting and preparing wild plants and fruits (Pieroni 2003), it is worth noting that in Portugal, the activity of collecting and preparing thistles is predominantly carried out by men (De Santayana 2008). The researchers conducted a study on individuals in the Albanian Alps who undergo a transformation from female to male. This unique form of transgenderism involves the integration of traditional knowledge related to the treatment of livestock ailments (ethnoveterinary), wild greens, and the use of weedy food and medicinal plants.

A total of 679 ethnobotanical surveys were conducted in Latin America up until the year 2012. Among the surveys analyzed, it was found that 41% of the ethnobotanical studies were conducted in Brazil, while 22% were carried out in Mexico. Peru accounted for 9% of these surveys, followed by Argentina with 8%, Bolivia with 6%, and the remaining countries of the Latin American continent accounted for 14% of the studies (Albuquerque et al. 2013). In addition to the notable emergence of ethnobiology as an academic discipline in Latin America, there appears to be a lag in the quantity of ethnobotanical research being disseminated to the scientific community in some nations. Based on the findings of research conducted on ethnobotanical knowledge in Latin America, with a particular focus on Mexico, Brazil, and Argentina, it is evident that there has been a notable increase in scientific investigations within the field of ethnobotany across the continent. The enhancement of this condition might be achieved through the allocation of more financial resources toward ethnobotanical research, along with investments in human capital and economic assets. Additionally, the establishment of thematic regions and the cultivation of ethnobotanical cultures, as exemplified in countries like Mexico and Brazil, would contribute to the improvement of this condition. The overarching aim of these theme areas is to facilitate and expand discussions within the disciplines, establish standardized procedures, and generate new insights to address the current knowledge gaps in many fields of ethnobotany. The promotion of ethnobotanical research can be facilitated by an increase in the availability of scientific conferences and journals for disseminating research findings. Additionally, the development of undergraduate and graduate programs can provide scholars with the necessary training to conduct high-quality research in this field. These methodologies have the potential to address the current gaps and deficiencies in ethnobotanical research and facilitate the integration of ethnobotanical knowledge across the Latin American region (Albuquerque et al. 2013).

Ethnobotany has gained recognition among the scientific community, particularly in the realm of financial studies. However, there remains a significant amount of work that still has to be undertaken in this discipline. There remains a substantial amount of ethnobotanical knowledge that has yet to be documented, particularly in various regions of Africa such as the ethnobotanical practices of the Khoi, Ndebele, and Swazi ethnic groups. This includes the examination of their traditional medicinal plant knowledge, the abundance of unrecorded information pertaining to the ethnobotany of these groups, as well as the dynamics between plants and human societies, folk classifications of plants, historical utilization practices of plants, the application of plants in the treatment of livestock ailments, and the utilization of medicinal plants prior to the colonial era. Furthermore, there are several applications of medicinal plants in the treatment of dental ailments (Van Wyk and Van

Staden 2002). Additionally, plant components are utilized in the production of cosmetics, perfumes, insect repellents, flowers for coloring, plants for yeast cultivation, textile production, materials for thatching, musical instruments, and technologies related to hunting, fishing, and other domains. The segregation of market-based research from indigenous knowledge-based research holds significant importance. Therefore, market-oriented research relies on the financial gains derived from the study in a reasonable timeframe. However, while such research may yield more immediate benefits, indigenous knowledge-based research holds greater urgency and significance due to its longstanding contribution to the comprehensive understanding of medicinal plant usage by traditional healers and native communities, particularly in Africa. A study has demonstrated that the process of urbanization, along with the strong influence of external cultural traditions, has significantly contributed to the substantial erosion of indigenous knowledge in South Africa. It is imperative to encourage scholars and researchers to actively engage in the opportunities presented by ethnobotanical research. They should diligently document the various facets of their own historical-linguistic knowledge and the indigenous traditions pertaining to medicinal plant usage among diverse ethnic communities (Van Wyk and Van Staden 2002). This endeavor is crucial for the preservation and transmission of this valuable information to future generations. It is evident that when traditional knowledge is thoroughly and properly documented, it remains unadulterated indefinitely. It is well acknowledged that individuals and generations have the capacity to comprehend and interpret indigenous knowledge from varying perspectives. The origins of contemporary technologies and scientific breakthroughs can be attributed to the indigenous and traditional indigenous knowledge held by our predecessors who may have lacked formal education. The prioritization of research on indigenous knowledge pertaining to the utilization of medicinal plants for the purpose of preventing and treating diseases in both humans and cattle is of particular significance, particularly within the African context. Therefore, it is not viable to provide oral indigenous knowledge to the scientific community without substantial and thoroughly documented traditional knowledge regarding the utilization of medicinal plants in Africa (Van Wyk and Van Staden 2002).

A study was done to investigate the ethnobotanical practices in northern Angola, revealing that a total of 2390 medicinal plants were documented as being utilized by traditional healers and the local people. The medical plants were classified into a total of 358 species belonging to 96 plant groups. Among these species, just three are exclusive to Angola, while around 20% are considered neophytes (Lautenschläger et al. 2018). As the distance from the collection site of medicinal plants rose, there was a corresponding increase in the number of citations documenting their usage. Based on the findings of this study, a significant percentage of women (83%) were engaged in the gathering and processing of medicinal plants sourced from both forested areas and cultivated gardens. The researchers of this study have identified previously unknown medicinal plants that are utilized by indigenous populations for the treatment of various ailments. One such plant is *Gardenia ternifolia*, which is believed to possess anti-measles properties. Additionally, the ethnobotanical knowledge and chemical constituents of the medicinal plant *Annona stenophylla* subsp. cuneata has not been previously identified or documented, particularly within the specific geographical region under study. Despite the enduring conflict in Angola, the utilization of medicinal plants by indigenous communities for the purpose of preventing and treating diseases in both humans and cattle continues to have significant importance as a fundamental aspect of their traditional heritage. Hence, plants play a crucial role in several aspects of human existence, particularly within the domain of health care. The aforementioned issue is further exacerbated by the persistently inadequate state of the medical infrastructure in Angola, particularly in rural regions (Lautenschläger et al. 2018).

The findings of a comprehensive examination of the utilization of medicinal plants in Ethiopia indicate that the nation possesses a significant level of biodiversity, estimated to encompass over 6000 distinct species of higher plants, of which around 10% are local to the region. The level of plant species biodiversity encompasses a significant portion of lower plant species as well. The extensive genetic diversity present within the diverse biotic composition of the country contributes significantly to its status as a prominent hub of biodiversity for plant species. Being one of the 12 Vavilovian centers of origin for cultivated plants and their wild counterparts, this nation boasts a diverse array of indigenous crops and genetic resources. Ethiopia has been recognized as a region of significant biodiversity for approximately five millennia, as evidenced by its historical use by ancient civilizations such as the Greeks, Egyptians, and Romans (Constant and Tshisikhawe 2018). These civilizations relied on Ethiopia as a primary source for exclusive commodities, including Myrrh, Frankincense, and various plant items, which were also utilized in the manufacturing of medicinal substances. Ethiopia boasts a significant abundance of plant species, encompassing a notable biodiversity. Notably, a substantial proportion of these plant species possess medicinal properties, thereby highlighting their potential therapeutic use. Similar to many other countries, a significant proportion of the Ethiopian population relies on the utilization of medicinal plants for the purpose of treating both human and cattle problems. Accordingly, it has been observed that around 95% of conventional medicinal compounds are derived from botanical sources. It can be argued that medicinal plants have historically served as a fundamental component of traditional medicines employed in Ethiopia for the purpose of preventing and treating various diseases affecting both humans and cattle. The process of preparing medicine using medicinal plants holds significant cultural importance within the traditional practices of the Ethiopian population. The preservation of Ethiopian indigenous healing knowledge is characterized by a strict veil of secrecy. However, it is also marked by a dynamic nature, as traditional healers actively strive to expand their knowledge base. This is achieved through the reciprocal transmission of traditional medicinal knowledge across generations, as well as through the study of traditional pharmacopeias. Obtaining indigenous medical knowledge from traditional practitioners poses challenges due to their assertion of ownership over such knowledge and their preference for transmitting it exclusively to their kin, particularly the eldest male offspring (Fenetahun and Eshetu 2017).

The findings of a research investigation on the ethnobotanical evaluation and physicochemical characteristics of frequently utilized medicinal plants in the southwestern region of Ethiopia revealed that a collective count of 72 plant species were identified by the participants as remedies for various human and livestock ailments. These species were further categorized into 61 genera and 39 families. The primary components of plant growth were herbs, which were followed by trees, shrubs, and climbers. The traditional healers in the study mostly utilized leaves as the most commonly employed plant parts for therapeutic purposes, followed by roots and seeds. The traditional healers employed a sequential approach in the production of medicinal herbs, which involved the processes of powdering, crushing, pounding, and pressing (Siraj et al. 2020).

5 Utilization of Ethnobotany in the Context of Community Conservation and Management of Medicinal Plant Resources

There is an increasing acknowledgment within a scholarly discourse that effective management of resources in a sustainable manner necessitates active involvement and engagement of local populations. Therefore, participatory methodologies have been devised as a means of engaging communities in research and development initiatives pertaining to agriculture, forestry, watersheds, and resource management. The utilization of medicinal plants holds significant importance within community development initiatives in rural Asian civilizations (Haji et al. 2020).

Ethnobotany is an academic discipline that examines the intricate interplay between a certain society and its surrounding environment, with a specific focus on the plant kingdom. These relationships can encompass various dimensions, including social, economic, ecological, symbolic, religious, commercial, or artistic aspects, among others. Ethnobotany is a field that integrates principles and methodologies from both the social and biological sciences, drawing upon a wide range of knowledge bases and research methods (Cowan and Ford 1978; Davis 1991; Jain 1987; Martin 1995). Initially, the field of ethnobotany predominantly relied on qualitative and descriptive methodologies, namely, including the documentation and categorization of plant species and their respective utilizations. The field of ethnobotany had additional advancement and expansion with the contributions of Conklin et al. (1980) and Cowan and Ford (1978), who incorporated anthropological methodologies to gain insights into knowledge systems. Researchers have also made advancements in developing quantitative ways to assess the quantities of goods or distributions of valuable species (Caballero 1992). In contemporary times, the field of ethnobiology has expanded beyond the confines of ethnobotanical research. It now encompasses a broader range of inquiries, focusing on the intricate interconnections between plants, animals, humans, soil, and climate, as well as the cultural practices that emerge from these interrelationships.

Ethnobiological approaches have been employed in several researches to investigate traditional management systems (Johnson 1980; Brush 1980; Alcorn 1984;

Altieri and Merrick 1987; Balée and Gély 1989; Aumeeruddy and Pinglo 1989; Shengji 1995). Since 1992, the People and Plants Initiative, a global program operated by WWF UNESCO-Kew Garden, has been implementing various field-based projects. These projects utilize ethnobotany as a means to collaborate with local communities in order to establish sustainable management systems for plant resources. The ethnobotanical advancements in the Hindu Kush-Himalayan region, specifically in India, China, Pakistan, Bangladesh, and Nepal, have primarily focused on traditional herbal medicine, the sustainable management of native plant resources, the cultivation of plants in home gardens, the preservation of traditional agrarian ecosystems, the cultural significance of plants in relation to conservation efforts, and the ethnobotanical practices specific to various ethnic groups. Studies on the ethnobotany of economically significant plant groups, such as bamboo, rattan, and edible plants, as well as the ethnobotany of rural development and biodiversity conservation, have been carried out with a notable focus on practical methodologies in the field (Jain 1986; Banik et al. 1998; Shrestha 1997; Rastogi et al. 1998; Shengji 1991; Pei 1998).

Ethnobotany, as an interdisciplinary field, holds the potential to safeguard the extensive traditional knowledge held by indigenous communities regarding their natural systems and the environment, encompassing the realm of therapeutic plants. This encompasses their understanding of the sustainable exploitation and preservation of various plant resources, ensuring their long-term viability while minimizing harm to their environments. Therefore, it is imperative to prioritize the documentation and incorporation of indigenous knowledge pertaining to land use, non-timber forest products, vegetation and forest management, medicinal plants, home gardens, agroforestry, swidden agriculture, and biodiversity. In the interim, it is imperative to sustain the planning process for the development of natural resources through the establishment of robust conversation and communication channels with indigenous communities, employing ethnobotanical methodologies. Additionally, the inclusion of local individuals in future management endeavors will serve to prevent any potential negative consequences on both the local populace and the environment, as highlighted by Pei (1998).

In particular, the objectives of ethnobotany in relation to the sustainable utilization of botanical resources are emphasized through the subsequent aspects:

- The objective is to evaluate the historical and current transformations and their influence on management through the application of ethnobotanical techniques, utilizing quantitative as well as qualitative methods.
- Enhancing comprehension of the direct interactions that occur between individuals and biotic components within a given environment can be facilitated by conducting ethnobotanical research. Such studies can contribute to the effective management of resources at the local level.
- To facilitate the transmission of reproducible indigenous techniques, practices, and knowledge to different locations within the same bioregion for the purpose of resource management. This transfer will be accomplished through a

farmer-to-farmer interaction, utilizing the biocultural "niche-transfer" idea that is inherent in indigenous environmental management.

- To enable the active involvement of community members in the process of gathering, recording, examining, and evaluating indigenous knowledge through on-site investigations.
- To enhance the involvement of local communities in the design of resource management initiatives for rural development and conservation projects. This will be achieved through the utilization of Rapid Rural Appraisal (RRA), Participatory Rural Appraisal (Adhikari and Paul 2018), and Participatory Biodiversity Assessment (PBA) methodologies.
- To conduct an inventory of plant species that are utilized by humans for various objectives such as medical applications, fodder for livestock, food sources, religious practices, swidden crops, non-timber forest products, home gardens, and other purposes. Typically, this process entails the participation of a substantial cohort of individuals residing in the vicinity, alongside a specific subset of informants who have been recognized within the local community. This type of inventory serves to record and preserve voucher specimens, as well as gather information from local communities regarding the surveyed species. The information collected encompasses both natural and human-use aspects, such as vernacular names, scientific names, habitat, utilization, properties, preparation, and maintenance.

6 Ethnobotanical Documentation and Inventory as a Repository for Details of Traditional Medicine

The area of ethnobotany has been extensively utilized in scientific research for the purpose of documenting indigenous knowledge of plant usage and compiling inventories of beneficial plants found in the local flora of Asian nations (Martin 1995; Jain 1995; Shengji 1988). Ethnopharmacology can be described as a scholarly discipline that focuses on investigating the therapeutic use of plants by indigenous societies, with the aim of scientifically substantiating its efficacy and potential adverse reactions. Ethnopharmacological approaches can be encouraged for usage in poor nations due to their cultural acceptance and potential to improve public health (Farnsworth 1993). Ethnobotany has been extensively utilized in Asian nations such as India, China, Bangladesh, and Nepal for the purpose of documenting traditional knowledge and compiling inventories of herbal remedies (Shengji 1988; Shah 1990; Manandhar 1980; Alam et al. 1996).

The acquisition of ethnobotanical knowledge pertaining to the therapeutic properties of plants and their utilization by indigenous societies serves multiple purposes, including the preservation of cultural heritage and ecological diversity, as well as the advancement of community healthcare and pharmaceutical research. For drug discovery purposes, this data is employed on the premise that a plant with a long history of use by native communities likely possesses nootropic properties (Farnsworth 1993). Ethnobotanical documentation and inventories are systematically structured to include local and scientific nomenclature, medicinal applications, and cultural interpretations, as well as ecological, botanical, phenological, distribution, harvesting, management, and conservation information pertaining to medicinal plants. The data is sourced from indigenous herbal practitioners, community healers, and herbal vendors operating within local marketplaces, in addition to community members who possess traditional expertise in this domain. The documentation and description of indigenous knowledge include a wealth of different and credible local experiences pertaining to the prevention, treatment, and maintenance of human health within the immediate surroundings. The integration of inventory and documentation can be effectively implemented within community resource management programs and local-level initiatives for biodiversity conservation. Additionally, this data can serve as valuable information for guiding the process of plant selection and sample collection in laboratory identification and pharmacological testing for the purpose of medication development. Therefore, it may be argued that the documentation and inventory serve as the repository of knowledge for traditional medicine.

The revised inventory and registry of TCM encompasses a total of 1581 animal species, 11,146 plant species, and 80 minerals. Among these, a range of 500–600 species are frequently employed as medicinal substances in TCM. The comprehensive and extensively researched knowledge of TCM has lately been consolidated into a six-volume series titled Chinese Medicinal Resources. This series was prepared by the Chinese National Corporation of Traditional and Herbal Medicine and published by Science Press in Beijing between 1994 and 1995. The series comprises a total of six volumes, which are as follows:

1. Chinese Medicinal Resources (287 pages)
2. Outline of Chinese Medicinal Resources (2069 pages)
3. Regionalization of Chinese Medicine (634 pages)
4. Commonly used Chinese Medicinal Materials (1138 pages)
5. Chinese Folk Prescriptions (1817 pages)
6. The Resource Atlas of Chinese Herbal Medicine (in English and Chinese, comprises 81 color maps).

The Atlas is a pharmaceutical document that holds scientific merit and carries practical importance. The series comprises a collection of materials pertaining to TCM, with an extensive inventory of 12,807 medicinal plants, minerals, and animals, as well as 20,000 Folk Prescriptions. Each entry in the dataset has information regarding resource distribution, harvesting practices, environmental conditions, historical utilization, and conservation efforts.

China is home to 55 distinct ethnic minority groups, including a population of over 100 million individuals. These ethnicities are dispersed among 18 provinces, collectively accounting for 64% of China's total land area. The ancient medical methods and herbal medicine expertise in China are commonly referred to as the "Ethno-Medicine" of China (EMC). According to surveys and documentation efforts conducted in the past two decades, the area encompasses more than 8000 species of plants and animals (Zheng 1997; Sheng-Ji 2001). As an illustration, the classified Tibetan Medicine encompasses a total of 1106 plant species, 448 animal species, and 840 naturally occurring minerals. The inclusion of ethnobotanical knowledge pertaining to Ethno-Medicine has been widely recognized as a valuable asset for promoting regional economic growth and facilitating the discovery of novel pharmaceuticals (Zheng 1997; Xiao 1999).

According to the World Wildlife Fund (WWF, 1990), traditional healers in India utilize over 2500 plant species for medicinal purposes. Additionally, Shah (1990) states that 100 plant species are consistently employed as medicinal resources in the region. According to ethnobotanists, over 800 therapeutic plants were documented in Nepal, while in Bangladesh, around 700 medicinal plants were reported (Banik et al. 1998). According to Saiju (Saiju 1990), it has been reported that around 50% of the pharmaceutical medications utilized in contemporary medicine in Pakistan are derived synthetically from raw ingredients based on petroleum chemicals. In recent times, there has been a growing inclination toward the documentation and inventorying of traditional herbal medicine in Himalayan nations through the utilization of ethnobotanical methodologies. International organizations, including WWF-International, IDRC, UNESCO, ICIMOD, and European financial agencies, have significantly contributed to the advancement and facilitation of many ethnobotanical initiatives in the countries of the Himalayan region.

7 Utilization of Ethnobotanical Knowledge in the Process of Pharmaceutical Development

The rapid progression and advancements in the fields of technology, science, and the global economy have had a profound impact on the global landscape and natural surroundings. The juxtaposition of significant advancements in human healthcare and the escalating degradation of the environment has led to an increasing need for natural goods and phytomedicine. Consequently, research and development efforts have switched toward the exploration of novel drugs. Numerous academic institutions and corporate entities within this domain have increasingly embraced the utilization of traditional medicine, particularly the exploration of botanical resources as a potential reservoir for novel pharmaceutical compounds. Ethnobotanical information is utilized by researchers as a valuable indicator for identifying plants that are highly suitable for subsequent screening and chemical analysis (Farnsworth 1993). This developmental trend encompasses not only Western nations but also developing nations in Asia.

Significant transformations have occurred in the realm of TCM since the establishment of the People's Republic of China half a century ago. The official confirmation of the legal status and scientific standing of TCM has been established. The constitutional law explicitly states the imperative to advance both modern medicine and traditional medicine. The overarching policy of combining CTM with Western Medicine has been effectively executed within the comprehensive national healthcare system. The establishment of the TCM medical care system through the utilization of TCM hospitals has effectively utilized and highlighted the distinctive features and benefits of TCM. With the progression of TCM education systems and research institutions, alongside enhancements in the standardization and legal framework of TCM, the industry has successfully constructed a sophisticated production system capable of fulfilling both domestic and foreign demands.

In the last five decades, there has been a significant focus on traditional medicine, encompassing TCM and ethnomedicine, for the purposes of drug discovery and manufacture. This attention has been driven by policy guidance and the nationwide advancement of TCM. The utilization of ethnobotanical knowledge in the process of drug creation is a crucial component of the technique, which is achieved through several approaches.

8 Conclusion

Ethnobotany is a branch of life science that focuses on the examination of the relationship between human beings and plant life. It encompasses the systematic exploration, observation, and identification of the various types of plant diversity utilized for the purpose of preventing and treating diseases in both humans and cattle. Additionally, this field of study examines the knowledge, beliefs, and practices of indigenous populations, particularly in relation to medicinal plants, encompassing cultural and religious aspects. Additionally, this study encompasses the processes by which individuals classify, separate, and connect plant species, as well as the interconnectedness between plants and humans. In order to get significant knowledge regarding the therapeutic use of plants, it is imperative for ethnobotanists to engage in discussions with native communities, thereby fostering an understanding and appreciation for their cultural practices and way of life. Ethnobotany can be regarded as a distinct subfield within the broader discipline of ethnobiology. Ethnobotanists employ several methodologies and resources in their ethnobotanical research, such as analyzing ancient texts, conducting surveys, engaging in dialogues with knowledgeable individuals, and conducting field investigations to explore the intricate connections between plants and human beings. In general, it is common for researchers to collaborate with indigenous individuals or traditional healers who possess expertise in plant knowledge. These collaborations serve the purpose of documenting the native biodiversity, particularly plants, as well as identifying the various botanical species. Additionally, these collaborations aim to document the specific plant parts that are utilized for treating ailments in both humans and livestock, along with the methods employed for their preparation and application.

Ethnobotanists are increasingly inclined to identify themselves as ethnobiologists or ethnoecologists due to the greater opportunities these fields offer for assessing the connection between human beings and their overall environment, as well as the interaction of societies with the external environment. This includes examining the impact of global trade on domestic economies and individual lives. Since 1992, the field of applied ethnobotany has emerged to encompass the study and implementation of collaborative approaches between humans and plants. This field focuses on actively engaging with indigenous communities and traditional practitioners to explore their knowledge and develop improved management strategies that shape specific utilization practices and social dynamics. The field of applied ethnobotanies strives to bridge the divide between indigenous knowledge and contemporary practices. It also acknowledges the correlation between indigenous practices and knowledge systems, as well as the influence of national and global trends in terms of approaches, strategies, and funding. In contemporary times, the neologism "ethnoecology" has emerged.

According to Martine, ethnoecology is a multidisciplinary field that incorporates various academic disciplines. The concept of ethnoecology encompasses various academic disciplines that examine the interactions between indigenous communities and their surrounding ecosystems. These subdisciplines include ethnobiology, ethnobotany, ethnoentomology, and ethnozoology. Ethnoecology is an academic subject that focuses on the understanding of the dynamic relationship between human people and the various components of their environment, such as plants, animals, and physical factors within a specific geographical location.

The significance of traditional medical knowledge extends beyond its potential impact on drug development and market prices, encompassing its relevance to health care throughout history and its potential implications for the future. The use of information banks pertaining to traditional medicine and ethnobotanical knowledge has served distinct functions in the realm of medication discovery in China and numerous other nations. Furthermore, it is anticipated that these resources will persist in their contributions to this discipline in the foreseeable future. The contemporary landscape is marked by prevalent issues such as the excessive extraction of medicinal plants, the deterioration of medicinal resources, and the erosion of traditional medical knowledge within the indigenous groups residing in areas where these resources are abundant. The international agenda has recently included discussions on several topics such as indigenous knowledge, intellectual property rights related to the creation of new drugs through bioprospecting and herbal plants, and the unregulated transboundary commerce of medicinal plants in certain border regions.

In the contemporary day, there exists a pressing imperative to prioritize the preservation of indigenous medical knowledge and the invaluable botanical resources associated with traditional medicine, particularly within the developing nations of Asia. The future prospects of ancient traditional medicine in the twenty-first century remain uncertain, contingent upon advancements in science and technology, the priorities of major industry stakeholders, and the regulatory frameworks governing cultural practices in developing nations. Furthermore, it is evident that currently, no medical system can be deemed flawless or comprehensive in its capacity to effectively address all types of ailments and medical conditions. This includes Western medicine, Eastern medicine, allopathic medicine, and homoeopathic medicine.

Hence, it is crucial for healthcare providers on a global scale to possess a comprehensive understanding and conduct a thorough examination of the medical traditions that are being implemented in other nations. Traditional medicines in Asia play a significant role in this context.

References

Adhikari PP, Paul SB (2018) History of Indian traditional medicine: a medical inheritance. History 11(1):421
Adjanohoun É-J (1980) Médecine traditionnelle et pharmacopée, contribution aux études ethnobotaniques et floristiques au Niger: rapport présenté à l'ACCT par E.-J. Adjanohoun [et al.]
Adnan M, Patel M, Snoussi M (2023) Ethnobotany and ethnopharmacology of medicinal and aromatic plants: steps towards drug discovery. CRC Press, Boca Raton
Alam M, Choudhury J, Hassan M (1996) Some folk formularies from Bangladesh. Bangladesh J Life Sci 8(1):49–63
Albuquerque UP, Silva JS, Campos JLA, Sousa RS, Silva TC, Alves RRN (2013) The current status of ethnobiological research in Latin America: gaps and perspectives. J Ethnobiol Ethnomed 9:1–9
Alcorn JB (1984) Development policy, forests, and peasant farms: reflections on Huastec-managed forests' contributions to commercial production and resource conservation. Econ Bot 38(4):389–406
Alexiades MN, Sheldon JW (1996) Selected guidelines for ethnobotanical research: a field manual. New York Botanical Garden, New York, USA, p 306. ISBN: 978-0-89327-404-7
Allison DB, Fontaine KR, Heshka S, Mentore JL, Heymsfield SB (2001) Alternative treatments for weight loss: a critical review. Crit Rev Food Sci Nutr 41(1):1–28
Altieri MA, Merrick L (1987) In situ conservation of crop genetic resources through maintenance of traditional farming systems. Econ Bot 41:86–96
Ansari N, Houlihan L, Hussain B, Pieroni A (2005) Antioxidant activity of five vegetables traditionally consumed by south-Asian migrants in Bradford, Yorkshire, UK. Phytother Res 19(10):907–911
Aumeeruddy Y, Pinglo F (1989) Phytopractices in tropical regions: a preliminary survey of traditional crop improvement techniques. Man and the Biosphere Programme, Paris, France, p 71
Aumeeruddy-Thomas Y, Shengji P, Pei S (2003) Applied ethnobotany: case-studies from the Himalayan region. WWF, Surrey
Balée W, Gély A (1989) Managed forest succession in Amazonia: the Ka'apor case. Adv Econ Bot:129–158
Banik R, Alam M, Pei S, Rastogi A (1998) Applied ethnobotany. In: Proceedings of the subregional training workshop on applied ethnobotany (December, 1997). Bangladesh Forest Research Institute (BFRI), Chittagong
Birhanu Z, Endale A, Shewamene Z (2015) An ethnomedicinal investigation of plants used by traditional healers of Gondar town, North-Western Ethiopia. J Med Plants Stud 3(2):36–43
Brush SB (1980) Potato taxonomies in Andean agriculture. In: Indigenous knowledge systems and development, pp 37–48
Caballero J (1992) Maya homegardens: past, present and future. Etnoecológica 1(1):35–54
Cheikhyoussef A, Shapi M, Matengu K, Mu Ashekele H (2011) Ethnobotanical study of indigenous knowledge on medicinal plant use by traditional healers in Oshikoto region, Namibia. J Ethnobiol Ethnomed 7:1–11
Chun-Lin L (2013) Modern ethnobotany: an introduction. Plant Divers 35(4):438
Conklin HC, Lupāih P, Pinther M (1980) Ethnographic atlas of Ifugao: a study of environment, culture, and society in northern Luzon. Yale University Press, vol-vii, p 115

Constant NL, Tshisikhawe MP (2018) Hierarchies of knowledge: ethnobotanical knowledge, practices and beliefs of the Vhavenda in South Africa for biodiversity conservation. J Ethnobiol Ethnomed 14:1–28

Cowan C, Ford RI (1978) The nature and status of ethnobotany, vol 67. University of Michigan Museum of Anthropology, Anthropological Papers, Ann Arbor, pp 263–288

Davidson-Hunt I (2000) Ecological ethnobotany: stumbling toward new practices and paradigms. MASA J 16(1):1–3

Davis W (1991) Towards a new synthesis in ethnobotany. Las Plantas y el Hombre Memorias del Primer Simposio Ecuatoriano de Etnobotánica y Botánica Económica ABYA-YALA Quito, Ecuador. pp. 339–357

De Santayana MP (2008) Estudios etnobotánicos en Campoo (Cantabria): conocimiento y uso tradicional de plantas. Editorial CSIC-CSIC Press, Madrid

de Santayana MP, Blanco E, Morales R (2005) Plants known as té in Spain: an ethno-pharmaco-botanical review. J Ethnopharmacol 98(1–2):1–19

Do Q-T, Bernard P (2006) Reverse pharmacognosy: a new concept for accelerating natural drug discovery. Adv Phytomed 2:1–20

Esha RT, Chowdhury MR, Adhikary S, Haque KMA, Acharjee M, Nurunnabi M et al (2012) Medicinal plants used by tribal medicinal practitioners of three clans of the Chakma tribe residing in Rangamati district, Bangladesh. Am Eur J Sustain Agr 6:74–84

Farnsworth NR (1993) Ethnopharmacology and future drug development: the North American experience. J Ethnopharmacol 38(2–3):137–143

Fenetahun Y, Eshetu G (2017) A review on ethnobotanical studies of medicinal plants use by agro-pastoral communities in Ethiopia. J Med Plants 5(1):33–44

Fokunang C, Ndikum V, Tabi O, Jiofack R, Ngameni B, Guedje N et al (2011) Traditional medicine: past, present and future research and development prospects and integration in the National Health System of Cameroon. Afr J Tradit Complement Altern Med 8(3):284

Francia S, Stobart A (2014) Critical approaches to the history of Western herbal medicine: from classical antiquity to the early modern period. Bloomsbury Academic, London

Gagnepain F (1950) Flore générale de l'Indochine [General flora of Indo-China]. 7 volumes+ suppl (1907). Masson & Cie, Paris, France

Ghorbani A, Naghibi F, Mosaddegh M (2006) Ethnobotany, ethnopharmacology and drug discovery. Iran J Pharm Sci 2(2):109–118

Haji L, Valizadeh N, Hayati D (2020) The role of local communities in sustainable land and forest management. In *Spatial Modeling in Forest Resources Management: Rural Livelihood and Sustainable Development. Springer*, 9, p 473–503

Heinrich M, Barnes J, Prieto-Garcia J, Gibbons S, Williamson EM (2017) Fundamentals of pharmacognosy and phytotherapy E-Book. Elsevier Health Sciences, Amsterdam

Hoft M, Barik S, Lykke A (1999) Quantitative ethnobotany, applications of multivariate and statistical analyses in ethnobotany, people and plants. UNESCO, Paris, p 3

Hong L, Guo Z, Huang K, Wei S, Liu B, Meng S et al (2015) Ethnobotanical study on medicinal plants used by Maonan people in China. J Ethnobiol Ethnomed 11(1):1–35

Hutchings A (1996) Zulu medicinal plants: an inventory. University of Natal Press, Pietermaritzburg

Idu M (2009) Current trends in ethnobotany. Editorial. Trop J Pharm Res 8(4):295–296

Inglis J (1994) Traditional knowledge in tropical environments—introduction. Parthenon Publishing Group Casterton, Hall, Carnforth, Lancashire, pp 3–4

Jahiel AR (1998) The organization of environmental protection in China. China Q 156:757–787

Jain S (1986) Ethnobotany. Interdiscip Sci Rev 11(3):285–292

Jain S (1987) Ethnobotany, its scope and various sub-disciplines. A manual of ethnobotany, pp 1–11

Jain S (1995) A manual of ethnobotany. Scientific Publishers, Jodhpur, p 179

Jaiswal YS, Williams LL (2017) A glimpse of Ayurveda—the forgotten history and principles of Indian traditional medicine. J Tradit Complement Med 7(1):50–53

Johnson A (1980) Ethnoecology and planting practices in a swidden agricultural system (Brazil). In: Brokensha D, Warren DM, Werner O (eds) Indigenous knowledge systems and development. University Press of America, New York

Judd WS, Campbell CS, Kellogg EA, Stevens PF, Donoghue MJ (1999) Plant systematics: a phylogenetic approach. Ecol Mediterr 25(2):215

Kankara SS, Ibrahim MH, Mustafa M, Go R (2015) Ethnobotanical survey of medicinal plants used for traditional maternal healthcare in Katsina state, Nigeria. S Afr J Bot 97:165–175

Kayser O (2018) Ethnobotany and medicinal plant biotechnology: from tradition to modern aspects of drug development. Planta Med 84(12/13):834–838

Keller K (1991) Legal requirements for the use of phytopharmaceutical drugs in the Federal Republic of Germany. J Ethnopharmacol 32(1–3):225–229

Kokni FK, Solanki HA, Patel DD (2016) Study of ethnomedicinal plants and its documentation of Waghai Forest, Gujarat. Life Sci Leafl 81:11–30

Kose LS, Moteetee A, Van Vuuren S (2015) Ethnobotanical survey of medicinal plants used in the Maseru district of Lesotho. J Ethnopharmacol 170:184–200

Kumar A, Kumar S, Komal, Ramchiary N, Singh P (2021) Role of traditional ethnobotanical knowledge and indigenous communities in achieving sustainable development goals. Sustain For 13(6):3062

Kumbhar KS (2022) Healing and harming: the "Noble Profession" of medicine in post-independence India, 1947–2015. Harvard University, Cambridge

Lautenschläger T, Monizi M, Pedro M, Mandombe JL, Bránquima MF, Heinze C et al (2018) First large-scale ethnobotanical survey in the province of Uíge, northern Angola. J Ethnobiol Ethnomed 14:1–73

Lev E, Amar Z (2000) Ethnopharmacological survey of traditional drugs sold in Israel at the end of the 20th century. J Ethnopharmacol 72(1–2):191–205

Li L (2000) Opportunity and challenge of traditional Chinese medicine in face of the entrance to WTO (World Trade Organization). Chin J Inform Tradit Chin Med 7:7–8

Liao Y (2011) Traditional Chinese medicine. Cambridge University Press, Cambridge

Lincoln JE (2012) Overview of the US FDA GMPS: good manufacturing practice (GMP)/quality system (QS) regulation (21 CFR part 820). J Valid Technol 18(3):17

Malla B, Gauchan DP, Chhetri RB (2015) An ethnobotanical study of medicinal plants used by ethnic people in Parbat district of western Nepal. J Ethnopharmacol 165:103–117

Manandhar N (1980) Medicinal plants of Nepal Himalaya. Ratna Pustak Bhandar, Bhotahity, Kathmandu, p 581

Manderson L, Jolly M (1997) Sites of desire/economies of pleasure: sexualities in Asia and the Pacific. University of Chicago Press, Chicago

Marcus J (1982) The plant world of the sixteenth-and seventeenth-century lowland Maya. In: Maya subsistence. Elsevier, Amsterdam, pp 239–273

Martin G (1995) Ethnobotany: a methods manual. Chapman & Hall, New York

Martin G (2004) Ethnobotany; a methods manual: WWF and IIED. Earth Scans, Camden, London

McClatchey W (2005) Medicinal bioprospecting and ethnobotany research. Ethnobot Res Appl 3:189–190

McClatchey WC, Mahady GB, Bennett BC, Shiels L, Savo V (2009) Ethnobotany as a pharmacological research tool and recent developments in CNS-active natural products from ethnobotanical sources. Pharmacol Ther 123(2):239–254

Mugomeri E, Chatanga P, Raditladi T, Makara M, Tarirai C (2016) Ethnobotanical study and conservation status of local medicinal plants: towards a repository and monograph of herbal medicines in Lesotho. Afr J Tradit Complement Altern Med 13(1):143–156

Mukherjee PK (2001) Evaluation of Indian traditional medicine. Drug Inf J 35(2):623–632

Mukherjee PK, Pitchairajan V, Murugan V, Sivasankaran P, Khan Y (2010) Strategies for revitalization of traditional medicine. Chin Herb Med 2(1):1–15

Naini V, Mamidala E (2013) An ethnobotanical study of plants used for the treatment of diabetes in the Warangal district, Andhra Pradesh, India. Biolife 1(1):24–28

Nebel S, Pieroni A, Heinrich M (2006) Ta chòrta: wild edible greens used in the Graecanic area in Calabria, Southern Italy. Appetite 47(3):333–342

Ochwang'i DO, Kimwele CN, Oduma JA, Gathumbi PK, Mbaria JM, Kiama SG (2014) Medicinal plants used in treatment and management of cancer in Kakamega County, Kenya. J Ethnopharmacol 151(3):1040–1055

Pardo-de-Santayana M, Tardío J, Blanco E, Carvalho AM, Lastra JJ, San Miguel E et al (2007) Traditional knowledge of wild edible plants used in the northwest of the Iberian Peninsula (Spain and Portugal): a comparative study. J Ethnobiol Ethnomed 3(1):1–11

Pei S (1998) Biodiversity conservation in the mountain development of Hindu Kush-Himalayas. In: Frontiers in biology; the challenge of biodiversity, biotechnology and sustainable agriculture. Academica Sinica, Taipei, pp 223–234

Peters CM (1996) Beyond nomenclature and use: a review of ecological methods for ethnobotanists. Adv Econ Bot 10:241–276

Petrovska BB (2012) Historical review of medicinal plants' usage. Pharmacogn Rev 6(11):1

Phillips O, Gentry AH (1993) The useful plants of Tambopata, Peru: I. Statistical hypotheses tests with a new quantitative technique. Econ Bot 47:15–32

Pieroni A (2000) Medicinal plants and food medicines in the folk traditions of the upper Lucca Province, Italy. J Ethnopharmacol 70(3):235–273

Pieroni A (2003) Women and plants case studies on gender relations in biodiversity management and conservation

Qiu J (2007) Traditional medicine: a culture in the balance. Nature 448(7150):126–129

Rahman IU, Afzal A, Iqbal Z, Ijaz F, Ali N, Shah M et al (2019) Historical perspectives of ethnobotany. Clin Dermatol 37(4):382–388

Rastogi A, Godbole A, Shengji P (1998) Applied ethnobotany in natural resource management. In: Traditional home gardens highlights of a training workshop held at Kohiima, Negaland, India (18–23 June 1997). International Centre for Integrated Mountain Development, Kathmandu

Saiju H (1990) Plant resources for economic sustenance of the mountain people: a review of the medicinal and aromatic plants. Commissioned position paper. ICIMOD, Kathmandu

Saito H (2000) Regulation of herbal medicines in Japan. Pharmacol Res 41(5):515–519

SarDesai DR (2018) Southeast Asian history: essential readings. Routledge, New York

Schulz V, Hänsel R, Tyler VE (2001) Rational phytotherapy: a physician's guide to herbal medicine. Psychology Press, London

Šeškauskaitė D, Gliwa B (2006) Some Lithuanian ethnobotanical taxa: a linguistic view on thorn apple and related plants. J Ethnobiol Ethnomed 2(1):1–4

Shah N (1990) Endangered medicinal and aromatic herbs of Indian Himalayas. In: 2nd international conference of ethnobiology. pp 22–26

Shale T, Stirk W, van Staden J (1999) Screening of medicinal plants used in Lesotho for antibacterial and anti-inflammatory activity. J Ethnopharmacol 67(3):347–354

Shengji P (1988) Ethnobotany and the exploitation of plant resources. Plant Divers 10(S1):1

Shengji P (1991) Conservation of biological diversity in templeyards and holy hills by the Dai ethnic minorities of China. Ethnobotany 3:27–35

Shengji P (1995) Regional study on biodiversity: concepts, frameworks, and methods. Yunnan University Press, Yunnan

Sheng-Ji P (2001) Ethnobotanical approaches of traditional medicine studies: some experiences from Asia. Pharm Biol 39(sup1):74–79

Shrestha KK (1997) Ethnobotany for conservation and community development. In: Proceedings of the national training workshop in Nepal:(January 6–13, 1997). Ethnobotanical Society of Nepal, Kathmandu

Siraj J, Belew S, Suleman S (2020) Ethnobotanical assessment and physicochemical properties of commonly used medicinal plants in Jimma zone, Southwest Ethiopia: traditional healers based cross-sectional study. J Exp Pharmacol 12:665–681

Sivasankari B, Anandharaj M, Gunasekaran P (2014) An ethnobotanical study of indigenous knowledge on medicinal plants used by the village peoples of Thoppampatti, Dindigul district, Tamilnadu, India. J Ethnopharmacol 153(2):408–423
Soejarto D, Gyllenhaal C, Riley M, Zhang H (2009) Ethnobotany of natural products. Encyclopedia of Life Support Systems (EOLSS), Developed under the Auspices of the UNESCO Eolss Publishers, Oxford. http://www.eolss.net
Thompson CM (2003) Medicine, nationalism, and revolution in Vietnam: the roots of a medical collaboration to 1945. East Asian Sci Technol Med 21:114–148
Tobyn GW (2017) English herbs for English bodies: the promotion of native plant remedies 1548–1659, Nicholas Culpeper and his antecedents. Lancaster University, Lancaster
Touwaide A (2004) Medieval herbals: the illustrative traditions. JSTOR, London
Tyler VE (2000) Herbal medicine: from the past to the future. Public Health Nutr 3(4a):447–452
Van Wyk B-E (2008) A review of Khoi-San and Cape Dutch medical ethnobotany. J Ethnopharmacol 119(3):331–341
Van Wyk B-E, Van Staden J (2002) A review of ethnobotanical research in southern Africa. S Afr J Bot 68(1):1–13
Van Wyk B-E, Wink M (2018) Medicinal plants of the world. Briza Publications, Pretoria
Watt JM, Brandwijk MG (1927) Suto (Basuto) medicines: part II. Bantu Stud 3(1):155–178
Wujisguleng W, Liu Y, Long C (2012) Ethnobotanical review of food uses of Polygonatum (Convallariaceae) in China. Acta Soc Bot Pol 81(4):239
Xiao P-G (1999) Century review and perspective on TCM. Research & Information on Traditional Chinese Medicine, New York, p 7
Zheng Y-L (1997) Greet the spring of exploitation and utilization of ethnodrugs. Chin J Ethnomed Ethnopharm 24:1–5

Evolution of Traditional Resources and Their Uses in the Field of Drug Discovery: Its Present-Day Impact in Society

Tilottoma Saha and Md Enamul Hoque

Contents

T. Saha
Department of Materials and Metallurgical Engineering, Bangladesh University of Engineering and Technology (BUET), Dhaka, Bangladesh

M. E. Hoque (✉)
Department of Biomedical Engineering, Military Institute of Science and Technology (MIST), Dhaka, Bangladesh
e-mail: enamul1973@gmail.com; enamul1973@bme.mist.ac.bd

© The Author(s), under exclusive license to Springer Nature Singapore Pte Ltd. 2024

A. Das Talukdar et al. (eds.), *Traditional Resources and Tools for Modern Drug Discovery*, Interdisciplinary Biotechnological Advances,
https://doi.org/10.1007/978-981-97-4600-2_2

Abstract Traditional resources, enriched with time-honored knowledge, have served as potent remedies for diverse diseases throughout history, igniting renewed interest in their integration into modern drug discovery. This chapter explores the evolution of traditional resources in drug development and their profound impact on society. Beginning with a global perspective on traditional medicine, it unearths healing practices across cultures. Scientific insights into the active compounds and mechanisms underpinning traditional remedies are examined, accompanied by evidence from preclinical studies and clinical research. The chapter investigates the opportunities and roadblocks presented by utilizing traditional resources in contemporary drug research, considering standardization, sustainability, and regulatory factors. Emphasizing their contemporary societal impact, it showcases traditional resource utilization in modern healthcare practices and their potential contribution to novel drug development. By embracing the confluence of ancient traditions and cutting-edge science, traditional resources offer a promising avenue for advancing integrative health care through the interdisciplinary approach of ethnopharmacology and sustainable sourcing practices.

Keywords Traditional resources · Drug discovery · Active compounds · Societal impact · Traditional medicine · Integrative health care · Ethnopharmacology · Sustainable sourcing

1 Introduction

1.1 Historical Significance of Traditional Resources

The search for new drugs and therapies has been a continuous endeavor in human history, rooted in the exploitation of traditional resources. Traditional resources, in this context, encompass a rich array of natural materials derived from plants, animals, minerals, and other sources that have been used for their medicinal properties across cultures and generations.

The historical significance of traditional resources in the field of drug discovery cannot be overstated. For millennia, indigenous communities and ancient civilizations have relied on their knowledge of local flora, fauna, and minerals to alleviate ailments and promote wellness. These time-honored practices, informed by empirical observation and passed down through oral traditions, have provided the foundation for modern pharmacology and drug development.

References to the historical use of traditional resources are abundant in the literature. For instance, the Ayurvedic texts of ancient India describe the therapeutic properties of various herbs and minerals, some of which are still investigated for their medicinal potential today (Sharma 2014). Traditional Chinese medicine has a similarly rich history of utilizing plant-based ingredients like ginseng and licorice in healing practices (Lin and Zhu 2002). Native American tribes have long used plants like echinacea and yarrow for their curative properties (Moerman 1996). Even in Western medicine, ancient texts such as "De Materia Medica" by Dioscorides have documented the use of plants like opium poppy and foxglove, which served as precursors to modern pharmaceuticals (Ali et al. 2020).

While these historical examples provide a glimpse into the deep-rooted relationship between traditional resources and medicine, the contemporary relevance of such practices is equally remarkable. Today, pharmaceutical companies and research institutions are increasingly turning to traditional resources as potential sources of novel drug compounds. The knowledge held by traditional healers and indigenous communities is seen as a valuable reservoir of untapped therapeutic agents, sparking a resurgence in ethnopharmacological research (Heinrich and Edwards 2007).

This chapter explores the evolution of traditional resources in the context of drug discovery, their historical significance, and their present-day impact on society (Fig. 1). We delve into the ways in which traditional knowledge is being integrated with modern scientific methodologies to develop innovative drugs, highlighting the potential benefits and challenges of this evolving field.

Moving forward, this chapter will detail the methodology behind the traditional resource-based drug discovery, the bioprospecting of plant and animal sources, and the ethical and legal considerations surrounding the utilization of traditional knowledge. Through a comprehensive examination of traditional resources, this chapter equips readers with a nuanced understanding of their multifaceted influence on contemporary drug discovery and their potential to shape the future of medicine.

1.2 Scope and Objectives

By comprehensively exploring the changing landscape of traditional resources, this chapter sheds light on their influence on drug discovery and their current impact on society. Our primary objectives are as follows:

1.2.1 Historical Exploration

We will delve deeper into the historical significance of traditional resources, offering a more in-depth analysis of their use in different cultures and time periods. This section will provide readers with a nuanced understanding of the diverse ways in which traditional resources have been employed for medicinal purposes throughout history (Balick and Cox 1996).

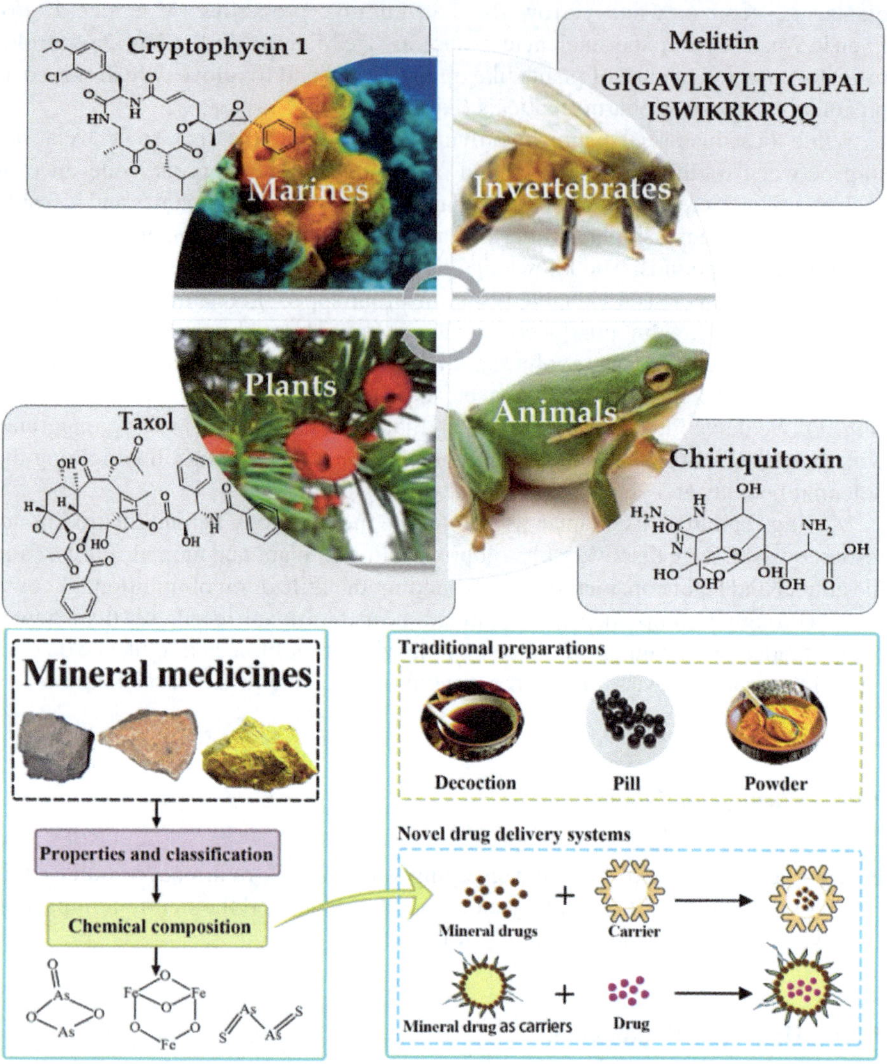

Fig. 1 Forging a synergy between contemporary and time-honored healing practices

1.2.2 Modern Drug Discovery

Our exploration will extend to the contemporary applications of traditional knowledge in drug discovery. This includes an examination of how traditional resources are being integrated into modern pharmaceutical research, offering insights into the development of novel therapeutic agents and potential breakthroughs in the field of medicine (Fabricant and Farnsworth 2001).

1.2.3 Methodology and Bioprospecting

We will discuss the methodologies employed in traditional resource-based drug discovery. This section will cover the techniques and approaches used to identify and extract bioactive compounds from traditional resources, shedding light on the scientific processes underpinning this innovative approach (Cragg and Newman 2005).

1.2.4 Ethical and Legal Considerations

An essential aspect of our discussion will revolve around the ethical and legal dimensions of utilizing traditional knowledge and resources in drug discovery. We will examine the challenges and responsibilities associated with respecting indigenous rights, protecting biodiversity, and ensuring fair and equitable sharing of benefits (Posey 1996).

1.2.5 Future Prospects

This chapter will conclude by addressing the future prospects of traditional resource-based drug discovery. By exploring current trends and ongoing research, we aim to provide readers with insights into the potential impact of traditional knowledge on the future of pharmaceuticals and health care (Heinrich et al. 2021).

In fulfilling these objectives, this chapter offers a comprehensive overview of the multifaceted influence of traditional resources on drug discovery, highlighting their historical importance and contemporary relevance. In an effort to bridge the gap between traditional and modern approaches, this study investigates the potential for traditional wisdom to enhance our understanding of modern science, ultimately benefiting society as a whole.

2 Traditional Medicine: Global Perspectives

2.1 Healing Practices in Different Cultures

Traditional medicine is a complex and diverse field deeply rooted in the cultural and ecological contexts of various societies. Understanding the nuances of these healing practices is essential for appreciating the full scope of traditional medicine's impact on drug discovery and modern health care.

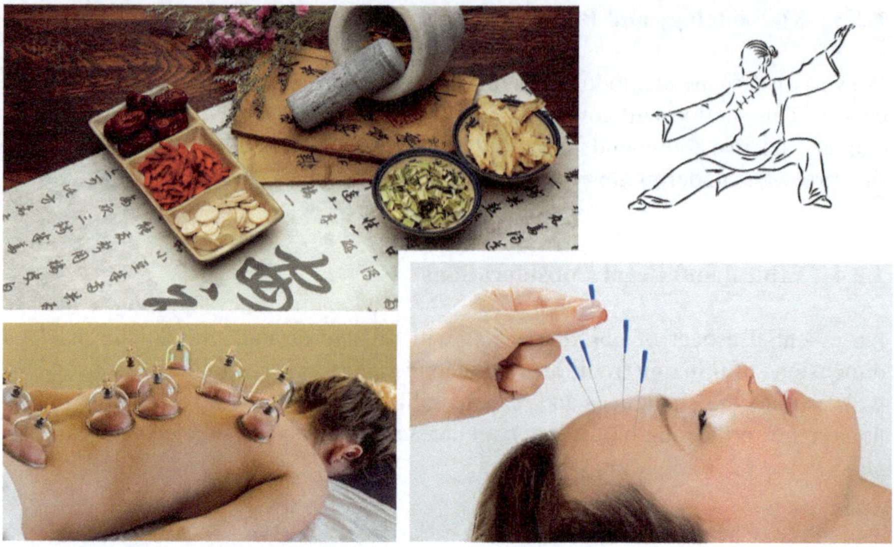

Fig. 2 Some traditional Chinese medicine (TCM)

2.1.1 Traditional Chinese Medicine (TCM)

Steeped in history, TCM stands as one of the world's most comprehensive traditional medical systems (Fig. 2). It revolves around the concept of balance, emphasizing the harmonious flow of vital energy, or Qi. TCM includes various modalities, such as the following:

- **Acupuncture**: In traditional Chinese medicine, it is a practice that restores balance in the body by inserting thin needles at specific locations.
- **Herbal Medicine**: A vast pharmacopeia of medicinal plants, including ginseng, astragalus, and licorice.
- **Cupping Therapy**: The use of heated glass cups to create suction on the skin, promoting circulation and alleviating various ailments.
- **Tai Chi and Qigong**: Mind-body practices that promote physical and mental well-being.

The integration of TCM into modern medicine and research has led to a deeper understanding of its components, with many herbal remedies gaining recognition for their potential therapeutic benefits (Bensky and Clavey 2015).

Fig. 3 Various aspects of ayurvedic tradition

2.1.2 Ayurveda

Ayurveda, the traditional medical system of India, is rooted in the concept of the three doshas—Vata, Pitta, and Kapha. It seeks to balance these life forces to maintain health and prevent disease (Fig. 3). Key components of Ayurveda include the following:

- **Herbal Medicine**: Utilizing an extensive range of herbs, such as turmeric, neem, and ashwagandha
- **Yoga and Meditation**: Practices that promote physical and mental well-being
- **Dietary Guidelines**: Prescribing specific diets to maintain doshic balance

Ayurvedic medicine's holistic approach has led to the exploration of its herbal remedies in modern pharmacology, with compounds like curcumin from turmeric gaining attention for their potential health benefits (Alam et al. 2023).

2.1.3 Native American Healing Traditions

Deeply connected to the land and the spiritual beliefs of Indigenous peoples (Fig. 4), Native American healing traditions incorporate various practices, including the following:

- **Herbal Medicine**: Utilizing local plants like echinacea, yarrow, and sage for their healing properties
- **Spiritual Ceremonies**: Rites and rituals that connect individuals with the spiritual world to promote healing

Fig. 4 Different types of Native American healing traditions

Fig. 5 Diverse varieties of African traditional healing practices

While these traditions have not been extensively studied in the context of modern drug discovery, they hold promise for unique therapeutic compounds and cultural preservation (Moerman 1996).

2.1.4 African Traditional Medicine

African traditional medicine is incredibly diverse, with practices varying from region to region (Fig. 5). Key aspects include the following:

Fig. 6 Diverse approaches within the Unani tradition of healing

- **Medicinal Plants**: A hallmark of traditional medicine is the use of a wide array of plants, animal parts, and minerals for therapeutic purposes.
- **Traditional Healers**: A central role played by healers in providing healthcare services and preserving indigenous knowledge.

The extensive knowledge of African traditional healers has prompted research into the therapeutic potential of these traditional remedies, particularly medicinal plants (Sofowora 1993).

2.1.5 Unani Medicine

Unani medicine has its roots in ancient Greece and was further developed in the Arab world. It follows a holistic approach to health, focusing on balancing the body's humors (Fig. 6). Unani medicine involves the following:

- **Herbal Remedies**: The use of herbs like licorice and black seed
- **Dietary Recommendations**: Suggesting dietary guidelines to maintain equilibrium within the body

Unani medicine's holistic approach has gained attention in the realm of natural and traditional medicine (Nisa and Kamili 2012).

This in-depth exploration of traditional healing practices in different cultures provides a foundation for understanding the rich heritage and potential of traditional resources in drug discovery and modern healthcare.

3 Scientific Basis of Traditional Resources

3.1 Active Compounds and Mechanisms

Understanding the mechanisms by which active compounds in traditional resources exert their therapeutic effects is vital for integrating traditional medicine into modern health care. Here, we explore the diverse mechanisms through which these compounds operate:

3.1.1 Plant-Derived Compounds

Curcumin

- **Anti-inflammatory**: Curcumin inhibits the activity of pro-inflammatory enzymes and signaling pathways, reducing inflammation (Shehzad et al. 2013).
- **Antioxidant**: With its free radical scavenging properties, it protects cells from the destructive consequences of oxidative damage (Nuge et al. 2022).
- **Immune Modulation**: Curcumin can modulate the immune response, potentially enhancing host defenses (Jagetia and Aggarwal 2007).

Artemisinin

- **Antimalarial Action**: Artemisinin disrupts the life cycle of the malaria parasite within red blood cells (O'Neill et al. 2010).
- **Selective Toxicity**: It exhibits greater toxicity to infected red blood cells than healthy ones, minimizing side effects (White 2008).

Quinine

- **Antimalarial Effect**: Quinine interferes with the parasite's ability to break down hemoglobin, essential for its survival (Phillips and Pasvol 1992).
- **Muscle Relaxant**: Quinine acts as a muscle relaxant, providing relief from cramps (Wise and Andrews 1996).

Morphine

- **Opioid Receptor Binding**: Through its interaction with opioid receptors distributed across the brain, spinal cord, and gastrointestinal tract, morphine effectively reduces pain perception (Inturrisi et al. 1987).
- **Analgesic Effect**: It modulates the perception of pain, providing relief (Gintzler 1982).

3.1.2 Animal-Derived Compounds

Hirudin

- **Anticoagulant Action**: Hirudin directly inhibits thrombin, an enzyme involved in blood clot formation (Stavrou and Schmaier 2003).

- **Blood Thinner**: It prevents excessive blood clotting and is used in various medical procedures (Hart and Gaffney 2011).

Conotoxins

- **Ion Channel Blockade**: Conotoxins target specific ion channels in nerve cells, affecting neurotransmission (Norton and Olivera 2006).
- **Pain Modulation**: Some conotoxins have analgesic properties and potential for pain management (Livett et al. 2004).

Snake Venom Compounds

- **Anticoagulant Activity**: Certain snake venom compounds can disrupt the blood clotting cascade, potentially preventing thrombosis (Kamiguti and Theakston 2003).
- **Neurological Effects**: Snake venoms contain neurotoxins that affect nerve cell function and could be of interest in neurology (Casewell et al. 2013).

3.1.3 Minerals and Inorganic Substances

Calcium Carbonate

- **Antacid Effect**: Calcium carbonate neutralizes stomach acid and provides relief from acid reflux and indigestion (Reimer and Smith 1998).

Sulfur

- **Antifungal Action**: Sulfur is effective against fungal infections by inhibiting fungal growth (Rosenthal 1975).
- **Keratolytic Properties**: It promotes the shedding of dead skin cells, aiding in the treatment of skin conditions (Scheman 2000).

Gold

- **Drug Delivery**: Gold nanoparticles can be used as drug carriers to enhance drug delivery and improve therapeutic outcomes (Dykman and Khlebtsov 2012).
- **Anti-inflammatory Potential**: Gold nanoparticles may modulate the immune response and inflammation (Bartneck et al. 2010).

Understanding these mechanisms allows researchers to explore the therapeutic potential of these traditional resources further and develop novel drugs based on their active compounds. This integration of traditional knowledge and modern science offers exciting prospects for advancing health care.

4 Integrating Traditional Resources into Modern Drug Discovery

4.1 Unlocking the Therapeutic Potential

The integration of traditional resources into modern drug discovery represents a dynamic and promising field of research. Traditional medicine, with its rich history and diverse array of remedies, holds the potential to yield novel treatments for a wide range of health conditions. To unlock this therapeutic potential, researchers engage in a multifaceted exploration of traditional resources.

4.1.1 Identifying Active Compounds

One of the primary steps in integrating traditional resources into modern drug discovery is the identification of active compounds within traditional remedies. Traditional healing practices often rely on a complex mixture of natural substances, such as herbs, roots, or animal-derived materials. Researchers aim to isolate and characterize the specific bioactive molecules responsible for the medicinal properties of these resources (Biswas et al. 2021).

By using various techniques, such as chromatography and mass spectrometry, scientists can pinpoint the active ingredients. For example, the compound curcumin, found in turmeric, has been isolated and studied for its anti-inflammatory and antioxidant properties, which have implications for various health conditions (Biswas et al. 2021).

4.1.2 Understanding Mechanisms

Understanding the mechanisms through which these active compounds exert their therapeutic effects is essential. This involves a comprehensive exploration of the molecular and physiological processes that underlie the impact of these compounds on the human body. Researchers seek to elucidate how these compounds interact with biological pathways and systems, ultimately leading to the observed health benefits (Biswas et al. 2021).

This knowledge is pivotal as it not only provides a scientific basis for the use of traditional resources but also informs the development of new drug candidates. For instance, artemisinin, derived from sweet wormwood, has a well-established mechanism of action against malaria parasites, making it a cornerstone in modern malaria treatment (Biswas et al. 2021).

4.1.3 Optimizing Formulations

Incorporating traditional remedies into modern health care often requires the development of standardized formulations and drug delivery systems. Traditional resources may have been used in diverse ways, such as teas, tinctures, or poultices. Researchers aim to optimize these preparations for modern use. This includes creating pharmaceutical dosage forms that enhance the bioavailability and efficacy of the active compounds while ensuring quality control throughout the production process (Biswas et al. 2021).

4.2 Enhancing Efficacy and Safety

While traditional resources offer a wealth of therapeutic potential, their integration into modern health care requires careful attention to efficacy and safety. Researchers address this challenge through several key strategies:

4.2.1 Standardizing Preparations

To ensure the consistent potency and quality of traditional resource-derived products, researchers work on the development of standardized formulations. This entails establishing rigorous quality control measures that guarantee the presence and concentration of active compounds within acceptable limits. Standardization ensures that patients receive a predictable and reliable dosage in each administration (World Health Organization 2003).

4.2.2 Toxicity Assessment

An essential aspect of enhancing the safety of traditional resource-derived drugs involves evaluating their toxicity. Researchers perform toxicological assessments, which involve evaluating potential adverse effects, are essential for establishing safe dosage levels. In cases where safety concerns arise, further research and modifications may be necessary to minimize side effects while preserving the therapeutic benefits of traditional remedies (World Health Organization 2003).

4.2.3 Clinical Trials

Conducting well-designed clinical trials is a critical step in establishing the safety and efficacy of traditional resource-derived drugs. Clinical trials allow researchers to gather robust data on the effectiveness of these remedies in treating specific health conditions and to monitor any potential side effects in a controlled environment.

Rigorous clinical trials provide the scientific evidence required for regulatory approval and widespread acceptance in mainstream health care (World Health Organization 2003).

The systematic enhancement of the efficacy and safety of traditional remedies not only ensures their reliability but also promotes their acceptance in modern healthcare systems. This integration bridges the gap between the wisdom of traditional medicine and the rigorous standards of contemporary medical practice, offering new avenues for innovative and effective treatments.

5 Impact on Society

5.1 Contemporary Use of Traditional Resources

Traditional resources continue to hold a significant place in modern society, playing a vital role in both health care and drug discovery. As of 2020, in a report by the World Health Organization (WHO), it is estimated that approximately 80% of the global population utilizes traditional medicine as their primary source of health care (World Health Organization 2020a). This reliance is particularly pronounced in low- and middle-income countries (LMICs), where access to modern health care often remains limited.

Beyond their direct use in health care, traditional resources have gained increasing importance in the realm of drug discovery. For instance, artemisinin, a potent antimalarial drug, was originally derived from the sweet wormwood plant, a staple in traditional Chinese medicine for centuries (Fabricant and Farnsworth 2001). Similarly, paclitaxel, a prominent cancer drug, has its origins in the Pacific yew tree, used in traditional Native American medicine (Fabricant and Farnsworth 2001).

The utilization of traditional resources in drug discovery brings forth several advantages. Firstly, these resources serve as a rich reservoir of novel chemical compounds with potential medicinal properties (Fabricant and Farnsworth 2001). Secondly, traditional knowledge systems provide invaluable insights into the potential therapeutic applications of these compounds (Fabricant and Farnsworth 2001). Thirdly, their use promotes equity and access to health care in LMICs, bridging gaps in healthcare accessibility (Fabricant and Farnsworth 2001).

Here are some noteworthy examples of the contemporary use of traditional resources in drug discovery:

- **Artemisinin**: Derived from the sweet wormwood plant, artemisinin is a powerful antimalarial drug that now forms the key ingredient in artemisinin-based combination therapies (ACTs), considered the gold standard for global malaria treatment (Fabricant and Farnsworth 2001).
- **Paclitaxel**: Originally sourced from the Pacific yew tree, paclitaxel is now one of the most widely used cancer drugs, with applications targeting a wide range of

cancers, including breast, ovarian, and lung cancers (Fabricant and Farnsworth 2001).

- **Vincristine and Vinblastine**: These two cancer drugs originate from the periwinkle plant, a traditional remedy in Europe and Asia. Vincristine and vinblastine are now instrumental in treating childhood cancers such as leukemia and lymphoma (Fabricant and Farnsworth 2001).
- **Etoposide**: Etoposide, a cancer drug, is derived from the mayapple plant, an integral component of traditional Native American medicine. It is utilized in cancer therapies targeting lymphoma, lung, and testicular cancers (Fabricant and Farnsworth 2001).

These examples showcase the remarkable contributions of traditional resources to the development of crucial modern drugs. As researchers delve deeper into the potential of traditional resources, we can anticipate the emergence of even more innovative medications in the years to come.

5.2 Societal Attitudes Toward Integrative Medicine

Integrative medicine, an approach to health care that amalgamates conventional and traditional medical practices, is gaining increasing popularity worldwide. People are seeking holistic and personalized approaches to their health, a trend reflected in the growing societal acceptance of integrative medicine (NCCIH 2018).

In 2018, a survey conducted by the National Center for Complementary and Integrative Health (NCCIH), a US government agency revealed that 68% of adults had utilized some form of complementary or alternative medicine (CAM) known as complementary and integrative health (CIH) previously (NCCIH 2018). Among CAM users, 44% had employed botanicals (plants or herbs) or dietary supplements, and 18% had undergone acupuncture (NCCIH 2018).

The burgeoning embrace of integrative medicine is indicative of the expanding recognition of the worth of traditional resources in health care (NCCIH 2018). It underscores the increasing demand for comprehensive and personalized health approaches that address not only physical symptoms but also the broader dimensions of wellness.

Several factors contribute to the escalating popularity of integrative medicine. One such factor is the mounting body of scientific evidence supporting traditional medicine practices (NCCIH 2018). Robust evidence now exists that confirms the effectiveness of acupuncture in treating chronic pain and highlights the efficacy of certain herbs and supplements in boosting the immune system and reducing inflammation (NCCIH 2018).

Another contributing factor is the growing availability of integrative medicine providers. Conventional healthcare practitioners are increasingly offering integrative medicine services, and a growing number of specialists in integrative medicine are emerging to meet the demand (NCCIH 2018).

The rising acceptance of integrative medicine holds promise for both individuals and society at large. Integrative medicine, by potentially improving health outcomes and reducing healthcare costs, this approach presents a positive development with far-reaching benefits (NCCIH 2018).

6 Challenges and Opportunities

While the use of traditional resources in drug discovery has many potential benefits, it also faces a number of challenges.

6.1 Standardization and Quality Control

One challenge is the standardization and quality control of traditional resources. Traditional resources can vary widely in their composition and potency, depending on a number of factors, such as the growing location, the time of year they are harvested, and the methods used to prepare them. This can make it difficult to develop reliable and reproducible drug discovery processes using traditional resources (World Health Organization 2022).

For example, a study published in the journal *Phytomedicine* found that the concentration of the active ingredient in a traditional herbal remedy for malaria varied by up to tenfold depending on the growing location of the plant. This variability can make it difficult to determine the correct dosage of the remedy and to ensure that it is safe and effective (World Health Organization 2022).

Another challenge is that traditional resources are often not cultivated in a controlled environment, which can make it difficult to ensure their quality and purity. For example, a study published in the journal *Drug Discovery Today* found that a traditional herbal remedy for cancer was contaminated with heavy metals. This contamination could potentially have serious health consequences for patients taking the remedy (World Health Organization 2022).

6.2 Sustainability and Ethical Considerations

Another challenge is the sustainability and ethical considerations of using traditional resources in drug discovery. It is important to ensure that the use of traditional resources does not lead to overexploitation or environmental damage. It is also important to respect the traditional knowledge of the communities that have used these resources for centuries (Heinrich and Jäger 2015).

For example, a study published in the journal *Conservation Biology* found that the overharvesting of wild medicinal plants was a major threat to biodiversity in some parts of the world. Additionally, a study published in the journal *Journal of Ethnopharmacology* found that some pharmaceutical companies were exploiting traditional knowledge without adequately compensating the communities that owned that knowledge (Heinrich and Jäger 2015).

Despite these challenges, the use of traditional resources in drug discovery has a number of promising opportunities.

Traditional resources can provide a rich source of new chemical compounds with potential medicinal properties. Highlighting the potential of traditional medicine, the antimalarial drug artemisinin was derived from the sweet wormwood plant, which has been used in traditional Chinese medicine for centuries. Similarly, the cancer drug paclitaxel was originally derived from the Pacific yew tree, which has been used in traditional Native American medicine (Ekor 2014).

Traditional knowledge systems can also provide valuable insights into the potential therapeutic uses of traditional resources. For example, traditional healers may know which plants are effective for treating specific diseases, and they may also know how to prepare and administer the plants in a way that is safe and effective (Ekor 2014).

Additionally, the use of traditional resources can help to promote equity and access to health care in LMICs. Traditional resources are often more affordable and accessible than conventional medicines, and they can be used to develop new drugs that are specifically tailored to the needs of LMICs. For example, a study published in the journal *The Lancet* found that a traditional herbal remedy was effective in treating malaria in children in Africa (Ekor 2014).

7 The Future of Traditional Resources

7.1 Collaborative Research and Knowledge Sharing

The future of traditional resources in health care and drug discovery holds promise and potential, underpinned by collaborative research and knowledge sharing.

7.1.1 A Collaborative Approach

Collaborative research that brings together traditional healers, local communities, scientists, and healthcare professionals has the potential to unlock new insights and innovations. By fostering an environment of mutual respect and understanding, it becomes possible to bridge the gap between traditional knowledge and modern science. Collaborative efforts can lead to the discovery of novel treatments and the development of standardized, safe, and effective healthcare products that integrate the wisdom of cultural heritage (World Health Organization 2014).

7.1.2 Knowledge Sharing

Efforts in knowledge sharing are pivotal for harnessing the wisdom of traditional medicine. Traditional healers, who possess a wealth of knowledge accumulated over generations, can work in partnership with researchers to document and share their expertise. This not only safeguards traditional knowledge but also contributes to the preservation of diverse healthcare practices. Digital platforms, ethnobotanical databases, and interdisciplinary conferences provide valuable channels for the exchange of information, benefiting both traditional practitioners and the broader scientific community (Leonti et al. 2020).

7.2 Potential Impact on Global Health

7.2.1 Addressing Healthcare Disparities

One of the remarkable opportunities that traditional resources offer is the potential to address healthcare disparities. In regions where access to modern health care is limited, traditional remedies can provide essential healthcare solutions. Integrating these resources into healthcare systems can lead to more equitable access to effective treatments, ultimately reducing disparities in health outcomes (World Health Organization 2020b).

7.2.2 Novel Drug Discovery

The integration of traditional resources into modern health care is leading to a renaissance in drug discovery. Traditional resources have been a source of inspiration for pharmaceutical research, offering a treasure trove of natural compounds with unique pharmacological properties. Here's a closer look at the potential impact of traditional resources on novel drug discovery:

7.2.2.1 Biodiversity as a Reservoir of Novel Compounds

Traditional resources, including plants, animals, and fungi, have evolved diverse chemical compounds over millions of years as part of their natural defense mechanisms. These compounds often possess remarkable bioactivity, which can be harnessed for therapeutic purposes. Scientists are increasingly turning to these natural reservoirs to discover novel drug candidates.

For instance, the rainforests of the Amazon, known for their unparalleled biodiversity, are a rich source of potential pharmaceuticals. Traditional Amazonian plant remedies, like the bark of the cinchona tree (Quina-Quina), led to the discovery of quinine, an essential treatment for malaria. Additionally, the venom of certain

Amazonian frogs contains compounds with powerful analgesic properties that may inspire future pain medications (Atanasov et al. 2021).

7.2.2.2 Bioprospecting and Drug Development

Bioprospecting is a systematic approach that involves the search for and development of new drugs or other products from natural sources, often guided by traditional knowledge. This approach has yielded several successful drugs (Ventola 2010). For instance:

- **Artemisinin**: In traditional Chinese medicine, sweet wormwood has a long history of use for treating fevers. This knowledge laid the foundation for the later discovery of artemisinin, a powerful antimalarial drug derived from the same plant. The identification and isolation of this compound from the plant revolutionized malaria treatment, leading to the development of artemisinin-based combination therapies (ACTs), which are now globally recognized as the gold standard for malaria treatment (Patwardhan and Mashelkar 2009).
- **Paclitaxel**: Derived from the Pacific yew tree, paclitaxel found applications in traditional Native American medicine. Widely used for its therapeutic effects on various cancers, this natural compound is particularly effective against breast, ovarian, and lung cancers (Patwardhan and Mashelkar 2009).
- **Vincristine and Vinblastine**: These cancer drugs, obtained from the periwinkle plant, have roots in European and Asian traditional medicine. Today, they are essential for the treatment of childhood cancers like leukemia and lymphoma (Patwardhan and Mashelkar 2009).

7.2.2.3 Understanding Mechanisms of Action

Traditional resources often come with an array of active compounds, each with its own mechanisms of action. These mechanisms can be explored and understood to develop novel drugs.

For example, curcumin, a compound found in turmeric, has antioxidant and anti-inflammatory properties, which have applications in various health conditions. The mechanism of how curcumin interacts with cellular pathways has inspired researchers to develop potential treatments for conditions like Alzheimer's disease and certain cancers (Patwardhan and Mashelkar 2009).

7.2.2.4 Combining Traditional Wisdom with Modern Science

Collaborations between traditional practitioners and modern scientists have the potential to uncover new drug candidates. By sharing their traditional knowledge and experiences, practitioners can guide researchers to resources with promising therapeutic properties. This collaborative approach not only respects traditional practices but also fast-tracks the discovery of new drugs.

The impact of traditional resources on novel drug discovery is ongoing and promising. As scientists continue to explore the vast diversity of traditional remedies, we can expect a steady stream of new and innovative drugs that have their roots in age-old healing traditions (Patwardhan and Mashelkar 2009).

7.2.3 Holistic Health Care

The integration of traditional resources aligns with the growing demand for holistic healthcare approaches. Beyond the treatment of specific ailments, traditional resources have the potential to foster a more integrative understanding of health, encompassing both conventional and complementary approaches. By combining traditional and modern practices, healthcare systems can offer a more holistic approach that addresses the physical, mental, emotional, and spiritual dimensions of well-being (Cragg and Newman 2013).

8 Conclusions and Future Directions

In the realm of drug discovery and health care, the influence of traditional resources is undeniable, extending from the annals of history to the forefront of modern science. This chapter has charted the significance of these resources, explored their scope, and illuminated their potential to shape our shared future.

8.1 Key Takeaways

- **Historical Significance**: Traditional resources, whether drawn from plants, animals, or minerals, have been central to the healing practices of diverse cultures throughout time. They've gifted us vital medicines, like quinine, paclitaxel, and artemisinin.
- **Scope and Objectives**: The incorporation of traditional resources into contemporary drug discovery is a multifaceted endeavor. It encompasses the exploration of new compounds, comprehension of their mechanisms of action, and the fusion of ancient wisdom with modern science to craft safer and more efficacious treatments.
- **Scientific Basis**: The bedrock of traditional resources is constituted by active compounds, each with their unique mechanism of action. These compounds have been the wellspring of numerous drugs and hold the promise of shaping the future of pharmacology.

- **Integrating Traditional Resources**: The utilization of traditional resources in modern drug development not only unlocks therapeutic potential but also elevates efficacy and safety standards. It presents opportunities for forging innovative drugs deeply rooted in traditional wisdom.
- **Impact on Society**: Despite their long history, traditional resources continue to be remarkably relevant in the present day. An estimated 80% of the world's population, especially in low- and middle-income countries (LMICs), utilize traditional medicine to address their primary healthcare needs. The ascent of integrative medicine, blending conventional and traditional modalities, reflects shifting societal attitudes.
- **Challenges and Opportunities**: The journey of assimilating traditional resources into modern health care is not devoid of challenges, including standardization, quality control, sustainability, and ethical considerations. However, these challenges are also gateways to scientific innovation and ethical practices.
- **The Future of Traditional Resources**: The future landscape of traditional resources is characterized by collaborative research, knowledge sharing, and the potential to revolutionize global health. This collaborative paradigm respects tradition while striving for healthcare equity and the creation of novel therapeutic candidates.

8.2 Future Directions

- **Research Collaborations**: Fostering collaborations between traditional healers, local communities, scientists, and healthcare professionals promises to unlock new frontiers in drug discovery. The foundation of such collaboration should rest on mutual respect and cross-cultural understanding.
- **Knowledge Sharing**: Preserving traditional wisdom is imperative. Initiatives aimed at documenting and sharing this knowledge will ensure its safeguarding for posterity. Digital platforms and interdisciplinary exchanges offer valuable avenues for knowledge dissemination.
- **Holistic Health Care**: The growing acceptance of integrative medicine, embracing both conventional and traditional practices, is a positive trend. It warrants support as it offers a holistic approach to health care that addresses not only physical but also mental, emotional, and spiritual well-being.
- **Ethical and Sustainable Practices**: Confronting the challenges of standardization, quality control, and the sustainability of traditional resources should be central. Ethical practices that honor cultural heritage and benefit local communities are paramount.
- **Innovative Drug Discovery**: The potential of traditional resources for drug discovery is far from exhausted. Researchers should persist in exploring the diverse compounds, their mechanisms of action, and the potential of collaborative approaches to craft groundbreaking drugs that carry the legacy of tradition into the future.

In summation, traditional resources wield an enduring and transformative influence on health care and drug discovery. Their integration into the fabric of modern medicine holds the promise of a future where health care is equitable, safe, and effective for all. As we advance, let us embrace these resources and their potential to shape a healthier world for generations to come.

References

Alam MR, Alimuzzaman S, Shahid MA, Karim F-E, Hoque ME (2023) Collagen/Nigella sativa/chitosan inscribed electrospun hybrid bio-nanocomposites for skin tissue engineering. J Biomater Sci Polym Ed 16:1–22. https://doi.org/10.1080/09205063.2023.2170139

Ali ME, Hoque ME, Hossain SS, Biswas MC (2020) Nanoadsorbents for wastewater treatment—next generation biotechnological solution. Int J Environ Sci Technol 17(9):4095–4132. https://doi.org/10.1007/s13762-020-02755-4

Atanasov AG et al (2021) Discovery and resupply of pharmacologically active plant-derived natural products: a review. Biotechnol Adv 39:107463

Balick MJ, Cox PA (1996) Plants, people, and culture: the science of ethnobotany. Scientific American Library, New York

Bartneck M et al (2010) Therapeutic and diagnostic applications of gold nanoparticles. Nanotechnol Sci Appl 3:1–10

Bensky D, Clavey S (2015) Chinese herbal medicine: Materia Medica. Eastland Press, Seattle

Biswas MC, Jony B, Nandy PK, Chowdhury RA, Halder S, Kumar D, Ramakrishna S, Hassan M, Ahsan MA, Hoque ME, Imam MA (2021) Recent advancement of biopolymers and their potential biomedical applications. J Polym Environ 30:51. https://doi.org/10.1007/s10924-021-02199-y

Casewell NR et al (2013) Solenodon paradox: a review of recent research. Open Quat 3(1):1–20

Cragg GM, Newman DJ (2005) Biodiversity: a continuing source of novel drug leads. Pure Appl Chem 77(1):7–24

Cragg GM, Newman DJ (2013) Natural products: a continuing source of novel drug leads. Biochim Biophys Acta Gen Subj 1830(6):3670–3695

Dykman LA, Khlebtsov NG (2012) Gold nanoparticles in biomedical applications: recent advances and perspectives. Chem Soc Rev 41(6):2256–2282

Ekor M (2014) The growing importance of traditional medicine in the management of infectious diseases. Expert Rev Anti-Infect Ther 12(5):601–610

Fabricant DS, Farnsworth NR (2001) The value of plants used in traditional medicine for drug discovery. Environ Health Perspect 109(Suppl 1):69–75

Gintzler AR (1982) Endorphin-mediated increase in the pain threshold elicited by intraventricular morphine in the rat. Eur J Pharmacol 86(2):179–180

Hart A, Gaffney A (2011) On the relevance of hirudin in contemporary medicine. J Am Acad Dermatol 65(2):467–468

Heinrich M, Edwards S (2007) Ethnopharmacological field studies: a critical assessment of their concept and methods. J Ethnopharmacol 124(1):1–17

Heinrich M, Jäger AK (2015) Traditional knowledge in drug discovery: the role of the WHO traditional medicine strategy 2014–2023. Ethnopharmacol 176:578–590

Heinrich M et al (2021) Traditional medicine—challenges and opportunities for new drug discovery. Medicines 8(8):45

Inturrisi CE et al (1987) Pharmacokinetics and pharmacodynamics of methadone in patients with chronic pain. Clin Pharmacol Ther 41(4):392–401

Jagetia GC, Aggarwal BB (2007) "Spicing up" of the immune system by curcumin. J Clin Immunol 27(1):19–35

Kamiguti AS, Theakston RD (2003) Snake venom components: their characteristics, interactions with platelets, and involvement in haemostasis. Toxicon 42(9):969–985
Leonti M et al (2020) Ethnopharmacology of medicinal plants used in North Kivu province (DR Congo). J Ethnopharmacol 251:112529
Lin J, Zhu L (2002) Historical perspective of traditional indigenous medical practices: the current renaissance and conservation of herbal resources. Environ Health Perspect 110(1):119–125
Livett BG et al (2004) Therapeutic applications of conotoxins that target the neuronal nicotinic acetylcholine receptor. Toxicon 44(3):259–274
Moerman DE (1996) An analysis of the food plants and drug plants of native North America. J Ethnopharmacol 52(1):1–22
NCCIH (2018) 2018 National Health Interview Survey: adult use of complementary health approaches. National Center for Complementary and Integrative Health, Bethesda, MD
Nisa SM, Kamili AN (2012) The Unani system of medicine: an analytical study. Res Rev J Bot Sci 1(2):22–30
Norton RS, Olivera BM (2006) Conotoxins down under. Toxicon 48(7):780–798
Nuge T, Liu X, Tshai KY, Lim SS, Nordin N, Hoque ME, Liu Z (2022) Accelerated wound closure: systematic evaluation of cellulose acetate effects on biologically active molecules release from amniotic fluid stem cells. Biotechnol Appl Biochem 69(3):906–919. https://doi.org/10.1002/bab.2162
O'Neill PM et al (2010) New antimalarials: a nature-derived quinine hybrid. Angew Chem Int Ed 49(17):3297–3299
Patwardhan B, Mashelkar RA (2009) Traditional medicine-inspired approaches to drug discovery: can Ayurveda show the way forward? Drug Discov Today 14(15–16):804–811
Phillips E, Pasvol G (1992) Anaemia of Plasmodium falciparum malaria. Baillieres Clin Haematol 5(2):315–330
Posey DA (1996) Cultural and spiritual values of biodiversity: a complementary contribution to the global biodiversity assessment. Intermediate Technology Publications
Reimer C, Smith RE (1998) Calcium carbonate antacid: comparison of acid neutralization capacity. J Am Pharm Assoc 38(6):767–769
Rosenthal SS (1975) Sulfur: its clinical uses. Am Fam Physician 11(5):102–106
Scheman AJ (2000) Topical treatment of common superficial tinea infections. Am Fam Physician 61(8):2378–2386
Sharma P (2014) Indigenous and traditional herbal medicine: ethnopharmacological importance. J Pharmacogn Phytochem 3(4):144–152
Shehzad A et al (2013) Curcumin: a potent modulator of multiple enzymes in multiple cancers. J Cell Biochem 114(1):21–30
Sofowora A (1993) Medicinal plants and traditional medicine in Africa. Wiley, Hoboken
Stavrou EX, Schmaier AH (2003) Factor XII: what does it contribute to our understanding of the physiology and pathophysiology of hemostasis & thrombosis. Thromb Res 112(5–6):321–327
Ventola CL (2010) The antibiotic resistance crisis: part 1: causes and threats. Pharm Ther 40(4):277–283
White NJ (2008) Qinghaosu (artemisinin): the Price of success. Science 320(5874):330–334
Wise R, Andrews JM (1996) The in vitro activity of quinine against erythrocytic and exoerythrocytic stages of Plasmodium falciparum. Trans R Soc Trop Med Hyg 90(4):429–431
World Health Organization (2003) Guidelines on safety monitoring of herbal medicines in pharmacovigilance systems. World Health Organization, Geneva
World Health Organization (2014) WHO traditional medicine strategy 2014–2023. World Health Organization, Geneva
World Health Organization (2020a) Global traditional and complementary medicine atlas. World Health Organization, Geneva
World Health Organization (2020b) Health equity through intersectoral action: an analysis of 18 country case studies. World Health Organization, Geneva
World Health Organization (2022) Traditional medicine: definitions, terminology and general considerations. World Health Organization, Geneva

Different Tools for Modern Drug Discovery Research

Srinivasan Kameswaran, B. Swapna, Mandala Ramakrishna, Bellamkonda Ramesh, and Manjunatha Bangeppagari

Contents

Abstract The least amount of pharmaceuticals that complete clinical studies makes it to market. Drug discovery, which includes drug creation and development, is a complex and expensive endeavor. Over the past three decades, technologies for finding and developing drugs using software played a significant role in the production of biologically active molecules. The pharmacokinetic and pharmacodynamic properties of drugs, as well as the structural interaction between the ligand and its target that determines how well they work together, can all be examined with the

S. Kameswaran (✉) · B. Swapna · M. Ramakrishna
Department of Botany, Vikrama Simhapuri University College, Kavali, Andhra Pradesh, India

B. Ramesh
Department of Internal Medicine, University of Nebraska Medical Center, Omaha, NE, USA

M. Bangeppagari
Department of Cell Biology and Molecular Genetics, Sri Devaraj Urs Academy of Higher Education and Research, Kolar, Karnataka, India

© The Author(s), under exclusive license to Springer Nature Singapore Pte Ltd. 2024

A. Das Talukdar et al. (eds.), *Traditional Resources and Tools for Modern Drug Discovery*, Interdisciplinary Biotechnological Advances,
https://doi.org/10.1007/978-981-97-4600-2_3

help of cutting-edge software-based techniques such as molecular modeling, "structure-based" online testing, a molecule interaction, "structure-based" drug design, and molecular simulation. Through the interactions of tiny compounds with structural macromolecules, computational techniques like docking enable hit detection and lead improvement. These techniques are quicker and more precise, and they offer insightful information on the results and workings of experiments. Additionally, the effective application of these strategies can lower the price of medication developing and development. These software are currently playing a crucial part in the various stages of drug development in the biomedical sciences. The most popular software for medication development and design is discussed in this chapter along with its successful uses.

Keywords Drug discovery · Pharmacokinetic · Docking · Drug designing · Pharmacodynamic · Bioactive compounds

1 Introduction

To find and develop new bioactive pharmaceuticals for a variety of therapeutic indications, the pharmaceutical industry now heavily relies on the use of software and model-based techniques. The process of finding new drugs has sped up, and various obstacles have been lessened with the proper usage of software and computer-based current technologies (Fig. 1). Modern medicinal chemistry techniques, like "molecular modeling," "structure-based drug design," virtual screening using structure, "ligand-based" modeling, and "molecular dynamics," offer helpful "tools" for understanding the pharmacokinetic and pharmacodynamic properties, as well as the structural activity relationship between ligands and their targets (Ferreira et al. 2015; Sliwoski et al. 2014). The use of simulations and modeling, like "ligand-based computer-assisted drug designing" ("CADD"), offers an effective model for modern therapeutic research conception and evaluation (Kimko and Pinheiro 2015; Gill et al. 2016). By using these methods, fewer animals will be required for research and preclinical drug discovery phases, handling large amounts of data will be easier, and study results will be more accurate (Hodos et al. 2016; Liu et al. 2016).

A single new drug's successful discovery, development, and introduction onto the market costs over $1 billion and takes almost 12 years to complete. The main challenges in developing new medications include prohibitively expensive, time-consuming, and inadequate risk, results-ambiguous, and using incredibly complicated processes. To address these challenges, novel and more cost-effective approaches to drug design and discovery—such as "molecular docking," software, and machine-learning drug design—must be used (Hodos et al. 2016; Wasko et al. 2015). This evaluation focuses on frequently used software and its possible applications in the field of novel medication development (Fig. 2).

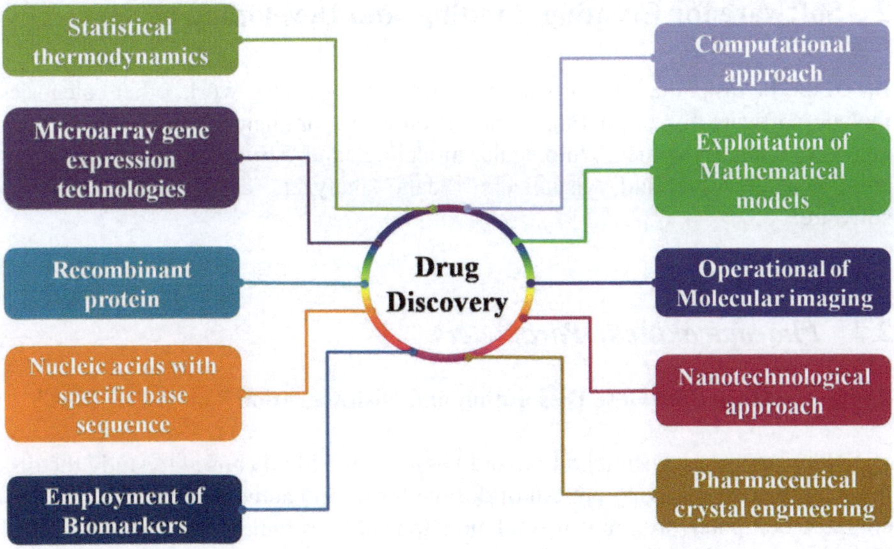

Fig. 1 Modern methods for finding new drugs

Fig. 2 Novel methods and approaches for drug discovery

2 Software for Creating, Finding, and Developing Drugs

Based on the tasks that the software performs and how they work, other categories include software for evaluating "pharmacological parameters," drug connections and "molecular dynamics," "molecular modeling" and "structural activity relationship," visualizations and "visualizers," "data" analyzers, and behavior "analysis software."

2.1 *Pharmacokinetic Parameters*

2.1.1 "Software for Dose Dissolution and Disintegration," or "DDD Plus"

DDDPlus ("Dosage Disintegration and Dissolution" Plus) is used to study the disintegrating and dissolving pattern of dosage forms and active compounds. Using a complex computer program, formulation specialists simulate the dissolution and disintegration of additives and "API," or "active pharmaceutical ingredients," in vitro under a range of experimental conditions. A single calibration experiment is often needed when formulating a new API. The rate of dissolution is then predicted by DDDPlus based on changes to the formulation or the testing setup. With this program, you may precisely determine the rate of dissolution and disintegration, eliminating the need for conventional "cut" and "try" methods to finish the composition model (Almukainzi et al. 2015; Uebbing et al. 2017; Kollipara et al. 2023).

With DDDPlus, you may choose from five different dosage forms and five different mathematical models that show how a chemical dissolves. The computational approaches used in the in vitro dissolving modeling describe the impact of the following parameters on dissolution:

- Physical and chemical characteristics of the formulation components being examined, including density, solubility, diffusion coefficient, and pKa values
- Producing qualities for dosage formulations with instantaneous release
- The distribution of particle sizes for every ingredient in the mixture
- Various fluid velocities and flow patterns for every experimental equipment
- The interaction between a mixture of ingredients and the drug's active component
- The reliance of the solubility, collapse, and "precipitation" on the pH of the microenvironment

The dissolution of micelles is aided by the addition of surfactants to the media.

2.1.1.1 Applications

- Uses the instrument speed and kind of apparatus to autonomously determine the liquid movement.
- DDDPlus features an optimization module that uses a single experimental data set to calibrate the dissolution rate of a medication.

2.1.2 Drug Development and Discovery Simulation Program, or GastroPlus

GastroPlus is software designed to simulate many absorption routes, including "intramuscular," "oral," "oral cavity," ocular, "intranasal," and "pulmonary," as well as "pharmacokinetic" and "pharmacodynamic" effects in animals and people. Its foundation is mechanistic theory.

It is possible to fit the model parameters to the data for one record at a time or for several records at once. Every time a record is entered, the program does a single simulation and modifies one or more model parameter values. Typically, for every hundred iterations, N simulations will be executed, where N is the total amount of data whose outcomes are compared to the measured results. The most popular weighting strategies are included in user-defined objective function weighting (Honório et al. 2013).

2.1.2.1 Applications

- Interactions between drugs based on transporters.
- Induction of metabolism and/or transporters.
- Associated with the "1-ranked dissolution/absorption (ACAT) model" in the industry.
- When appropriate, use physiological-based pharmacology or "compartmental pharmacokinetics."
- Employ competitively along with dependent on-time inhibitory kinetics for the "parent and/or metabolite."
- Either "simulate DDIs" for humans, beagles, rats, mice, rhesus monkeys, cynomolgus monkeys, rabbits, cats, or any other species.
- Take into account variations in the levels of enzyme expression in different populations.
- An integrated tool for quickly calculating the portion of in vitro experiments that are metabolized.

2.1.3 MapCheck

The results of the ion chamber are compared with the absolute dosage measurements of both systems using MapCheck. It contrasts the Portal Dosimetry from Varian and the MapCheck IMRT QA procedure from Sunnuclear.

The MapCheck system calibrates the diode array, generates an approval strategy for every area, and exports the estimated dose "map" ("Frontal") for every area before collecting data. Plan complexity raises the standard deviation. The average measured dose is not affected by the complexity of the strategy. It is an easy-to-use data analysis program that facilitates the commissioning process and produces thorough reports (Jursinic et al. 2010).

2.1.3.1 Applications

- MapCheck is utilized for verifying IMRT.
- MLC is detected by tiny sensors.
- The use of MapCheck for dose-based EPID IMRT quality assurance.

2.2 Interactions Between Ligands and Molecular Dynamics

2.2.1 AutoDock

An automated tool called AutoDock is used to forecast interactions between ligands and proteins, or biomacromolecular targets. Bimolecular X-ray crystallography is constantly developing, which aids in the provision of structural data for complex biomolecules including proteins along with genetic materials. In addition to being utilized as goals for innovative medications to treat ailments in people, animals, and plants, such frameworks may also be employed to understand fundamental biological principles.

There are several steps involved in AutoDock calculations:

- Using AutoDock tools to prepare coordinate files
- Atomic affinities are precalculated with AutoGrid
- Using AutoDock for ligand docking
- Utilizing AutoDock Tools for outcome analysis

2.2.1.1 Organize the Preparation of the File

AutoDock4.2 has optimized the protein and ligand model to include polarized "hydrogen atoms" rather than "hydrogen atoms" bonded to "carbon atoms." An enhanced "PDB format" called "PDBQT" is used to store collaborative data, including atom charge partials and atomic types. Currently, the "AutoDock force field" for the most common atoms uses multiple atomic kinds: various varieties of "aliphatic" and "aromatic carbon atoms," and a variety of "polar atoms" that form bonds with hydrogen as well as the ones that do not. The "PDBQT files" additionally contain the "torsional degrees of freedom." When some side chains in the peptide are thought to be flexible, a separate "PDBQT file" is additionally created for those chain locations. With "AutoDock Tools," the visual user experience for "AutoDock," and "PDBQT" files can be generated from standard PDB files.

2.2.1.2 AutoGrid Calculation

Precalculated "atomic affinity" possibilities are used to forecast fast energy evaluation for each type of atom throughout the ligands component getting coupled. Within this three-dimensional protein-embedded grid, an inspection atom appears at every "grid point." An AutoGrid affinity grid has been constructed for each kind of element inside the molecule; these networks primarily comprise electromagnetic and dissolution possibilities as well as carbon, oxygen, nitrogen, and hydrogen. The AutoDock computation uses the grid values to evaluate the kinetics of a particular "ligand" combination.

2.2.1.3 "Using AutoDock for Docking"

For docking, there are numerous techniques. One of the most effective techniques among them is the Lamarckian genetic algorithm (LGA). Genetic algorithms and simulated annealing traditionally accomplish docking. The optimal solution is found by combining the study of the anticipated energy and consistency of data obtained from running AutoDock multiple times, which produces multiple docked conformations for typical systems.

2.2.1.4 AutoDock Tools Are Used for Analysis

Tools for "visualizing confirmations," "ligand-protein interactions," "affinity potentials" produced by "AutoGrid," and grouping results based on conformational similarity are all included in the AutoDock Tools suite of methods for evaluating the outcomes of docking simulations (Rizvi et al. 2013).

2.2.1.5 Applications

- Aromatic ring identification.
- This applies to the investigation of a flexible ligand's conformational states; the ligand-protein interaction is assessed at every stage of the docking simulation using the maps produced by AutoGrid.

2.2.2 Schrodinger

Most of the problems these biomolecules will present can be resolved by using the extensive variety of applications available in Schrodinger software. It focuses on specific developments in biologics, ligand-receptor docking, molecular dynamics, and molecular modeling that were made to address these issues. The "hydrophobicity" and change in conformation of structure can be analyzed with this software, along with other structure-based properties of molecules.

Macrocysteine confirmation for biomolecular systems is carried out using an advanced virtual reality system which includes rapidity and precision. This program provides information about the atomic movements of macrocycles, which is utilized to gain a deeper understanding of energies, form, and durability. For setting up systems, executing simulations, and examining trajectories, Schrodinger offers robust and user-friendly graphical interfaces.

Molecular dynamics simulation software is used to study an array of stabilized folded "α-helical peptides" at various temperatures. The circular dichroism melting curves seen experimentally and the expected α-helical propensities generated from the simulations agreed well. Different stapled peptide binding affinities to MDM2 may be associated with the local flexibility of critical residues. These simulations investigate novel methods for creating "α-helical folded peptides" and creating strong inhibitors of α-helical protein-protein interactions (Kumar et al. 2010).

2.2.2.1 Applications

- Studies using molecular dynamics simulations
- Quantum mechanics
- Estimating binding affinity

2.2.3 Genomics-Oriented "Ligand Docking" or "GOLD"

"GOLD" ("Genetic Optimisation for Ligand Docking") is a genetic approach that makes it easier to dock adaptable ligands with a substance that possesses dynamic "hydroxyl groups." The Cambridge Structural Database's favorable conformations and empirical findings about weak chemical interactions serve as the foundation for the score function used by this software. The balance between GOLD's speed and the accuracy of its forecasts are managed by varying the genetic algorithm's parameter settings. It provides accurate atom typing for both ligands and proteins with dependable findings.

Gold is a component of the GOLD Suite program, which also consists of Hermes and GoldMine. GOLD offers all the capability needed to dock ligands into protein binding sites from prepared input files. The docking results are visualized, descriptors are calculated, and input files are prepared for docking with GOLD using the Hermes visualization. Protein residues' tautomeric states and appropriate ionization can be defined by adding hydrogen atoms, which can be retrieved from Hermes along with other input data. Additionally, the binding site and constraint settings are defined using the Hermes visualizer in the interactive docking setup. An analytical and post-processing tool for docking findings is called Gold Mine. The creation and editing of initial models will probably be done using GOLD in tandem with a modeling program (Yurieva et al. 2011).

2.2.3.1 Applications

- The genetic algorithm is employed for protein-ligand docking.
- Predicting binding modes.

2.2.4 BioSuite

Algorithms, chemoinformatics, macromolecular sequence analysis, and structural analysis are all used by BioSuite to support drug discovery. Across a large range of "bioinformatics" usage, it provides most of the few all-inclusive packages available. There are 79 distinct programs inside each of the four major sections that comprise it. "Molecular Dynamics Simulations," "Drug Design," "3D Modelling Structural Analysis," and "Genome and Proteome Sequence Analysis" are four of the core parts. Each has an intuitive graphics-user interface, ample documentation, and tutorials.

The BioSuite "Genome and Proteome Sequence Analysis module" handles applications for analyzing protein and nucleic acid sequences, including those of entire genomes and proteomes as well as single molecules. This module would make it possible to annotate genomes, forecast the secondary structures of proteins, determine the evolutionary relationships between animals, and compare two genomes to look for gene or protein similarities.

Building, analyzing, and forecasting the "3D structures of macromolecules" and "macromolecular complexes" are all possible with the 3D modeling and analysis module. The "Simulations" package successfully simulates a molecule's behavior in terms of its 3D form (NMITLI-BioSuite Team 2007; Bhattacharya et al. 2023). The following structural functionalities are available through the drug design module:

(a) To forecast the biological functioning of unidentified compounds, use QSAR.
(b) Discovery of pharmacophores with physiologically interacting compounds.
(c) Molecular set superimposition in three dimensions using alignment.
(d) Docking is used to identify the ligand's posture in three dimensions when it binds to a target.

2.2.4.1 Applications

- Studying sequences and genomes.
- Comparative genomics, 3D modeling, simulation, structural alterations, drug design, route modeling, and SNP analysis.

2.3 Structure-Activity Connection and Molecular Modeling

2.3.1 Maestro

A fully functional molecular visualization program is available for free called Maestro. For managing, sharing, and analyzing the outcomes of computer experiments, Maestro is an effective tool. Creating, displaying, and exchanging three-dimensional chemical models are made easier with its aid.

The foundation of Schrodinger's computational technology is Maestro. It is an effective and adaptable tool for computational chemistry molecular modeling. It oversees the gathering, organizing, and analysis of data. Calculations are simple and easy to set up because of Maestro's user-friendly interface. Projects requiring additional research automatically receive the computed results back. Gaining knowledge of detailed intermolecular interactions and molecular characteristics is made feasible by Maestro's extensive visualization options (Sabitha and Habeeb 2011; Hussain et al. 2023).

2.3.1.1 Applications

- Analysis of structures quantitatively.
- Molecular properties, electron density, and vibrational modes can all be visualized.

2.3.2 ArgusLab

The genetic modeling, imagery, and formulation program "ArgusLab" are available for Windows operating systems. Using an MS window-operated system PC, a kinematic study, like a geometrical optimization study, was performed using ArgusLab.

This program helps with the calculation of possible energy sources, molecular shapes, geometrical improvement, "atom coordinate vibration frequencies," "bond length," "bond angle," and "reaction pathways." It is based on the ideas of quantum mechanics. ArgusLab uses the geometry convergence function to get the least potential energy (Naz et al. 2009; Mathew et al. 2012).

2.3.2.1 Applications

- Calculations for molecular docking.
- It is utilized in the building of compounds.
- Molecules are constructed utilizing a model design.
- For the package of molecular modeling.

2.3.3 "Global Range Molecular Matching" or "GRAMM"

Proteins docking are performed using GRAMM software. By utilizing the two molecules' atomic coordinates, it makes structural predictions. It generates a list of high-score (low-energy) ligand locations that can be used directly or subsequently processed using different methods. Instead of using statistical sampling, this software conducts a thorough search to find any complex configuration that has a high-score steric fit.

Through the molecules' relative translations and rotations, this software does a thorough six-dimensional search. The molecular pairings could consist of two trans-membrane helices, two proteins, or a protein plus a smaller molecule. When there are significant conformational changes, it is employed to obtain high-resolution molecules for erroneous structures.

The intermolecular energy function is empirically smoothed using this method by varying an array of "atom-to-atom" abilities. The method determines the region with the world's lowest level of "intermolecular energy" for structures with different levels of precision. Prediction accuracy determines prediction quality. "High-resolution" models with few structural modifications can be docked to generate accurate predictions, as "ultralow-resolution" models will only generate the overall properties of the complex (Kundrotas and Vakser 2010; Devi and Mikilli 2013).

2.3.3.1 Applications

- It is employed in protein-ligand and protein-protein docking.

2.3.4 "SYBYL-X Suite"

"SYBYL-X" offers knowledge to help comprehend and assess different SARs to every one of the requirements that a suitable medicine candidate must meet. Using the analytical powers of the novel "Molecular Data Explorer" ("MDE") in "SYBYL-X," it quickly gains insights into data by examining and visualizing connections between various properties. The unmatched is extended while offering fresh perspectives on life science molecular discovery efforts.

SYBYL-X explores multiple aspects of the drug interaction mechanism with its receptor, hoping to uncover novel interaction mechanisms that could result in step leaps in potency or ways to enhance ADME or physical qualities without disrupting critical receptor connections (Dubey and Kalra 2013; Xu et al. 2014, 2023).

SYBYL-X enables anyone to:

- Create a three-dimensional homologous or skeletal model of the intended receptor.
- Define the features of the target protein's cavities and the surface where the protein interacts with the ligand.
- Use the docking software Surflex-Dock to anticipate and rationalize possible drug interactions with its receptor.
- Build libraries of snippets so that you may virtually filter them to locate useful bits. After such pieces are identified, either construct forms out of them or build a database of forms that can be inserted into an immune system aperture.
- To identify possible lead candidates, use "Surflex Dock" to virtually scan data that comprise proprietary or widely accessible compounds.

2.3.4.1 Applications

- Lead optimization to molecular modeling from the sequence.
- Design based on ligands.
- Design based on structure.
- A protein model can also be constructed using the SYBYL-X Suite.

2.3.5 "Sanjeevini"

The goal of this computer programming project is to provide an automated computing path for result designing. It makes use of a candidate medication and a bimolecular (protein) target. Potential active sites are located, docked, and scored, and four alternative medicine structures attached to peptide targets in addition to attaching liberated energies are returned using the software.

This software is used to upload the target protein and pharmaceutical molecule. The software opens a window with the results of several crucial pretests that were conducted based on the specifications required for the medication and protein files to be in an acceptable format when the file is uploaded. The software includes modules for "protein-ligand" complex synthesis, drug formulation, docks and assessing, and other functions (Jayaram et al. 2012; Singh et al. 2021; Alavi and Sharma 2023).

2.3.5.1 Applications

- It serves as a tool for creating drugs.
- It forecasts the affinity for binding.
- Estimating the affinity of protein-ligand binding.

2.3.6 Prediction of a Substance's Activity Spectrum, or PSAS

By comparing a database of existing structures, this software forecasts the potential biological activities of a novel pharmacological compound including a lead molecule. Averaging almost 95% of its predictions, PSAS predicts 4366 distinct kinds of biological activity. The biological activity prediction in PSAS is described in a qualitative "yes or no" or "active or inactive" manner. To find possible biological activities, a novel compound's structure is converted into two-dimensional structural equations. The molecular structure in PSAS is represented by the set of unique "MNA descriptors." Compounds with an identical array of "MNA descriptors" are considered similar in PSAS. Bayesian estimations of the likelihood of molecules falling into the groups of active and inert substances, respectively, form the basis of the biological activity spectrum prediction system known as PSAS. A "molecule's structural formula" can be found in a MOL file, which is used to do PSAS predictions (Pramely and Raj 2012; Azhaguraj et al. 2012; Islam et al. 2022; Talebi et al. 2023).

2.3.6.1 Applications

- Disclose novel effects and modes of action for compounds that are well-known in business and private databases.
- Find fresh leads based on particular biologically active features among the compounds in commercial and internal databases.
- Select the compounds with the most potential for a rapid screening process from the given samples.

2.4 Visualizers and Image Analysis

2.4.1 "A Medical Image Data Examiner" or "AMIDE"

The goal of AMIDE's development was to offer multimodality volumetric medical picture analysis. To allow a limitless amount of such components to be exhibited, changed, and evaluated at the same time, regions of interest (ROIs) and data sets (such as PET, CT, and MRI) are logically placed inside the framework of trees.

A tree abstraction made up of a series of items, including data sets and ROIs, serves as the foundation for AMIDE's data structure. This native data structure is located about the world data structure, and every item in "AMIDE" has an "Euclidean space" given to it (Loening and Gambhir 2012; You et al. 2022). The aforementioned types of objects are executed in AMIDE.

- Analyze the "AMIDE" root object, which functions to monitor variables that affect the entire study and arranges a set of related ROIs and medical images within an organized body.

- This data collection, which is used to encapsulate volumetric medical images, contains the raw picture data as well as the information needed to analyze it.
- A portion of a region designated for statistical computations is known as a "region of interest object" (ROI). Currently, ellipsoids, boxes, cylinders, and contours (two or three dimensions) are employed as ROIs.
- Fiducial reference markers: These are employed for rigid body registration of data sets, and they only encode a place in space.

2.4.1.1 Applications

- Offers the community studying molecular imaging research multimodality medical picture analysis
- Creates complex medical imaging techniques (such as factor analysis and heart polar maps) with interactive "wizard" interfaces

2.4.2 Visualizer for Discovery Studio

With "Discovery Studio Visualizer" ("DS Visualizer"), one may view, exchange, and analyze data related to proteins and small molecules. It is available for free and can be applied to both small and macromolecule applications.

It makes data analysis and transfer possible in a variety of formats, including images, 3D structures, SMILES, and sequences. You can get the necessary sequence and structures from PDB or NCBI. Through structure modification and computational work, molecular characteristics can be investigated (Rizvi et al. 2013; Sharma et al. 2022; Baroroh et al. 2023).

2.4.2.1 Applicability: Visual Aid

- Sophisticated visuals of molecules
- Graphics of publication quality

Macromolecule design:

- Editing protein sequences with many domains, such as antibodies.
- Secondary structure prediction.
- Modify and overlay protein structures.

Ligand-oriented architecture:

- Tools for sketching and assembling pieces
- Pharmacophore creation
- Adaptable cover ligands

Design based on structure:

- Identify, highlight, and modify ligand-binding sites.
- Draw 2D interaction graphs between ligands and receptors.
- Keep an eye on interactions that do not include bonds, such as satisfied, unsatisfied, and favorable encounters.

Exhibit several different molecular surface characteristics, such as:

- Charge, solvent accessibility, lipophilicity, ionizability, aromaticity, and H-bonds.
- Use ActiveX Control to see 3D molecules on websites and in Microsoft Office©.
- Record and distribute scenes using "Storyboard."
- Get access to heat maps, 2D and 3D charts, and more.

2.4.3 SCGE-Pro Imaging Software

Comet test and single-cell gel electrophoresis are two common uses for SCGE-Pro. We are working with the Computer Division to develop "imaging software" for "cytogenetic" and DNA repair analysis. The gene toxicity of factors from the environment such as chemicals, medicines, elevated and decreased "LET" radiation, and carcinogens is investigated using the comet assay. One imaging method that quantifies "loss of heterozygosity" ("LOH") for a "single gene" or "gene-specific repair" concerning "total DNA" is "fluorescence in situ hybridization" ("FISH"). These experiments investigate the process of "eukaryotic cell" healing as well as "intracellular DNA damage" with different types of cells. The influence of "3.3 MeV proton beams" on "DNA damage" in rat "leukocytes from the peripheral blood" is one study that uses the neutral comet test (Chaubey et al. 2001; Qian et al. 2021).

2.4.3.1 Applications

- Clinical applications include genetic instability, diabetes, cancer susceptibility, prenatal diagnostics, and DNA repair deficiency syndrome.
- Human biomonitoring: nutrition and aging.
- Aquatic or terrestrial environmental biomonitoring settings.
- Assessing the genotoxicity of chemicals and radiation using animal and human models.
- Agricultural sciences, radiation biology, clinical, and molecular epidemiology.

2.4.4 Software for Xenogen Living Images

"Wave Metrics IGOR Pro1," an efficient statistical analysis and computer device, is used by "Xenogen Living Image Software." The program creates a unique environment that is used for data collection and analysis. The software functions on both

Macintosh® and Windows®. Among the software's functional elements are a lab book window, a visual projection and interpretation window, an entire system condition and communication window, and an image acquisition control panel. The "Living Image" and IGOR Pro software tools are situated in the upper section of the "menu bar." To prevent interface clutter and confusion, the remaining menu items supporting Living Image software stay dormant while Igor Pro software operates.

A distinct image window display is displayed by software when opening or acquiring a Living Image data collection. The operable portion of the window overlays a luminous or fluorescent pseudo-color image on top of a photographic image. Unless otherwise noted, the software operates similarly for fluorescent and luminous images. The analysis tool that controls how the image is shown and measured is located at the top of the image. The image's right side features a color bar that illustrates the correlation between the image's pseudo-colors and its numerical data values. At the bottom of the window display, labeling information produced by the imaging system and the user together describes how to utilize each of the various controls in the image window (Xie et al. 2007).

2.4.4.1 Applications

- Applications for in vivo imaging
- Imaging at low light levels, including binning and sensitivity, calibrations and measurements, light sources in the background, and "dark charge management"

2.5 *"Data Analysis"*

2.5.1 GeneSpring

GeneSpring offers an overview of the info and analytical models that are accessible in the program, as well as definitions of terms related to important organizational features in the user interface.

Utilizing arrays to test hypotheses related to a certain scientific subject, this software represents a set of samples. Out of a chosen project, a fresh experiment is made in this. In a new experiment, specimens of a specific method are loaded and put through several common preliminary processing steps, including normalization, summarization, and baseline transformation, to retrieve the raw data in a format that is ready for analysis. The experiment that gave rise to it contained several samples, multiple interpretations that categorized the samples according to user-specified experimental criteria, and all additional objects that were produced as a consequence of different analysis stages in the experiment (Dey et al. 2023).

The three components of the software are the file system, database, and user interface layer. All objects are physically stored in the file system. These are kept in the installation folder's app/data subdirectory. All annotations (searchable attributes such as names, notes, and other details) connected to different file system objects are stored in an SQL database, which facilitates quick searches. Ultimately, pertinent objects are arranged into projects, experiments, analyses, etc. via the UI layer (Trivedi et al. 2023).

2.5.1.1 Applications

- Correction for batch effects
- Segmenting binary circles
- Filters for region of allelic imbalance and copy-neutral LOH events identification
- Find recurring patterns in a collection of samples

2.5.2 "QSARPro"

Using reliable statistical modeling, this program determines the relationship between a molecular activity or property and architectural factors, examines the association, and generates predictions quickly. Approximately 1000 molecular descriptors are evaluated with it, including physicochemical, electro-topological, topological, information theory-based, electrostatic, quantum, hydrophobic properties, MMFF atom types, etc.

Common tasks related to QSAR modeling include characteristic collection and estimation, empirical assessment for the calculated descriptions, assigning test and training groups, regression analysis, and outcomes analysis. To find the solution most suited for a given project, it evaluates several alternatives for test set, regression approach, linear or nonlinear regression, and classes of descriptors (Hung et al. 2014).

2.5.2.1 Applications

- Learn about and practice a range of variable selection and regression technique combinations.
- Creating a base for the ligand's position by lining up a specific group of atoms in the peptide that are active in the "co-crystal" molecule.
- Protein-protein interaction studies.

2.5.3 REST 2009 Software

REST 2009 Software is one program that can be used to analyze data on gene expression using qRT polymerase chain reaction investigations. Gene expression of reference genes is employed in compared gene activity analysis or quantification to standardize the activity levels of "genes of interest" (GOI) across different samples. This technique makes it possible to modify quantitative PCR data, for instance, to account for changes brought on by variations in sample loading.

The PCR efficiency of the reference genes and the gene of interest vary, and this software accounts for this variation using a mathematical model. When one reference gene is compared to several reference genes for normalization, the reliability of the results is increased (Ismail et al. 2011).

2.5.3.1 Applications

- Analyze whether there is a discernible difference between the samples and the controls.

2.6 Behavioral Study

2.6.1 Ethowatcher

Changes in behavior are thought to be a key indicator for diagnosing a variety of diseases. An experimental animal's complex behaviors are linked to changes in its morphology and physiology. These changes are commonly seen in the lab or "free-ranging animals" for a variety of biochemical or biological research-related causes, such as "ecology," "physiology," "neurosciences," "psychology," "genetics," "pharmacology," and disease. Advanced automated techniques capture specific behaviors indirectly by detecting their impacts through the use of "pressure" or "infrared" sensor activation or "image processing" techniques like those employed in the "video-tracking analysis." It is easy to collect a variety of data using video-tracking analysis, including variations in mobility action, including the duration and frequency of walking, movement velocity, and horizontal position.

C++ is the foundation of Ethowatcher, which runs in environments with C++ Builder 5.0. With the help of this integrated application, users can create and save behavioral changes for "offline" behavioral recordings (from digital video files) or "real-time" behavioral scoring (from analog video files or events occurring in the environment). The obtained digital video file can be processed using digital image processing techniques to automatically extract parameters (the length traveled, direction, speed, estimate target region, monitor plot), along with enabling subject (biological) monitoring. The software provides reports that are time-segmented and synchronized with identical timing sources. The reports cover the activity-related indices, sequence, length, rate, and length among the measured behaviors (Junior et al. 2012).

2.6.1.1 Applications

- Approval of an instrument for analyzing behavior in experimental animals.
- Analysis of laboratory animal video-tracking.

2.6.2 "Multimodal Animal Rotation System" or "MARS"

The entire 360° range of motion of an experiment animal is recorded via the Multimodal Animal Rotation System or MARS. Because of the software's clever design, tracking all the pertinent chemical and anatomical data of the experimental animal involves simply rotating a mouse to the appropriate locations or angles. Optical signals produced by the experimental orientation are also captured.

With this program, multimodal and multispectral data sets may be automatically co-registered and captured from all angles of acquisition. The program measures the ideal image or exports the whole rotation movements or video, increasing the resulting signal sensitivity. A co-registration program, multimodal visualization, and animal rotation device are all included in this software (Buehlmann et al. 2020).

2.6.2.1 Applications

- Cell monitoring
- Activities of enzymes
- Illness of the bones
- Infectious illness
- Monitoring and delivery of nanoparticles

3 Conclusion

The numerous software-based approaches that are currently being widely used in drug design and discovery have been addressed in this chapter. Without favoring known hits or leads, the effective application of software-based approaches made it possible to identify biologically active compounds in vitro. The various mechanisms underlying the complex target-ligand interaction are also being unlocked with the aid of novel techniques like docking. Drug discovery has benefited from the significant advancements and ongoing use of new software in the fields of pharmacokinetics and pharmacodynamics. The many biochemical industries' financial problems and drug discovery are improved by this intern. The most compelling proof that programmable and computer-aided methods will be employed to assist the costly, intricate, and very challenging process of designing and discovering new drugs will come from several previous drug examples, including the HIV viral enzyme inhibitor indinavir, which was discovered through software-based drug discovery.

References

Alavi A, Sharma V (2023) Role of docking in anticancer drug discovery. Lett Drug Des Discov 20(10):1490–1511

Almukainzi M, Okumu A, Wei H, Löbenberg R (2015) Simulation of in vitro dissolution behavior using DDDPlus™. AAPS PharmSciTech 16:217–221

Azhaguraj R, Milton MJ, Ganesh J, Ramakrishnan M, Antony S (2012) Prediction of biological activity spectra for secondary metabolites from marine macroalgae *Caulerpa* spp (Chlorophyta–Caulerpals). Int Res J Pharm 5:320–323

Baroroh U, Biotek M, Muscifa ZS, Destiarani W, Rohmatullah FG, Yusuf M (2023) Molecular interaction analysis and visualization of protein-ligand docking using Biovia Discovery Studio Visualizer. Indones J Comput Biol (IJCB) 2(1):22–30

Bhattacharya D, Li AS, Paul B, Dastidar UG, Santhakumar V, Sarkar D, Chau I, Li F, Ghosh T, Vedadi M, Talukdar A (2023) Development of selective class I protein arginine methyltransferase inhibitors through fragment-based drug design approach. Eur J Med Chem 260:115713

Buehlmann C, Mangan M, Graham P (2020) Multimodal interactions in insect navigation. Anim Cogn 23:1129–1141

Chaubey RC, Bhilwade HN, Rajagopalan R, Bannur SV (2001) Gamma ray induced DNA damage in human and mouse leucocytes measured by SCGE-Pro: a software developed for automated image analysis and data processing for Comet assay. Mutat Res Genet Toxicol Environ Mutagen 490(2):187–197

Devi NL, Mikilli S (2013) Homology modelling, molecular dynamics simulation and docking studies of HSP 105 of cystic fibrosis with the target protein HSP 70. Biomirror 4(12):102–108

Dey D, Shrivastava V, Joshi D, Singal CM, Tyagi S, Bhat MA, Jaiswal P, Sharma JB, Palanichamy JK, Sinha S, Seth P (2023) Hypoxia induces early neurogenesis in human fetal neural stem cells by activating the WNT pathway. Mol Neurobiol 60(5):2910–2921

Dubey A, Kalra SS (2013) Computational comparative modeling and visualization for HIV1 and HIV2 proteins via the software SYBYL-X. Int J Sci Res Publ 3(10):2250–3153

Ferreira LG, Dos Santos RN, Oliva G, Andricopulo AD (2015) Molecular docking and structure-based drug design strategies. Molecules 20(7):13384–13421

Gill SK, Christopher AF, Gupta V, Bansal P (2016) Emerging role of bioinformatics tools and software in evolution of clinical research. Perspect Clin Res 7(3):115

Hodos RA, Kidd BA, Shameer K, Readhead BP, Dudley JT (2016) *In silico* methods for drug repurposing and pharmacology. Wiley Interdiscip Rev Syst Biol Med 8(3):186–210

Honório TD, Pinto EC, Rocha HV, Esteves VS, dos Santos TC, Castro HC, Rodrigues CR, de Sousa VP, Cabral LM (2013) *In vitro–in vivo* correlation of efavirenz tablets using GastroPlus®. AAPS PharmSciTech 14:1244–1254

Hung TC, Lee WY, Chen KB, Chan YC, Lee CC, Chen CY (2014) In silico investigation of traditional Chinese medicine compounds to inhibit human histone deacetylase 2 for patients with Alzheimer's disease. Biomed Res Int 2014:769867

Hussain A, Hussain A, Sabnam N, Verma CK, Shrivastava N (2023) Insilico exploration of the potential inhibitory activity of DrugBank compounds against CDK7 kinase using structure-based virtual screening, molecular docking, and dynamics simulation approach. Arab J Chem 16(2):104460

Islam S, Hosen MA, Ahmad S, ul Qamar MT, Dey S, Hasan I, Fujii Y, Ozeki Y, Kawsar SM (2022) Synthesis, antimicrobial, anticancer activities, PASS prediction, molecular docking, molecular dynamics and pharmacokinetic studies of designed methyl α-D-glucopyranoside esters. J Mol Struct 1260:132761

Ismail Y, McCormick S, Hijri M (2011) A fungal symbiont of plant-roots modulates mycotoxin gene expression in the pathogen *Fusarium sambucinum*. PLoS One 6(3):e17990

Jayaram B, Singh T, Mukherjee G, Mathur A, Shekhar S, Shekhar V (2012) Sanjeevini: a freely accessible web-server for target directed lead molecule discovery. BMC Bioinformatics 13:1–13

Junior CF, Pederiva CN, Bose RC, Garcia VA, Lino-de-Oliveira C, Marino-Neto J (2012) ETHOWATCHER: validation of a tool for behavioral and video-tracking analysis in laboratory animals. Comput Biol Med 42(2):257–264

Jursinic PA, Sharma R, Reuter J (2010) MapCHECK used for rotational IMRT measurements: step-and-shoot, TomoTherapy, RapidArc. Med Phys 37(6Part1):2837–2846

Kimko H, Pinheiro J (2015) Model-based clinical drug development in the past, present and future: a commentary. Br J Clin Pharmacol 79(1):108–116

Kollipara S, Bhattiprolu AK, Boddu R, Ahmed T, Chachad S (2023) Best practices for integration of dissolution data into physiologically based biopharmaceutics models (PBBM): a biopharmaceutics modeling scientist perspective. AAPS PharmSciTech 24(2):59

Kumar DB, Kumar PV, Bhubaneswaran SP, Mitra A (2010) Advanced drug designing softwares and their application in medical research. Int J Pharm Sci 2:16–18

Kundrotas PJ, Vakser IA (2010) Accuracy of protein-protein binding sites in high-throughput template-based modeling. PLoS Comput Biol 6(4):e1000727

Liu T, Lu D, Zhang H, Zheng M, Yang H, Xu Y, Luo C, Zhu W, Yu K, Jiang H (2016) Applying high-performance computing in drug discovery and molecular simulation. Natl Sci Rev 3(1):49–63

Loening AM, Gambhir SS (2012) AMIDE: a free software tool for multimodality medical image analysis. Mol Imaging 2(3):131–137

Mathew B, Mathew GE, George S, Shamnas M, Raj VB (2012) Isolation, in silico design and anti-inflammatory activity of spirosolenol from the roots of *Solanum anguvi*. Chem Sci J 2012:83

Naz A, Bano K, Bano F, Ghafoor NA, Akhtar N (2009) Conformational analysis (geometry optimization) of nucleosidic antitumor antibiotic showdomycin by Arguslab 4 software. Pak J Pharm Sci 22(1):78–82

NMITLI-BioSuite Team (2007) BioSuite: a comprehensive bioinformatics software package (a unique industry—academia collaboration). Curr Sci 10:29–38

Pramely R, Raj LS (2012) Prediction of biological activity spectra of a few phytoconstituents of *Azadirachta indicia* A. Juss. J Biochem Technol 3(4):375–379

Qian Q, Chen Y, Wang JQ, Yang DQ, Jiang C, Sun J, Dong J, Li GC (2021) Use of the alkaline comet assay for monitoring genotoxic effects of arsenic in human populations. Mutat Res Genet Toxicol Environ Mutagen 867:503368

Rizvi SM, Shakil S, Haneef M (2013) A simple click by click protocol to perform docking: AutoDock 4.2 made easy for non-bioinformaticians. EXCLI J 12:831–857

Sabitha K, Habeeb SKM (2011) Molecular modeling and drug discovery of potential inhibitor for anticancer target gene melk (Maternal Embryonic Leucine Zipper Knase). Int Res J Pharm 2:141–145

Sharma A, Garg S, Bhatt AK, Sharma S, Gupta U, Garg MS (2022) Molecular docking studies on the anti-viral activity of cellulose acetate phthalate (CAP) as an adjuvant against SARS-CoV-2 by protease Mpro. NeuroQuantology 20(15):6481–6490

Singh A, Shekhar S, Jayaram B (2021) CADD: some success stories from Sanjeevini and the way forward. In: Innovations and implementations of computer aided drug discovery strategies in rational drug design, pp 1–18

Sliwoski G, Kothiwale S, Meiler J, Lowe EW (2014) Computational methods in drug discovery. Pharmacol Rev 66(1):334–395

Talebi A, Salehi M, Khaleghian A, Kubicki M (2023) Evaluation of anticancer activities, apoptosis, molecular docking, and antioxidant studies of new Ni (II), VO (IV), Cu (II) and Co (III) Schiff base complexes. Inorg Chim Acta 546:121296

Trivedi TS, Bhadresha KP, Patel MP, Mankad AU, Rawal RM, Patel SK (2023) Identification of hub genes associated with human cystic fibrosis: a meta-analysis approach. Hum Gene Ther 35:201139

Uebbing L, Klumpp L, Webster GK, Löbenberg R (2017) Justification of disintegration testing beyond current FDA criteria using in vitro and in silico models. Drug Des Devel Ther 11:1163–1174

Wasko MJ, Pellegrene KA, Madura JD, Surratt CK (2015) A role for fragment-based drug design in developing novel lead compounds for central nervous system targets. Front Neurol 6:197

Xie X, Xia W, Li Z, Kuo HP, Liu Y, Li Z, Ding Q, Zhang S, Spohn B, Yang Y, Wei Y (2007) Targeted expression of BikDD eradicates pancreatic tumors in noninvasive imaging models. Cancer Cell 12(1):52–65

Xu G, Chen Y, Shen K, Wang X, Li F, He Y (2014) The discovery of potentially selective human neuronal nitric oxide synthase (nNOS) inhibitors: a combination of pharmacophore modelling, CoMFA, virtual screening and molecular docking studies. Int J Mol Sci 15(5):8553–8569

Xu Q, Zhao Z, Liang P, Wang S, Li F, Jin S, Zhang J (2023) Identification of novel nematode succinate dehydrogenase inhibitors: virtual screening based on ligand-pocket interactions. Chem Biol Drug Des 101(1):9–23

You A, Kim JK, Ryu IH, Yoo TK (2022) Application of generative adversarial networks (GAN) for ophthalmology image domains: a survey. Eye Vis 9(1):1–9

Yurieva E, Agostinoa M, Ramsland PA (2011) Challenges and advances in computational docking: 2009 in review. J Mol Recogn 24:149–164

Importance of Traditional Knowledge in Modern Drug Discovery

Shafiq Ur Rahman, Salman Khan, Umair Sayed, Haider Ali, Kifayat Ullah, and Himayat Khan

Contents

Abstract In spite of advancement in medical technologies, traditional knowledge regarding medications still remains important and the key pillar for drug discoveries particularly in developing countries and in general for the entire globe. It's important to mention that traditional knowledge-based drug discoveries play a major role in the form of active pharmaceuticals in human therapy. The common traditional knowledge-based medicines include Unani, Islamic, Chinese, Ayurveda, Siddha, Irani, African, and Korean which are transmitted from ancient time from generation to generation. Ethnopharmacological researches on natural resources (plants, animals, microorganism, and minerals products) are of great importance in the development of active drugs as well as herbal medicines. On one side, these studies document the therapeutic uses of natural resources but, on the other hand, also safeguard the preservation of cultural heritage.

Because of the vast diversity in chemical nature/structures and functional groups of plants and microbial secondary metabolites, and also the development in the modern technologies for identification and characterization, they represent principal

S. U. Rahman (✉) · S. Khan · U. Sayed · H. Ali · K. Ullah · H. Khan
Department of Pharmacy, Shaheed Benazir Bhutto University, Sheringal, Dir (Upper), Khyber Pakhtunkhwa, Pakistan
e-mail: shafiq@sbbu.edu.pk

© The Author(s), under exclusive license to Springer Nature Singapore Pte Ltd. 2024

A. Das Talukdar et al. (eds.), *Traditional Resources and Tools for Modern Drug Discovery*, Interdisciplinary Biotechnological Advances,
https://doi.org/10.1007/978-981-97-4600-2_4

sources for new active pharmaceuticals. The demand in the world market for natural products (active pharmaceuticals, fragrances, colors, and flavors) is increasing day by day, and it is reported that about 80% of the world population depends on traditional medicine-based products for their health.

For the discovery of modern drugs, most of the researchers across the globe focused on traditional knowledge-based nontoxic plants and other related products. Therefore, significance of traditional knowledge regarding medications either directly from the communities or reported in books and journals should not be underestimated and ignored. This chapter highlights the importance and traditional knowledge-based latest discoveries of natural drugs.

Keywords Traditional knowledge · Natural products · Drug discoveries

1 Introduction

An effective process for finding new drugs can be found in traditional knowledge of herbal medicine. Indigenous pharmacopoeias profoundly rely on plant-based remedies, which are a major source of medication for the population (Garg et al. 2012). Since the beginning of time, numerous ailments have been successfully treated while using natural resources, primarily medicinal plants. For hundreds of years, plants have long been incredibly significant in the daily lives and been the primary source of medicines (Akbar 2020; Zhang et al. 2020). In fact, numerous discoveries of the important drugs and dosage form are based on traditional knowledge, and an overview is given in Fig. 1.

Numerous advantages have led to the usage of herbal remedies throughout history, including their affordability, accessibility, and lack of evidence of either effectiveness or resistance to whole plant extracts (Heinrich 2000). With time, the type of traditional medical knowledge has been consistently used in households and passed down from one generation to the next (Singh et al. 2012). For many people in both rural and urban areas today, plants as a form of treatment are still highly significant (Patwardhan et al. 2005). Through botanical surveys, and laboratory biological tests on animal models, scientific research has recently more focused on the use of plants in traditional medicine to treat a variety of diseases. It is an effort to identify specific species that have medicinal properties that could replace certain side-effect-producing chemical drugs (Tabassum and Ahmad 2011; Surh 2003).

Pharmacological substances from plants are being isolated and characterized at an ever-increasing rate. Many phytochemicals are isolated from plants like *Achyranthes aspera*, *Cleome viscosa* L., and *Cyperus rotundus* Linn used to cure various diseases. Tribal healers frequently employ plants to treat a variety of illnesses, including diabetes, ulcers, asthma, snakebites, respiratory problems,

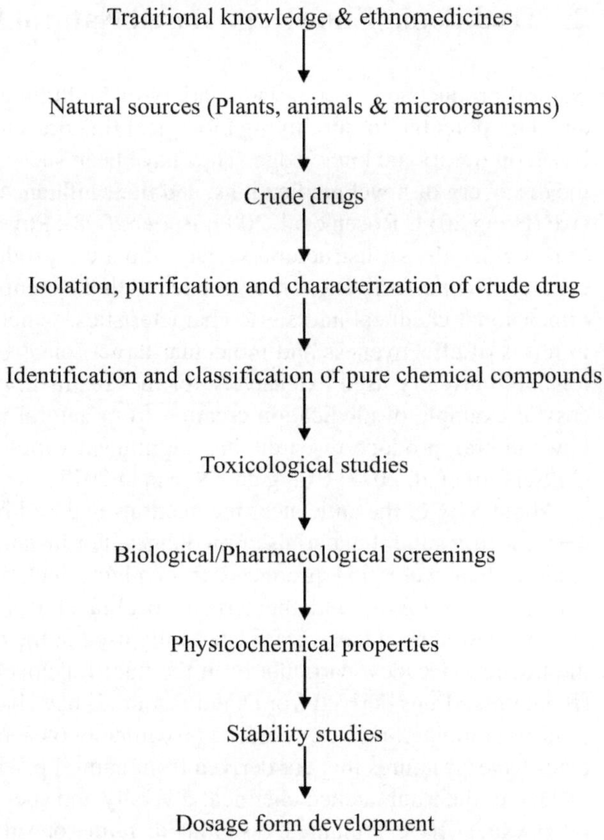

Fig. 1 An approach from drug discovery to dosage form development

leucorrhea, dandruff, and stomach disorders. The most common ways to make plant-based medicines are as powders, decoctions, pastes, and fluids (Groninger et al. 2002). The development of drugs from plants will be more comprehensive and can assist the population as much as possible. Saponins A and B from *Achyranthes aspera*, sterol from *Adhatoda vasica*, aspartate aminotransferase from *Andrographis paniculata*, and alkaloids like atropine from *Datura metal* are just a few of the phytochemicals that have been extracted from plants (Ganesh et al. 2021). Natural products and related structures continue to be crucial components of pharmacopoeias despite the significant synthetic diversity brought about by the advent of combinatorial chemistries and high-throughput screening methods over the past few decades. Going future, it's expected that natural products and similar structures will play an even bigger role in the creation of new and improved pharmaceutical drugs (Rasineni et al. 2010).

2 Traditional Knowledge-Based Natural Products

Natural products come in a wide variety of multidimensional chemical structures, and their potential for modifying biological functions has garnered a lot of attention based on traditional knowledge. They have been successfully used in the search for the discovery of novel medications, and their influence on chemo-biology is extensive (Hong 2011; Rosén et al. 2009; Butler 2008). Physical chemistry has been able to recognize the vast structural variety of natural products throughout the previous century. Their effectiveness is correlated with the complexity of their ordered three-dimensional chemical and steric characteristics, which provide numerous benefits in terms of effectiveness and molecular target selectivity. Artemisinin and its analogs are currently used extensively for the treatment of malaria, serving as a successful example of medication creation from natural materials. This demonstrates how natural product research has significantly aided medication development (Muschietti et al. 2013; Cragg and Newman 2013).

About 54% of the anticancer medications licensed between 1940 and 2002 were derived from natural materials or medicines. For instance, effective anticancer medications that were first produced from plants include the vinca alkaloids from *Catharanthus roseus* and the terpene paclitaxel from *Taxus baccata* (Li-Weber 2009). Natural products were successfully used in the creation of new medicines in the last few decades, particularly in the hunt for novel chemical structures. In the recent pass, drugs derived from natural sources have had a huge effect. It is actually true for antihypertensives, where approximately 64% of currently developed medicines have structures that are derived from natural products (Newman et al. 2003).

Due to their unmatched chemical diversity and special mood of action, the natural products have remained essential in numerous therapeutic development and research initiatives. With the passage of time, those natural compounds which have the capacities to interact with a broad range of various biological targets have undergone interesting and important developments, and several have appeared as the most important medications in the healthcare system (Zhu et al. 2012; Winter and Tang 2012; Li and Vederas 2009). For example, 13 natural products or their derivatives were among the 69 small-molecule novel medications accepted between 2005 and 2007 worldwide (Newman et al. 2003).

Over the past 50 years, many new medications have been evolved by using high-throughput screening techniques, and combinative chemistry, and these natural products and their chemical derivatives have remained vital elements in pharmacopoeias. Currently 250,000–500,000 plant species have been skillfully examined for bioactivities (Ngo et al. 2013). Therefore, there is a lot of possibilities for novel scientific discoveries that come from plants and other natural substances, which may be useful to learn a lot about distinctive chemical structures and their advance modes of action in the perspective of producing new drugs.

There is a vast application of herbal goods and herbal medication worldwide as shown in Fig. 2 with the trust that herbal medications are always harmless and carry no hazards because they are from natural sources. The World Health Organization

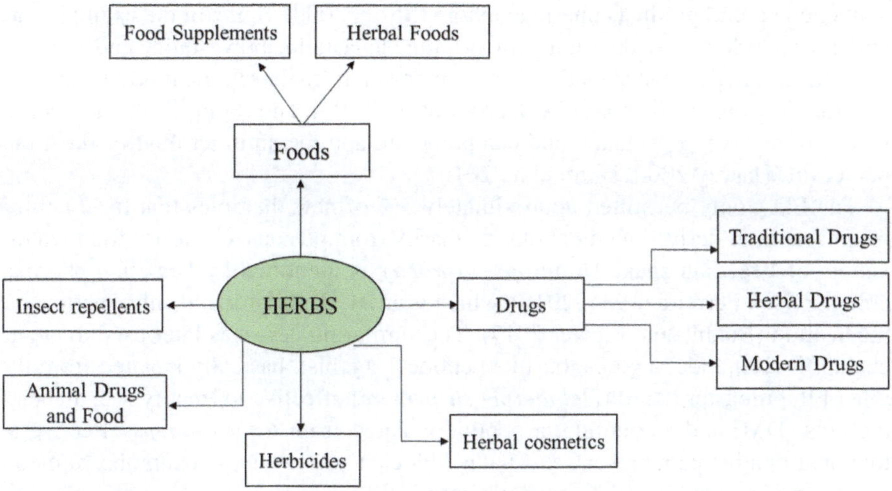

Fig. 2 Several uses and importance of plant herbs

(WHO) guidelines support developed countries to supplement their health programs with traditional herbal preparations provided they are nontoxic. For the large parts of the earth population, medicinal plants remain to show a dominant role in the healthcare system, and this is mainly genuine in developing countries, where herbal medicine has a continuous history of long use.

3 Development of Natural Products into New Medicines

Approximately 35% of the annual market demanded medicines have a link directly or indirectly to natural sources, such as plants (25%), microbes (13%), and animals (about 3%). In the research and production of novel medications, naturally derived substances are an important source for multinational pharmaceutical corporations. They act as a key source of therapeutic agents (both as pure medicines and phytomedicines), a base of raw materials for the production of complex, semisynthetic therapeutics, precursors for lead molecule design, and taxonomic markers for the finding of novel compounds, among other purposes.

In the world, one-third of the top-selling medications are natural substances or their derivatives. Approximately 60–80% of antibiotics and antineoplastic medications are derived from natural substances (Cragg et al. 1997).

Currently, Newman and Cragg described how natural ingredients were used in FDA-approved medications in the last four decades. A total of 1562 medications were approved by the FDA during the time, of which 64 (4%) were natural products that had not been altered, 141 (9.1%) were plant based drugs (mixtures), 320 (21%) were derivatives of natural products, and 61 (4%) were synthetic drugs that contained natural products pharmacophore (Strohl 2000). Some of the examples are anticholesterolemic medications simvastatin, lovastatin, pravastatin, and atorvastatin; the antihypertensive medications captopril and enalapril; the immunosuppressive medications cyclosporine A, tacrolimus (FK506), and ramipril; the anticancer medications taxol, docetaxel, and campothecin; and the antimicrobial medications penicillin (Harvey 2000; Dutra et al. 2016).

The FDA only permitted approximately 3% of new therapies that have animal origins, animal-derived medications, primarily from poisons. Captopril found in the venom of Brazilian snake (*Bothrops jararaca*), is identified by Brazilian pharmacologist S. H. Ferreira (1934–2016), which is an ACE inhibitor and antihypertensive medication (Bakhle and Flower 2017). The similar process was later used to create enalapril. Exenatide, a glucagon-like peptide 1 agonist, basically isolated from the gila of the monster lizard (*Heloderma suspectum*) effective to treat type 2 diabetes mellitus (DM) and ziconotide, a peptide isolated from *Conus magus* effective to treat neuropathic pain and is Ca^{++} channel blocker, are two more examples of medications made from animal sources (Calixto 2000).

The utilization of natural items in the process of finding and developing new drugs has some definite benefits. When contrasted to other sources, they represent chemical novelties and can generate lead drug candidates for intricate targets. The chemical diversity of organically produced elements is also unsurpassed by any collection of synthesized chemicals; it has complex two- and three-dimensional structures while still being capable to be absorbed and metabolized by the body (Calixto 2005).

The approach to natural environments physically as well as the procedures required to extract/isolate, purify, characterize, and classify active chemicals are expensive, slow, and challenging. The majority of pharmaceutical companies lament that some natural chemicals are more challenging to assay in high-throughput drug discovery programs (screening) than synthesized drugs.

Finally, it's critical to highlight how difficult it is to synthesize similar lead compounds because of the high structural complexity of natural molecules (Li and Vederas 2009; Newman and Cragg 2016; Smith and Vane 2003). Without a doubt, the progress of novel drugs for the treatment of specific diseases (chronic) and the identification of new therapeutic targets involved in numerous important chronic disorders have both greatly benefited the global pharmaceutical industry over the last two centuries in biodiversity-rich nations (Walsh and Fischbach 2010). The discovery of new medications and the creation of their dosage forms for the treatment of diverse diseases were both greatly aided by plants. In reality, following their conventional use has helped the development of numerous medications. Approximately 67% of the effective medicines used in cancer chemotherapy come from natural sources.

Many of these medications are in-depth evaluated in the latest sections of natural antineoplastic agents. There are numerous substances that act as precursors to the semisynthetic medications topotecan (hycamptin) and irinotecan (camptosar), such as camptothecin (topoisomerase inhibitor) isolated from the bark of *Camptothecin acuminate*. To fulfill the raising need of antineoplastic medication, taxol and its variants are now being prepared. Different plant components may work together synergistically in therapies, while occasionally only one plant component can be effective (Wangkheirakpam and Laitonjam 2016).

The discovery of taxol is used here as an example of a drug that came from conventional medicine. The European yew tree, *T. baccata*, was used as a phytotherapy for heart treatment (Tekol 2007). In the Central Himalayas, *T. baccata* is additionally employed as a traditional medicine to treat breast and ovarian tumor (Singh et al. 2020). A research team's examination of the species revealed specific substances present in yew bark had anticancer effects (Li et al. 2023). Subsequently, the compound paclitaxel (Taxol) was obtained from the bark of *T. baccata* plant. This elucidates the progression of a drug from traditional and vernacular medicines.

The discovery of a new drug, which has its origin in traditional medicine, starts with its conventional use documented in a specific system in the form of books or currently databases and portals. The development of the anticancer medication taxol from folk and conventional therapies began with traditional knowledge. First, the clinical trials were performed in rat models and then in complex models in the form of a solitary molecule. Subsequently, it underwent additional human testing. A competent body approved it for the curing of ovarian tumor, and further research on breast, colon, and lung malignancies has yielded encouraging results. These pharmacological and chemical studies led to the isolation and identification of taxol. Taxol became an anticancer medicine as a result of the positive clinical trials.

Instead, being bioactive, the medications may have numerous drawbacks, such as impenetrability or unwanted effects. Therefore, additional investigation was done to create and improve taxol compounds that were more effective. The very low degree of water solubility of paclitaxel is one disadvantage. A number of investigation teams created derivatives that are water soluble, such as smart taxanes, which are taxol derivatives created for pointed therapy of tumor, to get around this and boost its effectiveness. The first paclitaxel derivative that targets tumors was developed and reported to increase Taxol's pharmacological effectiveness through the synthesis of a water-soluble conjugation that targets tumors (Safavy et al. 1999).

Although the cancer drug taxol was a significant advancement, several more treatments have been developed that employ conventional medicine as a target. A significant development in the field of drug discovery was the identification of the antimalarial medication artemisinin from CTM. Compared to the paths taken by many other phytochemicals discovered during the drug development process, the discovery of artemisinin was quick. Artemisinin was developed by phytochemical evolution, giving it its distinctive sesquiterpene lactone (Kajani et al. 2014). TCM has inspired the creation of numerous other medicinal substances. Clinical investigations have shown that the antiquated Chinese medicine medication arsenic is a relatively safe and efficient treatment for acute promyelocytic leukemia (APL). APL is now treated with arsenic trioxide as a first option (Colin et al. 1989).

For the memory impairment, a powerful source of Chinese herb *Huperzia serrata* is used known is huperzine A. In both Europe and the United States, a huperzine A derivative is under the clinical trials to treat Alzheimer's disease. In Chinese traditional medicine, the molecules chuangxiongol and paeoniflorin are effectively used for the prevention of restenosis (Wittman and Kadow 1996). Clinical trials have provided evidence for the therapeutic effectiveness of similar Chinese medical techniques aiming at promoting blood flow in the controlling of cardiac ischemia/reperfusion injury and the therapy of ischemic disorders. Workout (to enhance the volume of blood flow) combined with extract shenlian (Chinese medicine) appears to have potential for preventing atherosclerotic plaque. A substance obtained from the root of *Salvia miltiorrhiza*, known as salvianolic acid B, has been found as a potential method to explain how endothelial cells' functions are affected by increased shear stress (Tu 2011).

Around the world, there are numerous illnesses that are lethal. Therefore, there is a huge demand for investigations in phytochemistry and pharmacology to develop cures for each a solitary bioactive or mixture (extract) with many compounds, even though researchers are continually trying to find effective medications to treat common illnesses, new illnesses, and microbial infections. Some of the plants used by the traditional people along with their constituents and pharmacological actions are given in Table 1.

Table 1 Some of the traditionally used medicinal plants with potential bioactivities

Plant	Family	Ethnomedical uses	Active constituents	Bioactivities	References
Adhatoda vasica	Acanthaceae	Colds, coughs, chronic bronchitis, and asthma	Vasicine, β-sitosterol, kaempferol, adhatodic acid	Abortifacient, anti-asthmatic, antitussive, antispasmodic, uterotonic	Garg et al. (2012)
Abutilon indicum	Malvaceae	Urinary diseases	Abulitin-A, 4-hydroxybenzamide, sesquiterpene, alantolactone	Wound healing, analgesic, antidiabetic	Akbar (2020)
Boerhavia diffusa	Nyctaginaceae	Body tonic and better general health	Punarnavine, ursolic acid, liirodendrin, steroids	Antibacterial, anti-inflammatory, anticonvulsant	Kumar et al. (2018)
Cleome viscosa	Capparidaceae	Pain	7-Phenoxy coumarin, lactone, cleomeolide	Antipyretic, antidiarrheal	Heinrich (2000)
Calotropis procera	Asclepiadaceae	Rheumatoid arthritis, filariasis edema, atopic dermatitis	Triterpenes, oleanolic acid	Anthelmintic	Singh et al. (2012)
Datura stramonium	Solanaceae	Acne and bronchitis	Hyoscyamine, scopolamine, allantoin, vitamin C, atropine	Anti-asthmatic, anticholinergic	Patwardhan et al. (2005)
Euphorbia thymifolia	Euphorbiaceae	Wound healing	Isomallotinic acid, 12-deoxyphorbol -13, 12-diacetate, 1-hexacosanol	Antioxidant, antiviral, antimicrobial	Tabassum and Ahmad (2011)
Hemidesmus indicus	Asclepiadaceae	Gastric aliments and sciatica	Hemidescine, rutin, indicine, denicunine, heminine, desinine	Antimicrobial, anticarcinogenic, anti-ulcerogenic, anti-inflammatory, antioxidant	Surh (2003)
Lantana indica Linn.	Verbenaceae	Food and general tonic	β-Sitosterol, 3-hydroxy-22 β-12 oleanen-28-oic acid	Antimycobacterial, antibacterial	Groninger et al. (2002)
Phyllanthus niruri	Euphorbiaceae	Hyperbilirubinemia and malaria	Vitamin C, robigenin, 12-hydroxyoctadec-9-enoic acid, 3,4,5-trihydroxybenzoic acid, saponin	Anti-HIV, antimalarial, anticancerous, anti-hepatitis, antidiabetic, hepatoprotectant, antihypertensive	Ganesh et al. (2021)

(continued)

Table 1 (continued)

Plant	Family	Ethnomedical uses	Active constituents	Bioactivities	References
Phyllanthus amarus	Euphorbiaceae	Jaundice	Lignans, flavonoids, polyphenols, triterpenes, steroids	Antimicrobial, antidiabetic, hypolipidemic, antioxidant, hepatoprotective	Rasineni et al. (2010)
Papaver somniferum	Papaveraceae	Smooth muscle relaxant, sedative analgesic and antitussive	Codeine, morphine, noscapine, papaverine	Analgesic, antitussive	Tavl (2022)
Sida cordifolia	Malvaceae	Breast-feeding mothers to provide strength and energy	β-Phenethylamines, quinazolines, carboxylated tryptamines	Anti-inflammatory, antioxidant, antibacterial	Chatterjee et al. (2006)
Solanum nigrum	Solanaceae	Diabetes mellitus and skin infections	Solanocarpidine, campesterol, caffeic acid	Antioxidant, antihyperlipidemic, antimicrobial, anticancer, antiulcer, neuroprotective	Akbar and Akbar (2020)
Strychnos nuxvomica	Loganiaceae	Snake bites	Brucin, loganin, struxine, strychnine	Anti-snake venom	Gayathri and Kannabiran (2009)
Tribulus terrestris	Zygophyllaceae	Tonic for heartburn	4-Hydroxy-3-methoxybezaldehyde, esters of fatty acids, hydroxybenzoic acid, *β*-sitosterol, xanthosine	Diuretic, contractile	Mahboubi (2022)
Vetiveria zizanioides	Poaceae	Headache	Benzoic acid, vetiverol, furfural, β-vetivone, vetivene	Wound healing, antioxidant	Alves and Rosa (2007)

4 Conclusion

Traditional knowledge-based medicine, in spite of its cryptic nature, is too valuable to be disregarded in the study and making of modern-day medications. In general, traditional medicine contains a single herb/part or formula comprised of varieties of phytochemical components, including but not limited to the class of flavonoids, alkaloids, glycosides, steroids, saponins and terpenoids, etc. The point that many plant-extracted medications used in clinical practice in modern era were produced from traditional medicine is remarkable, and it has been recognized that many useful medicines originated from plants were found through their use in traditional medicine.

References

Akbar S (2020) Handbook of 200 medicinal plants: a comprehensive review of their traditional medical uses and scientific justifications. Springer, Cham

Akbar S, Akbar S (2020) *Strychnos nux-vomica* L. (Loganiaceae) (Syn.: S. spireana Dop.). In: Handbook of 200 medicinal plants: a comprehensive review of their traditional medical uses and scientific justifications, pp 1689–1697

Alves RR, Rosa IM (2007) Biodiversity, traditional medicine and public health: where do they meet. J Ethnobiol Ethnomed 3:1–9

Bakhle YS, Flower RJ (2017) Sergio Henrique Ferreira (1934–2016). Br J Pharmacol 174(4):341

Butler MS (2008) Natural products to drugs: natural product-derived compounds in clinical trials. Nat Prod Rep 25(3):475–516

Calixto JB (2000) Efficacy, safety, quality control, marketing and regulatory guidelines for herbal medicines (phytotherapeutic agents). Braz J Med Biol Res 33:179–189

Calixto JB (2005) Twenty-five years of research on medicinal plants in Latin America: a personal view. J Ethnopharmacol 100(1–2):131–134

Chatterjee I, Chakravarty AK, Gomes A (2006) *Daboia russellii* and *Naja kaouthia* venom neutralization by lupeol acetate isolated from the root extract of Indian *sarsaparilla hemidesmus indicus* R. Br J Ethnopharmacol 106(1):38–43

Colin M, Guenard D, Gueritte-Voegelein F, Potier P, Rhone Poulenc Sante SA (1989) Taxol derivatives, their preparation and pharmaceutical compositions containing them. US Patent 4,814,470

Cragg GM, Newman DJ (2013) Natural products: a continuing source of novel drug leads. Biochim Biophys Acta (BBA) Gen Subj 1830(6):3670–3695

Cragg GM, Newman DJ, Snader KM (1997) Natural products in drug discovery and development. J Nat Prod 60(1):52–60

Dutra RC, Campos MM, Santos AR, Calixto JB (2016) Medicinal plants in Brazil: pharmacological studies, drug discovery, challenges and perspectives. Pharmacol Res 112:4–29

Ganesh SS, Rao PS, Nandal DH, Kunkulol R (2021) A review on pharmacological and phytochemical constituent of *Achyranthes aspera* L. Linn. Int J Pharmacogn 8(8):58–64

Garg V, Dhar VJ, Sharma A, Dutt R (2012) Facts about standardization of herbal medicine: a review. J Chin Integr Med 10(10):1077–1083

Gayathri M, Kannabiran K (2009) Antimicrobial activity of *Hemidesmus indicus*, Ficus bengalensis and Pterocarpus marsupium roxb. Indian J Pharm Sci 71(5):578

Groninger E, Boer MD, De Graaf SSN, Kamps WA, De Bont ESJM (2002) Vincristine induced apoptosis in acute lymphoblastic leukaemia cells: a mitochondrial controlled pathway regulated by reactive oxygen species. Int J Oncol 21(6):1339–1345

Harvey A (2000) Strategies for discovering drugs from previously unexplored natural products. Drug Discov Today 5(7):294–300
Heinrich M (2000) Ethnobotany and its role in drug development. Phytother Res 14(7):479–488
Hong J (2011) Role of natural product diversity in chemical biology. Curr Opin Chem Biol 15(3):350–354
Kajani AA, Bordbar AK, Esfahani SHZ, Khosropour AR, Razmjou A (2014) Green synthesis of anisotropic silver nanoparticles with potent anticancer activity using *Taxus baccata* extract. RSC Adv 4(106):61394–61403
Kumar R, Gautam S, Singh KD, Kumar P, Haque N, Yadav V, Kumar R, Diwakar RP, Rishikant Kumar SB (2018) Pharmacological properties of *Boerhavia diffusa*: a review. Int J Chem Stud SP4:72–80
Li JWH, Vederas JC (2009) Drug discovery and natural products: end of an era or an endless frontier. Science 325(5937):161–165
Li Y, Ma NL, Chen H, Zhong J, Zhang D, Peng W, Lam SS, Yang Y, Yue X, Yan L, Wang T (2023) High-throughput screening of ancient forest plant extracts shows cytotoxicity towards triple-negative breast cancer. Environ Int 181:108279
Li-Weber M (2009) New therapeutic aspects of flavones: the anticancer properties of Scutellaria and its main active constituents Wogonin, Baicalein and Baicalin. Cancer Treat Rev 35(1):57–68
Mahboubi M (2022) *Tribulus terrestris* and its efficacy in the treatment of urinary calculi. Nat Prod J 12(7):2–10
Muschietti L, Vila R, Setzer W (2013) Tropical protozoan diseases: natural product drug discovery and development. Evid Based Complement Alternat Med 2013:404250
Newman DJ, Cragg GM (2016) Natural products as sources of new drugs from 1981 to 2014. J Nat Prod 79(3):629–661
Newman DJ, Cragg GM, Snader KM (2003) Natural products as sources of new drugs over the period 1981–2002. J Nat Prod 66(7):1022–1037
Ngo LT, Okogun JI, Folk WR (2013) 21st century natural product research and drug development and traditional medicines. Nat Prod Rep 30(4):584–592
Patwardhan B, Warude D, Pushpangadan P, Bhatt N (2005) Ayurveda and traditional Chinese medicine: a comparative overview. Evid Based Complement Alternat Med 2:465–473
Rasineni K, Bellamkonda R, Singareddy SR, Desireddy S (2010) Antihyperglycemic activity of *Catharanthus roseus* leaf powder in streptozotocin-induced diabetic rats. Pharm Res 2(3):195
Rosén J, Gottfries J, Muresan S, Backlund A, Oprea TI (2009) Novel chemical space exploration via natural products. J Med Chem 52(7):1953–1962
Safavy A, Raisch KP, Khazaeli MB, Buchsbaum DJ, Bonner JA (1999) Paclitaxel derivatives for targeted therapy of cancer: toward the development of smart taxanes. J Med Chem 42(23):4919–4924
Singh A, Singh GS, Singh PK (2012) Medico-ethnobotanical inventory of Renukoot forest division of district Sonbhadra, Uttar Pradesh, India
Singh YD, Panda MK, Satapathy KB (2020) Ethnomedicine for drug discovery. In: Advances in pharmaceutical biotechnology: recent progress and future applications. Springer, Singapore, pp 15–28
Smith CG, Vane JR (2003) The discovery of captopril. FASEB J 17(8):788–789
Strohl WR (2000) The role of natural products in a modern drug discovery program. Drug Discov Today 5(2):39–41
Surh YJ (2003) Cancer chemoprevention with dietary phytochemicals. Nat Rev Cancer 3(10):768–780
Tabassum N, Ahmad F (2011) Role of natural herbs in the treatment of hypertension. Pharmacogn Rev 5(9):30
Tavl OF (2022) *Papaver somniferum* L. In: Novel drug targets with traditional herbal medicines: scientific and clinical evidence. Springer, Cham, pp 479–494
Tekol Y (2007) The medieval physician Avicenna used an herbal calcium channel blocker, *Taxus baccata* L. Phytother Res 21(7):701–702

Tu Y (2011) The discovery of artemisinin (qinghaosu) and gifts from Chinese medicine. Nat Med 17(10):1217–1220

Walsh CT, Fischbach MA (2010) Natural products version 2.0: connecting genes to molecules. J Am Chem Soc 132(8):2469–2493

Wangkheirakpam SD, Laitonjam WS (2016) Studies on the uses of some plants for medicinal and dyeing properties. Int J Chem 5(1):93–102

Winter JM, Tang Y (2012) Synthetic biological approaches to natural product biosynthesis. Curr Opin Biotechnol 23(5):736–743

Wittman MD, Kadow JF (1996) Bristol Myers Squibb Co. Amino acid derivatives of paclitaxel. U.S. Patent 5,489,589

Zhang L, Song J, Kong L, Yuan T, Li W, Zhang W, Hou B, Lu Y, Du G (2020) The strategies and techniques of drug discovery from natural products. Pharmacol Ther 216:107686

Zhu F, Ma XH, Qin C, Tao L, Liu X, Shi Z, Zhang CL, Tan CY, Chen YZ, Jiang YY (2012) Drug discovery prospect from untapped species: indications from approved natural product drugs. PLoS One 7(7):e39782

The Importance of Traditional Resources in Ethnomedicine

Hadia Hemmami, Mohammed Messaoudi, Barbara Sawicka, Wafa Zahnit, Nadjet Osmani, Mokhtar Benmohamed, and Abdelkrim Rebiai

Contents

H. Hemmami · A. Rebiai
Department of Chemistry, Hamma Lakhdar El-Oued University, El Oued, Algeria
e-mail: rebiai-abdelkrim@univ-eloued.dz

M. Messaoudi · N. Osmani
Nuclear Research Centre of Birine, Djelfa, Algeria

B. Sawicka (✉)
Department of Plant Production Technology and Commodities Science, University of Life Sciences in Lublin, Lublin, Poland
e-mail: barbara.sawicka@up.lublin.pl

W. Zahnit
Faculty of Mathematics and Matter Sciences, Laboratory of Valorization and Promotion of Saharan Resource (VPRS), University of Ouargla, Ouargla, Algeria

M. Benmohamed
Laboratory of Fundamental Sciences, University Amar Télidji of Laghouat, Laghouat, Algeria

© The Author(s), under exclusive license to Springer Nature Singapore Pte Ltd. 2024

A. Das Talukdar et al. (eds.), *Traditional Resources and Tools for Modern Drug Discovery*, Interdisciplinary Biotechnological Advances,
https://doi.org/10.1007/978-981-97-4600-2_5

Abstract The chapter "The Importance of Traditional Resources in Ethnomedicine" focuses on the research and use of traditional plant, animal, and microbial resources in ethnomedicine in the context of modern drug discovery. Ethnomedicine refers to traditional medicine practiced by various cultures and communities, using natural resources of plants, animals, and microorganisms to treat various conditions and diseases. The chapter analyzes the importance of these traditional resources as potential sources of new drugs and the possibility of using modern scientific tools in the identification and study of biological activities of substances from ethnomedicine. Various examples of plants, animals, and microorganisms that play an important role in ethnomedicine and have the potential to discover new drugs will be presented.

Keywords Bioactive compounds · Endophytic microorganisms · Therapeutics · Human health

1 Introduction to Ethnomedicine

Ethnomedicine is the study of conventional medical methodologies and remedies that have been developed and passed down through generations in specific cultural and ethnic groups. It encompasses the use of plants, minerals, animals, and different additional resources from nature for the treatment, prevention, and cure of diseases. Ethnomedicine is firmly grounded in the cultural and social ideologies of these communities and often involves rituals and traditional practices that address not only physical healing but also spiritual and holistic well-being. This interdisciplinary field explores the traditional medical and healing practices of different cultural groups, examining the knowledge, practices, and beliefs related to treatment, health, and disease within ethnic communities. These practices are often based on local traditions, mythology, beliefs, and intergenerational transmission of experiences (Aliev et al. 2023; Megersa et al. 2023). Health is a critical attribute of human existence that is influenced by a multitude of sociocultural and biological variables.

Health is an outcome characterized by optimal mental, social, and physical health. When any of these aspects of well-being is compromised or when the body fails to operate normally due to certain factors, the resulting state is referred to as disorder or, at times, sickness. The knowledge of this is grounded in scientific truths. However, conceptual differences have been identified within tribal communities or among locals regarding wellness, disease, and sickness. Their ideology and perspective diverge to some extent from the scientific method. It is a common misconception that diseases and illnesses are controlled and caused by God, supernatural forces, or alchemy. A physician cannot remedy a patient's problems with the prescription of medication alone. The function of an anthropologist in medicine is to investigate the etiology and management of illness and disease from the perspective of indigenous peoples, which may vary across tribal communities (Helman 2007). Anthropologists endeavor to ascertain the perspectives of people in the community concerning the causes of illness, wellness, and bad health; the methods by which they self-treat such conditions; the therapies and medicines they employ to remedy physical and mental conditions; any misconceptions they may have about contemporary medicine; and whether they harbor any fears or taboos concerning disease or the current medical system. In many cases, medical anthropologists suggest solutions that align with the entire community's culture and mentality based on the gathered data. A medicinal anthropologist functions in this manner as a social doctor, distinct from a medical practitioner. Tribal communities have upheld and conserved their traditional and native understandings of the medicinal properties of animals and vegetation since the dawn of time. Each tribal group possesses a distinct and particular understanding of ethnomedical practices that sets them apart from one another. Ethnomedicine is a form of traditional medicine observed within tribal societies, wherein members employ medicinal flora and animals to regale various ailments (Souza et al. 2018; Aliev et al. 2023).

Several parallels and distinctions exist among biomedical science, ethnomedicine, and Ayurveda. These later are united by the conventional understanding and application of herbal plants and certain animal-based goods (e.g., urine from cows, manure from cows, ghee, honey, and milk). The distinction is that Ayurveda is documented, universal, and worldwide; it abstains from consuming flesh from animals and meat, as well as religious rituals involving sorcery, in its approach to treating illnesses. The system of Ayurveda employs herbal remedies, meditation, yoga, and philosophy based on nature to prevent and treat illness. Conversely, ethnomedicine operates on the premise that knowledge is transmitted orally rather than in writing and is restricted to a select group of specialized, indigenous medical professionals known as "bargain" or "shamans" (Shrungeswara and Unnikrishnan 2019; Aliev et al. 2023).

Ethnomedicine is characterized by its tribal, mystical, rather than scientific nature (Erickson 2007). It examines the therapeutic methods employed by different cultural communities and personal encounters with diseases and illnesses. The field of ethnomedicine encompasses the utilization of medicinal products made from animals and plants, and magical effects, for the purpose of illness treatment. Biomedicine is a contemporary medical system that employs bio-products to address a range of conditions, including cold, cough, fever, gastrointestinal issues, and serious chronic illnesses. The utilization of advanced electromagnetic treatments and equipment achieves this. The approach is grounded in rigorous scientific knowledge and methodologies. Ethnomedicine is limited in its efficacy for life-threatening conditions such as diabetes, cancer, and hypertension. Its application is limited to treating common cold cough, stomach or liver issues, joint pain, and other related conditions. These deficiencies and constraints pose a significant risk to this traditional knowledge.

1.1 Definition and Scope of Ethnomedicine

Ethnomedicine can be defined as the science dealing with traditional healthcare practices, beliefs and concepts of ethnic communities that have evolved in isolation with the cultural evolution of races and experiences passed on by different generations. According to Foster, ethnomedicine comprises of those beliefs and practices relating to disease that are the product of indigenous cultural development and are not explicitly derived from the frame work of contemporary health care. In short, ethnomedicine, utilizing minerals, plants, and animal-derived goods for therapeutic and wellness applications not specified in the conventional medicine utilization of the distinct cultures, is a well-being-based concept founded upon the principles of biodiversity and is transmitted between decades within the ethnic communities of a particular nation or region. Acharya Charaka, one of the great scholars and physicians of Ayurveda, was the first to describe the importance and concept of ethnomedical studies including its applied aspects in his famous treatise *Charakasamhita*.

Charaka emphasized the significance of accurate recognition of plants as vital for the selection of correct ingredients of a drug either alone or in combination for the purpose of treatment. According to Charaka, it is insufficient to confirm a plant's presence for effective treatment. Instead, it is crucial to possess a comprehensive understanding of its morphological characteristics. To gain a comprehensive understanding of plants and their therapeutic applications, it is imperative to engage in reciprocal interactions with the hill tribe, cowherds, and sages residing in the forest. The perspectives put out by Charaka have garnered significant recognition and are often regarded as the fundamental principles of ethnomedical investigation, even in contemporary times.

1.2 History and Importance of Traditional Resources in Medicine

Herbalism is a folk medicine and traditional therapeutic practice that involves the application of extracts from plants and vegetation. Herbs constitute a significant portion of traditional medicine worldwide and are among the most prevalent forms of organisms on the planet. According to estimates, the total number of plant species, encompassing bryophytes, seed plants, and ferns, is approximately 350,000. As of 2004, 287,655 species have been officially identified (Gardens 2023). HM is also referred to as phytotherapy, botanical medicine, or herbal medicine and encompasses a range of substances, including herbal ingredients, preparations, and completed herbal items. These substances predominantly consist of active components derived from plants or other substances (Sim et al. 2013). Herbal therapy utilizes various plant components such as roots, seeds, berries, leaves, bark, fruits, flowers, or entire plants. Humans primarily relied on unrefined botanical substances for their medical requirements to maintain vitality and treat illnesses (Jack 1997). Prior to the discovery of aspirin, which originated from *Spiraea ulmaria*, Egyptian papyri recorded its use for fever and edema, and Greek Hippocrates advised its application for both fever and pain.

Despite written documents regarding MP currently over 5000 years to the Sumerians, who documented caraway, laurel, and thyme as having confirmed medicinal applications (Falodun 2010), archaeological research indicates that the utilization of HM has its origins in Iraq and China, dating back 8000 and 60,000 years, respectively (Leroi-Gourhan 1975; Rao et al. 2019). Therapists of mainstream medicine have challenged herbal medicine since the development of medical science in the West (or "traditional medicine") in the last century as a result of the absence of scientific evidence in contemporary medicine, notwithstanding its extensive history of effective utilization. Curiously, circumstances evolve with time. Recently, there has been a notable resurgence in the utilization of herbal remedies. The negative consequences of chemical medications and the lack of viable current remedies for different chronic conditions can be identified as contributing factors to this phenomenon, the phenomenon of microbial resistance's rise, and the remarkable allocation of resources toward drug development and research (R&D) (Pan et al. 2010). The approval of novel pharmaceuticals by the US Food and Drug Administration (FDA) has amounted to 1200 since 1950 (Munos 2009). Consequently, the utilization of herbs for medicinal intentions has experienced a surge in global prominence throughout the last four decades, encompassing nations that are industrialized as well as developing (Humber 2002). Furthermore, multinational pharmaceutical corporations, equipped with cutting-edge science and technology as well as innovative concepts, have initiated a reevaluation of botanicals as possible sources of new medicinal products and refocused their approaches on the development and exploration of natural product drugs (Seidl 2002; Li and Zhang 2008; Corson and Crews 2007; Schmidt et al. 2007).

Source of Drug Discovery: Many pharmaceuticals and modern medicines have their origins in traditional remedies. Scientists have discovered valuable compounds and active ingredients in plants, animals, and minerals used in ethnomedicine. These resources have functioned as a wellspring of incentive for drug discovery and the evolution of new medications. This historical context underscores the valuable contributions of traditional medicine to the advancement of healthcare practices (Schmidt et al. 2007; Alives et al. 2023).

In contemporary times, numerous "conventional" medicine practitioners are unhesitant to advise their patients on the efficacy of alternative and complementary therapies, herbal remedies, or botanicals for treating particular illnesses (Suzuki and Medicine 2004; Manek et al. 2010). A 2007 survey found that approximately 40% of adults and 11% of children engaged in complementary and alternative medicine (CAMT). Among the adult participants, white people accounted for 43.1%, and black adults accounted for 25.5% (Barnes et al. 2008). Furthermore, individuals with elevated academic achievement and money levels tend to utilize complementary and alternative medicine (CAM) and herbal remedies more frequently (Bishop et al. 2010; Moran et al. 2013). In this regard, a 2012 study found a significant correlation between a higher educational level and CAM use, with patients younger than 40 with cancer of the breast exhibiting a pattern toward increased use (Chung et al. 2014). The public's acceptance of CAM/CAMT remains unaffected by the fact that the precise facts and processes underlying the majority of traditional treatments and/or how they avoid illness are only partially understood at this time. Despite the global diversity of CAM/CAMT, all of them can be classified into two main categories: CAM/CAMT that is drug-based and CAM/CAMT that is not drug-based (Pan et al. 2012).

1.2.1 Cultural Significance

Traditional resources are deeply intertwined with the culture and identity of various communities. They often hold symbolic and spiritual value, and their use in healing reflects a profound connection with the natural world. Preserving these traditions is essential for maintaining cultural diversity:

1. **Local and Sustainable Treatment**: Traditional resources are often more accessible and affordable in areas with limited resources, where modern medicine may not be readily available. Moreover, these resources can sometimes be harvested sustainably, contributing to the protection of local ecosystems.
2. **Holistic Approach**: Ethnomedicine often embraces a holistic approach to health, considering not only physical symptoms but also emotional and spiritual aspects of well-being. Such an approach can complement modern medicine, which tends to focus more on the physical aspects of health (Megersa et al. 2023).
3. **Healthcare in Indigenous and Remote Communities**: In many indigenous and remote communities, traditional healing practices serve as a primary or sole

form of health care. They are essential for maintaining the health and well-being of these populations.

In summary, the significance of traditional resources in ethnomedicine lies in their historical importance, drug discovery potential, cultural relevance, and their role in providing health care in resource-limited or remote areas. The synergy of traditional and modern medicine can lead to more comprehensive healthcare solutions. However, ensuring the sustainable use and protection of these resources and respecting the cultural aspects of ethnomedicine are crucial.

2 Plants in Ethnomedicine

2.1 Selected Plants with Proven Medicinal Properties in Traditional Medicine

Traditional practitioners consider forests to be a vital source of medicinal flora. Schippmann et al. report that 52,885 species among the 422,000 species of blossoming plants worldwide are utilized for therapeutic purposes (Schippmann et al. 2002). Tropical rainforests are renowned for their exceptional biodiversity (Shah and Bhat 2019) and housing approximately half of the world's species (Myers 1988); consequently, these ecosystems possess tremendous potential as therapeutic resources. Nevertheless, the destruction of their habitats, deforestation, and degradation of forests diminish the accessibility of medicinal herbs (Shanley and Luz 2003). From 14% to approximately 6%, the global area covered by tropical rainforests has diminished. Under the current rate of deforestation, the remaining rainforests may cease to exist within less than four decades (Taylor 2005). Approximately 137 species of animals, insects, and plants are believed to be lost globally on a daily basis. The annual devastation of rainforests results in the loss of approximately 50,000 species (Shah and Bhat 2019). Recently, the drug sector has acknowledged the therapeutic value of tropical rainforest flora in developing new drugs (Pistorius and van Wijk 1993). Present-day pharmaceuticals derived from rainforest flora comprise 25% of the global supply. According to researchers, an estimated 70% of rainforest vegetation possesses anticancer abilities (Shah and Bhat 2019). Even with this immense potential, pharmacological analysis of tropical forest plants had examined less than 1% of the plant species as of 1995 (Roberson 2008). Simultaneously, scientists are expressing significant apprehensions regarding the devastation of tropical rainforests, which directly jeopardizes the identification of chemical components that could be included in traditional pharmaceutical systems. The potential extinction of plant species and the cultural traditions responsible for their discovery could preclude their official recognition for their medicinal properties. Many prominent pharmaceutical compounds utilized in contemporary medicine have been sourced from the conventional wisdom of native populations residing in tropical forests. The compound quinine, employed in treating malaria, is derived from the tree of cinchona. This tree is indigenous to the tropical rainforests of the Andes region. The Quench Indians have historically utilized quinine as a traditional medicinal remedy (Frost and Griffiths 2001). Primarily utilized as an ordeal poison by the indigenous peoples, physostigmine (Somani and Dube 1989), which is obtained from Calabar legumes in the tropical forests of Africa, is a treatment for glaucoma (Proudfoot 2006). Native Americans used cortisone, the active component in contraception tablets, originating from wild yams that were uncovered in South American tropical rainforests (Pariser 1997). Tubocurarine, a drug that relaxes muscles used in surgical procedures, is sourced from the curare lianas endemic to the Amazon jungles. Its historical usage can be traced back to indigenous communities' application of it as a toxic substance for coating arrow shafts (Place 2001). Treatments for pediatric leukemia and Hodgkin's disease, such as vincristine and vinblastine, are isolated from the pink periwinkle, a plant indigenous to the rainforests of Madagascar that is also utilized by traditional remedies (Halpern and Weverka 2003). One notable pharmaceutical breakthrough that faced significant challenges in the current century involved the identification of Calanolide A, an antihuman immunodeficiency virus (HIV) chemical. This substance was produced from a scarce tree species known as *Calophyllum lanigerum* var. *austrocoriaceum*, indigenous to Sarawak, Malaysia's tropical jungles. Upon the researchers from the National Cancer Institute in the United States' return to get samples from the tree, it was observed that the local inhabitants had already felled the tree. However, thankfully, specimens were successfully retrieved from the Singapore Botanical Gardens. Calanolide B, a further anti-HIV component, has been isolated from *Calophyllum teysmannii.* And the clinical studies for Calanolide A are underway, and Calanolide B continues to undergo preclinical testing (Butler 2005). Approximately 80% of the global population relies on botanical herbal remedies (Ekor 2014). Over a thousand companies are involved in the provision of medicinal plant-based goods, which constitute an annual worldwide market valued at 60$ billion (Newmaster et al. 2013). There is a rising requirement for herbal medicine in developed and developing nations (Ganie et al. 2015). Communities in the Pacific Islands have utilized traditional medicine since antiquity; oral transmission has been observed across successive generations. Progressively eroding, nevertheless, is conventional wisdom regarding the therapeutic properties of tropical forests and plants that are medicinal (Voeks and Leony 2004). The degradation of traditional medicinal knowledge can be ascribed to a multitude of noteworthy variables, including the growing prevalence of Western medicine, cultural shifts, and limited availability of medicinal plant habitats (Alves and Rosa 2007). The identification and economic significance of phytochemicals can be enhanced by utilizing traditional knowledge possessed by indigenous groups (Cox and Balick 1994; Balick et al. 1996). Approximately 86% of the botanical specimens utilized by conventional healers in Samoa have exhibited noteworthy pharmacological activities within experimental evaluations (Cox and Balick 1994). In contrast to the exceedingly low achievement rate of 1% observed in random plant assessment, Shaman Pharmaceuticals achieved a 50% success rate in developing drugs by employing indigenous understanding of traditional healers or Shamans in Africa, Asia, and

Latin America. Prior to testing a plant, Shaman Pharmaceuticals confirmed its medicinal utility in at least two distinct communities (Kumar and Tarui 2004). The initial laboratory experiments for HIV were carried out by Cox and Balick (1994) which revealed that the number of positive results associated with plant extracts utilized was 4× greater than that of randomly selected specimens (Cox and Balick 1994).

2.2 *Methods for Identifying Active Plant Ingredients*

When developing novel drugs, ethnobotanical data on plants with medicinal properties is frequently used to inform chemical examinations. The initial screening selection was based on established clinical efficacy and safety records of traditional botanicals (Fig. 1).

Following the identification and collection of plant materials based on ethnobotanical data and phytochemical examination, a screening process was initiated in collaboration with indigenous medicinal plant consumers. To evaluate the identified critical components of herbal remedies, phytochemistry, studies on animals in the pharmacy, and research studies are utilized. Subsequently, they may be incorporated into the formulation of novel pharmaceuticals in accordance with governmental regulations pertaining to advancements in medicine. Based on a comprehensive analysis of 400 scholarly articles published in China between October 1987 and October 1988, about 150 novel substances were identified and documented in relation to the chemical properties of medicinal plants employed in ethnomedicine. A minimum of 20 novel pharmaceuticals were invented and commercialized in the 1980s on the basis of conventional medicine. A considerable number of these novel pharmaceuticals were formed from chemical extracts derived from various MPs. Pure and biologically active substances derived from MP are occasionally employed

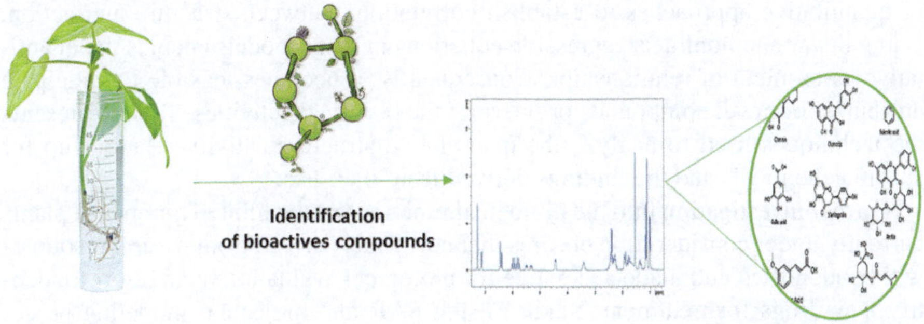

Fig. 1 Identification of active plant components

in formulating novel traditional Chinese medicine (TCM) medications albeit less frequently than the concentrated substances derived from medicinal herbs. The 1980s witnessed the establishment of several cooperation projects between Chinese research institutes and international pharmaceutical corporations, with the aim of conducting investigations on MP. As an illustration, the Chinese Academy of Sciences' Yunnan Institute of Tropical Botany (YITB) and Yamanouchi Pharmaceutical Ltd., a Japanese company, collaborated from 1986 to 1991 to investigate and explore the active ingredients in herbal remedies. The collaborative endeavor was executed employing an ethnobotanical methodology. An assessment was conducted on 155 candidate species of plants using ethnobotanical data collected from the available areas. Following the collection of plant materials consistent with ethnobotanical data, YITB scientists made samples. Additional pharmacological and chemical investigations were carried out in Japan. From 1988 to 1991, seven inventions were received in connection with this collaborative research endeavor between the two parties involved in the initiative.

2.3 The Use of Modern Tools for the Study of Plant Biological Activities

Using contemporary medicinal chemistry techniques, such as combined chemistry and molecular simulation, the identified naturally occurring bioactive substances may serve as principal compounds for optimizing structural characteristics in order to create new, more potent analogs. Moreover, natural substances exist as a group of structurally similar molecules together with other chemicals. As a result, several homologs can be derived from a single source, which can offer information relating to structure-activity relationships (SAR). The utilization of SAR is constrained by the absence of quantitative activity forecasting for novel compounds. Conversely, quantitative structure-activity (QSAR/QSPR) correlations enable the development of quantitative approaches to establish correlations between structure and action. Using linear and nonlinear regression equations in these models enables the quantitative assessment of relationships. Consequently, it becomes feasible to anticipate attributes for novel compounds or interpret the observed activities. Table 1 presents the techniques used to analyze the quantitative structure-activity relationship for specific chemicals and the findings derived from their use.

Further investigation into the biological functions of the limited number of plants currently under consideration ensures that new leads derived from natural resources will be identified and made accessible for biological evaluation in an effort to identify new drugs for treatment. Studies using SAR and molecular modeling procedures are applied to newly isolated substances with acceptable biological activity in modern natural product drug discovery to formulate and acquire analogs with greater resistance, fewer harmful impacts, as well as improved pharmacokinetic characteristics. Also, research may demonstrate that the pharmacological activity of

Table 1 Methodologies for quantitative structure-activity analysis for some bioactive compounds

Component	Descriptors	Chemometric method	Results	References
Naringin/ (-)-catechin	Steric, hydrogen bond donor fields, log P, electrostatic fields	CoMSIA, CoMFA, 3D-QSAR	The active sites for antibacterial action against *E. coli* include bulky unfavorable sites at C-3′ and C-4, H-bond donors at C-4, C-3, C-8, and C-3′, electronegative bonds at C-3 and C-4, and H-bond acceptors at C-4′	Fang et al. (2016)
Quercetin/ rutin	D/Dr05 RCI, Mor27u, BIC1, JGI1, Mor30m IC,	MLR, 3D-QSAR	The inhibition of the H1N1 virus is contingent upon various factors, including the electric charges, polarizabilities, masses of the atoms inside the molecule, and the conformation of the molecular structure	Mercader and Pomilio (2010)
Flavone	Surface tension, E_{HOMO}, molar refractivity, dipole moment	PCA, MLR, QSAR, MNLR, ANN,	GABA (A) receptor binding affinity constants (−log Ki) of flavone and its derivatives	Ghamali et al. (2017)
Naringenin	Electrostatic fields, log P	CoMFA, 3D-QSAR,	The chromane scaffold was seen to have sterically unfavorable areas located below the 3-position and beyond the 2-position	Jiang et al. (2017)
Naringenin Catechin Quercetin	R4u+, PJI2 Mor27m, G1e	PLS QSAR	Descriptors pertain to the interaction between flavonoids and a binding site, as well as their ability to penetrate the membrane of the bacteria of *Salmonella typhimurium* TA98	de Melo et al. (2010)

the compounds under investigation can be influenced by their interactions with specific enzymes. Analogs possessing optimal drug ability can be synthesized and assessed in the laboratory through various in vitro/in vivo biological experiments (Kitchen et al. 2004). The following is a succinct overview of the procedure involved in the design and development of analogs derived from a lead chemical that is naturally separated.

2.3.1 In Silico Ligand Construction and Preparation

This procedure necessitates PDB-formatted optimized three-dimensional models of ligands. PubChem, ZINC, and natural compound databases are dependable resources for obtaining the structures of recognized natural product in various acceptable forms, including mol., SDF., PDB, mol2, and others (Sorokina and

Steinbeck 2020). To minimize energy consumption, the geometry of the generated structures must be optimized. In advance of docking, energy reduction can be conducted using docking software such as Discovery Studio and AutoDock Vina. Alternatively, independent software tools like Chem 3D Ultra, Chimaera, Avogadro, and others can be employed for structure development and optimization purposes (Liao et al. 2011; Chen et al. 2020).

2.3.2 Target Preparation

The process involves generating or obtaining the three-dimensional configurations (PDB) of the molecules being studied, including enzymes (e.g., human serum albumin), receptors (e.g., PPAR-γ), and proteins (e.g., topoisomerase II, cyclooxygenase). These structures are then refined in terms of energy and geometry, defining the binding site and calculating standard scores for available natural ligands.

2.3.3 Docking

Docking software is utilized to compare the three-dimensional structures of natural products to the desired structure, and the binding energy determines the resulting rankings. Docking applications such as AutoDock, Discovery Studio, Flex X, AutoDock Vina, and MDock are widely utilized (Chen et al. 2020). Docking requires the PDB-formatted optimized three-dimensional structures of the ligands and targets. It employs a variety of search algorithms, including the Lamarckian genetic algorithm, to determine the optimal ligand-binding conformation. Conducting post-docking investigations is crucial to examining intermolecular interactions and providing support for the findings derived from docking.

2.3.4 Identification of Hit Molecule

Following the docking modeling, the obtained findings are subjected to analysis in order to identify the most significant interactions, as determined by the energy scores. It is customary to conduct molecular dynamics (MD) simulation studies on the best ten scores. The method of MD simulation can be applied to two systems in this regard: The first category is an Apo (unfinished) protein/receptor system, and the second category is a protein/receptor system linked with an interacting substance as predicted by docking analysis. Hit molecules exhibiting a strong affinity for the target are identified in accordance with the classification of ligands and their interactions with the target.

2.3.5 Optimization of Hits

Hit optimization involves identifying analogs of hits that exhibit superior affinity for the natural product. The analogs with the highest affinity are then developed, and QSAR software is used to analyze a variety of drug-like properties, such as pharmacodynamics, stability, as well as pharmacokinetic characteristics (Devillers 2013; Sullivan et al. 2014).

In contemporary drug discovery, as an alternative to extracting and testing natural substances, virtual screening or in silico analysis is utilized to evaluate an abundance of molecules derived from natural resources that exhibit different chemical structures. Educational institutions and research centers maintain libraries containing potentially millions of these structurally diverse chemicals. By employing virtual screening, prospective results for a specific pharmacological activity target can be detected, and an effective SAR can be generated via a lead optimization procedure. Virtual screening can effectively reduce the variety of substances required for the actual bioassay assessment (Stockwell 2004; Bindseil et al. 2001). To validate the anticipated novel biological properties, it is imperative that the hit molecules be physically accessible and subjected to appropriate bioassays. These compounds can potentially be synthesized in a laboratory or, if commercially available, purchased. ChemSpider, DNP, Phyto Pure, Natural Product Alert, Tim Tech, and others are examples of widely accessible natural product databases and libraries (OBrien and Haskins 2007).

3 Animals in Ethnomedicine

According to the World Health Organization (WHO), up to 80% of the world's six billion inhabitants depend predominantly on remedies derived from plants and animals (WHO 1993). The natural pharmacopeia used by conventional populations of humans encompasses a diverse array of species of wild plants and animals. Ingredients derived from indigenous flora and fauna are not just employed in conventional medicinal practices. Still, they are progressively esteemed as primary constituents in formulating contemporary pharmaceuticals and MP (Kang and Phipps 2003). Animal-derived goods and substances obtained from various animal organs have been utilized as therapeutic agents in diverse cultures since antiquity. These applications continue to persist in traditional medicine (Lev 2003; Adeola 1992; Angeletti et al. 1992). Zootherapy is the practice of applying therapeutic substances derived from or acquired from animals to treat human disorders (Costa-Neto 2005). "Every human culture with a structured medical system will utilize animals as medicines," according to Marques (1994). An extensive geographic reach and profound historical roots characterize zootherapy. Zootherapy is a significant alternative therapy that is widely performed in contemporary societies alongside other established therapeutic approaches. Using wild and domestic animals, together with their by-products such as hooves, skins, bones, feathers, and tusks, holds significant value in developing medicinal compounds that possess therapeutic, protective, and preventive properties (Adeola 1992; Angeletti et al. 1992). For instance, over 1500 animal species have been documented in traditional Chinese medicine (TCM) as having some medicinal applicability (Traditional and Medicine 1995). Approximately 15–20% of the Ayurvedic medicine in India comprises compounds derived from animals (Shrungeswara and Unnikrishnan 2019). Throughout the northeastern state of Bahia, Brazil, more than 180 species of medicinal plants have been documented (Costa-Neto 2004).

3.1 *Traditional Use of Various Animal Species for Medical Purposes*

The utilization of natural sources for therapeutic purposes, palliatives, curatives for various ailments, or recreational purposes has been of interest since ancient times (Spainhour 2005). Within this framework, the utilization of natural products for medicinal purposes likely predates the documented history of humanity by several millennia (Ji et al. 2009). In contrast to the consistent use of animal, vegetable, and mineral compounds by compounders of medicines in past centuries, animal compounds are less frequent than herbs and more frequent than minerals, according to surveys of previously utilized medications (MacKinney 1946). Various components of the formula medicinal products, including animal parts, whole animals, and products derived from animals, have been utilized medicinally in multiple areas of the world (Alakbarli 2006). Early documentation regarding the medicinal applications of natural products was composed in cuneiform on clay tablets in Mesopotamia around 2600 B.C. (Cragg and Newman 2001). Egypt also produced some of the earliest documented accounts of the use of medicinal products for healing. It is widely acknowledged that between 3000 and 6000 years ago, the ancient Egyptians formulated a diverse array of potent pharmacological substances derived from natural sources (Halberstein 2005). "By far the most prevalent form of treatment recommended in the medical papyri was the use of drugs, derived from a vast array of vegetable, animal, and mineral compounds and delivered in various ways," according to Nunn (2002). Our knowledge of Egyptian pharmacology and medicine is derived from inscriptions on tombs, monuments, and papyrus cylinders, with Papyrus Ebers being the most significant. Its approximate date of composition is 1550 B.C. (Lidgard 2005) and containing at least 800 remedies for a variety of minerals, herbs, and animals, of which a significant number include substances obtained from insects, fish, and untamed and domesticated birds, and animals. A diverse array of medicinal products was obtained from these animals, including but not limited to the raw materials (e.g., eggs, milk), excrements, bones, flesh, marrow, fat, and hide of carcasses; in specific instances, distinctive materials such as tortoise shells, feathers of birds, snake sloughs, and dirt extracted from beneath human fingernails were also utilized. In instances involving worms, insects, and small animals, it was common practice to utilize the entirety of their bodies. Conversely, only certain pieces were employed for regular animals, such as the head, hair, brain, horns, eyes, jawbone, teeth, heart, spleen, liver, gall, uterus, testicles, vulva, legs, etc. Undoubtedly, the most preferred substance was grease, or fat, precisely that of the goose and bovine. Grain-based substances were utilized in over a hundred recipes, albeit occasionally as a vehicle instead of the active component. Subsequent in significance were the excrements used in approximately 50% of the medications (MacKinney 1946).

3.2 *Biologically Active Compounds Derived from Animals as Potential Drugs*

Animals have also served as a reservoir of intriguing chemicals that possess potential therapeutic applications. The potency of epibatidine, derived from the dermal tissue of an Ecuadorian poison frog, is tenfold more significant than that of morphine (Spande et al. 1992). Venoms and poisons derived from animals have been instrumental in developing numerous therapeutic interventions for various ailments.

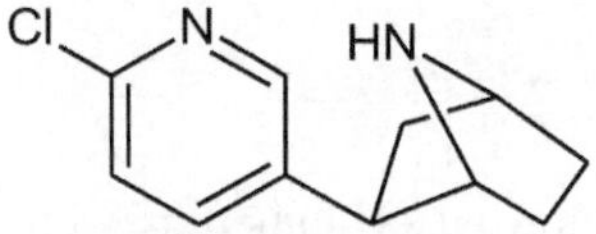

Epibatidine

The synthesis of morphine, a crystalline compound, from opium in a laboratory test tube was accomplished by Friedrich Wilhelm Sertürner in 1803/1804. This achievement marked the beginning of drug discovery incorporating natural products (Klockgether-Radke 2002); in 1826, E. Merck introduced a medicine of significant commercial importance (Kinghorn 2002).

3.3 *Ethical and Ecological Aspects of Using Animals in Medicine*

Differences abound in our treatment and attitudes toward animals, and there needs to be more clarity regarding the optimal conditions to utilize and maintain them. Initially, it is imperative to take into account the ecological consequences of all animal systems of management on a global and local scale while also designing an environment that adheres to the biospherical model. The interdependence of organisms and the environment, the self-sustaining character of ecosystems, and the significance of biodiversity in maintaining and ensuring the stability of ecosystems are the primary factors to consider. These factors may and ought to be considered when evaluating animal care. In addition to companion/urban canines, examples of farm, zoo, and circus animals illustrate them. Regarding the ethological requirements of the animals, the environment has to be considered. Two potential methodologies exist to address this matter: (1) a reductionist strategy, as exemplified by selecting experimental evaluations, and (2) a comprehensive evolution of strategy that prioritizes the identification of anguish and the extent of behavioral limitation (Megersa et al. 2023). In determining the etiological validity of an environment, the evaluation of an animal's ethological demands must consider both the species and the individual's requirements (interaction system, species-specific brain receptor features, and cognition) and experience, respectively. The impact of domestication on behavior and methods for evaluating distress. There are varying ethical perspectives about animals and their treatment, with proponents asserting that the ethical acceptability of human utilization of animals is contingent upon their placement in ecologically and ethically adequate habitats. The relationship between animals and humans should be defined by symbiosis, which entails mutual advantage, as opposed to a parasitic or exploitative dynamic, when the employer assumes the employee's role and the master assumes the role of the slave.

4 Microorganisms in Ethnomedicine

Pharmaceutical Substances Obtained from Fungi: Comprised of numerous well-known fungi, the fungal kingdom encompasses organisms that hold significant ecological and economic importance, including morels, yeasts, puffballs, rusts, truffles, and molds. Numerous illnesses of both animals and plants have been linked to fungi. However, the investigation of fungal metabolites has yielded multiple compounds that have demonstrated efficacy as antibiotics, antifungals, immunosuppressants, and cholesterol-lowering drugs.

Fleming's 1928 development of penicillin from *Penicillium notatum*, Chain, Florey, and colleagues' re-isolation and clinical investigations at the beginning of the 1940s and the subsequent commercialization of synthesized penicillin transformed drug discovery exploration (Sengupta et al. 2013). In an attempt to discover new antibiotics, pharmaceutical companies and scientific organizations assembled enormous collections of microorganisms on a global scale in response to the tremendous triumph of penicillin (Wainwright 1990). The initial years were extraordinarily fruitful, with the discovery of chloramphenicol, streptomycin, cephalosporin C, chlortetracycline, vancomycin, and erythromycin, among others. Presently, each of these chemicals or their derivatives is used as a pharmaceutical agent (Sengupta et al. 2013; Butler 2004).

Embryophytic fungi cryptoperiod quercetin, which was extracted from the medicinal plant *Tripterygium wilfordii* (native to Eurasia), has demonstrated antifungal properties against *Candida albicans* and *Trichophyton mentagrophyte*, two significant human fungal pathogens (Guo et al. 2000). As a bioactive concept, the new peptide cryptocyanin remains under consideration for its potential application in treating fungal illnesses affecting the epidermis and nails (Guo et al. 2000).

A study was conducted to examine the antibiotic properties of an unnamed endophytic fungus found in Costa Rica, resulting in the identification of Guanacaste Pene A. The study demonstrated the presence of antibiotic action against methicillin-sensitive and methicillin-resistant *Staphylococcus aureus* (MSSA and MRSA) as well as vancomycin-resistant *Enterococcus faecalis* (VRE) (Brady et al. 2000). It has been postulated that membrane disruption is the primary mechanism of bactericidal effect.

Tricyanic acid, extracted from an endophyte discovered in the imperiled tree *Torreya taxifolia*, exhibited cytotoxicity five to ten times higher in cancer cell lines sensitive to protein kinase C, the enzyme responsible for inducing apoptosis-mediated cell death (Dias et al. 2012).

4.1 Microorganisms in Traditional Folk Medicine Practices

Antibiotic research commenced with isolating streptomycin and penicillin, both of which originated from naturally existing organisms. The golden era of antibiotic research occurred between 1950 and 1960 when approximately half of all antibiotics presently employed were discovered (Davies 2006). As antibiotic resistance developed and industrial production of chemical constituents became more prevalent, the golden age ended. The annual toll from antimicrobial-resistant organisms is 700,000, projected to increase to ten million by 2050, resulting in an economic output loss of $100 trillion (O'Neill 2016). Recently, the World Health Organization (WHO) urgently demanded new antibiotics to combat the danger of antimicrobial resistance (WHO 2017). The World Health Organization, in its Strategy Agenda for Antimicrobial Resistance in the Region of the Western Pacific, urged its member nations to assist in discovering novel antimicrobial medications (WHO 2015). As a result, antimicrobial resistance has been designated a national priority in Fiji. Accelerating the search for novel antimicrobial substances is imperative due to the escalating resistance to antiquated antibiotics; the variety of plant chemicals presents a distinctive opportunity to tackle this pressing matter. Drug discovery reached its lowest point in 24 years in 2004, when a mere 25 new substances, referred to as New Chemical Entities, were introduced (Newman and Cragg 2007). This occurred as a result of the introduction of innovative drug discovery instruments, which caused drug manufacturers to shift their focus from sources that are natural.

Nevertheless, with the increasing reliance of contemporary medicine on natural product-derived compounds, there is a revitalized enthusiasm for exploring plant-based drug development. Half of the anticancer medications that have been developed since the 1940s are derived from natural substances (Newman and Cragg 2016). Natural products are 118 of the most popular 150 prescription medications sold in the United States (Daily 1997).

Seventy percent of the newly approved drugs in the United States over the past quarter century originated from natural products, regardless of the use of cutting-edge methodologies for developing products in laboratories (Steenhuysen 2007). Addressing the immobilization of antimicrobial drug discovery underscores the necessity of reorienting the process of drug discovery toward natural goods.

4.2 *Bioprospection of Microorganisms to Discover New Bioactive Compound*

Recently, bioprospecting for microorganisms containing bioactive substances of medicinal significance has emerged as a method for acquiring new and intriguing natural substances (Manganyi et al. 2019). As the development of drugs has progressed, medicinal plants have become an increasingly significant source of naturally occurring bioactive compounds (Toghueo and Boyom 2019). In contrast, the escalating scarcity and interest in plant-derived products, in conjunction with the endangerment of numerous ethnopharmacological plants, have led to a surge in the utilization of synthetic substances in the pharmaceutical industry (Rajamanikyam et al. 2017). Presently accessible synthetic substances employed in the realm of illness management and development of drugs exhibit diminished therapeutic advantages and engender intolerable side effects, including anemia and anorexia in patients who have cancer (Collins 2020). Hence, it is imperative to prioritize novel treatment approaches that prioritize the utilization of natural and secure biological products and technology (Ek-Ramos et al. 2019). A bioprocessing method is one of these strategies, in which beneficial bioactive substances with antifungal, antibacterial, antioxidant, anticancer, and antitumor properties are synthesized by microbial endophytes (Fig. 2) (Gupta et al. 2020).

Endophytes refer to microorganisms that reside within the tissues of plants without inducing any recognized adverse effects or illnesses in the host plant (Gouda et al. 2016). Over the course of time, there has been a gradual development of a

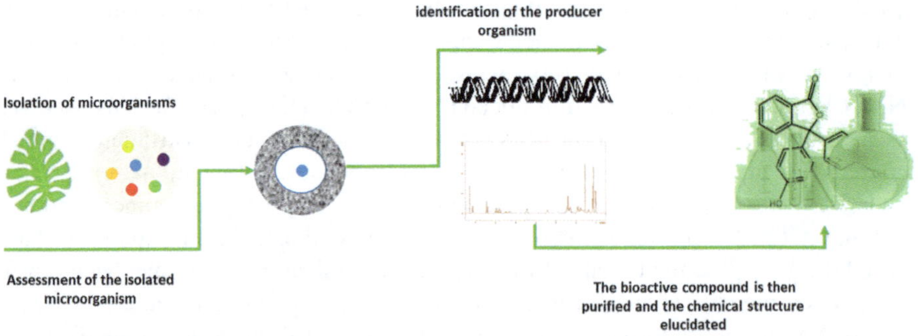

Fig. 2 General procedure of bioprospection of microorganisms for the discovery of biologically active natural compounds

mutually advantageous association between endophytes and host plants, which has developed over time (Adeleke et al. 2019). Comprehending the principles underlying endophyte-plant association can be used to effectively generate a wide range of advantageous bioactive chemicals (Venugopalan and Srivastava 2015). One potential approach to improving host plant production is the manipulation of growing circumstances to affect the efficient colonization by endophytes (Faeth and Fagan 2002). Moreover, the introduction of a specific bacterial or fungal endophyte into a host plant has the potential to enhance the generation of secondary metabolites in both the host plant and the companion endophyte (Firáková et al. 2007). This approach has the potential to offer helpful insights regarding the direct synthesis of therapeutically significant bioactive chemicals through the application of bioengineering techniques (Jia et al. 2016). The estimated number of endophytes present in different plant spheres exceeds one million (Blackwell 2011). However, the research conducted on these endophytes is limited, indicating a substantial potential for the discovery of endophytes with wide-ranging applications that have yet to be identified. Extensive research has been propelled by the capacity of endophytes to produce bioactive substances that are on par with those of their host plants and the immense diversity of life that those microbes support (Harrison and Griffin 2020). Endophytes contain several bioactive chemicals, such as phenolics, alkaloids, polyketides, flavonoids, and terpenoids, which have been identified as potential medicines for treating illnesses in humans like cancer, tumors, and infections caused by bacteria and fungi (Gupta et al. 2020). Therefore, the investigation and identification of previously unidentified endophytes containing bioactive substances that hold significant medicinal value have the potential to transform the administration of human well-being fundamentally. The advantageous uses of endophytes in sustainable agriculture have been the subject of extensive study. Key benefits that were reported included the biocontrol of phytopathogens (Sim et al. 2019), enhancement of stress tolerance in plants (Ferus et al. 2019), and enhancement of plant growth concerning nutrition solubilization, nitrogen fixation, and phytohormone synthesis (Mashiane et al. 2018). Nevertheless, there needs to be more research evaluating endophytes for bioactive substances, primarily emphasizing various therapeutic uses.

Fungi are recognized for their ability to induce various diseases in both plants and animals. However, extensive research on fungal metabolites has yielded numerous substances that have demonstrated efficacy, such as cholesterol-lowering, antibiotics, immunosuppressive agents, and antifungal agents (Table 2 and Fig. 3) (Gurnani et al. 2014).

Table 2 Active components derived from fungi

Isolates	Microorganism	Medicinal practices	References
Penicillin	*Penicillium notatum*	Antibiotic	Wainwright (1990)
Streptomycin	*Streptomyces griseus*	Antibiotic	Sengupta et al. (2013), Butler (2004)
Vancomycin	*Amycolatopsis orientalis*	Antibiotic	Sengupta et al. (2013), Butler (2004)
Cephalosporin C	*Cephalosporium acremonium*	Antibiotic	Sengupta et al. (2013), Butler (2004)
Chlortetracycline	*Streptomyces aureofaciens*	Antibiotic	Sengupta et al. (2013), Butler (2004)
Chloramphenicol	*Streptomyces venezuelae*	Antibiotic	Sengupta et al. (2013), Butler (2004)
Erythromycin	*Saccharopolyspora erythraea*	Antibiotic	Sengupta et al. (2013), Butler (2004)
β-Lactamase	*Streptomyces clavuligerus*	Frontline antibiotic	Van der Sar (2006)
Clavulanic acid	*Streptomyces clavuligerus*	Antibiotic	Van der Sar (2006)
Mevinolin	*Aspergillus terreus*	Cholesterol-lowering agent	Alberts et al. (1980)
Cholecystokinin (CCK)	*Aspergillus alliaceus*	Antagonist	Goetz et al. (1985)
Human cytomegalovirus (hCMV)	*Endophytic fungus*, *Cytonaema* sp.	Protease inhibitors	Herranz (2003)
Cryptosporiopsis quercina	*Tripterygium wilfordii*	Antifungal effect	Guo et al. (2000)
Sequoiatones A and B	*Aspergillus parasiticus*	Inhibition of human tumor cells	Gurnani et al. (2014), Stierle et al. (1999)
Guanacastepene A	*Unidentified endophytic fungus*	Antibiotic agent	Brady et al. (2000)
Anthracycline/doxorubicin	*Streptomyces peucetius*	Acute leukemia Soft tissue Bone sarcomas Lung cancer Thyroid cancer Hodgkins Non-Hodgkins lymphomas	Dewick (2002)

(continued)

Table 2 (continued)

Isolates	Microorganism	Medicinal practices	References
Torreyanic acid	*Torreya taxifolia*	Medical intervention for cancer cell lines exhibiting susceptibility to protein kinase C	Dias et al. (2012)

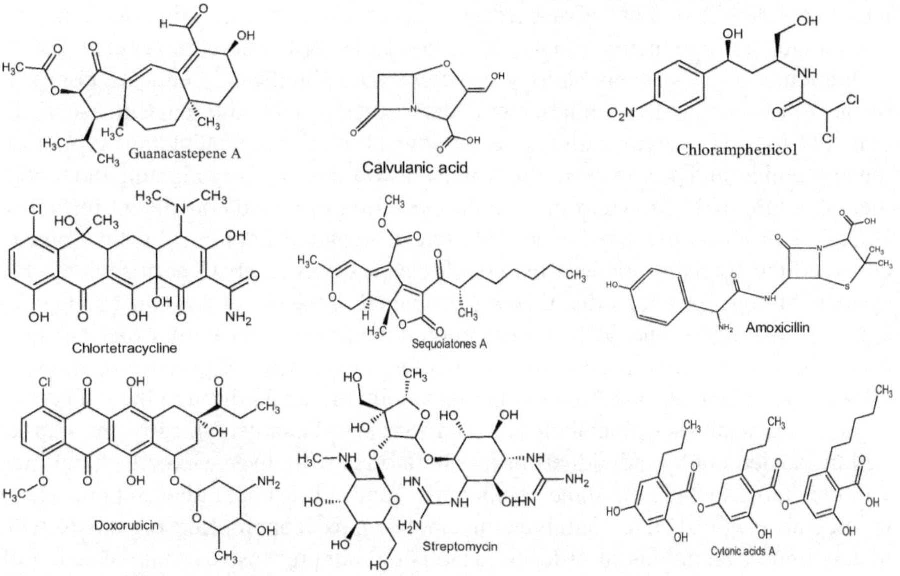

Fig. 3 Active components derived from fungi

4.3 The Use of Biotechnological Technologies in the Study of Microorganisms

In an emerging discipline within the life sciences, systems biology investigates the interrelationships among various constituents (such as regulatory networks and molecular pathways) inside an organism in order to uncover pertinent physiological processes and behavior (Kumar and Tarui 2004). The primary objective of this expanding discipline is to examine the intricacies of several regulatory, genetic, and metabolic activities within a cell, intending to comprehend the intricate nature of cellular networks (Van Seters et al. 1997). The conventional reductionist methodology for developing medications may lack the capability to identify activity when many active substances are present, as well as the synergistic effects of phytochemicals and prodrugs. The methodology is employed to investigate the efficacy of established targets through the utilization of receptor-binding tests. Utilizing an approach based on systems biology could enhance the effectiveness of drug development in holistic medicine, specifically when focusing on a medicinal plant with a well-established body of data (Megersa et al. 2023).

In order to establish the beneficial effects of herbal medications, clinical assessment is an indispensable method. Identifying the plant species and their active principles, followed by the synthesis of substantial quantities to meet the potentially enormous demand, is just one of the many stages involved in discovering herbal drugs derived from Indian traditional medicine. Diverse obstacles and concerns have been identified in the advancement of herbal medicines in India; these may be surmounted through international cooperation and coordination (Setzer et al. 2001).

Implementing a systems biology strategy would significantly enhance competitiveness vis-à-vis modern Chinese botanical medicine and traditional Chinese medicine (TCM). The organization's technological platforms, including genomics, metabolomics, and proteomics, offer robust instruments for investigating the fundamental nature of TCM syndrome and the operation of herbal constituent formulas. Botanical products that have undergone scientific and technological validation can be expedited by using modern methodologies such as reverse pharmacology and systems biology, provided that they are grounded in the principles underlying conventional medicine. The field of conventional medicine is currently experiencing a transformative process as communities explore novel ways of practice (Catalogue 2018). The techniques used to conduct metabolic modeling through the collection, storage, and analysis of metabolomic data vary significantly. Typically, these tasks will be carried out by individuals or in laboratories with diverse expertise, but they will inevitably include the same compounds. Hence, it is imperative and opportune to integrate established or contingent metabolic maps of appropriate organisms with metabolome evaluations in order to establish a comprehensive comprehension of metabolite concentrations and metabolic fluxes in these organisms, as well as their dynamic patterns under varying circumstances (Environment 2014; Whiffin and Kikkawa 1992).

5 Challenges and Perspectives in Drug Discovery from Ethnomedicine

5.1 The Need to Preserve the Knowledge of Traditional Medicine

The field of ethnobotany has been extensively employed in the scientific realm to chronicle indigenous understandings of plant utilization and to compile a comprehensive inventory of valuable plant species found in the local flora of Asian nations (Martin and Ethnobotany 1995; Ford 1994). Ethnopharmacology can be characterized as a scholarly domain that investigates the medical applications of plants within indigenous societies, with the aim of scientifically substantiating their efficacy and potential adverse reactions. Due to their cultural acceptability and ability to improve health, ethnopharmacological approaches can be recommended for use in poor nations (Farnsworth 1993). Nevertheless, ethnobotany has been extensively utilized in Asian nations such as India, China, Nepal, and Bangladesh to document traditional knowledge and compile inventories of herbal remedies (Pei 1998). The ethnobotanical knowledge of herbal remedies and their use by indigenous societies serves multiple purposes, including preserving cultural practices and biodiversity and advancing community medical care and pharmaceutical research. The provided material serves as a reference for the purpose of drug research, operating on the premise that a plant that has been historically employed by indigenous civilizations may possess allopathic properties (Farnsworth 1993). The documentation and inventory of ethnobotany are systematically structured, incorporating both local and scientific nomenclature, medical applications, cultural interpretations, as well as comprehensive details pertaining to the botany, ecology, phenology, harvesting practices, distribution patterns, management strategies, and conservation efforts associated with medicinal plants. The data is obtained from indigenous herbal practitioners, community healers, and herbal vendors at different local markets, as well as knowledgeable members of the community. The documented and detailed indigenous knowledge encompasses a wide range of local experiences that are rich, diversified, and frequently dependable in terms of preventing, treating, and preserving the health of individuals within the immediate vicinity. The incorporation of inventory and recording into community management of resources programs and preserving biodiversity at the local level can prove to be highly beneficial. Additionally, it can serve as a valuable resource for informing the process of plant choice and collecting samples for laboratory confirmation and pharmacological testing in the context of the development of drugs. Therefore, the records and inventory may be regarded as the repository of knowledge pertaining to traditional medicine. Traditional Chinese medicine (TCM) documentation and inventory has been revised to include 80 minerals, 11,146 plant species, and 1581 animal species. Between 500 and 600 species are frequently utilized as medicinal agents in TCM. A set of Chinese Medicinal Resources, published by Science Press, Beijing, from 1994 to 1995 and assembled into six volumes by the Chinese National Corporation of Traditional & Herbal Medicine, contains an exhaustive and well-documented compilation of TCM information. The series comprises six volumes:

- Chinese Medicinal Resources (287 pp.)
- Outline of Chinese Medicinal Resources (2069 pp.)
- Regionalization of Chinese Medicine (634 pp.)
- Commonly used Chinese Medicinal Materials (1138 pp.)
- Chinese Folk Prescriptions (1817 pp.)
- The Resource Atlas of Chinese Herbal Medicine (in Chinese and English, contains 81 color maps)

The Atlas paper holds scientific merit and practical importance within the field of pharmaceuticals. The series comprises a collection of articles on traditional Chinese medicine (TCM), encompassing a comprehensive list of 12,807 medicinal plants, rocks, and animals, as well as 20,000 traditional prescription drugs. The provided information encompasses the resource distribution, ecological conditions, extraction methods, historical utilization, and preservation data for every entry. China has a diverse population of 55 ethnic minority groups, including around 100 million individuals. These ethnicities are dispersed among 18 provinces, collectively accounting for 64% of China's total land. The assemblage of botanical medicine expertise and conventional medical systems comprise the "Ethno-Medicine" of China (EMC). The region is home to more than 8000 species of plants and animals, as determined by numerous inventories and recording efforts conducted within the past 20 years (Pei 1998). To illustrate, as documented, Tibetan Medicine comprises 840 natural minerals, 448 fauna, and 1106 plants. For developing novel pharmaceuticals and regional economies, ethnobotanical data on ethnomedicine has been regarded as a priceless asset. According to the WWF, traditional healers in India utilize a reported 2500 plant species. Additionally, Shah describes 100 kinds of plants routinely utilized as medication sources. Ethnobotanists documented approximately 800 herbal remedies in Nepal, whereas in Bangladesh, they documented 700 medicinal species. Reportedly, approximately 50% of the pharmaceuticals used in contemporary Pakistan are synthetically formulated from petroleum-chemical-based basic materials. Recently, there has been a growing inclination toward the documentation and inventorying of traditional herbal remedies in Himalayan nations through the utilization of ethnobotanical methodologies. WWF International, UNESCO, IDRC, ICIMOD, and European financial agencies have significantly contributed to the advancement and assistance of various ethnobotanical initiatives in the Himalayan nations.

5.2 Sustainable Use of Ethnomedicine Resources and Protection of Biodiversity

There is an increasing acknowledgment that the effective management of resources necessitates active involvement from the local populations. Therefore, community-based study and development programs in the fields of crops, forests, waterways, and management of resources have led to the development of participatory methodologies. The utilization of medicinal plants holds significant importance within social improvement programs implemented in rural Asian civilizations. Ethnobotany is a scientific field that investigates the interconnection between a certain society and its surrounding environment, with a special focus on the plant kingdom. These relationships can encompass several aspects, such as economic, social, ecological, figurative, religious, advertisement, or artistic considerations, among others. Ethnobotany is a field that encompasses elements from both the social and biological sciences. It entails a comprehensive approach that integrates many knowledge foundations and methodologies (Martin and Ethnobotany 1995). Initially, ethnobotany primarily relied on qualitative and descriptive methodologies, including compiling plant information and their applications. In the research conducted by

Conklin et al. in 1980 and Ford in 1994 (Conklin et al. 1980; Ford 1994), ethnobotany advanced and incorporated an appreciation of systems of knowledge by employing anthropological methodologies. Researchers also devised quantification techniques to determine quantities of items or distributions of valuable species (Caballero 1992). Presently, ethnobiology encompasses a broader domain than ethnobotanical research, as it investigates human perspectives on the interconnections between plants, animals, soil, and climate, as well as the practices that emerge from these perspectives. Traditional systems of management have been the focus of a multitude of ethnobiological investigations (Pei and Sajise 1995). Since 1992, the Humans and Plants Initiative, an international program overseen by WWFUNESCO-Kew Garden, has been implementing various field-based assignments. These projects seek to utilize ethnobotany as a means to collaborate with the local population in order to establish sustainable systems of management for the resources associated with plants. The current advancements in ethnobotany within the Hindu Kush-Himalayan region, encompassing India, China, Nepal, the nation of Pakistan, and the Republic of Bangladesh, have primarily focused on traditional medicinal uses of herbs, indigenously cultivated plant resources, domestic gardens, conventional agricultural systems, cultural understandings of the plant realm in relation to conservation efforts, and the ethnobotany specific to specific ethnic communities. The study of ethnobotany pertaining to economically significant plant groups such as rattan, bamboo, and plants that are edible, as well as the study of ethnobotany in relation to rural development and the preservation of biodiversity, has been undertaken with a significant focus on practical methodologies within the area (Pei 1991, 1998). The field of ethnobotany, being an interdisciplinary knowledge, has the capacity to safeguard the extensive traditional knowledge held by indigenous communities regarding their natural processes and environment, which encompasses therapeutic plants. This encompasses their understanding of the effective utilization and sustainable maintenance of various plant resources over an extended period while ensuring the preservation and conservation of their natural environments. Therefore, it is imperative to exert significant endeavors in order to record and incorporate indigenous knowledge pertaining to land utilization, vegetation, fur administration, non-timber forest resources, medicinal flora, agroforestry practices, residence, swidden agriculture, gardens, and biodiversity. Simultaneously, the process of developing natural assets should persist by fostering extensive interaction and communication with indigenous populations through the utilization of ethnobotanical methodologies. This will additionally guarantee the involvement of residents in future governance, thereby preventing any negative consequences on both the local population and the environment that could otherwise arise (Pei 1998). Further elaborating on the objectives of ethnobotany in relation to the long-term sustainability of plant resources, the subsequent topics are underscored:

- Apply qualitative and quantitative ethnobotanical approaches to evaluate historical and current developments and their influence on management.
- Ethnobotanical research and the facilitation of resource management agreements at the local level can help achieve a more comprehensive comprehension of the

close relationships between individuals and living organisms within a given ecosystem.

- The objective is to facilitate the transfer of reproducible indigenous methods, practices, and expertise to other regions within an identical bioregion to manage resources. This will be achieved through farmer-to-farmer trade, following the biocultural idea of "niche-transfer" in native management of the environment.
- To enhance the involvement of community members in the process of gathering, recording, examining, and evaluating traditional information via field research.
- The objective is to enhance the inclusion of rural growth and preservation projects in participatory resource management planning by utilizing rapid rural appraisal (RRA), participatory rural appraisal (PRA), and participatory biodiversity assessment (PBA) methodologies.
- To compile a comprehensive list of plant species, people utilize for medical, food, swidden crops, non-timber forest products, gardens at home, spiritual reasons, and other purposes. This often entails a substantial population of individuals residing in the local area, as well as a specific number of informants who have been identified from the local community. This category of inventories encompasses the documentation of information and the storage of voucher specimens. These inventories depend on indigenous knowledge derived from local people, encompassing both natural and human-use components of the species under survey. This includes vernacular names, scientific names, habitat, utilization, preparation, attributes, and maintenance strategies (Pei 1998; Alive et al. 2023).

5.3 *Integration of Traditional Knowledge with Modern Methods of Drug Discovery*

The utilization of ethnobotanical knowledge in the swift advancement of pharmaceuticals, coupled with the progress in technology, science, and the global economy, has significantly transformed the planet and its surroundings. Due to significant advancements in health care for humans and environmental degradation, there is an increasing need for natural goods and phytomedicine. This has led to a shift in efforts to conduct research and development toward producing new drugs. In recent years, numerous research organizations and companies have increasingly embraced TM and specified the utilization of botanical resources as a means to develop novel pharmaceuticals. Ethnobotanical information is employed by researchers as a means to identify plants that are highly suitable for subsequent chemical examination (Farnsworth 1993). The aforementioned evolution tendency extends beyond Western countries and encompasses developing nations in Asia. Traditional Chinese medicine has seen significant transformations considering the establishment of the People's Republic of China half a century ago. TCM's legal validity and scientific standing have been formally verified. The constitutional provision explicitly mandates the advancement of both contemporary and TM. The successful integration of traditional Chinese medicine (TCM) and Western medicine has been successfully implemented within the current national health care. Implementing TCM system of health care and utilizing TCM clinics have effectively harnessed the inherent qualities and benefits of TCM. With the advancement of the traditional Chinese medicine (TCM) educational system and research institutes, as well as the enhancement of standardization and legal frameworks, the TCM sector has successfully constructed a sophisticated production system that effectively caters to both domestic and foreign demands. In previous instances 5 years, there has been a significant focus on TM for medication development as well as manufacture, achieved through nationwide policy guidance and TCM development. Incorporating ethnobotanical data into drug development processes through several methodologies is a fundamental component of the technique.

6 Summary

6.1 The Potential of Traditional Resources in Ethnomedicine as a Source of Modern Drugs

In the last five decades, there has been a significant focus on traditional medicine for medication development and manufacture, achieved through nationwide policy guidance and TCM development. Incorporating ethnobotanical data into drug development processes through several methodologies is fundamental to the technique (King 1992). The ethnopharmacological method is grounded in botany, pharmacology, and chemistry and encompasses the processes of observation, proof of identity, outline, and experimental inquiry. Nevertheless, it is important to acknowledge that other academics have also made significant contributions to this field. Ethnopharmacology is the scientific investigation of chemicals with biological activity that have been traditionally used or observed by humans, considering various factors. The primary aims of ethnopharmacology encompass the preservation and documentation of significant cultural material before its potential loss, as well as the examination and assessment of the substances utilized. Therefore, it assumes a significant role in assessing natural goods, including herbal medications derived from elements of traditional and folklore. Ethnopharmacology encompasses domain investigations, illustrations of the use and impacts of traditional treatments, botanical identification, and phytochemical and pharmacological investigations (Cordell and Colvard 2005; Patwardhan 2005).

Ethnopharmacology transcends being an academic discipline of the past that relies on outdated methods. The scientific foundation for advancing active therapies derived from traditional remedies of diverse ethnic groups remains intact. The primary objective of this endeavor is to authenticate conventional formulations, either by isolating the active constituents or by examining diverse pharmacological observations (Mukherjee 2005). The utilization of Ayurvedic medications and preparations has consistently been a fundamental substance in managing different afflictions throughout distinct populations throughout India. The investigation of Ayurvedic medicines has led to the recognition of a multitude of potential drugs that have gained widespread acceptance in commercial markets. Numerous botanical specimens exhibiting promising therapeutic properties were initially extensively employed as Ayurvedic remedies (Shrungeswara and Unnikrishnan 2019).

In addition to multiple regulations, the majority of European and Asian governments have implemented successful measures to enhance the quality, safety, efficacy, and application of medicinal plants. Several contemporary pharmaceuticals employed to treat serious illnesses were derived from herbal remedies. The aforementioned compounds encompass reserpine derived from *Rauvolfia serpentina* (L.) Benth. ex-Kurz; curcumin is sourced from *Curcuma longa* L., along with a nuclide obtained from *Withania somnifera Dunal.*; sennoside is derived from *Cassia angustifolia*; quinine is separated from *Cinchona officinalis* L.; psoralen is isolated from *Ruta graveolens* L.; and glycyrrhizin is obtained from *Glycyrrhiza glabra* L. (Mukherjee et al. 2010). Furthermore, some lead compounds have been successfully obtained in the laboratory from ISM, including immunomodulatory agent botulinic acid (derived from *Nelumbo nucifera*), AChE inhibitor β-asarone (sourced from *Acorus calamus* L.), hepatoprotective methenamine (derived from *Murraya koenigii*), and tallianine (derived from *Semecarpus anacardium*), all of which have had their claimed therapeutic properties validated (Mukherjee et al. 2009).

As pharmaceutical research and manufacturing from Ayurvedic plants progress, major pharmaceutical companies conduct extensive pharmacologic screening of botanicals. The "Sushruta-Samhita" records that *Commiphora mukul* Hook's botanical species exhibited potential therapeutic properties for obesity and similar conditions. Historically, Ashwagandha (*Withania somnifera*) has been employed as an adaptogen. This is commonly referred to as "Indian Ginseng," which can enhance the immune system and decrease inflammation. The integration of traditional alternative therapies and research systems, such as Ayurveda, with the computational capabilities of combined scientific disciplines and high-capacity screening, holds promise for the development of structure-activity libraries. This integration holds the potential to facilitate the identification of active molecules. Further exploration of this topic can be conducted through the implementation of clinical trials, diverse pharmacokinetics, pharmacological research, herbal therapies, and herbal drug surveillance. Recent advancements in chemistry, genetics, chemical technology, molecular biology, spectroscopy, crystallography, and chromatography have the potential to impact the development of highly effective lead compounds in conventional medicine (Shrungeswara and Unnikrishnan 2019).

6.2 Challenges and Benefits of Using Traditional Knowledge in the Field of Drug Discovery and Future Prospects

Several published articles by Cambie and Ash (1994), Weiner (1984), Parham (1943), WHO (1998), and Chand et al. (2018) declared significant medicinal plants in Fiji were described. In contrast, evidence for each MP should include agreement among sources from local communities regarding the conventional applications of the species for managing a range of medical conditions. Around 50,000–70,000 medicinal plants are utilized globally, according to Schippmann et al. (2006), around 15,000 of these plants are imperiled by elimination as a result of fragmentation of habitat, excessive use, and climate change in plants for medicinal purposes. The critical nature of studying anthropogenic pressure and the disruption to native habitats arise from the significant threat that habitat fragmentation presents to threatened plants. This hazard manifests in the form of reduced population size and increased geographic isolation. According to data presented by the Royal Botanic Gardens at Kew, United Kingdom, an estimated 20% of plant species worldwide are currently threatened with extinction. The presence of endangered species within the majority of plant groups utilized for medication manufacturing is a subject of significant apprehension. Nevertheless, there needs to be more data regarding the precise geographical range and frequency of medicinal herbs in Fiji. An evaluation of MP obtained by indigenous groups from forest environments is necessary to collect data on their ecological condition. This research would additionally contribute to the classification of plant protection status, namely, identifying uncommon, endangered, and threatened plants and formulating strategies for their preservation and sustainable administration. The screening of plants, specifically indigenous biodiversity, for ability bioactive chemicals has the potential to aid in developing innovative drugs for modern medicine, considering concerns over antimicrobial resistance and the pharmaceutical sector's heavy reliance on nature-derived products. Although herbal treatments are widely utilized, only a small fraction, fewer than 1%, of medicinal plant species is actively cultivated for commercial purposes. This is mainly owing to concerns about the lower efficacy and secondary metabolites found in farmed stock (Schippmann et al. 2006). The majority of medicinal species rely heavily on wild stock, which underscores the imperative to acquire data on harvesting practices' sustainability. The limited availability of data regarding the diversity, number, and geographic distribution of MP obtained from natural habitats in Fiji is a significant constraint on implementing conservation strategies. This knowledge has the potential to inform the development of sustainable cultivation and oversight strategies aimed at reducing the impact of human activities on herbal assets. Moreover, there needs to be more knowledge regarding the trading and marketing of therapeutic plants. Research on economic viability, value enhancement, and advertising techniques can potentially contribute to poverty alleviation and strategic rural development. This includes cultivating valuable MP in herbal gardens or integrating them into agricultural systems, thereby enhancing the economic prosperity and overall well-being of rural populations.

6.3 Safety Studies

The notion that modern medications are hazardous foreign chemicals with adverse reactions, while herbal remedies are natural and risk-free, is widely held among the general public. Improper use of certain botanicals can result in fatal illnesses and other potentially fatal complications. Specialist knowledge and expertise are required to effectively tackle the intricacies of herbal drug formulations and comprehend the literature regarding their effectiveness and safety, which are informed by years of experience. Additionally, it is possible for consumers to inadvertently combine herbal items with prescription medication without the awareness of their therapist, potentially resulting in herb-drug interactions mediated by cytochrome enzymes. The isoenzymes of cytochrome P450 (CYPs) are mostly localized in the liver, although they can also be found in other organs such as the gastrointestinal tract, lungs, the liver, and the brain. The metabolism of medically important medications is primarily influenced by many isoforms, namely, CYP2C9, CYP1A2, CYP3A4, and CYP2D6. The existing body of scientific research contains numerous findings indicating that the concurrent utilization of phyto-compounds in conjunction with prescription drugs or over-the-counter (OTC) treatments might potentially lead to significant clinical adverse responses. This is attributed to the latter's capacity to modify human drug absorption and pharmacokinetics. The study of dietary supplements by the manufacturer for toxicological, preclinical, and clinical findings is optional and not currently required to meet standard drug safety guidelines (Cox and Balick 1994).

Regulations about the safety and evaluation of prospective reactions among phytoconstituents and traditional drugs have been established by multiple nations (Balick et al. 1996). Before granting licenses, regulatory bodies now mandate recording interactions among herbal medicines and CYP isoforms. Several phytoconstituents have been found to inhibit or stimulate cytochromes, leading to herb-drug interactions, through the use of in vitro, in vivo, and in silico approaches. Likewise, there have been documented instances of interactions between drugs involving phytoconstituents and traditional drugs. Publications have documented instances of active component interactions, such as those involving quercetin, allicin, and silymarin substances.

It is imperative for herbal experts, scientists, and producers to proactively ensure the effective functioning of pharmacovigilance, similar to how doctors record responses to modern medications. Individuals need to acknowledge that disclosing possible adverse responses to their medicinal products is a demonstration of bravery, ultimately leading to an enhancement rather than a decline in their regard for their profession. The use of herbal pharmacovigilance is not detrimental but rather impartial, particularly when it detects a novel and significant herbal health hazard. Pharmacovigilance can offer reassurance by presenting proof that some herbal health issues are either absent or insignificant. Therefore, it can also contribute to enhancing one of phytotherapy's primary characteristics, specifically its comparative safety in relation to conventional medications (Cox and Balick 1994).

7 Conclusion

Natural products have served as a catalyst for numerous advancements in the field of medication research. Numerous historical instances exist when natural products have not only served as therapeutic agents but have also contributed to the exploration of innovative avenues for drug separation. Screening mixtures of chemicals to separate and discover the active lead, not only from plant extracts but also from microorganisms, remains an appealing approach because of the time-consuming and costly nature of creating huge libraries of separated and chemically characterized natural products. Plants are highly proficient in combinatorial chemistry and continue to offer valuable insights into their medicinal capabilities, which can help alleviate humanity from life-threatening illnesses. The current understanding of traditional medicines has yielded numerous advancements in health care and pharmaceutical research, serving as a model for further exploration. The increasing pace of modern technology poses a significant threat to preserving ethnic knowledge utilized in health care. Therefore, it is suitable to cultivate and record conventional knowledge and health care, so facilitating the future development of appropriate medications for diverse conditions. In order to advance the field of medication development derived from natural substances, it is imperative to employ modern methodologies such as genomics, strategies for extracting novel compounds from unexplored settings, and screening methods. These approaches necessitate extensive exploration through both national and international cooperation and collaboration.

References

Adeleke RA, Nunthkumar B, Roopnarain A, Obi L (2019) Applications of plant–microbe interactions in agro-ecosystems. In: Microbiome in plant health disease: challenges opportunities, pp 1–34

Adeola MO (1992) Importance of wild animals and their parts in the culture, religious festivals, and traditional medicine, of Nigeria. Environ Conserv 19:125–134

Alakbarli F (2006) Medical manuscripts of Azerbaijan. Heydar Aliyev Foundation, Baku (1):274

Alberts A, Chen J, Kuron G, Hunt V, Huff J, Hoffman C, Rothrock J, Lopez M, Joshua H, Harris E (1980) Mevinolin: a highly potent competitive inhibitor of hydroxymethylglutaryl-coenzyme A reductase and a cholesterol-lowering agent. Proc Natl Acad Sci U S A 77:3957–3961

Aliev AM, Murtazaliev RA, Vagabova FA, Guseynova ZA, Ramazanova BA, Islamova FI, Alibegova AN, Osmanov RM, Mallaliev MM, Mamalieva MM, Radzhabov G, Anatov DM, Bussmann RW (2023) Ethnobotany of Dagestan. Ethnobot Res Appl 26:1–63. https://ethnobotanyjournal.org/index.php/era/article/view/5663

Alves RR, Rosa IM (2007) Ethnomedicine. Biodiversity, traditional medicine and public health: where do they meet? J Ethnobiol 3:1–9

Angeletti LR, Agrimi U, French D, Curia C, Mariani-Costantini R (1992) Healing rituals and sacred serpents. Lancet 340:223–225

Balick MJ, Elisabetsky E, Laird SA (1996) Medicinal resources of the tropical forest: biodiversity and its importance to human health. Columbia University Press, New York

Barnes PM, Bloom B, Nahin RL (2008) Complementary and alternative medicine use among adults and children: United States, 2007. Natl Health Stat Report 10(12):1–2
Bindseil KU, Jakupovic J, Wolf D, Lavayre J, Leboul J, Van Der Pyl D (2001) Pure compound libraries; a new perspective for natural product based drug discovery. Drug Discov Today 6:840–847
Bishop FL, Lewith GT, Medicine A (2010) Who uses CAM? A narrative review of demographic characteristics and health factors associated with CAM use. Evid Based Complement 7:11–28
Blackwell M (2011) The fungi: 1, 2, 3… 5.1 million species? Am J Bot 98:426–438
Brady SF, Singh MP, Janso JE, Clardy J (2000) Guanacastepene, a fungal-derived diterpene antibiotic with a new carbon skeleton. J Am Chem Soc 122:2116–2117
Butler MS (2004) The role of natural product chemistry in drug discovery. J Nat Prod 67:2141–2153
Butler R (2005) Anti-HIV drug from rainforest almost lost before its discovery. Mongabay com
Caballero J (1992) Maya homegardens: past, present and future. Etnoecológica 1:35–54
Cambie RC, Ash J (1994) Fijian medicinal plants. CSIRO Publishing, Clayton
Catalogue M (2018) Catalogue of the plants of Madagascar. Missouri Botanical Garden, St. Louis, Antananarivo Madagascar
Chand RR, Devi SS, Kumari SS, Kumar SS, Goundar NN, Naranyan N, Chandra P (2018) Traditional use of medicinal plants among selected villages in Fiji Islands: a review perspective. Pac Med Stud J 1:10–20
Chen G, Seukep AJ, Guo M (2020) Recent advances in molecular docking for the research and discovery of potential marine drugs. Mar Drugs 18:545
Chung VC, Ma PH, Lau CH, Wong SY, Yeoh EK, Griffiths SM (2014) Views on traditional Chinese medicine amongst Chinese population: a systematic review of qualitative and quantitative studies. Health Expect 17:622–636
Collins JM (2020) Cancer pharmacology. In: Abeloff's clinical oncology. Elsevier, Amsterdam
Conklin HC, Lupāih P, Pinther MJ (1980) Ethnographic atlas of Ifugao: a study of environment, culture, and society in northern Luzon. Yale University Press, New Haven
Cordell GA, Colvard MD (2005) Some thoughts on the future of ethnopharmacology. J Ethnopharmacol 100:5–14
Corson TW, Crews CM (2007) Molecular understanding and modern application of traditional medicines: triumphs and trials. Cell 130:769–774
Costa-Neto EM (2004) Implications and applications of folk zootherapy in the state of Bahia, Northeastern Brazil. Sustain Dev 12:161–174
Costa-Neto EM (2005) Animal-based medicines: biological prospection and the sustainable use of zootherapeutic resources. An Acad Bras Ciênc 77:33–43
Cox PA, Balick MJ (1994) The ethnobotanical approach to drug discovery. Sci Am 270:82–87
Cragg GM, Newman DJ (2001) Medicinals for the millennia: the historical record. Ann N Y Acad Sci 953:3–25
Daily GC (1997) Introduction: what are ecosystem services. In: Nature's services: societal dependence on natural ecosystems. Island Press, Washington, DC, pp 1–10
Davies J (2006) Where have all the antibiotics gone? Can J Infect Dis Med Microbiol 17:287–290
de Melo EB, Martins JPA, Jorge TCM, Friozi MC, Ferreira MMC (2010) Multivariate QSAR study on the antimutagenic activity of flavonoids against 3-NFA on Salmonella typhimurium TA98. Eur J Med Chem 45(10):4562–4569
Devillers J (2013) Methods for building QSARs. In: Computational toxicology, vol II. Humana, Totowa, NJ, pp 3–27
Dewick PM (2002) Medicinal natural products: a biosynthetic approach. Wiley, New York
Dias DA, Urban S, Roessner U (2012) A historical overview of natural products in drug discovery. Metabolites 2(2):303–336
Ekor M (2014) The growing use of herbal medicines: issues relating to adverse reactions and challenges in monitoring safety. Front Pharmacol 4:177
Ek-Ramos MJ, Gomez-Flores R, Orozco-Flores AA, Rodríguez-Padilla C, González-Ochoa G, Tamez-Guerra P (2019) Bioactive products from plant-endophytic gram-positive bacteria. Front Microbiol 10:463

Environment DO (2014) Department of Environment, Government of Fiji. Fiji's Fifth National Report to the United Nations Convention on Biological Diversity. SPREP LIBRARY, Government of Fiji Publication, Fiji. Physical 102p.: ill. (col.)
Erickson PI (2007) Ethnomedicine. Waveland Press, Long Grove
Faeth SH, Fagan WF (2002) Fungal endophytes: common host plant symbionts but uncommon mutualists. Integr Comp Biol 42:360–368
Falodun A (2010) Herbal medicine in Africa-distribution, standardization and prospects. Res J Phytochem 4:154–161
Fang Y, Lu Y, Zang X, Wu T, Qi X, Pan S, Xu X (2016) 3D-QSAR and docking studies of flavonoids as potent Escherichia coli inhibitors. Sci Rep 6(1):23634
Farnsworth NR (1993) Ethnopharmacology and future drug development: the North American experience. J Ethnopharmacol 38:137–143
Ferus P, Barta M, Konôpková J (2019) Endophytic fungus Beauveria bassiana can enhance drought tolerance in red oak seedlings. Trees 33:1179–1186
Firáková S, Šturdíková M, Múčková M (2007) Bioactive secondary metabolites produced by microorganisms associated with plants. Biologia 62:251–257
Ford RI (1994) The nature and status of ethnobotany. U of M Museum Anthro Archaeology, Ann Arbor
Frost L, Griffiths A (2001) Plants of Eden. Alison Hodge Publishers
Ganie SH, Upadhyay P, Das S, Sharma MP (2015) Authentication of medicinal plants by DNA markers. Plant Gene 4:83–99
Gardens (2023) Peoples and plants for sustainable future, Global Botanical Garden Congress (BGCI), BGjournal 21(1). January 2024, ISSN 1811-8712
Ghamali M, Chtita S, Aouidate A, Ghaleb A, Bouachrine M, Lakhlifi T (2017) Combining DFT and QSAR computation to predict the interaction of flavonoids with the GABA (A) receptor using electronic and topological descriptors. J Taibah Univ Sci 11(3):422–433
Goetz M, Lopez M, Monaghan R, Chang R, Lotti V, Chen T (1985) Asperlicin, a novel non-peptidal cholecystokinin antagonist from Aspergillus alliaceus. Fermentation, isolation and biological properties. J Antibiot 38(12):1633–1637
Gouda S, Das G, Sen SK, Shin H-S, Patra JK (2016) Endophytes: a treasure house of bioactive compounds of medicinal importance. Front Microbiol 7:1538
Guo B, Dai J-R, Ng S, Huang Y, Leong C, Ong W, Carté BK (2000) Cytonic acids A and B: novel tridepside inhibitors of hCMV protease from the endophytic fungus Cytonaema species. J Nat Prod 63:602–604
Gupta S, Chaturvedi P, Kulkarni MG, Van Staden J (2020) A critical review on exploiting the pharmaceutical potential of plant endophytic fungi. Biotechnol Adv 39:107462
Gurnani N, Mehta D, Gupta M, Mehta B (2014) Natural products: source of potential drugs. Afr J Basic Appl Sci 6:171–186
Halberstein RA (2005) Medicinal plants: historical and cross-cultural usage patterns. Ann Epidemiol 15:686–699
Halpern GM, Weverka P (2003) The healing trail: essential oils of Madagascar. Basic Health Publications, Inc., Laguna Beach
Harrison JG, Griffin EA (2020) The diversity and distribution of endophytes across biomes, plant phylogeny and host tissues: how far have we come and where do we go from here? Environ Microbiol 22:2107–2123
Helman C (2007) Culture, health and illness. CRC Press, Boca Raton
Herranz R (2003) Cholecystokinin antagonists: pharmacological and therapeutic potential. Med Res Rev 23:559–605
Humber JM (2002) Alternative medicine, ethics. JAMA 288:1655–1656
Jack DB (1997) One hundred years of aspirin. Lancet 350:437–439

Ji HF, Li XJ, Zhang HY (2009) Natural products and drug discovery: can thousands of years of ancient medical knowledge lead us to new and powerful drug combinations in the fight against cancer and dementia? EMBO Rep 10:194–200

Jia M, Chen L, Xin H-L, Zheng C-J, Rahman K, Han T, Qin L-P (2016) A friendly relationship between endophytic fungi and medicinal plants: a systematic review. Front Microbiol 7:906

Jiang W-J, Daikonya A, Ohkawara M, Nemoto T, Noritake R, Takamiya T, Kitanaka S, Iijima H (2017) Structure-activity relationship of the inhibitory effects of flavonoids on nitric oxide production in RAW264. 7 cells. Bioorg Med Chem 25(2):779–788

Kang S, Phipps M (2003) A question of attitude: South Korea's traditional medicine practitioners and wildlife conservation. A Traffic Azja Report. Published by TRAFFIC East Asia © 2003 in collaboration with Ministry of Health and Welfare, Republic of Korea Association of Korean Oriental Medicine Korea Oriental Drug Association

King S (1992) Medicines that changed the world. Pac Discov 45:23–31

Kinghorn AD (2002) Pharmacognosy in the twenty-first century. J Pharm Pharmacol 53:135–148

Kitchen DB, Decornez H, Furr JR, Bajorath J (2004) Docking and scoring in virtual screening for drug discovery: methods and applications. Nat Rev Drug Discov 3:935–949

Klockgether-Radke AFW (2002) Sertürner and the discovery of morphine. 200 years of pain therapy with opioids. Anasthesiol Intensivmed Notfallmed Schmerzther 37:244–249

Kumar P, Tarui N (2004) Identifying the contribution of indigenous knowledge in bioprospecting for effective conservation strategy. In: Bridging scales and epistemelgioes conference, Alexandria, Ezypt March, 2004. Citeseer, 17–20

Leroi-Gourhan A (1975) The flowers found with Shanidar IV, a Neanderthal burial in Iraq. Science 190:562–564

Lev E (2003) Traditional healing with animals (zootherapy): medieval to present-day Levantine practice. J Ethnopharmacol 85:107–118

Li X-J, Zhang H-Y (2008) Western-medicine-validated anti-tumor agents and traditional Chinese medicine. Trends Mol Med 14:1–2

Liao C, Sitzmann M, Pugliese A, Nicklaus MC (2011) Software and resources for computational medicinal chemistry. Future Med Chem 3:1057–1085

Lidgard H (2005) Quackery or complementary medicine—a historical approach to the present situation. Lund University, Lund

Mackinney LC (1946) Animal substances in materia medica: a study in the persistence of the primitive. J Hist Med Allied Sci 1:149–170

Manek NJ, Crowson CS, Ottenberg AL, Curlin FA, Kaptchuk TJ, Tilburt JC, Medicine A (2010) What rheumatologists in the United States think of complementary and alternative medicine: results of a national survey. BMC Complement 10:1–8

Manganyi MC, Tchatchouang C-DK, Regnier T, Bezuidenhout CC, Ateba CN (2019) Bioactive compound produced by endophytic fungi isolated from *Pelargonium sidoides* against selected bacteria of clinical importance. Mycobiology 47:335–339

Marques JGW (1994) A fauna medicinal dos índios Kuna de San Blas (Panamá) ea. hipótese da universalidade zooterápica. Anais da 46a Reunião Anual da SBPC (324)

Martin GJ, Ethnobotany A (1995) A methods manual. Chapman & Hill, London, pp 1–251

Mashiane AR, Adeleke RA, Bezuidenhout CC, Chirima GJ (2018) Community composition and functions of endophytic bacteria of Bt maize. S Afr J Sci 114:88–97

Megersa M, Nedi T, Belachew S (2023) Ethnobotanical study of medicinal plants used against human diseases in Zuway Dugda District, Ethiopia. Evid Based Complement Alternat Med 2023:5545294., 22 pp. https://doi.org/10.1155/2023/5545294

Mercader AG, Pomilio AB (2010) QSAR study of flavonoids and biflavonoids as influenza H1N1 virus neuraminidase inhibitors. Eur J Med Chem 45(5):1724–1730

Moran MS, Ma S, Jagsi R, Yang T-IJ, Higgins SA, Weidhaas JB, Wilson LD, Lloyd S, Peschel R, Gaudreau B (2013) A prospective, multicenter study of complementary/alternative medicine (CAM) utilization during definitive radiation for breast cancer. Int J Radiat Oncol Biol Phys 85:40–46

Mukherjee P (2005) Exploring green resources for drug development through ethnobotany. In: Chemistry for green environment. Narosa Publishing House, New Delhi
Mukherjee PK, Sahoo AK, Narayanan N, Kumar NS, Ponnusankar S (2009) Lead finding from medicinal plants with hepatoprotective potentials. Expert Opin Drug Discov 4:545–576
Mukherjee PK, Venkatesh M, Gantait A (2010) Ayurveda in modern medicine: development and modification of bioactivity
Munos B (2009) Lessons from 60 years of pharmaceutical innovation. Nat Rev Drug Discov 8:959–968
Myers N (1988) Tropical forests and their species: going, going. Biodiversity 1:28–35
Newman DJ, Cragg GM (2007) Natural products as sources of new drugs over the last 25 years. J Nat Prod 70:461–477
Newman DJ, Cragg GM (2016) Natural products as sources of new drugs from 1981 to 2014. J Nat Prod 79:629–661
Newmaster SG, Grguric M, Shanmughanandhan D, Ramalingam S, Ragupathy S (2013) DNA barcoding detects contamination and substitution in North American herbal products. BMC Med 11:1–13
Nunn JF (2002) Ancient Egyptian medicine. University of Oklahoma Press, Norman
O'neill J (2016) Tackling drug-resistant infections globally: final report and recommendations
Obrien P, Haskins J (2007) High content screening: a powerful approach to systems cell biology and drug discovery. In: Taylor (ed) pp 415–425
Pan SY, Pan S, Yu Z-L, Ma D-L, Chen S-B, Fong W-F, Han Y-F, Ko K-M (2010) New perspectives on innovative drug discovery: an overview. J Pharm Pharm Sci 13:450–471
Pan S-Y, Gao S-H, Zhou S-F, Tang M-K, Yu Z-L, Ko K-M (2012) New perspectives on complementary and alternative medicine: an overview and alternative therapy. Altern Ther Health Med 18:20–36
Parham HR (1943) Fiji native plants with their medical and other uses. [Continued]. J Polyn Soc 52:1–156
Pariser HS (1997) Explore Belize. Harry S. Pariser
Patwardhan B (2005) Ethnopharmacology and drug discovery. J Ethnopharmacol 100:50–52
Pei S (1991) Managing for biological diversity conservation in temple yards and holy hills: the traditional practices of the Xishuangbanna Dai Community southwest China. In: Ethics, religion biodiversity. White Horse Press, Cambridge, pp 118–132
Pei S (1998) Biodiversity conservation in the mountain development of Hindu Kush-Himalayas. In: Frontiers in biology the challenge of biodiversity, biotechnology sustainable agriculture. Academica Sinica, Taipei, pp 223–234
Pei S, Sajise P (1995) Regional study on biodiversity. Yunnan Univ. Press, Kunming
Pistorius R, Van Wijk J (1993) Biodiversity prospecting: commercializing genetic resources for export. Biotechnol Dev Monit 15:12–15
Place SE (2001) Tropical rainforests: Latin American nature and society in transition. Rowman & Littlefield, Lanham
Proudfoot A (2006) The early toxicology of physostigmine: a tale of beans, great men and egos. Toxicol Rev 25:99–138
Rajamanikyam M, Vadlapudi V, Upadhyayula SM (2017) Endophytic fungi as novel resources of natural therapeutics. Braz Arch Biol Technol 60
Rao J, Liu Y, Cheng L (2019) The development of the spinning wheel in ancient China. De Redactie 120
Roberson E (2008) Medicinal plants at risk: nature's pharmacy, our treasure chest. Center for Biological Diversity, Tucson
Schippmann U, Leaman DJ, Cunningham AB (2002) Impact of cultivation and gathering of medicinal plants on biodiversity: global trends and issues. In: Biodiversity the ecosystem approach in agriculture, forestry fisheries. Food and Agriculture Organization of the United Nations (FAO), Rome
Schippmann U, Leaman D, Cunningham A (2006) A comparison of cultivation and wild collection of medicinal and aromatic plants under sustainability aspects. Fronties 17:75–95

Schmidt BM, Ribnicky DM, Lipsky PE, Raskin I (2007) Revisiting the ancient concept of botanical therapeutics. Nat Chem Biol 3:360–366
Seidl PR (2002) Pharmaceuticals from natural products: current trends. An Acad Bras Ciênc 74:145–150
Sengupta S, Chattopadhyay MK, Grossart H-P (2013) The multifaceted roles of antibiotics and antibiotic resistance in nature. Front Microbiol 4:47
Setzer MC, Setzer WN, Jackes BR, Gentry GA, Moriarity DM (2001) The medicinal value of tropical rainforest plants from Paluma, North Queensland, Australia. Pharm Biol 39:67–78
Shah S, Bhat JA (2019) Ethnomedicinal knowledge of indigenous communities and pharmaceutical potential of rainforest ecosystems in Fiji Islands. J Integr Med 17:244–249
Shanley P, Luz L (2003) The impacts of forest degradation on medicinal plant use and implications for health care in eastern Amazonia. Bioscience 53:573–584
Shrungeswara AH, Unnikrishnan MK (2019) Evolution of dietary preferences and the innate urge to heal: drug discovery lessons from Ayurveda. J Ayurveda Integr Med 10:222–226
Sim TF, Sherriff J, Hattingh HL, Parsons R, Tee LB (2013) The use of herbal medicines during breastfeeding: a population-based survey in Western Australia. BMC Complement Alternat Med 13:1–10
Sim CSF, Cheow YL, Ng SL, Ting ASY (2019) Biocontrol activities of metal-tolerant endophytes against Ganoderma boninense in oil palm seedlings cultivated under metal stress. Biol Control 132:66–71
Somani S, Dube S (1989) Physostigmine an overview as pretreatment drug for organophosphate intoxication. Int J Clin Pharmacol Ther Toxicol 27:367–387
Sorokina M, Steinbeck C (2020) Review on natural products databases: where to find data in 2020. J Cheminform 12:20
Souza EN, Williamson EM, Hawkins JA (2018) Which plants used in ethnomedicine are characterized? Phylogenetic patterns in traditional use related to research effort. Front Plant Sci 9:834
Spainhour CB (2005) Natural products. In: Drug discovery handbook, pp 11–72
Spande TF, Garraffo HM, Edwards MW, Yeh HJ, Pannell L, Daly JW (1992) Epibatidine: a novel (chloropyridyl) azabicycloheptane with potent analgesic activity from an Ecuadoran poison frog. J Am Chem Soc 114(9):3475–3478
Steenhuysen J (2007) Mother nature still a rich source of new drugs. Reuters Limited, London
Stierle AA, Stierle DB, Bugni T (1999) Sequoiatones A and B: novel antitumor metabolites isolated from a redwood endophyte. J Organ Chem 64:5479–5484
Stockwell BR (2004) Exploring biology with small organic molecules. Nature 432:846–854
Sullivan K, Manuppello J, Willett C (2014) Building on a solid foundation: SAR and QSAR as a fundamental strategy to reduce animal testing. SAR QSAR Environ Res 25:357–365
Suzuki N, Medicine A (2004) Complementary and alternative medicine: a Japanese perspective. Evid Based Complement Alternat Med 1(2):113–118
Taylor L (2005) The healing power of rainforest herbs: a guide to understanding and using herbal medicinals. Square One Pub, New Hope
Toghueo RMK, Boyom FF (2019) Endophytes from ethno-pharmacological plants: Sources of novel antioxidants—a systematic review. Biocatal Agric Biotechnol 22:101430
Traditional CNCO, Medicine H (1995) Materia medica commonly used in China. Science Press, Beijing
Van Der Sar S (2006) Studies in the chemistry of fungal natural products. PhD Theses/Dissertations. http://hdl.handle.net/10092/1333, https://doi.org/10.26021/8925
Van Seters A, Bodeker G, Bhat KKS, Burley J, Vantomme P (eds) (1997) Forest based medicines in traditional and cosmopolitan health care. In: Medicinal plants for forest conservation. Health care, non-wood forest products. pp 5–11
Venugopalan A, Srivastava S (2015) Endophytes as in vitro production platforms of high value plant secondary metabolites. Biotechnol Adv 33:873–887
Voeks RA, Leony AJEB (2004) Forgetting the forest: assessing medicinal plant erosion in eastern Brazil. Econ Bot 58:S294–S306

Wainwright M (1990) Miracle cure: the story of penicillin and the golden age of antibiotics. Blackwell, Oxford

Weiner MA (1984) Secrets of Fijian medicine. Edition: View all formats and editions. Ministry of Health, Fiji. ISBN-10: 0912845023; ISBN-13: 978-0912845029

Whiffin T, Kikkawa JJOTFS (1992) The status of forest biodiversity in Oceania. pp 155–172

WHO (1993) Guidelines on the conservation of medicinal plants. WHO, IUCN & WWF. All requests for reproduction should be made to the Publications Officer, IUCN, Gland. ISBN 2-8317-0136-8

World Health Organization (WHO) (1998) Medicinal plants in the South Pacific: information on 102 commonly used medicinal plants in the South Pacific. World Health Organization, Geneva

World Health Organization (WHO) (2015) Action agenda for antimicrobial resistance in the Western Pacific Region. World Health Organization, Geneva

World Health Organization (WHO) (2017) Antibacterial agents in clinical development: an analysis of the antibacterial clinical development pipeline, including tuberculosis. World Health Organization, Geneva

Role of Ethnomedicinal Resources to Cure Metabolic Diseases

Shamee Bhattacharjee and Ankur Banerjee

Contents

Abstract The rising incidences and prevalence of metabolic diseases, creating enormous health and economic burden worldwide, are a global healthcare chal-

S. Bhattacharjee (✉) · A. Banerjee
Department of Zoology, West Bengal State University, Berunanpukhuria, West Bengal, India
e-mail: shamee.zoology@wbsu.ac.in

© The Author(s), under exclusive license to Springer Nature Singapore Pte Ltd. 2024

A. Das Talukdar et al. (eds.), *Traditional Resources and Tools for Modern Drug Discovery*, Interdisciplinary Biotechnological Advances,
https://doi.org/10.1007/978-981-97-4600-2_6

lenge. Such conditions, which can be inherited or acquired, are exemplified by type II diabetes, high blood pressure, dyslipidemia, non-alcoholic fatty liver disease, obesity, etc. and are caused by dysregulated metabolism, i.e. defects in the body's energy processing system. Increase in mortality and years of healthy life lost due to metabolic diseases calls for urgent attention to manage these diseases. Current treatment options largely involve prolonged allopathic medication which is disadvantageous because of associated adverse health effects. Moreover, these diseases often occur as a constellation of maladies requiring complex treatment regimens with multiple drugs culminating in adverse drug-drug interactions. Therefore, considering the limitations of existing therapeutic options, there is a renewed and growing interest in the use of ethnomedicines for the management of metabolic diseases. Ethnomedicine offers advantage over allopathy or chemical drugs, because these are part of indigenous traditional knowledge system and hence are time-tested, safer, economically affordable, accessible and hence sustainable in the long run. In this chapter, we discuss about ethnomedicines which are used in the treatment and management of diabetes, obesity, dyslipidemia, hypertension, fatty liver diseases, etc. in various parts of the world. We also discuss about the scientific investigations that have been performed to validate the therapeutic potential of widely used ethnomedicines against metabolic diseases. Finally, clinical trials conducted on ethnomedicinal resources for the cure of metabolic disorders and associated drawbacks will be reported.

Keywords Metabolic disease · Metabolic syndrome · Ethnomedicinal plants · Ethnozoological resources · Folk medicine

1 Introduction

Metabolic diseases are characterized by aberrations in metabolism which refer to those enzyme-catalysed reactions by which our body synthesizes energy and other essential molecules from carbohydrates, fat and proteins that we consume as food. These metabolic diseases can be broadly classified into two groups: (i) those that result from gene mutation causing inborn errors of metabolism and (ii) those that are acquired due to changing lifestyle and environmental factors. In case of genetic metabolic diseases, gene for an enzyme or cofactor involved in the metabolic pathway is mutated disrupting the metabolic pathway resulting in build-up of toxic by-products of metabolism or depletion of essential metabolites (Clemente-Suárez et al. 2023). Examples of some inborn errors of metabolism are phenylketonuria (PKU), cystic fibrosis, lysosomal storage disease (e.g. Tay Sach's disease), mitochondrial disorders, etc. These diseases occur due to mutation in specific genes. According to estimates, inborn errors of metabolism are prevalent in 50.9 cases per 100,000 live births which are relatively rare (Waters et al. 2018). In recent years, however, there is a growing concern regarding the increased incidences of metabolic diseases worldwide. Particularly, prevalence of obesity, hypertension, type II diabetes mellitus (T2DM), hyperlipidaemia and non-alcoholic fatty liver disease (NAFLD) has been rising at an astonishing rate in all countries creating enormous clinical and economic burden on the society. More often, these diseases, which share common risk factors, coexist in individuals and increase risks of many chronic diseases including cancer. Such multifactorial diseases involving hypertension, hyperglycaemia, dyslipidemia and obesity which increase the risk factor for cardiovascular diseases are often referred to as metabolic syndrome (MetS) (Alberti 2005). Urban lifestyle including changes in dietary habits and sedentary life is mainly attributed to the increased prevalence of such metabolic diseases (Chew et al. 2023) as depicted in Fig. 1.

At present, the first-line management strategy of metabolic diseases is largely aimed at reducing the risk factors through changes in dietary and lifestyle pattern. Therapy to reduce certain conditions such as high blood pressure, triglycerides or blood sugar is recommended if the risk factors are very high or lifestyle modifications fail to normalize the risk factors (Grundy et al. 2005). However, such therapeutic interventions usually involve multi-drug approach to tackle the complex pathophysiology of metabolic diseases (Lillich et al. 2021). Multiple treatments include drugs to lower high lipid, blood glucose, blood pressure, etc. But polypharmacy has a risk of adverse drug-drug interaction, poor patient compliance and toxic side effects such as hypoglycaemia, myalgia, weight gain, liver dysfunction, cardiovascular damage, etc. In addition, the economic cost of such treatment is also very high which creates a huge financial burden on the healthcare system across the

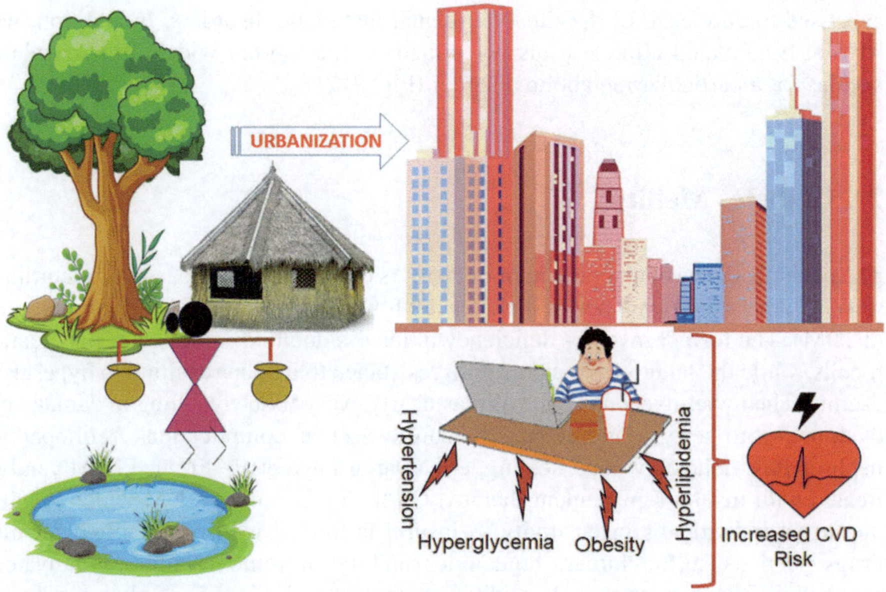

Fig. 1 Prevalence of metabolic diseases in urban lifestyle

globe. Therefore, the quest for an effective alternative treatment regimen to cure metabolic diseases is of utmost importance.

In this regard, ethnomedicine can be a valuable resource to tackle the global burden of metabolic diseases. Ethnomedicines are natural botanical and zoological resources traditionally used by ethnic communities to cure or manage various ailments and hence are envisaged to have limited toxicity. In addition, considering the chemical diversity of natural products, such folk medicines have the added advantage of polypharmacological or multi-target effect. Accumulating scientific evidences also support the clinical relevance of these ethnomedicinal resources in the management of metabolic disorders. The bioactive constituents of many such ethnomedicinal resources have been characterized which might serve as lead compounds in drug discovery.

This chapter is an endeavour to collate information regarding the usage of ethnomedicinal resources in different ethnic groups worldwide to combat metabolic diseases, especially those whose prevalence is on the rise due to adoption of a Western lifestyle and urbanization. Considering the huge number of articles available on ethnobotanical surveys, it is beyond the scope of this chapter to discuss each ethnomedicinal plant in detail. Therefore, we have limited our discussion on the ethnopharmacological and experimental validation of only the top three widely reported ethnomedicinal plant species across the world and provided a mechanistic insight wherever possible. As opposed to ethnobotanical resources, relatively scanty information is available on ethnozoological resources, and their mechanism of action is also not well investigated. In this chapter, we have also included information on the use of ethnozoological resources to treat metabolic disorders. Finally, we report clinical trials that have been conducted on these top three mostly utilized plant species used to cure each of the lifestyle-related metabolic disorders. In addition, we have also tabulated clinical trials performed on the top ten widely utilized plant species for a particular metabolic disorder (Fig. 2).

2 Diabetes Mellitus

The most pervasive characteristic of diabetes is chronic hyperglycaemia. According to WHO, there are two types of diabetes mellitus (DM): type I (T1DM) and type II (T2DM). The former involves deficiency in insulin due to destruction of pancreatic β-cells, while the latter is associated with resistance to insulin resulting in hyperglycaemia. The hyperglycaemic state affects nearly every tissue resulting in damage or dysfunction to several organs such as cardiovascular complications, retinopathy, nephropathy, delayed wound healing, etc. (Darenskaya et al. 2021). T1DM can be treated with insulin replacement therapy. On the other hand, T2DM which results due to an individual's insensitivity to insulin is treated with oral hypoglycaemic drugs such as sulfonylureas, biguanides and peroxisome proliferator-activated receptor-γ (PPARγ) agonists. In addition, a few new classes of anti-diabetic drugs are now used to treat T2DM such as α-glucosidase inhibitors, the inhibitors for

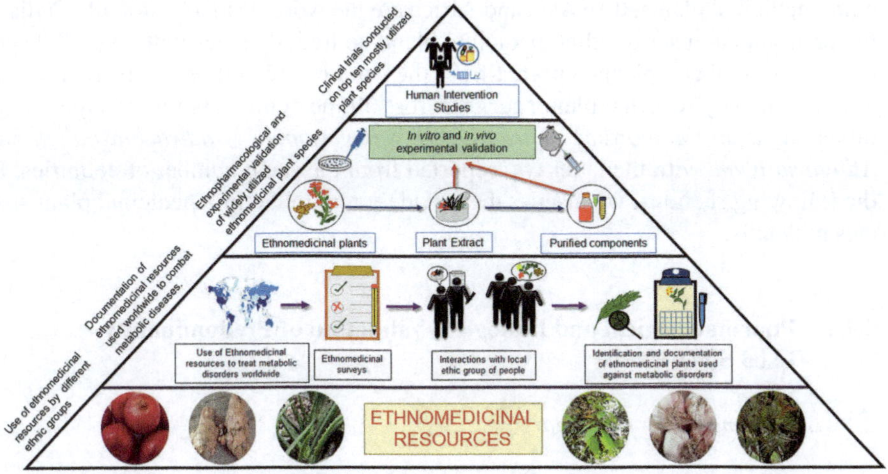

Fig. 2 Scheme of the chapter: ethnopharmacological, experimental and clinical validation of the top three widely reported ethnomedicinal plant species

dipeptidyl peptidase 4 (DPP4), agonists for glucagon-like peptide 1 (GLP1) receptor and SGLT2 inhibitors. Severe hypoglycaemia, weight gain, limited therapeutic efficacy and toxic side effects are the major drawbacks of these hyperglycaemics (Padhi et al. 2020). The problem is further compounded by improper access to medication and cost of modern anti-diabetic medicines in certain parts of the world compromising the optimum management of diabetes mellitus across the world (Fralick et al. 2022). At present, diabetes has become a pandemic, and its prevalence is rocketing sky high globally, especially in developing countries. It is estimated that 40% of the total diabetic population resides in India and China (Fralick et al. 2022). Thus, optimizing proper dose regimens to maintain optimum blood glucose level and preventing long-term complications associated with diabetes and with anti-diabetic drugs is a major challenge. In such a scenario, there is an increased tendency to resort to traditional ethnomedicinal resources as alternative treatment option for managing diabetes because of comparable safety, sustainability and cultural acceptability (Rutebemberwa et al. 2013).

2.1 Anti-diabetic Ethnomedicinal Plants

Majority of anti-diabetic ethnomedicines are botanicals that have been identified by ethnobotanical surveys among various ethnic groups of different countries. According to a recent comprehensive review, 2004 plants across 92 countries are reported to be traditionally used to treat diabetes (Aumeeruddy and Mahomoodally 2021a). Among these, majority of the plants belong to the family Leguminosae followed by Compositae and Lamiaceae. If we look at the continent-wise usage of ethnomedicinal plants, then Asia and Africa are the world leaders. Globally, India is the predominant user of ethnomedicinal plants to treat diabetes with over 300 species of anti-diabetic plants reported from the country. According to this review, the top three mostly reported plant species across all the continents for managing high blood sugar are *Momordica charantia*, *Syzygium cumini*, *Azadirachta indica* and *Allium sativum*, with the latter two reported from the same number of countries. In the following sections, we discuss these widely reported ethnomedicinal plant species in details.

2.1.1 Pharmacological and Biological Validation of Predominantly Used Species

2.1.1.1 *Momordica charantia*

Commonly known as bitter gourd or bitter melon, *M. charantia* is a vine widely cultivated in tropical countries. It has a long history of usage in traditional medicine for controlling blood glucose levels.

The major phytochemical constituents of bitter gourd fruits and seeds include cucurbitane-type triterpenoids, known as momordicosides, phytosterol glycosides named charantin, peptides such as the hypoglycaemic polypeptide-P, *M. charantia lectin* (MCL), *M. charantia* polysaccharide IIa (MCPIIa), *M. charantia* insulin receptor-binding protein (mcIRBP), etc.; flavonoids; and polyphenols (Çiçek 2022).

M. charantia has been widely studied for its hypoglycaemic effect in various in vitro and in vivo models (Peter et al. 2020). Its active constituents have been found to exert hypoglycaemic effect through several mechanisms from regulating glucose uptake, glucose absorption, insulin secretion and so on. Among the phytochemical constituents of bitter gourd, charantin, polypeptide-P and another polypeptide MCPIIa were identified as anti-diabetic principles. In addition, the cucurbitane-type triterpenoids were demonstrated to increase glucose uptake by enhancing translocation of the glucose transporter GLUT4 mediated by adenosine monophosphate kinase (AMPK) phosphorylation. mcIRBP has been demonstrated to stimulate glucose uptake resulting in significant reduction in blood glucose level. The triterpenoid glycosides were also reported to potently inhibit the enzymes α-glucosidase and α-amylase thereby delaying or decreasing the intestinal digestion and absorption of carbohydrates. Several bitter gourd triterpenoids were also demonstrated to inhibit glucose production by inhibiting gluconeogenesis and also stimulate insulin secretion. Thus, *M. charantia* has pleiotropic mechanism of action in controlling blood glucose levels, and all these effects have been recently reviewed in several articles (Xu et al. 2022; Çiçek 2022; Richter et al. 2023).

2.1.1.2 *Syzygium cumini*

This plant had been traditionally used in India as an astringent and anti-diarrhoeal and was introduced into Western medicine in 1880s. Commonly referred to as jambul or black plum, scientific investigations into the beneficial effect of this ethnomedicinal plant began quite early, and its beneficial effects against diabetes became known even before the discovery of insulin (Helmstädter 2008).

Various parts of *Syzygium* including the leaves, fruits, bark and seeds have therapeutic potential and have been traditionally used to treat many disorders especially gastric problems and diabetes. All these plant parts are rich in bioactive components, and each has a characteristic phytochemical composition which is different from the other.

The purple colour of jambolan fruits is due to the presence of anthocyanins such as delphinidin, petunidin and malvidin, and its astringent flavour arises from the large number of different types of tannins. In addition, the fruit pulp also contains phenolic acids such as gallic acid, caffeic acid and flavanols such as catechin, epicatechin, epigallocatechin and coumarins. The leaves, on the other hand, are characterized by the presence of terpenoids such as betulinic acid, α-pinene, camphor, etc. and flavonoids such as quercetin, myricetin, myricitrin and alkanes. Seeds are rich in tannins and fatty acids. For a detailed description of phytochemical composition of each part, readers are encouraged to read review articles available on this topic (Qamar et al. 2022; Ayyanar and Subash-Babu 2012). In addition, a characteristic component of this plant is a series of phenolipids, of which the major phenolipid identified in *S. cumini* is jambone E which has also been shown to have hypoglycaemic effect (Wang et al. 2023).

Several pre-clinical studies have investigated the anti-diabetic potential of various parts of *S. cumini* including fruit pulp, seed and leaves. Water and alcohol extracts of fruit pulp could reduce blood glucose level in a variety of diabetic animal models, and the reduction was comparable to standard anti-diabetic drugs. Similarly, seed powder or alcoholic and water extracts of seed administered to alloxan- or streptozotocin-induced diabetic animals via various routes were found to have significant anti-diabetic effect. *S. cumini* leaves and barks were also demonstrated to exert hypoglycaemic effect in experimental animals (Qamar et al. 2022; Zulcafli et al. 2020).

Mechanism of action includes inhibitory effect on α-glucosidase and α-amylase enzymes, activation of phosphatidyl inositol-3-kinase (PI3K)-mediated uptake of glucose by GLUT4 transporter and activation of PPARγ which transactivate *GLUT4* gene expression (Mahindrakar and Rathod 2021; Anandharajan et al. 2006). Jambone E has been demonstrated to activate Akt-mediated insulin signaling pathway thereby inhibiting gluconeogenesis and activating glycogen synthesis (Wang et al. 2023). Another bioactive component of *S. cumini*, myricetin, was found to inhibit the aggregation of human islet amyloid polypeptide (hIAPP) hormone and restored the function of pancreatic β-cells (Dubey et al. 2021).

2.1.1.3 *Azadirachta indica*

Commonly referred to as neem, this tree is extensively found in south-east Asia. It has been used in folklore medicine through ages and is believed to be a "Divine Tree" because of its health benefits.

Several hundred phytochemical constituents have been identified from various parts of neem including leaves, flowers, fruits and bark. Two major classes of phytochemicals constitute the different parts of neem tree: (i) isoprenoids including triterpenes (e.g. Nimbin), limonoids, etc. and (ii) non-isoprenoids including polyphenolics, coumarins, tannins, aliphatic compounds, etc. The neem leaves also contain characteristic glycoproteins, known as neem leaf glycoproteins (NLGP), which show immunomodulatory activity. Several glycosides such as nimbanene, nimbandiol, nimbolide, etc. have been reported in the aqueous extract of neem leaves. The primary bioactive component of neem, azadirachtin, is a limonoid derived from its seed (Islas et al. 2020).

A substantial number of studies, both in vitro and in vivo, have been conducted on anti-diabetic property of neem which reveal that all parts of neem exhibit hypoglycaemic activity. However, majority of the studies till date have been conducted using neem leaves. Leaves have been shown to exhibit more potent hypoglycaemic activity than seeds. Neem leaves were found to be more potent in controlling blood glucose level in T2DM than in T1DM. A few authors have suggested that alleviation of hyperglycaemia by neem extract is possibly due to inhibition of serotonin, which is known to interfere with insulin secretion thereby raising blood glucose level, while a few other authors have proposed that plant extracts improve the morphology of pancreatic islets leading to improve insulin secretion. The latter is supported by another study showing protective effect of rutin, a flavonoid glucoside, on Langerhans and β-cells of pancreas by inhibiting damage induced by oxidative stress, hence increasing insulin production. Significant number of studies have proven the inhibitory effect of neem extracts and bioactive components especially limonoids on carbohydrate-digesting enzymes present in liver, pancreas and intestine such as α-amylase and glucosidase. Neem extract and its components such as β-sitosterol were also shown to increase glucose uptake possibly by upregulating GLUT4 expression. In addition to its effect on blood glucose level per se, neem has been demonstrated to ameliorate oxidative stress and other complications arising from diabetes and has ameliorated complication such as diabetic nephropathy or diabetic retinopathy (Patil et al. 2022).

2.1.1.4 *Allium sativum*

Garlic or *Allium sativum* is a pungent smelling herbaceous spice which has been traditionally used in cooking and folk medicine since ancient times. It has described as a super food due to its immense health benefits (Van Den Driessche et al. 2018). According to various processing methods, different garlic products can be formed like fresh garlic, aged garlic extract (AGE), black garlic, garlic oil, garlic extract and so on.

Allicin [*S*-(2-propenyl)-2-propene-1-sulfinothioate], the most active component of garlic cloves which imparts its characteristic taste and smell, is formed from its precursor alliin by the enzyme alliinase when garlic is cut or crushed. In addition, garlic also contains other phytochemicals such as sulphur-containing compounds ajoenes, vinyldithiins, allyl sulphides and polyphenols such as kaempferol, luteolin, apigenin, flavonoids, saponins, tannins and so on (Ansary et al. 2020).

The pharmacological and therapeutic properties of garlic including its hypoglycaemic effect are well investigated in different experimental and clinical studies (Ribeiro et al. 2021). Significant reduction in blood glucose level and improved glucose tolerance has been demonstrated by treatment with aqueous and alcoholic garlic extracts in experimental models of T1DM and T2 DM.

Garlic products and their components target multiple mechanisms to control blood glucose level such as reducing energy intake in obese rats on high-fat/high-sucrose diet by regulating the brain-gut hormone hunger/satiety axis, stimulating insulin secretion, protecting pancreatic β-cells by inhibiting their degeneration, improving insulin resistance by activating AMPK- and GLUT4-mediated glucose uptake, regulating lipid metabolism and many other potential mechanisms which have been recently reviewed (Jiang et al. 2022).

2.2 *Ethnozoological Resources*

A literature search revealed that there are mentions of a few ethnozoological resources which are used to treat hyperglycaemia in folklore. For example, a survey in the Eastern Rif in Northern Morocco documented the use of various ethnozoological resources by the indigenous inhabitants, the Berbers, for health care. According to the survey, they use cow (*Bos taurus*) fat to cure hyperglycaemia (Budjaj et al. 2021). Camel fat is also reported to be used by Berbers of Western Africa to treat diabetes (Budjaj et al. 2021). Another survey carried out in Assam, a Northeastern state of India, revealed that the indigenous inhabitants use the fish Assamese snake head (*Channa stewartii*) to treat diabetes. The whole body is consumed after boiling the fish in water (Borah and Prasad 2017). A decoction prepared with *Limnephilus* sp. in combination with neem and Guinea pepper is reported to be used by the indigenous population of the Republic of Benin, Africa, to treat diabetes. They also use a decoction made with *Eudrilus eugeniae* and some plants to lower blood sugar level (Loko et al. 2019).

2.3 *Human Intervention Studies with Some Anti-diabetic Ethnomedicinal Plants*

2.3.1 Bitter Gourd

A number of clinical trials have been conducted with bitter melon to assess its hypoglycaemic potential in humans. A recent meta-analysis including ten clinical trials with 1045 T2DM patients documented lowering of plasma glucose levels in patients administered *M. charantia* monoherbal formulation as compared to the placebo group (Peter et al. 2019).

In a randomized placebo controlled study involving 90 volunteers with T2DM, significant reduction in blood glucose and glycated haemoglobin (Hb1Ac) levels were recorded in patients receiving bitter gourd extract for 12 weeks (Kim et al. 2020).

A randomized control trial tested the anti-hyperglycaemic effect of a polyherbal formulations (GlyCare-II) containing *M. charantia* in diabetic and pre-diabetic patients against metformin. After 120 days of twice-a-day treatment, the reduction blood glucose and Hb1Ac levels by GlyCare were comparable to that of metformin in diabetic and pre-diabetic patients. Interestingly, in newly diagnosed diabetic patients, treatment with GlyCare yielded more promising result than metformin (Majeed et al. 2021).

Another polyherbal formulation containing *M. charantia* was investigated for its ability to prevent diabetes in pre-diabetic patients in combination with lifestyle management. This intervention caused significant reduction in conversion of pre-diabetic to diabetic patients in comparison to placebo plus lifestyle management (Nakanekar et al. 2019).

2.3.2 Black Plum

Despite encouraging results in pre-clinical models, clinical trials with black plum have been largely disappointing. One randomized trial reported no antihyperglycemic effect of tea prepared from *S. cumini* leaves in patients with T2DM (Teixeira et al. 2006). Another randomized control trial tested the hypoglycaemic effect of tea prepared from leaves of *S. cumini* in non-diabetic volunteers but did not produce any encouraging results (Teixeira et al. 2000).

2.3.3 Neem

In a double-blinded randomized placebo-controlled clinical trial with 94 subjects, aqueous extract of neem leaves and twigs was investigated for hypoglycaemic effect in diabetic patients in India. Significant reduction in both fasting and postprandial glucose levels and that Hb1Ac levels was noted (Pingali et al. 2020). Similar improvement in serum glycaemic indices was obtained from other two randomized control clinical trials in volunteers with metabolic syndrome and with diabetes (Pingali et al. 2021; Kumari 2010).

2.3.4 Garlic

Many clinical trials have proven the potential of garlic as a potent anti-hyperglycaemic agent. In a study, 40 patients with metabolic syndrome were treated with 100 mg/kg body weight raw garlic extract twice a day for 4 weeks. The study revealed significant improvement of several components of metabolic syndrome including reduction in fasting glucose level in the patients following treatment. An earlier double-blinded placebo-controlled clinical trial in 60 T2DM patients treated with garlic powder tablet, Allicor, (300 mg twice a day) for 4 weeks had also revealed significant reduction in fasting glucose level in the diabetic patients. Similarly, consumption of garlic clove for 30 days also resulted in marked reduction in blood glucose levels in T2DM patients. In yet another human intervention study, Kyolic AGE administration 1.2 g/day for 12 weeks was demonstrated to induce anti-hyperglycaemic effect in patients with metabolic syndrome as compared to patients receiving placebo. This was accompanied by an increase in adiponectin levels which is known to enhance glucose uptake and reduce hepatic gluconeogenesis, hence lowering blood glucose level. All these clinical trials have been recently reviewed in an article (Ansary et al. 2020).

In addition to these clinical trials, human intervention studies conducted on the rest of the top ten ethnomedicinal plants reported to be used by majority of the countries against hyperglycaemia (according to the review by Aumeeruddy and Mahomoodally 2021a) has been provided in Table 1.

3 Hypertension

High blood pressure or hypertension is the leading and modifiable risk factor for cardiovascular ailments (Mills et al. 2020). It is characterized by a rise in blood pressure with systole >140 mmHg and diastole >90 mmHg. It is estimated that worldwide 40% of population above 25 years of age suffer from hypertension. Lifestyle factors such as unhealthy diet including high sodium intake and lack of physical activity are contributing to the increased prevalence of hypertension globally. Drug therapy for treatment of hypertension includes anti-hypertensive drug such as diuretics (act by inhibiting reabsorption of sodium and chloride), β-blockers (blocks the β-1 adrenergic receptors thereby reducing heart rate), calcium channel blockers (prevents entry of Ca^{2+} into vascular smooth muscles resulting in vasodilation) and blockers of the rennin-angiotensin system (inhibits the conversion of angiotensinogen to angiotensin I or that of the latter to angiotensin II). However,

Table 1 Human intervention studies with some widely reported anti-diabetic ethnomedicinal plants

Sl no.	Name of the plant	Study design	Outcome	References
1.	*Olea europaea* (Olive)	1. 330 mL olive leaf tea administered to pre-diabetic individuals thrice daily for 12 weeks 2. Two meta-analysis performed with the following objectives: (i) To assess relation between olive oil intake and risk of developing T2DM (ii) Analysing randomized control trials to investigate the effect of olive oil intake on glycaemic parameters of T2DM patients	1. Decreased fasting glucose level 2. Risk of T2DM reduced with increasing intake of olive oil 3. Treatment with olive oil cause marked decrease in Hb1Ac and blood glucose level in T2DM patients	Araki et al. (2019)
2.	*Trigonella foenum-graecum* (Fenugreek)	1. 12 patients with uncontrolled diabetes on metformin received either 2 g fenugreek or glibenclamide daily for 12 weeks 2. Meta-analysis on pharmacological effect of fenugreek including 21 trials with a total 1440 individuals assessed blood glucose level	1. Increased fasting insulin level 2. Reduction in Hb1Ac and blood glucose level	1. Najdi et al. (2019) 2. Fakhr et al. (2023)
3.	*Aloe vera*	1. Meta-analysis of eight trials involving pre-diabetic and T2DM patients (total 470) 2. Randomize control trial involving 50 T2DM patients with control group receiving conventional hypoglycaemic drug and the intervention group receiving polyherbal formulation containing 300 mg *Aloe vera* leaf gel	1. (i) In pre-diabetic patients *Aloe vera* significantly improved fasting glucose level without any effect on Hb1Ac (ii) In T2DM patients, marked reduction in Hb1Ac and marginal effect on fasting glucose level 2. The intervention group showed significant reduction in Hb1Ac level but no effect on fasting glucose	1. Suksomboon et al. (2016) 2. Ghorbani et al. (2019)

(continued)

Table 1 (continued)

Sl no.	Name of the plant	Study design	Outcome	References
4.	*Allium cepa* (Onion)	1. Meta-analysis of 14 randomized control trials to investigate impact of onion supplementation on parameters of metabolic syndrome 2. Randomized triple-blind control trial in 56 breast cancer patients who have completed second chemotherapy cycle received either high-dose or low-dose onion for 8 weeks	1. Intervention did not alter glycaemic parameters significantly 2. High-dose onion significantly reduced fasting blood glucose level and ameliorated insulin resistance in breast cancer patients	1. Hejazi et al. (2023) 2. Jafarpour-Sadegh et al. (2017)
5.	*Gymnema sylvestre* (Gurmar)	1. Meta-analysis of 10 studies with a total of 419 T2DM patients to investigate the effect of Gymnema sylvestre (GS) supplementation on glycaemic parameters	Significant reduction in fasting blood glucose, postprandial glucose and Hb1Ac	Devangan et al. (2021)

due to side effects from long-term use and associated costs, management of hypertension by using ethnomedicines is a viable alternative.

3.1 Anti-hypertensive Ethnomedicinal Plants

Several studies have documented the use of herbal remedies to treat hypertension in folklore medicine. Across the world, many ethnobotanical surveys have been conducted on the use of plant species to manage hypertension.

An article on the use of anti-hypertensive ethnomedicinal plants in Shiraz, Iran, has documented 27 such plants (Baharvand-Ahmadi et al. 2016). The top three species based on the frequency of recommendation by the traditional healers are *Borago officinalis* (51.85%), *Berberis vulgaris* (51.58%) and *Anethum graveolens* (48.14%). In a recent review article, herbal medicines used for hypotensive effects in various parts of Iran such as Sistan and Baluchestan province (*Nigella sativa*), in Mobarakeh of Isfahan (*Rumex crispus*, *Ziziphus jujuba* and *Olea europaea*), Kazeran (*Silybum marianum*, *Achillea tenuifolia*, *Cichorium intybus*, *Berberis vulgaris*, *Capsella bursa-pastoris*, *Equisetum arvense*, *Juglans regia* and *Melilotus indicus*), Lorestan (*Falcaria vulgaris*, *Crocus haussknechtii*, *Berberis integerrima*, *Allium ursinum*, *Tragopogon porrifolius* and *Anethum graveolens*) and Ilam Province (*Rheum ribes*, *Paliurus spina-christi*) have been discussed (Kamyab et al. 2021).

A similar study was conducted by another group in Northern Pakistan. The study, based on an interview of 250 traditional healers, has reported 192 plant species which are used to treat hypertension by indigenous people (Malik et al. 2018). Based on their relative frequency of citation (RFC) value, the top three plants species are *Bidens biternata*, *Marrubium vulgare* and *Stevia rebaudiana*. For the first time, the study has reported the utilization of 81 plant species for treating hypertension including *Ficus palmata*, *Senna tora*, *Teucrium polium*, *Valeriana officinalis* and *Ziziphus mauritiana*. Traditional use of anti-hypertensive ethnomedicinal plants in Lebanon has been documented in a study in which the anti-hypertensive potential of three of the plant species showing the highest RFC values (*Mentha longifolia*, *Viola odorata* and *Urtica dioica*) were also clinically tested (Samaha et al. 2019).

An ethnobotanical survey conducted among indigenous people of in Cameroon, Africa, belonging to 58 sociocultural groups has documented 28 species of plants for the treatment of hypertension (Tsabang et al. 2017). The predominantly used plants used in Cameroon according to this survey are *Moringa oleifera*, *Aloe buettneri* and *Aloe barteri*. Another survey carried out in South Africa around Northern Maputaland has reported 28 plant species which are used for treating hypertension.

A recent review article has comprehensively documented the use of ethnomedicinal plants against hypertension in 90 countries across the world. The country reporting use of maximum number of anti-hypertensive plants is Pakistan followed by Mexico (Aumeeruddy and Mahomoodally 2020). The species reported to be used in majority of the countries is *Allium sativum* (reported in 41 countries), followed by *Allium cepa* in 18 countries and *Olea europaea* in 17 countries. The following section provides a review of the pharmacological and biological validation of these three predominantly reported plant species used to cure hypertension in folklore.

3.1.1 Pharmacological and Biological Validation of Predominantly Used Species

3.1.1.1 *Allium sativum*

The phytochemical constituents of *Allium sativum* or garlic have been discussed in the preceding section on anti-diabetic ethnomedicinal plants.

There are enormous number of studies demonstrating the ability of garlic extracts and its bioactive components to reduce both systolic and diastolic pressure in experimental animals and human subjects. The mechanism of action includes reduction in prostaglandin levels, inhibition of thrombosis, stimulation of nitric oxide (NO) and hydrogen sulfide (H_2S) production leading to vasodilation and inhibition of angiotensin-converting enzyme (ACE) (El-Saber Batiha et al. 2020). Another form of garlic which is prepared by natural ageing of garlic for more than 10 months, known as AGE, is also reported elicit similar effects in experimental animals and human subjects (Matsutomo 2020). Furthermore, pre-biotic properties of garlic are also proposed to contribute to its anti-hypertensive effect (Ried 2019).

3.1.1.2 *Allium cepa*

Allium cepa, commonly known as onion, is a bulbous herbaceous plant which is the third most important crop plant after potato and tomato. Onion has a distinct flavour and the bulbs, either roasted or fried, are used as an important culinary ingredient across the world or eaten raw. Moreover, onion has also been used in traditional medicine to alleviate many disorders since ancient times (Teshika et al. 2019). It has a wide range of phytochemical constituents including phytoestrogens such as coumestrol and isoflavones and flavonoids such as quercetin, kaempferol, resveratrol, apigenin and organosulfur compounds such as S-methyl cysteine sulfoxide, thiosulfinates, etc. (Kianian et al. 2021).

Several in vivo studies have validated the use of *Allium cepa* as a potential anti-hypertensive agent. The mechanisms of action identified so far are inhibition of the enzyme ACE thereby reducing conversion of angiotensin I to angiotensin II, amelioration of oxidative stress, inhibition of Ca^{2+} influx into vascular smooth muscle cell and reduction in heart rate (Galavi et al. 2021). Quercetin, one of the major constituents of onion, has been demonstrated to reduce blood pressure by multiple mechanisms including increased nitric oxide synthase (NOS) and NO and decrease in angiotensin II production (Kianian et al. 2021).

3.1.1.3 *Olea europaea*

Commonly known as olive tree, *Olea europaea* is a rich source of polyphenols, and its fruits and extracted oil are essential components of the healthy Mediterranean diet. Olive leaves are reported to contain higher polyphenol content than oil or fruits, and there are many commercially available olive leave extracts marketed as food supplements and neutraceuticals (Saibandith et al. 2017). Olive has been used for its therapeutic potential since ancient times. It is rich in phenolics such as iridoids and secoiridoids, flavonoids, triterpenes, sterols and glycosides which are found in almost all parts of the tree. In addition, olive oil also contains mono- and polyunsaturated fatty acid as major constituents. Oleuropein is a secoiridoid which is a major constituent of olive and is found in higher concentrations in leaves than in fruits and oil (Hashmi et al. 2015).

The ability of olive leaf extract to reduce blood pressure has been experimentally proven in several studies. For instance, olive leaf extracts have been found to reduce systolic blood pressure and heart beat in a concentration-dependent manner in rabbits. It is also found to effectively exert anti-hypertensive effect in spontaneously hypertensive rat. Various mechanisms have been proposed for the observed hypotensive or anti-hypertensive effect of olive including inhibition of ACE, blocking of β-adrenergic receptors and antagonizing Ca^{2+} ion channels (Ivanov et al. 2018). Similar blood pressure-lowering effect has also been demonstrated for olive oil in many epidemiological, animal and clinical studies. Inhibition of oxidative stress, increased eNOS activity, reduction in plasma angiotensin II, etc. have been demonstrated as potential mechanism of action of olive oil. In addition to the above-mentioned mechanisms, oleic acid which is a component of olive oil has been proposed to be responsible for the hypotensive effect of olive oil by inhibiting G-protein-coupled receptors (Massaro et al. 2020).

3.2 *Anti-hypertensive Ethnozoological Resources*

Not much information is available on the use of faunal resources to treat hypertension. One study suggests the use of honey derived from *Apis mellifera* to treat hyper blood pressure (Hassan et al. 2023). The use of *Hirudo medicinalis* or leech is documented as a vasodilator. Earthworms are also used as anti-hypertensive agents in Chinese traditional medicine. Secretions from granular glands of amphibians possess many pharmacological properties including hypotensive effect (Costa-Neto 2005). Tangsa tribe, inhabitants of Arunachal Pradesh, India, use cooked brain of Assamese macaque (*Macaca assamensis*) and whole cooked body of spiny eel (*Mastacembelus armatus*) to treat blood pressure (Jugli et al. 2020). Snakehead fish, locally known as "haruan" or "shol", is used in south-east Asia, including North East India and Malaysia, to treat various ailments including cardiovascular ailments. Its skin extract, called shol fish skin extract (SFSE), contains an active compound, cardiotoxic factor II (CTF-II), with hypotensive effect (Karmakar et al. 2002). In the Republic of Benin, topical application of a powder made from *Limicolaria aurora is reported to be used to regulate blood pressure* (Loko et al. 2019).

3.3 *Human Intervention Studies with Some Anti-hypertensive Ethnomedicinal Plants*

3.3.1 Clinical Trials with Garlic

A meta-analysis of 9 randomized controlled trials (RCT) including 482 individuals suggested that various concentrations of garlic preparations were effective in significantly reducing systolic and diastolic blood pressure as compared to the placebo group in hypertensive individuals and were well tolerated. However, adverse gastrointestinal effects at higher doses partially limit the efficacy (Rohner et al. 2015).

An earlier meta-analysis including 20 clinical trials in more than 900 individuals suggested that garlic supplements significantly reduced systolic blood pressure (mean difference 5.1 ± 2.2 mmHg) and diastolic blood pressure (mean difference 2.6 ± 1.6 mmHg). In the trials included in this meta-analysis, 13 trials had used standard garlic supplement powder, 5 trials used AGE, 1 used garlic oil and 1 used egg yolk-enriched garlic powder (Ried 2016).

An updated meta-analysis of 12 trials with 553 participants with hypertension was performed to investigate the effect of garlic on hypertension. Of these 12 trials,

8 used garlic powder while 4 trials investigated the effect of Kyolic AGE. The meta-analysis confirmed the blood pressure-lowering effect of garlic supplements. Systolic pressure was reduced by an average of 8.3 ± 1.9 mmHg and diastolic pressure by 5.5 ± 1.9 mmHg which is comparable to standard anti-hypertensive allopathic medication. The study suggested that vitamin B status is an important determinant of anti-hypertensive response to garlic treatment.

The effect of Kyolic garlic on gut microbiome was assessed in another clinical trial, the GarGIC trial. The study demonstrated a pre-biotic effect Kyolic garlic exemplified by increased microbial diversity and growth of beneficial bacterial species, e.g. *Lactobacillus* and *Clostridia* species (Ried et al. 2018). In a recent triple-blinded placebo-controlled randomized trial, a modified form of garlic called aged black garlic (ABG), with different organosulfur and polyphenol composition than fresh garlic and optimized with *S*-allyl cysteine (SAC) was administered at low doses in volunteers with grade I hypertension. Significant reduction in systolic and diastolic blood pressure was noted after 12 weeks of treatment accompanied by increased antioxidant capacity, rise in NO and inhibition of ACE activity (Serrano et al. 2023).

3.3.2 Clinical Trial with Onion

In a double-blinded randomized placebo-controlled clinical trial, 70 overweight to obese participants with stage I hypertension or prehypertension were assigned to receive either onion skin extract powder or placebo treatment for 6 weeks separated by 6 weeks washout period. The study revealed that onion skin extract powder lowered ambulatory blood pressure in subjects with hypertension. Plasma quercetin levels increased noticeably in subjects receiving onion skin extract powder suggesting the blood pressure-lowering effect was due to quercetin (Brüll et al. 2015).

3.3.3 Clinical Trials with Roselle

A meta-analysis of 13 randomized control trials with 1205 participants was performed to investigate the effect of *H. sabdariffa* in patients with hypertension or hypertension with metabolic syndrome. Results suggested that there was a significant drop in systolic and diastolic blood pressure of individuals suffering from mild to moderate hypertension. But in patients having both hypertension and metabolic syndrome, the decrease in blood pressure was not significant (Abdelmonem et al. 2022).

3.3.4 Clinical Trial with Olive

In a double-blinded placebo-controlled trial, 61 volunteers with prehypertension or grade I hypertension were randomized to receive either commercially available liquid olive leaf extract or placebo for 6 weeks separated by a 4-week washout period. Olive leaf extract induced significant reduction in both daytime and 24 h systolic and diastolic pressure as compared to control (Lockyer et al. 2017). The anti-hypertensive potential of extra virgin olive oil (rich in monounsaturated fatty acid) was tested and compared with that of sunflower oil (polyunsaturated fatty acid) in a randomized controlled trial involving 23 hypertensive subjects. Treatment with the extra virgin olive oil but not with sunflower oil was demonstrated to significantly lower the requirement of daily anti-hypertensive medications in the volunteers (Ferrara et al. 2000).

A randomized control study was designed to investigate the effect of a combination of roselle and olive in Egyptian patients with grade I hypertension and compare the effect with that of captopril. The results suggested a significant drop in blood pressure with both low- and high-dose combination of roselle and olive. This reduction was found to be comparable to that mediated by captopril (Elkafrawy et al. 2020). A similar study was performed to investigate the blood pressure-lowering effect and tolerability of olive leaf extract alone in comparison to captopril in subjects with stage I hypertension. After 8 weeks of treatment, 500 mg olive leaf extract administered twice daily was found to exhibit hypotensive effect comparable to that of captopril (Susalit et al. 2011).

In a 12-week double-blinded placebo-controlled human intervention trial, volunteers were randomized to receive a novel formulation of olive leaf extract (500 mg), green coffee bean extract (100 mg) and beet powder (150 mg) or a matching placebo treatment twice daily for 6 weeks followed by a 6-week washout period. The study suggested no significant blood pressure-lowering effect of the formulation (Wong et al. 2014). Table 2 includes information on the human intervention studies conducted with some additional ethnomedicinal plants reported from majority of the countries (according to Aumeeruddy and Mahomoodally 2020).

4 Obesity

According to WHO, obesity is defined as excessive build-up of body fat for a given height that imposes significant health risk (Bray 2003). Since last few decades, it has become a significant public health concern. According to estimates, more than 30% of the world population is obese (The GBD 2015 Obesity Collaborators 2017). It is caused by a combination of environmental, lifestyle and genetic factors. Obesity increases the predisposition to a lot of other metabolic disorders and enhances the complication of diabetes, hepatic steatosis, osteoarthritis, dyslipidemia, etc. (Mayoral et al. 2020). Obesity management is aimed at decreasing calorie intake and increasing calorie expenditure thereby reducing body weight. Lifestyle

Table 2 Human intervention studies with some widely reported anti-hypertensive ethnomedicinal plants

Sl no.	Plant	Study design	Outcome	References
1.	*Persea americana* (Avocado)	1. Crossover randomized placebo-controlled trial. Sixteen volunteers received either 600 mg avocado pulp or placebo following which the participants were subjected to treadmill exercise 2. Review of five clinical trials out of which three conducted on elderly hypertensive patients receiving avocado leave decoction	1. Intervention group showed improved recovery of systolic blood pressure and heart rate after exercise 2. Significant reduction in blood pressure was reported in all the studies	1. Sousa et al. (2020) 2. Ichsan et al. (2023)
2.	*Citrus aurantiifolia* (Lime)	1. Three studies conducted to investigate the effect of *C. aurantiifolia* on anxiety levels in patients with different medical conditions 2. 100 acute myocardial infarction patients were randomized to receive either lemon inhalation therapy or placebo for 3 consecutive days	1. All the studies reported a decrease in anxiety levels accompanied by reduction in blood pressure 2. Lemon aroma therapy reduced systolic BP and anxiety in patients	1. Mannucci et al. (2018) 2. Rambod et al. (2020)
3.	*Moringa oleifera*	1. Randomized parallel placebo-controlled study in which 40 diabetic patients were randomized to receive *M. oleifera* leaves or placebo daily for 14 days 2. Prospective placebo-controlled randomized clinical study with 41 healthy volunteers. Intervention group received 120 g cooked *Moringa* leaves while the control group did not. Blood pressure was measured before meal and at regular intervals for 24 h for the two groups 3. 20 male stage I hypertensive patients were treated with 150 mL of *Moringa* leaf juice twice a day for 30 days	1. Mean systolic blood pressure decreased at one particular dose of *M. oleifera* (40 g leaves), but the effect was not dose dependent 2. 2 h postprandial systolic and diastolic blood pressure showed reduction in the intervention group 3. Significant reduction in both systolic and diastolic blood pressure was noted; post-intervention as compared to pre-intervention	1. Afiaenyi et al. (2023) 2. Chan Sun et al. (2020) 3. Sailesh et al. (2018)

(continued)

Table 2 (continued)

Sl no.	Plant	Study design	Outcome	References
4.	*Cymbopogon citratus* (Lemongrass)	1. 105 healthy volunteers received lemon grass tea prepared from 2, 4 and 8 g of lemongrass powder for 30 days 2. 29 healthy volunteers randomized to receive lemongrass oil massage once a week for 3 weeks. Control group did not receive the massage 3. Single-blinded clinical trial of 60 healthy women volunteers who either inhaled an oil mixture (prepared from lemon grass, cananga and patchouli extracts) or lavender oil. Control group did not inhale either of the oils	1. Significant reduction in systolic and diastolic blood pressure 2. Significant reduction in diastolic blood pressure but no change in systolic blood pressure 3. The mixture essential oil containing lemongrass was more effective in reducing blood pressure than lavender oil	Silva and Bárbara (2022)

interventions, anti-obesity medications and bariatric surgery are recommended for the management of obesity. Four structurally related anti-obesity medications, namely, phentermine, phendimetrazine, diethylpropion and benzphetamine, act to increase satiety by reducing food cravings thereby limiting calorie intake. At present, these medications are recommended for short-term use because of associated adverse effects such as insomnia, dryness of mouth, constipation, etc. In addition, five medications are available for long-term treatment. These are orlistat, lorcaserin and liraglutide and two combination therapies phentermine/topiramate and naltrexone/bupropion. However, adverse gastrointestinal events, marginal efficacy, diarrhoea and cognitive impairment limit the use of these drugs (Gadde et al. 2018). Therefore, due to the negligible benefit of these conventional treatment strategies, ethnomedicinal resources are gaining popularity for long-term weight management as a more viable alternative.

4.1 Anti-obesity Ethnomedicinal Plants

Several authors have documented the practice of using traditional ethnomedicinal plants in the amelioration of obesity, and a few review articles are also available. One recent review article, compiling information from 150 articles published till 2020, has recorded the use of 451 plant species which are traditionally used to treat obesity. According to the study, the top three countries using the maximum number of species are Mexico, Brazil and Burkina Faso. The five most popular species that were reported to be used are *Carica papaya*, *Zingiber officinale*, *Hibiscus sabdariffa*, *Foeniculum vulgare* and *Zea mays* (Aumeeruddy and Mahomoodally 2021b). According to an earlier review article which had compiled data on the ethnomedicinal plants used for obesity management from 1995 to 2015, 23 species had been biologically validated for their use as anti-obesity medicinal plants. However, strangely among those 23 plants, except *Carica papaya*, none of the other predominantly used ethnomedicinal plants, as mentioned above, were described (de Freitas Junior and de Almeida 2017). In yet another comprehensive review, a meta-analysis was performed for all data from 2000 to 2018 on the use of natural agents to treat obesity and their mode of action (Karri et al. 2019).

4.1.1 Pharmacological and Biological Validation of Predominantly Used Species

4.1.1.1 *Carica papaya*

Ethnobotanical surveys have revealed that various parts of this plant such as fruits, leaves and seeds are consumed in different countries. The fruit is consumed raw, whereas decoction of leaves and seeds is prepared for consumption.

Ethnopharmacological study using papaya seed extract has revealed that the predominant bioactive component is benzyl isothiocyanate. The extracts were also found to contain glucosinolates, α and δ-tocopherol, polyphenols, carotenoids, monounsaturated fatty acids, especially oleic acid and the enzyme papain (Kermanshai et al. 2001). The leaves of *Carica papaya* are also reported to contain fibres and polyphenols such as lycopene, tocopherols, benzyl isothiocyanate, etc. (Vuong et al. 2013).

A study was conducted to examine the effect of papaya seed oil on mice on high-fat diet. Results revealed that the calorie intake, adipocyte area, serum total cholesterol and weight gain were reduced in groups receiving papaya seed oil as compared to the control group (Santana et al. 2022).

Water extract of papaya leaves was found to reduce weight in male New Zealand rabbits. However, this was accompanied by increased levels of serum markers of hepatotoxicity (Omonkhua and Onoagbe 2011).

Similarly, papaya fruit juice was also demonstrated to reduce body and adipose tissue weight and lower LDL cholesterol and triglycerides in mice fed with high-fat diet. Pancreatic lipase was inhibited by papaya juice which probably decreased dietary fat absorption. Obesity-related inflammatory markers such as IL-6 and TNF-α are significantly reduced in papaya juice-treated group (Od-Ek et al. 2020). Very recently, another similar study has further validated the hypolidaemic and anti-obesity activity of papaya juice in male Wistar rats fed with high-fat diet (Matsuane et al. 2023).

The cysteine protease papain, which is a major component of papaya, was found to alleviate obesity by reducing body weight, adipose tissue weight and serum cholesterol of obese mice fed with high-fat diet. Mechanistically, papain was found to decrease the expressions of transcription factors related to adipogenesis such as peroxisome proliferator-activated receptor-γ (PPARγ), CCAAT/enhancer-binding protein (C/EBPα) and sterol regulatory binding protein-1 (SREBP-1) via inhibiting

Akt. Differentiation of pre-adipocytes to adipocytes was also suppressed by papain treatment. Moreover, papain also inhibited leptin and increased adiponectin by activating AMPK, the master regulator of energy metabolism (Kang et al. 2021).

4.1.1.2 *Zingiber officinale*

Several studies have documented the role of *Zingiber officinale*, commonly known as ginger, in reducing obesity. The rhizhome or the underground stem is a popular culinary spice used for its pungent flavour and taste. It is also used in various folk medicines across the world. Ginger is reported to contain more than 400 bioactive components belonging to diarylheptanoids, gingerol analogues (mainly 6-, 8- and 10-ginerol in fresh ginger and shogaols in dried gingers), monoterpenoid glucosides, sulfonates and steroids (Unuofin et al. 2021).

Multiple studies have demonstrated that ethanolic and aqueous extracts of ginger reduced lipid accumulation, body weight and cholesterol and triacyl glycerides in murine and rodent models. Markers of adipogenesis and lipogenesis, PPARγ and C/EBPα, were also modulated by ginger extract (Unuofin et al. 2021). Studies with active components of ginger such as gingerol and shogaols were also found to have anti-obesity effect via multiple mechanisms. Both gingerol and shogaol inhibited adipogenesis by preventing differentiation of pre-adipocytes to adipocytes thereby preventing lipid accumulation. The active components were also demonstrated to alter the expression of lipogenic markers such as fatty acid synthase, acetyl CoA carboxylase, lipoprotein lipase, etc. In addition, gingerol and shogaol also increase leptin and HDL cholesterol and activated the AMPK and PPAR δ pathways (Wang et al. 2017; Salaramoli et al. 2022).

4.1.1.3 *Hibiscus sabdariffa*

The red calyces of *Hibiscus sabdariffa* or roselle, as it is commonly known, are used in various parts of the world to prepare a popular drink, roselle drink. In addition, the shoot and fruits are also used in soups and sauces or eaten raw in salads (Salami and Afolayan 2021). The predominant bioactive components identified in roselle are organic acids such as hydroxy citric acid, dimethyl hibiscus acid, anthocyanins such as delphinidin 3-sambibioside (hibiscin), cyanidin 3-sambubioside (gossip cyanidin), flavonoids such as quercetin, luteolin and phenolic acids such as chlorogenic acid (Ojulari et al. 2019).

Several authors have documented a decrease in body weight in obese mice following treatment with *H. sabdariffa* extract. The active components including polyphenols and organic acids derived from *H. sabdariffa* are reported to contribute to its anti-obesity effect (Ojulari et al. 2019). Consumption of *H. sabdariffa* extract has been correlated with reduction in serum lipid levels such as cholesterol, triglycerides, LDL and VLDL in hyperlipidemic mice. This was accompanied by activation of AMPK and inhibition of fatty acid synthase. Moreover, the constituent polyphenols are also demonstrated to regulate transcription factors related to lipid and glucose metabolism, e.g. PPARs and SREBP-1c in hyperlipidemic mice model.

Studies have also pointed toward a potential epigenetic regulation of genes involved in lipid metabolism by roselle. For instance, quercetin inhibited histone acetyl transferase (HAT) activity in the promoters of inflammatory genes. Roselle polyphenols also regulated the expression of microRNAs (miRNAs) to modulate lipid metabolism in mice fed with high-fat diet (Herranz-López et al. 2017). A recent study has shown that *H. sabdariffa* extract reversed fibroblast growth factor 21 (FGF21) resistance in mice by inhibiting miR34a thereby increasing FGF receptor (FGFR1) expression. FGF21 plays important role in browning and thermogenesis of adipose tissue (Kartinah et al. 2022).

In addition to the above-mentioned mechanisms, *H. sabdariffa* extract has also been shown to ameliorate the alterations in gut microbiota induced by high-fat diet leading to decreased weight gain in mice. Ratio of *Firmicutes/Bacteroidetes* was found to improve by *H. sabdariffa* extract administration (Diez-Echave et al. 2020).

4.2 Ethnozoological Resources to Treat Obesity

Consumption of raw milk of *Bos taurus* (cow), roasted meat of *Hystrix cristata* (crested porcupine) and *Ovis aries* (sheep) is reported to be used for mitigating weight gain among the indigenous people of Ethiopia (Wendimu and Tekalign 2023). Consumption of the water bird *Sterna hirundo* meat has been documented to be used in East Pakistan to treat obesity (Rahman et al. 2022).

4.3 Human Intervention Studies with Some Anti-obesity Ethnomedicinal Plants

4.3.1 Clinical Trial with Ginger

A meta-analysis of 14 randomized clinical trials conducted till 2017 was performed to summarize the impact of ginger on weight reduction among 473 overweight or obese participants. The study concluded that ginger intake caused significant reduction in body weight and hip-to-waist ratio which was accompanied by an increase in serum HDL cholesterol level. Triglycerides and LDL cholesterol levels, however, remained unchanged with ginger supplementation (Maharlouei et al. 2019). According to one randomized controlled trial, the anti-obesity effect of ginger was more noticeable in individuals with AA genotype for the uncoupling protein and Trp64Trp genotype for β3 adrenergic receptor gene (Ebrahimzadeh Attari et al. 2015).

One double-blinded placebo-controlled randomized clinical trial (NCT02570633) involving 66 female volunteers with high body adiposity concluded that ginger intake (600 mg extract daily) for 3 months neither changed the resting energy expenditure nor did it alter anthropometric measurements of the participants (Braga Tibaes et al. 2022).

4.3.2 Clinical Trial with Roselle

A very recently published meta-analysis of 7 randomized clinical trials (5 conducted in Spain and two in Asia) with 332 participants examined the effect of *Hibiscus sabdariffa* or roselle in combination with other plant extracts on parameters associated with metabolic syndrome. The study revealed roselle and other plant extracts caused significant improvement in all anthropometric parameters as evident by substantial reduction in BMI, body weight and body fat percentage (García-Muñoz et al. 2023). One of the randomized clinical trials was conducted to assess the effect of a combination of *H. sabdariffa* with *Lippia citriodora* on satiety and hunger in obese individuals. A decrease in appetite sensation was observed in participants receiving the plants extracts which led to decreased calorie intake after a full meal (Serna et al. 2022).

A triple-blinded randomized clinical trial (IRCT201109122306N2) conducted in Iran on 72 obese adolescent volunteers reported amelioration of dyslipidemia in the group receiving 2gms of powdered roselle calices per day for 1 month as compared to the group receiving placebo (Sabzghabaee et al. 2013).

Apart from the above-mentioned ethnomedicinal plants, the other plants which appear among the top ten mostly reported to be used in folklore for treating obesity have not been thoroughly studied till date. Till date, there are no human intervention studies reported with papaya which is used to treat obesity in majority of the countries (according to the review by Aumeeruddy and Mahomoodally 2021b). However, a few clinical trials have been conducted with some plant species used in folklore in majority of the countries which we have tabulated in Table 3. Apart from this, the effect of garlic on body weight has been investigated in a few human intervention studies with one study, conducted with obese women on low calorie diet, reporting weight reduction in the garlic-treated group as well as in the placebo group. In another randomized control trial, raw crushed garlic was administered to diabetic patients at 100 mg/kg body weight with diet for 4 weeks. Significant reduction in waist circumference and serum triglyceride levels was noted in groups receiving garlic-supplemented diet. However, there was no significant change in BMI after garlic consumption (Choudhary et al. 2018). Thus, the effect of garlic on weight reduction remains inconclusive.

Table 3 Human intervention studies with some widely reported anti-obesity ethnomedicinal plants

Sl no.	Name of the plant	Study design	Outcome	References
1.	*Foeniculum vulgare* (Fennel)	1. 47 postmenopausal women were randomized to receive either 100 mg fennel capsules or placebo 2. Randomized control trial 64 obese women with polycystic ovarian syndrome, investigated the effect of fennel supplementation with a hypocaloric diet	1. No significant change in body weight in the intervention group 2. Fennel supplementation did not cause reduction in body weight	1. Saghafi et al. (2017) 2. Nadjarzadeh et al. (2021)
2.	*Caralluma* sp.	1. Randomized controlled trial with 43 adults having BMI ≥ 25 received the *Caralluma fimbriata* extract or placebo for 12 weeks 2. Meta-analysis of 7 clinical trials to investigate the effect of *Caralluma fimbriata* in appetite suppression and obesity control	1. Significant reduction in body weight, BMI, waist circumference was noted 2. The review found no significant changes in metabolic parameters	1. Astell et al. (2013) 2. Jayawardena et al. (2021)
3.	*Citrus* sp.	1. Meta-analysis of 13 clinical trials involving a total of 921 participants investigated the effect of citrus or its extracts on weight loss 2. 60 healthy volunteers were randomized to receive a combination of citrus polyphenol α-glucosyl hesperidin and tea polyphenol epigallocatechin-3-*O*-gallate or placebo for 12 weeks	1. Significant reduction in body weight observed in the intervention group 2. Significant reduction in body weight and BMI was noted in the intervention group	1. Wang et al. (2020) 2. Yoshitomi et al. (2021)

(continued)

Table 3 (continued)

Sl no.	Name of the plant	Study design	Outcome	References
4.	*Prunus avium* (Dark sweet cherry)	60 obese participants were stratified into 2 groups with one group receiving dark sweet cherry concentrated juice and the other receiving placebo for 30 days	Although body weight reduction was not observed, dark sweet cherry prevented the increase in body in weight and BMI as observed in the placebo group during the intervention period. Moreover, dark sweet cherry reduced systolic and diastolic blood pressure more effectively in individuals with high BMI	Arbizu et al. (2023)

5 Non-alcoholic Fatty Liver Disease (NAFLD)

NAFLD has become a global epidemic of chronic liver diseases with a prevalence ranging from *23% to 32%*. In NAFLD, the cause of excessive hepatic fat accumulation (≥5% fat infiltration in histology) cannot be identified. It includes a wide spectrum of liver diseases including non-alcoholic fatty liver (NAFL) which is the mildest form to more serious forms such as non-alcoholic steatohepatitis (NASH) leading to cirrhosis and liver failure. NAFL is characterized by hepatic steatosis, while in NASH, steatosis is accompanied by hepatocyte ballooning and lobular inflammation. Other components of the metabolic syndrome such as diabetes, dyslipidemia, hypertension, etc. often coexist in NAFLD patients. NAFLD progresses in two steps: (i) fat accumulation in the liver leading to insulin resistance and (ii) oxidative stress, oxidation of fatty acids and associated cellular and molecular changes. Increased mortality from other complications such as cardiovascular disease or cancer has been reported in NAFLD patients as compared to non-NAFLD patients (Pouwels et al. 2022).

Lifestyle modifications such as weight loss, healthy diet and physical activity play the most important role in the management of NAFLD. In addition, pharmacotherapy for NAFLD is recommended for patients with significant fibrosis (stage 2, i.e. with pericellular and portal fibrosis) and to patients with less severe NAFLD but who are at high risk of progression due to comorbidities. At present, there are only two treatment options for NAFLD patients: vitamin E and the PPAR-γ ligand Pioglitazone. However, there are limitations in the use of both these medications. For example, vitamin E has been found to be ineffective in NAFLD patients with diabetes. Also, vitamin E has not been found to be effective in alleviating fibrosis. Moreover vitamin treatment is also associated with a lot of adverse consequences such as prostate cancer, heart failure, etc. Thus, vitamin E is recommended for use only in selected NASH patients with caution. Treatment with the other drug, pioglitazone, also has yielded inconsistent outcome with one study (PIVENS trial recruiting non-diabetic individuals) finding it ineffective against fibrosis, while another study with pre-diabetic and diabetic patients demonstrated pioglitazone to be effective in alleviating NASH and also improve fibrosis score. Both the studies, however, revealed significant weight gain among the participants. Pioglitazone also has other side effects such as increased risk of bladder cancer, bone loss and fluid retention and is contraindicated in patients with heart failure. In addition to these two treatments, there are other medications which are under clinical trials but yet to be prescribed to patients (Paternostro and Trauner 2022).

Although there is no definite efficacious drug treatment for NAFLD, several ethnobotanical resources have been traditionally used worldwide for treating liver diseases. These have huge scope for the prevention and/or treatment of fatty liver disease.

5.1 Ethnomedicinal Plants Used to Treat NAFLD

Across the world, ethnobotanical surveys have identified several plant species used by traditional healers and local communities of a particular region to cure liver diseases. In one such ethnobotanical study in Africa, 575 traditional healers were interviewed throughout Burkina Faso which revealed the healers could identify various liver ailments such as hepatitis, cirrhosis, jaundice, etc., and they have herbal remedies for each condition. According to the study, the most common plant species based on number of mentions by traditional healers in descending order are *Cochlospermum tinctorium*, *Cassia sieberiana*, *Anogeissus leiocarpa*, *Tamarindus indica*, *Terminalia avicennioides*, *Carica papaya*, *Cochlospermum planchonii*, *Combretum micranthum*, *Terminalia macroptera* and so on. A total of 212 different plant species were identified in this survey (Tibiri et al. 2020). A similar ethnobotanical survey conducted in Multan, Pakistan, interviewed 43 herbal practitioners and documented 69 different plant species which are used against 15 different liver disorders such as jaundice, fatty liver, hepatitis, cirrhosis, etc. The most frequently used plant species are *Cichorium intybus*, *Artemisia absinthium*, *Solanum nigrum*, *Berberis vulgaris*, *Borago officinalis*, *Prunus domestica*, *Curcuma longa*, *Rheum emodi* and *Fumaria officinalis*.

According to a review article, Iranians residing in different regions of Iran and belonging to different cultures use 26 different plant species to treat and prevent liver diseases. These include some common species as documented in Pakistan such as *Cichorium intybus* and *Berberis vulgaris* (Moradi et al. 2016). A few other species documented in this study which has been clinically tested for their effect on liver function are *Silybum marianum*, *Cynara scolymus* and *Camellia sinensis* (Fakhri et al. 2022).

A recent article has reviewed the global usage of traditional medicinal plants in NAFLD and evaluated the pharmacological and biological studies done with these plant species and (Karadağ and Demírci 2022). We have selected the following plant species from this review article for a detailed discussion in the next section, based on their ethnobotanical use and total number of mentions against NAFLD in the literature: *Cichorium intybus*, *Olea europea* and *Vitis vinifera*. In addition, we have included information on the clinical trials conducted with these medicinal plants which have been documented to be used to treat NAFLD in ethnobotanical surveys.

5.1.1 Pharmacological and Biological Validation of Predominantly Used Species

5.1.1.1 *Cichorium intybus*

This plant, commonly referred to as chicory, is a herbaceous plant with a fleshy tap root and large leaves and is mainly distributed in Europe and Asia. It has been historically used as coffee substitute and as a remedy for various ailments in folklore. The plant is cultivated because of its numerous applications: the taproot is used to prepare coffee substitute; leaves are used as vegetables; the seeds, roots and leaves are used as forage, to produce witloof chicory (for etiolated buds or chicon); and its popularity as a medicinal plant is also growing steadily (Street et al. 2013).

The plant is a rich source of phytochemicals such as inulin which is a fructose polymer with a terminal glucose residue, chicoric acid, aliphatic compounds and their derivatives, coumarins, flavonoids, anthocyanins and volatile compounds (Street et al. 2013).

Hepatoprotective effects of chicory and its components have been demonstrated in many experiments (Ahmed et al. 2003; Landmann et al. 2014). Chicory seeds extract was shown to prevent hepatic steatosis caused by early- and late-stage diabetes in rats as evident by improved histological features (Ziamajidi et al. 2013). Polysaccharide fractions from chicory root also have been found to exert hepatoprotective effects in a NAFLD model in rat induced by high-fat diet (Ziamajidi et al. 2013).

Mechanisms identified till date are downregulation of SREBP-1 and PPAR-α genes. Chicory also acts as an agonist of PPAR-α (Ziamajidi et al. 2013). Serum metabolomic study revealed that chicory polysaccharides regulate several metabolic pathways such as amino acid metabolism, fatty acid biosynthesis and β-oxidation of long-chain fatty acids which might be potential mechanisms of intervention by chicory in NAFLD (Zhu et al. 2019).

5.1.1.2 *Olea europaea*

This plant and its phytoconstituents have been described in the above section as one of the top three widely reported plants against hypertension.

A large body of evidences point towards the health benefits of olive in NAFLD patients. Olive leaf powder supplementation in high-fat and high-cholesterol fed rats caused significant reduction in hepatic inflammation, fibrosis and hepatic cholesterol impeding NASH progression (Omagari et al. 2021). Olive leaves extracts also prevented development of spontaneous NASH (Omagari et al. 2010).

A recent study has shown that olive leaf extract reduced lipid and H_2O_2 accumulation and activate lipophagy in hepatic cells treated with high concentration of free fatty acids (Tacconi et al. 2023).

Oleuropein, a secoiridoid derived from olive, has been found promising in inhibiting progression of NASH to liver fibrosis in a high-fat diet-induced murine NASH model (Kim et al. 2014).

Mediterranean diet, characterized by the consumption of olive and olive oil, is famous for its health-promoting effects. Studies have shown that Mediterranean diet supplemented with extra virgin oil protects against hepatosteatosis by improving fat deposition and insulin resistance. Numerous studies have experimentally and clinically proven the beneficial effect of olive oil in NAFLD (Abenavoli et al. 2019).

5.1.1.3 *Vitis vinifera*

This is the common grapevine, also known as European grapes. Grapes have been used to make wines since ancient civilization. Grapes are one of the widely consumed fruits across the world. Apart from the fruit, grape skin, stalk and seeds, which are formed as by-products of wine-making process, are also utilized as nutritional supplements and cosmetic additives.

Grapes including its fruits, seeds and peels are rich in polyphenolic antioxidants and have been experimentally demonstrated to have protective effect against metabolic syndrome. Major phytochemicals that have been identified in grapes are flavonoids such as catechin, epicatechin, quercetin, resveratrol, stilbenes, proanthocyanidin, vitamin C, proteins and fats (Majeed et al. 2023).

Dietary grape consumption has been shown to alleviate symptoms of fatty liver such as high-fat diet-induced lipid accumulation in mice liver and also enhanced mice longevity (Dave et al. 2022). Polymerized anthocyanins from grape skin, proanthocyanidin from grapeseed and grape polyphenols rich in B-type proanthocyanidins have all been found to have protective effects against high-fat diet-induced hepatic steatosis (Fan et al. 2019; Sun et al. 2022; Mezhibovsky et al. 2021).

Grapes have been demonstrated to modulate genes of the lipid metabolism pathway including lipid synthesis, lipid degradation and lipid oxidation. Evidences also suggest activation of AMPK by grape extract which activate fatty acid oxidation and inhibit fatty acid synthesis by downregulation of the transcription factor SREBP-1. In addition, antioxidant activity of grapes also contributes to improved mitochondrial function facilitating reduced lipid accumulation in liver (Santos et al. 2017).

5.2 Ethnozoological Resources Used for Liver Diseases

Indigenous population of Pakistan use milk, meat and urine of camel (*Camelus dromedarius*) to treat jaundice and hepatitis. Bone, liver and spleen of *Canis lupus* is also reported to cure liver diseases (Ullah et al. 2020). In north-east India, the local tribes use many ethnozoological resources to cure liver ailments (Hussain and Tynsong 2021). The boiled head of *Wallago attu* is documented to be consumed by Karbi tribe of Assam, India, to improve liver function (Teronpi et al. 2012). A decoction of the snail *Archachatina marginata* is used to treat liver dysfunction by ethnic population of the Republic of Benin (Loko et al. 2019). The Sukuma tribe of Tanzania consumes the ashes after burning *Loxodonta africana* skin to treat hepatitis (Vats and Thomas 2015). Consumption of powdered meat and bones of *Cynopterus sphinx* mixed with oil and pepper is used for curing liver diseases in Ethiopia (Abebe et al. 2022).

5.3 Human Intervention Studies with Some Ethnomedicinal Plants Used Against NAFLD

5.3.1 Clinical Trials with Chicory

In a very recent meta-analysis of 5 clinical trials with 197 NAFLD patients, chicory was found to significantly reduce the levels of serum biomarkers of hepatotoxicity (Maleki et al. 2023).

A randomized placebo-controlled clinical trial was conducted with 92 NAFLD patients to assess the effect of turmeric, chicory seed, their combination or placebo on the risk factors of NAFLD. After 12 weeks of intervention, significant reduction in BMI and waist circumference was noted in the two groups receiving chicory. The combination group also showed decreased serum alkaline phosphatase level and increased serum HDL level. Thus, chicory supplementation proved beneficial reducing the risk factors in NAFLD patients (Ghaffari et al. 2019).

5.3.2 Clinical Trials with Olive

A meta-analysis of 6 randomized control trials with 344 individuals using olive oil in cooking or eating was performed. The individuals recruited for these studies were overweight or obese, NAFLD patients and women having polycystic ovarian syndrome (PCOS). All the studies reported improvement of NAFLD scoring with olive oil intervention. Before intervention, the percentage of patients with mild-to-moderate NAFLD was 58.8% which rose to 85% after the intervention. This was accompanied by reduction in serum alanine aminotransferase (ALT) and aspartate amino transferase (AST) in the intervention group (Ma et al. 2023).

A randomized double-blinded clinical trial with 66 NAFLD patients compared the impact of olive oil and sunflower oil on the severity of fatty liver. Results indicated that fatty liver grade and body fat percentage were significantly lesser in the olive oil-treated group as compared to sunflower group (Rezaei et al. 2019).

In a nested multicentre randomized clinical trial, 100 volunteers were randomized to receive Mediterranean diet supplemented with extra virgin olive oil (EVOO), a Mediterranean diet supplemented with mixed nuts or a control diet for 3 years. The results suggested that the prevalence of hepatic steatosis was significantly lesser in participants on Mediterranean diet supplemented with EVOO accompanied by a lower liver fat percentage (Pintó et al. 2019).

5.3.3 Clinical Trials with Grape

In a meta-analysis of 8 trials with a total of 291 volunteers suffering from metabolic syndrome, hypertension, T2DM or are healthy, intervention with grape products for 12 weeks or more improved liver function with a reduction in serum levels of ALT and AST (Ghaffar et al. 2022).

The effect of resveratrol which is a flavonoid component in grapes was investigated in a meta-analysis of 6 randomized control trials carried out from 2014 to 2019. Parallel controlled trials with a total of 266 volunteers who were NAFLD patients, obese or both revealed that resveratrol supplementation did not cause any significant improvement in anthropogenic parameters, liver function and lipid profile. However, it caused a significant reduction in TNFα and C-reactive protein indicating amelioration of inflammation (Rafiee et al. 2021).

However, as we could not find any comprehensive review article documenting the usage of ethnomedicinal plants against NAFLD in different parts of the world, the selection of most popular plant species used to manage NAFLD became difficult. However, we found a meta-analysis in the Cochrane database which included 77 randomized clinical trials with 6753 participants suffering from fatty liver disease investigating the effect of herbal medicines. According to this meta-analysis, species of plants which have been most commonly used in clinical trials are *Crataegus pinnatifida*, *Salvia miltiorrhiza*, *Alisma orientale*, *Bupleurum chinense*, *Cassia obtusifolia*, *Astragalus membranaceus* and *Rheum palmatum* (Liu et al. 2013).

6 Dyslipidemia

It is a chronic condition characterized by abnormally high levels of serum triglycerides, cholesterol or both. The most common and clinically relevant parameters to assess dyslipidemia are elevated total cholesterol, triglycerides, low-density lipoprotein (LDL) cholesterol, very low-density lipoprotein (VLDL) and decreased high-density lipoprotein (HDL) cholesterol in the serum. Dyslipidemia is a major risk factor for cardiovascular diseases (CVD), and one of the primary approaches to prevent and treat atherosclerotic CVD is lowering serum cholesterol and triglyceride levels. At present, the medications which are available to lower lipid levels include statins which inhibit 3-hydroxy-3-methylglutaryl-coenzyme A reductase (HMG-CoA reductase) which is the rate-limiting enzyme in cholesterol biosynthesis; ezetimibe that lowers LDL cholesterol; monoclonal antibody against proprotein convertase subtilisin/kexin type 9 (PSCK9) which inhibits PSCK9-mediated degradation of LDL receptor thereby increasing hepatocyte uptake of LDL cholesterol; and inclisiran which is an siRNA against PSCK9 mRNA. Triglyceride-lowering therapy primarily includes lifestyle modification and increase in exercise and dietary adjustments. In addition, a few medications such as fibrates which are agonists of PPAR-α affecting lipoprotein metabolism and omega-3-fatty acid supplementation are derived from fish intake (Ferraro et al. 2022).

Despite the available medications, the associated side effects and the increasing trend in the incidences and mortality associated with atherosclerotic CVD have prompted investigations for more effective and safer therapies to manage dyslipidemia. Several ethnobotanical studies have documented the use medicinal plants used in folklore to treat hyperlipidaemia. But our literature search did not yield any report of hypolipidemic ethnozoological resources used in folklore.

6.1 Ethnomedicinal Plants with Hypolipidemic Activity

In a review of 174 ethnobotanical surveys on the use of plants to treat hypercholesterolemia in folklore dating from 2001 to 2020, a total of 390 species were documented (Aumeeruddy and Mahomoodally 2022). According to this review, the top ten plant species in decreasing order of number of reports are *Allium sativum*, *Juglans regia*, *Olea europaea*, *Rosmarinus officinalis*, *Prunus dulcis*, *Citrus limon* (L.), *Linum usitatissimum*, *Allium cepa*, *Curcuma longa* and *Cynara scolymus*. The country reporting the use of maximum number of plant species is Turkey followed by Argentina and Brazil.

6.1.1 *Allium sativum*

An extensive body of experimental and clinical evidence exists on the ameliorating effect of garlic on hyperlipidaemia. In several experimental animal models, garlic has been demonstrated to reduce serum cholesterol and lipid levels. Mechanism of lipid-lowering activity of garlic includes inhibition of cholesterol and triglyceride synthesis by inhibiting enzymes such as fatty acid synthase, glucose-6-phosphate dehydrogenase and HMG-CoA reductase (Yeh and Liu 2001; Reuter et al. 1996). Recently, it was shown that attenuation of high-fat diet-induced hyperlipidaemia by garlic supplementation is accompanied by an increase in the α-diversity of the gut microbiome (Chen et al. 2019).

6.1.2 *Juglans regia*

It is a large deciduous tree, which produces the edible English or Persian walnuts. The Western Himalayan agroclimatic conditions are suitable for the production of high-quality walnuts. In addition to nutritional components, walnuts are also a rich source of non-nutritional bioactive phytochemicals which endow them with many health-promoting effects.

All parts of the tree including the nut, leaves, bark and fruits have been found to have therapeutic properties. Different parts of the plant display variations in their constituent phytochemicals. Walnuts are rich in γ-tocopherol or vitamin E and omega 3 fatty acids, phenolic compounds and phytosterols (Sharma et al. 2022). Leaves contain phenolic acids, tannins and flavonoids such as quercetin, quercetin-galactoside, caffeic acid, essential fatty acids, ascorbic acid and naphthalene derivatives such as 5-hydroxy-1-4-naphthoquinone or Juglone which is the most active component in walnut leaves and green husk (Delaviz et al. 2017).

Walnut oil as well as walnut leave extract have been shown to improve lipid profile by decreasing serum LDL cholesterol, triglycerides and increasing HDL cholesterol in both diabetic, hypercholesterolemic and normal experimental animals (Delaviz et al. 2017; Nergiz-Ünal et al. 2013).

Walnut oil increases cholesterol efflux from foam cells plausibly due to inhibition of the lipogenic enzyme stearoyl CoA desaturase$_1$ (SCD_1) through the activation of nuclear receptor farnesoid-X-receptor (FXR) pathway (Zhang et al. 2011). Walnuts have also been proposed to increase hepatic β-oxidation of fatty acids and enhance PPARα (Shimoda et al. 2009).

6.1.3 *Olea europaea*

Several in vitro and in vivo studies have investigated the hypolipidemic effect of olive extracts and its active component, oleuropein. Multiple experiments have demonstrated that olive leaf extract caused a significant reduction in total cholesterol, LDL cholesterol and triglycerides in rodent models. Moreover, increased serum HDL and amelioration of dyslipidemia were observed by supplementation of diet with olive leaf powder in diabetic rats and in cholesterol fed rats. Potential mechanisms proposed so far are inhibition of pancreatic lipase thereby delaying postprandial lipid absorption. Moreover, in cholesterol fed rat model, the hypoglycaemic effect of olive was found to reduce hepatic lipid accumulation (Saibandith et al. 2017; Acar-Tek and Ağagündüz 2020).

6.2 Human Intervention Studies on the Hypolipidemic Effect of Ethnomedicinal Plants

6.2.1 Clinical Trials with Garlic

A meta-analysis of 19 randomized clinical trials with 999 volunteers was performed to investigate the effect of garlic (garlic extract, garlic powder or processed garlic) supplementation on parameters of metabolic syndrome. The volunteers were suffering from either hyperlipidaemia, hypertension, NAFLD or metabolic syndrome. All the studies reported a significant reduction in serum triglycerides, total cholesterol and LDL levels with a slight increase in serum HDL level in the intervention group (Fu et al. 2023).

However, despite documentation of cholesterol-lowering effect of garlic in humans, there are also clinical trials which demonstrated no significant change in plasma lipid levels by garlic supplementation, especially those using commercially available garlic supplements. According to a study, garlic powder tablet at 0.6–1.2 g concentrations has not been found to be effective in ameliorating dyslipidemia (Gardner 2007).

6.2.2 Clinical Trial with Walnut

A meta-analysis of 13 randomized clinical trials conducted around Asia, Europe, the UK and the USA with 515 volunteers in the intervention group and 522 in the control group investigated the effect of walnut intake on serum lipid profile of the volunteers. The health status of the volunteers ranged from being healthy, diabetic, hypercholesterolemic, obese, having metabolic syndrome or with cardiovascular problems. The meta-analysis revealed significant improvement in the levels of serum total cholesterol, LDL cholesterol and triglycerides by walnut consumption. However, the HDL cholesterol level did not change significantly. The study also concluded that overweight/obese individuals (BMI > 25 kg/m^2) exhibited greater reduction serum triglycerides than those with BMI ≤ 25 kg/m^2. The beneficial effect of walnut consumption was pronounced in interventions lasting 12 weeks or lesser than that exceeding 12 weeks (Alshahrani et al. 2022). Comparable results were obtained from another meta-analysis of 26 randomized trials assessing the effect of walnut on serum lipids (Guasch-Ferré et al. 2018).

Contradictory to these results, a meta-analysis of four clinical trials to investigate the effect of walnut leaf extract on serum lipid profile of T2DM patients found no significant change in total cholesterol, LDL and triglyceride level post-intervention, though glycaemic indices showed significant improvement (Mirzababaei et al. 2022).

6.2.3 Clinical Trial with Olive

In a randomized clinical trial, 77 overweight/obese adults received 500 mg olive leaf extract or placebo for 8 weeks. No significant change in blood lipid profile was observed due to olive leaf intervention (Stevens et al. 2021). However, two separate clinical trials conducted on pre-hypertensive individuals or individuals with stage 1 hypertension who received 10 mL olive leaf extract concentrate (conatining136 mg oleuropein and 6 mg hydroxytyrosol) and olive leaf extract FLA(®)943 (containing 200 mg oleuropein), respectively, reported significant reduction in serum lipids post-intervention as compared to control (Lockyer et al. 2017; Susalit et al. 2011).

We have included information on the clinical trials conducted with a few more ethnomedicinal plants reported in majority of the countries (according to Aumeeruddy and Mahomoodally 2022) for lowering blood lipid in Table 4.

7 Discussion

The practice of using natural resources for healing purpose is as old as mankind. Evidence regarding the use of plants in preparing drugs dates back to thousands of years wherein as many as 250 plants have been mentioned (Kelly 2009). Natural resources have provided the raw material for discovery of many present-day pharmaceuticals. The discovery of morphine from *Papaver somniferum* in 1803 marked the onset of an unending quest for novel drug discovery against a variety of ailments. Since then, several plant-derived bioactive components have served as lead compounds in drug discovery programmes (Nasim et al. 2022). Even to this date, natural products continue to play a vital role in pharmaceutical research particularly because of the failure of synthetic molecules as lead compounds in many emerging diseases of the modern human civilization such as metabolic diseases (Mathur and Hoskins 2017).

Earlier, majority of the world population inhabited rural areas depending predominantly on agriculture, fishing and trade. However, over the time, in search of better economic opportunities, migration to urban areas as well as de novo urbanization began leading to shrinkage of rural areas. Urbanization is rapidly increasing globally, especially in the developing countries in Asia and Africa, with 54% of the world population residing in urban areas and which is projected to rise by 66% by 2050 (United Nations 2014). This rapid urbanization is accompanied by modification of the environment (e.g. deforestation, pollution, etc.) and resultant changes in the lifestyle and food habits which are exerting a negative impact on the metabolic health of human beings (Kurniawan et al. 2022). So, while increased urbanization has benefitted people with better economic development and access to improved healthcare facilities, at the same time, it has also put the world population at greater risk of developing chronic diseases including the burgeoning rise in metabolic disorders (Popkin 1999) (Fig. 1). Frequent occurrence of more than one metabolic disease in an individual, particularly the prevalence of MetS globally, necessitates

Table 4 Human intervention studies with some widely reported anti-hyperlipidemic ethnomedicinal plants

Sl no.	Name of the plant	Study design	Outcome	References
1.	*Rosmarinus officinalis* (Rosemary)	1. Randomized double-blinded study with 110 NAFLD patients receiving 4 g rosemary leaf or placebo along with a weight loss diet for 8 weeks 2. 48 adult volunteers received 2, 5 or 10 g of rosemary leaf powder 3. 45 T2DM patients and 15 non-diabetic volunteers received 3 g of rosemary per day for 4 weeks	1. Compared to baseline serum triglyceride decreased significantly in both the groups. No significant difference in the rosemary and placebo group after 8 weeks 2. Significant reduction in LDL cholesterol in all groups. Increase in HDL was significant only in the group receiving 10 g rosemary 3. Significant reduction in total cholesterol, LDL and triglycerides along with significant increase in serum HDL levels in both the groups	1. Akbari et al. (2022) 2. Labban et al. (2014) 3. Jamal (2014)
2.	*Prunus dulcis* (Almond)	1. Meta-analysis of 27 studies up to September 2020 with 1154 cases and 904 controls 2. Meta-analysis of 9 randomized double-blinded clinical trial to study the effect of almond on cardiometabolic parameters in T2DM patients	1. Significant reduction in triglyceride, total cholesterol and LDL. But insignificant effect on HDL level 2. Significant reduction in LDL cholesterol level with most pronounced effect with almond consumption >50 g/day	1. Asbaghi et al. (2021) 2. Moosavian et al. (2022)
3.	*Linum usitatissimum* (Flax)	1. Meta-analysis of 14 randomized control trials with a total of 1107 healthy and dyslipidemic subjects to investigate the effect of flaxseed on the lipid profile 2. A single-centre prospective pilot study with eight patients having severe dyslipidemia and receiving lipoprotein apheresis received flaxseed supplementation for 10 weeks at a dose of 28 g/day	1. Flaxseed significantly decreased total and LDL cholesterol and triglycerides in dyslipidemic patients while increasing HDL in healthy individuals 2. Significant reduction in total and LDL cholesterol in the intervention group	1. Sadat Masjedi et al. (2022) 2. Kanikowska et al. (2020)

(continued)

Table 4 (continued)

Sl no.	Name of the plant	Study design	Outcome	References
4.	*Allium cepa* (Onion)	1. Meta-analysis of 10 clinical trials with 446 subjects receiving onion supplementation or placebo	1. Significant reduction in LDL cholesterol and increase in HDL with no effect on triglyceride with respect to the control. Excepting one trial, the other nine trials showed significant reduction in total cholesterol as well	Huang et al. (2021)
5.	*Curcuma longa* (Turmeric)	1. Meta-analysis of 13 trials for LDL, 14 for total cholesterol, 12 for triglycerides and 16 for HDL to assess the effect of turmeric and curcuminoid on blood lipid profile in individuals with metabolic disease 2. Meta-analysis of 57 trials for LDL, 55 for total cholesterol, 58 for triglycerides and 59 trials for HDL to assess the effect of curcumin/turmeric supplementation	1. Turmeric and Curcuminoids significantly decreased blood total cholesterol, LDL and triglycerides while significantly increasing HDL level 2. Lipid profile improved significantly in the intervention group. There was not change in apolipoprotein A or B	1. Yuan et al. (2019) 2. Dehzad et al. (2023)

(continued)

Table 4 (continued)

Sl no.	Name of the plant	Study design	Outcome	References
6.	*Cynara scolymus* (Artichoke)	1. 75 hypercholesterolemic subjects were randomized to receive either 1280 mg of artichoke leaf extract or placebo for 12 weeks 2. 92 overweight volunteers with mild hypercholesterolemia received either 250 mg artichoke leaf extract twice a day or placebo for 8 weeks 3. 72 T2M patients with hypercholesterolemia either received 400 mg fibre-free artichoke leaf extract thrice a day or placebo for 8 weeks 4. 60 volunteers with hypercholesterolemia randomized to receive either 1 pill of Eufortyn® Colesterolo Plus or placebo (containing artichoke leaf extract, bergamot polyphenolic fraction, etc.) for 8 weeks 5. A pilot clinical trial with hypercholesterolemic patients tested the effect of treatment with a neutraceutical containing artichoke dry extract and berberine for 8 weeks	1. Modest reduction in total cholesterol but no effect on LDL, HDL or triglyceride 2. Significant increase in HDL and decrease in total cholesterol, LDL and LDL/HDL ratio as compared to placebo 3. The intervention group displayed significant reduction total cholesterol and LDL 4. Significant improvement in all the parameters of the blood lipid profile in the intervention group 5. Significant reduction in total cholesterol, LDL and triglycerides	1. Bundy et al. (2008) 2. Rondanelli et al. (2013) 3. Huseini et al. (2012) 4. Fogacci et al. (2022) 5. Cicero et al. (2019)

polypharmacological approach. However, such complex treatment regimen often leads to negative drug-drug interactions and is expensive as well. Moreover, as metabolic disorder is a chronic condition, it requires long-term intervention strategies. These factors result in undesirable toxic side effects causing poor management of metabolic disorder in the population at present (Lillich et al. 2021). Therefore, people are increasingly taking resort to traditional medicinal resources for their proven safety and time-tested health beneficial properties.

In this regard, the traditional knowledge of medicinal plants and other natural resources among the various ethnic communities serve as important source for exploring novel treatment strategies against metabolic disorder. From the preceding sections of this chapter, it is quite apparent that incredibly huge number of medicinal plants are used by various ethnic groups across the world to treat metabolic diseases. Scientific validation of the indicated use of the medicinal by the ethnic communities has been successfully done for many ethnomedicinal plants. The bioactive components responsible for their therapeutic potential have been identified, and mechanism of action has also been elucidated for several widely used ethnomedicinal plants (Fig. 3).

Resultantly, the anti-diabetic potential of bitter gourd, the anti-hypertensive property of garlic and the positive effect of olive in NAFLD, for example, are now widely explored globally for their proven efficacy. Data f rom pre-clinical and clinical studies largely agree with the reported ethnomedicinal use of the plants reported from maximum number of countries. However, exceptions also have been observed in some cases. For instance, *Carica papaya*, which is reported to be used to treat obesity in maximum number of countries, is not backed up by clinical studies as yet, though there are a lot of pre-clinical studies supporting the use of papaya against obesity. Similarly, there are no reports on the anti-obesity effect of *Bauhinia variegate*, which is among the top ten mostly reported plant species used for controlling obesity in ethnobotanical surveys. However, there are a lot of pre-clinical studies reported for another species of *Bauhinia*, *B. purpurea* against obesity. Recently, the active component of this species, namely, Bauhiniastatin-1, (BSTN1), has been isolated and found to have anti-obesity and lipid-lowering property (Karunakaran et al. 2021). Similarly, instead of the reported species of *Caralluma*, *C. tuberculata*, clinical trial for anti-obesity effect has been conducted on another species *Caralluma fimbriata*. Green tea (*Camellia sinensis*) reported to be used for weight reduction in

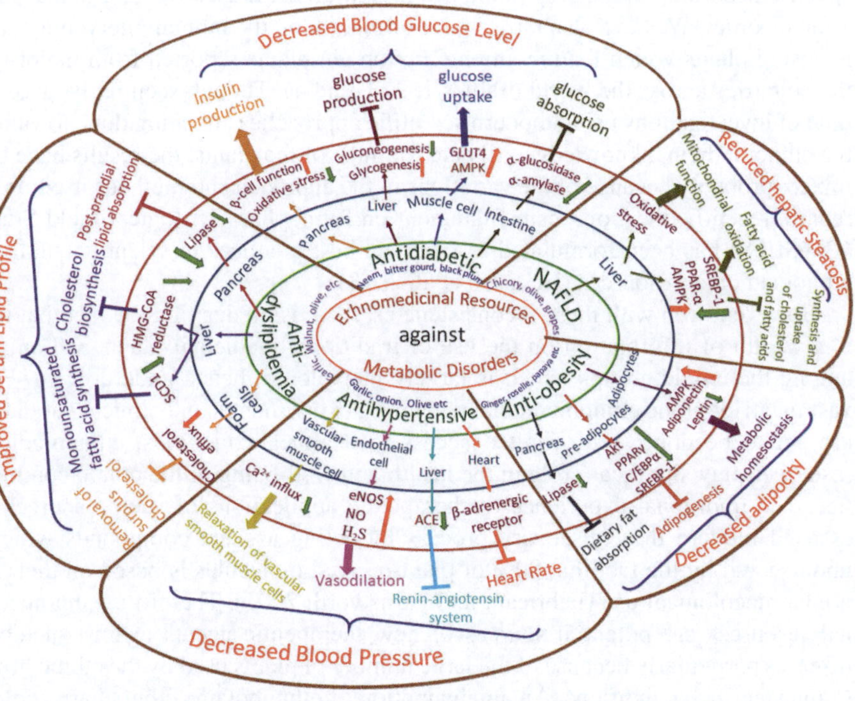

Fig. 3 Mechanism of action for several widely used ethnomedicinal plants

a few countries like Greece, Mexico and Iran has been widely investigated for its potent anti-obesity effect. Several clinical trials have also been performed which provide evidence toward its positive effect on obesity parameters (Lin et al. 2020).

There are also instances of plants reported for a particular use in folklore, but clinical trial has been conducted against different conditions. For example, *Annona muricata* is the fourth most reported ethnomedicinal plant reported for curing high blood pressure. Despite several pre-clinical studies investigating its anti-hypertensive potential, it has not been clinically tested for this effect. But a few clinical trials with *A. muricata have been conducted to* test its hypoglycaemic activity and anticancer activity (Coria-Téllez et al. 2018).

Clinical trials with some ethnomedicinal plants have yielded mixed results; for example, the hypolidaemic activity of walnut leaf extract is largely inconclusive because of contradictory results obtained from various clinical trials as mentioned in the above sections. This might be due to several reasons, the choice of volunteers, duration of the intervention, the phytochemical composition of the extracts used, the concentration of the extracts used and many others. All these factors have subdued the clinical evidences of these ethnomedicinal plants against metabolic disorders.

It is evident from this chapter that ethnobotanical surveys have documented a large number of plants and animals used by local communities out of which a very few have been pharmacologically validated. In this chapter, we have discussed the pre-clinical and clinical studies conducted on the three most widely reported plant species (according to recently published review articles) used for a particular metabolic disorder. We have also tabulated information on the human intervention studies using plants which feature among the top ten plants reported from majority of the countries across the world (Tables 1, 2, 3, and 4). Though seemingly large volume of investigations using modern scientific approaches are being done to validate the ethnomedicinal knowledge, till date for most of the plants, the results have been mixed, probably because of the variations in the approach and methods used. In this regard, recently, the Consensus Statement on Ethnopharmacological Field Studies (ConSEFS) has been formulated to facilitate best practices in scientific studies of ethnomedicinal resources (Heinrich et al. 2018).

However, even with these inconsistencies, there is no denying the fact that there is a wealth of information on the use of traditional medicinal plants and animals among the ethnic groups which is largely unexplored, hence underutilized. For a vast majority of the ethnomedicinal resources, experimental and clinical validations are yet to be done. With limited access to healthcare facilities, ethnomedicinal resources play significant role in the health and well-being of the ethnic communities. The traditional experience of therapeutic applications of such resources has assisted modern drug discovery process by acting as lead compounds which is underscored by the fact that 80% of plant-derived molecules is based on their ethnopharmacological use (Fabricant and Farnsworth 2001). Therefore, ethnomedicinal resources are potential sources of new therapeutic agents against metabolic diseases particularly because of the large number of plants used by the ethnic groups to manage such conditions. An amalgamation of ethnobotany, ethnopharmacology, analytical chemistry, suitable biological models, metabolomics and rigorous randomized placebo-controlled clinical trials is required for successful synthesis of new therapeutic drugs against the rising burden of metabolic diseases.

References

Abdelmonem M, Ebada MA, Diab S, Ahmed MM, Zaazouee MS, Essa TM, ElBaz ZS, Ghaith HS, Abdella WS, Ebada M, Negida A (2022) Efficacy of Hibiscus sabdariffa on reducing blood pressure in patients with mild-to-moderate hypertension: a systematic review and meta-analysis of published randomized controlled trials. J Cardiovasc Pharmacol 79:e64–e74. https://doi.org/10.1097/FJC.0000000000001161

Abebe D, Molla Y, Belayneh A, Kebede B, Getachew M, Alimaw Y (2022) Ethnozoological study of medicinal animals and animals' products used by traditional medicinal practitioners and indigenous people in Motta city administration and Hulet Eju Enessie District, East Gojjam, Northwest Ethiopia. Heliyon 8:e08829. https://doi.org/10.1016/j.heliyon.2022.e08829

Abenavoli L, Milanović M, Milić N, Luzza F, Giuffrè AM (2019) Olive oil antioxidants and non-alcoholic fatty liver disease. Expert Rev Gastroenterol Hepatol 13:739–749. https://doi.org/10.1080/17474124.2019.1634544

Acar-Tek N, Ağagündüz D (2020) Olive leaf (Olea europaea L. folium): potential effects on glycemia and lipidemia. Ann Nutr Metab 76:10–15. https://doi.org/10.1159/000505508

Afiaenyi IC, Ngwu EK, Okafor AM, Ayogu RN (2023) Effects of Moringa oleifera leaves on the blood glucose, blood pressure, and lipid profile of type 2 diabetic subjects: a parallel group randomized clinical trial of efficacy. Nutr Health 026010602311768. https://doi.org/10.1177/02601060231176873

Ahmed B, Al-Howiriny TA, Siddiqui AB (2003) Antihepatotoxic activity of seeds of Cichorium intybus. J Ethnopharmacol 87:237–240. https://doi.org/10.1016/S0378-8741(03)00145-4

Akbari S, Sohouli MH, Ebrahimzadeh S, Ghanaei FM, Hosseini AF, Aryaeian N (2022) Effect of rosemary leaf powder with weight loss diet on lipid profile, glycemic status, and liver enzymes in patients with nonalcoholic fatty liver disease: a randomized, double-blind clinical trial. Phytother Res 36:2186–2196. https://doi.org/10.1002/ptr.7446

Alberti G (2005) Introduction to the metabolic syndrome. Eur Heart J Suppl 7:D3–D5. https://doi.org/10.1093/eurheartj/sui021

Alshahrani SM, Mashat RM, Almutairi D, Mathkour A, Alqahtani SS, Alasmari A, Alzahrani AH, Ayed R, Asiri MY, Elsherif A, Alsabaani A (2022) The effect of walnut intake on lipids: a systematic review and meta-analysis of randomized controlled trials. Nutrients 14:4460. https://doi.org/10.3390/nu14214460

Anandharajan R, Jaiganesh S, Shankernarayanan NP, Viswakarma RA, Balakrishnan A (2006) In vitro glucose uptake activity of Aegle marmelos and Syzygium cumini by activation of Glut-4, PI3 kinase and PPARγ in L6 myotubes. Phytomedicine 13:434–441. https://doi.org/10.1016/j.phymed.2005.03.008

Ansary J, Forbes-Hernández TY, Gil E, Cianciosi D, Zhang J, Elexpuru-Zabaleta M, Simal-Gandara J, Giampieri F, Battino M (2020) Potential health benefit of garlic based on human intervention studies: a brief overview. Antioxidants 9:619. https://doi.org/10.3390/antiox9070619

Araki R, Fujie K, Yuine N, Watabe Y, Nakata Y, Suzuki H, Isoda H, Hashimoto K (2019) Olive leaf tea is beneficial for lipid metabolism in adults with prediabetes: an exploratory randomized controlled trial. Nutr Res 67:60–66. https://doi.org/10.1016/j.nutres.2019.05.003

Arbizu S, Mertens-Talcott SU, Talcott S, Noratto GD (2023) Dark sweet cherry (Prunus avium) supplementation reduced blood pressure and pro-inflammatory interferon gamma (IFNγ) in obese adults without affecting lipid profile, glucose levels and liver enzymes. Nutrients 15:681. https://doi.org/10.3390/nu15030681

Asbaghi O, Moodi V, Hadi A, Eslampour E, Shirinbakhshmasoleh M, Ghaedi E, Miraghajani M (2021) The effect of almond intake on lipid profile: a systematic review and meta-analysis of randomized controlled trials. Food Funct 12:1882–1896. https://doi.org/10.1039/D0FO02878A

Astell KJ, Mathai ML, McAinch AJ, Stathis CG, Su XQ (2013) A pilot study investigating the effect of Caralluma fimbriata extract on the risk factors of metabolic syndrome in overweight and obese subjects: a randomised controlled clinical trial. Complement Ther Med 21:180–189. https://doi.org/10.1016/j.ctim.2013.01.004

Aumeeruddy MZ, Mahomoodally MF (2020) Traditional herbal therapies for hypertension: a systematic review of global ethnobotanical field studies. S Afr J Bot 135:451–464. https://doi.org/10.1016/j.sajb.2020.09.008

Aumeeruddy MZ, Mahomoodally MF (2021a) Ethnomedicinal plants for the management of diabetes worldwide: a systematic review. CMC 28:4670–4693. https://doi.org/10.2174/0929867328666210121123037

Aumeeruddy MZ, Mahomoodally MF (2021b) Traditional herbal medicines used in obesity management: a systematic review of ethnomedicinal surveys. J Herb Med 28:100435. https://doi.org/10.1016/j.hermed.2021.100435

Aumeeruddy MZ, Mahomoodally MF (2022) Global use of folk medicinal plants against hypercholesterolemia: a review of ethnobotanical field studies. J Herb Med 32:100536. https://doi.org/10.1016/j.hermed.2022.100536

Ayyanar M, Subash-Babu P (2012) Syzygium cumini (L.) Skeels: a review of its phytochemical constituents and traditional uses. Asian Pac J Trop Biomed 2:240–246. https://doi.org/10.1016/S2221-1691(12)60050-1

Baharvand-Ahmadi B, Bahmani M, Tajeddini P, Rafieian-Kopaei M, Naghdi N (2016) An ethnobotanical study of medicinal plants administered for the treatment of hypertension. J Renal Inj Prev 5:123–128. https://doi.org/10.15171/jrip.2016.26

Borah MP, Prasad SB (2017) Ethnozoological study of animals based medicine used by traditional healers and indigenous inhabitants in the adjoining areas of Gibbon Wildlife Sanctuary, Assam, India. J Ethnobiol Ethnomed 13:39. https://doi.org/10.1186/s13002-017-0167-6

Braga Tibaes JR, Martins LB, Rodrigues AMDS, Amaral MHA, Teixeira AL, Ferreira AVM (2022) Ginger supplementation does not increase energy expenditure in female adults. Nutrition 103–104:111803. https://doi.org/10.1016/j.nut.2022.111803

Bray GA (2003) Evaluation of obesity: who are the obese? Postgrad Med 114:19–38. https://doi.org/10.3810/pgm.2003.12.1544

Brüll V, Burak C, Stoffel-Wagner B, Wolffram S, Nickenig G, Müller C, Langguth P, Alteheld B, Fimmers R, Naaf S, Zimmermann BF, Stehle P, Egert S (2015) Effects of a quercetin-rich onion skin extract on 24 h ambulatory blood pressure and endothelial function in overweight-to-obese patients with (pre-)hypertension: a randomised double-blinded placebo-controlled cross-over trial. Br J Nutr 114:1263–1277. https://doi.org/10.1017/S0007114515002950

Budjaj A, Benítez G, Pleguezuelos JM (2021) Ethnozoology among the Berbers: pre-Islamic practices survive in the Rif (northwestern Africa). J Ethnobiol Ethnomed 17:43. https://doi.org/10.1186/s13002-021-00466-9

Bundy R, Walker AF, Middleton RW, Wallis C, Simpson HCR (2008) Artichoke leaf extract (Cynara scolymus) reduces plasma cholesterol in otherwise healthy hypercholesterolemic adults: a randomized, double blind placebo controlled trial. Phytomedicine 15:668–675. https://doi.org/10.1016/j.phymed.2008.03.001

Chan Sun M, Ruhomally ZB, Boojhawon R, Neergheen-Bhujun VS (2020) Consumption of Moringa oleifera Lam leaves lowers postprandial blood pressure. J Am Coll Nutr 39:54–62. https://doi.org/10.1080/07315724.2019.1608602

Chen K, Xie K, Liu Z, Nakasone Y, Sakao K, Hossain MA, Hou D-X (2019) Preventive effects and mechanisms of garlic on dyslipidemia and gut microbiome dysbiosis. Nutrients 11:1225. https://doi.org/10.3390/nu11061225

Chew NWS, Ng CH, Tan DJH, Kong G, Lin C, Chin YH, Lim WH, Huang DQ, Quek J, Fu CE, Xiao J, Syn N, Foo R, Khoo CM, Wang J-W, Dimitriadis GK, Young DY, Siddiqui MS, Lam

CSP, Wang Y, Figtree GA, Chan MY, Cummings DE, Noureddin M, Wong VW-S, Ma RCW, Mantzoros CS, Sanyal A, Muthiah MD (2023) The global burden of metabolic disease: data from 2000 to 2019. Cell Metab 35:414–428.e3. https://doi.org/10.1016/j.cmet.2023.02.003

Choudhary PR, Jani RD, Sharma MS (2018) Effect of raw crushed garlic (Allium sativum L.) on components of metabolic syndrome. J Diet Suppl 15:499–506. https://doi.org/10.1080/19390211.2017.1358233

Cicero AFG, Fogacci F, Bove M, Giovannini M, Veronesi M, Borghi C (2019) Short-term effects of dry extracts of artichokeand berberis in hypercholesterolemic patients without cardiovascular disease. Am J Cardiol 123:588–591. https://doi.org/10.1016/j.amjcard.2018.11.018

Çiçek SS (2022) Momordica charantia L.—diabetes-related bioactivities, quality control, and safety considerations. Front Pharmacol 13:904643. https://doi.org/10.3389/fphar.2022.904643

Clemente-Suárez VJ, Martín-Rodríguez A, Redondo-Flórez L, López-Mora C, Yáñez-Sepúlveda R, Tornero-Aguilera JF (2023) New insights and potential therapeutic interventions in metabolic diseases. IJMS 24:10672. https://doi.org/10.3390/ijms241310672

Coria-Téllez AV, Montalvo-Gónzalez E, Yahia EM, Obledo-Vázquez EN (2018) Annona muricata: a comprehensive review on its traditional medicinal uses, phytochemicals, pharmacological activities, mechanisms of action and toxicity. Arab J Chem 11:662–691. https://doi.org/10.1016/j.arabjc.2016.01.004

Costa-Neto EM (2005) Animal-based medicines: biological prospection and the sustainable use of zootherapeutic resources. An Acad Bras Ciênc 77:33–43. https://doi.org/10.1590/S0001-37652005000100004

Darenskaya MA, Kolesnikova LI, Kolesnikov SI (2021) Oxidative stress: pathogenetic role in diabetes mellitus and its complications and therapeutic approaches to correction. Bull Exp Biol Med 171:179–189. https://doi.org/10.1007/s10517-021-05191-7

Dave A, Park E-J, Kumar A, Parande F, Beyoğlu D, Idle JR, Pezzuto JM (2022) Consumption of grapes modulates gene expression, reduces non-alcoholic fatty liver disease, and extends longevity in female C57BL/6J mice provided with a high-fat western-pattern diet. Food Secur 11:1984. https://doi.org/10.3390/foods11131984

de Freitas Junior LM, de Almeida EB (2017) Medicinal plants for the treatment of obesity: ethnopharmacological approach and chemical and biological studies. Am J Transl Res 9:2050–2064

Dehzad MJ, Ghalandari H, Amini MR, Askarpour M (2023) Effects of curcumin/turmeric supplementation on lipid profile: a GRADE-assessed systematic review and dose–response meta-analysis of randomized controlled trials. Complement Ther Med 75:102955. https://doi.org/10.1016/j.ctim.2023.102955

Delaviz H, Mohammadi J, Ghalamfarsa G, Mohammadi B, Farhadi N (2017) A review study on phytochemistry and pharmacology applications of Juglans Regia plant. Phcog Rev 11:145. https://doi.org/10.4103/phrev.phrev_10_17

Devangan S, Varghese B, Johny E, Gurram S, Adela R (2021) The effect of Gymnema sylvestre supplementation on glycemic control in type 2 diabetes patients: a systematic review and meta-analysis. Phytother Res 35:6802–6812. https://doi.org/10.1002/ptr.7265

Diez-Echave P, Vezza T, Rodríguez-Nogales A, Ruiz-Malagón AJ, Hidalgo-García L, Garrido-Mesa J, Molina-Tijeras JA, Romero M, Robles-Vera I, Pimentel-Moral S, Borras-Linares I, Arráez-Román D, Segura-Carretero A, Micol V, García F, Duarte J, Rodríguez-Cabezas ME, Gálvez J (2020) The prebiotic properties of Hibiscus sabdariffa extract contribute to the beneficial effects in diet-induced obesity in mice. Food Res Int 127:108722. https://doi.org/10.1016/j.foodres.2019.108722

Dubey R, Kulkarni SH, Dantu SC, Panigrahi R, Sardesai DM, Malik N, Acharya JD, Chugh J, Sharma S, Kumar A (2021) Myricetin protects pancreatic β-cells from human islet amyloid polypeptide (hIAPP) induced cytotoxicity and restores islet function. Biol Chem 402:179–194. https://doi.org/10.1515/hsz-2020-0176

Ebrahimzadeh Attari V, Asghari Jafarabadi M, Zemestani M, Ostadrahimi A (2015) Effect of Zingiber officinale supplementation on obesity management with respect to the uncoupling protein 1 -3826A>G and ß3-adrenergic receptor Trp64Arg polymorphism. Phytother Res 29:1032–1039. https://doi.org/10.1002/ptr.5343

Elkafrawy N, Younes K, Naguib A, Badr H, Kamal Zewain S, Kamel M, Raoof GFA, El-Desoky AM, Mohamed S (2020) Antihypertensive efficacy and safety of a standardized herbal medicinal product of Hibiscus sabdariffa and Olea europaea extracts (NW Roselle): a PHASE-II, randomized, double-blind, captopril-controlled clinical trial. Phytother Res 34:3379–3387. https://doi.org/10.1002/ptr.6792

El-Saber Batiha G, Magdy Beshbishy A, Wasef LG, Elewa YHA, Al-Sagan AA, Abd El-Hack ME, Taha AE, Abd-Elhakim YM, Prasad Devkota H (2020) Chemical constituents and pharmacological activities of garlic (Allium sativum L.): a review. Nutrients 12:872. https://doi.org/10.3390/nu12030872

Fabricant DS, Farnsworth NR (2001) The value of plants used in traditional medicine for drug discovery. Environ Health Perspect 109:69–75. https://doi.org/10.1289/ehp.01109s169

Fakhr L, Chehregosha F, Zarezadeh M, Chaboksafar M, Tarighat-Esfanjani A (2023) Effects of fenugreek supplementation on the components of metabolic syndrome: a systematic review and dose-response meta-analysis of randomized clinical trials. Pharmacol Res 187:106594. https://doi.org/10.1016/j.phrs.2022.106594

Fakhri M, Fakheri H, Azadbakht M, Moosazadeh M, Yousefi SS (2022) Effect of medicinal plants and natural products on liver enzymes in non-alcoholic fatty liver patients in Iran: a systematic review and meta-analysis. Int J Prev Med 13:87. https://doi.org/10.4103/ijpvm.IJPVM_313_20

Fan M, Choi Y-J, Tang Y, Bae SM, Yang HP, Kim E-K (2019) Efficacy and mechanism of polymerized anthocyanin from grape-skin extract on high-fat-diet-induced nonalcoholic fatty liver disease. Nutrients 11:2586. https://doi.org/10.3390/nu11112586

Ferrara LA, Raimondi AS, d'Episcopo L, Guida L, Dello Russo A, Marotta T (2000) Olive oil and reduced need for antihypertensive medications. Arch Intern Med 160:837. https://doi.org/10.1001/archinte.160.6.837

Ferraro RA, Leucker T, Martin SS, Banach M, Jones SR, Toth PP (2022) Contemporary management of dyslipidemia. Drugs 82:559–576. https://doi.org/10.1007/s40265-022-01691-6

Ferro Y, Pujia R, Mazza E, Lascala L, Lodari O, Maurotti S, Pujia A, Montalcini T (2022) A new nutraceutical (Livogen Plus®) improves liver steatosis in adults with non-alcoholic fatty liver disease. J Transl Med 20:377. https://doi.org/10.1186/s12967-022-03579-1

Fogacci F, Rizzoli E, Giovannini M, Bove M, D'Addato S, Borghi C, Cicero AFG (2022) Effect of dietary supplementation with Eufortyn® Colesterolo Plus serum lipids, endothelial reactivity, indexes of non- alcoholic fatty liver disease and systemic inflammation in healthy subjects with polygenic hypercholesterolemia: The ANEMONE Study. Nutrients 14:2099. https://doi.org/10.3390/nu14102099

Fralick M, Jenkins AJ, Khunti K, Mbanya JC, Mohan V, Schmidt MI (2022) Global accessibility of therapeutics for diabetes mellitus. Nat Rev Endocrinol 18:199–204. https://doi.org/10.1038/s41574-021-00621-y

Fu Z, Lv J, Gao X, Zheng H, Shi S, Xu X, Zhang B, Wu H, Song Q (2023) Effects of garlic supplementation on components of metabolic syndrome: a systematic review, meta-analysis, and meta-regression of randomized controlled trials. BMC Complement Med Ther 23:260. https://doi.org/10.1186/s12906-023-04038-0

Gadde KM, Martin CK, Berthoud H-R, Heymsfield SB (2018) Obesity. J Am Coll Cardiol 71:69–84. https://doi.org/10.1016/j.jacc.2017.11.011

Galavi A, Hosseinzadeh H, Razavi BM (2021) The effects of Allium cepa L. (onion) and its active constituents on metabolic syndrome: a review. Iran J Basic Med Sci 24:3. https://doi.org/10.22038/ijbms.2020.46956.10843

García-Muñoz AM, García-Guillén AI, Victoria-Montesinos D, Abellán-Ruiz MS, Alburquerque-González B, Cánovas F (2023) Effect of the combination of Hibiscus sabdariffa in combination with other plant extracts in the prevention of metabolic syndrome: a systematic review and meta-analysis. Foods 12:2269. https://doi.org/10.3390/foods12112269

Gardner CD (2007) Effect of raw garlic vs commercial garlic supplements on plasma lipid concentrations in adults with moderate hypercholesterolemia: a randomized clinical trial. Arch Intern Med 167:346. https://doi.org/10.1001/archinte.167.4.346

Ghaffar S, Naqvi MA, Fayyaz A, Abid MK, Khayitov KN, Jalil AT, Alsaikhan F, Hammid AT, Al-Gazally ME, Mohammadparast V, Jannat B, Nouri M (2022) What is the influence of grape products on liver enzymes? A systematic review and meta-analysis of randomized controlled trials. Complement Ther Med 69:102845. https://doi.org/10.1016/j.ctim.2022.102845

Ghaffari A, Rafraf M, Navekar R, Sepehri B, Asghari-Jafarabadi M, Ghavami S-M (2019) Turmeric and chicory seed have beneficial effects on obesity markers and lipid profile in non-alcoholic fatty liver disease (NAFLD). Int J Vitam Nutr Res 89:293–302. https://doi.org/10.1024/0300-9831/a000568

Ghorbani A, Zarvandi M, Rakhshandeh H (2019) A randomized controlled trial of a herbal compound for improving metabolic parameters in diabetic patients with uncontrolled dyslipidemia. EMIDDT 19:1075–1082. https://doi.org/10.2174/1871530319666190206213420

Grundy SM, Cleeman JI, Daniels SR, Donato KA, Eckel RH, Franklin BA, Gordon DJ, Krauss RM, Savage PJ, Smith SC, Spertus JA, Costa F (2005) Diagnosis and management of the metabolic syndrome: an American Heart Association/National Heart, Lung, and Blood Institute scientific statement. Circulation 112:2735–2752. https://doi.org/10.1161/CIRCULATIONAHA.105.169404

Guasch-Ferré M, Li J, Hu FB, Salas-Salvadó J, Tobias DK (2018) Effects of walnut consumption on blood lipids and other cardiovascular risk factors: an updated meta-analysis and systematic review of controlled trials. Am J Clin Nutr 108:174–187. https://doi.org/10.1093/ajcn/nqy091

Hashmi MA, Khan A, Hanif M, Farooq U, Perveen S (2015) Traditional uses, phytochemistry, and pharmacology of Olea europaea (olive). Evid Based Complement Alternat Med 2015:1–29. https://doi.org/10.1155/2015/541591

Hassan M, Haq SM, Amjad MS, Ahmad R, Bussmann RW, Pérez De La Lastra JM (2023) Invertebrates and herptiles for livelihoods—ethnozoological use among different ethnic communities in Jammu and Kashmir (Indian Himalayas). Front Pharmacol 13:1043155. https://doi.org/10.3389/fphar.2022.1043155

Heinrich M, Lardos A, Leonti M, Weckerle C, Willcox M, Applequist W, Ladio A, Lin Long C, Mukherjee P, Stafford G (2018) Best practice in research: consensus statement on ethnopharmacological field studies—ConSEFS. J Ethnopharmacol 211:329–339. https://doi.org/10.1016/j.jep.2017.08.015

Hejazi N, Ghalandari H, Nouri M, Askarpour M (2023) Onion supplementation and health metabolic parameters: a systematic review and meta-analysis of randomized controlled trials. Clin Nutr ESPEN 58:1–13. https://doi.org/10.1016/j.clnesp.2023.08.032

Helmstädter A (2008) Syzygium cumini (L.) SKEELS (Myrtaceae) against diabetes—125 years of research. Pharmazie 63:91–101

Herranz-López M, Olivares-Vicente M, Encinar J, Barrajón-Catalán E, Segura-Carretero A, Joven J, Micol V (2017) Multi-targeted molecular effects of Hibiscus sabdariffa polyphenols: an opportunity for a global approach to obesity. Nutrients 9:907. https://doi.org/10.3390/nu9080907

Huang W, Tang G, Zhang L, Tao J, Wei Z (2021) Effect of onion on blood lipid profile: a meta-analysis of randomized controlled trials. Food Sci Nutr 9:3563–3572. https://doi.org/10.1002/fsn3.2309

Huseini HF, Kianbakht S, Hajiaghaee R, Dabaghian FH (2012) Anti-hyperglycemic and anti-hypercholesterolemic effects of Aloe vera leaf gel in hyperlipidemic type 2 diabetic patients: a randomized double-blind placebo-controlled clinical trial. Planta Medica 78(4):311–316. https://doi.org/10.1055/s-0031-1280474

Hussain JF, Tynsong H (2021) Review: ethno-zoological study of animals-based medicine used by traditional healers of North-east India. Asian J Ethnobiol 4:1–22. https://doi.org/10.13057/asianjethnobiol/y040101

Ichsan B et al (eds) (2023) ICHWB 2022, AHSR 61, pp 236–243. https://doi.org/10.2991/978-94-6463-184-5_22

Iloon Kashkooli R, Najafi SS, Sharif F, Hamedi A, Hoseini Asl MK, Najafi Kalyani M, Birjandi M (2015) The effect of berberis vulgaris extract on transaminase activities in non-alcoholic fatty liver disease. Hepat Mon 15:e25067. https://doi.org/10.5812/hepatmon.25067

Islas JF, Acosta E, G-Buentello Z, Delgado-Gallegos JL, Moreno-Treviño MG, Escalante B, Moreno-Cuevas JE (2020) An overview of Neem (Azadirachta indica) and its potential impact on health. J Funct Foods 74:104171. https://doi.org/10.1016/j.jff.2020.104171

Ivanov M, Vajic U-J, Mihailovic-Stanojevic N, Miloradovic Z, Jovovic D, Grujic-Milanovic J, Karanovic D, Dekanski D (2018) Highly potent antioxidant Olea europaea L. leaf extract affects carotid and renal haemodynamics in experimental hypertension: the role of oleuropein. EXCLI J 17:Doc29.; ISSN 1611-2156. https://doi.org/10.17179/EXCLI2017-1002

Jacobs BP, Dennehy C, Ramirez G, Sapp J, Lawrence VA (2002) Milk thistle for the treatment of liver disease. Am J Med 113:506–515. https://doi.org/10.1016/S0002-9343(02)01244-5

Jafarpour-Sadegh F, Montazeri V, Adili A, Esfehani A, Rashidi M-R, Pirouzpanah S (2017) Consumption of fresh yellow onion ameliorates hyperglycemia and insulin resistance in breast cancer patients during doxorubicin-based chemotherapy: a randomized controlled clinical trial. Integr Cancer Ther 16:276–289. https://doi.org/10.1177/1534735416656915

Jamal A (2014) Effect of rosemary (Rosmarinus officinalis) on lipid profiles and blood glucose in human diabetic patients (type-2). Afr J Biochem Res 8:147–150. https://doi.org/10.5897/AJBR2014.0765

Jayawardena R, Francis TV, Abhayaratna S, Ranasinghe P (2021) The use of Caralluma fimbriata as an appetite suppressant and weight loss supplement: a systematic review and meta-analysis of clinical trials. BMC Complement Med Ther 21(1):279. https://doi.org/10.1186/s12906-021-03450-8

Jiang Y, Yue R, Liu G, Liu J, Peng B, Yang M, Zhao L, Li Z (2022) Garlic (Allium sativum L.) in diabetes and its complications: recent advances in mechanisms of action. Crit Rev Food Sci Nutr 64:1–51. https://doi.org/10.1080/10408398.2022.2153793

Jugli S, Chakravorty J, Meyer-Rochow VB (2020) Zootherapeutic uses of animals and their parts: an important element of the traditional knowledge of the Tangsa and Wancho of eastern Arunachal Pradesh, North-East India. Environ Dev Sustain 22:4699–4734. https://doi.org/10.1007/s10668-019-00404-6

Kamyab R, Namdar H, Torbati M, Ghojazadeh M, Araj-Khodaei M, Fazljou SMB (2021) Medicinal plants in the treatment of hypertension: a review. Adv Pharm Bull 11:601–617. https://doi.org/10.34172/apb.2021.090

Kang Y-M, Kang H-A, Cominguez DC, Kim S-H, An H-J (2021) Papain ameliorates lipid accumulation and inflammation in high-fat diet-induced obesity mice and 3T3-L1 adipocytes via AMPK activation. IJMS 22:9885. https://doi.org/10.3390/ijms22189885

Kanikowska D, Korybalska K, Mickiewicz A, Rutkowski R, Kuchta A, Sato M, Kreft E, Fijałkowski M, Gruchała M, Jankowski M, Bręborowicz A, Witowski J (2020) Flaxseed (Linum usitatissimum L.) supplementation in patients undergoing lipoprotein apheresis for severe hyperlipidemia—a pilot study. Nutrients 12:1137. https://doi.org/10.3390/nu12041137

Karadağ AE, Demírci FT (2022) Herbal preparations used against fatty liver disease: systematic review. J Lit Pharm Sci 11:177–188. https://doi.org/10.5336/pharmsci.2022-91392

Karmakar S, Dasgupta SC, Gomes A (2002) Pharmacological and haematological study of shol fish (Channa striatus) skin extract on experimental animal. Indian J Exp Biol 40:115–118

Karri S, Sharma S, Hatware K, Patil K (2019) Natural anti-obesity agents and their therapeutic role in management of obesity: a future trend perspective. Biomed Pharmacother 110:224–238. https://doi.org/10.1016/j.biopha.2018.11.076

Kartinah NT, Komara N, Noviati ND, Dewi S, Yolanda S, Radhina A, Heriyanto H, Sianipar IR (2022) Potential of Hibiscus sabdariffa Linn. in managing FGF21 resistance in diet-induced-obesity rats via miR-34a regulation. Vet Med Sci 8:309–317. https://doi.org/10.1002/vms3.653

Karunakaran RS, Lokanatha O, Muni Swamy G, Venkataramaiah C, Muni Kesavulu M, Appa Rao C, Badri KR, Balaji M (2021) Anti-obesity and lipid lowering activity of Bauhiniastatin-1 is mediated through PPAR-γ/AMPK expressions in diet-induced obese rat model. Front Pharmacol 12:704074. https://doi.org/10.3389/fphar.2021.704074

Kelly K (2009) History of medicine. Facts on File, New York, pp 29–50

Kermanshai R, McCarry BE, Rosenfeld J, Summers PS, Weretilnyk EA, Sorger GJ (2001) Benzyl isothiocyanate is the chief or sole anthelmintic in papaya seed extracts. Phytochemistry 57:427–435. https://doi.org/10.1016/S0031-9422(01)00077-2

Kianian F, Marefati N, Boskabady M, Ghasemi SZ, Boskabady MH (2021) Pharmacological properties of Allium cepa, preclinical and clinical evidences; a review. IJPR 20:107. https://doi.org/10.22037/ijpr.2020.112781.13946

Kim SW, Hur W, Li TZ, Lee YK, Choi JE, Hong SW, Lyoo K-S, You CR, Jung ES, Jung CK, Park T, Um S-J, Yoon SK (2014) Oleuropein prevents the progression of steatohepatitis to hepatic fibrosis induced by a high-fat diet in mice. Exp Mol Med 46:e92. https://doi.org/10.1038/emm.2014.10

Kim SK, Jung J, Jung JH, Yoon N, Kang SS, Roh GS, Hahm JR (2020) Hypoglycemic efficacy and safety of Momordica charantia (bitter melon) in patients with type 2 diabetes mellitus. Complement Ther Med 52:102524. https://doi.org/10.1016/j.ctim.2020.102524

Kumari DJ (2010) Hypoglycemic effect of Moringa oleifera and Azadirachta indica in type-2 diabetes. Bioscan 5:211–214. https://doi.org/10.4236/jamp.2017.58127

Kurniawan F, Manurung MD, Harbuwono DS, Yunir E, Tsonaka R, Pradnjaparamita T, Vidiawati D, Anggunadi A, Soewondo P, Yazdanbakhsh M, Sartono E, Tahapary DL (2022) Urbanization and unfavorable changes in metabolic profiles: a prospective cohort study of Indonesian young adults. Nutrients 14:3326. https://doi.org/10.3390/nu14163326

Labban L, Mustafa UE-S, Ibrahim YM (2014) The effects of rosemary (Rosmarinus officinalis) leaves powder on glucose level, lipid profile and lipid perodoxation. IJCM 05:297–304. https://doi.org/10.4236/ijcm.2014.56044

Landmann M, Kanuri G, Spruss A, Stahl C, Bergheim I (2014) Oral intake of chicoric acid reduces acute alcohol-induced hepatic steatosis in mice. Nutrition 30:882–889. https://doi.org/10.1016/j.nut.2013.11.015

Lillich FF, Imig JD, Proschak E (2021) Multi-target approaches in metabolic syndrome. Front Pharmacol 11:554961. https://doi.org/10.3389/fphar.2020.554961

Lin Y, Shi D, Su B, Wei J, Găman M, Sedanur Macit M, Borges Do Nascimento IJ, Guimaraes NS (2020) The effect of green tea supplementation on obesity: a systematic review and dose–response meta-analysis of randomized controlled trials. Phytother Res 34:2459–2470. https://doi.org/10.1002/ptr.6697

Liu ZL, Xie LZ, Zhu J, Li GQ, Grant SJ, Liu JP (2013) Herbal medicines for fatty liver diseases. Cochrane Database Syst Rev 2013. https://doi.org/10.1002/14651858.CD009059.pub2

Lockyer S, Rowland I, Spencer JPE, Yaqoob P, Stonehouse W (2017) Impact of phenolic-rich olive leaf extract on blood pressure, plasma lipids and inflammatory markers: a randomised controlled trial. Eur J Nutr 56:1421–1432. https://doi.org/10.1007/s00394-016-1188-y

Loko LEY, Medegan Fagla S, Orobiyi A, Glinma B, Toffa J, Koukoui O, Djogbenou L, Gbaguidi F (2019) Traditional knowledge of invertebrates used for medicine and magical–religious purposes by traditional healers and indigenous populations in the Plateau Department, Republic of Benin. J Ethnobiol Ethnomed 15:66. https://doi.org/10.1186/s13002-019-0344-x

Ma Y, Ding X, Gu J, Zhou S, Jiang Y (2023) Effects of olive oil on hepatic steatosis and liver enzymes: a systematic review. J Funct Foods 109:105815. https://doi.org/10.1016/j.jff.2023.105815

Maharlouei N, Tabrizi R, Lankarani KB, Rezaianzadeh A, Akbari M, Kolahdooz F, Rahimi M, Keneshlou F, Asemi Z (2019) The effects of ginger intake on weight loss and metabolic profiles among overweight and obese subjects: a systematic review and meta-analysis of randomized controlled trials. Crit Rev Food Sci Nutr 59:1753–1766. https://doi.org/10.1080/10408398.2018.1427044

Mahindrakar KV, Rathod VK (2021) Antidiabetic potential evaluation of aqueous extract of waste Syzygium cumini seed kernel's by in vitro α-amylase and α-glucosidase inhibition. Prep Biochem Biotechnol 51:589–598. https://doi.org/10.1080/10826068.2020.1839908

Majeed M, Majeed A, Nagabhusahnam K, Mundkur L, Paulose S (2021) A randomized, double-blind clinical trial of a herbal formulation (GlycaCare-II) for the management of type 2 diabetes in comparison with metformin. Diabetol Metab Syndr 13:132. https://doi.org/10.1186/s13098-021-00746-0

Majeed U, Shafi A, Majeed H, Akram K, Liu X, Ye J, Luo Y (2023) Grape (Vitis vinifera L.) phytochemicals and their biochemical protective mechanisms against leading pathologies. Food Chem 405:134762. https://doi.org/10.1016/j.foodchem.2022.134762

Maleki E, Sadeghpour A, Taherifard E, Izadi B, Pasalar M, Akbari M (2023) The effects of chicory supplementation on liver enzymes and lipid profiles in patients with non-alcoholic fatty liver disease: a systematic review and meta-analysis of clinical evidence. Clin Nutr ESPEN 55:447–454. https://doi.org/10.1016/j.clnesp.2023.04.025

Malik K, Ahmad M, Bussmann RW, Tariq A, Ullah R, Alqahtani AS, Shahat AA, Rashid N, Zafar M, Sultana S, Shah SN (2018) Ethnobotany of anti-hypertensive plants used in northern Pakistan. Front Pharmacol 9:789. https://doi.org/10.3389/fphar.2018.00789

Mannucci C, Calapai F, Cardia L, Inferrera G, D'Arena G, Di Pietro M, Navarra M, Gangemi S, Ventura Spagnolo E, Calapai G (2018) Clinical pharmacology of Citrus aurantium and Citrus sinensis for the treatment of anxiety. Evid Based Complement Alternat Med 2018:1–18. https://doi.org/10.1155/2018/3624094

Mansour-Ghanaei F, Hadi A, Pourmasoumi M, Joukar F, Golpour S, Najafgholizadeh A (2018) Green tea as a safe alternative approach for nonalcoholic fatty liver treatment: a systematic review and meta-analysis of clinical trials. Phytother Res 32:1876–1884. https://doi.org/10.1002/ptr.6130

Massaro M, Scoditti E, Carluccio MA, Calabriso N, Santarpino G, Verri T, De Caterina R (2020) Effects of olive oil on blood pressure: epidemiological, clinical, and mechanistic evidence. Nutrients 12:1548. https://doi.org/10.3390/nu12061548

Mathur S, Hoskins C (2017) Drug development: lessons from nature. Biomed Rep 6:612–614. https://doi.org/10.3892/br.2017.909

Matsuane C, Kiage BN, Karanja J, Kavoo AM, Rimberia FK (2023) Hypolipidaemic effects of papaya (Carica papaya L.) juice on rats fed on a high fat and fructose diet. J Nutr Sci 12:e76. https://doi.org/10.1017/jns.2023.61

Matsutomo T (2020) Potential benefits of garlic and other dietary supplements for the management of hypertension (review). Exp Ther Med 19:1479. https://doi.org/10.3892/etm.2019.8375

Mayoral L-C, Andrade G, Mayoral E-C, Huerta T, Canseco S, Rodal Canales F, Cabrera-Fuentes H, Cruz M, Pérez Santiago A, Alpuche J, Zenteno E, Ruíz H, Cruz R, Jeronimo J, Perez-Campos E (2020) Obesity subtypes, related biomarkers & heterogeneity. Indian J Med Res 151:11. https://doi.org/10.4103/ijmr.IJMR_1768_17

Mezhibovsky E, Knowles KA, He Q, Sui K, Tveter KM, Duran RM, Roopchand DE (2021) Grape polyphenols attenuate diet-induced obesity and hepatic steatosis in mice in association with reduced butyrate and increased markers of intestinal carbohydrate oxidation. Front Nutr 8:675267. https://doi.org/10.3389/fnut.2021.675267

Mills KT, Stefanescu A, He J (2020) The global epidemiology of hypertension. Nat Rev Nephrol 16:223–237. https://doi.org/10.1038/s41581-019-0244-2

Mirzababaei A, Daneshvar M, Abaj F, Daneshzad E, Hosseininasab D, Clark CCT, Mirzaei K (2022) The effect of walnut (Juglans regia) leaf extract on glycemic control and lipid profile in patients with type 2 diabetes mellitus: a systematic review and meta-analysis of randomized clinical trials. Clin Nutr Res 11:120. https://doi.org/10.7762/cnr.2022.11.2.120

Moosavian SP, Rahimlou M, Rezaei Kelishadi M, Moradi S, Jalili C (2022) Effects of almond on cardiometabolic outcomes in patients with type 2 diabetes: a systematic review and meta-analysis of randomized controlled trials. Phytother Res 36:1839–1853. https://doi.org/10.1002/ptr.7365

Moradi MT, Asadi-Samani M, Bahmani M, Shahrani M (2016) Medicinal plants used for liver disorders based on the ethnobotanical documents of Iran: a review. Int J PharmTech Res 9(5):407–415

Nadjarzadeh A, Ghadiri-Anari A, Ramezani-Jolfaie N, Mohammadi M, Salehi-Abargouei A, Namayande SM, Mozaffari-Khosravi H, Hosseini-Marnani E (2021) Effect of hypocaloric high-protein, low-carbohydrate diet supplemented with fennel on androgenic and anthropometric indices in overweight and obese women with polycystic ovary syndrome: a randomized placebo-controlled trial. Complement Ther Med 56:102633. https://doi.org/10.1016/j.ctim.2020.102633

Najdi RA, Hagras MM, Kamel FO, Magadmi RM (2019) A randomized controlled clinical trial evaluating the effect of Trigonella foenum-graecum (fenugreek) versus glibenclamide in patients with diabetes. Afr H Sci 19:1594. https://doi.org/10.4314/ahs.v19i1.34

Nakanekar A, Kohli K, Tatke P (2019) Ayurvedic polyherbal combination (PDBT) for prediabetes: a randomized double blind placebo controlled study. J Ayurveda Integr Med 10:284–289. https://doi.org/10.1016/j.jaim.2018.05.004

Nasim N, Sandeep IS, Mohanty S (2022) Plant-derived natural products for drug discovery: current approaches and prospects. Nucleus 65:399–411. https://doi.org/10.1007/s13237-022-00405-3

Nergiz-Ünal R, Kuijpers MJE, De Witt SM, Heeneman S, Feijge MAH, Garcia Caraballo SC, Biessen EAL, Haenen GRMM, Cosemans JMEM, Heemskerk JWM (2013) Atheroprotective effect of dietary walnut intake in ApoE-deficient mice: involvement of lipids and coagulation factors. Thromb Res 131:411–417. https://doi.org/10.1016/j.thromres.2013.01.003

Od-Ek P, Deenin W, Malakul W, Phoungpetchara I, Tunsophon S (2020) Anti-obesity effect of Carica papaya in high-fat diet fed rats. Biomed Rep 13:30. https://doi.org/10.3892/br.2020.1337

Ojulari OV, Lee SG, Nam J-O (2019) Beneficial effects of natural bioactive compounds from Hibiscus sabdariffa L. on obesity. Molecules 24:210. https://doi.org/10.3390/molecules24010210

Omagari K, Kato S, Tsuneyama K, Hatta H, Sato M, Hamasaki M, Sadakane Y, Tashiro T, Fukuhata M, Miyata Y, Tamaru S, Tanaka K, Mune M (2010) Olive leaf extract prevents spontaneous occurrence of non-alcoholic steatohepatitis in SHR/NDmcr-cp rats. Pathology 42:66–72. https://doi.org/10.3109/00313020903434389

Omagari K, Koba C, Nagata A, Ngo LCT, Yamasaki M, Fukuda A, Yuasa M, Suruga K, Inada N, Ichimura-Shimizu M, Tsuneyama K (2021) Olive leaf powder prevents nonalcoholic steatohepatitis in Sprague–Dawley rats fed a high-fat and high-cholesterol diet. Clin Nutr Open Sci 37:47–59. https://doi.org/10.1016/j.nutos.2021.04.002

Omonkhua AA, Onoagbe IO (2011) Long-term hepatotoxicity and hypoglycaemic study of aqueous extracts of Carica papaya leaves on normal rabbits. Glob J Pure Appl Sci 17:241–247

Padhi S, Nayak AK, Behera A (2020) Type II diabetes mellitus: a review on recent drug based therapeutics. Biomed Pharmacother 131:110708. https://doi.org/10.1016/j.biopha.2020.110708

Panahi Y, Kianpour P, Mohtashami R, Atkin SL, Butler AE, Jafari R, Badeli R, Sahebkar A (2018) Efficacy of artichoke leaf extract in non-alcoholic fatty liver disease: a pilot double-blind randomized controlled trial. Phytother Res 32:1382–1387. https://doi.org/10.1002/ptr.6073

Paternostro R, Trauner M (2022) Current treatment of non-alcoholic fatty liver disease. J Intern Med 292:190–204. https://doi.org/10.1111/joim.13531

Patil SM, Shirahatti PS, Ramu R (2022) Azadirachta indica A. Juss (neem) against diabetes mellitus: a critical review on its phytochemistry, pharmacology, and toxicology. J Pharm Pharmacol 74:681–710. https://doi.org/10.1093/jpp/rgab098

Peter EL, Kasali FM, Deyno S, Mtewa A, Nagendrappa PB, Tolo CU, Ogwang PE, Sesaazi D (2019) Momordica charantia L. lowers elevated glycaemia in type 2 diabetes mellitus patients: systematic review and meta-analysis. J Ethnopharmacol 231:311–324. https://doi.org/10.1016/j.jep.2018.10.033

Peter EL, Mtewa AG, Nagendrappa PB, Kaligirwa A, Sesaazi CD (2020) Systematic review and meta-analysis protocol for efficacy and safety of Momordica charantia L. on animal models of type 2 diabetes mellitus. Syst Rev 9:7. https://doi.org/10.1186/s13643-019-1265-4

Pingali U, Ali MA, Gundagani S, Nutalapati C (2020) Evaluation of the effect of an aqueous extract of Azadirachta indica (neem) leaves and twigs on glycemic control, endothelial dysfunction and systemic inflammation in subjects with type 2 diabetes mellitus—a randomized, double-blind, placebo-controlled clinical study. DMSO 13:4401–4412. https://doi.org/10.2147/DMSO.S274378

Pingali U, Vuppalanchi B, Nutalapati C, Gundagani S (2021) Aqueous Azadirachta indica (neem) extract attenuates insulin resistance to improve glycemic control and endothelial function in subjects with metabolic syndrome. J Med Food 24:1135. https://doi.org/10.1089/jmf.2020.4838

Pintó X, Fanlo-Maresma M, Corbella E, Corbella X, Mitjavila MT, Moreno JJ, Casas R, Estruch R, Corella D, Bulló M, Ruiz-Canela M, Castañer O, Martinez JA, Ros E (2019) A Mediterranean diet rich in extra-virgin olive oil is associated with a reduced prevalence of nonalcoholic fatty liver disease in older individuals at high cardiovascular risk. J Nutr 149:1920–1929. https://doi.org/10.1093/jn/nxz147

Popkin BM (1999) Urbanization, lifestyle changes and the nutrition transition. World Dev 27:1905–1916. https://doi.org/10.1016/S0305-750X(99)00094-7

Pouwels S, Sakran N, Graham Y, Leal A, Pintar T, Yang W, Kassir R, Singhal R, Mahawar K, Ramnarain D (2022) Non-alcoholic fatty liver disease (NAFLD): a review of pathophysiology, clinical management and effects of weight loss. BMC Endocr Disord 22:63. https://doi.org/10.1186/s12902-022-00980-1

Qamar M, Akhtar S, Ismail T, Wahid M, Abbas MW, Mubarak MS, Yuan Y, Barnard RT, Ziora ZM, Esatbeyoglu T (2022) Phytochemical profile, biological properties, and food applications of the medicinal plant Syzygium cumini. Foods 11:378. https://doi.org/10.3390/foods11030378

Rafiee S, Mohammadi H, Ghavami A, Sadeghi E, Safari Z, Askari G (2021) Efficacy of resveratrol supplementation in patients with nonalcoholic fatty liver disease: a systematic review and meta-analysis of clinical trials. Complement Ther Clin Pract 42:101281. https://doi.org/10.1016/j.ctcp.2020.101281

Rahman Q, Nadeem MS, Umair M, Altaf M, Ni J, Abbasi AM, Jameel MA, Pieroni A, Hamed MH, Ashraf S, Sadaf T (2022) Medicinal waterbirds in the traditional healthcare system: an assessment of biodiversity–cultural linkages in Eastern Khyber Pakhtunkhwa, Pakistan. J Ethnobiol Ethnomed 18:57. https://doi.org/10.1186/s13002-022-00554-4

Rahmani S, Asgary S, Askari G, Keshvari M, Hatamipour M, Feizi A, Sahebkar A (2016) Treatment of non-alcoholic fatty liver disease with curcumin: a randomized placebo-controlled trial: curcumin supplementation for NAFLD. Phytother Res 30:1540–1548. https://doi.org/10.1002/ptr.5659

Rambod M, Rakhshan M, Tohidinik S, Nikoo MH (2020) The effect of lemon inhalation aromatherapy on blood pressure, electrocardiogram changes, and anxiety in acute myocardial infarction patients: a clinical, multi-centered, assessor-blinded trial design. Complement Ther Clin Pract 39:101155. https://doi.org/10.1016/j.ctcp.2020.101155

Reuter HD, Koch HP, Lawson LD (1996) Therapeutic effects and applications of garlic and its preparations. In: Koch HP, Lawson LD (eds) Garlic: the science and therapeutic application of Allium sativum L and related species, 2nd edn. Lippincott Williams & Wilkins, Baltimore, MD, pp 135–212

Rezaei S, Akhlaghi M, Sasani MR, Barati Boldaji R (2019) Olive oil lessened fatty liver severity independent of cardiometabolic correction in patients with non-alcoholic fatty liver disease: a randomized clinical trial. Nutrition 57:154–161. https://doi.org/10.1016/j.nut.2018.02.021

Ribeiro M, Alvarenga L, Cardozo LFMF, Chermut TR, Sequeira J, De Souza Gouveia Moreira L, Teixeira KTR, Shiels PG, Stenvinkel P, Mafra D (2021) From the distinctive smell to therapeutic effects: garlic for cardiovascular, hepatic, gut, diabetes and chronic kidney disease. Clin Nutr 40:4807–4819. https://doi.org/10.1016/j.clnu.2021.03.005

Richter E, Geetha T, Burnett D, Broderick TL, Babu JR (2023) The effects of Momordica charantia on type 2 diabetes mellitus and Alzheimer's disease. IJMS 24:4643. https://doi.org/10.3390/ijms24054643

Ried K (2016) Garlic lowers blood pressure in hypertensive individuals, regulates serum cholesterol, and stimulates immunity: an updated meta-analysis and review. J Nutr 146:389S–396S. https://doi.org/10.3945/jn.114.202192

Ried K (2019) Garlic lowers blood pressure in hypertensive subjects, improves arterial stiffness and gut microbiota: a review and meta-analysis. Exp Ther Med 19:1472. https://doi.org/10.3892/etm.2019.8374

Ried K, Travica N, Sali A (2018) The effect of kyolic aged garlic extract on gut microbiota, inflammation, and cardiovascular markers in hypertensives: the GarGIC trial. Front Nutr 5:122. https://doi.org/10.3389/fnut.2018.00122

Rohner A, Ried K, Sobenin IA, Bucher HC, Nordmann AJ (2015) A systematic review and meta-analysis on the effects of garlic preparations on blood pressure in individuals with hypertension. Am J Hypertens 28:414–423. https://doi.org/10.1093/ajh/hpu165

Rondanelli M, Giacosa A, Opizzi A, Faliva MA, Sala P, Perna S, Riva A, Morazzoni P, Bombardelli E (2013) Beneficial effects of artichoke leaf extract supplementation on increasing HDL-cholesterol in subjects with primary mild hypercholesterolaemia: a double-blind, randomized, placebo-controlled trial. Int J Food Sci Nutr 64:7–15. https://doi.org/10.3109/09637486.2012.700920

Rutebemberwa E, Lubega M, Katureebe SK, Oundo A, Kiweewa F, Mukanga D (2013) Use of traditional medicine for the treatment of diabetes in Eastern Uganda: a qualitative exploration of reasons for choice. BMC Int Health Hum Rights 13:1. https://doi.org/10.1186/1472-698X-13-1

Saberi-Karimian M, Keshvari M, Ghayour-Mobarhan M, Salehizadeh L, Rahmani S, Behnam B, Jamialahmadi T, Asgary S, Sahebkar A (2020) Effects of curcuminoids on inflammatory status in patients with non-alcoholic fatty liver disease: a randomized controlled trial. Complement Ther Med 49:102322. https://doi.org/10.1016/j.ctim.2020.102322

Sabzghabaee A, Ataei E, Kelishadi R, Ghannadi A, Soltani R, Badri S, Shirani S (2013) Effect of Hibiscus sabdariffa calices on dyslipidemia in obese adolescents: a triple-masked randomized controlled trial. Mater Sociomed 25:76. https://doi.org/10.5455/msm.2013.25.76-79

Sadat Masjedi M, Mohammadi Pour P, Shokoohinia Y, Asgary S (2022) Effects of flaxseed on blood lipids in healthy and dyslipidemic subjects: a systematic review and meta-analysis of randomized controlled trials. Curr Probl Cardiol 47:100931. https://doi.org/10.1016/j.cpcardiol.2021.100931

Saghafi N, Ghazanfarpour M, Khadivzadeh T, Babakhanian M, Afiat M (2017) The effect of Foeniculum vulgare (fennel) on body composition in postmenopausal women with excess weight: a double-blind randomized placebo-controlled trial. J Menopausal Med 23:166. https://doi.org/10.6118/jmm.2017.23.3.166

Saibandith B, Spencer J, Rowland I, Commane D (2017) Olive polyphenols and the metabolic syndrome. Molecules 22:1082. https://doi.org/10.3390/molecules22071082

Sailesh KS, Jabir PK, Madhusudhan U, Archana R, Mukkadan JK (2018) Effect of Moringa oliefera leaves on blood pressure in hypertensive patients. Indian J Clin Anat Physiol 5:350–352. https://doi.org/10.18231/2394-2126.2018.0081

Salami SO, Afolayan AJ (2021) Evaluation of nutritional and elemental compositions of green and red cultivars of roselle: Hibiscus sabdariffa L. Sci Rep 11:1030. https://doi.org/10.1038/s41598-020-80433-8

Salaramoli S, Mehri S, Yarmohammadi F, Hashemy SI, Hosseinzadeh H (2022) The effects of ginger and its constituents in the prevention of metabolic syndrome: a review. Iran J Basic Med Sci 25:664–674. https://doi.org/10.22038/ijbms.2022.59627.13231

Samaha AA, Fawaz M, Salami A, Baydoun S, Eid AH (2019) Antihypertensive indigenous Lebanese plants: ethnopharmacology and a clinical trial. Biomol Ther 9:292. https://doi.org/10.3390/biom9070292

Santana LF, Do Espirito Santo BLS, Tatara MB, Negrão FJ, Croda J, Alves FM, De Oliveira Filiú WF, Cavalheiro LF, Nazário CED, Asato MA, De Faria BB, Do Nascimento VA, De Cássia Avellaneda Guimarães R, De Cássia Freitas K, Hiane PA (2022) Effects of the seed oil of

Carica papaya Linn on food consumption, adiposity, metabolic and inflammatory profile of mice using hyperlipidic diet. Molecules 27:6705. https://doi.org/10.3390/molecules27196705

Santos IB, De Bem GF, Cordeiro VSC, Da Costa CA, De Carvalho LCRM, Da Rocha APM, Da Costa GF, Ognibene DT, De Moura RS, Resende AC (2017) Supplementation with Vitis vinifera L. skin extract improves insulin resistance and prevents hepatic lipid accumulation and steatosis in high-fat diet–fed mice. Nutr Res 43:69–81. https://doi.org/10.1016/j.nutres.2017.05.007

Selvia A, Rahmawaty S (2023) The effect of giving avocado (Persea americana) leaf decoction on blood pressure in patients with hypertension: a review article. In: Ichsan B, Nursanto D, Sari M, Firmansyah, Porusia M, Hudiyawati D, Perdana SS (eds) Proceedings of the international conference on health and well-being (ICHWB 2022). Atlantis Press International BV, Dordrecht, pp. 236–243

Serna A, Marhuenda J, Arcusa R, Pérez-Piñero S, Sánchez-Macarro M, García-Muñoz AM, Victoria-Montesinos D, Cánovas F, López-Román FJ (2022) Effectiveness of a polyphenolic extract (Lippia citriodora and Hibiscus sabdariffa) on appetite regulation in overweight and obese grade I population: an 8-week randomized, double-blind, cross-over, placebo-controlled trial. Eur J Nutr 61:825–841. https://doi.org/10.1007/s00394-021-02678-x

Serrano JCE, Castro-Boqué E, García-Carrasco A, Morán-Valero MI, González-Hedström D, Bermúdez-López M, Valdivielso JM, Espinel AE, Portero-Otín M (2023) Antihypertensive effects of an optimized aged garlic extract in subjects with grade I hypertension and antihypertensive drug therapy: a randomized, triple-blind controlled trial. Nutrients 15:3691. https://doi.org/10.3390/nu15173691

Sharma M, Sharma M, Sharma M (2022) A comprehensive review on ethnobotanical, medicinal and nutritional potential of walnut (Juglans regia L.). Proc Indian Natl Sci Acad 88:601–616. https://doi.org/10.1007/s43538-022-00119-9

Shimoda H, Tanaka J, Kikuchi M, Fukuda T, Ito H, Hatano T, Yoshida T (2009) Effect of polyphenol-rich extract from walnut on diet-induced hypertriglyceridemia in mice via enhancement of fatty acid oxidation in the liver. J Agric Food Chem 57:1786–1792. https://doi.org/10.1021/jf803441c

Silva H, Bárbara R (2022) Exploring the anti-hypertensive potential of lemongrass—a comprehensive review. Biology 11:1382. https://doi.org/10.3390/biology11101382

Solhi H, Ghahremani R, Kazemifar AM, Hoseini Yazdi Z (2014) Silymarin in treatment of non-alcoholic steatohepatitis: a randomized clinical trial. Caspian J Intern Med 5:9–12

Sousa FH, Valenti VE, Pereira LC, Bueno RR, Prates S, Akimoto AN, Kaviani M, Garner DM, Amaral JAT, De Abreu LC (2020) Avocado (Persea americana) pulp improves cardiovascular and autonomic recovery following submaximal running: a crossover, randomized, double-blind and placebo-controlled trial. Sci Rep 10:10703. https://doi.org/10.1038/s41598-020-67577-3

Stevens Y, Winkens B, Jonkers D, Masclee A (2021) The effect of olive leaf extract on cardiovascular health markers: a randomized placebo-controlled clinical trial. Eur J Nutr 60:2111–2120. https://doi.org/10.1007/s00394-020-02397-9

Street RA, Sidana J, Prinsloo G (2013) Cichorium intybus : traditional uses, phytochemistry, pharmacology, and toxicology. Evid Based Complement Alternat Med 2013:1–13. https://doi.org/10.1155/2013/579319

Suksomboon N, Poolsup N, Punthanitisarn S (2016) Effect of Aloe vera on glycaemic control in prediabetes and type 2 diabetes: a systematic review and meta-analysis. J Clin Pharm Ther 41:180–188. https://doi.org/10.1111/jcpt.12382

Sun HY, Gu AX, Huang BY, Zhang T, Li JP, Shan AS (2022) Dietary grape seed proanthocyanidin alleviates the liver injury induced by long-term high-fat diets in Sprague Dawley rats. Front Vet Sci 9:959906. https://doi.org/10.3389/fvets.2022.959906

Susalit E, Agus N, Effendi I, Tjandrawinata RR, Nofiarny D, Perrinjaquet-Moccetti T, Verbruggen M (2011) Olive (Olea europaea) leaf extract effective in patients with stage-1 hypertension: comparison with captopril. Phytomedicine 18:251–258. https://doi.org/10.1016/j.phymed.2010.08.016

Tacconi S, Longo S, Guerra F, Moliteni C, Friuli M, Romano A, Gaetani S, Paradiso VM, Difonzo G, Caponio F, Lofrumento D, Vergara D, Bucci C, Dini L, Giudetti AM (2023) An aqueous olive leaf extract (OLE) ameliorates parameters of oxidative stress associated with lipid accumulation and induces lipophagy in human hepatic cells. Food Funct 14:5805–5819. https://doi.org/10.1039/D3FO00817G

Teixeira CC, Rava CA, Mallman Da Silva P, Melchior R, Argenta R, Anselmi F, Almeida CRC, Fuchs FD (2000) Absence of antihyperglycemic effect of jambolan in experimental and clinical models. J Ethnopharmacol 71:343–347. https://doi.org/10.1016/S0378-8741(00)00185-9

Teixeira CC, Fuchs FD, Weinert LS, Esteves J (2006) The efficacy of folk medicines in the management of type 2 diabetes mellitus: results of a randomized controlled trial of Syzygium cumini (L.) Skeels. J Clin Pharm Ther 31:1–5. https://doi.org/10.1111/j.1365-2710.2006.00700.x

Teronpi V, Singh H, Tamuli A, Teron R (2012) Ethnozoology of the Karbis of Assam, India: use of ichthyofauna in traditional health-care practices. Ancient Sci Life 32:99. https://doi.org/10.4103/0257-7941.118547

Teshika JD, Zakariyyah AM, Zaynab T, Zengin G, Rengasamy KR, Pandian SK, Fawzi MM (2019) Traditional and modern uses of onion bulb (Allium cepa L.): a systematic review. Crit Rev Food Sci Nutr 59:S39–S70. https://doi.org/10.1080/10408398.2018.1499074

The GBD 2015 Obesity Collaborators (2017) Health effects of overweight and obesity in 195 countries over 25 years. N Engl J Med 377:13–27. https://doi.org/10.1056/nejmoa1614362

Tibiri A, Boria S, Traoré TK, Ouédraogo N, Nikièma A, Ganaba S, Compaoré J-M, Ouédraogo I, Guissou IP, Carraz M (2020) Countrywide survey of plants used for liver disease management by traditional healers in Burkina Faso. Front Pharmacol 11:563751. https://doi.org/10.3389/fphar.2020.563751

Tsabang N, Yedjou CG, Tchounwou PB (2017) Phytotherapy of high blood pressure in three phytogeographic regions of Cameroon. Pharm Anal Acta 08:530. https://doi.org/10.4172/2153-2435.1000530

Ullah I, Shoukat A, Khan MF, Shah GM, Tabassam S, Sajid M, Siddique H, Badshah KD (2020) Indigenous knowledge of zootherapeutic use among the people of Hazara division Khyber-Pakhtunkhwa, Pakistan. IJTK 19. https://doi.org/10.56042/ijtk.v19i3.41452

United Nations (2014) World urbanization prospects: the 2018 revision produced by the Population Division of the UN Department of Economic and Social Affairs (UN DESA). https://www.un.org/development/desa/en/news/population/2018-revision-of-world-urbanization-prospects.html

Unuofin JO, Masuku NP, Paimo OK, Lebelo SL (2021) Ginger from farmyard to town: nutritional and pharmacological applications. Front Pharmacol 12:779352. https://doi.org/10.3389/fphar.2021.779352

Van Den Driessche JJ, Plat J, Mensink RP (2018) Effects of superfoods on risk factors of metabolic syndrome: a systematic review of human intervention trials. Food Funct 9:1944–1966. https://doi.org/10.1039/C7FO01792H

Vats R, Thomas S (2015) A study on use of animals as traditional medicine by Sukuma Tribe of Busega District in North-western Tanzania. J Ethnobiol Ethnomed 11:38. https://doi.org/10.1186/s13002-015-0001-y

Vuong QV, Hirun S, Roach PD, Bowyer MC, Phillips PA, Scarlett CJ (2013) Effect of extraction conditions on total phenolic compounds and antioxidant activities of Carica papaya leaf aqueous extracts. J Herb Med 3:104–111. https://doi.org/10.1016/j.hermed.2013.04.004

Wah Kheong C, Nik Mustapha NR, Mahadeva S (2017) A randomized trial of silymarin for the treatment of nonalcoholic steatohepatitis. Clin Gastroenterol Hepatol 15:1940–1949.e8. https://doi.org/10.1016/j.cgh.2017.04.016

Wang J, Ke W, Bao R, Hu X, Chen F (2017) Beneficial effects of ginger Zingiber officinale Roscoe on obesity and metabolic syndrome: a review. Ann N Y Acad Sci 1398:83–98. https://doi.org/10.1111/nyas.13375

Wang X, Li D, Liu F, Cui Y, Li X (2020) Dietary citrus and/or its extracts intake contributed to weight control: evidence from a systematic review and meta-analysis of 13 randomized clinical trials. Phytother Res 34:2006–2022. https://doi.org/10.1002/ptr.6673

Wang G, Xu J, Ma H, Mu Y, Xu W, Yan N, Liu W, Zheng D, Huang X, Li L (2023) Phenolipid JE improves metabolic profile and inhibits gluconeogenesis via modulating AKT-mediated insulin signaling in STZ-induced diabetic mice. Pharmacol Res 187:106569. https://doi.org/10.1016/j.phrs.2022.106569

Waters D, Adeloye D, Woolham D, Wastnedge E, Patel S, Rudan I (2018) Global birth prevalence and mortality from inborn errors of metabolism: a systematic analysis of the evidence. J Glob Health 8:021102. https://doi.org/10.7189/jogh.08.021102

Wendimu A, Tekalign W (2023) An ethnozoological study of traditional medicinal animals and their products from Wolaita, Southern Ethiopia. Heliyon 9:e12733. https://doi.org/10.1016/j.heliyon.2022.e12733

Wong R, Garg M, Wood L, Howe P (2014) Antihypertensive potential of combined extracts of olive leaf, green coffee bean and beetroot: a randomized, double-blind, placebo-controlled crossover trial. Nutrients 6:4881–4894. https://doi.org/10.3390/nu6114881

Xu B, Li Z, Zeng T, Zhan J, Wang S, Ho C-T, Li S (2022) Bioactives of Momordica charantia as potential anti-diabetic/hypoglycemic agents. Molecules 27:2175. https://doi.org/10.3390/molecules27072175

Yeh Y-Y, Liu L (2001) Cholesterol-lowering effect of garlic extracts and organosulfur compounds: human and animal studies. J Nutr 131:989S–993S. https://doi.org/10.1093/jn/131.3.989S

Yoshitomi R, Yamamoto M, Kumazoe M, Fujimura Y, Yonekura M, Shimamoto Y, Nakasone A, Kondo S, Hattori H, Haseda A, Nishihira J, Tachibana H (2021) The combined effect of green tea and α-glucosyl hesperidin in preventing obesity: a randomized placebo-controlled clinical trial. Sci Rep 11:19067. https://doi.org/10.1038/s41598-021-98612-6

Yuan F, Dong H, Gong J, Wang D, Hu M, Huang W, Fang K, Qin X, Qiu X, Yang X, Lu F (2019) A systematic review and meta-analysis of randomized controlled trials on the effects of turmeric and curcuminoids on blood lipids in adults with metabolic diseases. Adv Nutr 10:791–802. https://doi.org/10.1093/advances/nmz021

Zhang J, Grieger JA, Kris-Etherton PM, Thompson JT, Gillies PJ, Fleming JA, Vanden Heuvel JP (2011) Walnut oil increases cholesterol efflux through inhibition of stearoyl CoA desaturase 1 in THP-1 macrophage-derived foam cells. Nutr Metab (Lond) 8:61. https://doi.org/10.1186/1743-7075-8-61

Zhu H, Wang Z, Wu Y, Jiang H, Zhou F, Xie X, Wang R, Hua C (2019) Untargeted metabonomics reveals intervention effects of chicory polysaccharide in a rat model of non-alcoholic fatty liver disease. Int J Biol Macromol 128:363–375. https://doi.org/10.1016/j.ijbiomac.2019.01.141

Ziamajidi N, Khaghani S, Hassanzadeh G, Vardasbi S, Ahmadian S, Nowrouzi A, Ghaffari SM, Abdirad A (2013) Amelioration by chicory seed extract of diabetes- and oleic acid-induced non-alcoholic fatty liver disease (NAFLD)/non-alcoholic steatohepatitis (NASH) via modulation of PPARα and SREBP-1. Food Chem Toxicol 58:198–209. https://doi.org/10.1016/j.fct.2013.04.018

Zulcafli AS, Lim C, Ling AP, Chye S, Koh R (2020) Antidiabetic potential of Syzygium sp.: an overview. Yale J Biol Med 93:307–332

Ethnic Plants as Therapeutic Agents for Human Diseases

Battana Swapna, Mandala Ramakrishna, Srinivasan Kameswaran, and Pitchika Gopi Krishna

Contents

B. Swapna · M. Ramakrishna · S. Kameswaran (✉)
Department of Botany, Vikrama Simhapuri University College, Kavali, Andhra Pradesh, India

P. G. Krishna
Department of Zoology, Vikrama Simhapuri University College,
Kavali, Andhra Pradesh, India

© The Author(s), under exclusive license to Springer Nature Singapore Pte Ltd. 2024

A. Das Talukdar et al. (eds.), *Traditional Resources and Tools for Modern Drug Discovery*, Interdisciplinary Biotechnological Advances,
https://doi.org/10.1007/978-981-97-4600-2_7

Abstract Drug discovery and development is the necessity as global rise in disease prevalence among humans is a continuous challenge due to climate change. Drug discovery is the process that gives effective lead compounds for diverse targets such as cancer, neurodegenerative diseases, AIDS, cardiac diseases, and pain. The early steps of drug innovation entail identifying natural compounds. These compounds can be obtained through chemical synthesis or by isolating them from plants. Plants and their active ingredients act as sources for various new drugs. Identification of the accurate candidate plants for plant-based new drug discovery involves random screening of plants or evidence-based screening. Evidence-based screening includes application of either Ayurvedic knowledge, traditional documented use, ethnomedicine, or extensive literature search. Analysis of the components of the ancient formulations and their characteristics may provide an in-depth idea of the appropriate candidate plants. Ethnomedicine has been one of the important sources of lead identification in modern drug discovery. Ethnomedicine refers to the knowledge, skills, and practices that indigenous people and different cultures used to maintain health. This approach would lead to saving of time and cost. This chapter focuses on ethnic plants with therapeutic potential against different human diseases hoping to shed new light on discovery of novel therapeutic drugs from known plant natural sources. Ethnomedicinal resources will remain as key providers to development of effective drugs to curb global health challenges.

Keywords Ethnomedicine · Therapeutic agents · Drug development · Neurodegenerative diseases · Cancer

1 Introduction

Drug discovery is becoming increasingly difficult despite the introduction of advanced technologies. Traditional or ethnomedicine-inspired drug discovery strategy helps at this time. The use of medicinal plants to treat various diseases dates back thousands of years across different regions and cultures. Ethnobotany explores the intricate relationship between humans and plants, revealing how different societies use plants for various purposes, including medicine. Through the study of these conventional practices, scientists were proficient to isolate active constituents for contemporary medicine.

Dr. Pushpangadan obtained information from Kani tribe about *Trichopus zeylanicus* as energy booster which helped in the drug "Jeevani" development. This exemplifies the process of transforming ethnic knowledge and further laboratory observations into a commercial drug for global market (Padma 2005). Guggulipid was developed from *Commiphora mukul* taking the lead from Ayurveda is another example. It is being manufactured and marketed by Cipla.

Dysphania ambrosioides (L.) Mosyakin & Clemants from the Amaranthaceae family and *Clausena anisata* (Willd.) Hook. f. ex Benth. from the Rutaceae family are two plant species commonly used in Cameroon to repel insects. A study has shown that the essential oil blend of these two plants has a significant effect in killing mosquito larvae. The findings support the traditional use of these plants and suggest their potential for developing safe and effective herbal insecticides against the vector of filariasis (Pavela et al. 2018).

Combining ethnobotanical knowledge with phytochemistry can aid in the discovery of potential therapeutic compounds for new drug development.

2 Cancer

Cancer, one of the foremost causes of death, signifies a high healthcare challenge worldwide, incurring significant medical costs for those affected (Olatunde et al. 2021). The global incidence of cancers was estimated at 29.8 million in 2020, of which 19.3 million were recent cancer cases and 10 million mortalities due to cancer (Ferlay et al. 2015, 2019). In addition, "lung cancer (11.4%), breast (6.9%), colorectal (10.0%)" are the most commonly diagnosed cancers (Hyuna et al. 2021). A few hallmarks of cancer are apoptosis (Snellenberg et al. 2014), angiogenesis (Mar et al. 2015; Frink et al. 2016), production of growing signals (Courtnay et al. 2015), and metastasis. Due to these properties tumor cells are characterized by continuous growth, long survival times, and the possibility of invasion of normal cells. Furthermore, if left unchecked, cancer cells can ultimately be fatal for the patient (Shonia et al. 2019). Various approaches are currently used to treat cancer including chemotherapy, radiation therapy, and surgery. While chemotherapy can offer short-term relief and extend patients' lives, many of these drugs have side effects (Gao et al. 2013). Because of this, there is a growing interest in exploring natural products as potential anticancer agents. Numerous studies have confirmed the effectiveness of natural bioactive compounds derived from plants in fighting cancer. Some ethnic plants having therapeutic potential against different types of cancers are discussed below (Table 1).

2.1 Acorus calamus *L. Bauj*

Acorus calamus L. Bauj belonging to the Acoraceae family is used traditionally for gastrointestinal and nervous problems (Afzal et al. 2022). Studies on *Acorus calamus* rhizomes led to the isolation of compounds such as β-asarone. The cytotoxic effect of this bioactive compound was demonstrated in various human cancer cell lines (Rahamooz Haghighi et al. 2017; Li et al. 2018; Koca 2018).

Table 1 Ethnic plants as potential therapeutic agents for cancer

S. no.	Scientific name	Family	Plant parts used	Ethnobotanical use with reference	Anticancer potential with references
1.	*Andrographis paniculata* (Burm.f.) Wall. ex. Nees	Acanthaceae	Leaf	Fevers, jaundice, snake bite (Kumar et al. 2021a, b)	Malik et al. (2021), Shimura et al. (2021), Paul et al. (2021), Tundis et al. (2023)
2.	*Annona muricata* L.	Annonaceae	Fruit, leaf	Diabetes, breast cancer, head and neck cancer (Lutoti et al. 2023)	Silihe et al. (2023)
3.	*Alpinia galanga* L.	*Zingiberaceae*	Rhizome	inflammation, digestive disorders (Abubakar et al. 2018)	Samarghandian et al. (2014), Hadjzadeh et al. (2014), Lakshmi et al. (2019), Ahlina et al. (2020)
4.	*Artemisia annua* L.	Asteraceae	Leaf	Malaria, cold, cough (Septembre-Malaterre et al. 2020)	Lang et al. (2019), Xu et al. (2020)
5.	*Asclepias curassavica* L.	*Apocynaceae*	Aerial parts	Pain, inflammation (Ticktin and Dalle 2005; Alonso-Castro et al. 2021)	Zheng et al. (2019)
6.	*Carica papaya* L.	Caricaceae	Fruit, seeds	Diabetes, digestive problems, skin diseases (Nguyen et al. 2016)	Anilkumar and Bhanu (2022), Mahrous and Noseer (2023)
7.	*Decalepis hamiltonii* Wight & Arn	Periplocaceae	Tuberous roots	Appetizer, flatulence (Ashalatha et al. 2010)	Shathish and Guruvayoorappan (2014), Mallikarjuna and Dharmesh (2018), Shivashankarappa et al. (2022)
8.	*Phyllanthus urinaria* (L.)	Euphorbiaceae	Whole plant	Jaundice (Geethangili and Ding 2018; Orman et al. 2023)	Sayuti et al. (2020), Saahene et al. (2021), Abdel-Sattar et al. (2023)
9.	*Saussurea lappa* C.B. Clarke	Asteraceae	Roots	Asthma, digestive disorders, and skin conditions (Madhuri et al. 2012; Kumari et al. 2023)	Tian et al. (2017), Ansari et al. (2021), Okubo et al. (2021), Peng et al. (2022)
10.	*Simarouba glauca* DC	Simaroubaceae	Leaf	Cancer (Patil and Gaikwad 2011)	Jose et al. (2020), Vikas et al. (2022)

(continued)

Table 1 (continued)

S. no.	Scientific name	Family	Plant parts used	Ethnobotanical use with reference	Anticancer potential with references
11.	*Solanum nigrum* L.	Solanaceae	Fruit	Liver tonic, skin infections (Ahmed et al. 2018; Umair et al. 2017)	Al Sinani et al. (2016), Lai et al. (2016), Razali et al. (2016), Dong et al. (2021), Nawaz et al. (2021)
12.	*Urtica dioica*	Urticaceae	Leaf	Diabetes, hypertension (El Haouari and Rosado 2019; Jan et al. 2021)	Esposito et al. (2019), Wenyuan et al. (2021), Karakol et al. (2022)
13.	*Vitex negundo* L.	Verbenaceae	Leaf	Rheumatic pains (Yabesh et al. 2014)	Sichaem et al. (2019), Van Vo et al. (2022)
14.	*Ziziphus mauritiana* Lam.	Rhamnaceae	Leaf, seed	Anxiety, sleeping disorders (El Maaiden et al. 2020)	Mishra et al. (2011), Beg et al. (2016), Kabir et al. (2020), Prakash et al. (2022)
15.	*Zingiber officinale* Rosc	Zingiberaceae	Rhizome	Pain, constipation, digestive troubles, fever, cramps, hypertension, dementia, and infections (Singh et al. 2022)	Hsu et al. (2015), Rastogi et al. (2015), Liang et al. (2019), Bawadood et al. (2020), Hsu et al. (2023)

2.2 Aegle marmelos *(L.) Correa*

Leaves and fruits of *Aegle marmelos* (L.) Correa belonging to the Rutaceae family is used conventionally for diarrhea, gastric troubles, and constipation (Kala 2006). The ethanolic extract of *A. marmelos* fruit pulp showed suppression of proliferating activity by reducing the development of breast tumors in rats (Akhouri et al. 2020; Bhatti et al. 2013; Kumar et al. 2021a, b; Pynam and Dharmesh 2018).

2.3 Allium cepa *L.*

Allium cepa L. bulbs are used like condiment as well as medicine for cold, cough, and asthma traditionally (Teshika et al. 2019). Polyphenols from *Allium cepa* upregulated P^{53} protein and suppressed Akt activity (Lee et al. 2014). In studies involving H295R and SW-13 cell lines, quercetin from *Allium cepa* demonstrated a stronger cytotoxic response on adrenocortical carcinoma cells (Veiga et al. 2022).

2.4 Asparagus racemosus *Willd.*

Asparagus racemosus roots are used for deworming by Santhal tribe in Assam, India (Soren and Yadav 2021). Acemosin from *Asparagus* showed moderate cytotoxicity against HepG2 cancer cell lines (Quang et al. 2018). Shatavarins in *Asparagus* exhibited anticancerous properties in both in vitro and in vivo experimental models (Mitra et al. 2012).

2.5 Andrographis paniculata *(Burm. f.) Wall. ex Nees*

Leaves and aerial parts of *Andrographis paniculata* are traditionally used for fevers, jaundice, and snake bite (Kumar et al. 2021a, b). Andrographolide and stigmasterol inhibited breast cancer by suppressing COX-2 expression through inactivation of p300 signaling (Malik et al. 2021).

2.6 Catharanthus roseus

Catharanthus roseus native to India and China belongs to family Apocynaceae. *Catharanthus roseus* is used by folklore for treating diabetes, asthma, insomnia, leukemia, and cancer (Pooja 2017). The methanolic extracts of *C. roseus* showed remarkable anticancerous activity on Hep-2 cell line. Effect of *C. roseus* flower ethanolic extract was evaluated against cytotoxicity in "human epithelial cervical carcinoma cell line" using MTT assay (Harshini et al. 2020).

2.7 Centella asiatica

Centella asiatica, also known as brahmi, is a traditional medicinal plant that belongs to the Apiaceae family and is commonly found in India and China. Various extracts of *Centella* leaves have been studied for their effects on human lung epithelial carcinoma cell line (A-549) and have been found to have alkaloids with cytotoxic properties (Iwan et al. 2016). Additionally, *C. asiatica* juice repressed the proliferation of a malignant HepG2 cell line via apoptosis (Hussin et al. 2014).

2.8 Curcuma longa

Turmeric, also known as *Curcuma longa*, is part of the Zingiberaceae family and has extensive health benefits. It has been found to have anti-HIV, anti-inflammatory, antitumor, antioxidant, and antimicrobial properties (Ankit et al. 2019). Studies have shown that curcumin, a component of turmeric, may help prevent colon and breast cancer. Tests using SRB and MTT assays for cytotoxicity and cell viability on breast cancer cell lines (MCF-7) demonstrated the potential effectiveness of curcumin (Giordano and Tommonaro 2019). Curcumin exerts its antiproliferative properties against colorectal cancer cell lines (Ojo et al. 2022).

2.9 Euphorbia tirucalli

Euphorbia tirucalli L. belonging to Euphorbiaceae thrives well in Brazil. Interestingly, the latex from *E. tirucalli* has been consumed by those who have been diagnosed with cancer. In vitro studies have confirmed the antitumor properties of the latex, making it a promising treatment option for cancer patients. Although limited in vivo evaluations have been conducted, *E. tirucalli* has great potential to be used as an adjunct in cancer treatment (Martins et al. 2020).

2.10 Leucas aspera *L.*

Leucas aspera is traditionally taken as leafy vegetable and used as medicine for cold, cough, and respiratory problems. Ethanolic extract of *Leucas aspera* leaf and flower showed significant anticancerous activity (Madhu et al. 2019).

2.11 Tinospora cordifolia

Tinospora cordifolia, also known as Giloi or Guduchi, belongs to the Menispermaceae family and is distributed in Europe and East Asia (Pooja 2017). The extract of Giloi is used to treat various cancers, including brain, breast, intestine, head and neck, and prostate cancers. The stem extracts of *T. cordifolia* have been found to initiate programmed cell death and inhibit apoptosis. In vitro cytotoxicity of *T. cordifolia* stem ethanolic extract toward "murine monocyte/-macrophages (J-774-A-1), human melanoma (A-375), and human breast cancer (MCF-7) cell lines" was assessed by the colorimetric MTT assay and TBE method (Palmieri et al. 2019).

2.12 Withania somnifera L.

Withania somnifera (Ashwagandha) belonging to family Solanaceae consists of substances such as withanolides, "withananine, anaferine, withanine, β-sisterol, tropanol, chlorogenic acid, somniferiene, scopoletin and somniferimine" that bring about anticancerous and antidiabetes activities. The cytotoxic effects of various extracts of *W. somnifera* plant parts on Hep-2 cells were evaluated using "MTT assay and the TBE method" (Ahmed et al. 2018).

3 Neurodegenerative Diseases

Oxidative stress in the central nervous system is related to "neurodegenerative diseases" like Alzheimer's disease, Parkinson's disease, and Huntington's disease which cause severe health issues and financial burden among people around the world (Tsoukalas et al. 2021). The treatment option for these disorders includes use of natural antioxidants to counteract cellular oxidative stress. The understanding that plant extracts show other positive activities along with antioxidant activity led to further research on plants for their potential use in prevention or management of neurodegenerative diseases. Some ethnic plants having therapeutic potential against neurodegenerative diseases are discussed below (Table 2).

3.1 Bacopa monnieri (L.) Wettst.

B. monnieri, known as brahmi in India, has been used from ancient times in Ayurveda as brain tonic. Bacosides derived from *Bacopa monnieri* has anti-inflammatory and Aβ aggregation inhibitor properties contributing as important therapeutic agents for Alzheimer's disease (Abdul Manap et al. 2019; Chaudhari et al. 2017). *B. monnieri* showed neuroprotective effects by suppressing the level of pro-inflammatory cytokines and reducing α-synuclein levels in Parkinson's disease animal model (Singh et al. 2020).

3.2 Mucuna pruriens (L.) DC.

Mucuna pruriens at a dosage of 17.5 mg/kg, enhanced motor function and decreased dyskinesia in patients with advanced Parkinson's disease with fewer adverse effects as compared with the treatment of levodopa and carbidopa together (Cilia et al. 2017; Johnson et al. 2018). *Mucuna* attenuate the MPTP-induced neurotoxicity by NF-κB and pAkt pathway (Rai et al. 2017).

Table 2 Ethnic plants as potential therapeutic agents for neurodegenerative diseases

S. no.	Scientific name	Family	Plant parts used	Ethnobotanical use with reference	Neurodegerative disease therapeutic effect with references
1.	*Acalypha indica* L.	Euphorbiaceae	Aerial parts	Bronchitis, wound healing (Zahidin et al. 2017)	Uddin and Zidorn (2020)
2.	*Achyranthes aspera* L.	Acanthaceae	Leaves, seeds	Wound healing, piles, skin diseases and dental problems (Rajalakshmi et al. 2019a, b)	Viswanatha et al. (2017), Lin et al. (2019), Sharma et al. (2021)
3.	*Allium sativum* L.	Amaryllidaceae	Bulb	Cough, cold, and chest congestion (Devi et al. 2014)	Hazzaa et al. (2020), Nayak et al. (2021), Nadeem et al. (2021)
4.	*Azadirachta indica* L.	Meliaceae	All parts	Diabetes, wound healing, stomach disorders (Sujarwo et al. 2016)	Sandhir et al. (2021), Xiang et al. (2018)
5.	*Calotropis procera* (Ait.) R.Br.	Asclepiadaceae	Leaf	Skin diseases (Yaniv and Koltai 2018)	Verma et al. (2017), Yoneyama et al. (2017)
6.	*Glycyrrhiza glabra* Linn.	Fabaceae	Root, leaf	Cough, phlegm, dyspnea, spasms, and pain (Xiaoying et al. 2017)	Wu et al. (2020), Singh et al. (2016), Zulfugarova et al. (2023), Bayram and Hacimuftuoglu (2022), Karthikkeyan et al. (2020)
7.	*Moringa oleifera* Lam.	Moringaceae	Leaf	Paralysis, helminthiasis, sores, and skin infections (Popoola and Obembe 2013; Yabesh et al. 2014)	Giacoppo et al. (2017), Ghimire et al. (2021), González-Burgos et al. (2021), Mundkar et al. (2022)
8.	*Morus alba* L.	Moraceae	Fruit	Constipation, urinary infections, dizziness, insomnia (Rodrigues et al. 2019; Yadav et al. 2022)	Paudel et al. (2019), Xia et al. (2019), Kuk et al. (2017), Sood et al. (2023)
9.	*Ocimum basilicum* L.	Lamiaceae	Leaf	Cold, cough, constipation (Bhat et al. 2020)	Ayuob et al. (2018), Seyed et al. (2021), Simayi et al. (2022)
10.	*Papaver somniferum* L.	Papaveraceae	Seeds	Pain (Bulut et al. 2017)	Sharma et al. (2021), Altinoz et al. (2022), Sirin et al. (2022)
11.	*Phyllanthus emblica* L.	Euphorbiaceae	Fruit	Diarrhea, jaundice, inflammation (Panmei et al. 2019)	Jang et al. (2017), Rajalakshmi et al. (2019a, b), Husain et al. (2019), Wu et al. (2022)

(continued)

Table 2 (continued)

S. no.	Scientific name	Family	Plant parts used	Ethnobotanical use with reference	Neurodegerative disease therapeutic effect with references
12.	*Portulaca oleracea* L.	Portulacaceae	Leaf	Burns, headache (Paniagua-Zambrana et al. 2020)	Martins et al. (2016), Berezutsky et al. (2022), Fazelinejad and Keighobadi (2023)
14.	*Vitis vinifera* L.	Vitaceae	Fruit	Anemia, piles (Akgül et al. 2016)	Choi et al. (2023), Calapai et al. (2017), Rapaka et al. (2019), Duangjan et al. (2021)
15.	*Vernonia amygdalina* Del	Asteraceae	Leaf	Diabetes (Tugume et al. 2016)	Oladele et al. (2020), Afees et al. (2022), Oboh et al. (2022), Ajeleti et al. (2023)
16.	*Withania somnifera* (L.) Dun	Solanaceae	Root, fruit	Fever, rejuvenation (Afewerky et al. 2021)	Gupta and Kaur (2019), Crane et al. (2019)

3.3 Cyperus rotundus *L.*

Cyperus rotundus (Cyperaceae) tuber is used as traditional folk medicine for stomach disorders and as astringent (Peerzada et al. 2015). *C. rotundus* extract showed anti-convulsant activity by repairing spatial memory impairment in rats, through improved neurogenesis. The anti-convulsant effect can be attributed to the rich flavonoid content present in *C. rotundus* (Shakerin et al. 2020).

3.4 Apium graveolens *L.*

Apium graveolens (celery) is used traditionally for hypertension and gout. An aqueous extract of *A. graveolens* and DL-3-*n*-butylphthalide (DL-NBP) were studied for neurological disorders related with diabetes. DL-NBP showed neuroprotective effects through "normalization of hippocampal morphology, improving synaptic plasticity" and reduced apoptosis in neurons (Wang et al. 2021). Studies on extract of *A. graveolens* at doses ranging from 80 to 1000 mg/kg given intraperitoneally exhibited betterment of antiepileptic activity (Ye et al. 2018; Choupankareh et al. 2018).

There is promising research on the potential benefits of *Apium* in various neurological disorders. For instance, studies have shown that it could help with conditions such as "Parkinson's disease, neurotoxicity, chronic cerebral hypoperfusion, chronic intermittent hypoxia, hypercapnia, anxiolytics, and perinatal effects" (Abu-Taweel 2020; Chonpathompikunlert et al. 2018; Wongtawatchai et al. 2017; Tan et al. 2023).

3.5 Crocus sativus *L.*

Crocus sativus L. (Saffron) is a spice used in cosmetics, textile dyes, and medicine (Kothari et al. 2021). Saffron is used in ancient Greece, Rome, Egypt, South and Central Asia, India, and China (Krishnan et al. 2017; Rahmani et al. 2017). Saffron was utilized as folklore medicine (Mohtashami et al. 2021; Shah et al. 2017). *Crocus* is a popular natural cosmetic in Iranian and Indian medicine for improving of skin tone. Saffron has numerous therapeutic properties such as aphrodisiac, antimicrobial, antiseptic and, anti-inflammatory. It has anticancerous activity also (Siddique et al. 2020).

The results of a study have shown that the compounds crocin and crocetin extracted from saffron have demonstrated an ability to enhance learning and memory skills in rats and mice that were experiencing learning behavior impairments induced by ethanol. Furthermore, promising results were observed when saffron was orally administered as an adjuvant in the treatment of neurodegenerative disorders and related memory impairments (Roshanravan and Ghaffari 2022; Yang et al. 2023).

4 Conclusion

Ethnobotanical information about the various medicinal plants provides leads to explore potential therapeutic agents for combating diseases. Various clinical studies and preclinical results suggest that many molecular aspects and many phytochemicals are being explored to bring them from the laboratory to the bedside and that in the future several new and valuable therapeutic options based on ethnic plants will be available to the public for treating several diseases like cancer and neurodegenerative diseases. Furthermore, in order to comply with international standards, comprehensive standardization of methods for evaluating bioaccessability, effectiveness, safeness, quality, composition, manufacturing processes, and product regulation should be accomplished. This suggests that the abovementioned plants could be considered as therapeutic agents to fight against various types of human cancer and neurodegenerative disorders. Further research in this direction will pave way for producing an optimal preparation from the plants.

References

Abdel-Sattar OE, Allam RM, Al-Abd AM, Avula B, Katragunta K, Khan IA, El-Desoky AM, Mohamed SO, El-Halawany A, Abdel-Sattar E, Meselhy MR (2023) Cytotoxic and chemomodulatory effects of *Phyllanthus niruri* in MCF-7 and MCF-7ADR breast cancer cells. Sci Rep 13(1):2683

Abdul Manap AS, Vijayabalan S, Madhavan P et al (2019) *Bacopa monnieri*, a neuroprotective lead in Alzheimer disease: a review on its properties, mechanisms of action, and preclinical and clinical studies. Drug Target Insights 13:117739281986641

Abubakar IB, Malami I, Yahaya Y, Sule SM (2018) A review on the ethnomedicinal uses, phytochemistry and pharmacology of Alpinia officinarum Hance. J Ethnopharmacol 224:45–62

Abu-Taweel GM (2020) Celery ameliorating against neurobehavioral and neurochemical disorders of perinatal lipopolysaccharides exposure in mice offspring. J King Saud Univ Sci 32:1764–1771

Afees OJ, Olufunmilola EI, Oluwaseyi AK, Igbo EJ, Arietarhire L, Soremekun O, Ogunbiyi O, Bamidele R, Taiye AS, Yinka OS, Afolabi T (2022) Vernonia amygdalina alleviates neurodegenerative effects of high fat diet and streptozotocin-induced type 2 diabetes mellitus in the hippocampi of adult male Wistar rats. In: Endocrine abstracts, vol 86. Bioscientifica

Afewerky HK, Ayodeji AE, Tiamiyu BB, Orege JI, Okeke ES, Oyejobi AO, Bate PNN, Adeyemi SB (2021) Critical review of the *Withania somnifera* (L.) Dunal: ethnobotany, pharmacological efficacy, and commercialization significance in Africa. Bull Natl Res Cent 45:1–16

Afzal S, Ayoub M, Raja WY (2022) *Acorus calamus*: a review on its phytochemical and pharmacological profile. In: Masoodi MH, Rehman MU (eds) Edible plants in health and diseases. Springer, Singapore. https://doi.org/10.1007/978-981-16-4959-2_17

Ahlina FN, Nugraheni N, Salsabila IA, Haryanti S, Da'i M, Meiyanto E (2020) Revealing the reversal effect of galangal (*Alpinia galanga* L.) extract against oxidative stress in metastatic breast cancer cells and normal fibroblast cells intended as a co-chemotherapeutic and anti-ageing agent. Asian Pac J Cancer Prev APJCP 21(1):107–117

Ahmed W, Mofed D, Zekri AR, El-Sayed N, Rahouma M, Sabet S (2018) Antioxidant activity and apoptotic induction as mechanisms of action of *Withania somnifera* (Ashwagandha) against a hepatocellular carcinoma cell line. J Int Med Res 46(4):1358–1369

Ajeleti AO, Akunna GG, Ojo JH, Irozuoke A (2023) Neuro-protective and neuro-curative effect of Vernonia Amygdalina in rat model of neurodegenerative disease. J Trace Elem Miner 4:100062

Akgül G, Yılmaz N, Celep A, Celep F, Çakılcıoğlu U (2016) Ethnobotanical purposes of plants sold by herbalists and folk bazaars in the center of Cappadocica. Nevşehir, Turkey

Akhouri V, Kumari M, Kumar A (2020) Therapeutic effect of *Aegle marmelos* fruit extract against DMBA induced breast cancer in rats. Sci Rep 10(1):18016

Al Sinani SS, Eltayeb EA, Coomber BL, Adham SA (2016) Solamargine triggers cellular necrosis selectively in different types of human melanoma cancer cells through extrinsic lysosomal mitochondrial death pathway. Cancer Cell Int 16(1):1–12

Alonso-Castro AJ, Arana-Argáez V, Yáñez-Barrientos E, Torres-Romero JC, Chable-Cetz RJ, Worbel K, de Jesús Euan-Canto A, Wrobel K, González-Ibarra A, Solorio-Alvarado CR, del Carmen Juárez-Vázquez M (2021) Pharmacological activities of *Asclepias curassavica* L. (Apocynaceae) aerial parts. J Ethnopharmacol 281:114554

Altinoz MA, Guloksuz S, Ozpinar A (2022) Immunomodifying and neuroprotective effects of noscapine: implications for multiple sclerosis, neurodegenerative, and psychiatric disorders. Chem Biol Interact 352:109794

Anilkumar A, Bhanu A (2022) In vitro anticancer activity of "Methanolic extract of papaya blackseeds" (MPB) in Hep G2 cell lines and its effect in the regulation of bcl-2, caspase-3 and p53 gene expression. Adv Cancer Biol 4:100025

Ankit S, Robbie H, Sudeep C (2019) Traditional herbal knowledge among the inhabitants: a case study in Urgam Valley of Chamoli Garhwal, Uttarakhand, India. Evid Based Complement Alternat Med 2019:1–21

Ansari S, Hasan K, Bhat S (2021) Anticancer, antioxidant, and hepatoprotective activity of Saussurea lappa, CB clarke (qust) on human hepatoma cell line. J Cancer Res Ther 17(2):499–503

Ashalatha K, Venkateswarlu Y, Priya AM, Lalitha P, Krishnaveni M, Jayachandran S (2010) Anti-inflammatory potential of *Decalepis hamiltonii* (Wight and Arn) as evidenced by down regulation of pro inflammatory cytokines—TNF-α and IL-2. J Ethnopharmacol 130(1):167–170

Ayuob NN, El Wahab MGA, Ali SS, Abdel-Tawab HS (2018) *Ocimum basilicum* improve chronic stress-induced neurodegenerative changes in mice hippocampus. Metab Brain Dis 33:795–804

Bawadood AS, Al-Abbasi FA, Anwar F, El-Halawany AM, Al-Abd AM (2020) 6-Shogaol suppresses the growth of breast cancer cells by inducing apoptosis and suppressing autophagy via targeting notch signaling pathway. Biomed Pharmacother 128:110302

Bayram C, Hacimuftuoglu A (2022) Investigation of antioxidant efficacy of Glycyrrhiza glabra L. extract in glutamate toxicity-induced primary neuron culture. Anatol J Biol 3(1):18–24

Beg MA, Teotia U, Farooq S (2016) In vitro antibacterial and anticancer activity of Ziziphus. J Med Plants Stud 4(5):230–233

Berezutsky MA, Durnova NA, Sheremetyewa AS, Matvienko UA, Kurchatova MN (2022) Experimental studies of geroprotective and anti-aging effects of chemical compounds of Portulaca oleracea L. Adv Gerontol 12(2):122–127

Bhat GM, Majeed H, Islam MA, Rather TA, Bakshi M, Shahkhan FA, Bhat RA (2020) Indigenous medicinal practices of *Ocimum basilicum* L. in rural Kashmir. J Pharmacogn Phytochem 9(5):1876–1881

Bhatti R, Singh J, Saxena AK, Suri N, Ishar MPS (2013) Pharmacognostic standardisation and antiproliferative activity of *Aegle marmelos* (L.) Correa leaves in various human cancer cell lines. Indian J Pharm Sci 75(6):628

Bulut G, Haznedaroğlu MZ, Doğan A, Koyu H, Tuzlacı E (2017) An ethnobotanical study of medicinal plants in Acipayam (Denizli-Turkey). J Herb Med 10:64–81

Calapai G, Bonina F, Bonina A, Rizza L, Mannucci C, Arcoraci V, Laganà G, Alibrandi A, Pollicino C, Inferrera S, Alecci U (2017) A randomized, double-blinded, clinical trial on effects of a Vitis vinifera extract on cognitive function in healthy older adults. Front Pharmacol 8:776

Chaudhari KS, Tiwari NR, Tiwari RR, Sharma RS (2017) Neurocognitive effect of nootropic drug Brahmi (Bacopa monnieri) in Alzheimer's disease. Ann Neurosci 24(2):111–122

Choi J, Choi SY, Hong Y, Han YE, Oh SJ, Lee B, Choi CW, Kim MS (2023) The central administration of vitisin a, extracted from Vitis vinifera, improves cognitive function and related signaling pathways in a scopolamine-induced dementia model. Biomed Pharmacother 163:114812

Chonpathompikunlert P, Boonruamkaew P, Sukketsiri W, Hutamekalin P, Sroyraya M (2018) The antioxidant and neurochemical activity of *Apium graveolens* L. and its ameliorative effect on MPTP-induced Parkinson-like symptoms in mice. BMC Complement Altern Med 18:103

Choupankareh S, Beheshti F, Karimi S, Sadeghnia HR, Rakhshandeh H, Rezaeipour M, Hosseini M (2018) The effects of aqueous extract of *Apium graveolens* on brain tissues oxidative damage in pentylenetetrazole-induced seizures model in rat. Curr Nutr Food Sci 14:47–53

Cilia R, Laguna J, Cassani E, Cereda E, Pozzi NG, Isaias IU, Contin M, Barichella M, Pezzoli G (2017) Mucuna pruriens in Parkinson disease: a double-blind, randomized, controlled, crossover study. Neurology 89:1–7

Courtnay R, Ngo DC, Malik N, Ververis K, Tortorella SM, Karagiannis TC (2015) Cancer metabolism and the Warburg effect: the role of HIF-1 and PI3K. Mol Biol Rep 42:841–851

Crane EA, Heydenreuter W, Beck KR, Strajhar P, Vomacka J, Smiesko M, Mons E, Barth L, Neuburger M, Vedani A, Odermatt A (2019) Profiling withanolide A for therapeutic targets in neurodegenerative diseases. Bioorg Med Chem 27(12):2508–2520

Devi A, Rakshit K, Sarania B (2014) Ethnobotanical notes on Allium species of Arunachal Pradesh, India. Indian J Tradit Knowl 13(3):606–61213

Dong Y, Hao L, Fang K, Han XX, Yu H, Zhang JJ, Cai LJ, Fan T, Zhang WD, Pang K, Ma WM (2021) A network pharmacology perspective for deciphering potential mechanisms of action of Solanum nigrum L. in bladder cancer. BMC Complement Med Ther 21:1–14

Duangjan C, Rangsinth P, Zhang S, Gu X, Wink M, Tencomnao T (2021) Vitis vinifera leaf extract protects against glutamate-induced oxidative toxicity in HT22 hippocampal neuronal cells and increases stress resistance properties in Caenorhabditis elegans. Front Nutr 8:634100

El Haouari M, Rosado JA (2019) Phytochemical, anti-diabetic and cardiovascular properties of Urtica dioica L. (Urticaceae): a review. Mini Rev Med Chem 19(1):63–71

El Maaiden E, El Kharrassi Y, Qarah NA, Essamadi AK, Moustaid K, Nasser B (2020) Genus Ziziphus: a comprehensive review on ethnopharmacological, phytochemical and pharmacological properties. J Ethnopharmacol 259:112950

Esposito S, Bianco A, Russo R, Di Maro A, Isernia C, Pedone PV (2019) Therapeutic perspectives of molecules from Urtica dioica extracts for cancer treatment. Molecules 24(15):2753

Fazelinejad H, Keighobadi N (2023) Inhibitory effect of *Portulaca oleracea* L. extract against the lysozyme fibrillation and its possible role in ameliorate of neurodegenerative diseases. Exp Anim Biol 11(3):21–31

Ferlay J, Soerjomataram I, Dikshit R, Eser S, Mathers C, Rebelo M (2015) Cancer incidence and mortality worldwide: sources, methods and major patterns in GLOBOCAN 2012. Int J Cancer 136:E359–E386

Ferlay J, Colombet M, Soerjomataram I et al (2019) Estimating the global cancer incidence and mortality in 2018: GLOBOCAN sources and methods. Int J Cancer 144:1941–1953

Frink RE, Peyton M, Schiller JH, Gazdar AF, Shay JW, Minna JD (2016) Telomerase inhibitor imetelstat has preclinical activity across the spectrum of non-small cell lung cancer oncogenotypes in a telomere length dependent manner. Oncotarget 7:31639

Gao W, Xiang B, Meng TT, Liu F, Qi XR (2013) Chemotherapeutic drug delivery to cancer cells using a combination of folate targeting and tumor microenvironment-sensitive polypeptides. Biomaterials 34:4137–4149

Geethangili M, Ding ST (2018) A review of the phytochemistry and pharmacology of *Phyllanthus urinaria* L. Front Pharmacol 9:1109

Ghimire S, Subedi L, Acharya N, Gaire BP (2021) Moringa oleifera: a tree of life as a promising medicinal plant for neurodegenerative diseases. J Agric Food Chem 69(48):14358–14371

Giacoppo S, Rajan TS, De Nicola GR, Iori R, Rollin P, Bramanti P, Mazzon E (2017) The isothiocyanate isolated from Moringa oleifera shows potent anti-inflammatory activity in the treatment of murine subacute Parkinson's disease. Rejuvenation Res 20(1):50–63

Giordano A, Tommonaro G (2019) Curcumin and cancer. Nutrients 11(10):2376

González-Burgos E, Ureña-Vacas I, Sánchez M, Gómez-Serranillos MP (2021) Nutritional value of Moringa oleifera Lam. leaf powder extracts and their neuroprotective effects via antioxidative and mitochondrial regulation. Nutrients 13(7):2203

Gupta M, Kaur G (2019) Withania somnifera (L.) Dunal ameliorates neurodegeneration and cognitive impairments associated with systemic inflammation. BMC Complement Altern Med 19:1–18

Hadjzadeh MR, Ghanbari H, Keshavarzi Z, Afshari JT (2014) The effects of aqueous extract of alpinia galangal on gastric cancer cells (AGS) and L929 cells in vitro. Iran J Cancer Prev 7:142–146

Harshini D, Sheriff R, Jadon R, Ashwini B, Krupa BN (2020, November) Classification of ultrasound thyroid nodule images into benign and malignant lesions. In: In 2020 IEEE International Conference for Innovation in Technology (INOCON). IEEE, pp 1–4

Hazzaa SM, Abdelaziz SAM, Abd Eldaim MA, Abdel-Daim MM, Elgarawany GE (2020) Neuroprotective potential of allium sativum against monosodium glutamate-induced excitotoxicity: impact on short-term memory, gliosis, and oxidative stress. Nutrients 12(4):1028

Hsu YL, Hung JY, Tsai YM, Tsai EM, Huang MS, Hou MF, Kuo PL (2015) 6-Shogaol, an active constituent of dietary ginger, impairs cancer development and lung metastasis by inhibiting the secretion of CC-chemokine ligand 2 (CCL2) in tumor-associated dendritic cells. J Agric Food Chem 63(6):1730–1738

Hsu CM, Su HC, Yang MY, Tsai YT, Tsai MS, Yang YH, Wu CY, Chang SF (2023) 6-Shogaol is a potential treatment for head and neck squamous cell carcinoma. Int J Med Sci 20(2):238–246

Husain I, Zameer S, Madaan T, Minhaj A, Ahmad W, Iqubaal A, Ali A, Najmi AK (2019) Exploring the multifaceted neuroprotective actions of Emblica officinalis (Amla): a review. Metab Brain Dis 34:957–965

Hussin F, Eshkoor SA, Rahmat A et al (2014) The *Centella asiatica* juice effects on DNA damage, apoptosis and gene expression in hepatocellular carcinoma (HCC). BMC Complement Altern Med 14:32

Hyuna S, Jacques F, Rebecca L, Siegel ML (2021) Global cancer statistics 2020: GLOBOCAN estimates of incidence and mortality worldwide for 36 cancers in 185 countries. CA Cancer J Clin 71:209–249

Iwan SH, Ngakan MRW, Ratna D (2016) Anticancer Activity of Centella asiatica Leaves Extract in Benzo(a)pyrene-Induced Mice. International Journal of Pharmacognosy and Phytochemical Research 8(1):80–84

Jan HA, Abbasi AM, Turi MA, Bussmann RW, Paniagua-Zambrana NY (2021) *Urtica dioica* L. Urticaceae. In: Kunwar RM, Sher H, Bussmann RW (eds) Ethnobotany of the Himalayas. Ethnobotany of mountain regions. Springer, Cham. https://doi.org/10.1007/978-3-030-45597-2_252-1

Jang H, Srichayet P, Park WJ, Heo HJ, Kim DO, Tongchitpakdee S, Kim TJ, Jung SH, Lee CY (2017) Phyllanthus emblica L. (Indian gooseberry) extracts protect against retinal degeneration in a mouse model of amyloid beta-induced Alzheimer's disease. J Funct Foods 37:330–338

Johnson SL, Park HY, DaSilva NA, Vattem DA, Ma H, Seeram NP (2018) Levodopa-reduced *Mucuna pruriens* seed extract shows neuroprotective effects against Parkinson's disease in murine microglia and human neuroblastoma cells, *Caenorhabditis elegans*, and *Drosophila melanogaster*. Nutrients 10(9):1139

Jose A, Kannan E, Madhunapantula SV (2020) Anti-proliferative potential of phytochemical fractions isolated from Simarouba glauca DC leaf. Heliyon 6(4):e03836

Kabir SR, Asaduzzaman AKM, Amin R, Haque AT, Ghose R, Rahman MM, Islam J, Amin MB, Hasan I, Debnath T, Chun BS (2020) *Ziziphus mauritiana* fruit extract-mediated synthesized silver/silver chloride nanoparticles retain antimicrobial activity and induce apoptosis in MCF-7 cells through the Fas pathway. ACS Omega 5(32):20599–20608

Kala CP (2006) Ethnobotany and ethnoconservation of Aegle marmelos (L.) Correa. Indian Journal of Traditional Knowledge 5(4):537–540

Karakol P, Saraydin SU, Bozkurt M, Hepokur C, Inan ZDS, Turan M (2022) Anticancer effects of Urtica dioica in breast cancer. Asian Pac J Cancer Prev APJCP 23(2):673

Karthikkeyan G, Najar MA, Pervaje R, Pervaje SK, Modi PK, Prasad TSK (2020) Identification of molecular network associated with neuroprotective effects of Yashtimadhu (Glycyrrhiza glabra L.) by quantitative proteomics of rotenone-induced Parkinson's disease model. ACS Omega 5(41):26611–26625

Koca HB (2018) Effects of *Acorus calamus* plant extract on prostate cancer cell culture. Anatol J Bot 2(1):46–51

Kothari D, Thakur R, Kumar R (2021) Saffron (Crocus sativus L.): gold of the spices—a comprehensive review. Hortic Environ Biotechnol 62(5):661–677

Krishnan AR, Mathew J, Salam SA, Jiju V, Elessy A (2017) A review on saffron as an alternative therapy in medicine and dermatology. Eur J Pharm Med Res 4:283–286

Kuk EB, Jo AR, Oh SI, Sohn HS, Seong SH, Roy A, Choi JS, Jung HA (2017) Anti-Alzheimer's disease activity of compounds from the root bark of Morus alba L. Arch Pharm Res 40:338–349

Kumar PY, Parimalam M, Kumar D, Joseph E, David D, Vinolia R (2021a) Screening of phytochemicals, in vitro assessment of antioxidant, anti-inflammatory, Tlc profiling and anticancer activity of Aegle marmelos (L.) leaves. Ann Rom Soc Cell Biol 25(4):18061–18071

Kumar S, Singh B, Bajpai V (2021b) Andrographis paniculata (Burm. f.) Nees: traditional uses, phytochemistry, pharmacological properties and quality control/quality assurance. J Ethnopharmacol 275:114054

Kumari R, Negi M, Thakur P, Mahajan H, Raina K, Sharma R, Singh R, Anand V, Ming LC, Goh KW, Calina D (2023) Saussureacostus (Falc.) Lipsch.: a comprehensive review of its pharmacology, phytochemicals, ethnobotanical uses, and therapeutic potential. Naunyn Schmiedebergs Arch Pharmacol 397(3):1505–1524

Lai YJ, Tai CJ, Wang CW, Choong CY, Lee BH, Shi YC, Tai CJ (2016) Anti-cancer activity of Solanum nigrum (AESN) through suppression of mitochondrial function and epithelial-mesenchymal transition (EMT) in breast cancer cells. Molecules 21(5):553

Lakshmi S, Suresh S, Rahul BS, Saikant R, Maya V, Gopi M, Padmaja G, Remani P (2019) In vitro and in vivo studies of 5, 7-dihydroxy flavones isolated from Alpinia galanga (L.) against human lung cancer and ascetic lymphoma. Med Chem Res 28:39–51

Lang SJ, Schmiech M, Hafner S, Paetz C, Steinborn C, Huber R, El Gaafary M, Werner K, Schmidt CQ, Syrovets T, Simmet T (2019) Antitumor activity of an Artemisia annua herbal preparation and identification of active ingredients. Phytomedicine 62:152962

Lee ZW, Teo XY, Tay EW, Tan CH, Hagen T, Moore PK, Deng LW (2014) Utilizing hydrogen sulfide as a novel anti-cancer agent by targeting cancer glycolysis and p H imbalance. British journal of pharmacology 171(18):4322–4336

Li L, Yang Y, Wu M, Yu Z, Wang C, Dou G, He H, Wang H, Yang N, Qi H, Xu X (2018) β-Asarone induces apoptosis and cell cycle arrest of human glioma U251 cells via suppression of HnRNP A2/B1-mediated pathway in vitro and in vivo. Molecules 23(5):1072

Liang T, He Y, Chang Y et al (2019) 6-Shogaol a active component from ginger inhibits cell proliferation and induces apoptosis through inhibition of STAT-3 translocation in ovarian cancer cell lines (A2780). Biotechnol Bioproc E 24:560–567

Lin LW, Tsai FH, Lan WC, Cheng YD, Lee SC, Wu CR (2019) Steroid-enriched fraction of Achyranthes bidentata protects amyloid β peptide 1–40-induced cognitive dysfunction and neuroinflammation in rats. Mol Neurobiol 56:5671–5688

Lutoti S, Kaggwa B, Kamba PF, Mukonzo J, Sesaazi CD, Katuura E (2023) Ethnobotanical survey of medicinal plants used in breast cancer treatment by traditional health practitioners in Central Uganda. J Multidiscip Healthc 16:635–651

Madhu GC, Kannaiyan J, Paulraj B, Veeramani V (2019) Anti-diabetic, anti-cancer activity and associated toxicity of Leucas aspera extract in Wistar albino rats. Int J Pharm Sci Drug Res 11:387–392

Madhuri K, Elango K, Ponnusankar S (2012) *Saussurea lappa* (Kuth root): review of its traditional uses, phytochemistry and pharmacology. Orient Pharm Exp Med 12:1–9

Mahrous NS, Noseer EA (2023) Anticancer potential of Carica papaya Linn black seed extract against human colon cancer cell line: in vitro study. BMC Complement Med Ther 23(1):271

Malik Z, Parveen R, Parveen B, Zahiruddin S, Khan MA, Khan A, Massey S, Ahmad S, Husain SA (2021) Anticancer potential of andrographolide from *Andrographis paniculata* (Burm. f.) Nees and its mechanisms of action. J Ethnopharmacol 272:113936

Mallikarjuna SE, Dharmesh SM (2018) Swallow root (*Decalepis hamiltonii*) pectic oligosaccharide (SRO1) induces cancer cell death via modulation of galectin-3 and survivin. Carbohydr Polym 186:402–410

Mar AC, Chu CH, Lee HJ, Chien CW, Cheng JJ, Yang SH (2015) Interleukin-1 receptor type 2 acts with c-Fos to enhance the expression of IL-6 and VEGF-A in colon cancer cells and induce angiogenesis. J Biol Chem 290:22212–22224

Martins WB, Rodrigues SA, Silva HK, Dantas CG, Júnior WDL, Cardoso JC, Gomes MZ (2016) Neuroprotective effect of *Portulaca oleracea* extracts against 6-hydroxydopamine-induced lesion of dopaminergic neurons. An Acad Bras Cienc 88:1439–1450

Martins CG, Appel MH, Coutinho DS, Soares IP, Fischer S, de Oliveira BC, Fachi MM, Pontarolo R, Bonatto SJ, Fernandes LC, Iagher F (2020) Consumption of latex from *Euphorbia tirucalli* L. promotes a reduction of tumor growth and cachexia, and immunomodulation in Walker 256 tumor-bearing rats. J Ethnopharmacol 255:112722

Mishra T, Khullar M, Bhatia A (2011) Anticancer potential of aqueous ethanol seed extract of *Ziziphus mauritiana* against cancer cell lines and Ehrlich ascites carcinoma. Evid Based Complement Alternat Med 2011:765029

Mitra SK, Prakash NS, Sundaram R (2012) Shatavarins (containing Shatavarin IV) with anticancer activity from the roots of *Asparagus racemosus*. Indian J Pharmacol 44(6):732–736

Mohtashami L, Amiri MS, Ramezani M, Emami SA, Simal-Gandara J (2021) The genus *Crocus* L.: a review of ethnobotanical uses, phytochemistry and pharmacology. Ind Crop Prod 171:113923

Mundkar M, Bijalwan A, Soni D, Kumar P (2022) Neuroprotective potential of *Moringa oleifera* mediated by NF-kB/Nrf2/HO-1 signaling pathway: a review. J Food Biochem 46(12):e14451

Nadeem MS, Kazmi I, Ullah I, Muhammad K, Anwar F (2021) Allicin, an antioxidant and neuroprotective agent, ameliorates cognitive impairment. Antioxidants 11(1):87

Nawaz A, Jamal A, Arif A, Parveen Z (2021) In vitro cytotoxic potential of *Solanum nigrum* against human cancer cell lines. Saudi J Biol Sci 28(8):4786–4792

Nayak S, Kumar NA, Hegde A, Kini RD, Blossom V, Poojary R (2021) Neuroprotective role of *Allium cepa* and *Allium sativum* on hippocampus, striatum and cerebral cortex in Wistar rats. Res J Pharm Technol 14(5):2406–2411

Nguyen TT, Parat MO, Shaw PN, Hewavitharana AK, Hodson MP (2016) Traditional aboriginal preparation alters the chemical profile of *Carica papaya* leaves and impacts on cytotoxicity towards human squamous cell carcinoma. PLoS One 11(2):e0147956

Oboh G, Adedayo BC, Adetola MB, Oyeleye IS, Ogunsuyi OB (2022) Characterization and neuroprotective properties of alkaloid extract of *Vernonia amygdalina* Delile in experimental models of Alzheimer's disease. Drug Chem Toxicol 45(2):731–740

Ojo OA, Adeyemo TR, Rotimi D, Batiha GES, Mostafa-Hedeab G, Iyobhebhe ME, Elebiyo TC, Atunwa B, Ojo AB, Lima CMG, Conte-Junior CA (2022) Anticancer properties of curcumin against colorectal cancer: a review. Front Oncol 12:881641

Okubo S, Ohta T, Fujita H, Shoyama Y, Uto T (2021) Costunolide and dehydrocostuslactone from *Saussurea lappa* root inhibit autophagy in hepatocellular carcinoma cells. J Nat Med 75:240–245

Oladele JO, Oyeleke OM, Oladele OT, Olaniyan M (2020) Neuroprotective mechanism of *Vernonia amygdalina* in a rat model of neurodegenerative diseases. Toxicol Rep 7:1223–1232

Olatunde A, Nigam M, Singh RK, Panwar AS, Lasisi A, Alhumaydhi FA et al (2021) Cancer and diabetes: the interlinking metabolic pathways and repurposing actions of antidiabetic drugs. Cancer Cell Int 21:499

Orman E, Bekoe SO, Asare-Nkansah S, Kralisch I, Jato J, Spiegler V, Agyare C, Bekoe EO, Hensel A (2023) Towards the development of analytical monograph specifications for the quality assessment of the medicinal plant *Phyllanthus urinaria*. Phytochemistry 215:113854

Padma TV (2005) Ayurveda. Nature 436(7050):486–486

Palmieri A, Scapoli L, Iapichino A, Mercolini L, Mandrone M, Poli F, Giannì AB, Baserga C, Martinelli M (2019) Berberine and *Tinospora cordifolia* exert a potential anticancer effect on colon cancer cells by acting on specific pathways. Int J Immunopathol Pharmacol 33:2058738419855567

Paniagua-Zambrana NY, Bussmann RW, Romero C (2020) *Portulaca oleracea* L. Portulacaceae. In: Ethnobotany of the Andes. Springer, Cham, pp 1515–1520

Patil MS, Gaikwad D (2011) A critical review on medicinally important oil yielding plant laxmitaru (*Simarouba glauca* DC.). J Pharm Sci Res 3:1195–1213

Paudel P, Park SE, Seong SH, Jung HA, Choi JS (2019) Novel Diels–Alder type adducts from Morus alba root bark targeting human monoamine oxidase and dopaminergic receptors for the management of neurodegenerative diseases. Int J Mol Sci 20(24):6232

Paul S, Roy D, Pati S, Sa G (2021) The adroitness of andrographolide as a natural weapon against colorectal cancer. Front Pharmacol 12:731492

Pavela R, Maggi F, Lupidi G, Mbuntcha H, Woguem V, Womeni HM, Barboni L, Tapondjou LA, Benelli G (2018) *Clausena anisata* and *Dysphania ambrosioides* essential oils: from ethno-medicine to modern uses as effective insecticides. Environ Sci Pollut Res 25:10493–10503

Peerzada AM, Ali HH, Naeem M, Latif M, Bukhari AH, Tanveer A (2015) *Cyperus rotundus* L. traditional uses, phytochemistry, and pharmacological activities. J Ethnopharmacol 174:540–560

Peng Y, Zhou T, Wang S, Bahetjan Y, Li X, Yang X (2022) Dehydrocostus lactone inhibits the proliferation of esophageal cancer cells in vivo and in vitro through ROS-mediated apoptosis and autophagy. Food Chem Toxicol 170:113453

Pooja T (2017) Plants with anticancer properties: a review on traditional plants and herbs are used to evaluation for their anticancer potential. J Pharm Res 11(s):547–553

Popoola JO, Obembe OO (2013) Local knowledge, use pattern and geographical distribution of *Moringa oleifera* Lam. (Moringaceae) in Nigeria. J Ethnopharmacol 150(2):682–691

Prakash O, Usmani S, Gupta A, Jafri A, Ullah MF, Wahab S, Arshad M, Kumar S (2022) Bioactive extracts of *Ziziphus mauritiana* induces apoptosis in A549 human lung epithelial carcinoma cells through the generation of reactive oxygen species. Curr Cancer Ther Rev 18(1):57–68

Pynam H, Dharmesh SM (2018) Antioxidant and anti-inflammatory properties of marmelosin from Bael (*Aegle marmelos* L.); inhibition of TNF-α mediated inflammatory/tumor markers. Biomed Pharmacother 106:98–108

Quang DN, Nanthalath P, Khamko VA, Soulinhong X, Vidavone V (2018) Acemosin-a cytotoxic 20-norsteroid from *Asparagus racemosus*. Fitoterapia 131:221–224

Rahamooz Haghighi S, Asadi MH, Akrami H, Baghizadeh A (2017) Anti-carcinogenic and anti-angiogenic properties of the extracts of *Acorus calamus* on gastric cancer cells. Avicenna J Phytomed 7(2):145–156

Rahmani AH, Khan AA, Aldebasi YH (2017) Saffron (*Crocus sativus*) and its active ingredients: role in the prevention and treatment of disease. Pharm J 9(6):873

Rai SN, Birla H, Singh SS, Zahra W, Patil RR, Jadhav JP, Gedda MR, Singh SP (2017) Mucuna pruriens protects against MPTP intoxicated neuroinflammation in Parkinson's disease through NF-κB/pAKT signaling pathways. Front Aging Neurosci 9:421

Rajalakshmi S, Vijayakumar S, Arulmozhi P (2019a) Ethnobotanical survey of medicinal plants in Thanjavur and its surrounding (Tamil Nadu-India). Acta Ecol Sin 39(5):380–397

Rajalakshmi S, Vijayakumar S, Praseetha PK (2019b) Neuroprotective behaviour of *Phyllanthus emblica* (L) on human neural cell lineage (PC12) against glutamate-induced cytotoxicity. Gene Rep 17:100545

Rapaka D, Bitra VR, Vishala TC, Akula A (2019) *Vitis vinifera* acts as anti-Alzheimer's agent by modulating biochemical parameters implicated in cognition and memory. J Ayurveda Integr Med 10(4):241–247

Rastogi N, Duggal S, Singh SK, Porwal K, Srivastava VK, Maurya R, Bhatt ML, Mishra DP (2015) Proteasome inhibition mediates p53 reactivation and anti-cancer activity of 6-gingerol in cervical cancer cells. Oncotarget 6(41):43310

Razali FN, Sinniah SK, Hussin H, Abidin NZ, Shuib AS (2016) Tumor suppression effect of *Solanum nigrum* polysaccharide fraction on breast cancer via immunomodulation. Int J Biol Macromol 92:185–193

Rodrigues EL, Marcelino G, Silva GT, Figueiredo PS, Garcez WS, Corsino J, Guimarães RCA, Freitas KC (2019) Nutraceutical and medicinal potential of the *Morus* species in metabolic dysfunctions. Int J Mol Sci 20(2):301

Roshanravan N, Ghaffari S (2022) The therapeutic potential of *Crocus sativus* Linn.: a comprehensive narrative review of clinical trials. Phytother Res 36(1):98–111

Saahene RO, Agbo E, Barnes P, Yahaya ES, Amoani B, Nuvor SV, Okyere P (2021) A review: mechanism of *Phyllanthus urinaria* in cancers-NF-κB, P13K/AKT, and MAPKs signaling activation. Evid Based Complement Alternat Med 2021:4514342

Samarghandian S, Hadjzadeh MAR, Afshari JT et al (2014) Antiproliferative activity and induction of apoptotic by ethanolic extract of *Alpinia galanga* rhizhome in human breast carcinoma cell line. BMC Complement Altern Med 14:192

Sandhir R, Khurana M, Singhal NK (2021) Potential benefits of phytochemicals from *Azadirachta indica* against neurological disorders. Neurochem Int 146:105023

Sayuti M, Riwanto I, Boediono BP, Akbar TIS (2020) Anticancer activity of *Phyllanthus niruri* Linn extract in colorectal cancer patients: a phase II clinical trial. Syst Rev Pharm 11(10):313–317

Septembre-Malaterre A, Lalarizo Rakoto M, Marodon C, Bedoui Y, Nakab J, Simon E, Hoarau L, Savriama S, Strasberg D, Guiraud P, Selambarom J (2020) *Artemisia annua*, a traditional plant brought to light. Int J Mol Sci 21(14):4986

Seyed MA, Ayesha S, Azmi N, Al-Rabae FM, Al-Alawy AI, Al-Zahrani OR, Hawsawi Y (2021) The neuroprotective attribution of *Ocimum basilicum*: a review on the prevention and management of neurodegenerative disorders. Futur J Pharm Sci 7:1–14

Shah ZA, Mir R, Matoo JM, Dar MA, Beigh MA (2017) Medicinal importance of saffron: a review. J Pharmacogn Phytochem 6(5):2475–2478

Shakerin Z, Esfandiari E, Razavi S, Alaei H, Ghanadian M, Dashti G (2020) Effects of *Cyperus rotundus* extract on spatial memory impairment and neuronal differentiation in rat model of Alzheimer's disease. Adv Biomed Res 9:17

Sharma R, Garg N, Verma D, Rathi P, Sharma V, Kuca K, Prajapati PK (2021) Indian medicinal plants as drug leads in neurodegenerative disorders. In: Nutraceuticals in brain health and beyond. Academic Press, London, pp 31–45

Shathish K, Guruvayoorappan C (2014) *Decalepis hamiltonii* inhibits tumor progression and metastasis by regulating the inflammatory mediators and nuclear factor κB subunits. Integr Cancer Ther 13(2):141–151

Shimura T, Sharma P, Sharma GG, Banwait JK, Goel A (2021) Enhanced anti-cancer activity of andrographis with oligomeric proanthocyanidins through activation of metabolic and ferroptosis pathways in colorectal cancer. Scientific Reports 11(1):7548

Shivashankarappa A, Sanjay KR, Shah D, Tagat A (2022) *Decalepis hamiltonii* derived gold nanoparticles and photodynamic cytotoxic evaluation on skin melanoma (B16F10) cells as an effective drug delivery vehicle. J Drug Deliv Sci Technol 76:103766

Shonia S, Kanga RS, Ammu KR (2019) Bioactive compounds: natural defense against cancer? Biomol Ther 9:758

Sichaem J, Nguyen H-H, Nguyen V-H, Mac D-H, Mai D-T, Nguyen H-C, Tran T-N-M, Pham N-K-T, Nguyen H-H, Niamnont N, Duong T-H (2019) A new labdane-type diterpenoid from the leaves of *Vitex negundo* L. Nat Prod Res 35(14):2329–2334

Siddique HR, Fatma H, Khan MA (2020) Medicinal properties of saffron with special reference to cancer—a review of preclinical studies. Saffron 233–244

Silihe KK, Mbou WD, Ngo Pambe JC, Kenmogne LV, Maptouom LF, Kemegne Sipping MT, Zingue S, Njamen D (2023) Comparative anticancer effects of *Annona muricata* Linn (Annonaceae) leaves and fruits on DMBA-induced breast cancer in female rats. BMC Complement Med Ther 23(1):234

Simayi J, Nuermaimaiti M, Hailati S, Han M, Reheman Z, Wumaier A, Zhou W (2022) A network pharmacology-based study on the mechanism of dibutyl phthalate of *Ocimum basilicum* L. against Alzheimer's disease through the AKT/GSK-3β pathway. Biomed Res Int 2022:9494548

Singh P, Singh D, Goel RK (2016) Protective effect on phenytoin-induced cognition deficit in pentylenetetrazol kindled mice: a repertoire of *Glycyrrhiza glabra* flavonoid antioxidants. Pharm Biol 54(7):1209–1218

Singh B, Pandey S, Rumman M, Mahdi AA (2020) Neuroprotective effects of *Bacopa monnieri* in Parkinson's disease model. Metab Brain Dis 35:517–525

Singh P, Mishra G, Pottoo FH, Singh B, Zeleke MM (2022) *Zingiber officinale*: its ethanobotanical uses, phytochemistry, and pharmacology. In: Masoodi MH, Rehman MU (eds) Edible plants in health and diseases. Springer, Singapore

Sirin S, Dolanbay SN, Aslim B (2022) Role of plant derived alkaloids as antioxidant agents for neurodegenerative diseases. Health Sci Rev 6:100071

Snellenberg S, Cillessen SA, Van Criekinge W, Bosch L, Meijer CJ, Snijders PJ et al (2014) Methylation-mediated repression of PRDM14 contributes to apoptosis evasion in HPV-positive cancers. Carcinogenesis 35:2611–2618

Sood P, Singh V, Shri R (2023) Morus alba fruit diet ameliorates cognitive deficit in mouse model of streptozotocin-induced memory impairment. Metab Brain Dis 38(5):1657–1669

Soren AD, Yadav AK (2021) Studies on the anthelmintic potentials of the roots of *Asparagus racemosus* willd. (Asparagaceae). Clin Phytosci 7(1):1–8

Sujarwo W, Keim AP, Caneva G, Toniolo C, Nicoletti M (2016) Ethnobotanical uses of neem (*Azadirachta indica* A. Juss.; Meliaceae) leaves in Bali (Indonesia) and the Indian subcontinent in relation with historical background and phytochemical properties. J Ethnopharmacol 189:186–193

Tan TYC, Lim XY, Norahmad NA, Chanthira Kumar H, Teh BP, Lai NM, Syed Mohamed AF (2023) Neurological applications of celery (*Apium graveolens*): a scoping review. Molecules 28(15):5824

Teshika JD, Zakariyyah AM, Zaynab T, Zengin G, Rengasamy KR, Pandian SK, Fawzi MM (2019) Traditional and modern uses of onion bulb (*Allium cepa* L.): a systematic review. Crit Rev Food Sci Nutr 59(sup1):S39–S70

Tian X, Song HS, Cho YM, Park B, Song YJ, Jang S, Kang SC (2017) Anticancer effect of *Saussurea lappa* extract via dual control of apoptosis and autophagy in prostate cancer cells. Medicine 96(30):e7606

Ticktin T, Dalle SP (2005) Medicinal plant use in the practice of midwifery in rural Honduras. J Ethnopharmacol 96(1–2):233–248

Tsoukalas D, Zlatian O, Mitroi M, Renieri E, Tsatsakis A, Izotov BN, Burada F, Sosoi S, Burada E, Buga AM, Rogoveanu I (2021) A novel nutraceutical formulation can improve motor activity and decrease the stress level in a murine model of middle-age animals. J Clin Med 10(4):624

Tugume P, Kakudidi EK, Buyinza M, Namaalwa J, Kamatenesi M, Mucunguzi P, Kalema J (2016) Ethnobotanical survey of medicinal plant species used by communities around Mabira Central Forest Reserve, Uganda. J Ethnobiol Ethnomed 12:1–28

Tundis R, Patra JK, Bonesi M, Das S, Nath R, Das Talukdar A et al (2023) Anti-cancer agent: the labdane diterpenoid-andrographolide. Plants 12(10):1969

Uddin MJ, Zidorn C (2020) Traditional herbal medicines against CNS disorders from Bangladesh. Nat Prod Bioprospect 10(6):377–410

Umair M, Altaf M, Abbasi AM (2017) An ethnobotanical survey of indigenous medicinal plants in Hafizabad district, Punjab-Pakistan. PLoS One 12(6):e0177912

Van Vo G, Nguyen TP, Nguyen HT, Nguyen TT (2022) In silico and in vitro studies on the anti-cancer activity of artemetin, vitexicarpin and penduletin compounds from Vitex negundo. Saudi Pharmaceutical Journal 30(9):1301–1314

Veiga AA, Irioda AC, Mogharbel BF, Bonatto SJR, Souza LM (2022) Quercetin-rich extracts from onions (*Allium cepa*) play potent cytotoxicity on adrenocortical carcinoma cell lines, and quercetin induces important anticancer properties. Pharmaceuticals 15:754

Verma SVS, Varma RKVRK, Singh SSS (2017) Nootropic activity of *Calotropis gigantea* (Linn) root against scopolamine induce amnesia in albino rats. Int J Ind Herbs Drugs 2(4):13–17

Vikas B, Kunjiraman K, Rajam SSN, Anil S (2022) Caspase-dependent apoptosis induced by *Simarouba glauca* on human non-small-cell lung cancer, A549 cells. Asian Pac J Cancer Prev APJCP 23(6):1867

Viswanatha GL, Venkataranganna MV, Prasad NBL, Godavarthi A (2017) *Achyranthes aspera* attenuates epilepsy in experimental animals: possible involvement of GABAergic mechanism. Metab Brain Dis 32:867–879

Wang BN, Wu CB, Chen ZM, Zheng PP, Liu YQ, Xiong J, Xu JY, Li PF, Mamun AA, Ye LB et al (2021) DL-3-n-butylphthalide ameliorates diabetes-associated cognitive decline by enhancing PI3K/Akt signaling and suppressing oxidative stress. Acta Pharmacol Sin 42:347–360

Wenyuan JI, Shaoyin WEI, Wei LIU (2021) Effects of *Urtica dioica* extract on malignant biological behaviors of breast cancer cells and its possible mechanism. Chin J Cancer Biother 28(8)

Wongtawatchai T, Sarsutham K, Sukketsiri W, Tipmanee V, Chonpathompikunlert P (2017) Anti-stress effects of *Apium graveolens* on rats subjected to immobilization. Int Food Res J 24:1490

Wu G, Liu J, Li S, Gao W, Qiu M, Yang C, Ma Y, Song X (2020) Glycyrrhizic acid protects juvenile epileptic rats against hippocampal damage through activation of Sirtuin3. Brain Res Bull 164:98–106

Wu M, Cai J, Fang Z, Li S, Huang Z, Tang Z, Luo Q, Chen H (2022) The composition and anti-aging activities of polyphenol extract from *Phyllanthus emblica* L. fruit. Nutrients 14(4):857

Xia CL, Tang GH, Guo YQ, Xu YK, Huang ZS, Yin S (2019) Mulberry Diels-Alder-type adducts from Morus alba as multi-targeted agents for Alzheimer's disease. Phytochemistry 157:82–91
Xiang X, Wu L, Mao L, Liu Y (2018) Anti-oxidative and anti-apoptotic neuroprotective effects of *Azadirachta indica* in Parkinson-induced functional damage. Mol Med Rep 17(6):7959–7965
Xiaoying W, Han Z, Yu W (2017) Glycyrrhiza glabra (Licorice): ethnobotany and health benefits. In: Sustained energy for enhanced human functions and activity. Academic Press, London, pp 231–250
Xu C, Zhang H, Mu L, Yang X (2020) Artemisinins as anticancer drugs: novel therapeutic approaches, molecular mechanisms, and clinical trials. Front Pharmacol 11:529881
Yabesh JM, Prabhu S, Vijayakumar S (2014) An ethnobotanical study of medicinal plants used by traditional healers in silent valley of Kerala, India. J Ethnopharmacol 154(3):774–789
Yadav S, Nair N, Biharee A, Prathap VM, Majeed J (2022) Updated ethnobotanical notes, phytochemistry and phytopharmacology of plants belonging to the genus Morus (Family: Moraceae). Phytomed Plus 2(1):100120
Yang W, Qiu X, Wu Q, Chang F, Zhou T, Zhou M, Pei J (2023) Active constituents of saffron (*Crocus sativus* L.) and their prospects in treating neurodegenerative diseases. Exp Ther Med 25(5):1–14
Yaniv Z, Koltai H (2018) *Calotropis procera*, apple of Sodom: ethnobotanical review and medicinal activities. Israel J Plant Sci 65(1–2):55–61
Ye X, Rong Z, Li Y, Wang X, Cheng B, Cheng Y, Luo H, Ti Y, Huang X, Liu Z et al (2018) Protective role of L-3-n-butylphthalide in cognitive function and dysthymic disorders in mouse with chronic epilepsy. Front Pharmacol 9:734
Yoneyama T, Arai MA, Akamine R, Koryudzu K, Tsuchiya A, Sadhu SK, Ahmed F, Itoh M, Okamoto R, Ishibashi M (2017) Notch inhibitors from *Calotropis gigantea* that induce neuronal differentiation of neural stem cells. J Nat Prod 80(9):2453–2461
Zahidin NS, Saidin S, Zulkifli RM, Muhamad II, Ya'akob H, Nur H (2017) A review of *Acalypha indica* L. (Euphorbiaceae) as traditional medicinal plant and its therapeutic potential. J Ethnopharmacol 207:146–173
Zheng X, Xu Y, Liu B, Qi Y, Zhang Y, Li X, Zhang X, Pu X, Li S, Chen Z, Wan C (2019) Ethyl acetate extract of *Asclepias curassavica* induced apoptosis in human cancer cells via activating p38 and JNK MAPK signaling pathways. Evid Based Complement Alternat Med 2019:9076269
Zulfugarova P, Zivari-Ghader T, Maharramova S, Ahmadian E, Eftekhari A, Khalilov R, Turksoy VA, Rosić G, Selakovic D (2023) A mechanistic review of pharmacological activities of homeopathic medicine licorice against neural diseases. Front Neurosci 17:1148258

Biomarkers from Medicinal Plants

Amit Kumar Pradhan, Umakanta Chowra, Manabendra Nath, Saurov Jyoti Roy, Bikash Kalita, Bikash Kundu, Jinu Devi Rajkumari, and Bhaben Tanti

Contents

A. K. Pradhan
Department of Botany, Pragjyotish College, Guwahati, Assam, India
e-mail: akpradhan@pragjyotishcollege.ac.in

U. Chowra
Department of Botany, Guwahati College, Guwahati, Assam, India

M. Nath · B. Kalita · B. Kundu · B. Tanti (✉)
Department of Botany, Gauhati University, Guwahati, Assam, India
e-mail: btanti@gauhati.ac.in

S. J. Roy
Department of Botany, Rangia College, Rangia, Assam, India

J. D. Rajkumari
Department of Botany, Cotton University, Guwahati, Assam, India

© The Author(s), under exclusive license to Springer Nature Singapore Pte Ltd. 2024

A. Das Talukdar et al. (eds.), *Traditional Resources and Tools for Modern Drug Discovery*, Interdisciplinary Biotechnological Advances,
https://doi.org/10.1007/978-981-97-4600-2_8

Abstract With the growing interest in natural and holistic approaches to healthcare, the exploration of biomarkers derived from medicinal plants have been received too much attention. Biomarkers sourced from therapeutic plants comprise a broad range of molecular and biochemical indicators, including proteins, genetic markers, and secondary metabolites, that can play pivotal roles in multiple facets of healthcare. These biomarkers offer insights into the therapeutic properties, safety, and efficacy of plant-derived compounds, aiding in the identification of promising phytochemicals for drug development. However, the discovery of biomarker in medicinal plants undergoes major issues related to variability due to environmental factors and the need for robust validation. Numerous biomarkers remain unvalidated because of overlapping signals, with typically only the most intense signals being selected to represent the target metabolites, leaving the less intense ones overlooked. Moreover, there's uncertainty regarding whether the chosen biomarkers genuinely reflect typical characteristics. Thus, analysis metabolite profiles and selecting specific biomarkers continues to pose significant challenges. Modern statistical techniques like meta-analysis and computational systems biology, along with next-generation sequencing technology, have opened up fresh possibilities for overcoming biological complexity. To elucidate the pivotal molecular determinants involved in secondary metabolite biosynthesis, researchers have conducted meta-analyses on publicly accessible transcriptomics datasets derived from medicinal plants with significant economic value. These analyses aim to identify differentially expressed genes in shoot and root tissues, shedding light on the underlying mechanisms governing secondary metabolite production. This study has taken into account the recent intervention of contemporary omics technology, which provides an overview of the developing field of biomarker discovery in medicinal plants and highlights its possible ramifications for drug development, personalized medicine, and the broader understanding of plant-based therapeutics.

Keywords Biomarkers · Metabolites · Drug development · Therapeutic · Phytocompounds

1 Introduction

Throughout ancient times, people have utilized a wide variety of medicinal plants to improve health and cure ailments. Plants are truly amazing gifts from the natural world that have long been used as the foundation of conventional medicine to treat a wide range of illnesses and support human health (Rupani and Chavez 2018). However, the pervasive issue of adulteration and the substitution of genuine materials with spurious alternatives have emerged as significant concerns for both users and industry stakeholders. The imperative for authenticating medicinal plants has consequently risen to the forefront, driven by the paramount goals of ensuring safety and efficacy in traditional and modern medicinal practices (Willcox and Bodeker 2004; WHO 2013). With the recent upliftment in scientific technology, biomarkers have been used as potent indicators for the identification and prognosis of variable biological, pathogenic, or pharmacological processes. Molecular advancement with the incorporation of omics technology underpins traditional practices into a fascinating frontier towards the exploration of biomarkers from medicinal plants. Biomarkers act as chemical messengers, providing quantifiable and measurable indicators of a plant's biological state, its response to environmental stimuli, and its potential impact on human health. They include a wide variety of molecules, including proteins, enzymes, gene transcripts, and secondary metabolites like terpenes, alkaloids, and flavonoids. They help identify the precise bioactive substances that give rise to a plant's therapeutic qualities by translating the language contained in the putative biomarkers, which facilitates the creation of standardized and focused herbal formulations (Aina et al. 2024; Qiu et al. 2023). They are specified to perform as crucial markers in ensuring the authenticity, purity, and potency of herbal products, safeguarding consumer safety and efficacy. By trailing the impact of environmental factors on biomarker profiles, it also promotes sustainable harvesting practices and conservation efforts (Wang et al. 2023). Biomarkers harbour immense potential for tailoring herbal remedies to individual needs based on their unique genetic and metabolic profiles, which offer a powerful lens into understanding the intricate mechanisms by which plants exert their therapeutic effects (Goetz and Schork 2018) (Fig. 1).

Approximately 70,000 different medicinal plants are used in various systems; India alone is home to more than 2000 of these plants. Medicinal plants play a significant role in alternative medicine by offering all-natural cures for a wide range of illnesses. Since medicinal plants have diverse genetic compositions and produce a range of secondary metabolites, they have been utilized for medical research and treatment since ancient times (Pradhan et al. 2020). Remarkable chemical and biological properties can be found in the large reserves of secondary metabolites found in medicinal plants. The identification of particular metabolites in medicinal plants can yield important insights into the therapeutic qualities of the plant and contribute to the establishment of its medicinal efficacy. The collaborative application of metabolomics with transcriptomics and proteomics exhibits broad applicability across various domains, including chemotaxonomy, nutrition, pharmacology, and

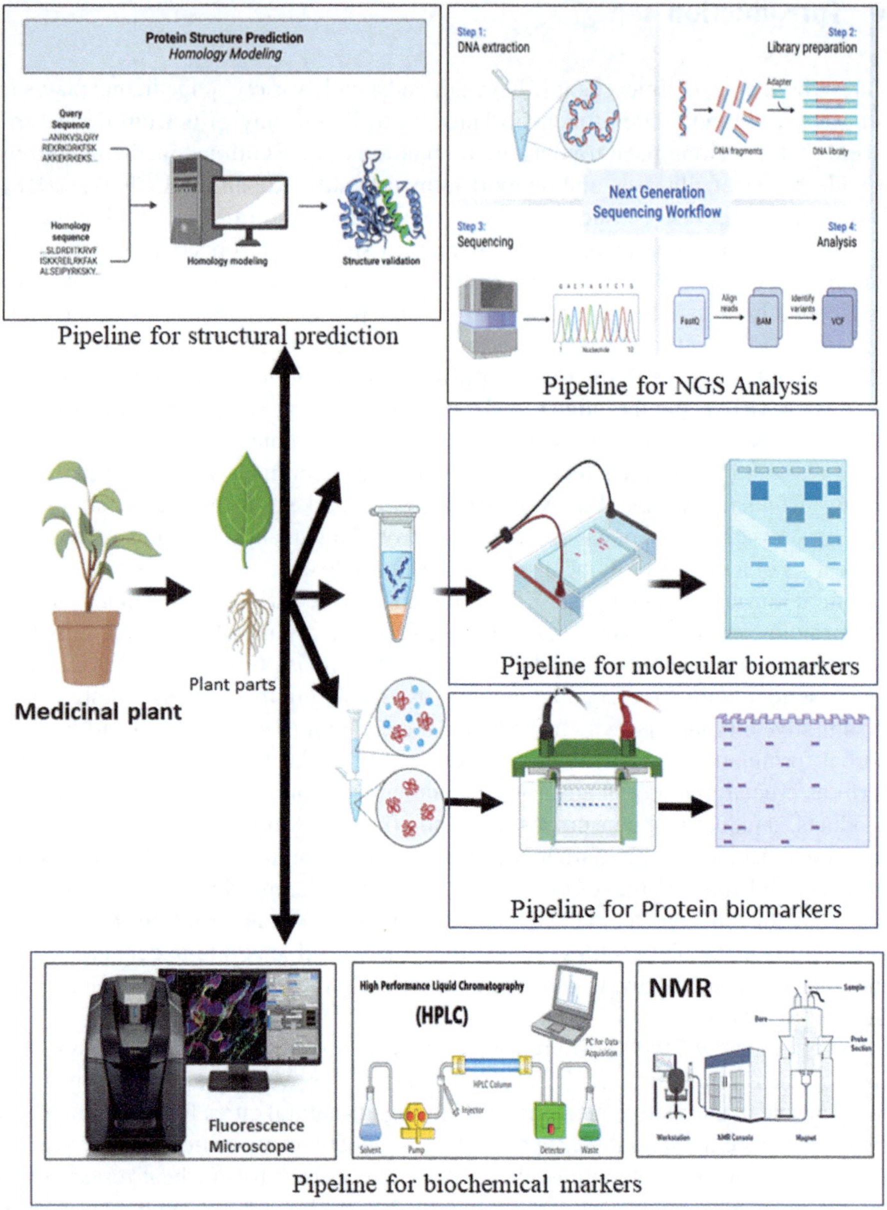

Fig. 1 Schematic representation to the application of advanced scientific techniques in identification and utilization of biomarkers for medicinal plants

personalized medicine (Elshafie et al. 2023). By integrating metabolomics into plant science research, scientists can delve deeper into the intricate biochemical pathways and metabolic interactions shaping the plant kingdom (Marchev et al.

2021). Additionally, explorations into the differences among wild, cultivated, and engineered plants provide valuable insights into the future of crop production (Pradhan et al. 2020) clarification of biosynthesis pathways, exploration of correlations between metabolic markers and the biological activity of medicinal plants, quality assurance in herbal preparations, and more (Marchev et al. 2020). The synthesis of a wide range of metabolites controlled by various genes and transcription factors has been made feasible in recent years by sophisticated high-throughput analytical tools like computational meta-analysis and next-generation sequencing (NGS) technology. These instruments are employed in plant metabolomics and metabolomics research, and they are also used in metabolite profiling (Pradhan et al. 2019). A range of secondary metabolites found in plants, including phenolics, alkaloids, saponins, tannins, and terpenes, are frequently investigated for their pharmacological effects and can function as biomarkers for the therapeutic potential of plants (Salam et al. 2023). The diverse chemical structures and biological activities of secondary metabolites in medicinal plants make them valuable biomarkers (Hussein and El-Anssary 2019; Seca and Pinto 2019).

This study delves into the exciting world of biomarkers from medicinal plants. The diverse array of biomarkers, their significance in therapeutic and clinical aspects of medical research, and the cutting-edge analytical techniques are employed to unlock their secrets. Additionally, this developing field's future directions and challenges are explored, emphasizing how it can transform many facets of comprehending and utilizing nature's pharmacy's healing power.

2 Metabolites as a Biomarker for Medicinal Plants and Their Efficiency in Drug Development

The plant kingdom is estimated to contain over 200,000 metabolites, encompassing primary metabolites essential for plant life maintenance and secondary metabolites that function in plant defence against stress conditions (Roy et al. 2024; Salam et al. 2023). Many Asian countries utilize medicinal herbs, rich in metabolites, for treating diseases, attributing the therapeutic effects to these compounds (Wink 2015). Small molecules called metabolites are produced during physiological processes and are crucial in defining particular phenotypic features. Although primary metabolites cannot directly function as traditional biomarkers, specific metabolites within these pathways can indicate specific physiological or pathological conditions (Zhong et al. 2023). Primary metabolites include amino acids, proteins, nucleic acids, carbohydrates, and polysaccharides, which are ubiquitous in all functioning cells and contribute to growth and evolution (Salam et al. 2023).

Originating from primary metabolic pathways, secondary metabolites, which do not involved in growth, are extensively used in traditional medicine due to their diverse biological effects. Terpenoids, which is known as the biggest course of plant metabolites with over 40,000 compounds, and phenolics, comprising about 8000 compounds, are noticeable among secondary metabolites (Elshafie et al. 2023; Pradhan et al. 2023). Plant secondary metabolites have many pharmacological activities, including antibacterial, anthelmintic, anticoagulant, stimulant, analgesic, anti-acne, and antioxidant properties (Zandavar and Babazad 2023). Medicinal plants which are in the same genus have been found similarities in their morphology and genetic DNA sequences along with chemical metabolism, due to comparable metabolic pathways (Fig. 1). Although the metabolites found in medicinal herbs from different genera are often diverse, those within the same genus tend to have significant similarities (Mahanta et al. 2022).

2.1 Alkaloids

Alkaloids are common nitrogenous essential organic compounds that exist widely in many kinds of plants, and they often possess pharmacological activities that can be beneficial for medicinal purposes. Alkaloids, constituting around 20% of plant-based secondary metabolites, are low-molecular-weight compounds, with approximately 12,000 variants identified across various plant genera (Dey et al. 2020). Notably, certain alkaloids hold significant therapeutic importance, such as reserpine, recognized for its antihypertensive properties in *Rauwolfia serpentina* (Sagi et al. 2016), and vinblastine and vincristine found in *Catharanthus roseus* (El-Sayed and Verpoorte 2007).

Beyond their well-known applications as anaesthetics, cardioprotective agents, and anti-inflammatory substances, alkaloids exhibit diverse pharmacological activities. Many possess antimicrobial, antifungal, CNS-stimulant, and antiviral properties. In the realm of medicinal plants, alkaloids play a crucial role as biomarkers, serving as measurable indicators to assess the presence or quantity of specific substances (Thawabteh et al. 2019). Alkaloids are categorized based on their plant sources, such as opium, vinca, belladonna, *Cinchona*, and ergot alkaloids, or according to their medicinal properties, including analgesic, stimulant, or anti-malarial alkaloids (Bhambhani et al. 2021). Among the alkaloids, reserpine, obtained from *Rauwolfia serpentine* (Sagi et al. 2016), outstands prominently for its pharmacological significance as antihypertensive and antitumor agents, respectively. Furthermore, a number of other alkaloids also exhibit noteworthy pharmacological activities encompassing antimicrobial, antifungal, CNS-stimulant, and antiviral properties. Several alkaloids that are frequently employed as biomarkers in medicinal plants are listed below:

2.1.1 Quinine

It is an alkaloid found in *Cinchona* that is utilized as a biomarker for anti-malarial activity. It is a member of the aryl amino alcohol class of medications (Ceravolo et al. 2021). During the eighteenth and nineteenth centuries, *Cinchona* bark and its derivative quinine gained widespread acceptance and replaced powdered bark as the standard treatment for intermittent fevers worldwide.

2.1.2 Morphine

It is isolated from *Papaver somniferum*, also called opium poppy. This well-known alkaloid has analgesic (pain-relieving) qualities, making it a useful biomarker (Vadhel et al. 2023).

2.1.3 Vinblastine and Vincristine

These alkaloids are utilized as biomarkers for their anticancer properties and are extracted from the leaves of the Madagascar periwinkle (*Catharanthus roseus*) (El-Sayed and Verpoorte 2007). Vinblastine and vincristine medications exhibit potent preventive effects on various types of cancer forms in different organs like the breast, liver, ovarian, head and neck, and testicular, solid sarcoma, and malignant melanoma, whether originating spontaneously or through transplantation and including genetic disorder like leukaemia (Skubnik et al. 2021).

2.1.4 Atropine and Scopolamine

Atropine isolated from dried leaves and flower tops of *Datura metel*. Scopolamine is another alkaloid found in plants that belong to Solanaceae family. Plants such as henbane (*Hyoscyamus niger*), angel's trumpets (*Brugmansia*), jimson weed (*Datura*), mandrake (*Mandragora officinarum*), deadly nightshade (*Atropa belladonna*) and corkwood (*Duboisia*). These alkaloids are used as biomarkers for their antimuscarinic effects (O'Connor 2012).

2.1.5 Ephedrine and Pseudoephedrine

These are alkaloids naturally found in plants like the ephedra herb, a conventional therapeutic plant with a history traversing thousands of years. It has been demonstrated that these alkaloids are useful in treating both moderate and severe cases of asthma. They have dilator properties for the bronchial tubes, stimulate the heart and central nervous system, and raise blood pressure. Leveraging their bronchodilator and decongestant properties, these alkaloids serve as valuable biomarkers in various applications. These alkaloids serve as imperative pointers within the distinguishing proof and standardization of therapeutic plants (Tang et al. 2023).

2.1.6 Caffeine

More than 60 plant species' leaves, fruits, and seeds contain the alkaloid caffeine, which has a potent psychoactive component and is the most widely consumed drug by people globally (Kim et al. 2019; Zulak et al. 2006). The main purpose of the caffeine that is extracted from *Coffea canephora* is to treat brain disorders as a stimulant medication (DePaula and Farah 2019).

2.2 Phenolics

The presence of phenolic compounds in medicinal plants can make them valuable biomarkers with broad applications in plant identification and medicinal plant quality evaluation. Chemical compounds with a hydroxyl group and one or more aromatic rings are known as phenolics (Sun and Shahrajabian 2023). This diverse group includes more than 8000 commonly classified into phenolic acids and analogues, flavonoids, stilbenes, tannins, coumarins, lignans, quinones, and curcuminoids (Cosme et al. 2020). The presence and quantity of specific phenolic compounds can be used for chemotaxonomic purposes, aiding in the identification and classification of medicinal plant species. Among them, phenolic acids constitute a major group of phenolic compounds widely present in the plant kingdom (Cai et al. 2004). Different species or varieties of plants may contain unique phenolic profiles, helping to ensure the accurate identification of plant material for medicinal use (Shan et al. 2005). To illustrate, gallic acid is widely present in medicinal herbs like *Barringtonia racemosa*, *Cornus officinalis*, *Cassia auriculata*, *Polygonum aviculare*, *Punica granatum*, *Rheum officinale*, *Rhus chinensis*, *Sanguisorba officinalis*, and *Terminalia chebula*. It also serves as a biomarker for these species, along with being found in dietary spices like thyme and clove.

For example, many medicinal herbs, such as *Cassia auriculata*, *Sanguisorba officinalis*, *Barringtonia racemosa*, *Rheum officinale*, *Polygonum aviculare*, *Rhus chinensis*, and *Terminalia chebula*, contain a significant amount of gallic acid. It also serves as a biomarker for these species, along with being found in dietary spices like thyme and clove (Sun and Shahrajabian 2023). Caffeic acid has been identified in ginger (*Zingiber officinale* Rosc.), sage (*Salvia officinalis* L.), and parsley (*Petroselinum crispum* L.).

In contrast, coumaric acid which is present in basil (*Ocimum basilicum* L.), thyme (*Thymus vulgaris* L.), and oregano (*Origanum vulgare* L.) (De la Rosa et al. 2019; Huang et al. 2009). The biomarkers caffeic acid, chlorogenic acid, and 3,5-caffeoyl quinic acid distinguish *Artemisia afra* and *Artemisia annua* (Xiao et al. 2022). Epigallocatechin gallate (EGCG), a well-studied flavonol, acts as a potential biomarker in green tea (*Camellia sinensis*) and grapes (*Vitis vinifera*) due to its rich phytochemical content (Zhao et al. 2022).

Certain phenolic compounds with well-established therapeutic effects are responsible for many plants' medicinal qualities. These compounds can act as biomarkers, revealing information about a plant's biological activity and possible health benefits. Capsaicinoids, which are common phenols found in *Capsicum* spp., are responsible for rubefacient, circulatory stimulant, and analgesic activity (Spille et al. 2008). Meanwhile, *Selaginella tamariscina*, also known as resurrection plant, contains active ingredients such as aminoflavones, hinoki flavones, isocryptomerin, and hevea flavones, which exhibit antibacterial and antifungal properties (Gechev et al. 2014).

2.3 Flavonoids

Flavonoids are plant secondary metabolites that contain a ring called benzopyrone with phenolic or polyphenolic groups located at different sites (Smith et al. 2023). These substances are found in large quantities in leaves, fruits, vegetables, and plants. They have potential uses in medicinal chemistry. Apart from their cardioprotective and neuroprotective properties, flavonoids also possess antiviral, anti-inflammatory, antioxidant, and anticancer properties. The kind, possible mechanism of action, and bioavailability of flavonoids determine the particular biological activities they display. Flavonoids are helpful biomarkers that can be applied to the identification and description of medicinal plants. The identity, quality, and possible therapeutic benefits of a plant can be ascertained by analysing the concentration and presence of particular flavonoids in plant extracts. In this sense, flavonoids can act as chemical markers or fingerprints for the identification and standardization of medicinal plants. It is crucial to remember that using flavonoids as biomarkers for medicinal plants necessitates a thorough examination and comprehension of the unique flavonoid profile of each species of plant (Ullah et al. 2020).

The most prevalent subclasses of flavonoids in nature include is oflavones, flavanonols, anthocyanins, flavones, flavonols, and flavanones. Plant flavonoids and other phenolics have a variety of benefits for human health, including the promotion of the immune system, the prevention of cancer, the treatment of diseases, the prevention of infections, and the protection of the skin from UV radiation (Sun and Shahrajabian 2023). Most prevalent flavones are found mostly in the aerial portions of *Artemisia annua*, inflorescences of *Chrysanthemum* and morifolium, and the roots of *Scutellaria baicalensis*. Flavonoid in *Cytisus multiflorus* is the derivative of chrysin; in *Malva sylvestris*, the main flavonoid is kaempferol-3-*O*-rutinoside; and in *Sambucus nigra*, the principal flavonoid is quercetin-3-*O*-rutinoside. *Carthamus tinctorius*, *Rhus verniciflua*, and *Caesalpinia sappan* are among the medicinal herbs that contain chalcones (phloretin, carthamin, sappan, chalcone, butein, etc.) (Cai et al. 2004). *Ginkgo biloba* is a well-known herbal remedy used for cognitive health, which contains flavonoids such as quercetin and kaempferol (Liu et al. 2015). Citrus fruits like oranges, lemons, and grapefruits are rich in flavonoids such as hesperidin and naringin (Alam et al. 2014).

2.4 Tannins

Tannins are naturally occurring polyphenolic compounds that range in molecular weight from 500 to 4000. They are soluble in water. Dietary plants and medicinal herbs contain a broad class of polyphenolics called tannins. In medicinal plants, tannins can serve as biomarkers or indicators of specific bioactive compounds that contribute to the plant's therapeutic properties (Miele et al. 2019). Gallotannins, ellagitannins, and proanthocyanidins are some important tannin that is found as potential biomarkers in some medicinal plants. *Glycyrrhiza glabra*, *Euphorbia hirta*, and *Rhus succedanea* have all been found to contain certain gallotannins. The peels of *P. granatum* and the fruits of *T. chebula* were found to contain several ellagitannins. Proanthocyanidins and leucoanthocyanidins were present in *Camellia sinensis* and *Areca catechu*, respectively (Huang et al. 2009). The traditional Chinese medicinal plants that have been tested and linked to anticancer properties include over 32 species that contain tannin constituents, such as gallotannins. Over 32 species of medicinal plants used in traditional Chinese medicine have been tested and found to contain tannin constituents such as ellagitannins, proanthocyanidins, and gallotannins, which are linked to anticancer properties (Cai et al. 2004). Ginsenosides are frequently employed as biomarkers to assess *Panax ginseng* quality. Chinese white ginseng (CWG), which is rich in ginsenoside Ro and chikusetsusaponin Iva, can be distinguished from Korean white ginseng (KWG) using biomarkers such as ginsenoside Rf and notoginsenoside R3 isomer (Chen et al. 2019).

2.5 Saponins

Plant saponins are a broad and varied class of secondary metabolites present in many different kinds of plants. Amphipathic compounds known as saponins are composed of a hydrophilic sugar moiety and a hydrophobic aglycone known as sapogenin. The biosynthesis of saponins is a complex process involving the mevalonate pathway for triterpenoid saponins and the mevalonate-independent pathway for steroid saponins (Moses et al. 2014). Numerous enzymes are essential to these metabolic processes which includes cytochrome P450 monooxygenases and glycosyltransferases (Haralampidis et al. 2002; Lizhu et al. 2016). Their defence systems against infections and herbivores are among their main functions. Cell lysis can result from saponin disruption of cell membranes, which can also form compounds with sterols. Plants are shielded from microbial invasions and herbivore feeding by this deterrent strategy (Sawai and Saito 2011; Mugford and Osbourn 2013). Because of their expectorant and emetic qualities, plants containing saponins have been employed in traditional medicine. Furthermore, saponins have the potential to function as allelopathic substances, impacting plant-plant interactions by preventing adjacent competing species from proliferating (Mugford and Osbourn 2013).

2.6 *Terpenes*

Terpenes are essential to plants defence mechanisms, interspecies signalling and a host of physiological processes (Toffolatti et al. 2021). Terpenes are divided into three groups based on the quantity of isoprene units they contain: diterpenes (which have four isoprene units), sesquiterpenes (which have three isoprene units), and monoterpenes (which have two isoprene units) (Pattanaik and Lindberg 2015). The sheer variety of terpenes results in a broad spectrum of aromatic profiles and functional properties. For instance, monoterpenes are often responsible for the characteristic scents of many essential oils, while sesquiterpenes contribute to the distinct aroma of plants like sage (*Salvia officinalis*) (Masyita et al. 2022) and chamomile (*Chamaemelum nobile*). The biosynthesis of terpenes occurs via the methylerythritol phosphate (MEP) pathway, also known as the non-mevalonate pathway (Bergman et al. 2019; Zhao et al. 2013). Terpenes have a variety of ecological functions, including defence mechanisms and interactions between plants and other abiotic stressors (Singh and Sharma 2015). The essential oils that are synthesized by terpenes have the ability to function as chemical defenxes against infections and herbivores. Certain terpenes possess anti-inflammatory, analgesic, and antimicrobial qualities, while others could repel or poison herbivores (Sharifi-Rad et al. 2017; Masyita et al. 2022).

3 Analytical Techniques Utilized in Identification of Metabolites to Be Used as Biomarkers

Novel metabolic biomarkers have been identified and characterized in recent decades due to technological advancements in metabolomics and metabolic profiling (Liu et al. 2021). The complete metabolomics profile of an organism can be simply provided by employing a number of different techniques. Utilizing high-throughput techniques, the primary objective of metricomics is to identify and measure every small molecule metabolite, both internal and external, within an organism (Chen et al. 2022). High-resolution mass spectrometry (HRMS) and nuclear magnetic resonance (NMR) spectroscopy are currently the two primary analytical methods utilized in the field of metabolomics. HRMS is frequently utilized in conjunction with the separation methods that are currently accessible, such as capillary electrophoresis (CE), gas chromatography (GC), and liquid chromatography (LC) (Pezzatti et al. 2020). The most widely used method among them all is definitely LC-HRMS, which normally uses an electrospray ionization (ESI) source (Table 1). The primary reason for this is that the ESI has a gentler ionization process than NMR analysis, making it easier to obtain structural information. It is also more sensitive than NMR analysis (Jaeger and Lisec 2022). These approaches, however, might use a range of analytical techniques, each with particular benefits and limitations. Therefore, no single analytical method can currently measure and identify every metabolite in a

Table 1 Metabolites as biomarkers with therapeutic application and techniques utilized for analysis

Sl. no.	Metabolites	Application as biomarkers	Techniques used for analysis	References
1	Carnosine	Antioxidant properties	HPLC	Park et al. (2005)
2	Proline betaine	Secondary metabolite	1H NMR	Heinzmann et al. (2010)
3	Sphingomyelins	A substantial part of the plasma and lens membranes in the eyes. It also contributes to brain myelination and cellular signalling	ESI-MS-MS	Altmaier et al. (2009), Schneider et al. (2019)
4	4-*O*-Methylgallic acid	Suppresses inflammation	HPLC-ESI-MS-MS	Mennen et al. (2006), Na et al. (2006)
5	Long-chain acylcarnitine's	Diagnostic biomarkers for Parkinson's disease	CE–TOF/MS	Saiki et al. (2017)
6	N8-acetylspermidine (N8AS)	Regulates cardiomyocyte cell death and cardiac dysfunction during ischemic injury	LC–FT spectroscopy MS	Nayak et al. (2020)
7	Total plasma tau	Biomarker for predicting dementia	1H NMR	Yilmaz et al. (2017), Pase et al. (2019)
8	Naringenin	Anti-diabetic, anti-inflammatory, hepatoprotective, nephroprotective, immunomodulatory, and antioxidant properties	HPLC-ESI-MS-MS	Mennen et al. (2006), Salehi et al. (2019)
9	Cotinine	Cotinine is a metabolite of nicotine that can be used as a biomarker for cigarette smoking exposure	High-performance liquid chromatography (HPLC)	Nan et al. (2001), Babhadiashar et al. (2014)
10	Lysophosphatidic acid	LPA enhances the tumorigenic attributes of cancer cells	Mass spectrometry	Sutphen et al. (2004), Chae et al. (2022)

single sample at the same time; as result the comprehensive metabolomic data must be evaluated by combining data from several platforms (Chen et al. 2022).

3.1 Nuclear Magnetic Resonance (NMR) Spectroscopy

Nuclear magnetic resonance (NMR) spectroscopy is a powerful tool for studying metabolites because it can yield quantitative, qualitative, and non-destructive data on a wide range of metabolites in a complex biological material. By examining the chemical shift of atoms in a molecule, this method provides structural information about metabolites and facilitates the identification and characterization of specific biomarkers. This spectroscopy technique involves subjecting a chemical to a magnetic field. Isomerization of the radiation is absorbed by the compound's isotopes, including ^{1}H, ^{13}C, ^{14}N, ^{15}N, and ^{17}O, which resonate at different frequencies depending on their location within the small molecule. The resulting NMR spectrum, which is a collection of peaks at different locations and intensities, shows a unique pattern for each molecule. An internal reference is used to compare the reported NMR chemical shifts (Nagana Gowda and Raftery 2022). Since NMR can reveal details about the more prevalent but challenging to identify metabolites, it can be viewed as an adjunct to GC and LC-MS methods. The primary disadvantage of NMR is that it is far less sensitive than MS (Emwas et al. 2019).

3.2 Mass Spectrometry

The utilization of mass spectrometry (MS) methods is a fundamental aspect of metabolite analysis for biomarker development. It is the most versatile, sensitive, and specific method available for detecting and quantifying a broad range of metabolites found in living organisms. Determining the mass-to-charge (m/z) ratios of the ions that remain after fragmenting the ionized parent chemical is the primary use of mass spectrometry. With data collection software, a full scan mass spectrum derived from target analyses can be obtained. Spectral libraries provide highly certain positive identification of target biomolecules. The advantages of MS-based methods include their connection to GC (gas chromatography) or LC (liquid chromatography), high sensitivity, speed, and excellent selectivity (Zhang et al. 2020).

4 Proteins as Biomarker for Medicinal Plants

Proteins are vital biomarkers found in medicinal plants that provide important information about the physiological reactions, biochemical activity, and therapeutic potential of these plants. They play diverse role, ranging from defence mechanisms against pathogens to the synthesis of bioactive compounds with medicinal properties. By analysing the expression, localization, and function of proteins in medicinal plants, promising candidates for drug discovery can be identified by researchers, optimize cultivation practices, and monitor plant responses to environmental stresses. Protein biomarkers can be applied to medicinal plants and herbal products to ensure quality control and authenticity. The identity, purity, and authenticity of medicinal plant samples can be confirmed, adulteration or contamination with foreign substances can be detected, and the consistency and effectiveness of herbal remedies can be guaranteed by researchers through analysing specific proteins' or protein profiles' abundance and composition. Protein-based assays can be used for target protein quantification and biomarker validation in medicinal plant extracts or formulations. Examples of these assays are mass spectrometry-based proteomics and enzyme-linked immunosorbent assays (ELISA) (Fig. 1).

Proteomics is used to examine proteins that are essential for cellular activity, revealing the entirety of expressed proteins in specific organisms or cell types at specific conditions. In addition to genome sequencing, many biological inquiries persist, as protein expression, localization, and activity fluctuate across conditions. The study of protein expression in diverse cellular contexts is crucial for decoding their biological significance. Proteomic analysis provides comprehensive assessment of cellular activities in clinical research, offering a comprehensive view of cellular dynamics across various diseases (Regon et al. 2022).

Plant proteomic analyses are currently applied to various species through protein characterization by MS and relationship assessments regarding biological plant functions. Originally, these analyses were conducted in the *Oryza* and *Arabidopsis* genera.

For instance, Nogueira et al. (2007) assessed *Vigna unguiculata* defence protein changes related to pathogenesis using 2DE-Maldi-TOF, while Datta et al. (2013) verified the comparable trial conducted on *Arabidopsis thaliana*. Protein identification has also been done using 2DE-LC-MS/MS in *Araucaria angustifolia* and *Manihot esculenta*, respectively, that are linked to metabolism, development, and biosynthesis, as well as those that control energetic metabolic pathways, translation processes, oxidative stress regulation, and cell signalling (Li et al. 2010; Balbuena et al. 2011).

Numerous plants have been studied for possible medical applications, especially in the management of chronic illnesses like diabetes, with the goal of developing more widely available forms of care. Within this framework, the purpose of this investigation was to characterize two plant proteins, "abajeru" (*Chrysobalanus icaco*) and "cow's paw" (*Bauhinia variegata* and *Bauhinia forficata*), that are utilized in traditional medicine as hypoglycaemic agents for the management of diabetes. Characterization of the proteome allowed for species differentiation. Trypsin was used to digest proteins in-solution through the filter-assisted sample preparation (FASP) technique. The next stage of the peptide analysis involved using liquid chromatography-tandem mass spectrometry (LC-MS/MS) to characterize the proteins. 131 proteins in total were found. Cellular respiration, transport, photosynthesis, and metabolism were the primary biological activities of these proteins.

Although proteins involved in regulating glucose levels were discovered, insulin-like proteins were not identified. The results are valuable for the proteomic characterization of phytotherapeutic plants and may serve as a guide for the assessed species in Brazil, where there seems to be a paucity of data in this field.

These methods have opened up new directions for the study of ethnobotany and phytomedicine, including the identification of species and biomarkers that point to the potential use of medicinal plants in the management of specific chronic illnesses. Medicinal plants with hypoglycaemic potential have been included in studies examining the therapeutic potential of ethanol-based medicinal plant extracts, like *Bauhinia forficate*, *Chrysobalanus icaco*, and *Bauhinia variegate*, which could be employed to treat diabetes (Salgueiro et al. 2016; Azevedo et al. 2006; White et al. 2016). Proteins linked to insulin may be functionally related to dependent pathways because protein-protein interactions are a sign of a functional partnership. Insulin-associated fragments are protein components that contain amino acid sequences that function as biomarkers. These sequences can be used by mass spectrometry to identify plants that show hypoglycaemic potential (Altan and Unal 2014; Hardikar et al. 2016; Stadlbauer et al. 2016). The possibility of other hypoglycaemic action mechanisms, such as new compounds or connections, cannot be completely ruled out. The distinct plant proteins from *C. icaco* were used to identify the enriched KEGG pathways, which showed a strong correlation with the biological process-related GO terms. According to the identified GO terms, the majority of the pathways were found to be connected to metabolism. Moreover, reactions to particular kinds of stress, like exposure to chemicals or abiotic stimuli, were linked to these pathways and GO terms. Pedrete et al. (2019) carried out identifying proteins with mass spectrometry on the assessed plants, producing intriguing results and a significant amount of identified proteins. Only a smaller percentage of these proteins were specific to each of the three species; the majority were shared by all. Remarkably, enzymes involved in the glucose metabolic pathway were found, but no proteins linked to insulin were found. Adjustments to the identification parameters are also proposed to improve the detection of a wider variety of proteins with different molecular roles. Moreover, they propose that investigating other biological characteristics of plants, like age and life cycle, may help differentiate proteins between species. These results signify a significant breakthrough in the proteomic profiling of plants used in phytomedicine and medicine. Subsequent research is anticipated to enhance comprehension of the detected proteins and aid in the identification of biomarkers crucial for regulating the effective management of diverse ailments.

5 Genomic Approaches to Develop Biomarkers for Medicinal Plants

Finding the connections between genetic regulations, metabolite phenotyping and biomarker identification are made possible by the combination of metabolomics and transcriptomics. The identification and characterization of plants with therapeutic potential for drug development depend heavily on molecular and genetic markers (Hao and Xiao 2020; Pan et al. 2013). These genetic and molecular markers help develop drugs by revealing genetic variability and variety, biosynthetic pathways, and potential medicinal applications of plants (Pandey et al. 2023).

As stated by Ganie et al. (2015), an identified gene or DNA sequence linked to specific genetic traits at a specific location on a chromosome is referred to as a genetic marker. Nucleotide changes or mutations within genomic loci may be the cause of this variation. Due to the species-specific genetic information that is unaffected by age, physiological state, or environmental factors, these markers are essential for improving medicinal plant species when used in conjunction with DNA-based molecular techniques, which are thought to be more reliable (Ganie et al. 2015; Cheng et al. 2010). The effectiveness of molecular marker approaches depends on how well they can detect polymorphisms within specific accessions (Paul et al. 2021). To find biomarkers from medicinal plants, numerous important genetic and molecular techniques have been employed. One of the first techniques for using restriction enzyme digestion to identify restrictions fragment length polymorphism (RFLP) was a variation in DNA sequence. The direct detection of single-nucleotide variations, or single-nucleotide polymorphisms (SNPs), is a more modern method. These developments have increased molecular markers' accessibility, which facilitates the creation of genetic maps and the development of treatments for drug discovery utilizing conventionally used medicinal plants. (Mahanta et al. 2021).

DNA barcoding markers, like the rbcL and matK regions, are significantly used to guarantee the accurate botanical identity for medication development from plant species (Gesto-Borroto et al. 2021; Hasan et al. 2021). Recently, CRISPR-Cas9 has been widely utilized to specifically modify genes and synthesis therapeutic compounds. Guo et al. (2022) reported the application of CRISPR in *Cannabis sativa* for cannabinoid biosynthesis for therapeutic uses.

Based on the variable genomic approaches and advancements in molecular marker techniques (Gupta et al. 2001), in recent decades, markers have been divided into three classes:

1. RFLPs, RAPDs, and their modified variants are examples of first-generation molecular markers.
2. Molecular markers of the second generation, including SSRs, AFLPs, and their modified versions.
3. SNPs as well as ESTs are considered third-generation molecular markers.

5.1 Restriction Fragment Length Polymorphism (RFLP)

Due to their crucial role in the advancement of molecular genetics, RFLPs are regarded as revolutionary findings in the field of genetic markers. This technique utilizes genetic variation arising from changes in the binding and cleavage sites of restriction enzymes. Moreover, RFLP is also caused by any rearrangements that occur within the genomic region that is surrounded by these restriction sites, altering their distribution and causing polymorphism. The size of the resulting DNA fragments varies after enzymatic digestion, requiring separation by Southern blot analysis and visualization by hybridization to particular probes, which may be either heterologous or homologous. Alternative methods to southern analysis in RFLPs are provided by polymerase chain reactions, especially in cases where the flanking nucleotide sequences are known. However, it is important to remember that size differences must be revealed by gel electrophoresis of PCR products when length polymorphism results from relatively large deletions or insertions (roughly 100 bp, depending on the size of the undigested PCR product) (Zhu et al. 2020). On the other hand, PCR products that exhibit length polymorphism due to base substitution at a restriction site need to be digested using a restriction enzyme in order to reveal the RFLP. The co-dominant nature of this marker is one of its notable advantages.

The usefulness of PCR-RFLP in the identification of medicinal plants is highlighted by a large body of literature, which also applies to other DNA-based markers. For instance, six plant species were identified by PCR-RFLP targeting the internal transcribed spacer (ITS) region using ITS1 (F) and ITS4 (R) primers: *Solanum indicum*, *Desmodium giganicum*, *Oroxylum indicum*, *Solanum xanthocarpum*, *Tribulus terrestris*, and *Aegle marmelos*. Similarly, PCR-RFLP was used to distinguish *Boerhavia diffusa* L. from other species by PCR-generated 700 bp ITS region and then restriction with Msp I, resulting in four distinct fragment (Biswas and Biswas 2013). Feng et al. (2010) successfully used ITS-specific primers to distinguish *Angelica sinensis* from seven different adulterant species of *Angelica*, amplifying 520 bp only from the adulterants' DNA and remained undetermined in *A. sinensis*. Jena et al. (2009) studied the physiology and taxonomy of Indian *Citrus* using PCR-RFLP with the trnD-trnT and rbcL-ORF regions, in addition to sequence analysis of the intragenic spacer region of the trnD-trnT gene. Fifty genotype-collected plant samples, comprising wild, semi-wild, and domesticated varieties, were examined by them. Some *Citrus* genotypes appeared to have uncertain origins, as evidenced by the moderate to low bootstrap support of trnL-trnF sequence-based trees, whereas PCR-RFLP analysis produced a phylogenetic tree with moderate to high bootstrap values. Nonetheless, *Citrus maxima*, *C. reticulata*, as well as *C. medica* were identified apart by the PCR-RFLP and trnL-trnF data. These examples underscore the versatility and applicability of PCR-RFLP in the field of biological sciences, particularly in the context of identifying medicinal plants.

RFLPs present a great challenge because of their low sensitivity for detection, particularly when attempting to obtain valid profiles from samples that have significantly degraded because of age or environmental factors or from trace biological evidence. Moreover, RFLP analysis is characterized by time-consuming procedures, involves the use of radioactivity, and requires labour-intensive efforts, making automation challenging. However, this can be mitigated through the adoption of PCR-based DNA typing systems, which offer faster and more efficient alternatives.

5.2 Random Amplification of Polymorphic DNA (RAPD)

Using polymerase chain reaction (PCR) and short synthetic oligonucleotide primers of random sequences (10–12 bp), RAPD amplifies genomic DNA at low annealing temperatures. Based on their sizes, the resultant amplicons are separated on agarose gels. The primers can anneal to complementary sequences in the genome because of their short length, which requires a low annealing temperature range of 28–38 °C. As a result, depending on the primer binding sites and nucleotide sequence homology, the amplified DNA displays size variations. A requirement for the arbitrary RAPD primer selection process is that the primers must have at least 40% GC content and no palindrome sequences.

Most RAPD fragments arise from the amplification of single loci, yielding two types of polymorphisms: presence (1) or absence (0) of bands. Different copy numbers or relative sequence abundances may cause bands to have varying intensities, which may help to differentiate homozygote dominant from heterozygote samples. However, inconsistencies in band intensity do not always correlate with copy number, potentially stemming from primer mismatch. Another drawback of RAPD markers is their low reproducibility, which can be addressed by optimizing PCR conditions, selecting appropriate DNA extraction protocols to remove contaminants, and scoring only reproducible DNA fragments.

RAPD has found extensive use in authenticating medicinal plant species. For instance, differentiation between two varieties of *Silybum marianum* was achieved using specific banding patterns generated by various primers. Similarly, distinct species of the parasite *Cuscuta* were identified using specific RAPD primers. Additionally, RAPD analysis aided in characterizing *Clitoria ternatea* and *Convolvulus pluricaulis* at inter-zonal levels. Laboratories with limited budgets favour RAPD due to its reliance on basic equipment like thermal cyclers and gel electrophoresis units. RAPD is effective at identifying taxa that are not at the species level because it examines both coding and non-coding sections of the genome.

In a research done by Mehrnia et al. (2005), 20 distinct plants of Iranian *Astragalus microcephalus* were selected, and they collected and used RAPD markers to analyse their fresh leaves. The findings supported a previous study's finding that RAPDs can be used to analyse the systematic relationships between closely related species and sub-specific taxa by Mehrnia et al. (2005). For *Podophyllum hexandrum* that was collected from high-altitude areas of the Northwestern Himalayas, a similar methodology was used. Random amplified polymorphic DNA

(RAPD) analysis of the 12 gathered accessions showed significant genetic diversity, which was linked to differences in climatic and geographic circumstances. This emphasizes how useful RAPD markers are as instruments for analysing genetic relationships and variation patterns among medicinal plant species that are prioritized and endangered (Tripathi et al. 2012).

5.3 *Amplified Fragment Length Polymorphism (AFLP)*

DNA fragments from a total genomic DNA can be selectively amplified using the AFLP technique, which was developed by Dorothea et al. in 1996. The dominant marker system AFLP is useful for molecular characterization and DNA fingerprinting of germplasm collections because it offers multi-locus and genome-wide profiles (Leela Tatikonda et al. 2009). Its advantages include higher reproducibility, resolution, and sensitivity compared to other techniques. AFLP can amplify 50–100 fragments simultaneously without prior sequence information, making it beneficial for studying diverse organisms such as bacteria, fungi, and plants (Hao et al. 2009). Additionally, AFLP markers can measure recombination rates, aiding in genetic map construction (Terashima et al. 2002).

High-density linkage maps were created in ryegrass (*Lolium perenne* L.) using AFLP, which facilitated marker-assisted breeding and the identification of quantitative trait loci (QTLs) for agronomic traits (Bert et al. 1999). AFLP analysis combined molecular markers, volatile oil composition, and morphological traits to help resolve taxonomic disputes in the *Ocimum* genus (Labra et al. 2004). Similarly, *Rosa damascena* genotypes were found to be identical due to vegetative propagation when AFLP markers were used to examine the species' genetic relationships (Baydar et al. 2004).

5.4 *Simple Sequence Repeats (SSR)*

These short DNA units, which range in length from two to five base pairs, are repeated several times at a particular genomic locus to form repetitive sequences (SSRs).

The complete co-dominance of these markers makes them useful for a variety of tasks, including population structure analysis, selfing and outcrossing research, kinship determination, fingerprinting, and marker-assisted selection. Microsatellites or SSRs are molecules that are distributed throughout the genome and are locally duplicated in tandem (Zhi et al. 2017). Therefore, it is essential to use primers specific to the locus to amplify a particular microsatellite in a locus-specific manner (Pei et al. 2023).

Primers need to be designed after microsatellites are identified. With the aid of instruments like microsatellite finders, this can be completed automatically or manually for small or limited sequence data. Depending on the objectives of the study, distinct flanking regions for co-dominant markers or the microsatellites themselves for dominant markers may be used in primer design. Sequence analysis and microsatellite-enriched library cloning are common steps in the discovery of microsatellites. It is also possible to recognize clones with repeats using fluorescently labelled probes. The microsatellite-containing DNA segment is then used for primer design after being sequence verified (Ganie et al. 2015).

The hyper-variability of microsatellite markers is ascribed to the sporadic insertion or deletion of one or two nucleotides as a result of replication slippage (Ganie et al. 2015). American and Oriental ginseng roots were analysed using 16 microsatellites, which allowed for the successful separation of Chinese and American ginseng samples. Microsatellites have also been utilized to analyse species of capsicum, which are widely used as spices. There were 5751 pairs of SSR primers used in a study with 800 lines that were taken from Central and South America.

There are 2245 *C. annuum* markers were found by successfully mapping these primers onto the tomato genome. A total of 96 markers covering the whole tomato genome were chosen for additional examination. *Capsicum* lines exhibited polymorphism in 60 markers, resulting in the clustering of 192 lines into five groups. *Capsicum* lines were further separated into three main groups by the plastid genes matK and rbcL. The genotype specificity of 19 marker loci to species and clusters was observed (Huaying et al. 2021; Ganie et al. 2015). Although SSR markers are useful in many ways, there are some restrictions associated with their use. The time and resources needed to isolate and characterize each SSR locus are a significant drawback, especially in cases where it is difficult to obtain the target plant species' genomic sequence. This procedure entails the distinct isolation and characterization of every SSR locus, which may be costly and time-consuming (Ahmad et al. 2018).

These alleles could result from a variety of causes, including inadequate DNA quantity or poor primer annealing brought on by nucleotide sequence divergence. Furthermore, null allele formation may also result from mutations in the primer binding site. Null alleles can make it more difficult to determine genotypic and allelic frequencies, which could result in an underestimation of heterozygosity (Silva et al. 2017).

5.5 Variable Number Tandem Repeats (VNTRs)

A class of genetic loci known as VNTRs is found in eukaryotes and is distinguished by short DNA sequences (15–100 bp) that are arrayed in tandem. These loci show allelic variation, with alleles varying in the sequence's number of repeats. VNTRs, which fall under the category of repetitive DNA, have shown to be very useful in DNA fingerprinting, especially for forensic applications. In the beginning, VNTRs were a significant source of RFLP markers for genome linkage analysis (Eslami Rasekh et al. 2021). However, with the advancement of genome sequencing technologies, VNTRs have become indispensable in forensic investigations, such as DNA fingerprinting and database management like CODIS (Yang et al. 2014).

Use of restriction endonucleases to separate VNTR sequences from a DNA sample and gene probes for identification via Southern blotting are two methods for analysing VNTR sequences. Alternative methods include amplifying VNTR alleles via PCR amplification, which is followed by gel electrophoresis for separation (Marwal and Gaur 2020). It is possible to compare the resulting patterns without the need for specific gene probes. The possibility of two people having the same sets of alleles for even a few of the VNTR loci is incredibly unlikely because of the high allelic diversity at these loci. As a result, DNA samples have a high degree of certainty when linked to particular people. By computing the total molecular weight of the repeated sequence, one can ascertain the number of repeats present in a VNTR sequence (Smittipat et al. 2005).

While there are certain similarities between VNTRs and STRs, the primary difference is the length of the repeated sequence. The repeated sequence in VNTRs usually has a length of between 10 and 100 bp (Fan and Chu 2007). Furthermore, Kim et al. (2019) identified the mitochondrial DNAs (mtDNAs) of sugar beet, and its closely related plants contain VNTR loci. Four unrelated tandem repeat loci (TR1, TR2, TR3, and TR4) were found in the mitochondrial genomes of beets, suggesting the presence of minisatellite-like VNTRs in plant mitochondrial genomes with this discovery. The TR1 locus was notably inserted into the rn26-repeats, a family of three recombination repeat sequences.

5.6 Single-Nucleotide Polymorphisms (SNPs)

Single-nucleotide polymorphisms, also known as variations in DNA sequence, are caused by modifications to a single nucleotide (A, T, G, and C) in the genome sequence. Because of their high density and mutational stability, SNPs are particularly useful DNA markers for population genetics and mapping candidate genes for abiotic stress tolerance and disease resistance (Nelson et al. 2004). For a variation to be considered an SNP, it must exist in at least 1% of the population (Allemailem et al. 2021). SNPs can appear in both the genome's coding and non-coding regions. Because genetic codes are degenerate, single-nucleotide polymorphisms (SNPs) located within coding sequences do not always alter the amino acid sequence of the resulting protein. Non-coding RNA sequence, transcription factor binding, and gene splicing can all be impacted by SNPs that are not found in protein-coding regions (Boussaha et al. 2012). Twelve different SNP genotyping approaches are available among different ways to find new SNPs at the moment. High-throughput capabilities and automation are features of many of these techniques. Their primary function is to differentiate between a precise match and a mismatch, specifically at the SNP locus, between the target DNA and a known-sequence probe. However, there are some exceptions, such as the recently created and used "invasive cleavage assay" and "reduced representation shotgun (RRS)" techniques, which typically yield the target DNA from a PCR product (Gupta et al. 2001). Coles et al. (2005) used a sequenced tag library that was expressed to find single-nucleotide polymorphisms (SNPs) in quinoa (*Chenopodium quinoa* Willd.). The SNPs identified by this study are crucial for characterizing gene expression and regulation related to seed development. Furthermore, these SNPs can be immediately useful for germplasm development, genetic mapping projects, and the exploration of evolutionary connections within the *Chenopodium* genus.

5.7 Short Tandem Repeats (STRs)

DNA segments known as STRs are defined by brief repeating units, usually lasting two to six base pairs. These repetitive DNA sequences in a genome come in different sizes and are categorized based on the length of the core repeat unit, the number of consecutive repeats, and/or the overall length of the repeat region (Fan and Chu 2007). STRs are typically found in the vicinity of the chromosomal centromere, which is a crucial structural location on chromosomes, and the beneficial characteristics of STRs are what make them particularly helpful for identification (Edelmann et al. 2010). Due to their small allele size, STR markers are more suitable for use in forensic applications. This is especially true in cases of damaged DNA, where smaller target product sizes improve the success of PCR amplification. A unique genetic profile of an individual can be determined by examining several STR loci and measuring the repeats of a particular STR sequence at each locus (Christophe et al. 2012). The length variations of simple sequence tandem repeats (STRs) are becoming more and more important in plant genetics. Across plant species, polymorphisms and Mendelian inheritance have been reported in certain STRs, especially those at the dinucleotide level (Wang et al. 2023).

5.8 Next-Generation Sequencing (NGS) Technology

The filtering, assembly, and sequence analysis processes in NGS technology mainly rely on complex algorithms. Commercially available tools such as DNAStar, CLCBio, and GeneSpring make it simple to analyse NGS data these days. The ability to quickly and cheaply analyse the transcriptomes of medicinal plant species has been revolutionized by the development of this method. RNA sequencing has replaced whole-genome profiling as the recommended method. Gene mapping, transcript secondary structure, and cell type-specific transcriptomics can all be explored with transcriptome data acquired by next-generation sequencing (NGS). These data provide priceless insights into how plants react to environmental stimuli and developmental stages (Hong et al. 2022; Sangwan 2014).

NGS has played a major role in the identification of molecular signatures linked to various physiological functions found in plant transcriptomes. For instance, the leaf transcriptome was sequenced using Illumina's reversible dye terminator sequencing technology of *Costus pictus*, revealing transcripts linked to this plant species' anti-diabetic characteristics (Annadurai et al. 2012). Yun et al. (2015) employed NGS and developed microsatellite markers for the endangered medicinal plant *Aconitum austrokoreense* using Roche 454 GS-FLX Titanium Platform (NGS). The same platform was used to clarify the components of *Dendrobium officinale* biosynthetic pathway through EST formation (Vaini et al. 2019). Luo et al. (2010) using the Roche/454 Titanium platform, the identification of genes involved in the biosynthesis pathway, and the transcriptomics regulation of those genes were made possible for *Huperzia serrata* and *Phlegmariurus carinatus*.

Additionally, sesquiterpenes (VoTPS1, VoTPS2) from the roots of *Valeriana officinalis* were identified using transcript reads (Pyle et al. 2012). Yu et al. (2022) found SSRs in the transcripts of sea buckthorn (*Hippophae rhamnoides* L.), with the majority of these being mononucleotide SSRs. For species phylogeny studies and samples of degraded or mixed species, Roche/454 pyrosequencing or Illumina NGS is preferred over traditional full-length sequencing, which is best suited for obtaining complete barcodes.

6 Imaging Biomarkers of Medicinal Plants

Imaging the biomarkers of medicinal plants involves using advanced imaging techniques to visualize the distribution, localization, and expression of specific biomolecules within plant tissues. This method offers insightful information about the physiological processes, metabolic pathways, and bioactive compounds present in medicinal plants, facilitating their characterization and utilization in various applications, including drug discovery, plant breeding, and agronomy. Metabolites serve as significant biomarkers in identifying and characterizing medicinal plants due to their broad range of chemical compositions and bioactive properties. These substances which are generated by metabolic processes in plants are essential for defence mechanisms, growth, and environmental adaption of plants (Ramakrishna and Ravishankar 2011). The primary driving force behind the application of mass spectrometry (MS) techniques such as liquid chromatography-mass spectrometry (LC-MS) and gas chromatography-mass spectrometry (GC-MS) is high levels of sensitivity and specificity (Pitt 2009; Krone et al. 2010). These methods are highly sensitive, even for low-abundance molecules, and they work well for metabolite detection and measurement across a broad spectrum. Their ability to distinguish between molecules is based on the ratio of mass-to-charge which makes it possible to identify different metabolites with good accuracy. The resurgence of many of these compounds in the middle of the twentieth century was made possible by advances in analytical methods such as chromatography, which laid the foundation for the field of phytochemistry to be established. The utilization of medicinal plants in traditional medicine has garnered scientific backing due to the demonstration of these secondary metabolites' diverse biological effects. Many techniques can be applied to meet the need for high-throughput analysis and separation of different metabolites, including capillary electrophoresis (CE), imaging mass spectrometry (IMS), and high-performance liquid chromatography (HPLC) (Patel et al. 2021). Metabolite profiles can be analysed in a robust and efficient manner by following the HPLC's adaptability in separating a wide range of compound which is compatibility with different detectors. Methods like capillary electrophoresis are great for separating charged molecules because they not only improve on current approaches but also provide additional information about the ionic substances. Also, spatio-temporal mapping of metabolites in tissues is made possible by imaging techniques such as MALDI-TOF MS (Sun et al. 2021) which combines two fundamental processes: mass spectrometry using time-of-flight and desorption/ionization. Using UPLC-ESIHRMSn for metabolomic analysis, 36 biomarkers were found in different sections of *Gentiana crassicaulis*. The results showed that triterpenes and flavonoids are mostly found in the aerial portions of the plant, whereas iridoids are mostly found in the roots. The biomarkers emodin-8-*O*-glucoside and gallic acid-3-*O*-glucoside in *Rheum palmatum* are used to differentiate between raw and processed products. Furthermore, it was discovered that acetate, glutamine, and glycocholate could be highly sensitive and specific biomarkers for the diagnosis of hepatocellular carcinoma (HCC), pancreatic cancer (PCC), and colorectal cancer (CRC) (Wronska et al. 2022; Guo et al. 2021). NMR (nuclear magnetic resonance) spectroscopy is another essential technique in the field of metabolomics. NMR uses radiofrequency pulses and a magnetic field to study the atomic nuclei in metabolites (Tognarelli et al. 2015). It provides quantitative, structural, and non-destructive information about metabolites.

In addition to profiling-based analysis, the recent utilization of in vivo imaging has gained a significant importance based on its capacity to track the spatial distribution and dynamics in real time within intact plant tissues (Kim et al. 2015). The in vivo imaging includes application of techniques like (a) fluorescence imaging which fluorescent probes or markers that specifically binds to target biomolecules, such as proteins, enzymes, or metabolites, allowing for their visualization under a fluorescence microscope; (b) positron emission tomography (PET) imaging, which involves administering radioactive tracers to plants and then using a PET scanner to detect their distribution and accumulation in different tissues; and (c) magnetic resonance imaging (MRI) that allows for extremely detailed imaging of plant tissues, which combines radio waves and strong magnetic fields to show internal structures, water content, and physiological changes brought on by illness or stress. Bioluminescence imaging (BLI) that involves introducing genetically encoded bioluminescent reporters into plant cells, which emit light when activated by specific biomolecular interactions. By monitoring bioluminescent signals emitted from living plants, the expression patterns of target genes or proteins over time can be tracked (Zhu and Hamachi 2020; Kim et al. 2015).

7 Challenges and Future Perspectives

Biomarkers derived from medicinal plants have a lot of potential, but there are a few obstacles that prevent them from being fully used in healthcare. The absence of standardized procedures for biomarker identification, validation, and analysis is a major problem. As per reports, only 37% of studies on medicinal plants include information on standardized methods for biomarker analysis. This inconsistency hampers the reproducibility and comparability of research findings, limiting the widespread adoption of biomarkers (Parveen et al. 2015).

Furthermore, understanding the complex interactions between plant constituents and human biology requires extensive research efforts. For example, as reported by Newman and Cragg (2016), merely 10% of identified therapeutic plant compounds have undergone thorough investigation to determine their pharmacological properties. This highlights the need for further research to comprehend the mechanisms behind the therapeutic benefits of medicinal plants. Translating biomarker findings from preclinical studies to clinical settings poses another challenge that necessitates robust clinical trials and validation studies. However, the investment in clinical research on medicinal plants remains disproportionately low compared to conventional pharmaceuticals. A World Health Organization (WHO 2013; 2019) report states that traditional medicine and medicinal plants receive less than 5% of funding for global health research (Adeosun and Loots 2024). This limited funding hinders the progress of clinical validation studies and impedes the integration of biomarkers into personalized medicine approaches (Abdallah et al. 2023).

Collaboration between researchers, physicians, and regulatory agencies is crucial to overcoming these obstacles. Developing standardized protocols and databases for biomarker research can enhance the consistency and reliability of findings. Furthermore, utilizing cutting-edge technologies like transcriptomics and metabolomics can offer deeper insights into the intricate relationships between plant components and human biology (Davis et al. 2020). However, substantial investments in rigorous clinical trials are crucial to validate the clinical utility of biomarkers and facilitate their integration into personalized medicine approaches (Ahmad et al. 2023).

8 Conclusion

Through the identification and examination of particular biomarkers linked to therapeutic outcomes, scientists can acquire significant understanding of the pathways and targets implicated. This information can direct the creation of more specialized and potent herbal remedies. The consistent quality and effectiveness of herbal products can be ensured by using biomarkers to identify and quantify the major active ingredients found in medicinal plants. This helps prevent adulteration and ensures patients receive the intended therapeutic benefits. The development of personalized medicine techniques is possible through the identification of biomarkers that forecast individual responses to particular plant-based therapies. This allows for tailoring treatments to each patient's unique needs and maximizing therapeutic outcomes.

References

Abdallah EM, Alhatlani BY, de Paula Menezes R, Martins CHG (2023) Back to nature: medicinal plants as promising sources for antibacterial drugs in the post-antibiotic era. Plants (Basel, Switzerland) 12(17):3077. https://doi.org/10.3390/plants12173077

Adeosun WB, Loots DT (2024) Medicinal plants against viral infections: a review of metabolomics evidence for the antiviral properties and potentials in plant sources. Viruses 16(2):218. https://doi.org/10.3390/v16020218

Ahmad A, Wang JD, Pan YB, Sharif R, Gao SJ (2018) Development and use of simple sequence repeats (SSRs) markers for sugarcane breeding and genetic studies. Agronomy 8(11):260. https://doi.org/10.3390/agronomy8110260

Ahmad A, Imran M, Ahsan H (2023) Biomarkers as biomedical bioindicators: approaches and techniques for the detection, analysis, and validation of novel biomarkers of diseases. Pharmaceutics 15(6):1630. https://doi.org/10.3390/pharmaceutics15061630

Aina O, Bakare OO, Fadaka AO, Keyster M, Klein A (2024) Plant biomarkers as early detection tools in stress management in food crops: a review. Planta 259(3):60. https://doi.org/10.1007/s00425-024-04333-1

Alam MA, Subhan N, Rahman MM, Uddin SJ, Reza HM, Sarker SD (2014) Effect of citrus flavonoids, naringin and naringenin, on metabolic syndrome and their mechanisms of action. Adv Nutr 5(4):404–417. https://doi.org/10.3945/an.113.005603

Allemailem KS, Almatroudi A, Alrumaihi F, Makki Almansour N, Aldakheel FM, Rather RA, Afroze D, Rah B (2021) Single nucleotide polymorphisms (SNPs) in prostate cancer: its implications in diagnostics and therapeutics. Am J Transl Res 13(4):3868–3889

Altan F, Unal R (2014) Focus on proteomics and bioinformatics in translational research and plant research of obesity and diabetes. MOJ Proteomics Bioinformatics 1(2):36–41

Altmaier E, Kastenmüller G, Römisch-Margl W, Thorand B, Weinberger KM, Adamski J, Suhre K (2009) Variation in the human lipidome associated with coffee consumption as revealed by quantitative targeted metabolomics. Mol Nutr Food Res 53(11):1357–1365

Annadurai RS, Neethiraj R, Jayakumar V, Damodaran AC, Rao SN, Katta MAVSK, Gopinathan S, Sarma SP, Senthilkumar V, Niranjan V, Gopinath A, Mugasimangalam RC (2012) De novo transcriptome assembly (NGS) of Curcuma longa L. rhizome reveals novel transcripts related to anticancer and antimalarial terpenoids. PLoS One 8(2):e56217. https://doi.org/10.1371/journal.pone.0056217

Azevedo CR, Maciel FM, Silva LB, Ferreira ATS, Da Cunha M, Machado OLT, Fernandes KVS, Oliveira AEA, Xavier-Filho J (2006) Isolation and intracellular localization of insulin-like proteins from leaves of *Bauhinia variegata*. Braz J Med Biol Res 39:1435–1444. https://doi.org/10.1590/S0100-879X2006001100007

Babhadiashar N, Sotoudeh M, Azizi E, Didevar R, Malekzadeh R, Ghahremani MH (2014) Correlation between cigarette smoking and urine cotinine level in gastric cancer patients. Iran J Pharm Res 13(1):313

Balbuena TS, Dias LL, Martins ML, Chiquieri TB, Santa-Catarina C, Floh EI, Silveira V (2011) Challenges in proteome analyses of tropical plants. Braz J Plant Physiol 23:91–104. https://doi.org/10.1590/S1677-04202011000200001

Baydar NG, Baydar H, Debener T (2004) Analysis of genetic relationships among Rosa damascena plants grown in Turkey by using AFLP and microsatellite markers. J Biotechnol 111(3):263–267. https://doi.org/10.1016/j.jbiotec.2004.04.014

Bergman ME, Davis B, Phillips MA (2019) Medically useful plant terpenoids: biosynthesis, occurrence, and mechanism of action. Molecules 24(21):3961. https://doi.org/10.3390/molecules24213961

Bert PF, Charmet G, Sourdille P, Hayward MD, Balfourier F (1999) A high-density molecular map for ryegrass (Lolium perenne) using AFLP markers. Theor Appl Genet 99:445–452. https://doi.org/10.1007/s001220051256

Bhambhani S, Kondhare KR, Giri AP (2021) Diversity in chemical structures and biological properties of plant alkaloids. Molecules 26(11):3374. https://doi.org/10.3390/molecules26113374

Biswas K, Biswas R (2013) Identification of medicinal plants using PCR-RFLP in Dasamula—an Ayurvedic drug. J Pharma Biosci 3:93–98

Boussaha M, Guyomard R, Cabau C, Esquerré D, Quillet E (2012) Development and characterisation of an expressed sequence tags (EST)-derived single nucleotide polymorphisms (SNPs) resource in rainbow trout. BMC Genomics 13:1–11. https://doi.org/10.1186/1471-2164-13-238

Cai Y, Luo Q, Sun M, Corke H (2004) Antioxidant activity and phenolic compounds of 112 traditional Chinese medicinal plants associated with anticancer. Life Sci 74(17):2157–2184. https://doi.org/10.1016/j.lfs.2003.09.047

Ceravolo IP, Aguiar AC, Adebayo JO, Krettli AU (2021) Studies on activities and chemical characterization of medicinal plants in search for new antimalarials: a ten year review on ethnopharmacology. Front Pharmacol 12:734263. https://doi.org/10.3389/fphar.2021.734263

Chae CS, Sandoval TA, Hwang SM, Park ES, Giovanelli P, Awasthi D et al (2022) Tumor-derived lysophosphatidic acid blunts protective type I interferon responses in ovarian cancer. Cancer Discov 12(8):1904–1921. https://doi.org/10.1158/2159-8290.CD-21-1181

Chen W, Balan P, Popovich DG (2019) Analysis of ginsenoside content (Panax ginseng) from different regions. Molecules 24(19):3491. https://doi.org/10.3390/molecules24193491

Chen Y, Li EM, Xu LY (2022) Guide to metabolomics analysis: a bioinformatics workflow. Metabolites 12(4):357. https://doi.org/10.3390/metabo12040357

Cheng HD, Lin CS, Gen XP, Yong P (2010) Authentication of medicinal plants by DNA-based markers and genomics. Plant Gene 4:83–99. https://doi.org/10.1016/j.plgene.2015.10.002

Coles ND, Coleman CE, Christensen SA, Jellen EN, Stevens MR, Bonifacio A, Rojas-Beltran JA, Fairbanks DJ, Maughan PJ (2005) Development and use of an expressed sequenced tag library in quinoa (Chenopodium quinoa Willd.) for the discovery of single nucleotide polymorphisms. Plant Sci 168(2):439–447. https://doi.org/10.1016/j.plantsci.2004.09.007

Cosme P, Rodríguez AB, Espino J, Garrido M (2020) Plant phenolics: bioavailability as a key determinant of their potential health-promoting applications. Antioxidants 9(12):1263. https://doi.org/10.3390/antiox9121263

Christophe VN, Filip VN, David VH, Dieter D (2012) Forensic STR analysis using massive parallel sequencing Forensic Science International: Genetics 6(6) 810–818. https://doi.org/10.1016/j.fsigen.2012.03.004

Datta R, Sinha R, Chattopadhyay S (2013) Changes in leaf proteome profile of Arabidopsis thaliana in response to salicylic acid. J Biosci 38:317–328. https://doi.org/10.1007/s12038-013-9308-9

Davis KD, Aghaeepour N, Ahn AH, Angst MS, Borsook D, Brenton A, Burczynski ME, Crean C, Edwards R, Gaudilliere B, Hergenroeder GW, Iadarola MJ, Iyengar S, Jiang Y, Kong JT, Mackey S, Saab CY, Sang CN, Scholz J, Segerdahl M, Tracey I, Veasley C, Wang J, Wager TD, Wasan AD, Pelleymounter MA (2020) Discovery and validation of biomarkers to aid the development of safe and effective pain therapeutics: challenges and opportunities. Nat Rev Neurol 16(7):381–400. https://doi.org/10.1038/s41582-020-0362-2

De la Rosa LA, Moreno-Escamilla JO, Rodrigo-García J, Alvarez-Parrilla E (2019) Phenolic compounds. In: Postharvest physiology and biochemistry of fruits and vegetables. Woodhead Publishing, Duxford, pp 253–271. https://doi.org/10.1016/B978-0-12-813278-4.00012-9

DePaula J, Farah A (2019) Caffeine consumption through coffee: content in the beverage, metabolism, health benefits and risks. Beverages 5(2):37. https://doi.org/10.3390/beverages5020037

Dey P, Kundu A, Kumar A, Gupta M, Lee BM, Bhakta T, Dash S, Kim HS (2020) Analysis of alkaloids (indole alkaloids, isoquinoline alkaloids, tropane alkaloids). In: Recent advances in natural products analysis, pp 505–567. https://doi.org/10.1016/B978-0-12-816455-6.00015-9
Dorothea M, Richard M, Brian GL, Vos P, Richard PO (1996) The use of AFLP finger printing for the detection of genetic variation in fungi. Mycol Res 100(9):1107–1111
Edelmann J, Hering S, Augustin C, Kalis S, Szibor R (2010) Validation of six closely linked STRs located in the chromosome X centromere region. Int J Legal Med 124:83–87. https://doi.org/10.1007/s00414-009-0328-9
El-Sayed M, Verpoorte R (2007) Catharanthus terpenoid indole alkaloids: biosynthesis and regulation. Phytochem Rev 6:277–305. https://doi.org/10.1007/s11101-006-9047-8
Elshafie HS, Camele I, Mohamed AA (2023) A comprehensive review on the biological, agricultural and pharmaceutical properties of secondary metabolites based-plant origin. Int J Mol Sci 24(4):3266. https://doi.org/10.3390/ijms24043266
Emwas AH, Roy R, McKay RT, Tenori L, Saccenti E, Gowda GN et al (2019) NMR spectroscopy for metabolomics research. Meta 9(7):123. https://doi.org/10.3390/metabo9070123
Eslami Rasekh M, Hernández Y, Drinan SD, Fuxman Bass JI, Benson G (2021) Genome-wide characterization of human minisatellite VNTRs: population-specific alleles and gene expression differences. Nucleic Acids Res 49(8):4308–4324. https://doi.org/10.1093/nar/gkab224
Fan H, Chu JY (2007) A brief review of short tandem repeat mutation. Genomics Proteomics Bioinformatics 5(1):7–14. https://doi.org/10.1016/S1672-0229(07)60009-6
Feng T, Liu S, He XJ (2010) Molecular authentication of the traditional Chinese medicinal plant Angelica sinensis based on internal transcribed spacer of nrDNA. Electron J Biotechnol 13(1):9–10
Ganie SH, Upadhyay P, Das S, Sharma MP (2015) Authentication of medicinal plants by DNA markers. Plant Gene 4:83–99. https://doi.org/10.1016/j.plgene.2015.10.002
Gechev TS, Hille J, Woerdenbag HJ, Benina M, Mehterov N, Toneva V et al (2014) Natural products from resurrection plants: potential for medical applications. Biotechnol Adv 32(6):1091–1101. https://doi.org/10.1016/j.biotechadv.2014.03.005
Gesto-Borroto R, Medina-Jiménez K, Lorence A, Villarreal ML (2021) Application of DNA barcoding for quality control of herbal drugs and their phytopharmaceuticals. Rev Bras 31:127–141. https://doi.org/10.1007/s43450-021-00128-7
Goetz LH, Schork NJ (2018) Personalized medicine: motivation, challenges, and progress. Fertil Steril 109(6):952–963. https://doi.org/10.1016/j.fertnstert.2018.05.006
Guo YX, Zhang Y, Gao YH, Deng SY, Wang LM, Li CQ, Li X (2021) Role of plant-derived natural compounds in experimental autoimmune encephalomyelitis: a review of the treatment potential and development strategy. Front Pharmacol 12:639651. https://doi.org/10.3389/fphar.2021.639651
Guo M, Chen H, Dong S, Zhang Z, Luo H (2022) CRISPR-Cas gene editing technology and its application prospect in medicinal plants. Chin Med 17(1):33. https://doi.org/10.1186/s13020-022-00584-w
Gupta PK, Roy JK, Prasad M (2001) Single nucleotide polymorphisms: a new paradigm for molecular marker technology and DNA polymorphism detection with emphasis on their use in plants. Curr Sci 80:524–535
Hao DC, Xiao PG (2020) Pharmaceutical resource discovery from traditional medicinal plants: pharmacophylogeny and pharmacophylogenomics. Chin Herb Med 12(2):104–117. https://doi.org/10.1016/j.chmed.2020.03.002
Hao DC, Chen SL, Xiao PG (2009) Authentication of medicinal plants based on molecular biology and genomics. Chin J Pharm Biotechnol 16:490–494
Haralampidis K, Trojanowska M, Osbourn AE (2002) Biosynthesis of triterpenoid saponins in plants. In: History and trends in bioprocessing and biotransformation, pp 31–49. https://doi.org/10.1007/3-540-44604-4_2
Hardikar MR, Varma ME, Kulkarni AA, Kulkarni PP, Joshi BN (2016) Elucidation of hypoglycemic action and toxicity studies of insulin-like protein from Costus igneus. Phytochemistry 124:99–107

Hasan N, Choudhary S, Naaz N, Sharma N, Laskar RA (2021) Recent advancements in molecular marker-assisted selection and applications in plant breeding programmes. J Genet Eng Biotechnol 19(1):128. https://doi.org/10.1186/s43141-021-00231-1

Heinzmann SS, Brown IJ, Chan Q, Bictash M, Dumas ME, Kochhar S, Nicholson JK (2010) Metabolic profiling strategy for discovery of nutritional biomarkers: proline betaine as a marker of citrus consumption. Am J Clin Nutr 92(2):436–443

Hong UVT, Tamiru-Oli M, Hurgobin B, Okey CR, Abreu AR, Lewsey MG (2022) Insights into opium poppy (Papaver spp.) genetic diversity from genotyping-by-sequencing analysis. Sci Rep 12(1):111. https://doi.org/10.1038/s41598-021-04056-3

Huang WY, Cai YZ, Zhang Y (2009) Natural phenolic compounds from medicinal herbs and dietary plants: potential use for cancer prevention. Nutr Cancer 62(1):1–20. https://doi.org/10.1080/01635580903191585

Huaying W, Baiming Y, Huan W, Hongxing X (2021) Impact of different numbers of microsatellite markers on population genetic results using SLAF-seq data for Rhododendron species Abstract Scientific Reports 11(1) https://doi.org/10.1038/s41598-021-87945-x

Hussein RA, El-Anssary AA (2019) Plants secondary metabolites: the key drivers of the pharmacological actions of medicinal plants. Herb Med 1(3):11–30. https://doi.org/10.5772/intechopen.76139

Jaeger C, Lisec J (2022) Towards unbiased evaluation of ionization performance in LC-HRMS metabolomics method development. Metabolites 12(5):426. https://doi.org/10.3390/metabo12050426

Jena SN, Kumar S, Nair NK (2009) Molecular phylogeny in Indian Citrus L. (Rutaceae) inferred through PCR-RFLP and trnL-trnF sequence data of chloroplast DNA. Sci Hortic 119(4):403–416. https://doi.org/10.1016/j.scienta.2008.08.030

Kim JE, Kalimuthu S, Ahn BC (2015) In vivo cell tracking with bioluminescence imaging. Nucl Med Mol Imaging 49:3–10. https://doi.org/10.1007/s13139-014-0309-x

Kim S, Song Y, Ha B, Moon YJ, Kim M, Ryu H, Ro HS (2019) Variable number tandem repeats in the mitochondrial DNA of Lentinula edodes. Genes 10(7):542. https://doi.org/10.3390/genes10070542

Krone N, Hughes BA, Lavery GG, Stewart PM, Arlt W, Shackleton CH (2010) Gas chromatography/mass spectrometry (GC/MS) remains a pre-eminent discovery tool in clinical steroid investigations even in the era of fast liquid chromatography tandem mass spectrometry (LC/MS/MS). J Steroid Biochem Mol Biol 121(3–5):496–504. https://doi.org/10.1016/j.jsbmb.2010.04.010

Labra M, Miele M, Ledda B, Grassi F, Mazzei M, Sala F (2004) Morphological characterization, essential oil composition and DNA genotyping of Ocimumbasilicum L. cultivars. Plant Sci 167(4):725–731. https://doi.org/10.1016/j.plantsci.2004.04.026

Leela Tatikonda LT, Wani SP, Seetha Kannan SK, Naresh Beerelli NB, Sreedevi TK, Hoisington DA et al (2009) AFLP-based molecular characterization of an elite germplasm collection of *Jatropha curcas* L., a biofuel plant. Plant Sci 176:505. https://doi.org/10.1016/j.plantsci.2009.01.006

Li X, Bonawitz ND, Weng JK, Chapple C (2010) The growth reduction associated with repressed lignin biosynthesis in Arabidopsis thaliana is independent of flavonoids. Plant Cell 22(5):1620–1632. https://doi.org/10.1105/tpc.110.074161

Liu XG, Wu SQ, Li P, Yang H (2015) Advancement in the chemical analysis and quality control of flavonoid in Ginkgo biloba. J Pharm Biomed Anal 113:212–225. https://doi.org/10.1016/j.jpba.2015.03.006

Liu R, Bao ZX, Zhao PJ, Li GH (2021) Advances in the study of metabolomics and metabolites in some species interactions. Molecules 26(11):3311. https://doi.org/10.3390/molecules26113311

Lizhu, Zhao Bin, Zang Wen, Qi Fangfang, Chen Haibo, Wang Yoshihiro, Kano Dan, Yuan (2016) Pharmacokinetic study of isocorynoxeine metabolites mediated by cytochrome P450 enzymes in rat and human liver microsomes Fitoterapia 11149–11157. https://doi.org/10.1016/j.fitote.2016.04.008

Luo H et al (2010) Transcriptome analysis of medicinal plant Salvia miltiorrhiza and identification of genes related to tanshinone biosynthesis. PLoS One 8(11):e80464. https://doi.org/10.1371/journal.pone.0080464

Mahanta S, Gogoi B, Chetia P, Tanti B, Handique PJ (2021) Identification of active phytochemical from traditional herbal knowledge-base targeting pantothenate synthetase for anti-tuberculosis therapy. Lett Drug Des Discov 18(9):859–871

Mahanta S, Naiya T, Biswas K, Changkakoti L, Mohanta YK, Tanti B et al (2022) Plant source derived compound exhibited in silico inhibition of membrane glycoprotein in SARS-CoV-2: paving the way to discover a new class of compound for treatment of COVID-19. Front Pharmacol 13:805344

Marchev AS, Koycheva IK, Aneva IY, Georgiev MI (2020) Authenticity and quality evaluation of different Rhodiola species and commercial products based on NMR-spectroscopy and HPLC. Phytochem Anal 31(6):756–769. https://doi.org/10.1002/pca.2940

Marchev AS, Vasileva LV, Amirova KM, Savova MS, Balcheva-Sivenova ZP, Georgiev MI (2021) Metabolomics and health: from nutritional crops and plant-based pharmaceuticals to profiling of human biofluids. Cell Mol Life Sci 78(19–20):6487–6503. https://doi.org/10.1007/s00018-021-03918-3

Marwal A, Gaur RK (2020) Molecular markers: tool for genetic analysis. In: Animal biotechnology. Academic Press, Oxford, pp 353–372. https://doi.org/10.1016/B978-0-12-811710-1.00016-1

Masyita A, Sari RM, Astuti AD, Yasir B, Rumata NR, Emran TB et al (2022) Terpenes and terpenoids as main bioactive compounds of essential oils, their roles in human health and potential application as natural food preservatives. Food Chem X 13:100217. https://doi.org/10.1016/j.fochx.2022.100217

Mehrnia M, Zarre S, Sokhan-Sanj A (2005) Intra-and inter-specific relationships within the Astragalus microcephalus complex (Fabaceae) using RAPD. Biochem Syst Ecol 33(2):149–158

Mennen LI, Sapinho D, Ito H, Bertrais S, Galan P, Hercberg S, Scalbert A (2006) Urinary flavonoids and phenolic acids as biomarkers of intake for polyphenol-rich foods. Br J Nutr 96(1):191–198

Miele S, Tegli S, Izquierdo CG, Cerboneschi M, Bargiacchi E (2019) Hydrolysable tannins in agriculture. In: Tannins-structural properties, biological properties and current knowledge, pp 79–91

Moses T, Papadopoulou KK, Osbourn A (2014) Metabolic and functional diversity of saponins, biosynthetic intermediates and semi-synthetic derivatives. Crit Rev Biochem Mol Biol 49(6):439–462. https://doi.org/10.3109/10409238.2014.953628

Mugford ST, Osbourn A (2013) Saponin synthesis and function. In: Isoprenoid synthesis in plants and microorganisms: new concepts and experimental approaches, pp 405–424. https://doi.org/10.1007/978-1-4614-4063-5_28

Na HJ, Lee G, Oh HY, Jeon KS, Kwon HJ, Ha KS et al (2006) 4-O-Methylgallic acid suppresses inflammation-associated gene expression by inhibition of redox-based NF-κB activation. Int Immunopharmacol 6(10):1597–1608

Nagana Gowda GA, Raftery D (2022) Quantitative NMR methods in metabolomics. In: Metabolomics and its impact on health and diseases. Springer, Cham, pp 143–164. https://doi.org/10.1007/164_2022_612

Nan HM, Kim H, Lim HS, Choi JK, Kawamoto T, Kang JW et al (2001) Effects of occupation, lifestyle and genetic polymorphisms of CYP1A1, CYP2E1, GSTM1 and GSTT1 on urinary 1-hydroxypyrene and 2-naphthol concentrations. Carcinogenesis 22(5):787–793

Nayak A, Liu C, Mehta A, Ko YA, Tahhan AS, Dhindsa DS, Quyyumi AA (2020) N8-Acetylspermidine: a polyamine biomarker in ischemic cardiomyopathy with reduced ejection fraction. J Am Heart Assoc 9(11):e016055

Nelson MR, Marnellos G, Kammerer S, Hoyal CR, Shi MM, Cantor CR, Braun A (2004) Large-scale validation of single nucleotide polymorphisms in gene regions. Genome Res 14(8):1664–1668. https://doi.org/10.1101/gr.2421604

Newman DJ, Cragg GM (2016) Natural products as sources of new drugs from 1981 to 2014. J Nat Prod 79(3):629–661. https://doi.org/10.1021/acs.jnatprod.5b01055

Nogueira FCS, Gonçalves EF, Jereissati ES, Santos M, Costa JH, Oliveira-Neto OB et al (2007) Proteome analysis of embryogenic cell suspensions of cowpea (Vigna unguiculata). Plant Cell Rep 26:1333–1343. https://doi.org/10.1007/s00299-007-0327-6

O'Connor SE (2012) Strategies for engineering plant natural products: the iridoid-derived monoterpene indole alkaloids of Catharanthus roseus. In: Methods in enzymology, vol 515. Academic Press, San Diego, pp 189–206. https://doi.org/10.1016/B978-0-12-394290-6.00009-4

Pan SY, Zhou SF, Gao SH, Yu ZL, Zhang SF, Tang MK et al (2013) New perspectives on how to discover drugs from herbal medicines: CAM's outstanding contribution to modern therapeutics. Evid Based Complement Alternat Med 2013:627375. https://doi.org/10.1155/2013/627375

Pandey Y, Chaturvedi T, Swaroop H, Gupta AK, Shanker K, Tiwari G (2023) Phytochemical and genetic marker (SCoT and CBDP) based study of genetic diversity and population structure in natural populations of Cannabis sativa L.: a high-value sustainable biodiversity of North-Indian Himalaya. Ind Crop Prod 200:116892. https://doi.org/10.1016/j.indcrop.2023.116892

Park YJ, Volpe SL, Decker EA (2005) Quantitation of carnosine in humans plasma after dietary consumption of beef. J Agric Food Chem 53(12):4736–4739

Parveen A, Parveen B, Parveen R, Ahmad S (2015) Challenges and guidelines for clinical trial of herbal drugs. J Pharm Bioallied Sci 7(4):329–333. https://doi.org/10.4103/0975-7406.168035

Pase MP, Beiser AS, Himali JJ, Satizabal CL, Aparicio HJ, DeCarli C, Seshadri S (2019) Assessment of plasma total tau level as a predictive biomarker for dementia and related endophenotypes. JAMA Neurol 76(5):598–606

Patel MK, Pandey S, Kumar M, Haque MI, Pal S, Yadav NS (2021) Plants metabolome study: emerging tools and techniques. Plants 10(11):2409. https://doi.org/10.3390/plants10112409

Pattanaik B, Lindberg P (2015) Terpenoids and their biosynthesis in cyanobacteria. Life 5(1):269–293. https://doi.org/10.3390/life5010269

Paul D, Mahanta S, Tag H, Das SK, Das Gupta D, Tanti B et al (2021) Identification of tyrosine kinase inhibitors from Panax bipinnatifidus and Panax pseudoginseng for RTK—HER2 and VEGFR2 receptors, by in silico approach. Mol Divers 26(4):1933–1955

Pedrete TA, Hauser-Davis RA, Moreira JC (2019) Proteomic characterization of medicinal plants used in the treatment of diabetes. Int J Biol Macromol 140:294–302. https://doi.org/10.1016/j.ijbiomac.2019.08.035

Pei D, Song S, Kang J, Zhang C, Wang J, Dong T et al (2023) Characterization of simple sequence repeat (SSR) markers mined in whole grape genomes. Genes 14(3):663. https://doi.org/10.3390/genes14030663

Pezzatti J, Boccard J, Codesido S, Gagnebin Y, Joshi A, Picard D et al (2020) Implementation of liquid chromatography–high resolution mass spectrometry methods for untargeted metabolomic analyses of biological samples: a tutorial. Anal Chim Acta 1105:28–44. https://doi.org/10.1016/j.aca.2019.12.062

Pitt JJ (2009) Principles and applications of liquid chromatography-mass spectrometry in clinical biochemistry. Clin Biochem Rev 30(1):19

Pradhan AK, Shandilya ZM, Lahkar L, Hasnu S, Kalita J, Borgohain D, Tanti B (2019) Comparative metabolomics approach towards understanding chemical variation in rice under abiotic stress. In: Advances in rice research for abiotic stress tolerance. Woodhead Publishing, Duxford, pp 537–550

Pradhan AK, Rehman M, Saikia D, Jyoti SY, Poudel J, Tanti B (2020) Biochemical and molecular mechanism of abiotic stress tolerance in plants. In: Plant ecophysiology and adaptation under climate change: mechanisms and perspectives I: general consequences and plant responses. Springer, Singapore, pp 825–853

Pradhan AK, Shandilya ZM, Sarma P, Bora RK, Regon P, Vemireddy LNR, Tanti B (2023) Concurrent effect of aluminum toxicity and phosphorus deficiency in the root growth of aluminum tolerant and sensitive rice cultivars. Acta Physiol Plant 45(2):33

Pyle BW, Tran HT, Pickel B, Haslam TM, Gao Z, MacNevin G et al (2012) Enzymatic synthesis of valerena-4, 7 (11)-diene by a unique sesquiterpene synthase from the valerian plant (Valeriana officinalis). FEBS J 279(17):3136–3146. https://doi.org/10.1111/j.1742-4658.2012.08692.x

Qiu S, Cai Y, Yao H, Lin C, Xie Y, Tang S, Zhang A (2023) Small molecule metabolites: discovery of biomarkers and therapeutic targets. Signal Transduct Target Ther 8(1):132. https://doi.org/10.1038/s41392-023-01399-3

Ramakrishna A, Ravishankar GA (2011) Influence of abiotic stress signals on secondary metabolites in plants. Plant Signal Behav 6(11):1720–1731. https://doi.org/10.4161/psb.6.11.17613. PMID: 22041989; PMCID: PMC3329344.

Regon P, Dey S, Rehman M, Pradhan AK, Chowra U, Tanti B et al (2022) Transcriptomic analysis revealed reactive oxygen species scavenging mechanisms associated with ferrous iron toxicity in aromatic Keteki Joha rice. Front Plant Sci 13:798580

Roy SJ, Regon P, Tanti B (2024) Morpho-physiochemical responses of Capsicum chinense Jacq. (*Bhut Jolokia*) under different abiotic stresses. Vegetos 1–16

Rupani R, Chavez A (2018) Medicinal plants with traditional use: ethnobotany in the Indian subcontinent. Clin Dermatol 36(3):306–309. https://doi.org/10.1016/j.clindermatol.2018.03.005

Sagi S, Avula B, Wang YH, Khan IA (2016) Quantification and characterization of alkaloids from roots of Rauwolfia serpentina using ultra-high performance liquid chromatography-photo diode array-mass spectrometry. Anal Bioanal Chem 408:177–190. https://doi.org/10.1007/s00216-015-9093-4

Saiki S, Hatano T, Fujimaki M, Ishikawa KI, Mori A, Oji Y et al (2017) Decreased long-chain acylcarnitines from insufficient β-oxidation as potential early diagnostic markers for Parkinson's disease. Sci Rep 7(1):7328

Salam U, Ullah S, Tang ZH, Elateeq AA, Khan Y, Khan J et al (2023) Plant metabolomics: an overview of the role of primary and secondary metabolites against different environmental stress factors. Life 13(3):706. https://doi.org/10.3390/life13030706

Salehi B, Fokou PVT, Sharifi-Rad M, Zucca P, Pezzani R, Martins N, Sharifi-Rad J (2019) The therapeutic potential of naringenin: a review of clinical trials. Pharmaceuticals 12(1):11

Salgueiro ACF, Folmer V, da Silva MP, Mendez ASL, Zemolin APP, Posser T et al (2016) Effects of *Bauhinia forficata* tea on oxidative stress and liver damage in diabetic mice. Oxidative Med Cell Longev 2016:8902954. https://doi.org/10.1155/2016/8902954

Sangwan RS (2014) Next-generation sequencing: a challenge to meet the increasing demand for plant genome sequencing. J Plant Biochem Biotechnol 45:232–245

Sawai S, Saito K (2011) Triterpenoid biosynthesis and engineering in plants. Front Plant Sci 2:25. https://doi.org/10.3389/fpls.2011.00025

Schneider N, Hauser J, Oliveira M, Cazaubon E, Mottaz SC, O'Neill BV, Deoni SC (2019) Sphingomyelin in brain and cognitive development: preliminary data. eNeuro 6(4):ENEURO.0421

Seca AM, Pinto DC (2019) Biological potential and medical use of secondary metabolites. Medicines 6(2):66. https://doi.org/10.3390/medicines6020066

Shan B, Cai YZ, Sun M, Corke H (2005) Antioxidant capacity of 26 spice extracts and characterization of their phenolic constituents. J Agric Food Chem 53(20):7749–7759. https://doi.org/10.1021/jf051513y

Sharifi-Rad J, Sureda A, Tenore GC, Daglia M, Sharifi-Rad M, Valussi M et al (2017) Biological activities of essential oils: from plant chemoecology to traditional healing systems. Molecules 22(1):70. https://doi.org/10.3390/molecules22010070

Silva FC, Torrezan GT, Brianese RC, Stabellini R, Carraro DM (2017) Pitfalls in genetic testing: a case of a SNP in primer-annealing region leading to allele dropout in BRCA 1. Mol Genet Genomic Med 5(4):443–447. https://doi.org/10.1002/mgg3.295

Singh B, Sharma RA (2015) Plant terpenes: defense responses, phylogenetic analysis, regulation and clinical applications. 3 Biotech 5(2):129–151. https://doi.org/10.1007/s13205-014-0220-2

Skubnik J, Pavlíčková VS, Ruml T, Rimpelová S (2021) Vincristine in combination therapy of cancer: emerging trends in clinics. Biology 10(9):849. https://doi.org/10.3390/biology10090849

Smith E, Lewis A, Narine SS, Emery RN (2023) Unlocking potentially therapeutic phytochemicals in Capadulla (Doliocarpus dentatus) from Guyana using untargeted mass spectrometry-based metabolomics. Metabolites 13(10):1050. https://doi.org/10.3390/metabo13101050

Smittipat N, Billamas P, Palittapongarnpim M, Thong-On A, Temu MM, Thanakijcharoen P et al (2005) Polymorphism of variable-number tandem repeats at multiple loci in Mycobacterium tuberculosis. J Clin Microbiol 43(10):5034–5043. https://doi.org/10.1128/jcm.43.10.5034-5043.2005

Spille F, Alves MK, Vieira SM, Carvalho TA, Leite CE, Lunardelli A et al (2008) Anti-inflammatory effects of red pepper (Capsicum baccatum) on carrageenan-and antigen-induced inflammation. J Pharm Pharmacol 60(4):473–478. https://doi.org/10.1211/jpp.60.4.0010

Stadlbauer V, Haselgrübler R, Lanzerstorfer P, Plochberger B, Borgmann D, Jacak J et al (2016) Biomolecular characterization of putative antidiabetic herbal extracts. PLoS One 11(1):e0148109

Sun W, Shahrajabian MH (2023) Therapeutic potential of phenolic compounds in medicinal plants—natural health products for human health. Molecules 28(4):1845. https://doi.org/10.3390/molecules28041845

Sun J, Shi H, Xue Y, Cheng W, Yu M, Ding C et al (2021) Releasing bacteria from functional magnetic beads is beneficial to MALDI-TOF MS based identification. Talanta 225:121968. https://doi.org/10.1016/j.talanta.2020.121968

Sutphen R, Xu Y, Wilbanks GD, Fiorica J, Grendys EC Jr, LaPolla JP et al (2004) Lysophospholipids are potential biomarkers of ovarian cancer. Cancer Epidemiol Biomark Prev 13(7):1185–1191

Tang S, Ren J, Kong L, Yan G, Liu C, Han Y et al (2023) Ephedraeherba: a review of its phytochemistry, pharmacology, clinical application, and alkaloid toxicity. Molecules 28(2):663. https://doi.org/10.3390/molecules28020663

Terashima K, Matsumoto T, Hayashi E, Fukumasa-Nakai Y (2002) A genetic linkage map of Lentinula edodes (shiitake) based on AFLP markers. Mycol Res 106(8):911–917. https://doi.org/10.1017/S0953756202006275

Thawabteh A, Juma S, Bader M, Karaman D, Scrano L, Bufo SA, Karaman R (2019) The biological activity of natural alkaloids against herbivores, cancerous cells and pathogens. Toxins 11(11):656. https://doi.org/10.3390/toxins11110656

Toffolatti SL, Maddalena G, Passera A, Casati P, Bianco PA, Quaglino F (2021) Role of terpenes in plant defense to biotic stress. In: Biocontrol agents and secondary metabolites. Woodhead Publishing, Duxford, pp 401–417. https://doi.org/10.1016/B978-0-12-822919-4.00016-8

Tognarelli JM, Dawood M, Shariff MI, Grover VP, Crossey MM, Cox IJ et al (2015) Magnetic resonance spectroscopy: principles and techniques: lessons for clinicians. J Clin Exp Hepatol 5(4):320–328. https://doi.org/10.1016/j.jceh.2015.10.006

Tripathi N, Chouhan DS, Saini N, Tiwari S (2012) Assessment of genetic variations among highly endangered medicinal plant Bacopa monnieri (L.) from Central India using RAPD and ISSR analysis. 3 Biotech 2:327–336

Ullah A, Munir S, Badshah SL, Khan N, Ghani L, Poulson BG et al (2020) Important flavonoids and their role as a therapeutic agent. Molecules 25(22):5243. https://doi.org/10.3390/molecules25225243

Vadhel A, Bashir S, Mir AH, Girdhar M, Kumar D, Kumar A et al (2023) Opium alkaloids, biosynthesis, pharmacology and association with cancer occurrence. Open Biol 13(5):220355. https://doi.org/10.1098/rsob.220355

Vaini JO, Mota KG, Ojeda AP, Barreiros JP, Moreira RG, Hilsdorf AWS (2019) Development and characterization of 20 polymorphic microsatellite markers for Epinephelus marginatus (Lowe, 1834) (Perciformes: Epinephelidae) using 454 pyrosequencing. Genet Mol Biol 42:74–79. https://doi.org/10.1590/1678-4685-GMB-2018-0067

Wang H, Chen Y, Wang L, Liu Q, Yang S, Wang C (2023) Advancing herbal medicine: enhancing product quality and safety through robust quality control practices. Front Pharmacol 14:1265178. https://doi.org/10.3389/fphar.2023.1265178

White PAS, Cercato LM, Batista VS, Camargo EA, De Lucca Jr W, Oliveira AS et al (2016) Aqueous extract of Chrysobalanus icaco leaves, in lower doses, prevent fat gain in obese high-fat fed mice. J Ethnopharmacol 179:92–100. https://doi.org/10.1016/j.jep.2015.12.047

Willcox ML, Bodeker G (2004) Traditional herbal medicines for malaria. Clin Rev 329(7475):1156–1159

Wink M (2015) Modes of action of herbal medicines and plant secondary metabolites. Medicines 2(3):251–286. https://doi.org/10.3390/medicines2030251

World Health Organization (2013) WHO traditional medicine strategy: 2014–2023. World Health Organization, Geneva

World Health Organization (2019) WHO global report on traditional and complementary medicine 2019. WHO Press, Geneva

Wronska N, Szlaur M, Zawadzka K, Lisowska K (2022) The synergistic effect of triterpenoids and flavonoids—new approaches for treating bacterial infections? Molecules 27(3):847. https://doi.org/10.3390/molecules27030847

Xiao Q, Mu X, Liu J, Li B, Liu H, Zhang B, Xiao P (2022) Plant metabolomics: a new strategy and tool for quality evaluation of Chinese medicinal materials. Chin Med 17(1):45. https://doi.org/10.1186/s13020-022-00601-y

Yang Y, Xie B, Yan J (2014) Application of next-generation sequencing technology in forensic science. Genomics Proteomics Bioinformatics 12(5):190–197. https://doi.org/10.1016/j.gpb.2014.09.001

Yilmaz A, Geddes T, Han B, Bahado-Singh RO, Wilson GD, Imam K et al (2017) Diagnostic biomarkers of Alzheimer's disease as identified in saliva using 1H NMR-based metabolomics. J Alzheimers Dis 58(2):355–359

Yu L, Diao S, Zhang G, Yu J, Zhang T, Luo H et al (2022) Genome sequence and population genomics provide insights into chromosomal evolution and phytochemical innovation of Hippophaerhamnoides. Plant Biotechnol J 20(7):1257–1273. https://doi.org/10.1111/pbi.13802

Yun YE, Yu JN, Nam GH, Ryu SA, Kim S, Oh K, Lim CE (2015) Next-generation sequencing identification and characterization of microsatellite markers in Aconitum austrokoreense Koidz., an endemic and endangered medicinal plant of Korea. Genet Mol Res 11;14(2):4812–4817. https://doi.org/10.4238/2015.May.11.13

Zandavar H, Babazad MA (2023) Secondary metabolites: alkaloids and flavonoids in medicinal plants. In: Herbs and spices-new advances. IntechOpen. https://doi.org/10.5772/intechopen.108030

Zhang XW, Li QH, Xu ZD, Dou JJ (2020) Mass spectrometry-based metabolomics in health and medical science: a systematic review. RSC Adv 10(6):3092–3104. https://doi.org/10.1039/c9ra08985c

Zhao L, Chang WC, Xiao Y, Liu HW, Liu P (2013) Methylerythritol phosphate pathway of isoprenoid biosynthesis. Annu Rev Biochem 82:497–530. https://doi.org/10.1146/annurev-biochem-052010-100934

Zhao LQ, Shan CM, Shan TY, Li QL, Ma KL, Deng WW, Wu JW (2022) Comparative transcriptomic analysis reveals the regulatory mechanisms of catechins synthesis in different cultivars of Camellia sinensis. Food Res Int 157:111375. https://doi.org/10.1016/j.foodres.2022.111375

Zhi ZC, Jing QJ, Agarwal G, Shan WJ, Zhen RX, Han X et al (2017) Genome-wide discovery of microsatellite markers from diploid progenitor species, Arachis duranensis and *A. ipaensis*, and their application in cultivated peanut (*A. hypogaea*). Front Plant Sci 8:1209. https://doi.org/10.3389/fpls.2017.01209

Zhong H, Duan BH, Du FM, Wang WM, Qiao H (2023) Identification of key genes, biological functions, and pathways of empagliflozin by network pharmacology and its significance in the treatment of type 2 diabetes mellitus. Ann Transl Med 11(2):123. https://doi.org/10.21037/atm-22-6406

Zhu H, Hamachi I (2020) Fluorescence imaging of drug target proteins using chemical probes. J Pharm Anal 10(5):426–433. https://doi.org/10.1016/j.jpha.2020.05.013

Zhu H, Zhang H, Xu Y, Laššáková S, Korabečná M, Neužil P (2020) PCR past, present and future. Biotechniques 69(4):317–325. https://doi.org/10.2144/btn-2020-0057

Zulak K, Liscombe D, Ashihara H, Facchini P (2006) Alkaloids. Plant secondary metabolism in diet and human health, vol 5. Blackwell Publishing, Oxford, pp 102–136. https://doi.org/10.1002/9780470988558.ch4

Role of Metabolomics in the Discovery of Modern Drugs from Traditional Resources

Shyamalima Saikia, Minakshi Puzari, and Pankaj Chetia

Contents

Abstract Metabolomics, the science of metabolome, is an emerging "omics" technology that deals with the characterization of metabolites and metabolic pathways in biological systems. It uncovers the intricate network of chemical reactions and pathways occurring within cells through analysis of the small molecules generated from these processes. Metabolomics provides valuable insights into the metabolic state of cells and the functioning of biological systems under various conditions, including health, disease, and environmental changes, identifies biomarkers for disease diagnosis, and facilitates the discovery of new therapeutics and targets. Moreover, metabolomics has given a new dimension to biomedicine, facilitating the

S. Saikia · M. Puzari · P. Chetia (✉)
Molecular Plant Taxonomy and Bioinformatics Research Laboratory, Department of Life Sciences, Dibrugarh University, Dibrugarh, Assam, India
e-mail: chetiapankaj@dibru.ac.in

© The Author(s), under exclusive license to Springer Nature Singapore Pte Ltd. 2024

A. Das Talukdar et al. (eds.), *Traditional Resources and Tools for Modern Drug Discovery*, Interdisciplinary Biotechnological Advances,
https://doi.org/10.1007/978-981-97-4600-2_9

discovery of active molecules from natural products by reducing both time and costs. Additionally, it offers enhanced reproducibility with minimal sample handling errors. Metabolomics employs both experimental and computational tools that significantly contribute to discovering probable drug-like compounds from traditional resources by analyzing chemical libraries and identifying novel drug targets through metabolic network analyses. Metabolomics techniques, characterized by their high-throughput, reduced sample size, increased sensitivity, expedited analysis, and comprehensive coverage of metabolites, have propelled drug discovery by identifying promising drug candidates with novel mechanisms of action. Within this context, the article aims to emphasize the role of metabolomics in drug discovery, particularly highlighting the recent technological advances that facilitate the identification of promising drug candidates from traditional resources. This article also sheds light on the potential future application of metabolomics in the field of drug discovery.

Keywords Chemical libraries · Drug discovery · Metabolomics · Natural products · NMR · Target identification

1 Introduction

Drug discovery and development is a multistep process, comprising the identification of novel lead molecules against a target associated with a particular disease of interest, ending with different phases of clinical trials, validation, and approval from regulating bodies such as the Food and Drug Administration (FDA). The traditional drug discovery process is expensive and time-consuming and possesses a high risk of failure in clinical trials, costing over $1–2 billion for approval of each new drug (Nicholson et al. 1999). This sector, therefore, demands new technologies that could predict the pharmacokinetic and molecular properties of drug-like compounds at the early stage, provides insight into the interaction of the compounds at the target sites, and reveals possible metabolic and cellular changes (Nicholson et al. 1999). The integration of multi-omics approaches with computational technologies has made many advances in this field by minimizing the time, cost, and failure rate (Wishart 2016). Over the past 25 years, advancements in omics science and high-throughput next-generation sequencing (NGS) technologies have provided remarkable advancements in biomedical research and provided new insights with its diverse application in medicine, ranging from identifying infection mechanisms and diagnosing diseases to discovering novel drug targets and potent drug molecules (Cuperlovic-Culf and Culf 2016). This has revolutionized the drug discovery process, offering a more detailed understanding of disease mechanisms and potential therapeutic interventions (Holmes et al. 2007).

Metabolomics, the science of metabolome, deals with the characterization of metabolites and metabolic pathways in biological systems. It uncovers the complex network of chemical reactions and pathways occurring within cells by analyzing the small molecules produced as a result of these processes (Nicholson et al. 1999). The metabolic profile of a living system serves as a direct signature of its phenotype and biochemical activity (Clish 2015). Even slight alternation of the metabolome may affect the biological system at the molecular, cellular, and physiological levels, including the genome (DNA), proteome (proteins), transcriptome (RNA), and epigenome (Lu and Chen 2017) (Fig. 1). Such metabolic abnormalities resulting from the accumulation or deficiency of metabolites and dysfunction of metabolic pathways are well-established hallmarks of diseases (Holmes et al. 2007). As metabolomics can provide insight into qualitative and quantitative changes of metabolites (in cells, tissue, or organisms), it successfully identified metabolic causes and biomarkers for various difficult-to-treat diseases, including diabetes, Alzheimer's, and cancer. It also contributed significantly to identifying novel therapeutic targets and lead compounds for treating several global health issues (Kumar et al. 2014).

Metabolomics, with bioinformatics, significantly contributes to drug discovery with the growing databases containing molecular and spectral information of alternate druggable targets, promising compounds, and leads (Lu and Chen 2017). It enhances the discovery of promising leads from natural resources by reducing toxicological screening costs, improving clinical trials, and shortening the time required for the overall development pipeline (Wishart 2016). Metabolomics plays a critical role in drug discovery as metabolome reflects genomic and protein alternations associated with the disease (Cuperlovic-Culf and Culf 2016). Analyzing the metabolites in biological fluids that are in circulation can frequently provide a quick and

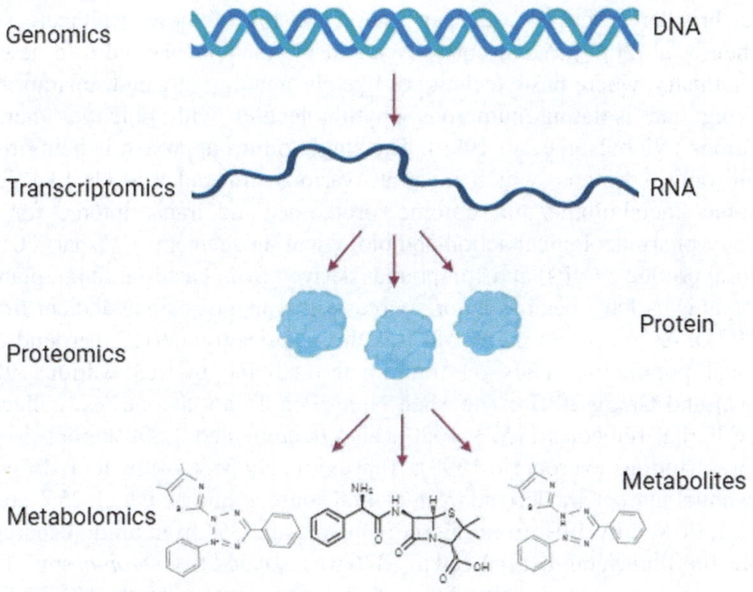

Fig. 1 Transfer of biological information through the four main biomolecules and their respective omics sciences

highly accurate assessment of the condition of the living system. Metabolomics focused on small-molecule metabolites has emerged as a specialized technique for metabolic biomarker and pathway research, elucidating putative processes underlying human illnesses and unlocking treatment opportunities (Wishart 2016; Rinschen et al. 2019). Numerous metabolic profiles have recently been produced to uncover the underlying mechanism and metabolic networks for investigating therapeutic targets (Kumar et al. 2014; Rinschen et al. 2019).

This article summarizes the growing role of metabolomics in the drug discovery sector, focusing on the recent technological advances for identifying novel drug candidates from traditional resources. Additionally, the current analytical approaches applied in metabolomics and their potential future application in the drug discovery field are briefly discussed in this chapter.

2 Metabolomics in Identification of Lead Molecules from Natural Products

Since the dawn of civilization, humans have relied on plants and plant parts for food and remedies to treat illnesses. The use of crude extracts of various plants to treat several diseases sowed the seeds for later-day phytochemistry (Wink 2015). Traditional medicine holds cultural theories, beliefs, and experiences unique to various populations, regardless of scientific variation. However, the limited availability of scientific investigation, empirical diagnosis, and rigorous therapeutic methods obstruct broader access to a deeper comprehension of these healthcare services (Rinschen et al. 2019). More curiosity about phytochemistry led to a new era in phytochemistry, where basic techniques like chromatography and separation aid in identifying and isolating numerous phytomolecules with potential therapeutic applications (Nicholson et al. 1999). The single omics approach is being replaced by multi-omics designs, which integrate various molecular levels like genome, epigenome, metabolome, microbiome, proteome, and transcriptome for deeper analysis of pharmacological action and biological mechanisms (Wishart 2016).

Natural products (NPs) and compounds derived from bacteria, fungi, plants, and other organisms have been a major source of therapeutics since ancient times (Li et al. 2021). According to the World Health Organization (WHO), around 60% of the global population relies on traditional medicine to treat various illnesses (Newman and Cragg 2007). The Shen Nung Pen Ts'ao, the earliest collection of Chinese herbal remedies (1573–1620), also documented 1,898 herbal drugs and 8,160 prescriptions (Verpoorte 1998). Approximately 35% of the total drugs in the global annual market are derived from natural sources, among which 25% are plant-derived, followed by 13% from microorganisms and 3% from animal sources. For example, morphine commercialized in 1826 was isolated from *Opium* spp. (Demain 1999). According to reports, out of the 1,562 drugs approved by the US FDA during 1981 and 2014, 64 were pure natural products, 320 drugs were derived from natural sources, 141 were herbal mixtures, and 61 were synthetic drugs formulated from pharmacophores found in natural products (Najmi et al. 2022). Metabolites, mainly of plant origin, are excellent drugs or drug precursors with many physiological effects and therapeutic properties (Gorlenko et al. 2020). The plant secondary metabolites possess anti-inflammatory, antimicrobial, antiviral, anticancer, and immunosuppressive properties (Verpoorte 1998). These metabolites can be identified, screened, optimized, isolated for drug discovery, and validated using computational techniques, high-throughput analytical techniques, and metabolomics (Rinschen et al. 2019). The application of metabolomics in different research areas is illustrated in Fig. 2.

NPs are chemical compounds or metabolites produced by living organisms. They are often investigated for therapeutic properties and considered as potential candidates for drug discovery and development (Newman and Cragg 2016). NPs have played a pivotal role in drug discovery, contributing significantly to treating various infectious diseases, cancer, multiple sclerosis (fingolimod), and cardiovascular diseases (statins) (Newman and Cragg 2007). Due to structural complexity and enormous diversity, NPs offer both advantages and challenges in drug discovery (Demain 1999). They possess high molecular weight (MW) and contain many sp3 oxygen and carbon atoms while having fewer halogen and nitrogen atoms. Additionally, they tend to have a high number of hydrogen bond (H-bond) donors and acceptors. NPs usually exhibit higher hydrophobicity and molecular rigidity compared to synthetic compounds (Atanasov et al. 2021; Sun and Wesolowski 2021). These properties are advantageous because NPs are the primary source of structurally optimized

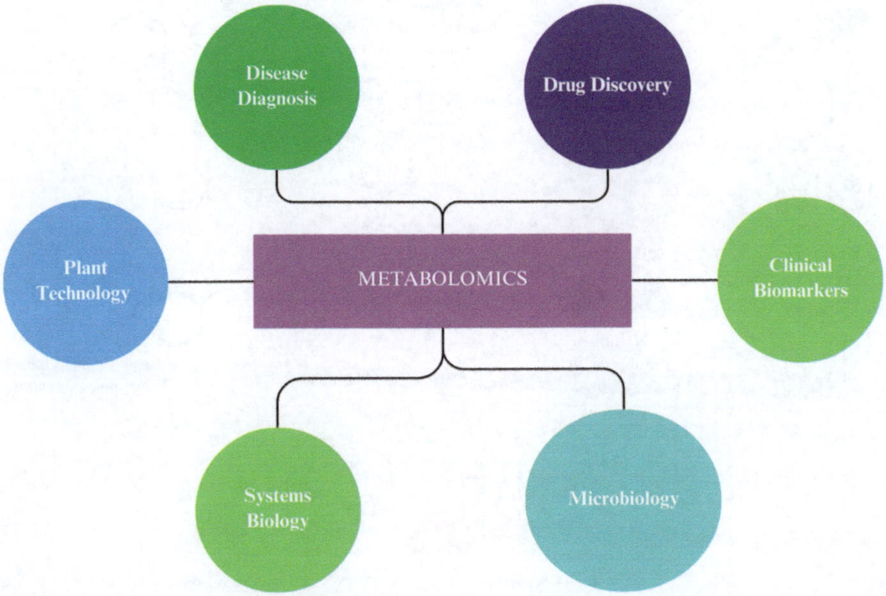

Fig. 2 Applications of metabolomics: an omics science in different research areas

orally administered drugs that adhere to 'Lipinski's rule of five,' ensuring safety and efficacy in traditional medicine practices (Lipinski et al. 1997). Natural product pools, rich in bioactive compounds from diverse sources, serve as both direct agents and structural scaffolds for synthesizing analogs with enhanced bioactivity and safety (Thomford et al. 2018). Despite the advantages and successful discoveries of lead compounds from NPs, various drawbacks have prompted pharmaceutical industries to reduce NP-based drug discovery programs (Newman and Cragg 2007; Harvey 2008). NP discovery deals with the identification of bioactive metabolites, whereas metabolomics focuses on comprehensive analysis, identification of all metabolites present in a biological sample, and exaction of a specific compound from a complex dataset of metabolic mixtures (Verpoorte 1998; Atanasov et al. 2021). However, both paradigms utilize similar analytical techniques and share the goal of identifying bioactive metabolites with therapeutic properties. As stated, metabolomics represents a holistic approach that attempts to analyze complex mixtures of metabolites utilizing high-throughput techniques to identify known and unknown novel compounds without prior isolation or purification (Rinschen et al. 2019). Different strategies for developing potent NP and NP analogs with bioactive properties are illustrated in Fig. 3.

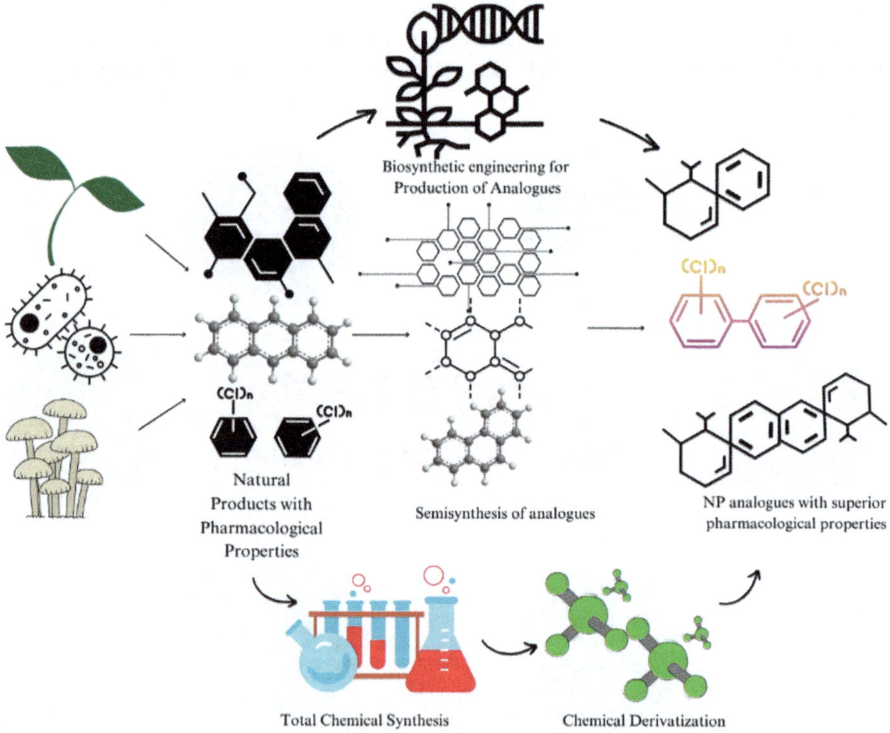

Fig. 3 Strategies for producing natural products and NP analogs with bioactive properties

Morphine was the first pharmacologically active compound, isolated from *Papaver somniferum* by Friedrich Sertürner around 200 years ago, which initiated the search to identify promising compounds from natural resources (Krishnamurti and Rao 2016). The discovery of penicillin during the Second World War is another historical event that expanded the screening of microorganisms in the search for drug molecules. Around 80% of the drugs till the 1990s originated from NPs or analogs inspired by them (Cook and Wright 2022). Several bioactive molecules serve as antibiotics, antiparasitics, immunosuppressive agents, growth promotors, herbicides, and insecticides, which are clinically essential for veterinary and agricultural applications (Abdel-Razek et al. 2020). Antibiotics such as penicillin, erythromycin, and tetracycline and antimalarials like quinine and artemisinin originated from NPs (Srivastav et al. 2020). Additionally, antiparasitics such as avermectin, anticancer drugs like doxorubicin, and lipid control agents like lovastatin are NP analogs drugs that saved millions of lives (Wink 2015; Keller 2019; Gorlenko et al. 2020). Traditionally, actinomycetes, fungi, and angiosperms have been the primary sources of drug discovery. In 2007, 91 compounds of plant origin, along with 17 compounds from cyanobacteria and 18 from marine organisms, were in clinical trials (Zhang et al. 2017; Abdel-Razek et al. 2020).

Natural products are categorized into five primary types according to their chemical composition: alkaloids, fatty-acid-derived substances, polyketides, terpenoids, and steroids, shikimate-derived compounds, and non-ribosomal polypeptides (Harvey 2008; Thomford et al. 2018; Atanasov et al. 2021). Secondary metabolites of plants possess great pharmacological importance because of their diverse pharmacological activities (Gorlenko et al. 2020). As plant-derived metabolites and bioactive compounds are easily available and effective against several pharmacological targets, they are more reliable and preferable as compared to synthetic compounds (Wink 2015). Plant secondary metabolites possess antioxidant, antimicrobial, anticancer, antidiabetic, antihypertensive, and anti-inflammatory activities (Verpoorte 1998). Terpenes are secondary metabolites of plants that are lipophilic in nature and thus can easily interact with cell membranes and modulate the activity of membrane proteins or ion channels (Sun and Wesolowski 2021). Therefore, many herbs with terpenes are used as relaxants against cramps. Several plants with sesquiterpene lactones are traditionally used because of antibiotic, antifungal, insecticidal, cytotoxic, and other properties (Harvey 2008; Wink 2015). Artemisinin is an excellent drug against *Plasmodium falciparum* isolated from *Artemisia annua* (Srivastav et al. 2020).

Among the microbes, bacteria are the treasure house of secondary metabolites with promising therapeutic properties (Abdel-Razek et al. 2020). Dr. William Colley pioneered the use of bacteria in cancer treatment, employing a combination of *Streptococcus pyogenes* and *Serratia marcescens* cultures against sarcomas (Sun and Wesolowski 2021). This remarkable accomplishment significantly impacts the field of bacteria-mediated cancer therapy and the development of bacteria-derived compounds for treating many human diseases, including cancer (Macalino et al. 2015; Abdel-Razek et al. 2020). Numerous clinically significant anticancer medications, including bleomycin, carfilzomib, ixabepilone, mitomycin C, rapamycin, romidepsin, pentostatin, actinomycin D, and doxorubicin, originate from bacterial sources (Abdel-Razek et al. 2020). Actinomycin D is derived from *Streptomyces antibioticus* and *S. parvulus* and was approved for treating choriocarcinoma, Ewing sarcoma, testicular cancer, rhabdomyosarcoma, solid tumors, and Wilms tumors alone or in combination with other drugs (Abdel-Razek et al. 2020). Actinomycin D was reported to possess a broad range of activity because of its toxicity in tissue necrosis, dermatotoxicity, gastrointestinal enterotoxicity, and myelosuppression (Abdel-Razek et al. 2020; Fouillaud and Dufossé 2022). Similarly, bleomycins, a group of glycopeptide-based anti-tumor antibiotics, was isolated from *S. verticillus* and used for the treatment of Hodgkin's and non-Hodgkin's lymphoma, melanoma, ovarian cancer, and squamous cell cancers. Many other anti-tumor drugs from the bleomycin family include phleomycins (from *Streptomyces verticillus*), tallysomycins, and zorbamycin isolated from *S. verticillus*, *Streptoalloteichus hindustanus*, and *S. flavoviridis*, respectively (El-Tarabily and Sivasithamparam 2006; Abdel-Razek et al. 2020; Sun and Wesolowski 2021). The conventional methods of discovering novel bioactive molecules from bacteria involve isolating and culturing bacteria in specific media, conducting phenotypic screening, purifying compounds, investigating pharmacological properties, and their mechanism of action using *in vitro* and *in vivo* models (Newman and Cragg 2007). The culture-independent metabolomics approach with metagenomics is more reliable compared to traditional methods for analyzing the complete metabolic profile of bacteria (Kumar et al. 2014). Several studies have reported new secondary metabolites having broad-spectrum activity from polyketide synthases (PKS) and non-ribosomal peptide synthases (NRPS) using the metagenomics approach (Abdel-Razek et al. 2020; Sun and Wesolowski 2021). For instance, bryostatin 1 is a polyketide produced by bryozoan *Bugula neritina* and has been evaluated for the treatment of cancers in many clinical trials (Abdel-Razek et al. 2020). Actinomycetes are considered a prolific source of antibiotics that produce various types of antibiotics belonging to glycopeptides, macrolides, β-lactams, and tetracyclines class. Examples of FDA-approved drugs derived from marine actinomycetes include salinosporamide A from *Salinispora tropica* and salinipyrones A and B from *Salinispora pacifica* (Abdel-Razek et al. 2020; Sun and Wesolowski 2021).

Various marine organisms, including algae, sponges, ascidians, bryozoans, mollusks, and soft corals, produce several secondary metabolites with potential therapeutic properties (Abdel-Razek et al. 2020). Sponges are the most extensively studied invertebrates, from which many promising bioactive compounds have been isolated in the last 50 years (Zhang et al. 2017). Of the over 5,000 natural products extracted from sponges, approximately 1,000 bioactive compounds were isolated in 2009 alone (Abdel-Razek et al. 2020; Lee et al. 2021). Sponges have been reported to produce over 3,300 antibiotics and other bioactive compounds (Abdel-Razek et al. 2020; Sun and Wesolowski 2021). Similarly, since the cutting-edge discovery of penicillin from *Penicillium notatum*, fungi have been recognized as significant sources of secondary metabolites and therapeutic agents (Keller 2019; Sun and Wesolowski 2021). Furthermore, many drugs, such as cyclosporine, fusidic acid, griseofulvin, and mycophenolic acid, were derived from fungal metabolites

(Abdel-Razek et al. 2020; Fouillaud and Dufossé 2022). Statins, another important class of antilipidemic drugs derived from microbial secondary metabolites, play a key role in treating cardiovascular diseases (Bills and Gloer 2016; Fouillaud and Dufossé 2022). Mevastatins, derived from *Penicillium citrinum*, and lovastatin, derived from *Aspergillus terreus*, are notable examples of this class. Taxol, a drug known for its anti-tumor and antifungal properties, is also derived from *Sorangium cellulosum* (Abdel-Razek et al. 2020).

Analysis of natural product mixture is the first step of metabolomics analysis, followed by structural optimization and deposition in online chemical databases (Thomford et al. 2018). The storage of structures and details of identified compounds isolated from natural resources in databases has smoothened the drug discovery and development process, rendering metabolomics a promising approach in this field (Berman 2005). One such database is the Dictionary of Natural Products (DNP), which contains over 1,70,000 unique plant secondary metabolites so far (Sorokina and Steinbeck 2020). Several online databases are available that store metabolomics datasets from both nuclear magnetic resonance (NMR) and mass spectroscopy (MS) for a variety of metabolites. Some examples include the Biological Magnetic Resonance Bank (BMRB), the Human Metabolome Database (HMDB), the Yeast Metabolomics Database (YMDB), and the Golm Metabolome Database (GMD) (Sorokina and Steinbeck 2020). MINE is a newly developed database, data of which are derived from the Kyoto Encyclopedia of Genes and Genomes (KEGG) (Jeffryes et al. 2015). This encompasses metabolites identified in LC-MS spectra that have not been previously observed but are likely to occur based on known metabolites and common biochemical reactions (Jeffryes et al. 2015; Sorokina and Steinbeck 2020). The KEGG Compound database holds around 5,71,000 compounds, of which approximately 93% are listed in the PubChem database (Sorokina and Steinbeck 2020). Some important natural product databases are depicted in Table 1.

3 Advanced Metabolomics Techniques for Lead Discovery and Target Identification

Unraveling the pathophysiological basis of a disease is the most crucial step of clinical research (Alarcon-Barrera et al. 2022). Metabolomics, with high-throughput analytical techniques, provides a comprehensive profile of the metabolites and helps to detect the tiny changes in small molecules involved in disease progression (Cuperlovic-Culf and Culf 2016). The broad-range physiological properties and complexity of the metabolome require the application of multiple standard analytical techniques for metabolite profiling (Kumar et al. 2014). Spectroscopic methods are the commonly used analytical techniques to study the compounds of interest and identification of a diverse group of metabolites. Among the techniques, nuclear magnetic resonance (NMR), gas chromatography (GC), mass spectrometry (MS)

Table 1 List of important natural product databases

Database	Description	Estimated size of compounds	Sources
3DMET	A database of three-dimensional structures and molecular properties of metabolites	18,248	Generalistic
BiGG models	Genome-scale metabolic network reconstruction database linked to KEGG, PubChem, and others	7,339	Metabolites
CEMTDD	CEMTDD stands for Chinese Ethnic Minority Traditional Drug Database for the identification of lead molecules and the elucidating relationship between herbs, compounds, biological targets, and other related diseases	4,060	Plants
CHDD	CHDD (Chinese Traditional Medicinal Herbs Database) is a database of small molecules obtained from Chinese medicinal plants	>30,000	Plants
ChEBI	ChEBI (Chemical Entities of Biological Interest) is a freely available chemical database of small compounds	15,736	Chemical compounds
ChEMBL	ChEMBL is a curated repository of drug-like bioactive molecules, including bioactivity, molecular, and genomic data	23,99,743	Small molecules
ChemDB	A chemical database of small molecules, founded in 2005	5 M commercially available small molecules	Plants
COCONUT	COCONUT (Collection of Open Natural Products) is a freely available database for the storage, search, and analysis of compounds isolated from NPs	4,07,270	Plants
Dr. Duke's Phytochemical and Ethnobotanical database	An online database of phytochemicals that stores biological activity and toxicity data. It was developed by James A. Duke for pharmaceutical and biomedical research	49,788	Plants
KEGG Compound Database	Database for chemical structures integrated with genomics	19,270	Metabolites
NPASS	NPASS (Natural Product Activity and Species Source) database is a freely accessible database that stores information on the source and biological activity of natural products	35,032	Animal, fungi, plant

(continued)

Table 1 (continued)

Database	Description	Estimated size of compounds	Sources
PhytoHub	Freely available database for phytochemicals and metabolites	1,200	Plants and animals
PubChem	PubChem is the World's largest accessible chemical database, which is maintained by the National Library of Medicine (NLM) and the National Institute of Health (NIH)		Generalistic
UNPD	Universal Natural Product Database	2,13,100	Generalistic
ZINC	ZINC is a freely available database of commercially available compounds developed by Irwin and Shoichet Laboratories, University of California	Contains over 750 million purchasable compounds and over 85,190 natural compounds	Generalistic

coupled with separation techniques liquid chromatography (LC), and supercritical fluid chromatography (SFC) employed widely in this field (Wang et al. 2010; Kumar et al. 2014; Cuperlovic-Culf and Culf 2016). These techniques could reveal a deeper understanding of changes in normal metabolite profiles under diseased conditions and help in the identification of biomarkers (Wang et al. 2010). Recent developments in metabolomics-based methods introduced several new techniques, including LC and GC combined with high-resolution mass spectrometry (HRMS) (Cuperlovic-Culf and Culf 2016; Guo et al. 2023). These new methods provide new ways to detect and identify biomarkers (e.g., biomarkers of effect) and exposure (relevant xenobiotic mixtures or emerging compounds) simultaneously in environmental or biological samples (Holmes et al. 2007). The choice of the analytical technique is determined by the sample matrix, the amount of the sample, and the properties and concentration of the metabolites (Wang et al. 2010). Nuclear magnetic resonance (NMR) and mass spectrometry (MS) are the two major techniques that are widely employed in metabolomics (Wang et al. 2010; Aminov 2022; Guo et al. 2023).

3.1 Role of Nuclear Magnetic Resonance in Drug Discovery

Metabolomics involves two sequential phases: firstly, selecting an analytical method to reveal the metabolite profile in a biological sample, followed by analyzing and interpreting the data generated (Cuperlovic-Culf and Culf 2016). The NMR is the most powerful technique for characterizing the metabolites found within biological samples (Lu and Chen 2017). It is a rapid and highly reliable spectroscopic method dependent on the absorption and reemission of energy by atomic nuclei, which results from the variation in the external magnetic field (Pellecchia et al. 2002). The metabolomics data generated depends on the type of atomic nuclei targeted by the applied magnetic field (Alonso et al. 2015). In addition to metabolite profile, NMR is one of the most suitable methods for analyzing the binding of ligands to their biological target. It also provides structural detail of the protein-ligand interactions and helps to optimize the weak-binding hits into high-affinity leads (Emwas et al. 2020). NMR can detect molecules even when present at a very low concentration, with a detection limit as low as one millimole per liter (1 mmol/L). This technique identifies the compounds through analysis of the chemical shifts or J-coupling patterns arising from the magnetic interaction between the nearby nuclei, which provides insight into the molecular conformation of the compounds (Cuperlovic-Culf and Culf 2016; Guo et al. 2023). NMR utilizes nuclei with nonzero spins, such as 1H (proton), 13C (carbon), 15N (nitrogen), and 31P (phosphorus), which absorb and reemit electromagnetic (EM) radiation at specific frequencies. It is a highly sensitive technique that can analyze complex mixtures without the need for prior separation of the components (Emwas et al. 2020). In NMR, a compound is introduced into a magnetic field, and the isotopes within the compounds, for example, 1H, 13C, 14N, 15N, and 17O absorb EM radiation and resonate at frequencies determined by their positions within the small molecule (Alonso et al. 2015; Emwas et al. 2020). Following this interaction, the resulting NMR spectrum, comprising peaks distributed at various positions and intensities, is unique for each compound. NMR is non-discriminating and can detect any compound containing protons, carbon, nitrogen, or oxygen. It typically demands minimal to no sample preparation (Pellecchia et al. 2002; Lu and Chen 2017). However, it is less sensitive compared to MS and thus often misses low-abundance metabolites. Hydrogen is the most frequently targeted nucleus (1H-NMR) in the analysis of biological samples due to the abundant presence of hydrogen nuclei, making it the nucleus of choice for such analyses (Pellecchia et al. 2002). Moreover, carbon (13C-NMR) and phosphorus (31P NMR) are less frequently targeted to acquire additional information on the

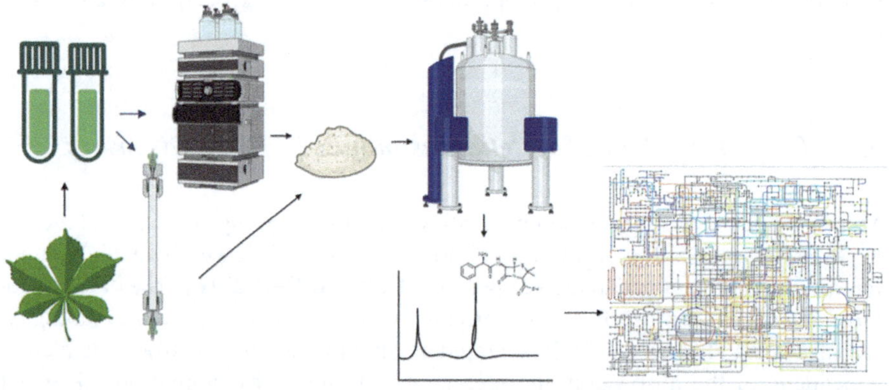

Fig. 4 Generalized workflow of metabolomics data using separation techniques and NMR

specific metabolites (Emwas et al. 2020). The resultant spectral data quantifies the concentration of the metabolite and provides its structural information (Fig. 4).

NMR has been integral to drug discovery since its discovery in 1946 by Felix Bloch and Edward Purcell (Emwas et al. 2020). Due to its numerous distinctive advantages, NMR has emerged as the gold standard in medical and pharmacological studies. It is highly sought in drug discovery because of its wide application in molecule identification and structural elucidation (Pellecchia et al. 2002). NMR is quantitative, nondestructive, and highly reproducible, allowing researchers to achieve measurements at an atomic level under various temperatures, time points, and concentrations (Pellecchia et al. 2002; Emwas et al. 2020). NMR provides crucial insights into the interaction between ligands and targets and the structural alterations ensuing from binding. This information is vital for assessing the effectiveness of drug-like compounds and their potential influence on host physiology. The utilization of NMR has proven highly advantageous in the realms of drug-protein interactions, stereochemistry analysis, isomer identification, and drug toxicity evaluation (Cuperlovic-Culf and Culf 2016; Emwas et al. 2020).

The protein-protein, protein-ligand, and protein-nucleic acid interactions are of particular interest in structural biology and drug discovery (Wang et al. 2010). NMR spectroscopy is one of the most powerful methods for describing macromolecular complexes with both high and low affinity in this context (Alonso et al. 2015). One or two frequency axes serve as reference points for spectral data obtained through NMR techniques. In one-dimensional NMR (1D-NMR), a single-frequency axis is utilized, where peaks of each molecule are positioned according to their respective resonant frequencies (Emwas et al. 2020). 1D-NMR is most widely employed for high-throughput metabolomics investigations. In contrast, 2D-NMR spectra, which utilize two-frequency axes, are primarily used for characterizing compounds that cannot be identified through 1D-NMR spectroscopy (Pellecchia et al. 2002). The additional dimension in 2D-NMR helps in distinguishing spectral peaks that might otherwise overlap, providing vital information about the chemical characteristics of the metabolite. Both the 1D and 2D frequency axes are referenced by the chemical shift, which is typically expressed in parts per million (ppm) (Pellecchia et al. 2002). The calculated chemical shift is determined by subtracting the resonance frequency from that of a reference substance and then dividing it by the operating frequency (OF) of the spectrometer (Emwas et al. 2020). NMR spectroscopy can be employed in investigating drug-target interactions through the fragment-based drug design (FBDD) approach, which can be classified further into two primary branches: the target-based (protein-based) approach and the ligand-based (drug-based) approach (Pellecchia et al. 2002). In the target-based approach, the changes due to the ligand binding on the targets are monitored using a method called structure-activity relationship (SAR) analysis via NMR (Lu and Chen 2017). Conversely, in the ligand-based approach, the binding affinity or nonbinding behavior of the drug or promising compounds is screened using techniques such as saturation transfer difference (STD) and other nuclear overhauser effect (NOE) type methods (Pellecchia et al. 2002). Macromolecular NMR is crucial to numerous ongoing structural and functional genomics projects across Europe, Japan, and the United States. Currently, over 2000 NMR structures have been registered in the Protein Data Bank (Emwas et al. 2020).

3.2 Role of Mass Spectrometry in Drug Discovery

Mass spectrometry (MS) is one of the widely employed, highly sensitive techniques for the separation of metabolites in biological systems (Wang et al. 2010). This technique requires the separation of the target samples using various separation techniques like GC, LC, high-pressure liquid chromatography (HPLC), ultra-HPLC, and capillary electrophoresis for sample analysis (Cuperlovic-Culf and Culf 2016; Guo et al. 2023). The choice of separation technique depends on the nature of the targeted metabolites. For instance, metabolic profiling of amyotrophic lateral sclerosis (ALS) and Parkinson's disease (PD) using MS utilizes LC and coulometric array detection for prior separation of the sample (Alseekh et al. 2021). Among the separation techniques used for MS-based detection of metabolites, LC combination with mass spectroscopy (LC-MS) is one of the high-resolution techniques employed for identifying, quantifying, and characterizing metabolites (Klont and Hopfgartner 2021). The metabolic profile is determined by analyzing the peaks recorded at different retention times. The accurate identification of the compounds based on the peaks generated is a major challenge (Meissner et al. 2022). Therefore, untargeted and targeted metabolomics, two different approaches, are used for the interpretation of metabolome data (Cuperlovic-Culf and Culf 2016; Aminov 2022). In untargeted metabolomics, a complete metabolite profile analysis is done using advanced multivariate analyses that identify known and unknown compounds (Meissner et al. 2022). Conversely, targeted metabolomics is quantitative and focuses on the characterization of known metabolites. Apart from LC-MS and GC-MS techniques, less commonly utilized metabolomics methodologies involving MS include ultra-high-performance liquid chromatography and mass spectrometry (UPLC-MS), high-performance liquid chromatography-diode array detection electrospray ionization tandem mass spectrometry (HPLC-DAD-ESI-MS), capillary electrophoresis-mass spectrometry (CE-MS), high-performance liquid chromatography-electrospray ionization mass spectrometry (HPLC-ESI-MS), matrix-assisted laser desorption ionization-time-of-flight mass spectrometry (MALDI-TOF-MS), ultra-high-performance liquid tandem chromatography quadrupole time-of-flight mass spectrometry (UHPLC-QTOF-MS), and inductively coupled plasma-mass spectrometry (ICP-MS) (Alseekh et al. 2021; Klont and Hopfgartner 2021; Guo et al. 2023). Moreover, Fourier transform infrared or Raman scattering spectroscopy is also used in metabolomics research. According to reports, LC-MS is the most commonly utilized technique in the field of metabolomics, accounting for 83% of applications, followed by GC-MS (30%), NMR (26%), imaging mass spectrometry (6%), MALDI-MS (4%), and CE-MS (3%) (Klont and Hopfgartner 2021; Meissner et al. 2022). The MS-based identification of metabolites typically comprises several stages. Firstly, the sample is prepared and extracted; secondly, metabolites are separated using column chromatography methods such as LC, GC, or EC. Thirdly, the metabolites are ionized using an ionization source, followed by separation of the sample based on mass-to-charge (*m/z*) ratio using a mass analyzer, detection of the peaks generated, and analysis (Alseekh et al. 2021; Klont and Hopfgartner 2021; Guo et al. 2023).

The understanding of changes in proteins or drug targets mediated by drugs is a challenge for pharmacological intervention (Alseekh et al. 2021). Through recent advances, MS-based proteomics allows rationalized profiling of complete proteome, their properties, and functional modulations on multiple levels within a few hours (Guo et al. 2023). The modern proteomics techniques with computational tools and databases have made easy characterization of proteomes to gene products, reveal their overall role in protein signaling networks, and produce comprehensive functional annotation on proteome-wide scale within a few seconds (Cuperlovic-Culf and Culf 2016; Aminov 2022).

4 Application of Metabolomics in Drug Discovery

The identification of promising compounds with novel structures and drug-like properties is the first step in drug discovery (Macalino et al. 2015). This is accomplished by high-throughput screening of chemical libraries containing thousands to millions of lead molecules (Van Drie 2007). Compounds exhibiting strong affinity for targets are considered high-affinity ligands, which are subjected to further analysis through structure-based drug design (SBDD) and cell-based assays to determine their potential as orally active drugs (Macalino et al. 2015). A ligand with high binding affinity may not be a true lead, as compounds having high MW and hydrophobicity result in high binding affinity, which are unfavorable drug-like characteristics. Compounds with poor physiological properties also lean toward providing deceptive biological activity, such as inhibition of target protein through induced protein aggregation, micelle formation, or precipitation (Aminov 2022). Selectivity is of equal importance to efficacy, which increases the binding affinity of promising drug-like compounds to their target. In addition to improving efficacy, a drug-like compound has to be optimized so that it can specifically fit in the active site of the target (Rinschen et al. 2019; Alarcon-Barrera et al. 2022). Moreover, for successful drug discovery, the similarity of lead compounds and drug targets is evaluated using various secondary assays, which identify functional homologs with substantial sequence and structural resemblance. Nevertheless, the absence of thorough toxicity analysis and efficiency evaluation often leads to the failure of potential leads during clinical trials (Lu and Chen 2017; Alseekh et al. 2021). Furthermore, assessing the bioavailability and stability of compounds represents a critical yet challenging aspect of drug discovery. Metabolomics thus attempts to assist in identifying and eliminating faulty drug leads before clinical trials. The introduction of new metabolomics approaches, coupled with advancements in computational technologies, has substantially broadened and enhanced the success of drug discovery (Kumar et al. 2014; Alonso et al. 2015).

4.1 Role of CADD in Drug Discovery and Development

Computational methods have been integral to the drug discovery process since the 1980s. The computer-aided drug design (CADD) approach is divided into ligand-based drug design (LBDD) and structure-based drug design (SBDD), based on the availability and utilization of the target structure (Gschwend et al. 1996; Ferreira and Andricopulo 2018). CADD involves the prediction of the structure of drug targets, identification of active sites for ligand binding, screening of compounds for drug-like properties, calculation of molecular properties and ADMET (absorption, distribution, metabolism, elimination, and toxicity) parameters, and QSAR (Quantitative Structure-Activity Relationship) analysis identifying the biological activity of identified leads (Elokely and Doerksen 2013; Wang and Zhu 2016; Chandrasekaran et al. 2018). In CADD, compounds with undesirable properties can be eliminated more easily to identify compounds with more chances for success in later phases of drug discovery and development (Van Drie 2007; Macalino et al. 2015).

In SBDD, the structure of the drug target is known, but the ligand structure is unknown. Conversely, in LBDD, the structure of the ligand is known, but the target structure is unknown (Meng et al. 2011). SBDD stands out as a more dependable approach for designing, identifying, and optimizing structures of promising drug candidates, rendering them suitable for clinical testing (Maia et al. 2020). However, for this approach, the three-dimensional structure of the drug target or protein is essential, which can be obtained through NMR, X-ray crystallography, cryo-electron microscopy, etc. (Acharya et al. 2011). Another advantage of CADD is the *in silico* prediction of structure using homology modeling, which enables the prediction of the structure of proteins or drug targets whose experimental 3D structures are not available (Almihyawi et al. 2022). The idea of employing protein structure in drug discovery for designing novel therapeutics originated in the 1960s, with the initial prediction of a protein structure through X-ray crystallography (Van Drie 2007). The pioneering research by John Kendrew and Max Perutz on the crystal structure of myoglobin and hemoglobin unveiled crucial insights into the oxygen-carrying functions of these proteins. Furthermore, their work shed light on the molecular mechanisms underlying sickle cell anemia, suggesting potential treatment options (Gurung et al. 2021). Similarly, the groundbreaking contributions of Fred Sanger and Dorothy Hodgkin in determining the amino acid sequence and three-dimensional structure of insulin, respectively, played a pivotal role in the development of synthetic insulin for the treatment of diabetes (Macalino et al. 2015; Lee et al. 2022). These novel studies epitomize how understanding protein structure and function, coupled with three-dimensional structural data, significantly influences the drug discovery process (Macalino et al. 2015). Additionally, the gradual increase of protein or target structures deposited in the Protein Data Bank (PDB) encourages the expansion of the CADD process (Macalino et al. 2015). Therefore, CADD is considered the most powerful approach for drug discovery and development as it eliminates complex compounds with molecular obesity, which is considered one of the prime reasons for the failure of drug candidates. As per Polanki's analysis, a positive correlation exists between the simplicity of the drug molecule and its market success (Van Drie 2007; Meng et al. 2011; Macalino et al. 2015; Maia et al. 2020). On the contrary, LBDD is an indirect approach that begins with a single compound or a set of compounds targeting a specific entity without prior knowledge of its three-dimensional structure (Attwood et al. 2020; Ajjarapu et al. 2022). This approach is based on the principle that molecules with analogous structures are likely to exhibit similar properties. Similarity-based search, QSAR (quantitative structure-activity relationship), and pharmacophore modeling are some of the techniques commonly used in this method (Maia et al. 2020).

4.2 CADD in Drug Target Identification

Drug targets are biological entities whose functionality is altered upon binding with a specific compound, leading to a physiological response (Cuperlovic-Culf and Culf 2016). The traditional approach of drug target identification, analysis of the interaction of drugs with the targets, and the physiological response via biological experiments are expensive and time-consuming. As a result, it is substituted with a computational approach from linear mode to network mode, i.e., from one drug, one target, one disease approach to multidrug, multi-target, multi-disease approach (Alonso et al. 2015; Floris et al. 2018). The identification of probable drug targets is the initial stage of the drug discovery and development. G protein-coupled receptors (GPCRs) and kinases have served as common drug targets for several years (Alhosaini et al. 2021). Nevertheless, with recent technological advances, new targets showing promising specificity have been identified in the last few years to cure many diseases. One such notable success was "Sotorasib" the first selective KRAS (Ki-ras2 Kristen rat sarcoma viral oncogene homolog) inhibitor, approved for the treatment of *KRAS* G12C mutant tumors in 2021 (Rask-Andersen et al. 2011; Zdrazil et al. 2020).

A target is considered druggable if it plays a vital role in the disease, and inhibiting or modifying its activity can mitigate the adverse effects caused by the disease (Attwood et al. 2020; Almihyawi et al. 2022). Genetic-driven drug target identification and validation is a promising tool that started in 2003. This approach revealed the molecular basis of autosomal dominant hyperglycemia (ADH). Gain of function (GOF) of the PCSK9 gene encoding subtilisin protein results in ADH, while its loss of function (LOF) reduces coronary heart disease (CHD) by decreasing low-density lipoprotein (LDL) cholesterol (Floris et al. 2018; Zdrazil et al. 2020). Therefore, inhibition of the *PCSK9* gene or PCSK9 protein could drop LDL concentration and prevent CHD. Genome-wide association studies have contributed to an increase in the drug discovery process, rising from 2% to 8.2% among approved drugs. Rask-Andersen et al. (2011) identified 435 effective drug targets in the human genome modulated by 989 unique drugs through 2242 drug-target interactions (Rask-Andersen et al. 2011). According to studies, many cell surface receptors and extracellular proteins have been identified as suitable drug targets for treating various diseases. Numerous antibodies have been developed to treat cancer and inflammatory diseases, including antibody-drug conjugates that specifically target extracellular proteins and cell surface receptors (Rask-Andersen et al. 2011; Attwood et al. 2020). The characterization of drug target properties can be accomplished by comparing RNA (transcriptomics) or protein expression (proteomics) in target tissues with those in healthy tissues.

The G protein-coupled receptors (GPCRs) consist of approximately 800 genes, representing about 2% of the human genome. As of 2020, among the 1500 drugs approved by the FDA, approximately 460 drugs target GPCRs. Among these, around 94% of the drugs target Class A GPCR, 4% target Class B GPCR, and 2% target Class C and Class F GPCRs (Alhosaini et al. 2021). For instance, drugs such as albiglutide, pramlintide, and sitagliptin are approved for treating type 1 and type 2 diabetes and target GPCR receptors (Alhosaini et al. 2021). Similarly, α- and β-androgenic receptors are established targets for most cardiac and antiplatelet drugs. Mogamulizumab for treating leukemia also targets GPCRs. Network-based approaches methods are widely used to predict novel drug targets (Murthy et al. 2013). Thermal proteome profiling is another new technique for drug target identification that measures the changes in the proteome using quantitative MS (Guo et al. 2023). Validation of the identified drug target is crucial as it ensures the involvement of the target in the disease mechanism (Wang et al. 2010). This can be achieved through structure-activity relationship analysis, genetic manipulation, mutant generation, altering signaling pathways downstream of the presumed target, knockout study, and overexpression analysis (Baxevanis and Ouellette 2004). Besides, genomics, functional genomics, machine learning, and artificial intelligence (AI) are also used for the identification of novel drug targets.

4.3 Metabolic Pathway Databases for Drug Discovery

Metabolic pathways are interlinked reactions of metabolites or reactants, products of which are utilized in various cellular activities (Nicholson et al. 1999). Understanding the mechanism of a drug target inhibitor or metabolite through mapping the cellular and molecular interactions is challenging for biologists. However, in-depth analysis of the signaling and metabolic pathways, reaction participants, and interactions of genes, proteins, and enzymes are of much importance in the drug discovery process (Wittig 2001; Clish 2015; Rinschen et al. 2019). In metabolomics, metabolites or proteins are screened for desirable changes in gene expression, signal transduction, or inhibition of enzyme activity. Thus, analyses of genetic and metabolic pathways aid in delineating the target's involvement in the disease and identify crucial intermediates that could function as alternative drug targets (Sun and Wesolowski 2021). For instance, various interconnected pathways are linked to carbohydrate metabolism, including glycolysis, Krebs cycle, gluconeogenesis, glycogenesis, and others, through several interconnected intermediates (Al-Masaudi et al. 2017). Von Gierke's disease is an autosomal disorder resulting from a deficiency of glucose-6-phosphatase (G6Pase). This enzyme breaks down stored liver glycogen into glucose during starvation. Mutations in G6Pase disrupt this process, causing hypoglycemia by inhibiting the final step of gluconeogenesis (Zhang and Tang 2018). Conversely, cancer cells primarily rely on ATP produced through glycolysis as their primary energy source, a phenomenon referred to as the Warburg effect. P53, a tumor suppressor gene (TSG), is activated by stress signals, including DNA damage, oxidative stress, and activation of oncogenes (Shulaev 2006). Activation of P53 leads to activation of downstream regulatory genes, thereby causing cellular senescence, cell cycle arrest, and apoptosis (Zhang and Tang 2018). Mutations arising from the LOF of TSGs or the GOF of oncogenes disrupt the well-balanced environment regulated by P53. This disruption leads to increased glucose uptake and uncontrolled cell-growth in the body (Zhang and Tang 2018). Therefore, P53 can elucidate the metabolic phenotype of cancer, and the interconnected intermediates associated with P53 expression can potentially serve as potential targets for anticancer drugs. Some of the widely used pathway databases in computational drug discovery include KEGG, Pathway Commons, MetaCyc, etc. (described in Table 2) (Wittig 2001; Baxevanis and Ouellette 2004; Likić 2006).

On the contrary, proteome interaction is another novel approach in drug discovery, which focuses on the interactions among proteins, unveiling their associations with other cellular proteins that could serve as alternative drug targets (Kleiner 2019; Meissner et al. 2022). The Biomolecular Interaction Network Database (BIND) is an example of a freely accessible database that archives protein-protein, protein-DNA, and genetic interactions in yeast (*Saccharomyces cerevisiae*), *Caenorhabditis elegans*, and *Drosophila melanogaster* (Baxevanis and Ouellette 2004). The STRING is another notable database for predicting protein-protein interaction and is widely used for drug target identification (Szklarczyk et al. 2019). Some of the important databases available for protein-protein or gene-gene interactions are depicted in Table 2.

4.4 *In Silico Screening of Metabolites or Compounds for Physicochemical and Pharmacokinetic Properties*

The physicochemical and molecular properties of a drug molecule determine its potential therapeutic properties. Among these properties, boiling point (BP), bioconcentration factor (BCF), melting point (MP), partition coefficient (LogP), and water solubility are essential in determining the success rate of a drug candidate (Jambhekar and Breen 2013). The lipophilicity of a drug-like compound is one of

Table 2 Metabolic pathway and interaction analysis databases widely used in the CADD approach

Database type	Name of the database	Description
Metabolic pathway databases	KEGG Pathway Database	Integrated database for metabolic and regulatory pathways. It is used for mapping several metabolic pathways involved in cellular processes, human diseases, genetic information processing, and drug development
	MetaCyc	A curated metabolic database encompassing metabolic pathways from various life domains. It includes pathways associated with both primary and secondary metabolism. Presently, it contains 3,105 pathways, 18,566 reactions, and 18,973 metabolites
	PlantCyc	Database consists of metabolic pathways for approximately 500 plants
	Pathway Commons	Integrated resource of a publically available database of biological pathways and molecular interactions. It integrates data from 22 public databases
	HumanCyc	Curated database for human metabolic pathways, human genome, enzymatic reactions, and human metabolites
	Reactome	A freely accessible online repository of biological pathways, with bioinformatics tools for analyzing and visualizing the metabolites and small molecules implicated in these pathways
	BioCyc	A database comprising 20,043 metabolic pathways from eukaryotes and microbes. It contains data from 1,46,000 publications
Interaction analysis databases	BIND	The Biomolecular Interaction Network Database for molecular interactions, pathways, and protein interactions. Also allow to submit new records
	STRING	A repository containing known and predicted protein-protein interactions derived from genomic context prediction, laboratory experiments, automated text mining, co-expression analyses, and existing knowledge stored in databases
	MINT	Public and open-source database for prediction of protein-protein interactions
	BioGRID	The Biological General Repository for Interaction Datasets (BioGRID) is an openly accessible database for analyzing genetic and protein interactions in model organisms and humans. It currently contains over 17,40,000 interactions

the main parameters that determine absorption, distribution, permeability, and excretion within biological systems (Blake 2000; Malan and Chetty 2002). The lipophilicity of a drug determines the rate of passive diffusion of the molecule across the hydrophobic cell membrane. It influences the metabolism, solubility, and toxicity of the compound (Blake 2000; Malan and Chetty 2002; Chandrasekaran et al. 2018). Aqueous solubility is another crucial property linked to the effectiveness of drug candidates, determining their absorption, distribution, and elimination from the body. Drugs with poor solubility get eliminated before entering the bloodstream (Malan and Chetty 2002). Besides molecular size, hydrogen-bond acceptors,

H-bond donors, and the charge of the compound are also crucial for the efficient binding of the drug candidates to their targets. The "rule of five" (RO5), formulated by Christopher A. Lipinski in 1997, evaluates the drug-like characteristics of a compound and predicts whether a compound or metabolite with pharmacological attributes can be administered orally as a drug (Lipinski et al. 1997). Consideration of this rule is crucial during the optimization of lead molecules to enhance their activity and selectivity. According to Lipinski's RO5, a compound qualifies as an orally active drug if it meets specific criteria: a molecular mass of under 500 Da, no more than five H-bond donors, no more than ten H-bond acceptors, and a calculated partition coefficient (log P) value not exceeding five (Lipinski et al. 1997). Apart from the molecular properties described by Lipinski, the oral bioavailability of drug-like compounds is also influenced by the number of rotatable bonds and the polar surface area (PSA) (Veber et al. 2002). Therefore, the bioactive compounds identified from different sources are screened and filtered for drug-like properties using various computational tools prior to molecular docking (Veber et al. 2002).

Optimization of ADME properties and toxicity analysis of drug candidates is the most challenging part of drug discovery (Ferreira and Andricopulo 2019). However, the calculation of ADMET properties of a small molecule of the metabolite is essential to predict its behavior inside the living body (Chandrasekaran et al. 2018). In addition to demonstrating efficacy, a drug molecule should exhibit suitable ADMET properties at therapeutic doses (Ferreira and Andricopulo 2019). The calculation of ADMET properties of a potential drug candidate emphasizes several factors such as absorption site, distribution rate, time taken for the drug to metabolize, mode of action, rate, and route of excretion and its toxic effects on the body (Guan et al. 2019; Ferreira and Andricopulo 2019). Various tools are used for the prediction of ADMET properties in the CADD approach, including ADMET predictor, ADME suite, pkCSM, QikProp, DataWarrior, ADMEWORKS predictor, StarDrop, MetaTox, ADMETlab, AdmetSAR, MetaSite, and Tox suite for toxicity prediction (Chandrasekaran et al. 2018).

Molecular docking (MD) is the final step in the process of identifying an active drug candidate for a target with an available three-dimensional structure (Wang and Zhu 2016). The initial algorithm for MD was developed in 1980 by Kuntz et al. (Salmaso and Moro 2018). It is a computational technique that uses mathematical tools to identify the most suitable inhibitor of the target with the highest binding affinity (Meng et al. 2011). Molecular docking can also predict top binding conformations of the compound and structure of the protein-ligand complex (Raval and Ganatra 2022). Scoring functions are utilized to arrange the potential interactions between small molecules and targets, which aid in predicting the most probable structure found in nature (Meng et al. 2011). Molecular dynamic simulation is a more robust and realistic technique for the drug discovery approach, which simulates docked complex artificial systems that resemble certain parameters of the biological system (Almihyawi et al. 2022). It is more powerful compared to molecular docking for successful drug discovery and development.

5 Conclusion

Metabolomics is a growing field with great potential to transform the entire public health scenario from disease diagnosis to treatment. Metabolomics-based approaches have demonstrated greater efficacy compared to conventional methods in drug metabolism and toxicology research. It has diverse applications across all four phases of drug discovery and development, from target identification and lead molecule discovery to preclinical testing and clinical trials. The complete integration of metabolomics into drug discovery poses several challenges, including limited familiarity of the researchers with metabolomics technology, inadequate tools for data annotation and structural elucidation, and insufficient research on the metabolic pathways of different biomolecules. Metabolomics alone cannot reveal the molecular mechanisms underlying diseases, disease biomarkers, and novel infection pathways or expedite the drug discovery process due to the complexity of metabolic pathways, metabolite interactions, and various other factors. Hence, employing a focused and functional analysis strategy alongside a multi-omics approach with advanced imaging techniques and computational tools can help to overcome these obstacles. However, the ability to unite chemistry and biology through multi-omics integrative analysis holds significant promise in making metabolomics an invaluable tool for advancing drug discovery and development in the future.

Acknowledgments None.

Declaration of Competing Interest The author declares no competing interests relevant to this article's content.

References

Abdel-Razek AS, El-Naggar ME, Allam A et al (2020) Microbial natural products in drug discovery. Processes 8:470. https://doi.org/10.3390/pr8040470

Acharya C, Coop A, Polli JE, MacKerell AD (2011) Recent advances in ligand-based drug design: relevance and utility of the conformationally sampled pharmacophore approach. CAD 7:10–22. https://doi.org/10.2174/157340911793743547

Ajjarapu SM, Tiwari A, Ramteke PW et al (2022) Ligand-based drug designing. In: Bioinformatics. Elsevier, Amsterdam, pp 233–252

Alarcon-Barrera JC, Kostidis S, Ondo-Mendez A, Giera M (2022) Recent advances in metabolomics analysis for early drug development. Drug Dis Today 27:1763–1773. https://doi.org/10.1016/j.drudis.2022.02.018

Alhosaini K, Azhar A, Alonazi A, Al-Zoghaibi F (2021) GPCRs: the most promiscuous druggable receptor of the mankind. Saudi Pharm J 29:539–551. https://doi.org/10.1016/j.jsps.2021.04.015

Al-Masaudi S, El Kaoutari A, Drula E et al (2017) A metagenomics investigation of carbohydrate-active enzymes along the gastrointestinal tract of Saudi sheep. Front Microbiol 8:666. https://doi.org/10.3389/fmicb.2017.00666

Almihyawi RAH, Naman ZT, Al-Hasani HMH et al (2022) Integrated computer-aided drug design and biophysical simulation approaches to determine natural anti-bacterial compounds for Acinetobacter baumannii. Sci Rep 12:6590. https://doi.org/10.1038/s41598-022-10364-z

Alonso A, Marsal S, Juliã A (2015) Analytical methods in untargeted metabolomics: state of the art in 2015. Front Bioeng Biotechnol 3:23. https://doi.org/10.3389/fbioe.2015.00023

Alseekh S, Aharoni A, Brotman Y et al (2021) Mass spectrometry-based metabolomics: a guide for annotation, quantification and best reporting practices. Nat Methods 18:747–756. https://doi.org/10.1038/s41592-021-01197-1

Aminov R (2022) Metabolomics in antimicrobial drug discovery. Exp Opin Drug Discov 17:1047–1059. https://doi.org/10.1080/17460441.2022.2113774

Atanasov AG, Zotchev SB, Dirsch VM, Supuran CT (2021) Natural products in drug discovery: advances and opportunities. Nat Rev Drug Discov 20:200–216. https://doi.org/10.1038/s41573-020-00114-z

Attwood MM, Jonsson J, Rask-Andersen M, Schiöth HB (2020) Soluble ligands as drug targets. Nat Rev Drug Discov 19:695–710. https://doi.org/10.1038/s41573-020-0078-4

Baxevanis AD, Ouellette BFF (2004) Bioinformatics: a practical guide to the analysis of genes and proteins. Wiley, New York

Berman HM (2005) Structural databases of biological macromolecules, 1st edn. Wiley, New York

Bills GF, Gloer JB (2016) Biologically active secondary metabolites from the fungi. Microbiol Spectr 4. https://doi.org/10.1128/microbiolspec.funk-0009-2016

Blake J (2000) Chemoinformatics—predicting the physicochemical properties of "drug-like" molecules. Curr Opin Biotechnol 11:104–107. https://doi.org/10.1016/S0958-1669(99)00062-2

Chandrasekaran B, Abed SN, Al-Attraqchi O et al (2018) Computer-aided prediction of pharmacokinetic (ADMET) properties. In: Dosage form design parameters. Elsevier, London, pp 731–755

Clish CB (2015) Metabolomics: an emerging but powerful tool for precision medicine. Cold Spring Harb Mol Case Stud 1:a000588. https://doi.org/10.1101/mcs.a000588

Cook MA, Wright GD (2022) The past, present, and future of antibiotics. Sci Transl Med 14:eabo7793. https://doi.org/10.1126/scitranslmed.abo7793

Cuperlovic-Culf M, Culf AS (2016) Applied metabolomics in drug discovery. Expert Opin Drug Discov 11:759–770. https://doi.org/10.1080/17460441.2016.1195365

Demain AL (1999) Pharmaceutically active secondary metabolites of microorganisms. Appl Microbiol Biotechnol 52:455–463. https://doi.org/10.1007/s002530051546

Elokely KM, Doerksen RJ (2013) Docking challenge: protein sampling and molecular docking performance. J Chem Inf Model 53:1934–1945. https://doi.org/10.1021/ci400040d

El-Tarabily KA, Sivasithamparam K (2006) Non-streptomycete actinomycetes as biocontrol agents of soil-borne fungal plant pathogens and as plant growth promoters. Soil Biol Biochem 38:1505–1520. https://doi.org/10.1016/j.soilbio.2005.12.017

Emwas A-H, Szczepski K, Poulson BG et al (2020) NMR as a "gold standard" method in drug design and discovery. Molecules 25:4597. https://doi.org/10.3390/molecules25204597

Ferreira LLG, Andricopulo AD (2018) Editorial: chemoinformatics approaches to structure- and ligand-based drug design. Front Pharmacol 9:1416. https://doi.org/10.3389/fphar.2018.01416

Ferreira LLG, Andricopulo AD (2019) ADMET modeling approaches in drug discovery. Drug Dis Today 24:1157–1165. https://doi.org/10.1016/j.drudis.2019.03.015

Floris M, Olla S, Schlessinger D, Cucca F (2018) Genetic-driven druggable target identification and validation. Trend Genet 34:558–570. https://doi.org/10.1016/j.tig.2018.04.004

Fouillaud M, Dufossé L (2022) Microbial secondary metabolism and biotechnology. Microorganisms 10:123. https://doi.org/10.3390/microorganisms10010123

Gorlenko CL, Kiselev HY, Budanova EV et al (2020) Plant secondary metabolites in the battle of drugs and drug-resistant bacteria: new heroes or worse clones of antibiotics? Antibiotics 9:170. https://doi.org/10.3390/antibiotics9040170

Gschwend DA, Good AC, Kuntz ID (1996) Molecular docking towards drug discovery. J Mol Recognit 9:175–186. https://doi.org/10.1002/(SICI)1099-1352(199603)9:2<175::AID-JMR260>3.0.CO;2-D

Guan L, Yang H, Cai Y et al (2019) ADMET-score—a comprehensive scoring function for evaluation of chemical drug-likeness. Med Chem Commun 10:148–157. https://doi.org/10.1039/C8MD00472B

Guo S, Qiu S, Cai Y et al (2023) Mass spectrometry-based metabolomics for discovering active ingredients and exploring action mechanism of herbal medicine. Front Chem 11:1142287. https://doi.org/10.3389/fchem.2023.1142287

Gurung AB, Ali MA, Lee J et al (2021) An updated review of computer-aided drug design and its application to COVID-19. Biomed Res Int 2021:1–18. https://doi.org/10.1155/2021/8853056

Harvey A (2008) Natural products in drug discovery. Drug Discov Today 13:894–901. https://doi.org/10.1016/j.drudis.2008.07.004

Holmes E, Loo RL, Cloarec O et al (2007) Detection of urinary drug metabolite (xenometabolome) signatures in molecular epidemiology studies via statistical total correlation (NMR) spectroscopy. Anal Chem 79:2629–2640. https://doi.org/10.1021/ac062305n

Jambhekar SS, Breen PJ (2013) Drug dissolution: significance of physicochemical properties and physiological conditions. Drug Dis Today 18:1173–1184. https://doi.org/10.1016/j.drudis.2013.08.013

Jeffryes JG, Colastani RL, Elbadawi-Sidhu M et al (2015) MINEs: open access databases of computationally predicted enzyme promiscuity products for untargeted metabolomics. J Cheminform 7:44. https://doi.org/10.1186/s13321-015-0087-1

Keller NP (2019) Fungal secondary metabolism: regulation, function and drug discovery. Nat Rev Microbiol 17:167–180. https://doi.org/10.1038/s41579-018-0121-1

Kleiner M (2019) Metaproteomics: much more than measuring gene expression in microbial communities. mSystems 4:e00115-19. https://doi.org/10.1128/mSystems.00115-19

Klont F, Hopfgartner G (2021) Mass spectrometry based approaches and strategies in bioanalysis for qualitative and quantitative analysis of pharmaceutically relevant molecules. Drug Dis Today Technol 40:64–68. https://doi.org/10.1016/j.ddtec.2021.10.004

Krishnamurti C, Rao SC (2016) The isolation of morphine by Serturner. Indian J Anaesth 60:861–862. https://doi.org/10.4103/0019-5049.193696

Kumar B, Prakash A, Ruhela RK, Medhi B (2014) Potential of metabolomics in preclinical and clinical drug development. Pharma Rep 66:956–963. https://doi.org/10.1016/j.pharep.2014.06.010

Lee Y-J, Cho Y, Tran HNK (2021) Secondary metabolites from the marine sponges of the genus Petrosia: a literature review of 43 years of research. Mar Drugs 19:122. https://doi.org/10.3390/md19030122

Lee JW, Maria-Solano MA, Vu TNL et al (2022) Big data and artificial intelligence (AI) methodologies for computer-aided drug design (CADD). Biochem Soc Transact 50:241–252. https://doi.org/10.1042/BST20211240

Li G, Peng X, Guo Y et al (2021) Currently available strategies for target identification of bioactive natural products. Front Chem 9:761609. https://doi.org/10.3389/fchem.2021.761609

Likić VA (2006) Databases of metabolic pathways. Biochem Mol Biol Educ 34:408–412. https://doi.org/10.1002/bmb.2006.494034062680

Lipinski CA, Lombardo F, Dominy BW, Feeney PJ (1997) Experimental and computational approaches to estimate solubility and permeability in drug discovery and development settings. Adv Drug Del Rev 23:3–25. https://doi.org/10.1016/S0169-409X(96)00423-1

Lu Y, Chen C (2017) Metabolomics: bridging chemistry and biology in drug discovery and development. Curr Pharmacol Rep 3:16–25. https://doi.org/10.1007/s40495-017-0083-4

Macalino SJY, Gosu V, Hong S, Choi S (2015) Role of computer-aided drug design in modern drug discovery. Arch Pharm Res 38:1686–1701. https://doi.org/10.1007/s12272-015-0640-5

Maia EHB, Assis LC, de Oliveira TA et al (2020) Structure-based virtual screening: from classical to artificial intelligence. Front Chem 8:343. https://doi.org/10.3389/fchem.2020.00343

Malan SF, Chetty DJ (2002) Physicochemical properties of drugs and membrane permeability. S Afr J Sci 98(7):385–391

Meissner F, Geddes-McAlister J, Mann M, Bantscheff M (2022) The emerging role of mass spectrometry-based proteomics in drug discovery. Nat Rev Drug Discov 21:637–654. https://doi.org/10.1038/s41573-022-00409-3

Meng X-Y, Zhang H-X, Mezei M, Cui M (2011) Molecular docking: a powerful approach for structure-based drug discovery. CAD 7:146–157. https://doi.org/10.2174/157340911795677602

Murthy D, Attri KS, Gokhale RS (2013) Network, nodes and nexus: systems approach to multitarget therapeutics. Curr Opin Biotechnol 24:1129–1136. https://doi.org/10.1016/j.copbio.2013.02.009

Najmi A, Javed SA, Al Bratty M, Alhazmi HA (2022) Modern approaches in the discovery and development of plant-based natural products and their analogues as potential therapeutic agents. Molecules 27(2):349. https://doi.org/10.3390/molecules27020349

Newman DJ, Cragg GM (2007) Natural products as sources of new drugs over the last 25 years. J Nat Prod 70:461–477. https://doi.org/10.1021/np068054v

Newman DJ, Cragg GM (2016) Natural products as sources of new drugs from 1981 to 2014. J Nat Prod 79:629–661. https://doi.org/10.1021/acs.jnatprod.5b01055

Nicholson JK, Lindon JC, Holmes E (1999) "Metabonomics": understanding the metabolic responses of living systems to pathophysiological stimuli via multivariate statistical analysis of biological NMR spectroscopic data. Xenobiotica 29:1181–1189. https://doi.org/10.1080/004982599238047

Pellecchia M, Sem DS, Wüthrich K (2002) NMR in drug discovery. Nat Rev Drug Discov 1:211–219. https://doi.org/10.1038/nrd748

Rask-Andersen M, Almén MS, Schiöth HB (2011) Trends in the exploitation of novel drug targets. Nat Rev Drug Discov 10:579–590. https://doi.org/10.1038/nrd3478

Raval K, Ganatra T (2022) Basics, types and applications of molecular docking: a review. IJCAAP 7:12–16. https://doi.org/10.18231/j.ijcaap.2022.003

Rinschen MM, Ivanisevic J, Giera M, Siuzdak G (2019) Identification of bioactive metabolites using activity metabolomics. Nat Rev Mol Cell Biol 20:353–367. https://doi.org/10.1038/s41580-019-0108-4

Salmaso V, Moro S (2018) Bridging molecular docking to molecular dynamics in exploring ligand-protein recognition process: an overview. Front Pharmacol 9:923. https://doi.org/10.3389/fphar.2018.00923

Shulaev V (2006) Metabolomics technology and bioinformatics. Brief Bioinform 7:128–139. https://doi.org/10.1093/bib/bbl012

Sorokina M, Steinbeck C (2020) Review on natural products databases: where to find data in 2020. J Cheminform 12:20. https://doi.org/10.1186/s13321-020-00424-9

Srivastav VK, Egbuna C, Tiwari M (2020) Plant secondary metabolites as lead compounds for the production of potent drugs. In: Egbuna C, Kumar S, Ifemeje JC et al (eds) Phytochemicals as lead compounds for new drug discovery. Elsevier, Amsterdam, pp 3–14

Sun S, Wesolowski SS (2021) Biologically active metabolites in drug discovery. Bioorgan Med Chem Lett 48:128255. https://doi.org/10.1016/j.bmcl.2021.128255

Szklarczyk D, Gable AL, Lyon D et al (2019) STRING v11: protein–protein association networks with increased coverage, supporting functional discovery in genome-wide experimental datasets. Nucleic Acid Res 47:D607–D613. https://doi.org/10.1093/nar/gky1131

Thomford NE, Senthebane DA, Rowe A et al (2018) Natural products for drug discovery in the 21st century: innovations for novel drug discovery. Int J Mol Sci 19:1578. https://doi.org/10.3390/ijms19061578

Van Drie JH (2007) Computer-aided drug design: the next 20 years. J Comput Aided Mol Des 21:591–601. https://doi.org/10.1007/s10822-007-9142-y

Veber DF, Johnson SR, Cheng H-Y et al (2002) Molecular properties that influence the oral bioavailability of drug candidates. J Med Chem 45:2615–2623. https://doi.org/10.1021/jm020017n

Verpoorte R (1998) Exploration of nature's chemodiversity: the role of secondary metabolites as leads in drug development. Drug Dis Today 3:232–238. https://doi.org/10.1016/S1359-6446(97)01167-7
Wang G, Zhu W (2016) Molecular docking for drug discovery and development: a widely used approach but far from perfect. Future Med Chem 8:1707–1710. https://doi.org/10.4155/fmc-2016-0143
Wang JH, Byun J, Pennathur S (2010) Analytical approaches to metabolomics and applications to systems biology. Semin Nephrol 30:500–511. https://doi.org/10.1016/j.semnephrol.2010.07.007
Wink M (2015) Modes of action of herbal medicines and plant secondary metabolites. Medicines 2:251–286. https://doi.org/10.3390/medicines2030251
Wishart DS (2016) Emerging applications of metabolomics in drug discovery and precision medicine. Nat Rev Drug Discov 15:473–484. https://doi.org/10.1038/nrd.2016.32
Wittig U (2001) Analysis and comparison of metabolic pathway databases. Brief Bioinform 2:126–142. https://doi.org/10.1093/bib/2.2.126
Zdrazil B, Richter L, Brown N, Guha R (2020) Moving targets in drug discovery. Sci Rep 10:20213. https://doi.org/10.1038/s41598-020-77033-x
Zhang Z, Tang W (2018) Drug metabolism in drug discovery and development. Acta Pharm Sin B 8:721–732. https://doi.org/10.1016/j.apsb.2018.04.003
Zhang H, Dong M, Chen J et al (2017) Bioactive secondary metabolites from the marine sponge genus Agelas. Mar Drugs 15:351. https://doi.org/10.3390/md15110351

Reverse Pharmacology: Transdisciplinary Approach of Drug Discovery from Ethnomedicine

Kiran R. Giri, Kamlesh Palandurkar, Anshuman Trigunayat, and Reena Giri

Contents

K. R. Giri (✉) · A. Trigunayat
Department of Pharmacology, Institute of Medical Sciences, Banaras Hindu University, Varanasi, India

K. Palandurkar
Department of Biochemistry, Institute of Medical Sciences, Banaras Hindu University, Varanasi, India

R. Giri
Department of Pharmacology, Government Medical College, Akola, India

© The Author(s), under exclusive license to Springer Nature Singapore Pte Ltd. 2024

A. Das Talukdar et al. (eds.), *Traditional Resources and Tools for Modern Drug Discovery*, Interdisciplinary Biotechnological Advances,
https://doi.org/10.1007/978-981-97-4600-2_10

1 Introduction

Reverse pharmacology is the process of drug discovery that starts with traditional medicine and works backward in search of active compounds and understanding their mechanism of action. The attention of academicians in the late twentieth century was raised in the direction of reverse pharmacology as their pharmaceutical collaborations explored novel methods for medication discovery. Due to the emergence of drug resistance, there was a need for new molecules contending resistance. The cost of overall treatment had been raised by drug resistance. The cost of resistance includes the direct expenses for alternative new therapies, deaths, and health issues produced by the increasing incidence of antibiotic resistance. Switching from cheap chloroquine therapy to artemisinin combination treatment for malaria which is now recommended by the WHO had resulted in a \$300–\$500 million increase in cost (Howard and Scott 2005). In conventional drug development, molecular targets are identified, and then molecular libraries are searched for molecules interacting with them. In reverse pharmacology, the bed-to-laboratory approach starts with clinical observations or effects of herbal or traditional medications, and then the laboratory finds the underlying mechanism and prospective therapeutic target molecules (Fig. 1). This method establishes a link between the therapeutic impact of a medicine or chemical and the precise target or mechanism of action. Artemisinin extracted from the plant *Artemisia annua* which has been used for generations to cure fevers in traditional Chinese medicine. Tu Youyou, a Chinese researcher,

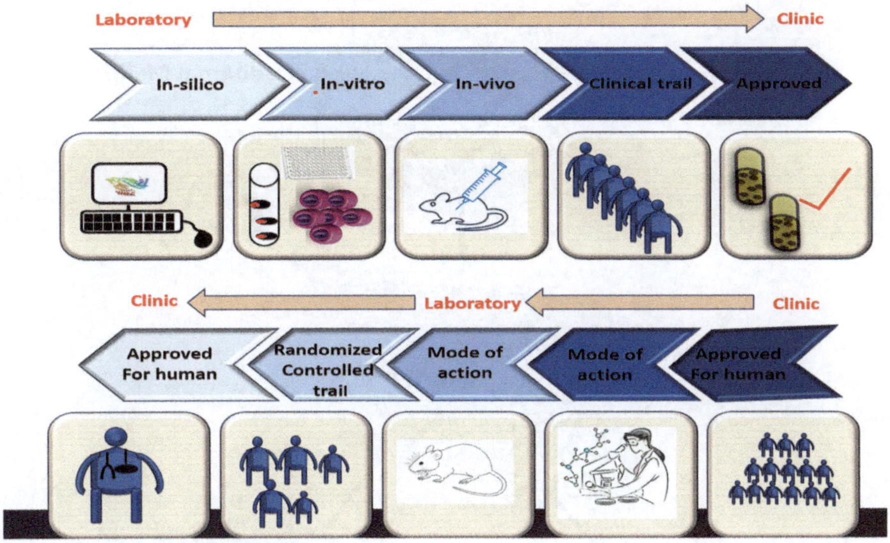

Fig. 1 Drug discovery process: conventional vs reverse

directed the efforts to identify new antimalarial medications during the Vietnam War. She identified a chemical and named it artemisinin which proved its efficiency against malaria parasites. Tu Youyou's revolutionary discovery won her Nobel Prize in Physiology or Medicine in 2015. Artemisinin and its derivatives are currently the world's first-line malaria treatments (Awari et al. 2022). In reverse pharmacology for the discovery of drug targets and the creation of novel therapeutic approaches, various clinical investigation processes include techniques like extraction and isolation of active compounds, in vitro testing using cell cultures, cytokine levels, target identification, animal studies, biomarker identification, clinical trials, patient samples analysis, target validation, optimization and drug development, regulatory approval, and post-marketing surveillance (Fig. 2).

The bed-to-laboratory strategy promises the interlinking of laboratory findings and clinical outcomes, highlighting the relevance of research to real-world therapeutic applications. The drug development processes can be accelerated by focusing on herbal and other sources of molecules with proven medicinal promise (Fig. 3). The approach is highly dependent on patient inputs regarding the efficacy. On the other hand, this technique connects traditional medicine with current scientific evidence-based techniques. The history of reverse pharmacology in India is indistinguishably associated with the country's rich heirloom of traditional health systems like Ayurveda and notable efforts to integrate traditional knowledge into contemporary drug research. In this series of discussions, it will be noteworthy to mention the names of two stalwarts, Professor Ramnath Chopra and Professor Gana Nath Sen, who established the fields of pharmacology and reverse pharmacology, respectively. Professor Sen and Professor Bose Rauwolfia's serpentina modification led us to reverse pharmacology.

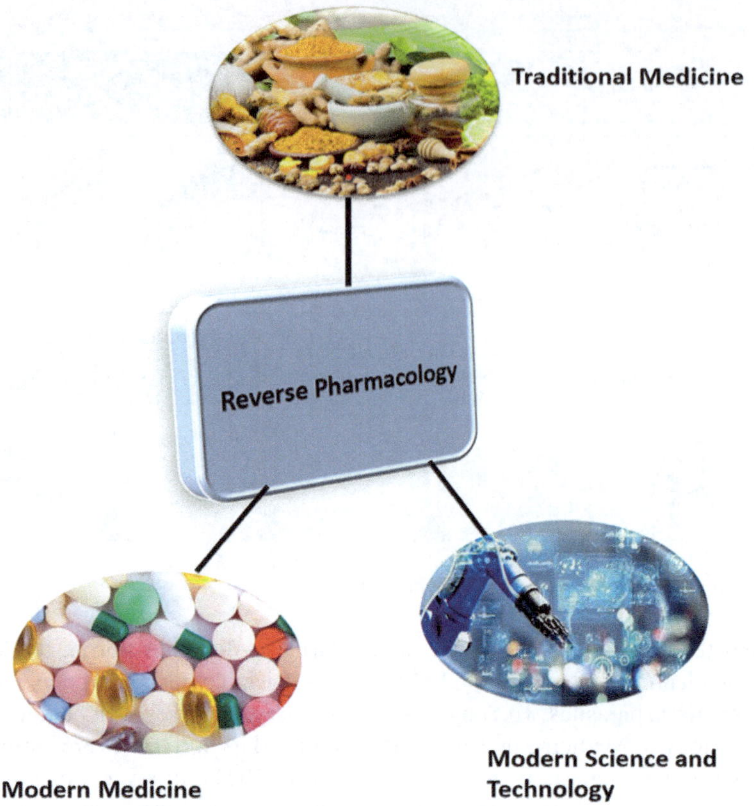

Fig. 2 Reverse pharmacology is amalgamation of traditional medicine, modern medicine, and modern science and technology

2 Traditional Medicine

Reverse pharmacology is an incorporation of ethnomedicine with modern knowledge and techniques. Reverse pharmacology not only speeds up the process of drug discovery but also decreases the cost (Patwardhan et al. 2008). It is also a milestone to validate our exemplary traditional knowledge. Drug development is costly, and wide resources are used for it. The investigational new drug stage costs are around \$10–\$15 million in India. Recently 32% of the small molecules which are approved by the regulatory agencies belong to the category of natural products and their derivatives (Huang and Zhang 2022). According to WHO, 40% of newly discovered drugs are of natural origin. 60% of cancer-targeting medications and 75% of anti-infective drugs are derived from natural resources.

Nuno Sales Craveiro et al.'s work "Drug Withdrawal Due to Safety: A Review of the Data Supporting Withdrawal Decision" investigates the causes of drug withdrawal owing to safety concerns (Craveiro et al. 2020). The research looked at 133

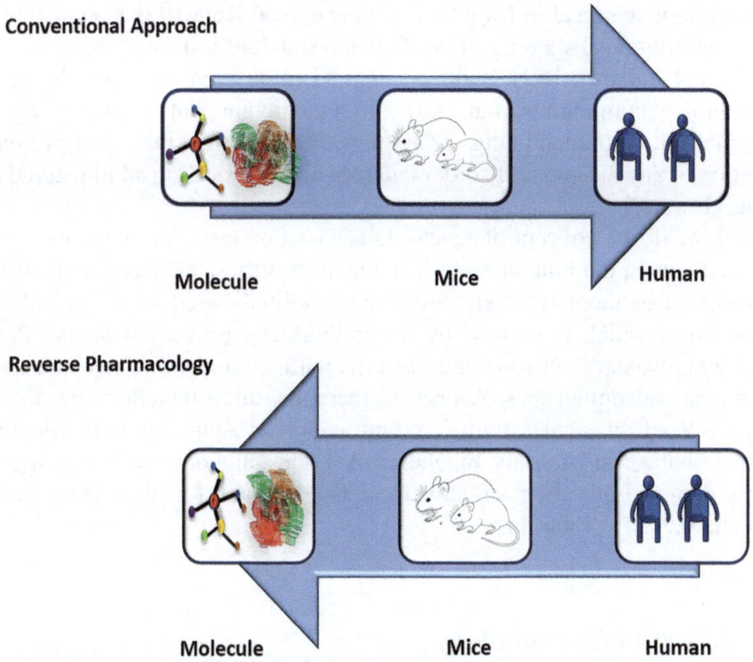

Fig. 3 Target-based drug discovery

medications that were pulled from the market between 1990 and 2010 in different World Health Organization member nations. The most prevalent sources of information leading to drug discontinuation were spontaneous reports and case reports, with clinical trials providing data in 18.0% of instances. Among the different most common causes of medication withdrawal, hepatotoxicity, cardiac problems, hypersensitivity, and nephrotoxicity were responsible for 69.2% of medicine withdrawals. As per the research, geographical differences in the medication in different regions' medicine safety assessments and withdrawal are different. The American continent has the lowest average withdrawal time 17.5 years and that of Africa has the longest of 23.0 years.

In the bedside-to-beachside reverse pharmacology drug development process, withdrawals are minimized, as safety is too well-known due to years of usage of medicine in that region as a traditional drug. These medicines or their raw forms as a herb have been used for centuries and are built on a comprehensive knowledge of health and sickness.

Ayurveda is the most traditional and comprehensive medicinal systems in the world, comprising a massive variety of natural and herbal treatments and lifestyle practices. Researchers and scientists have started a careful investigation of traditional medicines to find the active chemicals responsible for the reported therapeutic benefits which was also mentioned as phytopharmaceuticals.

Phytopharmaceutical in New Drug Clinical Trial Rule 19 defines a "phytopharmaceutical drug means a drug of purified and standardised fraction, assessed qualitatively and quantitatively with a defined minimum of four bio-active or phytochemical compounds of an extract of a medicinal plant or its part, for internal or external use on human beings or animals, for diagnosis, treatment, mitigation or prevention of any disease or disorder but does not include drug administered through parenteral route."

The three-dosha concept of Ayurveda is based on three basic energies or principal ingredients in the human body. The balance of these doshas is critical for optimal health. Treatment regimens are often modified based on a person's Prakriti (constitution), which is defined by the individual's primary dosha(s). Ayurvedic medicine emphasizes nutrition and lifestyle, with recommendations based on a person's dosha and imbalances. Ayurvedic therapies often include yoga, meditation, and specialized physical activities. Combination of Ayurveda with contemporary medicine, using clinical trials, biochemical studies, and other scientific methodologies to demonstrate their effectiveness is explored for new drug candidates (Eranezhath 2011) (Table 1).

2.1 *Reverse Pharmacology*

The process drug discovery approach focuses on identifying and characterizing molecular targets in traditional medicinal practices by matching proteins or receptors associated with disease processes. It involves designing and screening compounds to modify their activity and achieve desired therapeutic effects. This systematic and standardized approach allows documentation of lead compounds with anticipated pharmacological properties for further testing for safety and efficacy as therapeutic agents.

2.2 *Ethnomedicine*

Ethnomedicine is an interdisciplinary approach to studying traditional medical practices and knowledge from a specific culture, ethnicity, or community. It incorporates traditional healing methods, including medicinal plants, animals, rituals, and natural ingredients. By embracing the cultural and social framework of healthcare, ethnomedicine offers fresh insights into diverse cure approaches and contributes to the conservation of traditional medicinal knowledge.

Table 1 Compare and contrast between Ayurveda and reverse pharmacology

	Ayurveda	Reverse pharmacology
Standardization and quality control	The lack of standardized production processes leads to variability in composition and potency	Emphasizes precision and standardization, ensuring consistency in drug development. This is crucial for reproducibility and quality control
Scientific validation and evidence base	Rooted in traditional knowledge but may lack rigorous scientific validation and a robust evidence base	Relies on modern scientific methodologies, emphasizing evidence-based practices
Isolation of active compounds	Consists of complex mixtures of herbs and minerals, making it challenging to identify and isolate specific active compounds	Focuses on isolating and understanding the specific active compounds responsible for therapeutic effects. This allows for targeted drug development and precise mechanisms of action
Individualized vs generalized treatment	Emphasizes individualized treatments based on a person's constitution (Prakriti) and imbalances (Vikriti). However, this personalized approach may lack generalizable evidence	Strives for a more generalized approach, identifying drug candidates that can have broad applications across diverse populations. This is essential for mass healthcare interventions
Time-tested vs rapid drug development	Ayurveda relies on time-tested formulations developed over centuries, which can be perceived as a strength but may not align with the rapid pace of modern drug development	Reverse pharmacology aims for efficient drug development through targeted research, utilizing contemporary technologies to accelerate the discovery and development process

2.3 *Ethnopharmacology*

Ethnopharmacology is an interdisciplinary field that studies traditional knowledge and practices of indigenous groups worldwide, focusing on the use of medicinal plants, animal parts, and natural substances for health improvement and treatment. It aims to document, analyze, and understand the detection of bioactive compounds of therapeutic importance in modern medicine.

3 Ethnopharmacological Surveys and Documentation

These are taken up for the following objectives:

- Identification of traditional medicinal practices and remedies used and tested by various cultures.
- Selection of traditional medications and therapies demonstrating promising therapeutic properties.

- Identification and isolation of specific active compounds by employing different techniques of molecular separation like HPLC, to extract compounds. The chemical structure of isolated and purified compounds can be determined by spectroscopic and crystallographic methods such as nuclear magnetic resonance (NMR) and mass spectrometry,
- Screening and bioassays: The activity of purified compounds against particular molecular targets can be evaluated in appropriate disease models. The selected compounds demonstrating reassuring pharmacological properties can be subjected to further research.
- Identification of the molecular mechanisms responsible for the activity of selected compounds: Employment of cellular and molecular biology methods to find out how compounds change the upstream and downstream activity in signaling pathways and how they change physiological processes by interaction with biological targets.
- Compound optimization for leads: Perform in silico analysis of the compounds for research of the structure-activity relationship (SAR) among the biosimilars to improve the selectivity and potency of lead compounds. Evaluate the safety features of the selected optimized compounds derived from in silico analysis.
- Validation in preclinical and clinical settings: Undertake animal model studies for the preclinical investigations for the assessment of the pharmacokinetics, toxicity, and efficacy of the selected lead compounds. Lead compounds that have been deemed safe may advance to human trials to assess their safety, dose, and effectiveness. These studies also provide valuable data on possible adverse reactions, interactions with other drugs, and eventual consequences.
- Standardization and formulation: Articulate suitable delivery systems for the active compounds for enhancing the biological effects (e.g., extracts, capsules, tablets). To guarantee uniformity in the composition and potency of the product quality control, protocols and benchmarks shall be implemented. The next step in production management shall be dealing with concerns regarding product stability and batch-to-batch variations.
- Market access and regulatory approval: For the approval of the compound as a drug formulation, assemble and submit regulatory documents with information from preclinical and clinical studies to the authorities. Create collaborations with regulatory bodies for efficient confrontation of safety and effectiveness issues. Complete the regulatory approval process to commercialize the drug.
- Post-marketing surveillance is to be implemented for the detection and justification of any unanticipated negative consequences. Pragmatic evidence needs to be gathered regarding medication efficacy and evaluated outcomes detected in patients. Labeling or treatment guidelines will be modified by continuous monitoring as and when required.

Phase I Clinical Trials

Phase I trials are open-label studies. These studies are aim to determine the optimal dose range and safety profile of a product under investigation in healthy volunteers.

Phase II Clinical Trials
Phase II trials are randomized controlled trials (RCTs) that involve a larger, more diverse group of participants (diseased population). The aim of phase II trial is to find the optimal dosage and regimen for a treatment. The primary endpoints in phase II trials are related to efficacy, while secondary endpoints may include safety, pharmacokinetics, and exploratory measures.

Phase III Clinical Trials
The objective of phase III trials is to collect supplementary data about the safety and efficacy of an investigational product and validate previous conclusions. Extensive information collected help in preparation for regulatory approval. These trials frequently employ randomized controlled trials (RCTs) and double-blind designs on a more extensive and heterogeneous patient population which are the measures to reduce bias (Fig. 4).

Post-marketing Surveillance (Phase IV)
Monitor the drug's safety and efficacy after market approval.

3.1 Hit to Lead Process

The hit-to-lead process comprises of identification of a molecular target associated with a disease, screening of a large library of compounds, confirmation and characterization of hits, selection of leads based on activity, optimization of leads for pharmacological properties, and ADME-Tox assessments. Following an extensive series of testing and optimization iterations, a lead compound is ultimately chosen as the candidate drug. The candidate should possess the pharmacokinetic properties, safety profiles, and therapeutic effects that are desirable. Preclinical development evaluates safety and efficacy through in vitro and in vivo investigations. In the event of success, the drug candidate advances to human subjects in clinical trials. Risks may be mitigated in favor of positive outcomes and benefits, which may lead to regulatory approval being granted. This collaborative and iterative procedure encompasses the participation of multidisciplinary scientific teams. One such drug development is the medicine for psoriasis developed by the Lupin and Department of Science and Technology, Government of India. This is a novel plant-based formulation named Desoris. Desoris (LLL 3348) is a lyophilized aqueous extract prepared from the leaves of *Argemone mexicana* to treat chronic stable plaque-type psoriasis (Fig. 5).

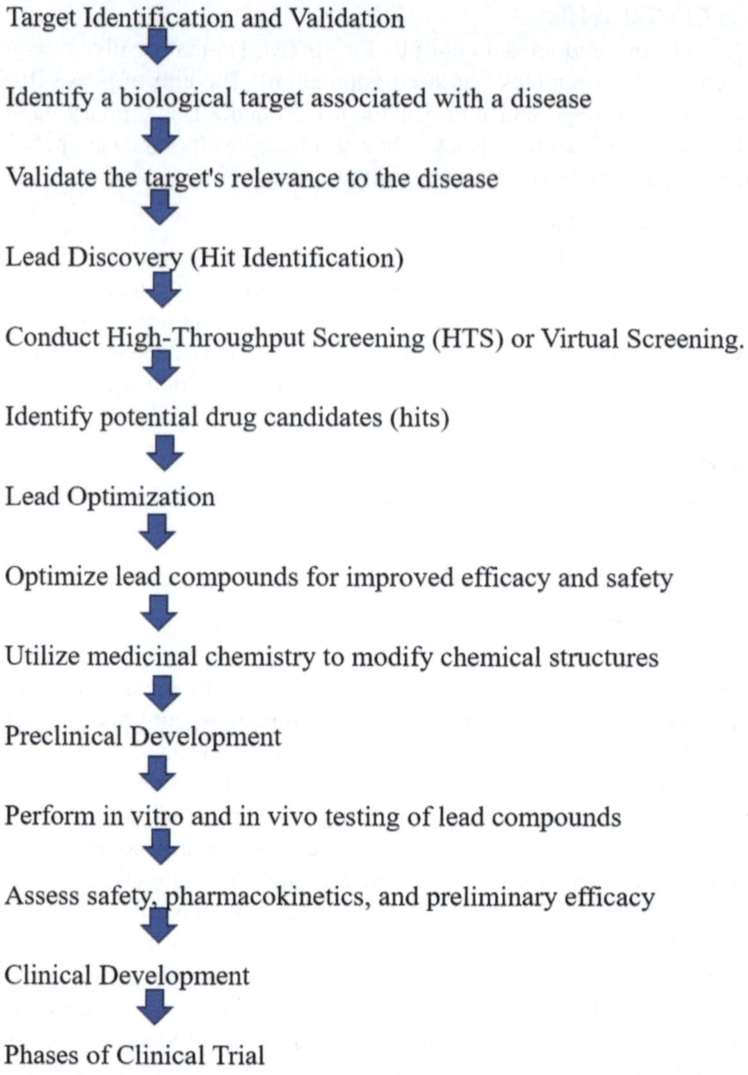

Fig 4 Key steps in a conventional drug discovery process

4 Tools and Technologies in Reverse Pharmacology

4.1 Genomics and Proteomics in Target Identification

Proteomics tools and computational approaches have revolutionized research by identifying therapeutic targets, detecting illness biomarkers, and developing more effective medications. Proteins involved in signal transduction pathways and regulatory enzymes are potential targets, with protein kinases accounting for 25% of

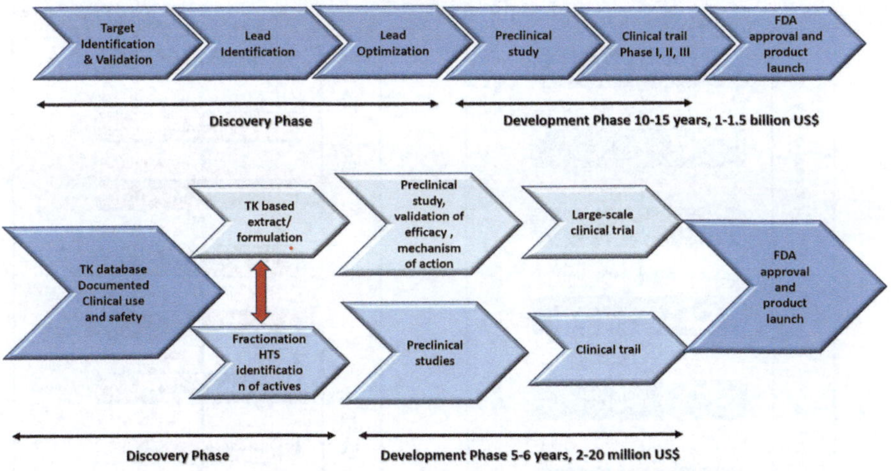

Fig. 5 Drug development process

biotechnology and pharmaceutical research efforts. New chemical techniques detect oxidized cysteine in proteins, which is crucial for medication specificity and selectivity. Metabolite profiling methods complement genomes and proteomics to track disease progression, drug metabolism, and chemical toxicity. Diagnostic and discovery glycan arrays assess changes in surface receptors and understand drug resistance and tolerance. Mechanisms behind host-pathogen interactions are investigated using comparative genomes and proteomics methods for expression and functional analysis. Paclitaxel, a reverse pharmacology medication, targets the tubulin subunit in microtubules and has anticancer capabilities. It is widely used in cancer therapy due to its natural origins and sustainable source and is often used in personalized cancer treatment regimens based on disease type and genetic features (Otimenyin 2021) (Fig. 6).

4.2 *High-Throughput Screening Methods*

High-throughput screening (HTS) has been a crucial aspect of pharmaceutical research for over 25 years, connecting pharmacology, computational chemistry, and medicinal chemistry. The automation and miniaturization of in vitro assays have made HTS methods more repeatable and sustainable, resulting in higher data quality and consistency. Various detection technologies, such as absorbance, fluorescence, luminescence, and radiometric approaches, are used to detect biological, chemical, or physical occurrences. HTS is evolving to address complex pathobiology for unmet medical needs, with integrated campaigns using various screening modalities becoming common. Current screening challenges include limited use of

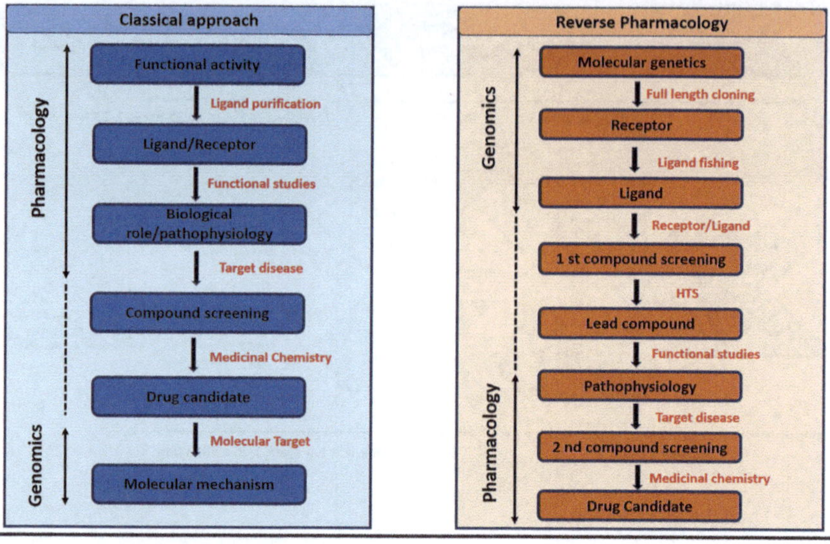

Fig. 6 Classical approach of drug discovery vs reverse pharmacology

physiologically relevant cells at scale. Future prospects include internal chemical collections, fragment screening, ligand-based design, and in silico approaches.

Lovastatin was initially isolated from the fungus *Aspergillus terreus*. In the late 1970s, researchers at Merck conducted high-throughput screening of microbial extracts for compounds with cholesterol-lowering potential. In that, HTS lovastatin, extracted from *Aspergillus terreus*, prominently established its ability to inhibit HMG-CoA reductase. Lovastatin, one of the statin group of drugs, lowers cholesterol levels by blocking the enzyme HMG-CoA reductase, a rate-limiting step in cholesterol synthesis. Lovastatin and other statins are prescribed for managing hypercholesterolemia and reducing the risk of atherosclerotic diseases like myocardial infarction and stroke (Mulder et al. 2015).

4.3 Systems Biology and Computational Approaches

Systems biology is an interdisciplinary field that comprehends and examines complex biological systems using mathematical and computer modeling methods. It makes use of probability models so that similar targets, pathways, and chemicals can be assigned to functions or pathology. Reverse pharmacology makes use of network analysis in the identification of biological pathway regulatory steps that might be potential drug targets. It examines and makes simulations of biological pathways and their interconnections and route analysis to make calculated predictions to identify linkages between pathways and illnesses of interest. This method builds a systematic picture of the molecular landscape and explores its links to illnesses and possible therapeutic targets, and it is assisted by integration of heterogeneous omics data. Even drug reactions can be predicted with the use of dynamic modeling tools, as similar compounds may have similar effects and side effects which is called the temporal behavior of biological systems. When we rank important nodes according to their status in biological networks, reverse pharmacology becomes more effective and targeted. The calculated process of drug development and experimental validation is nowadays guided by data-driven hypothesis generation, predictive modeling, multiscale modeling, and personalized medicine techniques to increase the efficiency of the process. Torcetrapib, which is being used to decrease cholesterol, has been associated with higher mortality and cardiovascular risk and was identified by systems biology research. The research identified biological processes and signaling pathways utilized by torcetrapib using a human-reconstructed signaling network and microarray gene expression data based on the predictive model. It discovered enhanced signaling pathways and 388 upregulated genes potentially linked to the detrimental cardiovascular effects of torcetrapib. The research highlights the importance of understanding intricate relationships between target molecules and cellular pathways and using in silico techniques to forecast unintended consequences arising from the use of a potential molecule (Fan et al. 2012) (Tables 2, 3, 4, and 5).

5 Applications of Reverse Pharmacology

- Reverse pharmacology is playing a key role in identifying novel drug sources and targets, repurposing existing drugs and herbs, and optimizing personalized medicine.
- It empowers scientists by investigating the observed clinical effects of traditional medicines and understanding their underlying mechanisms. It helps especially in complex and multifactorial diseases.
- It helps in the identification of different biomarkers associated with disease progression, patient prognosis, or treatment response.

Table 2 Compare and contrast between conventional and reverse pharmacology

Conventional pharmacology	Reverse pharmacology
Begins with a known target	Begins with an observed phenotype or clinical effect
More rigid as it starts with a specific target	More flexibility, exploring multiple targets based on observed effects
Failure risk is high especially if the chosen target is not validated	Reduce risk by starting with observed clinical effects
Often used for well-understood diseases with known molecular targets	Particularly useful for complex diseases where the underlying biology is not well understood
Costly	Cost involved is comparatively less
The time involved is more	Less time involved

Table 3 Disease-wise pharmacological potential correlates from Ayurveda

Disease	Ayurvedic drug	Origin plant	Active molecules	References
Hypertension	*Convolvulus pluricaulis* (shankhpushpi)	*Convolvulus pluricaulis*	Scopoletin, 4-hydroxycinnamic acid, kaempferol, quercetin, and ayapanin	Somarathinam K et al. (2023)
	Boerhavia diffusa (punarnava)	*Boerhavia diffusa*	Punarnavoside	Nayak S et al. (2016)
	Acorus calamus (vacha)	*Acorus calamus*	Alpha-asarone, beta-asarone	Shah AJ et al. (2009)
Diabetes	*Syzygium cumini* (jamun)	*Syzygium cumini*	Jambosine, ellagic acid, anthocyanins, glucoside, ellagic acid, isoquercetin, kaemferol and myrecetin.	Perera PRD et al. (2017)
	Gymnema sylvestre (gurmar)	*Gymnema sylvestre*	Gymnemic acids	Kanetkar P et al. (2007)
	Phyllanthus emblica (amla)	*Phyllanthus emblica*	Emblicanin A and B, gallic acid	D'souza JJ et al. (2014)
CNS disorders	*Nardostachys jatamansi* (jatamansi)	*Nardostachys jatamansi*	Jatamansone, jatamansic acid	Panara K et al. (2020)
	Valeriana wallichii (tagara)	*Valeriana wallichii*	Valerenic acid, valtrate	Yoo DY et al. (2015)
	Mucuna pruriens (kapikachhu)	*Mucuna pruriens*	L-DOPA (levodopa)	Zahra W (2022)
Skin disorders	*Psoralea corylifolia* (bakuchiol)	*Psoralea corylifolia*	Bakuchiol	Alam F et al. (2017)
	Rubia cordifolia (manjistha)	*Rubia cordifolia*	Alizarin, purpurin	Wen M et al. (2022)
	Saraca indica (ashoka)	*Saraca indica*	Tannins, flavonoids	Rathod CP et al. (2021)
Carcinoma	*Terminalia arjuna* (arjuna)	*Terminalia arjuna*	Arjunic acid, arjunolic acid	Ahmad MS et al. (2014)
	Tinospora cordifolia (guduchi)	*Tinospora cordifolia*	Tinosporaside, cordifolide	Kundu et al. (2021)
	Aloe vera	*Aloe vera*	Aloin, aloe-emodin	Eshun and He (2004)

(continued)

Table 3 (continued)

Disease	Ayurvedic drug	Origin plant	Active molecules	References
Arthritis	*Zingiber officinale* (ginger)	*Zingiber officinale*	Gingerol, shogaol	Mashhadi et al. (2013)
	Triphala (combination of *Emblica officinalis*, *Terminalia bellirica*, and *Terminalia chebula*)	*Emblica officinalis*, *Terminalia bellirica*, *Terminalia chebula*	Tannins, gallic acid, chebulinic acid	Baliga et al. (2012)
	Curcuma longa (turmeric)	*Curcuma longa*	Curcumin	Hewlings and Kalman (2017)
	Curcuma longa (turmeric)	*Curcuma longa*	Curcumin	Hewlings and Kalman (2017)

Table 4 Established drugs which are derived from plants

Drug	Mechanism of action	Origin plant	Origin plant use	Drug use
Aspirin	Irreversibly inhibits COX-1 and COX-2 enzymes, reducing prostaglandin synthesis	Willow bark (*Salix* spp.)	Traditional remedy for pain and fever	Analgesic, anti-inflammatory, antipyretic, antiplatelet
Morphine	Binds to mu-opioid receptors in the CNS, reducing pain perception	Opium poppy (*Papaver somniferum*)	Traditional analgesic in various cultures	Analgesic, anesthetic
Quinine	Disrupts *Plasmodium falciparum* parasite's ability to break down hemoglobin	Cinchona tree (*Cinchona* spp.)	Traditional treatment for fevers	Antimalarial
Paclitaxel (Taxol)	Stabilizes microtubules, inhibiting cell division	Pacific yew tree (*Taxus brevifolia*)	Traditional medicine for various conditions	Antineoplastic
Artemisinin	Generates free radicals, damaging the *Plasmodium* parasite	Sweet wormwood (*Artemisia annua*)	Traditional Chinese medicine for fevers	Antimalarial
Digoxin	Inhibiting the activity of the sodium-potassium ATPase pump in cardiac myocytes	Foxglove	Dropsy	Heart failure, atrial flutter, fibrillation

- Its integration with systems biology enables an inclusive understanding of biological systems through different algorithms.
- Repurposing expedites the drug development process because of the established pharmaceutical safety dosage and already established clinical effects, for

Table 5 Disease-wise list of modern medicine drugs derived from plants

Drug	Origin plant	Origin plant use	Derivation process	Active molecules
Anticancer				
Paclitaxel (Taxol)	Pacific yew tree (*Taxus brevifolia*)	Traditional medicine for various conditions	Semi-synthesized from precursor 10-DAB	Paclitaxel
Vincristine	Periwinkle (*Catharanthus roseus*)	Traditional medicine for diabetes	Extracted from the periwinkle plant	Vincristine
Camptothecin	Happy tree (*Camptotheca acuminata*)	Traditional medicine for various ailments	Isolated from the bark and wood of the happy tree	Camptothecin
Podophyllotoxin	Mayapple (*Podophyllum peltatum*)	Traditional medicine for various ailments	Isolated from the rhizome of the mayapple plant	Podophyllotoxin
Respiratory conditions				
Ephedrine	*Ephedra sinica*	Traditional medicine for respiratory conditions	Extracted from the stems of *Ephedra sinica*	Ephedrine
Theophylline	Tea (Camellia sinensis)	Traditional beverage	Isolated from tea leaves or synthesized	Theophylline
Digestive disorders				
Alosetron	*Aloe vera*	Traditional remedy for skin conditions	Semi-synthesized from anthraquinones in aloe vera	Alosetron
Neurological disorders				
Galantamine	Snowdrop (*Galanthus nivalis*)	Traditional remedy for various ailments	Extracted from the bulbs of the snowdrop plant	Galantamine
Cardiovascular health				
Reserpine	Indian snakeroot (*Rauwolfia serpentina*)	Traditional remedy for snake bites	Extracted from the root of the Indian snakeroot	Reserpine
Digoxin	Foxglove (*Digitalis purpurea*)	Traditional remedy for heart conditions	Extracted from the leaves of foxglove	Digoxin
Anti-inflammatory				
Colchicine	Autumn crocus (*Colchicum autumnale*)	Traditional remedy for joint pain	Isolated from the seeds of the autumn crocus	Colchicine
Curcumin	Turmeric (*Curcuma longa*) (Fig. 7)	Traditional medicine for digestive issues	Extracted from the rhizomes of turmeric	Curcumin

(continued)

Table 5 (continued)

Drug	Origin plant	Origin plant use	Derivation process	Active molecules
Quercetin	Various plants, e.g., onions, apples, tea	Found in various traditional foods	Extracted from plant sources or synthesized	Quercetin

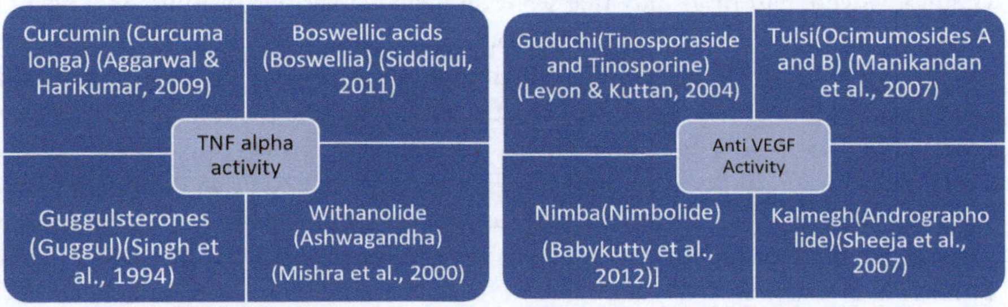

Fig. 7 Natural molecules with active compound showing anti-inflammatory and antiangiogenic activity by modulating TNF-α activity and VEGF activity

example, in Ayurvedic medicine, *Rauwolfia serpentina* has been used for centuries for the treatment of a various range of ailments, including mental disorders, insomnia, and hypertension. Its calming properties were distinguished by traditional physicians, who used it to treat illnesses associated with the nervous system.

- We can also use it to more efficiently evaluate drug safety in toxicology and drug safety investigations. It is still frequently used in partnership with conventional target-based methodologies to achieve a more holistic understanding of disease biology and to formulate more efficient therapeutic modalities.

6 Government Initiatives

Government Initiatives: For scientific validation of traditional treatments, the Government of India has offered funds and research assistance. Collaborative approaches from traditional medicine practitioners, university researchers, and pharmaceutical firms help in documenting traditional knowledge linked to medicinal plants, formulations, and treatment practices and way a path for reverse pharmacology research.

6.1 "New Millennium Indian Technology Leadership Initiative" (NMITLI) Arthritis

The goal of the "New Millennium Indian Technology Leadership Initiative" (NMITLI) Arthritis Project was to establish a standard for Ayurvedic treatments for knee osteoarthritis. Two hundred and forty-five patients with knee osteoarthritis symptoms were part of a study that was conducted at various institutions. Five oral formulations, each of which contained four plant extracts, were examined. The substances were discovered to possess immunomodulatory, analgesic, and anti-inflammatory properties. There was evidence that multiple plant formulations were more effective than single medicines, suggesting a synergistic clinical effect. Candidates for eligibility screening were between 40 and 70 years of age who were walking, had OA knees diagnosed, and were receiving analgesics and/or NSAIDs to ease their pain. During the trial, patients were told not to include any new therapy. Following 5 years, an IND and NDA file was submitted, and two Indian and PCT patent applications were filed through this initiative (Chopra et al. 2010.)

6.2 Traditional Knowledge Digital Library (TKDL)

The Indian government has established a Traditional Knowledge Digital Library (TKDL) to study traditional medicinal plants. The library, eventually categorized as a Traditional Knowledge Resource Classification (TKRC), aims to improve scientific innovation, particularly in developing nations like India, by translating ancient Sanskrit texts into international acceptable resources (Vaidya and Devasagayam 2007).

6.3 Foundation for the Revitalisation of Local Health Traditions (FRLHT)

The Foundation for the Revitalisation of Local Health Traditions (FRLHT) is an Indian nonprofit organization dedicated to preserving and promoting traditional health systems, particularly Ayurveda and indigenous medical knowledge. They focus on research, documentation, medicinal plant conservation, education, and biodiversity preservation and participate in projects like medicinal plant gardens and sustainable harvesting methods (Patwardhan 2012).

6.4 Banaras Hindu University and the Ministry of AYUSH Joint Initiative

An organized initiative of Banaras Hindu University and the Ministry of AYUSH, in which data on Ayurvedic medicine and the diseases in which that medicine is useful will be collected, in which the Department of Kriya Sharir, Faculty of Ayurveda, Banaras Hindu University, has been chosen as the nodal center. It has been said that this type of data will prove to be very useful in the future for reverse pharmacology in which we can know which medicines are proving to be more useful for which diseases, which will help us in finding out their active ingredients via using reverse pharmacology techniques.

6.5 Ayusoft and Ayugenomic

AyuSoft, a collaboration between India's Centre for Development of Advanced Computing and the University of Pune, aims to transform traditional Ayurvedic books into interactive information libraries. The database, which contains 5 lakh entries from nine books, uses a decision support system, data-mining tool, and digitized searchable texts for precise information exploration.

Ayugenomics is a drug development platform that combines Ayurvedic knowledge with current genomics, proteomics, and pharmacogenetics. Ayugenomics® is a legally registered trademark that intends to generate novel drug discovery methodologies by merging old Ayurvedic knowledge with current scientific technology. Initial research linked certain HLA alleles to Prakriti kinds, providing a genetic foundation for the Ayurvedic idea of unique Prakriti types (Patwardhan et al. 2008).

7 Challenges and Limitations

7.1 Validation of Target

The robust validation tools like pharmacophore modeling, quantitative structure-activity relationship (QSAR) analysis, virtual screening, molecular dynamic simulation, and fragment-based drug designing are required for authentic target identification.

7.2 Complex Biological Structures

Complex pathway analysis is important to prevent inaccurate predictions and obstructions; sometimes off-target interactions of a drug can lead to unexpected side effects. Unanticipated pathway cross talk, for example, is notable cross talk between the EGFR pathway and the mitogen-activated protein kinase (MAPK) pathway which can lead to drug resistance as seen with erlotinib and gefitinib.

7.3 Biomarkers and Its Correlation

Precise biomarkers and its correlation with disease are important to observe the actual effects of the compounds.

7.4 Data Integration

Accurate prediction models and robust calculation is required to integrate genomics, proteomics, and clinical information.

7.5 Translation of Identified Target

Preclinical studies help in translating results into successful drug development, alternatives to the animal models like synthetic and tissue culture models needed to be explored for evaluation.

7.6 Regulatory and Ethical Considerations

The problems in the long documentation processes and time delays from the regulatory bodies and ethical committees can be extremely time-consuming and delay the drug development process.

7.7 *Time and Financial Expenditure*

As there can be thousands of candidates in the mixture of complex biological matrix, identification of single and effective content can be an expensive and time-consuming endeavor.

7.8 *Durability and Reproducibility*

The reproducibility and robustness of experimental findings to prevent massive waste of resources and the ultimate failure of drug development enterprises are of utmost importance.

7.9 *Integration with Computational Models*

Computational model can predict interaction, along with computational models; experimental validation is required for accurate molecular identification and prediction of drugs action and efficacy.

7.10 *Heterogeneity of Patients*

The genetic and environmental impacts on individual patients are unpredictable. They can result in diverse responses to medications. Development of effective medication for wide range of population is crucial in drug development.

8 Future Directions

8.1 *Personalized Medicine*

Reverse pharmacology can offer personalized medicine by considering individual molecular, genetic, and physiological characteristics, including allergies, in drug prescribing.

8.2 Synchronization of Artificial Intelligence and Big Data

Artificial intelligence and big data analytics can help process large amounts of biological and clinical data, leading to sophisticated computational models that identify novel targets and improve drug development processes.

8.3 The Pharmacology of Networks

Advances in biological networks and signaling pathways will enable the integration of network pharmacology approaches, enabling the differentiation of multitarget pharmaceuticals and understanding of interventions' impacts on interconnected pathways.

8.4 Functional Genomics and the Technology of CRISPR

Functional genomics and CRISPR technology can be used to examine the impacts of genetic modifications on disease phenotypes and drug efficacy in patients' cultured cells, validating the relevance of targets.

8.5 Single-Cell Examination

Single-cell examination techniques can recognize treatment efficacy in disease-relevant cell populations and investigate cellular response to drug treatment.

8.6 Systems Biology and Metabolic Biology

Systems biology and metabolic biology can be applied to study metabolic alterations linked to diseases and drug reactions.

8.7 Gene Therapies and Immunotherapy

Reverse pharmacology knowledge can be used to develop novel immunotherapies and gene therapies.

8.8 Patient-Derivates Organoids

Patient-derived options like xenografts and organoids can detect the individual effects of drugs with greater precision, making drug discovery pipelines more acceptable

8.9 Therapies Involving Drug Repurposing and Combination

Drug repurposing involves the documentation of novel applications for conventional medications. The understanding of disease pathways can help us to investigate already marketed or used drugs and their combination therapies for improving efficacy and overcoming resistance.

8.10 Monitoring in Real-Time and Biomarkers

To evaluate drug responses and modify treatment protocols integration of real-time monitoring methodologies is essential determination of reliable biomarkers for early detection of treatment efficacy or adverse events.

8.11 International Cooperation

International cooperation is crucial for the translation of scientific findings into clinical applications.

8.12 Ethical and Regulatory Considerations

Ethical and regulatory considerations include establishing standard operating procedures (SOPs) and adhering to good clinical practice guidelines.

8.13 Documentation and Data Integrity

Adherence to the regulatory requirements regarding data documentation and integrity is quintessential to the acceptance by authorities.

8.14 Pharmacovigilance Entails and Post-marketing Surveillance

Pharmacovigilance ensures post-marketing surveillance of drugs and helps identify rare complications and side effects.

8.15 Working in Partnership with Regulatory Authorities

Partnerships with regulatory authorities foster long-term collaboration and compliance with regulatory requirements.

8.16 The Governance of Genetic Research

Genetic research, with particular prominence on reverse pharmacology applications that include genomics, gene editing, or other genetic interventions, has been a great debate on ethical perspectives, and adhering to national and international regulations is important to ensure passing through ethical reviews.

9 Ethical Consideration

Research on reverse pharmacology may face disparities in benefits, but collaborating globally with diverse populations can improve generalizability and inclusivity. Pressures from the market might force an accelerated translation, jeopardizing safety evaluations. Efficiency and moral obligations must be balanced. Standardizing ethical norms may improve international cooperation. Intellectual property protection is also important ethical concern. Resource-constrained environments may be addressed by funding ethical training and capacity-building initiatives.

10 Conclusion

A major contribution of reverse pharmacology is that it has integrated traditional medicine and contemporary scientific methods into a systematic and standardized process of drug discovery. This not only speeds up the process but also reduces market withdrawal, and it follows and validates the broader heritage of traditional healing systems such as Ayurveda. The drug development process involves several steps, which includes everything from target discovery to monitoring of the drug after it is approved for sale in the market. The comparison between Ayurveda and reverse pharmacology reveals disparities in terms of standardization, evidentiary foundation, therapeutic methodology, and rate of progress. Proteomics, genomics, and high-throughput screening are important tools in reverse pharmacology. The study of ethnomedicine, ethnopharmacology, and conducting ethnomedicinal surveys has a critical role in the identification, isolation, and recognition of active molecules. The worldwide influence of India's traditional medicine highlights the incomparable potential of reverse pharmacology in defining the future of drug development.

References

Aggarwal BB, Harikumar KB (2009) Potential therapeutic effects of curcumin, the anti-inflammatory agent, against neurodegenerative, cardiovascular, pulmonary, metabolic, autoimmune and neoplastic diseases. Int J Biochem Cell Biol 41(1):40–59. https://doi.org/10.1016/j.biocel.2008.06.010

Ahmad MS, Ahmad S, Gautam B, Arshad M, Afzal M (2014) Terminalia arjuna, a herbal remedy against environmental carcinogenicity: An in vitro and in vivo study. Egypt J Med Hum Genet. 15(1):61–67, ISSN 1110-8630. https://doi.org/10.1016/j.ejm

Mishra LC, Singh BB, Dagenais S (2000) Scientific basis for the therapeutic use of Withania somnifera (ashwagandha): a review. Altern Med Rev 5(4):334–346

Alam F, Khan GN, Asad MHHB (2018) Psoralea corylifolia L: Ethnobotanical, biological, and chemical aspects: A review. Phytother Res. 32(4):597–615. https://doi.org/10.1002/ptr.6006. Epub 2017 Dec 15. PMID: 29243333; PMCID: PMC7167735.

Awari D, Awari D, Wankhede S, Ganjiwale R, Rathi L (2022) Reverse pharmacology: an innovative approach to the study of herbs. Res J Pharm Technol 15(12):5875–5878. https://doi.org/10.52711/0974-360X.2022.00991

Babykutty S, S PP, J NR, Kumar MA, Nair MS, Srinivas P, Gopala S (2012) Nimbolide retards tumor cell migration, invasion, and angiogenesis by downregulating MMP-2/9 expression via inhibiting ERK1/2 and reducing DNA-binding activity of NF-κB in colon cancer cells. Mol Carcinog 51(6):475–490. https://doi.org/10.1002/mc.20812. Epub 2011 Jun 15. PMID: 21678498

Baliga MS, Meera S, Mathai B (2012) Scientific validation of the ethnomedicinal properties of the Ayurvedic drug Triphala: a review. Chin J Integr Med 18(12):946–954. https://doi.org/10.1007/s11655-012-1299-x

Chopra A, Saluja M, Tillu G, Venugopalan A, Sarmukaddam S, Raut AK et al (2010) A randomized controlled exploratory evaluation of standardized Ayurvedic formulations in symptomatic osteoarthritis knees: a Government of India NMITLI project. Evid Based Complement Alternat Med 2011:724291. https://doi.org/10.1155/2011/724291

Craveiro NS, Lopes BS, Tomás L, Almeida SF (2020) Drug withdrawal due to safety: a review of the data supporting withdrawal decision. Curr Drug Saf 15(1):4–12. https://doi.org/10.2174/1574886314666191004092520

D'souza JJ, D'souza PP, Fazal F, Kumar A, Bhat HP, Baliga MS (2014) Anti-diabetic effects of the Indian indigenous fruit Emblica officinalis Gaertn: active constituents and modes of action. Food Funct. 5(4):635–644. https://doi.org/10.1039/c3fo60366k. PMID: 24577384

Eranezhath S (2011) Ayurveda poised for change, so is the 'ancient science of life'. Ancient Sci Life 30(3):63. PMID: 22557429; PMCID: PMC3336257

Eshun K, He Q (2004) Aloe vera: a valuable ingredient for the food, pharmaceutical and cosmetic industries—a review. Crit Rev Food Sci Nutr 44(2):91–96. https://doi.org/10.1080/10408690490424694

Fan S, Geng Q, Pan Z, Li X, Tie L, Pan Y, Li X (2012) Clarifying off-target effects for torcetrapib using network pharmacology and reverse docking approach. BMC Syst Biol 6:1–12. https://doi.org/10.1186/1752-0509-6-152

Hewlings SJ, Kalman DS (2017) Curcumin: a review of its' effects on human health. Foods 6(10):92. https://doi.org/10.3390/foods6100092

Howard DH, Scott RD (2005) The economic burden of drug resistance. Clin Infect Dis 41(Supplement_4):S283–S286. https://doi.org/10.1086/430792

Huang B, Zhang Y (2022) Teaching an old dog new tricks: drug discovery by repositioning natural products and their derivatives. Drug Discov Today 27(7):1936–1944. https://doi.org/10.1016/j.drudis.2022.02.007

Kundu M, Majumder R, Das CK, Mandal M (2021) Natural products based nanoformulations for cancer treatment: current evolution in Indian research. Biomed Mater 16(4):044101. https://doi.org/10.1088/1748-605X/abe8f2

Somarathinam K, Gunalan S, Sailapathi A, A DAM, Pannerselvam B, Mohanasundari M, Balaraman AK, Kothandan G (2023) Antihypertensive effects of pentacyclic triterpenoid from Convolvulus pluricaulis and its plausible mechanism of action hypothesizing its selectivity targeting Mineralocorticoid receptor of RAAS. Phytomedicine Plus, 3(1):100408, ISSN 2667–0313, https://doi.org/10.1016/j.phyplu.2023.100408

Kanetkar P, Singhal R, Kamat M (2007) Gymnema sylvestre: A memoir. J Clin Biochem Nutr. 41(2):77–81. https://doi.org/10.3164/jcbn.2007010. PMID: 18193099; PMCID: PMC2170951

Leyon PV, Kuttan G (2004) Effect of Tinospora cordifolia on the cytokine profile of angiogenesis-induced animals. Int Immunopharmacol 4(13):1569–1575. https://doi.org/10.1016/j.intimp.2004.06.015. PMID: 15454110

Manikandan P, Vidjaya Letchoumy P, Prathiba D, Nagini S (2007) Proliferation, angiogenesis and apoptosis-associated proteins are molecular targets for chemoprevention of MNNG-induced gastric carcinogenesis by ethanolic Ocimum sanctum leaf extract. Singapore Med J 48(7):645–651. PMID: 17609827

Mashhadi NS et al (2013) Anti-oxidative and anti-inflammatory effects of ginger in health and physical activity: review of current evidence. Int J Prev Med 4(Suppl 1):S36. PMID: 23717767

Mulder KC, Mulinari F, Franco OL, Soares MS, Magalhães BS, Parachin NS (2015) Lovastatin production: from molecular basis to industrial process optimization. Biotechnol Adv 33(6):648–665. https://doi.org/10.1016/j.biotechadv.2015.04.001

Nayak S*, Nayak S, Prasad Dash D, Das B (2016) A clinical study on the effect of boerhaavia diffusa (Punarnava) in essential hypertension. Ayushdhara [Internet]. 2016 Apr.3 [cited 2024 Jul.18];2(6). Available from: https://ayushdhara.in/index.php/ayushdhara/article/view/24

Otimenyin SO (2021) Ethnobotanical survey: the foundation to evidence-based validation of medicinal plants. In: Evidence based validation of traditional medicines: a comprehensive approach, pp 441–466. https://doi.org/10.1007/978-981-15-8127-4_22

Patwardhan B (2012) The quest for evidence-based Ayurveda: lessons learned. Curr Sci 102:1406–1417. https://www.jstor.org/stable/24107798

Patwardhan B, Vaidya AD, Chorghade M, Joshi SP (2008) Reverse pharmacology and systems approaches for drug discovery and development. Curr Bioact Compd 4(4):201–212. https://doi.org/10.2174/157340708786847870

Kalpesh P, Mukeshkumar N, Nishteswar K (2020) Central nervous system depressant activity of Jatamansi (Nardostachys jatamansi DC.) rhizome. AYU (An international quarterly journal of research in Ayurveda) 41(4):250–254. https://doi.org/10.4103/ayu.AYU_251_20

Perera PRD, Ekanayake S, Ranaweera KKDS (2017) Antidiabetic compounds in syzygium cumini decoction and ready to serve herbal drink. Evid Based Complement Alternat Med. 2017:1083589. https://doi.org/10.1155/2017/1083589. Epub 2017 May 9. PMID: 28572825; PMCID: PMC5440793.

Sabu MC, Smitha K, Kuttan R (2002) Anti-diabetic activity of green tea polyphenols and their role in reducing oxidative stress in experimental diabetes. J Ethnopharmacol 83(1–2):109–116. https://doi.org/10.1016/S0378-8741(02)00217-9

Sharma B, Salunke R, Balomajumder C, Daniel S, Roy P (2010) Anti-diabetic potential of alkaloid rich fraction from Capparis decidua on diabetic mice. Journal of ethnopharmacology. 127(2):457–462. https://doi.org/10.1016/j.jep.2009.10.013

Shah AJ, Gilani AH (2009) Blood pressure-lowering and vascular modulator effects of Acorus calamus extract are mediated through multiple pathways. J Cardiovasc Pharmacol. 54(1):38–46. https://doi.org/10.1097/FJC.0b013e3181aa5781. PMID: 19528816.

Sheeja K, Guruvayoorappan C, Kuttan G (2007) Antiangiogenic activity of Andrographis paniculata extract and andrographolide. Int Immunopharmacol 7(2):211–221. https://doi.org/10.1016/j.intimp.2006.10.002. Epub 2006 Nov 2. PMID: 17178389

Siddiqui MZ (2011) Boswellia serrata, a potential antiinflammatory agent: an overview. Indian J Pharm Sci 73(3):255–261. https://doi.org/10.4103/0250-474X.93507

Singh RB, Niaz MA, Ghosh S (1994) Hypolipidemic and antioxidant effects of Commiphora mukul as an adjunct to dietary therapy in patients with hypercholesterolemia. Cardiovasc Drugs Ther 8(4):659–664. https://doi.org/10.1007/BF00877420

Rathod CP, Ghante MH (2021) Pharmacological importance of saraca asoca: A review. Research Journal of Pharmacognosy and Phytochemistry. 13(3):131–135. https://doi.org/10.52711/0975-4385.2021.00022

Vaidya AD, Devasagayam TP (2007) Current status of herbal drugs in India: an overview. J Clin Biochem Nutr 41(1):1. https://doi.org/10.3164/jcbn.2007001

Wen M, Chen Q, Chen W, Yang J, Zhou X, Zhang C, Wu A, Lai J, Chen J, Mei Q, Yang S, Lan C, Wu J, Huang F, Wang L (2022) A comprehensive review of Rubia cordifolia L.: Traditional uses, phytochemistry, pharmacological activities, and clinical applications. Front Pharmacol. 13:965390. https://doi.org/10.3389/fphar.2022.965390. PMID: 36160419; PMCID: PMC9500525.

Yoo DY, Jung HY, Nam SM, Kim JW, Choi JH, Kwak YG, Yoo M, Lee S, Yoon YS, Hwang IK (2015) Valeriana officinalis extracts ameliorate neuronal damage by suppressing lipid peroxidation in the gerbil hippocampus following transient cerebral ischemia. J Med Food. 18(6):642–647. https://doi.org/10.1089/jmf.2014.3295. Epub 2015 Mar 18. PMID: 25785762; PMCID: PMC4458752.

Zahra W, Birla H, Singh SS, Rathore AS, Dilnashin H, Singh R, Keshri PK, Gautam P, Singh SP (2022) Neuroprotection by mucuna pruriens in neurodegenerative diseases. Neurochem Res. 47(7):1816–1829. https://doi.org/10.1007/s11064-022-03591-3. Epub 2022 Apr 5. PMID: 35380400

Role of Flavonoids as Ethnomedicine for the Treatment of Complex Neurodegenerative Diseases

Sourav Das and Anupam Nath Jha

Contents

Abstract The role of flavonoids as ethnomedicine for treating complex diseases is an area of interest in scientific research. The utilization of flavonoid-rich plants for medicinal purposes traces back to ancient civilizations, where indigenous communities passed down knowledge through generations. Flavonoids, recognized for their diverse health benefits, find applications in ethnomedicine across cultures. This chapter explores the historical and cross-cultural use of flavonoid-rich plants in traditional healing practices, showcasing their roles in treating various ailments. In response to contemporary health challenges, scientists have isolated thousands of flavonoids, exploring their potential medicinal effects. Flavonoids, with their diverse biological activities, emerge as potential candidates in combating neurodegenerative disorders. Epidemiological evidence suggests a correlation between flavonoid-

S. Das
Life Sciences Division, Institute of Advanced Study in Science and Technology, Guwahati, Assam, India

A. N. Jha (✉)
Department of Molecular Biology and Biotechnology, Tezpur University, Tezpur, India
e-mail: anjha@tezu.ac.in

© The Author(s), under exclusive license to Springer Nature Singapore Pte Ltd. 2024

A. Das Talukdar et al. (eds.), *Traditional Resources and Tools for Modern Drug Discovery*, Interdisciplinary Biotechnological Advances,
https://doi.org/10.1007/978-981-97-4600-2_11

rich diets and reduced risk of neurodegenerative conditions. The chapter systematically reviews the intricate landscape of complex diseases such as Alzheimer's, Parkinson's, and Huntington's, underscoring their multifactorial nature. Specific flavonoids and their mechanisms take center stage, providing valuable insights into potential therapeutic interventions. This exploration serves not only to understand the traditional wisdom embedded in ethnomedicine but also to facilitate a seamless connection between this ancient knowledge and the advancements of modern scientific exploration. The bridging of this gap holds promise for innovative approaches to treating and preventing complex diseases, rooted in the natural compounds found in flavonoid-rich plants.

Keywords Ethnomedicine · Flavonoids · Alzheimer's · Parkinson's · Huntington's

1 Introduction

Nature, with its diverse natural products, serves as a potent defense against various diseases, including those caused by microbes. In the face of emerging diseases, drug resistance, and the limitations of synthetic drugs, scientists, in collaboration with ethnomedicine, are turning to nature's treasure troves. By exploring the natural metabolites of plants, especially for managing viral infections, researchers aim to find effective and sustainable solutions to health challenges. This shift underscores the importance of harnessing the therapeutic potential found in the richness of nature's offerings, guided by traditional knowledge and practices in ethnomedicine. Scientifically, ethnomedicine involves the examination of traditional healing methods that rely on the bioactive substances found in plants and animals. This field of study focuses on healthcare practices within diverse ethnic groups, particularly those with limited access to western medicinal approaches (Mahapatra et al. 2019). Ethnomedicine and traditional healing practices are ancient systems of healthcare that have been passed down through generations in various cultures around the world (Sabiha et al. 2022). It is an expansive interdisciplinary field, encompassing the knowledge of using natural pharmaceuticals and understanding the ethnic origins of pharmacologically active ingredients. Ethnomedicine has been utilized in a wide array of ancient civilizations, including traditional Chinese medicine (TCM), Ayurveda (India), African multipractices, and Unani healing traditions during the Mughal era in India. Practitioners, guided by traditions and beliefs, may lack awareness of modern treatment theories. Nevertheless, their approach, rooted in generations of successful outcomes, focuses on relieving suffering and uncovering the underlying causes with a belief in "healing from within." These practices are deeply rooted in the knowledge, beliefs, and traditions of specific communities, and they often involve the use of medicinal plants, rituals, and spiritual elements to treat and prevent illnesses. Ethnomedicine and traditional healing play a significant role in the healthcare landscape, particularly in regions where modern western medicine may not be readily accessible or culturally accepted.

Flavonoids, a subset of polyphenols, have captivated the attention of scientists, nutritionists, and health enthusiasts for their multifaceted roles in human health (Manzoor et al. 2023). Flavonoids constitute a category of natural molecules that are found in various plant-derived sources, including cereals, fruits, bark, root crops, stems, flowers, tea, dark chocolate, and wine (Das et al. 2020). The consumption of flavonoids is primarily through the plant-based foods we eat. For instance, quercetin, a flavonol, can be found in apples, onions, and red grapes, while epicatechin, a flavanol, is abundant in cocoa products (Panche et al. 2016; Kesavan et al. 2018). Anthocyanins, responsible for the vibrant colors of blueberries and blackberries, are another prominent subgroup of flavonoids. The diversity of flavonoid-rich foods allows for a balanced and colorful diet, which is associated with numerous health benefits. The term "flavonoid" encompasses a wide range of chemical structures, but all share a common backbone, a structure that imparts a spectrum of colors and a host of biological activities. They form an extensive category of polyphenolic compounds distinguished by a benzo-γ-pyrone structure, and they are commonly found throughout various plant species. The basic structure of flavonoids, as illustrated in Fig. 1 (Kumar and Pandey 2013; Das et al. 2018), comprises a 15-carbon skeleton featuring two benzene rings joined by a 3-carbon bridging chain.

Flavonoids can be categorized into various groups, including flavonols, flavones, flavanones, flavanols (or catechins), anthocyanins, and isoflavones. This categorization is determined by factors such as molecular composition, oxidation level, and the unsaturation of the connecting chain (C3) (Fig. 1) (Kumar and Pandey 2013; Shen et al. 2022). These variations in structure lead to differences in color, solubility, and bioactivity. Moreover, flavonoids exist in both free aglycone and glycoside-bound forms, with the latter being predominant in the diet for flavones and flavonols.

The intake of flavonoids has been associated with a myriad of health benefits (Ullah et al. 2020). One of the most recognized benefits is their potent antioxidant activity (Kozłowska and Szostak-Wegierek 2014). Flavonoids help combat oxidative stress by neutralizing harmful free radicals in the body, which is crucial in preventing chronic diseases and cellular damage. Furthermore, flavonoids exhibit anti-inflammatory properties, which play a pivotal role in mitigating chronic inflammatory conditions. Flavonoids have also been associated with cardiovascular health (Rees et al. 2018). Certain flavonoids, such as quercetin and resveratrol, have been shown to play a role in lowering cholesterol levels and blood pressure, consequently reducing the risk of heart disease (Larson et al. 2012; Akbari et al. 2020). Additionally, these compounds have the potential to enhance blood vessel function and promote healthy blood flow. Cognitive health is another area where flavonoids make a positive impact. These phytochemicals have been studied for their neuroprotective effects, possibly lowering the susceptibility to neurodegenerative conditions like Parkinson's and Alzheimer's disease (Maan et al. 2020). Emerging research has also suggested that flavonoids may play a role in cancer prevention. Their anti-inflammatory and antioxidant properties can help protect against DNA damage and the development of cancerous cells.

This chapter highlights the profound connection to ethnomedicine, where flavonoids, renowned for their anti-inflammatory, antioxidant, and anti-cancer attributes,

(a) Basic skeletal framework of flavonoid

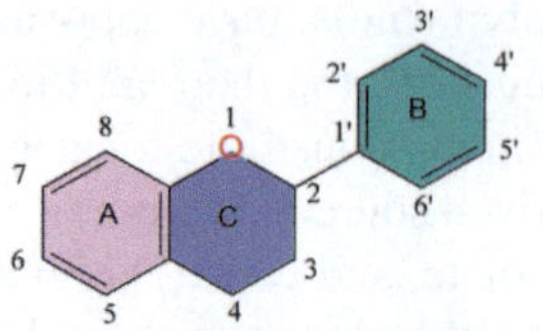

(b) Different classes of flavonoid

Flavanol	Flavanone	Flavonol

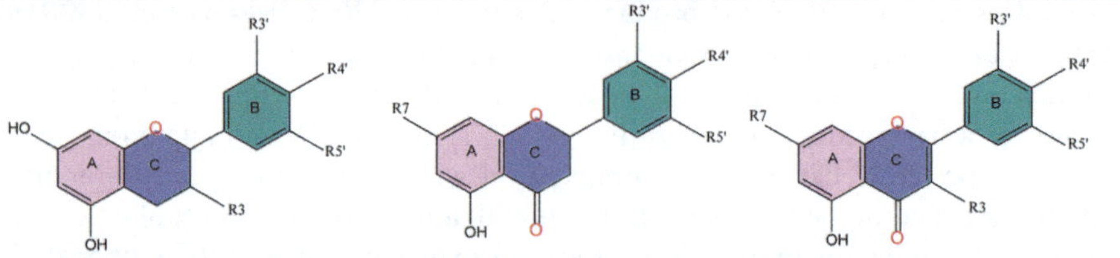

Flavanol	R3	R3'	R4'	R5'
Catechin	OH	-	OH	OH
Epicatechin	OH	OH	OH	-
Epigallocatechin	OH	OH	OH	OH
Epigallocatechin gallate	O	OH	OH	OH

Flavanone	R7	R3'	R4'	R5'
Naringenin	OH	H	OH	H
Hesperetin	OH	H	OCH_3	OH
Eriodictyol	OH	OH	OH	H
Pinocembrim	OH	H	H	H

Flavonol	R3	R7	R3'	R4'	R5'
Quercetin	OH	OH	OH	OH	H
Rutin	Rut	OH	OH	OH	H
Kaempferol	OH	OH	H	OH	H
Myricetin	OH	OH	OH	OH	OH

Flavone	Isoflavone	Anthocyanin

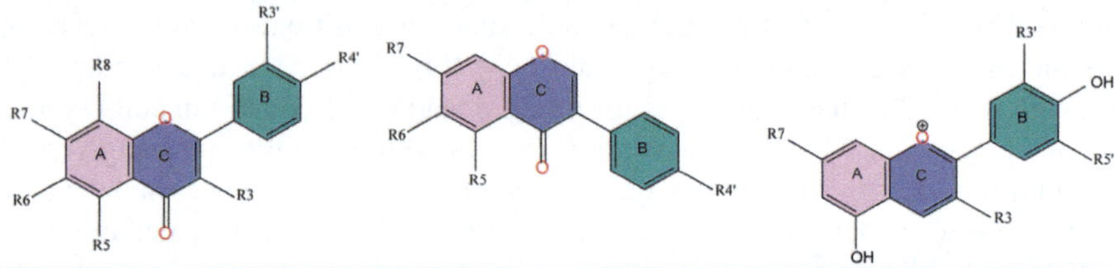

Flavone	R5	R6	R7	R8	R3'	R4'
Luteolin	OH	H	OH	H	OH	OH
Apigenin	OH	H	OH	H	H	OH
Tangeretin	OH	H	OH	H	H	OH
Baicalein	OH	OH	OH	H	H	H

Isoflavone	R5	R6	R7	R4'
Genistein	OH	H	OH	OH
Diadzein	H	H	OH	OH
Biochanin A	OH	H	OH	OCH_3
Glycitein	H	OCH_3	OH	OH

Anthocyanin	R3	R7	R3'	R5'
Cyanidin	OH	OH	OH	H
Malvidin	OH	OH	OCH_3	OCH_3
Petunidin	OH	OH	OH	OCH_3
Delphinidin	OH	OH	OH	OH

Fig. 1 (**a**) The basic chemical structure of flavonoid. (**b**) Classes of flavonoids along with few flavonoids under each class

emerge as integral components of indigenous healing traditions. The historical tapestry of ethnomedicine weaves a narrative of the enduring role played by flavonoids in fostering health and well-being. The chapter further shows the intricate interplay between flavonoids and neuroprotection, with a focus on combating neurodegenerative diseases. Further, this chapter underscores flavonoids as valuable allies in navigating the complex terrain of combating neurodegenerative disorders. Their polypharmacological actions and interactions with disease-associated proteins present promising avenues for drug development, offering hope for effective interventions in the challenging realm of ethnomedicine and neurodegenerative diseases.

2 Flavonoids in Ethnomedicine: Unraveling Nature's Healing Tapestry

The utilization of plants rich in flavonoids for medicinal use can be traced back to ancient civilizations. Indigenous communities, with a deep connection to their natural surroundings, have passed down knowledge of medicinal plants through generations. In ethnomedicine, flavonoids are often recognized for their potential health advantages, encompassing anti-cancer, anti-inflammatory, antioxidant, and antiviral benefits. Indigenous communities have historically utilized plants containing flavonoids to formulate remedies for conditions ranging from infections to chronic diseases. The use of flavonoid-rich herbs and plants in ethnomedicine reflects the long-standing observation and application of natural compounds for maintaining health and addressing medical concerns within specific cultural contexts. Many of these plants, rich in flavonoids, have been used for treating wounds, infections, and various health conditions. The ethnomedicinal practices of Indigenous peoples often involve a holistic approach, considering the interconnectedness of the environment, spirituality, and well-being. For example, the ancient Egyptians used chamomile, which contains about 50 flavonoids, to treat various ailments such as fever and inflammation (Dai et al. 2022). The ancient Greeks used juice extracted from willow bark, which contains flavonoids, to relieve pain, inflammation, and fever (Sulima and Przyborowski 2019). Traditional Chinese medicine, Ayurveda in India, traditional African medicine, Kampo medicine, Siddha medicine, Amazonian traditional medicine, Unani medicine, Maori traditional medicine, western herbalism, and indigenous healing practices in various parts of the world incorporated plants containing flavonoids to address a wide range of health issues (Singh 2016; Li et al. 2019; Achika et al. 2023). The historical texts of these traditions document the use of specific plants for their flavonoid content, emphasizing their roles in promoting health and treating ailments. The rise of novel infectious, chronic, and drug-resistant diseases has led scientists to explore medicinal plants as potential agents for treating and preventing illnesses. Hence, based on the ethnomedicinal values of flavonoids, thousands of them have been isolated from natural sources and screened for their medicinal effects (Tahir et al. 2021). For example, six flavonoids were isolated from ethnomedicine "dragon's blood" which refers to a resin obtained from various plant species, and it has been used in traditional medicine for various purposes (Yi et al. 2012; Lang et al. 2020). High concentrations of catechin have been found in in *Withania somnifera* (ashwagandha) which possess anti-microbial effects (Alam et al. 2011). Roots of medicinal plant *Achyranthes bidentata* contain quercetin and kaempferol which have significant role in osteoarthritis treatment (Ma et al. 2019), while *Mangifera indica* (mango) is known to be a good source of quercetin and rutin inhibits α-amylase and α-glucosidase activities (Gondi and Prasada Rao 2015). Hence, the historical and cross-cultural use of flavonoid-rich plants in medicinal practices reflects an enduring bond between indigenous communities and nature. From ancient civilizations to modern times, the ethnomedicinal wisdom surrounding flavonoids persists as a holistic approach to health, acknowledging the interconnectedness of the environment and spirituality. The documented utilization of specific plants for their flavonoid content across diverse traditions underscores their pivotal role in promoting well-being. As contemporary challenges prompt a renewed focus on natural remedies, the isolation and screening of thousands of flavonoids exemplify a bridge between traditional knowledge and scientific exploration, offering a promising avenue for the development of effective, culturally rooted therapeutic interventions.

3 Flavonoids in Addressing Intricate Neurodegenerative Conditions

Complex diseases, also known as multifactorial diseases, are disorders that result from the contributions of multiple genes and environmental factors. The onset of these diseases is influenced by a blend of environmental, hereditary, and lifestyle elements, rendering them more challenging to address in comparison to straightforward genetic conditions. Complex neurological diseases encompass a diverse range of disorders affecting the nervous system, posing intricate challenges for both diagnosis and treatment (Siuly and Zhang 2016). These conditions involve abnormalities in the structure and function of the spinal cord, brain, and peripheral nerves, often resulting in profound impacts on cognition, movement, sensory perception, and other essential functions. Disorders like Alzheimer's disease, marked by gradual memory impairment and cognitive deterioration, exemplify the complexity of neurodegenerative conditions (Breijyeh et al. 2020). Similarly, Parkinson's disease, marked by tremors, rigidity, and impaired motor control, underscores the intricate nature of these disorders (Hill et al. 2023). Additionally, autoimmune diseases like multiple sclerosis involve the immune system erroneously attacking the central nervous system, resulting in diverse range of neurological symptoms (Akyuz et al. 2023). The intricate interplay of genetic, environmental, and immunological factors further complicates our understanding of diseases like amyotrophic lateral sclerosis (ALS), where the deterioration of motor neurons leads to muscle weakness and paralysis (Taylor et al. 2016). Successfully addressing these complex neurological diseases necessitates ongoing research, multidisciplinary approaches to patient care, and a comprehensive understanding of the intricate mechanisms underlying each disorder. Complex diseases do not follow a clear inheritance pattern, and their genetic component may be small compared to the influence of environment, lifestyle, age, and diet. The specific factors contributing to the emergence of complex diseases are not yet fully understood, and researchers often use statistical models to study the risk factors and their interactions (Saha et al. 2023; Bora et al. 2020). Understanding the causes of complex diseases is a challenging task, as the effects of individual factors may be obscured or confounded by other contributing factors. However, cross-disciplinary studies and multidisciplinary efforts are being conducted to better monitor and elucidate disease development. The identification of major contributing genes for complex disorders is an ongoing area of research (Hojaisa and Jha 2023), and the development of more effective treatments and prevention strategies is a priority for researchers in the field of genetics and medicine (Saha and Nath Jha 2023).

As previously mentioned, owing to their biological activity, structural variations, bioavailability, metabolism, and distribution, diverse flavonoids can manifest unique effects on human health. Among the vast array of secondary metabolites derived from plants, flavonoids have garnered significant attention in epidemiological investigations, emphasizing their beneficial role in preventing neurodegenerative diseases (Solanki et al. 2015; Maher 2019). Historically, flavonoids earned recognition primarily for their ability to neutralize free radicals and antioxidant properties (Jovanovic et al. 1994). These characteristics hinted at their potential in combating oxidative stress, a process linked to aging and various diseases, including neurodegenerative conditions. However, recent studies have expanded our understanding, revealing that flavonoids possess a diverse array of activities that go beyond their antioxidative roles. Figure 2 emphasizes the protective effects of flavonoids against different conditions associated with neurodegenerative disorders. It should be noted that, flavonoids possess polypharmacological effects, i.e., these molecules can simultaneously interact with multiple targets or pathways within biological systems, enhancing their therapeutic potency (Li et al. 2014; Baby et al. 2016). In the realm of neuroprotection, flavonoids may play a crucial role, showcasing their diverse pharmacological potential. Emerging evidence suggests that these compounds may be especially potent in inhibiting the pathways of toxicity associated with age-related neurodegenerative diseases (Jones et al. 2012). This broader

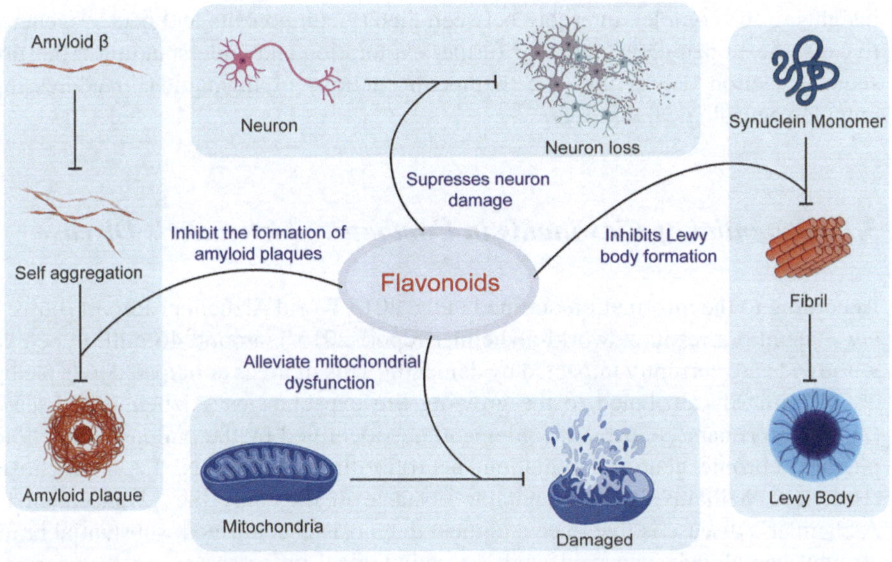

Fig. 2 Protective actions of flavonoids against neurodegenerative diseases

spectrum of activities positions flavonoids as multifaceted players in the complex landscape of neuroprotection, promising not only antioxidant defense but also potential interventions in the intricate processes underlying age-related neurodegenerative disorders (Tahir et al. 2021). Numerous epidemiological investigations have suggested a correlation between a diet rich in flavonoids and a reduced risk of Parkinson's and Alzheimer's diseases (De Andrade Teles et al. 2018). The anti-inflammatory and antioxidative attributes of flavonoids are believed to play a crucial role in alleviating oxidative stress and inflammation, processes associated with the pathogenesis of neurodegenerative conditions. Though additional research is required to establish causal relationships and elucidate specific mechanisms, the expanding body of evidence underscores the potential of flavonoids as valuable dietary components for promoting brain health and potentially impeding the initiation or advancement of neurodegenerative disorders. Several years ago, a groundbreaking epidemiological study, encompassing a massive cohort of approximately 130,000 individuals tracked over a span of 20–22 years, sought to unravel the connection between flavonoid intake and the risk of developing Parkinson's disease (Gao et al. 2012). The study meticulously examined the dietary habits of participants, focusing on five primary flavonoids sources—berries, tea, red wine, orange, and apples. The investigation utilized a comprehensive database of food compositions and detailed surveys on the frequency of food consumption to assess flavonoid intake. Intriguingly, the results uncovered significant association between higher flavonoid intake and a remarkable 40% reduction in the risk of Parkinson's among men who fell within the highest quintile of flavonoid consumption in comparison to those in the lowest quintile. Nonetheless, notably, this significant association was not mirrored in women, emphasizing potential gender-specific variations in the impact of flavonoids on Parkinson's disease risk. This study contributes valuable insights to the complex interplay between dietary components and neurodegenerative diseases, opening avenues for further exploration and understanding. The subsequent section delves into the therapeutic actions of flavonoids in addressing complex neurological diseases.

3.1 Potential of Flavonoids in Combating Alzheimer's Disease

According to the information outlined in the 2015 World Alzheimer Report (https://www.alzint.org/resource/world-alzheimer-report-2015/), around 46 million people worldwide are currently impacted by dementia. This figure is anticipated to triple by 2050, primarily attributed to the growing life expectancies. Alzheimer's disease (AD), the primary contributor to dementia, is identified by the buildup of amyloid plaques, chronic neuroinflammation, neurofibrillary tangles (NFTs), and gliosis (Hole and Williams 2020). A notable obstacle in impeding the advancement of Alzheimer's disease is that once a clinical diagnosis is confirmed, substantial brain atrophy has already occurred, and the pathological processes are well underway (Holtzman et al. 2011). Flavonoids display significant therapeutic promise in tackling Alzheimer's disease and are considered promising candidates for future drug development (Hole and Williams 2020). The results suggest that the primary mechanism by which flavonoids contribute to Alzheimer's treatment involves diminishing the production of β-amyloid protein and reactive oxygen species (ROS). Examination of around 127 flavonoids for their anti-Alzheimer's activity uncovered that their efficacy is associated with their role as inhibitors of acetylcholinesterase and butyrylcholinesterase (Abou Baker 2022). Scientific results have revealed a myriad of ways in which these natural compounds intervene, offering a promising narrative in the relentless fight against cognitive decline. Firstly, flavonoids increase the activity of α-secretase, a protective enzyme that helps stop the formation of Aβ plaques (Baptista et al. 2014). This is like a secret weapon to disrupt Alzheimer's plans. Secondly, flavonoids inhibit β-secretase, another key player in the formation of Aβ plaques (Baptista et al. 2014). It's like creating a strong shield to block the first step of the disease. Flavonoids also interfere with fibrillation, disrupting the aggregation of β-protein that contribute to Alzheimer's (Uddin et al. 2020). They showcase metal-chelating activity, preventing the clumping together of beta amyloid proteins. In a powerful move, flavonoids boost blood flow in the brain. This not only feeds the brain but also serves as a protective barrier against the detrimental effects of β-amyloid proteins. Flavonoids work to decrease the levels of β-amyloid proteins, countering their damaging effects on the brain. They also play a role in stopping factors that harm nerves, like nitric oxide (NO), reactive oxygen species (ROS), and acetyl choline esterase (AChE) (Abou Baker 2022). For instance, in both neuron-like cells carrying the human Swedish mutant APP transgene and primary neurons, it was discovered that luteolin, a flavonoid derived from citrus, has the capacity to diminish the production of amyloid precursor protein (APP) by influencing the amyloidogenic γ-secretase activity. This action of luteolin results in a reduction in the production of Aβ, a key component associated with Alzheimer's disease (Rezai-Zadeh et al. 2009). Further, there is support for the positive impact of epigallocatechin-3-gallate, a tea polyphenol in lowering brain Aβ levels by regulating the processing of APP (Rezai-Zadeh et al. 2005). Interestingly, EGCG induces an increase in the nonamyloidogenic processing of APP by augmenting α-secretase cleavage (Obregon et al. 2006). Apigenin, a flavonoid, has shown potential in alleviating learning and memory deficits associated with AD. This beneficial effect is achieved through its ability to inhibit the deposition of β-amyloid and reduce insoluble Aβ concentrations (Zhao et al. 2013). The function of some of the flavonoids as anti-Alzheimer agents are listed in Table 1. Therefore, the intricate and diverse mechanisms through which flavonoids exhibit therapeutic promise against Alzheimer's disease provide a hopeful panorama in the realm of neuroprotection. As we navigate the complexities of their actions, from disrupting plaque formation to enhancing blood flow and addressing key enzymes, the potential for these natural compounds as allies in the battle against cognitive decline is increasingly evident. The ongoing exploration of flavonoids, exemplified by specific compounds like luteolin, EGCG, and other flavonoids, fuels optimism for future drug development and underscores the importance of continued research in unlocking novel solutions for Alzheimer's and related neurodegenerative disorders.

Table 1 List of flavonoids showing anti-Alzheimer effect

Flavonoids	Mechanism of action
Silibinin (Song et al. 2016)	Inhibition of inflammatory reaction Reduction in reactive oxygen species (ROS) generation
Epicatechin (Jin Heo and Lee 2005)	Diminishes neurotoxic effects induced by Aβ
Hesperidin (Abou Baker et al. 2020)	Facilitates neural differentiation Reduction in β-amyloid plaques Inhibits AChE
Wogonin (Huang et al. 2017)	Inhibition of amyloidogenic pathway
Baicalein (Im et al. 2005)	Elevates dopaminergic levels
Quercetin (Paula et al. 2019)	Diminishing β-amyloidosis and mitigating tauopathy in the amygdala and hippocampus
Naringin (Sachdeva et al. 2014)	Enhances conditions of oxido-nitrosative stress and inflammatory surge induced by mitochondrial dysfunction
Myricetin (Ramezani et al. 2016)	Safeguards pyramidal neurons in the hippocampus and enhances cognitive function by ameliorating learning and memory impairments

3.2 *Potential of Flavonoids in Combating Parkinson's Disease*

The prevalence of Parkinson's disease (PD), ranking as the second-most widespread neurodegenerative disorder after AD, is on the rise globally as the population ages. PD is a complex and challenging condition that impacts millions of people around the world. Two main characteristics of Parkinson's disease (PD) include the continual and advancing decline of dopaminergic neurons and the existence of intracellular Lewy bodies within the substantia nigra (SN). The lesions that develop in the substantia nigra lead to a reduction in dopamine (DA) levels and a decline in the functioning of DA nerve pathways. It unfolds as a puzzle, affecting not just movement but also the way people think and behave (Gopalakrishna and Alexander 2015). At its core, Parkinson's is marked by a gradual breakdown of dopaminergic neurons, crucial players in our brain's communication system (MacMahon Copas et al. 2021). As these neurons fade away, they take with them vital functions, giving rise to a cascade of troublesome symptoms affecting movement, thinking, and behavior. Beyond the typical motor symptoms, individuals suffering from PD also display disruptions in involuntary bodily functions, cognitive functions, and psychological aspects. In the pathogenesis of PD, a distinct pattern emerges, marked by the aberrant folding and clustering of proteins, accompanied by mitochondrial dysfunction and resulting oxidative stress. Thus, in recent years, there has been a growing interest in natural antioxidants as potential means to reduce the risk of PD (Sandoval-Avila et al. 2019). As a result, both researchers and the pharmaceutical industry have redirected their focus toward flavonoids, drawn to their exceptional antioxidative capabilities and other therapeutic attributes (Costas and Faro 2021). Studies have shown diet rich in fruits and vegetables containing flavonoids may increase survival rates among individuals with PD (Zhang et al. 2022). Flavonoids have demonstrated the ability to influence several pathological pathways and display protective effects on the nervous system in animal models and cellular models conducted in vivo and in vitro, as well as in human studies related to PD (Ullah et al. 2020). Research demonstrated that flavonoids, including quercetin, significantly improved the coordination and motor balance of mice under PD induction. Additionally, remarkable improvements were detected in the effectiveness of various antioxidants, encompassing glutathione peroxidase and superoxide dismutase (Lv et al. 2012). Quercetin administration in neurons mitigates microglia activation, which is an early indication of PD development, and also inhibits cell death (Bournival et al. 2012). Further, studies have shown that flavonoids can offer advantages in averting and managing neurodegenerative conditions such as PD by reducing neuroinflammation, oxidative stress, mitochondrial dysfunction, and improved support of neurotrophic factors (Jung and Kim 2018; Meng-zhen et al. 2022). Baicalein, a naturally occurring flavonoid extracted from *Scutellaria baicalensis* Georgi, mitigated the decline of dopaminergic neurons in PD (Hung et al. 2016). It also possesses anti-oxidative, neuroprotective, antiapoptotic, and anti-inflammatory properties against PD (Zhu et al. 2019). Similarly, citrus flavonoid and naringenin have shown neuroprotective effects in PD. It is suggested that naringenin holds promise as a potential treatment for PD, since it decreases α-synuclein expression and neuroinflammation (Mani et al. 2018). Kaempferol is suggested to play a protective role in PD by preventing the degeneration of dopamine neurons. This protective effect is attributed to its ability to inhibit lipid peroxidation-mediated mitochondrial damage (Han et al. 2021). Specifically, kaempferol is believed to promote lipophagy, a process involving the removal of damaged or excess lipids through autophagy, which is essential for maintaining mitochondrial health. By preventing mitochondrial damage and supporting lipophagy, kaempferol may offer a potentially innovative therapeutic approach in the ongoing battle against PD and other neurodegenerative diseases. This mechanism highlights the intricate interplay between cellular processes and the potential of natural compounds like kaempferol in addressing the underlying factors contributing to neurodegeneration in PD and related disorders. The effects of some of the flavonoids having potential subsiding effect on PD are listed in Table 2.

Therefore, understanding the role of flavonoids in the context of PD reveals a promising landscape for potential therapeutic interventions. From the complex interplay of dopaminergic neuron loss to the multifaceted impact on oxidative stress, neuroinflammation, and mitochondrial dysfunction, flavonoids emerge as natural compounds with significant neuroprotective potential. The evidence presented, from quercetin's positive effects on motor coordination to baicalein and naringenin demonstrating neuroprotective properties, underscores the diverse pathways through which flavonoids can influence and potentially alleviate the burden of PD. As we unravel the complexities of PD, the role of flavonoids stands out as a beacon of hope, offering not only a better understanding of the fundamental mechanisms but also potential avenues for therapeutic strategies. Further research in this

Table 2 List of flavonoids showing anti-Parkinson's effect

Flavonoids	Mechanism of action
Morin (Wang et al. 2023)	Protects dopaminergic neurons from induced neurotoxicity
Tangeretin (Datla et al. 2001)	Protects against the decline of dopamine, reduces behavioral deficits and dopaminergic neuron damage
Luteolin (Chen et al. 2008)	Safeguards dopaminergic neurons from inflammation-induced injury by hindering microglial activation
Rutin (De Araújo et al. 2023)	Raises nerve-derived neurotrophic factor mRNA and elevates glia-derived neurotrophic factor levels. Additionally, augments anti-inflammatory activity
Epigallocatechin-3-gallate (Mandel et al. 2004)	Inhibits aggregation of α-synuclein and Aβ fibrillation. Protects depletion of DA and loss of neurons
Nobiletin (Jeong et al. 2014)	Anti-oxidation, prevents DA neuron damage, reduces motor impairment and cognitive deficits
Daidzein (De Rus Jacquet et al. 2021)	Increases motor performance and antioxidant activity and decreases the loss of DA neurons

field holds the promise of advancing our understanding and bringing us closer to effective treatments for Parkinson's and related neurodegenerative disorders.

3.3 *Potential of Flavonoids in Combating Huntington's Disease*

Huntington's disease (HD) is an uncommon, inherited neurodegenerative disorder marked by the progressive decline of nerve cells in the brain. It is an inherited genetic disorder with an autosomal dominant pattern, implying that if an individual is affected, there is a 50% likelihood of passing the mutated gene to their offspring (Tabrizi et al. 2020). The root cause of HD lies in a mutation within the HTT gene, which is responsible for providing instructions for the synthesis of a protein called huntingtin. HD is a genetic disorder inherited in an autosomal dominant manner. It is caused by a genetic mutation distinguished by an erratic increase of a trinucleotide repeat (CAG) in the DNA (Young 2003). The elongated CAG repeats lead to the incorporation of an unusually long polyglutamine tract in the huntingtin protein. The existence of this altered protein is a pivotal factor in the progression of HD. The mutated form of the huntingtin protein (mHTT) leads to the degeneration of neurons, particularly in the basal ganglia. The symptoms of HD usually become apparent in mid-adulthood, often manifesting around the ages of 30–50; however, onset can occur at any age. Early symptoms may include mood swings, irritability, and clumsiness. As the disease progresses, individuals with HD experience involuntary movements (chorea), which can include jerking or writhing motions. Motor control and coordination deteriorate, leading to difficulties with balance and coordination. HD also affects cognitive function, leading to memory loss, impaired judgment, and difficulties with concentration. Behavioral changes, such as impulsivity and mood disorders, are common. HD is progressive, and the severity of symptoms worsens over time. The rate of progression can vary among individuals. Research on the role of flavonoids in HD is an evolving area with potential implications for neuroprotection. Flavonoids are known to mitigate oxidative stress implicated in neurodegenerative conditions. While investigations into flavonoids have primarily cantered on disorders like Alzheimer's and Parkinson's, their general neuroprotective features suggest a broader relevance. These compounds exhibit anti-inflammatory effects and may modulate signaling pathways crucial for cell survival and neuroplasticity. Although direct evidence of the impact of flavonoids on HD is limited, the published results indicate that certain flavonoids may hold promise for potential clinical use. Flavonoids such as hesperidin and naringin have shown positive effects toward HD in a study (Kumar and Kumar 2010). When compared with a group that received 3-nitropropionic acid (3-NP), a compound associated with neurotoxicity, the group that received naringin and hesperidin exhibited enhanced locomotor activity, muscle grip strength, and a decrease in body weight loss. Additionally, these flavonoids contributed to a decrease in lipid peroxidation and nitrite concentration, indicators of oxidative stress. Furthermore, the crucial antioxidant enzymes, superoxide dismutase and catalase, had their activities restored. These findings suggest that hesperidin and naringin may have protective properties, counteracting the negative effects induced by 3-NP in aspects such as body weight, muscle function, and oxidative stress. The compound fisetin has been found to activate the extracellular signal-regulated kinase (ERK) and diminish the impact of mutant huntingtin (HTT) (Maher et al. 2011). This suggests that fisetin may have a potential therapeutic effect in mitigating the consequences associated with the mutated form of the huntingtin protein, which is implicated in conditions like HD. The activation of ERK, a signaling pathway involved in cell regulation and survival, may contribute to the protective mechanisms exerted by fisetin against the negative effects of mutant HTT. In another study, rutin treatment substantially restored multiple factors, including biochemical, behavioral, and histological aspects (Suganya and Sumathi 2017). This included the reversal of reduced body weight, improvements in locomotor activities, memory enhancement, and restoration of antioxidant levels. Additionally, rutin treatment resulted in a decline in elevated levels of lipid peroxides, nitrite, the activity of acetylcholine esterase, and glial fibrillary acidic protein. Epigallocatechin-3-gallate (EGCG) has been found to hinder the clumping of mutant huntingtin exon 1 (HTT exon 1) protein in vitro (Ehrnhoefer et al. 2006). This implies that EGCG may play a potential role in preventing the formation of protein aggregates associated with mutant huntingtin, which is a key pathological feature in HD. Inhibiting protein aggregation is of interest in neurodegenerative disorders, as the accumulation of misfolded proteins can contribute to the progression of HD. Myricetin has been discovered to engage with the CAG motif of RNA, and this interaction results in preventing the synthesis and retention of muscleblind-like-1 (MBNL1) splicing factor (Khan et al. 2018). This finding suggests that myricetin may play a role in modulating RNA function and addressing the dysregulation associated with certain conditions, particularly those involving trinucleotide repeat expansions, such as myotonic dystrophy. By preventing the translation and sequestration of MBNL1, myricetin may contribute to the regulation of RNA processing, offering a potential avenue for therapeutic intervention in conditions where RNA dysfunction plays a role. Although further research is required to fully understand the mechanisms and therapeutic potential of flavonoids in the context of HD, the existing findings suggest that these natural compounds could have positive effects in mitigating the impact of the disease. Therefore, their potential to interact with proteins associated with the disease, such as the mutant huntingtin protein, warrants exploration. As research progresses, a deeper understanding of the specific mechanisms through which flavonoids may offer therapeutic benefits in the context of HD is anticipated, holding promise for future interventions in this challenging neurodegenerative disorder.

4 Conclusion

In conclusion, this book chapter highlights the promising avenue of flavonoids in the realm of neuroprotection, particularly in addressing complex neurodegenerative diseases such as Huntington's, Alzheimer's, and Parkinson's. With a historical foundation deeply rooted in ethnomedicine, flavonoids, renowned for their antioxidant, anti-inflammatory, and anticancer properties, have played a crucial role in indigenous healing practices. Epidemiological studies highlight a potential link between flavonoid-rich diets and a reduced risk of neurodegenerative conditions. Specific flavonoids, including luteolin, EGCG, myricetin, and others, demonstrate their capacity to modulate key pathways associated with disease progression. These compounds exhibit diverse mechanisms, from inhibiting plaque formation to enhancing blood flow, showcasing their multifaceted nature in neuroprotection. As research progresses, this book chapter emphasizes that flavonoids stand out as valuable allies in the intricate landscape of combatting Alzheimer's, Parkinson's, and Huntington's diseases. The polypharmacology of flavonoids and their interactions with disease-associated proteins open new avenues for drug development, offering hope for effective interventions in the complex realm of neurodegenerative disorders.

References

Abou Baker DH (2022) An ethnopharmacological review on the therapeutical properties of flavonoids and their mechanisms of actions: a comprehensive review based on up to date knowledge. Toxicol Rep 9:445–469. https://doi.org/10.1016/J.TOXREP.2022.03.011

Abou Baker DH, Ibrahim BMM, Hassan NS et al (2020) Exploiting Citrus aurantium seeds and their secondary metabolites in the management of Alzheimer disease. Toxicol Rep 7:723–729. https://doi.org/10.1016/j.toxrep.2020.06.001

Achika JI, Yusuf AJ, Ayo RG, Liman DU (2023) Flavonoids from Nigerian indigenous medicinal plants and their bioactivities: a review. Phytomed Plus 3:100405. https://doi.org/10.1016/J.PHYPLU.2023.100405

Akbari M, Tamtaji OR, Lankarani KB et al (2020) The effects of resveratrol on lipid profiles and liver enzymes in patients with metabolic syndrome and related disorders: a systematic review and meta-analysis of randomized controlled trials. Lipids Health Dis 19:25. https://doi.org/10.1186/s12944-020-1198-x

Akyuz E, Celik BR, Aslan FS et al (2023) Exploring the role of neurotransmitters in multiple sclerosis: an expanded review. ACS Chem Neurosci 14:527–553. https://doi.org/10.1021/ACSCHEMNEURO.2C00589

Alam N, Hossain M, Khalil MI et al (2011) High catechin concentrations detected in Withania somnifera (ashwagandha) by high performance liquid chromatography analysis. BMC Complement Altern Med 11:65. https://doi.org/10.1186/1472-6882-11-65

Baby B, Antony P, Al Halabi W et al (2016) Structural insights into the polypharmacological activity of quercetin on serine/threonine kinases. Drug Des Devel Ther 10:3109–3123. https://doi.org/10.2147/DDDT.S118423

Baptista FI, Henriques AG, Silva AMS et al (2014) Flavonoids as therapeutic compounds targeting key proteins involved in Alzheimer's disease. ACS Chem Neurosci 5:83–92. https://doi.org/10.1021/CN400213R

Bora N, Jha AN (2020) In silico metabolic pathway analysis identifying target against leishmaniasis - a kinetic modeling approach Front Genet 11:179. https://doi.org/10.3389/fgene.2020.00179

Bournival J, Plouffe M, Renaud J et al (2012) Quercetin and sesamin protect dopaminergic cells from MPP+-induced neuroinflammation in a microglial (N9)-neuronal (PC12) coculture system. Oxid Med Cell Longev 2012:921941. https://doi.org/10.1155/2012/921941

Breijyeh Z, Karaman R, Muñoz-Torrero D, Dembinski R (2020) Comprehensive review on Alzheimer's disease: causes and treatment. Molecules 25:5789. https://doi.org/10.3390/MOLECULES25245789

Chen HQ, Jin ZY, Wang XJ et al (2008) Luteolin protects dopaminergic neurons from inflammation-induced injury through inhibition of microglial activation. Neurosci Lett 448:175–179. https://doi.org/10.1016/J.NEULET.2008.10.046

Costas C, Faro LRF (2021) Do naturally occurring antioxidants protect against neurodegeneration of the dopaminergic system? A systematic revision in animal models of Parkinson's disease. Curr Neuropharmacol 20:432–459. https://doi.org/10.2174/1570159X19666210421092725

Dai YL, Li Y, Wang Q et al (2022) Chamomile: a review of its traditional uses, chemical constituents, pharmacological activities and quality control studies. Molecules 28:133. https://doi.org/10.3390/MOLECULES28010133

Das S, Khanikar P, Hazarika Z et al (2018) Deciphering the interaction of 5,7-dihydroxyflavone with hen-egg-white lysozyme through multispectroscopic and molecular dynamics simulation approaches. ChemistrySelect 3:4911–4922. https://doi.org/10.1002/slct.201800280

Das S, Hazarika Z, Sarmah S et al (2020) Exploring the interaction of bioactive kaempferol with serum albumin, lysozyme and hemoglobin: a biophysical investigation using multi-spectroscopic, docking and molecular dynamics simulation studies. J Photochem Photobiol B Biol 205:111825. https://doi.org/10.1016/j.jphotobiol.2020.111825

Datla KP, Christidou M, Widmer WW et al (2001) Tissue distribution and neuroprotective effects of citrus flavonoid tangeretin in a rat model of Parkinson's disease. Neuroreport 12:3871–3875

De Andrade Teles RB, Diniz TC, Costa Pinto TC et al (2018) Flavonoids as therapeutic agents in Alzheimer's and Parkinson's diseases: a systematic review of preclinical evidences. Oxid Med Cell Longev 2018:7043213. https://doi.org/10.1155/2018/7043213

De Araújo FM, Frota AF, de Jesus LB et al (2023) Protective effects of flavonoid rutin against aminochrome neurotoxicity. Neurotox Res 41:224–241. https://doi.org/10.1007/S12640-022-00616-1

De Rus Jacquet A, Ambaw A, Tambe MA et al (2021) Neuroprotective mechanisms of red clover and soy isoflavones in Parkinson's disease models. Food Funct 12:11987–12007. https://doi.org/10.1039/D1FO00007A

Ehrnhoefer DE, Duennwald M, Markovic P et al (2006) Green tea (−)-epigallocatechin-gallate modulates early events in huntingtin misfolding and reduces toxicity in Huntington's disease models. Hum Mol Genet 15:2743–2751. https://doi.org/10.1093/hmg/ddl210

Gao X, Cassidy A, Schwarzschild MA et al (2012) Habitual intake of dietary flavonoids and risk of Parkinson disease. Neurology 78:1138–1145. https://doi.org/10.1212/WNL.0b013e31824f7fc4

Gondi M, Prasada Rao UJS (2015) Ethanol extract of mango (Mangifera indica L.) peel inhibits α-amylase and α-glucosidase activities, and ameliorates diabetes related biochemical parameters in streptozotocin (STZ)-induced diabetic rats. J Food Sci Technol 52:7883–7893. https://doi.org/10.1007/s13197-015-1963-4

Gopalakrishna A, Alexander SA (2015) Understanding Parkinson disease: a complex and multifaceted illness. J Neurosci Nurs 47:320–326. https://doi.org/10.1097/JNN.0000000000000162

Han X, Zhao S, Song H et al (2021) Kaempferol alleviates LD-mitochondrial damage by promoting autophagy: implications in Parkinson's disease. Redox Biol 41:101911. https://doi.org/10.1016/j.redox.2021.101911

Hill DR, Huters AD, Towne TB et al (2023) Parkinson's disease: advances in treatment and the syntheses of various classes of pharmaceutical drug substances. Chem Rev 123(23):13693–13712. https://doi.org/10.1021/ACS.CHEMREV.3C00479

Hojaisa S, Jha AN (2023) Databases and tools for microbial genome and human microbiome studies. Springer, Singapore

Hole KL, Williams RJ (2020) Flavonoids as an intervention for Alzheimer's disease: progress and hurdles towards defining a mechanism of action 1. Brain Plast 6:167–192. https://doi.org/10.3233/BPL-200098

Holtzman DM, Morris JC, Goate AM (2011) Alzheimer's disease: the challenge of the second century. Sci Transl Med 3:77sr1. https://doi.org/10.1126/SCITRANSLMED.3002369

Huang DS, Yu YC, Wu CH, Lin JY (2017) Protective effects of wogonin against Alzheimer's disease by inhibition of amyloidogenic pathway. Evid Based Complement Altern Med 2017:3545169. https://doi.org/10.1155/2017/3545169

Hung KC, Huang HJ, Wang YT, Lin AMY (2016) Baicalein attenuates α-synuclein aggregation, inflammasome activation and autophagy in the MPP+-treated nigrostriatal dopaminergic system in vivo. J Ethnopharmacol 194:522–529. https://doi.org/10.1016/J.JEP.2016.10.040

Im HI, Joo WS, Nam E et al (2005) Baicalein prevents 6-hydroxydopamine-induced dopaminergic dysfunction and lipid peroxidation in mice. J Pharmacol Sci 98:185–189. https://doi.org/10.1254/JPHS.SC0050014

Jeong KH, Jeon M-T, Kim HD et al (2014) Nobiletin protects dopaminergic neurons in the 1-methyl-4-phenylpyridinium-treated rat model of Parkinson's disease. J Med Food 18:409–414. https://doi.org/10.1089/jmf.2014.3241

Jin Heo H, Lee CY (2005) Epicatechin and catechin in cocoa inhibit amyloid β protein induced apoptosis. J Agric Food Chem 53:1445–1448. https://doi.org/10.1021/JF048989M

Jones QRD, Warford J, Rupasinghe HPV, Robertson GS (2012) Target-based selection of flavonoids for neurodegenerative disorders. Trends Pharmacol Sci 33:602–610. https://doi.org/10.1016/J.TIPS.2012.08.002

Jovanovic SV, Steenken S, Tosic M et al (1994) Flavonoids as antioxidants. J Am Chem Soc 116:4846–4851. https://doi.org/10.1021/ja00090a032

Jung UJ, Kim SR (2018) Beneficial effects of flavonoids against Parkinson's disease. J Med Food 21:421–432. https://doi.org/10.1089/jmf.2017.4078

Kesavan P, Banerjee A, Banerjee A et al (2018) An overview of dietary polyphenols and their therapeutic effects. In: Polyphenols: mechanisms of action in human health and disease. Elsevier, London, pp 221–235

Khan E, Tawani A, Mishra SK et al (2018) Myricetin reduces toxic level of CAG repeats RNA in Huntington's disease (HD) and spino cerebellar ataxia (SCAs). ACS Chem Biol 13:180–188. https://doi.org/10.1021/acschembio.7b00699

Kozłowska A, Szostak-Wegierek D (2014) Flavonoids—food sources and health benefits. Ann Natl Inst Hyg 65:79–85. https://doi.org/10.1145/313451.313558

Kumar P, Kumar A (2010) Protective effect of hesperidin and naringin against 3-nitropropionic acid induced Huntington's like symptoms in rats: possible role of nitric oxide. Behav Brain Res 206:38–46. https://doi.org/10.1016/j.bbr.2009.08.028

Kumar S, Pandey AK (2013) Chemistry and biological activities of flavonoids: an overview. Sci World J 2013:162750

Lang G-Z, Li C-J, Gaohu T-Y et al (2020) Bioactive flavonoid dimers from Chinese dragon's blood, the red resin of Dracaena cochinchinensis. Bioorg Chem 97:103659. https://doi.org/10.1016/j.bioorg.2020.103659

Larson AJ, Symons JD, Jalili T (2012) Therapeutic potential of quercetin to decrease blood pressure: review of efficacy and mechanisms. Adv Nutr 3:39–46. https://doi.org/10.3945/an.111.001271

Li B, Xiong M, Zhang H-Y (2014) Elucidating polypharmacological mechanisms of polyphenols by gene module profile analysis. Int J Mol Sci 15:11245–11254

Li J, Bai L, Li X et al (2019) Antidiabetic potential of flavonoids from traditional Chinese medicine: a review. Am J Chin Med 47:933–957. https://doi.org/10.1142/S0192415X19500496

Lv C, Hong T, Yang Z et al (2012) Effect of quercetin in the 1-methyl-4-phenyl-1,2,3,6-tetrahydropyridine-induced mouse model of Parkinson's disease. Evid Based Complement Altern Med 2012:928643. https://doi.org/10.1155/2012/928643

Ma D, Yu T, Peng L et al (2019) PIM1, CYP1B1, and HSPA2 targeted by quercetin play important roles in osteoarthritis treatment by *Achyranthes bidentata*. Evid Based Complement Altern Med 2019:1205942. https://doi.org/10.1155/2019/1205942

Maan G, Sikdar B, Kumar A et al (2020) Role of flavonoids in neurodegenerative diseases: limitations and future perspectives. Curr Top Med Chem 20:1169–1194

MacMahon Copas AN, McComish SF, Fletcher JM, Caldwell MA (2021) The pathogenesis of Parkinson's disease: a complex interplay between astrocytes, microglia, and T lymphocytes? Front Neurol 12:666737. https://doi.org/10.3389/FNEUR.2021.666737/BIBTEX

Mahapatra AD, Bhowmik P, Banerjee A et al (2019) Ethnomedicinal wisdom: an approach for antiviral drug development. Academic Press, Cambridge, MA

Maher P (2019) The potential of flavonoids for the treatment of neurodegenerative diseases. Int J Mol Sci 20:3056. https://doi.org/10.3390/ijms20123056

Maher P, Dargusch R, Bodai L et al (2011) ERK activation by the polyphenols fisetin and resveratrol provides neuroprotection in multiple models of Huntington's disease. Hum Mol Genet 20:261–270. https://doi.org/10.1093/hmg/ddq460

Mandel S, Maor G, Youdim MBH (2004) Iron and α-synuclein in the substantia nigra of MPTP-treated mice. J Mol Neurosci 24:401–416. https://doi.org/10.1385/JMN:24:3:401

Mani S, Sekar S, Barathidasan R et al (2018) Naringenin decreases α-synuclein expression and neuroinflammation in MPTP-induced Parkinson's disease model in mice. Neurotox Res 33:656–670. https://doi.org/10.1007/S12640-018-9869-3/FIGURES/8

Manzoor MA, Sabir IA, Shah IH et al (2023) Flavonoids: a review on biosynthesis and transportation mechanism in plants. Funct Integr Genomics 23:1–14. https://doi.org/10.1007/S10142-023-01147-4/

Meng-zhen S, Ju L, Lan-chun Z et al (2022) Potential therapeutic use of plant flavonoids in AD and PD. Heliyon 8:e11440. https://doi.org/10.1016/J.HELIYON.2022.E11440

Obregon DF, Rezai-Zadeh K, Bai Y et al (2006) ADAM10 activation is required for green tea (–)-epigallocatechin-3-gallate-induced α-secretase cleavage of amyloid precursor protein*. J Biol Chem 281:16419–16427. https://doi.org/10.1074/jbc.M600617200

Panche AN, Diwan AD, Chandra SR (2016) Flavonoids: an overview. J Nutr Sci 5:e47. https://doi.org/10.1017/jns.2016.41

Paula PC, Maria SGA, Luis CH, Patricia CGG (2019) Preventive effect of quercetin in a triple transgenic Alzheimer's disease mice model. Molecules 24:2287. https://doi.org/10.3390/MOLECULES24122287

Ramezani M, Darbandi N, Khodagholi F, Hashemi A (2016) Myricetin protects hippocampal CA3 pyramidal neurons and improves learning and memory impairments in rats with Alzheimer's disease. Neural Regen Res 11:1976–1980. https://doi.org/10.4103/1673-5374.197141

Rees A, Dodd GF, Spencer JPE (2018) The effects of flavonoids on cardiovascular health: a review of human intervention trials and implications for cerebrovascular function. Nutrients 10:1852. https://doi.org/10.3390/nu10121852

Rezai-Zadeh K, Shytle D, Sun N et al (2005) Green tea epigallocatechin-3-gallate (EGCG) modulates amyloid precursor protein cleavage and reduces cerebral amyloidosis in Alzheimer transgenic mice. J Neurosci 25:8807–8814. https://doi.org/10.1523/JNEUROSCI.1521-05.2005

Rezai-Zadeh K, Douglas Shytle R, Bai Y et al (2009) Flavonoid-mediated presenilin-1 phosphorylation reduces Alzheimer's disease β-amyloid production. J Cell Mol Med 13:574–588. https://doi.org/10.1111/J.1582-4934.2008.00344.X

Sabiha S, Serrano R, Hasan K et al (2022) The genus Cynometra: a review of ethnomedicine, chemical, and biological data. Plants 11:3504. https://doi.org/10.3390/plants11243504

Sachdeva AK, Kuhad A, Chopra K (2014) Naringin ameliorates memory deficits in experimental paradigm of Alzheimer's disease by attenuating mitochondrial dysfunction. Pharmacol Biochem Behav 127:101–110. https://doi.org/10.1016/j.pbb.2014.11.002

Saha D, Nath Jha A (2023) Computational multi-target approach to target essential enzymes of Leishmania donovani using comparative molecular dynamic simulations and MMPBSA analysis. Phytochem Anal 34:842–854. https://doi.org/10.1002/pca.3213

Saha D, Khataniar A, Singh AK, Jha AN (2023) Chapter 5—Review of methods for encapsulation of nutraceutical compounds. In: Inamuddin, Altalhi T, Neves Cruz JBT-N (eds) Nutraceuticals. Academic Press, Cambridge, MA, pp 127–156

Sandoval-Avila S, Diaz NF, Gómez-Pinedo U et al (2019) Neuroprotective effects of phytochemicals on dopaminergic neuron cultures. Neurologia 34:114–124. https://doi.org/10.1016/J.NRL.2016.04.018

Shen N, Wang T, Gan Q et al (2022) Plant flavonoids: classification, distribution, biosynthesis, and antioxidant activity. Food Chem 383:132531. https://doi.org/10.1016/j.foodchem.2022.132531

Singh AK (2016) Exotic ancient plant introductions: part of Indian 'Ayurveda' medicinal system. Plant Genet Resour 14:356–369. https://doi.org/10.1017/S1479262116000368

Siuly S, Zhang Y (2016) Medical big data: neurological diseases diagnosis through medical data analysis. Data Sci Eng 1:54–64. https://doi.org/10.1007/s41019-016-0011-3

Solanki I, Parihar P, Mansuri ML, Parihar MS (2015) Flavonoid-based therapies in the early management of neurodegenerative diseases. Adv Nutr 6:64–72. https://doi.org/10.3945/an.114.007500

Song X, Zhou B, Zhang P et al (2016) Protective effect of silibinin on learning and memory impairment in LPS-treated rats via ROS–BDNF–TrkB pathway. Neurochem Res 41:1662–1672. https://doi.org/10.1007/S11064-016-1881-5/FIGURES/7

Suganya SN, Sumathi T (2017) Effect of rutin against a mitochondrial toxin, 3-nitropropionicacid induced biochemical, behavioral and histological alterations—a pilot study on Huntington's disease model in rats. Metab Brain Dis 32:471–481. https://doi.org/10.1007/s11011-016-9929-4

Sulima P, Przyborowski JA (2019) Purple willow (Salix purpurea L.) and its potential uses for the treatment of arthritis and rheumatism. Academic Press, Cambridge, MA

Tabrizi SJ, Flower MD, Ross CA, Wild EJ (2020) Huntington disease: new insights into molecular pathogenesis and therapeutic opportunities. Nat Rev Neurol 16:529–546. https://doi.org/10.1038/s41582-020-0389-4

Tahir MS, Almezgagi M, Zhang Y et al (2021) Mechanistic new insights of flavonols on neurodegenerative diseases. Biomed Pharmacother 137:111253. https://doi.org/10.1016/j.biopha.2021.111253

Taylor JP, Brown RH, Cleveland DW (2016) Decoding ALS: from genes to mechanism. Nature 539:197–206. https://doi.org/10.1038/nature20413

Uddin MS, Kabir MT, Niaz K et al (2020) Molecular insight into the therapeutic promise of flavonoids against Alzheimer's disease. Molecules 25:1267. https://doi.org/10.3390/MOLECULES25061267

Ullah A, Munir S, Badshah SL et al (2020) Important flavonoids and their role as a therapeutic agent. Molecules 25:5243. https://doi.org/10.3390/MOLECULES25225243

Wang Z, Cui J, Li D et al (2023) Morin exhibits a neuroprotective effect in MPTP-induced Parkinson's disease model via TFEB/AMPK-mediated mitophagy. Phytomedicine 116:154866. https://doi.org/10.1016/J.PHYMED.2023.154866

Yi T, Tang Y, Zhang J et al (2012) Characterization and determination of six flavonoids in the ethnomedicine "dragon's blood" by UPLC-PAD-MS. Chem Cent J 6:116. https://doi.org/10.1186/1752-153X-6-116

Young AB (2003) Huntingtin in health and disease. J Clin Invest 111:299–302. https://doi.org/10.1172/JCI17742

Zhang X, Molsberry SA, Yeh T-S et al (2022) Intake of flavonoids and flavonoid-rich foods and mortality risk among individuals with Parkinson disease. Neurology 98:e1064. https://doi.org/10.1212/WNL.0000000000013275

Zhao L, Wang JL, Liu R et al (2013) Neuroprotective, anti-amyloidogenic and neurotrophic effects of apigenin in an Alzheimer's disease mouse model. Molecules 18:9949–9965. https://doi.org/10.3390/MOLECULES18089949

Zhu Q, Zhuang X, Lu J (2019) Neuroprotective effects of baicalein in animal models of Parkinson's disease: a systematic review of experimental studies. Phytomedicine 55:302–309. https://doi.org/10.1016/J.PHYMED.2018.09.215

Nutraceuticals of Natural Origin and Their Significance in Modern Medicine

Mukesh Singh Sikarwar and Vaibhav Kumar Rathi

Contents

M. S. Sikarwar (✉)
College of Pharmacy, Teerthanker Mahaveer University, Moradabad, Uttar Pradesh, India
e-mail: mukesh.pharmacy@tmu.ac.in

V. K. Rathi
Department of Pharmaceutical Sciences, Quantum University, Roorkee, Uttrakhand, India

© The Author(s), under exclusive license to Springer Nature Singapore Pte Ltd. 2024

A. Das Talukdar et al. (eds.), *Traditional Resources and Tools for Modern Drug Discovery*, Interdisciplinary Biotechnological Advances,
https://doi.org/10.1007/978-981-97-4600-2_12

Abstract Modern drugs are popular due to their role in emergency medicine however their roles are limited in chronic diseases due to their long-term usage and associated side effects. Various research have been conducted on nutraceuticals for prophylactic and therapeutic use. Significant role of micronutrients in diseases management has also been identified. Nutraceuticals are supplements with nutritional and medicinal value having proper dosage forms. Nutraceuticals work in synergy and support the pharmacokinetics, for example, vitamin D with K2, vitamin D with calcium, curcumin in combination with boswellic acid, and many such other combinations. Ayurveda nutraceuticals are also now emerging and making contribution in Asian market. Nutraceuticals are not tightly regulated in most part of the world; however, some countries like Japan have regulation in place to regulate nutraceuticals under Food for Specified Health Uses (FOSHU), whereas the USA regulates it under Dietary Supplement Health and Education Act of 1994. Currently nutraceutical is being utilized as adjuvant to existing therapies to reduce the reactive oxygen species in cancer, diabetes, rheumatoid arthritis, hypertension, and other diseases having significant inflammation and oxidative stress.

Keywords Nutraceuticals · Clinical nutraceuticals · Synergistic nutraceuticals

1 Introduction to Nutraceuticals

A portmanteau of "nutrition" and "pharmaceuticals", the term "nutraceuticals" refers to the dynamic relationship between food and health. Technological breakthroughs and changing consumer tastes have caused a significant upheaval in the worldwide health and wellness environment in recent years. This change has ushered in a time where cutting-edge technology and new trends complement and occasionally contradict conventional approaches to healthcare. This introduction sheds light on the critical role that nutraceuticals will play in determining the direction of preventive healthcare in the future by examining its changing landscape in comparison to the newest trends and technology (Kalra 2003).

Nutraceuticals are a broad category of items that include functional foods, herbal products, and dietary supplements. They have become well-known due to their potential health advantages that go beyond simple nutrition. The need for natural and comprehensive approaches to well-being has led to an exponential expansion in the nutraceutical sector. The effectiveness of several nutraceutical chemicals in promoting health, avoiding disease, and improving general quality of life has been validated by scientific studies (Violi 2010).

On the other hand, new technologies have sparked a revolution in healthcare by bringing in creative solutions that make use of data analytics, artificial intelligence, and personalised medicine. Technology is changing the way people approach and maintain their health. Examples include wearables that track physiological data and genetic testing that provides individualised nutrition recommendations. The fusion of technology and nutraceuticals has ushered in a new era of "smart nutrition", where dietary decisions are guided by data-driven insights to maximise health results (Al-Okbi et al. 2023).

The scientific foundations of nutraceutical efficacy, technology interventions in healthcare, and the opportunities and challenges that occur at this crossroads will all be covered in this investigation of nutraceuticals and the newest technologies. Aiming for a balanced and evidence-based approach to holistic health, it is imperative that we take into account the possible advantages and difficulties presented by these opposing forces as we traverse this dynamic terrain (Zuchi et al. 2010).

2 Current Market of Nutraceuticals

2.1 Market Size and Growth

The market for nutraceuticals was valued at USD 291.33 billion globally in 2022; it is projected to expand at a compound annual growth rate (CAGR) of 9.4% between 2023 and 2030. The main reasons fuelling the market's expansion are the growing prevalence of lifestyle-related illnesses, preventative healthcare, and growing consumer interest in diets that promote health (Grandview 2023).

US market was 77.7 billion dollars in 2021 and predicted to have 7.4% CAGR, majorly focusing on functional foods (Grandview 2023).

The value of the nutraceutical market in India was 110.5 billion rupees in 2012 and 260 billion Indian rupees in 2017. This was predicted to increase to 650 billion rupees by 2022, suggesting that the industry has significant room for expansion (Statista 2018).

Key points of the market growth are:

- The global nutraceutical market is expected to reach a value of over $780 billion by 2030.
- The Asia-Pacific proven to be fastest-growing market for nutraceuticals, with a CAGR of over 9% from 2022 to 2030, followed by North America and Europe (Allied Market Research 2022).

- Vitamins and minerals are the largest category of nutraceuticals, followed by herbs and dietary supplements.
- The demand for personalized nutraceuticals is growing, as consumers are increasingly looking for products that are tailored to their individual needs.
- The use of technology in the nutraceutical industry is growing, as companies are using technology to develop new products and to deliver them to consumers in a more convenient way (Markets and Markets 2022).

2.2 *Key Segments and Categories*

In 2022, probiotics held 26.47% of market share which is largest in categories of nutraceuticals; probiotics are also projected to be fastest-growing category of these with annual growth rate of 14.0%. Following categories were found to be promising:

2.2.1 Functional Foods and Beverages

The nutraceutical market's largest and most rapidly expanding category is projected to have a worldwide value of $330 billion by 2022. Functional foods and beverages are meals and beverages that have been improved with extra nutrients or other components that have been proven to offer health advantages. Functional foods and beverages including fortified cereals, probiotic-rich yoghurt, and energy drinks.

2.2.2 Dietary Supplements

The nutraceutical market's second-largest category has a worldwide value of $160 billion in 2022. Dietary supplements are items designed to enhance the diet by providing additional vitamins, minerals, herbs, or other substances. They are accessible in many forms, such as pills, capsules, powders, and liquids.

2.2.3 Personal Care Products

The nutraceutical market's smallest category has a worldwide value exceeding $20 billion in 2022. Personal care goods are items designed to enhance the aesthetic or physical condition of the skin, hair, or nails. These products may include components that have been scientifically proven to have positive effects on health, such as antioxidants, vitamins, and minerals (PRNewswire 2023).

2.2.3.1 Outlook of Nutraceuticals

See Table 1.

Table 1 Outlook of nutraceuticals

Ingredients	Product	Application	References
Aloe vera	Dietary supplements	Allergy and intolerance	Amar Surjushe et al. (2008)
Amino acids	Functional beverages	Overall nutrition	Wu (2013)
Botanical ingredients • Ashwagandha • Curcumin • Ginseng • Hemp	Functional foods	Various medicinal effects such as anticarcinogens, antioxidants, anti-inflammatory	Gulati and Ottaway (2008)
Cannabidiol (CBD)		Energy and endurance	Walker et al. (2020)
Carnitine		Heart health	Zhao et al. (2022)
Carotenoids • Astaxanthin • Lutein • Lycopene • Zeaxanthin • Beta-carotene		Antioxidant, skin Eye health Heart and prostate Blue light protection Antioxidant and immunity	Honda (2020)
Spirulina		Personalised nutrition	Priyanka et al. (2023)
Collagen		Post pregnancy and reproductive health	Edgar et al. (2018)
Colostrum		General health	Playford and Weiser (2021)
Dairy ingredients		Balanced nutrition	Górska-Warsewicz et al. (2019)
Enzymes		Allergy and intolerance	Agyei et al. (2019)
Essential oils		GI and mental health	Matera et al. (2023)
Fats and oils		General Well-being	Dhara and Chakrabarti (2020)
Glucosamine/chondroitin		Sports nutrition and joint health	Deal and Moskowitz (1999)
Isoflavones		Oestrogen-mimicking effects	Křížová et al. (2019)
Krill		Antioxidants, heart, and joint health	Joob and Wiwanitkit (2015)
Minerals • Calcium • Iron • Magnesium • Selenium		Overall health	Alagawany et al. 2020)
Omega-3s • Marine derived • Plant derived		Antioxidants and joint health	Chandra et al. (2022)
Prebiotics and probiotics		Gut health	Damián et al. (2022)
Proteins		Overall health	Alagawany et al. (2020)
Stevia		Antidiabetic and sweetener	Chughtai et al. (2020)
Vitamins (all categories)			Alagawany et al. (2020)
Whey		Health supplement	Patel (2015)

2.3 Consumer Trends and Preferences

Consumers are increasingly demanding nutraceuticals that are science-backed and clinically proven, made with natural and high-quality ingredients, tailored to their individual needs, convenient and easy to use, tasteful, and enjoyable.

Some of the most popular nutraceutical ingredients include prebiotics and probiotics, omega-3 fatty acids, vitamins and minerals, and herbal extracts.

2.4 Data and Statistics

Key findings of the preferences:

- According to a survey conducted in 2021 by Kerry, 72% of customers worldwide are prepared to pay a higher price for nutraceuticals that have scientific proof supporting their effectiveness.
- A 2022 report by "Grand View Research" predicts that the global probiotics market will reach USD 62.3 billion by 2028.
- A 2023 survey by Nutra Ingredients found that 57% of consumers are taking nutraceuticals to improve their overall health and well-being (Avinash 2021; Grandview research 2022).

2.5 Major Market Players

Some of the major companies that dominates nutraceutical space include SDSM, Amway, Pfizer Inc., Nestle, The Kraft Heinz Company, The Hain Celestial Group, Inc., Nature's Bounty, General Mills Inc., Danone, and Tyson Foods.

3 Nutraceuticals from Natural Sources

3.1 Plant Based

3.1.1 Curcumin

Curcumin obtained from Turmeric (*Curcuma longa*) is a one of the most important nutraceuticals used in Indian cuisine. It contains curcumin, a compound with powerful anti-inflammatory and antioxidant properties. Turmeric has been studied for its potential to treat a variety of health conditions, including arthritis, heart disease, and cancer. Sources of turmeric include turmeric powder, turmeric capsules, and turmeric tea (Hewlings and Kalman 2017).

3.1.2 Resveratrol

Resveratrol is a polyphenol found in grapes, red wine, and berries. It has been shown to have antioxidant, anti-inflammatory, and anticancer properties. Resveratrol is also being studied for its potential to improve heart health and cognitive function. Sources of resveratrol include red wine, grape juice, berries, and resveratrol supplements (Dhingra et al. 2021).

3.1.3 Quercetin

Quercetin is a flavonoid, a type of polyphenol found in many fruits, vegetables, and grains. Specially found in citrus fruits. It is a potent antioxidant and has a wide range of potential health benefits (Boots et al. 2011).

3.1.4 Gingerol

Gingerol is a resin obtained from ginger. Ginger is a therapeutic spice with a lengthy historical background. It is frequently employed for the treatment of nausea and vomiting. Ginger possesses both anti-inflammatory and antioxidant characteristics (Mashhadi et al. 2012).

3.1.5 Allicin

Allicin is a sulphur-containing compound found in garlic. It is responsible for the characteristic smell and taste of garlic. Allicin has a number of potential health benefits including antimicrobial, antioxidant, anti-inflammatory, cardioprotective, and anticancer (Rahman and Lowe 2006).

3.1.6 *Ginko biloba*

Ginkgo biloba is a tree that has been utilised in traditional Chinese medicine for an extended period of time. This tree is among the most ancient live trees that have been discovered. *Ginkgo biloba* comprises many chemicals, such as flavonoids, terpenoids, and ginkgolides. These substances possess several potential health advantages, such as enhancing cognitive processes, alleviating anxiety, enhancing blood circulation, acting as antioxidants, and promoting overall health (Birks and Grimley Evans 2007).

3.1.7 Withanolides

Withania somnifera, also known as ashwagandha, is a woody shrub that has been used in traditional Indian medicine for centuries. It is considered an adaptogen, a substance that helps the body adapt to stress and maintain balance. Ashwagandha is a rich source of withanolides which have numerous health benefits including stress antianxiety, anti-inflammatory, cognitive enhancement, antioxidants, sleep regulation, antidiabetic, and reproductive health (Chandrasekhar et al. 2012; Rathi et al. 2021).

3.2 Animal Based

3.2.1 Omega-3 Fatty Acids

Polyunsaturated fat (PUFA) found in fish (salmon, mackerel, and tuna) and in some plant oils (flaxseeds, chia seeds, walnuts, soybean, and canola). Omega-3 fatty acids are essential fatty acids as body does not produce them and supplemented through diet. Omega-3 fatty acids are essential for body processes like inflammation, heart health, and brain function. Omega-3 fatty acids are known to reduce risk of dementia and Alzheimer's disease, reduce the risk of heart diseases, reduce risk of cancer, and have remarkable anti-inflammatory activity as well (Kris-Etherton et al. 2002).

3.2.2 Collagen Peptides

The most prevalent protein in the human body, collagen, is the source of collagen peptides, a class of nutraceuticals. The skin, bones, and cartilage are examples of connective tissues that contain collagen. It gives these tissues stability and structure. Collagen peptides improve skin health, joint health, and bone and gut health (Lim et al. 2019).

3.2.3 Coenzyme Q10

Every cell in the body contains coenzyme Q10 (CoQ10), a naturally occurring molecule that resembles a vitamin. It is a crucial part of the chain that transports electrons, which is necessary for the mitochondria to produce energy. It is fund in organ meats, beef, pork, fish oil, nuts, seeds, and fruits. Additionally, CoQ10 has potent antioxidant, improved heart health, and anti-inflammatory and also shown to enhance athletic performance properties that shield cells from harm from free radicals (Liu et al. 2011).

3.2.4 Creatine

Muscle cells naturally contain a chemical called creatine. It can increase muscular strength, power, and endurance and aid in the body's energy production during high-intensity activity. Creatine improves muscle strength and power from 5% to about 20% in clinical study. It has shown to reduce muscle fatigue and speed up recovery from exercise (McMorris et al. 2007).

3.2.5 Probiotics

When ingested, probiotics, which include live bacteria and yeasts, can have positive effects on one's health. Fermented foods like yoghurt, kefir, and sauerkraut are common sources of them. It is also possible to consume probiotics as supplements. Probiotics shows a wide range of health benefits like improved gut health, improves symptoms of IBS, improves mental health, enhances immune functions, and improves skin health (Hemarajata and Versalovic 2013).

3.3 Marine Based

3.3.1 Astaxanthin

Powerful antioxidant astaxanthin is present in many different types of creatures in nature, such as fish, crustaceans (shrimps and lobster), and algae (*Haematococcus pluvialis*). It's a carotenoid, a kind of pigment that imparts colours to plants and animals. The most powerful known natural antioxidant is astaxanthin, which has been demonstrated to offer a variety of possible health advantages like reduced oxidative stress, improved cognitive function, reduced inflammation, improved eye health, and reduced risk of chronic diseases (Ambati et al. 2014).

3.3.2 Marine Collagen

Marine collagen comes from marine sources, such fish skin. Amino acids, the building blocks of protein, are abundant in it. Numerous possible health advantages of marine collagen have been demonstrated, including improved skin health, joint health, bone health, muscle health, and gut health (Lim et al. 2019).

3.3.3 Fucoidan

Fucoidan is a sulphated polysaccharide that is found in brown seaweed (*Macrocystis pyrifera*). It has been shown to have a number of potential health benefits, like anticancer, anti-inflammatory, antiviral, and antioxidant properties (Luthuli et al. 2019).

4 Ayurveda Nutraceuticals and Their Significance in Modern Medicine

The concept of *Aahara Kalpanas* can directly be correlated with nutraceuticals in Ayurveda; it literally conceptualises medicated food. Other Ayurveda terms like *Hita*, *Ahita*, *Pathypathya*, *satmya*, and *Asatmya* can directly be associated with nutraceuticals. *Ajasrika Rasyana* can be consumed on regular basis to maintain the health (Kumar and Patil 2022).

Evidence of the therapeutic value of food in the treatment and prevention of disease comes from the ancient Indian culture. Ayurveda has made explicit reference to the therapeutic advantages of Ahara. Ahara and Vihara are topics that are well-documented in Ayurvedic literature. The Charak Samhita and more recent masterpieces like Bhavaprakash both discuss nutritional guidelines. It has been stated that men who maintain a healthy diet will not require medication. Similar to this, even the best medication won't help if someone isn't adhering to dietary restrictions (Ujjaliya et al. 2018).

The combination of herbs and nutraceuticals effectively mitigated several chronic illnesses and enhanced individuals' overall well-being. Salicin, found in willow bark (*Salix nigra*), has been proven in clinical trials to have analgesic, anti-inflammatory, febrifuge, antiarthritic, and astringent properties. Psoralen, a flavonoid that is found in parsley (*Petroselinum crispum*), has advantageous properties like febrifuge, diuretic, and carminative. The terpenoids in peppermint (*Mentha piperita*), particularly the bioactive compound menthol, help to treat colds and the flu. Tannins found in lavender (*Lavandula angustifolia*) aid in lowering blood pressure and stress levels and are beneficial for respiratory conditions including asthma (Chanda et al. 2019).

The authors of this paper focus on the topic of expanding complementary and alternative (CAM) approaches. The main objective of CAM is to explore and study complementary and alternative healing methods using rigorous scientific methods. CAM aims to promote advanced research, educate researchers, inform the public about effective healing methods, and provide scientific evidence for new discoveries. CAM is dedicated to researching and financing therapies that have robust ethical justifications, urgent public health requirements, and first evidence. Due to meticulous research, the adverse effects of natural goods are likewise minimised, making them a sought-after alternative (Cooper 2005).

Nutrition has always been prime focus of Ayurveda from ancient times; it is quoted in Ayurveda:

> Tat cha nityam prayunjeet svasthyam yen anuvartate
> Ajaatanam vikaranam anuttpattikaram cha yat.
> (Charaka Samhita: Sutra Sthana: 5)
> (The diet which besides providing the basic nutrition to the body, helps to maintain the healthy state of the body and prevents the occurrence of diseases should be consumed) (Sinha 2011).

Another Ayurveda book *Vaidhya Jeevana: 1/10* quotes "If wholesome diet is given in a planned way, then there is no need to administer medicines and if wholesome diet is not being used then too there is no need to give medicine, as this is not going to cure the disease in the absence of wholesome diet" (Shah et al. 2021).

Ayurveda describes herbs and minerals found in nature as Rasayanas; they can be administered as single or compound. There are various categories of Rasayanas; one of them is called Ahar-Rasayana, which directly correlated with nutraceuticals. Table 2 of Rasayanas is given below which are meant to be taken on regular basis along with usual food to maintain health (Singh et al. 2014).

Authors have mentioned the traditional classification of nutraceuticals; different classes and their summarized introduction including examples are given in Table 3 (Kumar Jha et al. 2021).

Some plants that have nutraceuticals potentials and are used traditionally are given in Table 4 along with their potential uses (Pandey et al. 2013).

Table 2 Rasayanas and their general effects

S. no.	Name of Rasayana	Effect
1.	Chayvanprash	General health
2.	Brahma Rasayana	Mental illness
3.	Narasimha Rasayana	Health and vigour
4.	Arjuna Rasayana	Cardiovascular system
5.	Brhami Ghrit	Increased memory in young and old

Table 3 Classifications of nutraceuticals based on traditional system

Type of classification	Summary	Examples
Chemical	Primary metabolites that have well-defined role in metabolism; this includes products from plants and animals	Carbohydrates, amino acids, fatty acids, organic acids
Herbals	Nutraceuticals with herb that can prevent various chronic diseases and improve the quality of life	*Salix nigra* is a very good analgesic, antipyretic, astringent due to presence of salicin
Phytochemicals	Classified based on phytochemicals found in the plants	Flavonoids present in various plant-based food have very good antioxidant activity which can be utilised as anticancer, antidiabetic, in heart diseases and as overall health tonic
Probiotic	Microorganisms that have symbiotic relation with humans and make digestion better and maintain healthy environment inside the body	*Bacillus bulgaricus*

Table 4 Plants and their potential use as nutraceuticals

S. no.	Plant names	Common names	Uses
1.	*Asparagus racemosus* Willd	Shatavari	Potent rejuvenator, strengthen immune system, female reproductive system
2.	*Garcinia cambogia* Dr	Garcinia	This plant is known to have hydroxycitric acid which inhibit synthesis of lipids, thus preventing formation of LDL and tri-glycerides. It also contains vitamin C and traditionally used as heart tonic
3.	*Melia azadirachta* L.	Neem	Used as general health tonic and prevents various diseases. The extract also shows antispasmodic activity
4.	*Commiphora mukul* Engl.	Guggul	Joint pain, immune-modulation, lowers cholesterol level, increases WBC count, anti-inflammatory
5.	*Glycyrrhiza glabra* L.	Liquorice	It has mild laxative and antispasmodic properties; traditionally it is used as expectorant and also possess good antimutagenic properties
6.	*Momordica charantia* L.	Karela	It contains a polypeptide compound called gurmarin which has similar structure to bovine insulin thus has strong sugar regulation effect
7.	*Moringa pterygosperma* Gaertn	Shigru	The pterygospermin found in this plant has antibiotic like activity
8.	*Nardostachys jatamansi* DC.	Jatamansi	Jatamansi is very effective for mental health and can be used for regulating the sleep
9.	*Piper longum* L.	Peepli	It has stimulating effect on both digestive system and respiratory system thus can be used as tonic for both
10.	*Piper nigrum* L.	Black pepper	It enhances the absorbance of other nutrients and aids in the digestion
11.	*Terminalia chebula* Retz.	Haritaki	It is safe and effective purgative, tonic, and expectorant; it is also a part of famous Triphala powder of India
12.	*Withania somnifera* (L.) Dunal	Ashwagandha	It is called Indian ginseng because of similar properties; it is said to be one of the best health tonic and rejuvenator
13.	*Tinospora cordifolia* Miers	Guduchi	It enhances immune system and inhibit bacterial growth and shown to enhance growth of macrophages. It is also a rich source of vitamin C
14.	*Zingiber officinale* Rosc	Ginger	It enhances absorption of nutrients and used as adjuvant in many traditional formulations. It also relaxes bowel muscles

5 Clinical Nutraceuticals and Their Significance in Modern Medicine

Traditional medicine has strong origins in the idea of utilizing food as a type of medicine. In order to address memory impairment brought on by Alzheimer's disease and cure depression, researchers are actively looking into innovative applications for well-established drugs. These problems have grown to be a significant social and economic challenge as the number of aged people increases globally.

There are several unintended side effects that frequently follow contemporary medical procedures like radiation and chemotherapy. As a result, there is a sizable amount of study examining the ability of herbal supplements to lessen these negative effects. This whole scenario has led to clinical research on nutraceuticals which in turn also lead to development and exploration of new age nutraceutical medicines based on clinical data (Williamson et al. 2019).

Table 5 shows different clinical nutraceuticals products, their sources, and general summary.

Table 5 Different clinical nutraceuticals products, their sources, and summary

S. no.	Name of nutraceutical	Source	Summary
1.	Astaxanthin (xanthophyll carotenoid)	*Haematococcus pluvialis* (micro-algae), yeast *Phaffia rhodozyma*	The compound known as AXE incorporates both *cis-* and *trans*-isomers of 3,30-dihydroxy-β, β-carotene-4,40-dione. The *trans* isomer is the only one that is present in the synthetic counterpart. By activating peroxisome proliferator-activated receptors, it serves as an anti-inflammatory agent and promotes programmed cell death. It boasts a powerful potential to counteract the effects of oxidation and functions by activating these receptors. In clinical studies, AXE has been shown to have either minimal or no adverse effects (Ambati et al. 2014; Fakhri et al. 2018)
2.	Cannabidiols (non-euphoric phytocanabinoids)	*Cannabis sativa*	CBD has low affinity to CB1 and CB2 receptors; thus it is not euphoric like THC (tetrahydrocannabinols); CBD also shows various effects on humans like analgesic, anti-inflammatory, and behaviour management. Clinically CBD has proved to show effect on severe early onset of epilepsy, antipsychotic, and antidepressant effects. CBD also shows anxiolytic effect possibly via the activation of 5-HT1A receptor. CBD has shown to improve mood and amnesia, treatment of neuropathic pain in multiple sclerosis. Larger amount decreases symptoms of Crohn's disease and ulcerative colitis. CBD is very safe up to dose of 600 mg; higher amount has only shown to decrease appetite and cause nausea (Crippa et al. 2018; White 2019)

(continued)

Table 5 (continued)

S. no.	Name of nutraceutical	Source	Summary
3.	Monk fruit	*Siraitia grosvenorii* a.k.a. *Momordica grosvenorii*	The fruit includes a group of chemical compounds called mogroside, which are cucurbitane triterpene glycosides. These compounds have been revealed to have numerous beneficial effects like anti-asthmatic, anti-tussive, glucose-lowering, and immunomodulator activity. The clinical use of mogroside IIIE effectively reduces the severity of acute lung damage via modulating the TLR4/MAPK/NF-κB pathway through the activation of AMPK. The typical daily consumption of monk fruit does not exhibit any signs of toxicity or negative consequences (Engels and Brinckmann 2017)
4.	Ginseng products	*Panax ginseng*	Most active ingredient in ginseng is called a group of compounds called ginsenosides. They are dammarane saponins classified into series Ra to Rs and two main categories that are derived from protopanaxadiol (PPD) and protopanaxatriols (PPT). Ginsenoside RG1 is shown to have memory enhancement and neuroprotective effects. GRB1 improves spatial memory; GRH1 and RH2 have anti apoptotic, anti-angiogenic, and antioxidant properties. Both also have direct effect on the memory. The formulations and extracts are generally found to be safe and also showed no drug interactions. Apart from mild effect in antihypertensive drugs like amlodipine and losartan (Geng et al. 2010; Mancuso and Santangelo 2017)
5.	Nigella seed	*Nigella sativa* (black cumin)	The main chemical constituent is thymoquinone (30–48%); this plant is very famous in unani medicine for it being a "Panacea", but clinical studies show only mild improvement in inflammation, respiratory illness, and arthritis. Though it is shown to have a good paediatric seizure control compound. Nigella doesn't show any concerning adverse effects (Gholamnezhad et al. 2016)

Table 5 (continued)

S. no.	Name of nutraceutical	Source	Summary
6.	Omega-3 fatty acids	Salmon, tuna, herring, sardines, mackerel, soybeans, etc.	Alpha-linolenic acid (ALA) is found in plant seeds, EPA is found in sea algae and fish, and DHA is found in sea algae and fish. This category improves cardiovascular health by changing cell lipid composition and membrane fluidity. By inhibiting nuclear factor-kappaB, it has anti-inflammatory actions (Mozaffarian and Wu 2011). Omega-3 polyunsaturated fatty acids (PUFA) have been found to have a positive impact on depression and bipolar illness, according to a controlled clinical trial. The reference is from the study (Grosso et al. 2014) A multicentre, randomised, double-blind research included 202 knee osteoarthritis (OA) and chronic knee pain patients. Participants were randomly assigned to receive either 4.5 g omega-3 fatty acids or 15 mL/day of a 1:9 blend of fish oil and sunola oil with 0.45 g omega-3s. Low-dose PUFA decreased pain more than higher dose on the Western Ontario and McMaster Universities Arthritis Index (WOMAC) pain score at 3, 6, 12, and 24 months and cartilage volume change at 24 months. But cartilage volume didn't alter (Hill et al. 2016)
7.	Glucosamine and chondroitin	Cartilage of animals, seashells of sea animals, and synthetic preparation	The GAIT study also assessed chondroitin sulphate's knee osteoarthritis treatment efficacy. The results showed a relatively helpful reduction in knee pain, notably for those people with moderate-to-severe knee pain (Clegg et al. 2006) Based on data demonstrating its effectiveness, the Osteoarthritis Research Society International (OARSI) recommendations (2020) suggest chondroitin sulphate for the treatment of osteoarthritis in the knee (Bannuru et al. 2019)

(continued)

Table 5 (continued)

S. no.	Name of nutraceutical	Source	Summary
8.	Melatonin	Tart berry (*Prunus cerasus*), goji berry (*Lycium chinense*), eggs, milk, fish, nuts (especially pistachios)	The pineal gland produces melatonin (*N*-acetyl-5-methoxytryptamine). It is essential for sleep regulation and circadian rhythm. Main intracellular proteins like quinone reductase 2, calmodulin, calreticulin, and tubulin affects the melatonin's physiological effects. Melatonin also binds membrane MT1 and MT2 receptors. Melatonin selectively interacts with organ receptor subtypes, causing varied physiological effects. Melatonin is neuroprotective (Pandi-Perumal et al. 2008). The correlation between reproductive activity and seasonal changes in day length is influenced by the fluctuation in the daily release of melatonin, which is determined by the number of hours each day. These discoveries ignited the inquiry into the role of melatonin and the pineal gland in human reproduction (Venkatramanujam Srinivasan Spence and Brzezinski 2009) The study's findings suggest that melatonin possesses strong antioxidant properties and might potentially be used to treat inflammation, oxidative stress, and cellular apoptosis. Melatonin has been shown to effectively reduce the toxicity of methamphetamine and detrimental prescription medications. Additionally, it might be beneficial in the treatment or delay of several age-related ailments. Humans readily tolerate melatonin, and it is considered to have a high level of safety. The study's authors hypothesise that melatonin's ability to specifically target mitochondria may be the underlying reason for its numerous good effects (Reiter et al. 2016)

6 Synergistic Nutraceuticals

Many current researches show how combining nutraceuticals can lead to improved bioavailability, enhanced biological effects, and reduced side effects. The article also provides examples of synergistic combinations for a variety of health conditions, including cardiovascular disease, cancer, and cognitive decline. Overall, the research on synergistic nutraceuticals is still in its early stages, but the findings are promising. Synergistic nutraceuticals have the potential to provide more effective and safer ways to prevent and treat a variety of health conditions (Image 1).

Nutraceuticals synergism works by different mechanisms that are described in Table 6.

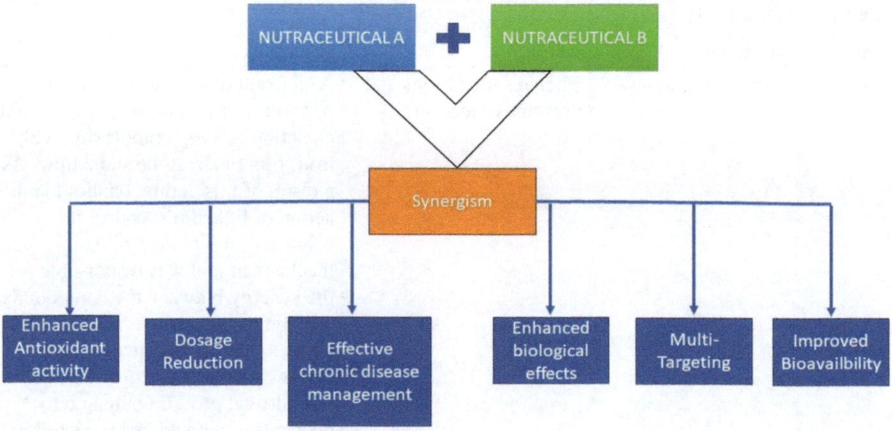

Image 1 A general diagram showing different effects of synergism

Table 6 Mechanism of synergism and their actions

S. no.	Mechanism	Action	Examples
1.	Act on a new/ multiple mechanism	Act on new targets or multiple new pathways or activate new pathways	Combination of fish oil and medium chain triglycerides can decrease *N*-acyl turines which is not activated by one of these; this combination decreases the chances of diabetic onset (Jackson and Jewell 2020)
2.	Mutual interaction/ protection	Increased solubility, stability, and bioavailability	Proanthocyanidins and quercetin can combine to increase quercetin's stability and solubility, which has a synergistic impact. Benefits of quercetin include better absorption and increased antioxidant activity in vivo (Zhao et al. 2015)
3.	Sacrificial oxidation	Weaker one traps peroxy radicals and protects stronger antioxidant	In a combination of lycopene and β-carotene, lycopene functions as a protective shield by rapidly undergoing isomerisation and degradation, thanks to its superior scavenging abilities. Lycopene, when mixed with β-carotene, slows down the process of isomerisation until β-carotene is completely used up (Heymann et al. 2015)

(continued)

Table 6 (continued)

S. no.	Mechanism	Action	Examples
4.	Metal chelation	Chelates metal ions and prevents oxidation	As a result of its interaction with peroxide, curcumin has shown metal chelating activity complexing with iron from haem at the same time. As a result of this, a direct antioxidant action of linked resveratrol is enhanced. It was determined that the mechanism that was responsible for the synergy between the compounds was the occurrence of metal chelation when it was investigated Oxyresveratrol has a greater amount of hydroxyl groups compared to resveratrol, which is why it exhibits stronger metal chelation and strong and straight antioxidant effects. This indicates a synergistic relationship between the two compounds (Aftab and Vieira 2010; Nan Aftab and Vieira, 2010)
5.	Regeneration	Scavenge ROS and regenerate antioxidant	Weaker antioxidants replenish the antioxidant capacity of the stronger one. Resveratrol was discovered to be more effective in the process of reinforcing curcumin in this particular instance. Therefore, the curcumin and resveratrol have antioxidant action The combined effect resulted in a 15.5% increase compared to the average of separate effects. Additionally, there was a fourfold increase in antioxidant synergy (Aftab and Vieira 2010)
6.	Spatial distribution	Increased lipid/water interface and scavenge aqueous peroxyl radials	α-Tocopherol in combination with rosmarinic acid is synergized by physical localisation of antioxidants in emulation droplets (Altunkaya et al. 2016)

7 Synergistic Food for Common Diseases

7.1 Immunomodulators

Nutrition, genetics, age, gender, socioeconomic level, and eating habits, among other factors, all have a direct effect on the immune system. Oxidation reduces immune system function, but foods such as green tea and black pepper increase the bioavailability of epigallocatechin gallate (EGCG), which is contained in green tea.

EGCC is an extremely effective scavenger of free radicals. The absorption was shown to be ten times higher than when tea was drunk alone. Vitamin C has also been shown to increase catechin absorption by fivefold (Tewari et al. 2000; Lambert et al. 2004; Majchrzak et al. 2004).

7.2 Prevention of Infection

Lack of nutrition causes immune system to become weak especially in children; elderly and people with comorbid conditions are strongly affected; thus food and optimum nutrition plays a very important role in the prevention of infection. The skin of rats treated with curcumin and ginger extract showed rapid healing as compared to single item and control (Bhagavathula et al. 2009).

Authors studied the interaction between curcumin and PUFA, DHA, and EPA derived from fish oil; the combination showed potent anti-inflammatory by reducing the oxidative stress (Saw et al. 2010).

7.3 Cardiovascular Diseases

Combination of garlic and honey reduces blood cholesterol level and triglyceride level; garlic with the combination of fish oil (omega-3) synergistically boosts immunity and improved cardiovascular health (Morcos 1997).

7.4 Diabetes

The paper reveals combination of 80% apple cider vinegar and 20% whole blue berry juice as potential hypoglycaemic properties (Boath et al. 2012). Black currant which is rich in anthocyanins when combined with rowan berry improve sugar level by acting synergistically. The combination was found to maintain glycaemic level for type 2 diabetes (Agustinah et al. 2016).

7.5 Cancer

Food such as broccoli and tomatoes and antitumour and antiangiogenic properties lower the growth of tumour cells (Canene-Adams et al. 2005); apple and berry juice in the presence of vitamins, minerals, and fibres are found to prevent growth of cancer cells by recuing the oxidative stress (Jaganathan et al. 2014); red meat exposes the consumer to oxidative stress and have carcinogenic effects; it is also found that adding rosemary to red meat hinders the formation of carcinogens and reduces oxidative stress as well (Tsen et al. 2006).

8 Clinical Findings of Popular Synergistic Combinations

8.1 Vitamin D with Vitamin K

Clinical investigations on animals and people reveal that vitamin D increases vitamin K-dependent bone protein concentrations, promoting bone growth in vitro. Vitamin D enhances osteoblast-specific gene expression and regulates osteocalcin transcription through the 1,25(OH)D-responsive element in the gene promoter. Vitamin K and 1,25(OH)D stimulate mineralisation differently than vitamin K alone.

Furthermore, in rats, the passage suggests that 1,25(OH)D receptor binding can undergo gamma-carboxylation in the presence of vitamin K. This implies that vitamin K has the potential to modify the intrinsic biochemical properties of the nuclear receptors, affecting their binding to DNA and thereby influencing biological processes related to bone health (Van Ballegooijen et al. 2017).

- In human osteoblast cell cultures, glycoxidation restricts the maturation of osteoblasts, potentially impacting various bone markers. However, the negative effects of glycoxidation on osteoblast maturation can be reversed by the addition of vitamins D and K. The combined effect of these vitamins appears to induce beneficial biochemical changes in the bone, and this reversal of glycoxidation may have therapeutic implications for bone metabolic diseases like osteoporosis. In summary, the passage highlights the potential positive impact of vitamins D and K on counteracting the adverse effects of glycoxidation in osteoblasts, offering a potential avenue for therapeutic intervention in bone-related conditions (Gigante et al. 2014).

8.2 N-*Acetyl Cysteine with Glutathione*

Different health conditions have been studied in clinical research for synergistic effects of *N*-acetylcysteine (NAC) and glutathione (GSH).

- Acetaminophen Toxicity: The research looked at the effect of NAC and GSH in treating acetaminophen poisoning as one entity. According to the findings, combination therapy significantly lessened liver damage and increased survival rates compared to GSH alone or NAC alone (Heard 2008).
- COPD: The aim of the present study was to examine the combined effects of NAC and GSH on COPD patients. Results obtained show that combined therapy significantly improved lung function than when using NAC or GSH only.

Besides, oxidative stress markers were decreased and exercise capacity enhanced (Sadowska et al. 2006).

- Parkinson's Disease: In order to better understand the potential benefits of *N*-acetylcysteine (NAC) and glutathione (GSH) together for the treatment of Parkinson's disease (PD), authors carried out pilot research. For a period of 12 weeks, a group of 12 PD patients were randomly randomized to receive 600 mg/day of NAC, 1000 mg/day of GSH, or a combination of 600 mg/day of NAC and 1000 mg/day of GSH (Tenório et al. 2021).
- The study's findings demonstrated that combining NAC and GSH improved motor function and slowed the course of Parkinson's disease in people with PD better than either NAC or GSH alone. Additionally, the combined treatment greatly decreased inflammatory indicators and increased antioxidant capacity (Martinez-Banaclocha 2023).
- Cardiovascular Disease: The impact of *N*-acetylcysteine (NAC) and glutathione (GSH) on the cardiotoxicity caused by doxorubicin (DXR). It goes over how GSH and NAC affect DXR-induced cardiotoxicity in vitro and in vivo. The key findings are that whereas oral NAC given 1 h before and 2 h after DXR is ineffective, intravenous GSH given 10 min before and 1 h after DXR inhibits the development of DXR-induced cardiotoxicity. On isolated rat atria, DXR produces a negative inotropic effect that is inhibited by both GSH and NAC. A strong relationship exists between the two medicines' capacity to raise the myocardium's non-protein sulfhydryl group content and their cardioprotective actions (Villani et al. 1990).

8.3 NAC with Vitamin C

- The researchers studied the effect of NAC and vitamin C combination on aluminium chloride ($AlCl_3$)-induced haemolysis effect on erythrocytes. Various assays were conducted to measure the impact of $AlCl_3$ on haemolysis, annexin V-PE binding, forward scatter, intracellular Ca^{2+} levels, reactive oxygen species (ROS) production, and glutathione levels in erythrocytes. The results showed that $AlCl_3$ induced haemolysis, PS externalisation, membrane blebbing, and cell shrinkage in erythrocytes, as well as increased intracellular Ca^{2+} levels and ROS production. Vitamin C alone was not very effective and NAC shown to exert some antioxidant effect on the eryptosis. The combination effect of NAC and vitamin C was not clear in this study (Zangeneh et al. 2021).

8.4 NAC with Ginkgo

- The aim of the study was to evaluate the preventive properties of EGb761, an extract from *Ginkgo biloba*, and *N*-acetylcysteine (NAC) against oxidative DNA damage in the liver. Four groups (sham-operation, control, NAC, and EGb761) comprised 40 rats. In comparison to the sham-operated group, hepatic ischaemia lasting 90 min and then reperfusion led to elevated levels of thiobarbituric acid-reactive compounds, serum ALT, AST, and LDH activity as well as 8-OHdG, a marker of oxidative DNA damage. Nonetheless, lipid peroxidation and the production of 8-OHdG were decreased by pretreatment with both NAC and EGb761. According to the study, antioxidant compounds like EGb761 and NAC may help shield hepatic tissue from postischaemic reperfusion harm (Keles et al. 2008).

8.5 *Curcumin with Bosmeric Acid*

- The combination effect was studied on a group of 201 patients for osteoarthritis pain. The combination reduced WOMAC total score from baseline at 84 days compared to placebo. Researchers have also suggested to continue use of at least 3 months and supposed to reduce the use of NSAIDS (Sethi et al. 2022).

8.6 *Curcumin with Piperine*

- (Shoba et al. 1998) Investigated influence of piperine on the pharmacokinetics of curcumin in rats and humans. When curcumin was given to rats in the dose of 2 g/kg, only moderate concentration was observed in the serum, but when it was given with 20 mg/kg piperine, the bioavailability increased to 154% and total time taken to achieve maximum serum concentration also reduced from 4 to 1 h.

- In humans, total 2 g curcumin was undetectable in serum even after 4 h, but when curcumin was given at 20 mg with curcumin, the high concentration was observed within 1 h of administration and bioavailability was 2000%.

8.7 *S-Adenosyl-L methionine and* Gingko

- While the synergetic effect of both SAMe and *Gingko* was not studied, individually, SAMe and *Gingko* both were studied for dementia and cognitive impairment.
- SAMe was given 400 mg intravenously, 200 mg intramuscularly, and 400 mg twice day for 40 days. The MMSE and SCAG were given at baseline (time 0) and on Days 20 and 60 of treatment. At Day 60, MMSE and SCAG total scores showed significant changes ($P < 0.01$) compared to initial assessments. Four of 19 SCAG items rose considerably on Day 20. On Day 60, 13 of 19 things improved from the first (Fontanari et al. 1994).

- *Ginkgo* (dosage over 200 mg/day) had benefits at 24 weeks (207/276 vs. 178/273, OR 1.66, 95% CI 1.12–2.46, $P = 0.001$), but not at lower doses. *Ginkgo* improves cognition after 12 weeks in any dose. Higher doses had no effect (Birks and Grimley Evans 2009).

8.8 Vitamin A, Vitamin A Plus Zinc, and Multiple Micronutrients

Around 293 students were finally included in the study, who were divided into three groups.

- Vitamin A group: Children in this group received vitamin A supplementation once every 14 days.
- Vitamin A plus zinc group: Children in this group were given zinc in the form of zinc gluconate tablets 5 days a week, from Monday to Friday.
- Vitamin A in combination with multiple micronutrients group: In this group children received vitamin A plus multiple micronutrients in the form of chewable tablets five times every 14 days.
- The mixed micronutrients provided 100% of the Chinese dietary reference intakes (DRIs) for each supplementation agent, demonstrating that supplementation of micronutrients, including vitamin A alone, vitamin A combined with zinc, and vitamin A plus multiple micronutrients, effectively enhanced height gains and reduced the prevalence of anaemia in preschool children.
- The addition of zinc in combination with vitamin A was shown to be more efficacious in enhancing the height and height-for-age z-scores (HAZ) of children, but the combination of vitamin A with various micronutrients was more beneficial in boosting haemoglobin levels and weight in preschool children. After a period of 6 months, the occurrence of anaemia, sub-clinical vitamin A insufficiency, and zinc deficiency showed a notable decrease in all three groups receiving supplementation (Chen et al. 2012).

9 Profile of some Nutraceuticals Currently Used in Modern Medicine

9.1 Curcuma *(Turmeric)*

See Image 2.

- **Biological source**: the cleaned, dried, and ground rhizome of natural strains of *Curcuma longa* L. rhizome contains about 40% curcumin. It belongs to the family Zingiberaceae (Kendrick 2012).
- **Chemical constituents**: The turmeric rhizome is naturally comprised of 69.4% carbohydrates, 6.3% protein, 5.1% fat, and 3.5% minerals. The primary chemical

Image 2 *Curcuma*

constituents found in the rhizome are essential oil and curcuminoids. The scent of the substance is attributed to the essential oil, while the bright yellow colour is attributed to the curcuminoids, specifically curcumin and its counterparts (Dosoky et al. 2019). More than 100 different chemical constituents were isolated from different genera of *Curcuma*, some of the main chemical constituents that were found to be responsible for its nutraceutical and pharmaceutical activities are oleo-resin named curcuminoids (curcumin, demethoxycurcumin, and bisdemethoxycurcumin), turmerone, and zingiberene to name a few (Dosoky and Setzer 2018).

- **Uses:** *Curcuma* is used in wound healing and arthritis as antioxidant; in IBD (inflammatory bowel disease), antimutagenic and anticarcinogenic, antimicrobial, antidiabetic, antiangiogenic, hepatoprotective, dementias, and tardive dyskinesia; in general respiratory and gastric health; and in cosmetics (Nair 2019).

9.2 Ginger

See Image 3.

- **Biological source**: Ginger is derived from a flowering plant known as *Zingiber officinale* Roscoe, whose root and rhizome are widely used as a spice and for its medicinal properties. It belongs to the Zingiberaceae family.
- **Chemical constituents**: Gingerol, the primary bioactive component found in ginger, is accountable for its distinctive taste and therapeutic qualities. It possesses both anti-inflammatory and antioxidant properties. Zingerone exhibits antioxidant potential, while shogaol is produced from gingerol via the drying process after storage. Shogaol, in addition to its anti-inflammatory and antioxidant capabilities, is far more fragrant and stronger than gingerol (Semwal et al. 2015).

Image 3 Ginger tubers

- **Uses**: Ginger has been recognised for its ability to decrease the manufacture of prostaglandins by blocking the activity of cyclooxygenase-1 and cyclooxygenase-2 enzymes, making it an effective anti-inflammatory agent (Grzanna et al. 2005). Ginger is known for its digestive properties, helping to alleviate nausea and indigestion (Hu et al. 2011). Ginger is also studied for its antioxidant and antidiabetic effect and found to be significantly effective against type 2 diabetes (Mahluji et al. 2013).

9.3 *Black Pepper*

See Image 4.

- **Biological source**: Black pepper is known as *Piper nigrum* which belongs to the family Piperaceae; though it is native to Southern India, it is grown in other tropical countries as well (Srinivasan 2007).
- **Chemical constituents**: Piperine is the main chemical constituents of this drug, which is responsible for its pungent taste and heat. It is proven to have antioxidant and anti-inflammatory activity. Pepper is also known to have essential oils like beta-caryophyllene, limonene, pinene, and sabinene which produces its district aroma. It contains various minerals such as manganese, magnesium, and iron. Pepper is also found to contain peptidase in small amount so it might aid in digestion. Piperidine alkaloids are also found in pepper (Butt et al. 2013).
- **Uses**: Black pepper is famous as spice; it enhances the flavour of dishes, but medicinally piperine has been studied for enhancing bioavailability of some other nutrients such as curcumin. Piperine has also been shown to possess anti-inflammatory and antioxidant properties, and it stimulates the production of digestive enzymes. Pepper also relieves respiratory symptoms such as cough and congestion (Meghwal and Goswami 2013).

Image 4 Black pepper

Image 5 Garlic bulbs

9.4 Garlic

See Image 5.

- **Biological sources**: Garlic is obtained from bulbs of *Allium sativum* which is member of Liliaceae family. It is native to central Asia (Omar 2013).
- **Chemical constituents**: Well-known bioactive constituents named allicin makes garlic's characteristic odour, which is shown to possess antimicrobial properties. Garlic also contains sulphur compounds like diallyl sulphide and diallyl trisulfide which contribute to its health benefits. Allium contains an enzyme called allinase which converts alliin into allicin when garlic is crushed or chopped. It also contains flavonoids which contributes to antioxidant and other health-

improving properties. Vitamins C and B6 as well as manganese and selenium minerals are also found in garlic.

- **Uses**: Allicin showed antimicrobial properties in various studies. It also shows anti-inflammatory and antioxidant activities. Allicin is also shown to lower blood pressure and cholesterol level. Traditionally garlic is used to boost immune system especially in cold and flu season (Ankri and Mirelman 1999; Nantz et al. 2012; Bayan et al. 2014).

9.5 *Tomato*

See Image 6.

- **Biological sources**: Tomato is a rich source of lycopene, which is derived from ripen fruits of *Solanum lycopersicum*, which belongs to the family Solanaceae. Tomatoes are native to western South America but are now cultivated globally.
- **Chemical constituents**: Tomato contains a red colour carotenoid pigment called lycopene which is known to be potent antioxidant. It also contains various flavonoids like quercetin and kaempferol. Tomatoes are rich in vitamins such as vitamin C and A and various B vitamins, minerals such as potassium and folate, and carbohydrates such as fructose and glucose. Citric acid and malic acids among other organic acids are also found in tomatoes, which contributes to its taste.
- **Uses**: Tomatoes are high in lycopene, which has been researched for possible health advantages. These benefits include antioxidant activity and a possible lower risk of developing certain disease like cardiovascular, cancer, and diabetes. Tomatoes also contains vitamin C which prevents skin diseases. Tomatoes have various vitamins and minerals which make them a very good nutraceutical candidate (Giovannucci 2002; Story et al. 2010; Nasir et al. 2014).

Image 6 Tomato fruit

Image 7 Green tea powder

9.6 *Green Tea*

See Image 7.

- **Biological source**: The *Camellia sinensis* (Theaceae) plant's leaves are used to make green tea. Green tea, in contrast to black tea, is processed with less oxidation, maintaining its original aroma and colour. The main varieties of green tea have different qualities, such as sencha, matcha, gyokuro, and others.
- **Chemical constituents**: Green tea contains flavonoid-like catechins such as epicatechin gallate (ECG), epigallocatechin (EGC), and epicatechin (EC), all of which have antioxidant characteristics. Green tea contains lower levels of caffeine compared to black tea or coffee, but it still provides a mild stimulant effect. An amino acid name theanine contributes to the unique flavour of green tea and has calming effects. Tea contains various vitamins like B and C; it also has minerals like potassium, calcium, and magnesium.
- **Uses**: Green tea has various types of phenolics and exert strong antioxidant effects and prevents various diseases like cancer and diabetes; it also enhances metabolic rate and promotes fat oxidation thus supporting weight management. Combination of theanine and caffeine enhances cognitive benefits and functions. Regular consumption of green tea shows reduced cholesterol level and improving cardiac functions. Green tea is consumed as beverages (Cabrera et al. 2006; Hursel et al. 2009; Tang et al. 2019).

9.7 *Soy Isoflavones*

See Image 8.

Image 8 Soybeans

- **Biological source**: Natural substances called soy isoflavones can be found in soy products and *Glycine max* belonging to family Fabaceae. Isoflavones are a kind of phytoestrogen, which are substances obtained from plants that have structural characteristics with oestrogen from humans.
- **Chemical constituents**: Genistein, found in high amounts in soybeans, is the most studied soy isoflavone and has various health benefits like cancer prevention, improved cardiac functions, and alleviation of menopausal symptoms (Goh et al. 2022). Another well-known soy isoflavone, Daidzein, is transformed by certain gut bacteria into equal in the intestines. The third main isoflavone in soy is glycitein, albeit it is often present in lesser concentrations than genistein and daidzein.
- **Uses**: Isoflavone known as genistein alleviate menopausal symptoms due to their oestrogens like effects. Soy isoflavones also shown to have some bone-protective effects and reduced risk of osteoporosis. It is also associated with improved cardiac health, prevents cancer, improves cognitive functions, and has beneficial effects on the skin. Soy is also a very good nutraceutical because of high content of protein and various elements (Taku et al. 2012; Rizzo and Baroni 2018).

10 Conclusion

In conclusion, studies on natural origin nutraceuticals seem to be a promising field in contemporary medicine with a lot of therapeutic potential. Empirical evidence has demonstrated that naturally occurring bioactive compounds provide advantageous properties for human well-being, encompassing the mitigation and remediation of a diverse array of ailments. This review highlights the importance of nutraceuticals in enhancing overall health, with a special focus on their antioxidant, anti-inflammatory, and immune-regulating properties.

The research that has been given highlights the potential of nutraceuticals to supplement traditional medical methods by offering a thorough and well-rounded foundation for healthcare. Furthermore, natural substances are attractive candidates for preventative treatments due to their relatively low risk of side effects, particularly when it comes to treating chronic illnesses. In light of the growing prevalence of lifestyle-related illnesses, incorporating nutraceuticals into traditional medicine has potential for improving patient outcomes and treating health issues.

Nonetheless, it is critical to acknowledge the necessity of more study to clarify the modes of action, ideal dose, and long-term impacts of nutraceuticals. A more sophisticated understanding of their safety and effectiveness characteristics will be facilitated by comprehensive research and rigorous clinical trials. In addition, it is imperative to tackle issues pertaining to standardisation, quality control, and regulatory frameworks to guarantee uniformity and dependability in the manufacturing and dissemination of nutraceuticals.

References

Aftab N, Vieira A (2010) Antioxidant activities of curcumin and combinations of this curcuminoid with other phytochemicals. Phytother Res 24:500–502. https://doi.org/10.1002/ptr.2960

Agustinah W, Sarkar D, Woods F, Shetty K (2016) Apple and blueberry synergies for designing bioactive ingredients for the management of early stages of type 2 diabetes. J Food Qual 39:370–382. https://doi.org/10.1111/jfq.12206

Agyei D, Akanbi TO, Oey I (2019) Chapter 9—enzymes for use in functional foods. In: Kuddus M (ed) Enzymes in food biotechnology. Academic Press, London, pp 129–147

Alagawany M, Elnesr SS, Farag MR, Tiwari R, Yatoo MI, Karthik K, Michalak I, Dhama K (2020) Nutritional significance of amino acids, vitamins and minerals as nutraceuticals in poultry production and health—a comprehensive review. Vet Q 41(1):1–29. https://doi.org/10.1080/01652176.2020.1857887

Allied market research (2022), Nutraceuticals Market by Type, (Functional Beverage, Functional Food, and Dietary Supplements), Form (Capsules and Tablets, Liquid, Powder, Others), and Sales Channel (Hypermarkets/Supermarkets, Specialty Stores, Pharmacies, and Online Channels): Global Opportunity Analysis and Industry Forecast, 2021–2030. https://www.alliedmarketresearch.com/nutraceuticals-market. Accessed on 10-Nov-2023

Al-Okbi SY, Ali O, Aly AS et al (2023) Management of metabolic syndrome by nutraceuticals prepared from chitosan and ferulic acid with or without beta-sitosterol and their nanoforms. Sci Rep 13:12176. https://doi.org/10.1038/s41598-023-38837-9

Altunkaya A, Gökmen V, Skibsted LH (2016) pH dependent antioxidant activity of lettuce (L. sativa) and synergism with added phenolic antioxidants. Food Chem 190:25–32. https://doi.org/10.1016/J.FOODCHEM.2015.05.069

Ambati RR, Phang S-M, Ravi S, Aswathanarayana RG (2014) Astaxanthin: sources, extraction, stability, biological activities and its commercial applications—a review. Mar Drugs 12:128–152

Ankri S, Mirelman D (1999) Antimicrobial properties of allicin from garlic. Microbes Infect 1:125–129. https://doi.org/10.1016/S1286-4579(99)80003-3

Avinash Lal (2021), The Future of Food: Top Trends in 2021, https://www.kerry.com/insights/kerrydigest/2021/food-trends-in-2021.html##:~:text=A%20rise%20in%20home%20cooks.popular%20in%20the%20coming%20year. Accessed on 9-Nov-2023.

Bannuru RR, Osani MC, Vaysbrot EE et al (2019) OARSI guidelines for the non-surgical management of knee, hip, and polyarticular osteoarthritis. Osteoarthr Cartil 27:1578–1589. https://doi.org/10.1016/j.joca.2019.06.011

Bayan L, Koulivand P, Gorji A (2014) Garlic: a review of potential therapeutic effects. Avicenna J Phytomed 4:1–14

Bhagavathula N, Warner RL, DaSilva M et al (2009) A combination of curcumin and ginger extract improves abrasion wound healing in corticosteroid-impaired hairless rat skin. Wound Repair Regen 17:360–366. https://doi.org/10.1111/j.1524-475X.2009.00483.x

Birks J, Grimley Evans J (2007) Ginkgo biloba for cognitive impairment and dementia. Cochrane Database Syst Rev. https://doi.org/10.1002/14651858.CD003120.pub2

Birks J, Grimley Evans J (2009) Ginkgo biloba for cognitive impairment and dementia. Cochrane Database Syst Rev. https://doi.org/10.1002/14651858.CD003120.pub3

Boath AS, Stewart D, McDougall GJ (2012) Berry components inhibit α-glucosidase in vitro: synergies between acarbose and polyphenols from black currant and rowanberry. Food Chem 135:929–936. https://doi.org/10.1016/J.FOODCHEM.2012.06.065

Boots AW, Drent M, de Boer VCJ et al (2011) Quercetin reduces markers of oxidative stress and inflammation in sarcoidosis. Clin Nutr 30:506–512. https://doi.org/10.1016/j.clnu.2011.01.010

Butt MS, Pasha I, Sultan MT et al (2013) Black pepper and health claims: a comprehensive treatise. Crit Rev Food Sci Nutr 53:875–886. https://doi.org/10.1080/10408398.2011.571799

Cabrera C, Artacho R, Giménez R (2006) Beneficial effects of green tea—a review. J Am Coll Nutr 25:79–99. https://doi.org/10.1080/07315724.2006.10719518

Canene-Adams K, Campbell JK, Zaripheh S et al (2005) The tomato as a functional food. J Nutr 135:1226–1230

Chanda S, Tiwari RK, Kumar A, Singh K (2019) Nutraceuticals inspiring the current therapy for lifestyle diseases. Adv Pharmacol Sci 2019:6908716

Chandra S, Saklani S, Kumar P et al (2022) Nutraceuticals: pharmacologically active potent dietary supplements. Biomed Res Int 2022:2051017. https://doi.org/10.1155/2022/2051017

Chandrasekhar K, Kapoor J, Anishetty S (2012) A prospective, randomized double-blind, placebo-controlled study of safety and efficacy of a high-concentration full-spectrum extract of Ashwagandha root in reducing stress and anxiety in adults. Indian J Psychol Med 34:255–262. https://doi.org/10.4103/0253-7176.106022

Chen L, Liu Y-F, Gong M et al (2012). Effects of vitamin A, vitamin A plus zinc, and multiple micronutrients on anemia in preschool children in Chongqing, China. In Asia Pac J Clin Nutr 21:3–11.

Chughtai MF, Pasha I, Zahoor T et al (2020) Nutritional and therapeutic perspectives of Stevia rebaudiana as emerging sweetener; a way forward for sweetener industry. CyTA J Food 18:164–177. https://doi.org/10.1080/19476337.2020.1721562

Clegg DO, Reda DJ, Harris CL et al (2006) Glucosamine, chondroitin sulfate, and the two in combination for painful knee osteoarthritis. N Engl J Med 354:795–808. https://doi.org/10.1056/NEJMoa052771

Cooper EL (2005) CAM, eCAM, bioprospecting: the 21st century pyramid. Evid Based Complement Alternat Med 2:125–127

Crippa JA, Guimarães FS, Campos AC, Zuardi AW (2018) Translational investigation of the therapeutic potential of cannabidiol (CBD): toward a new age. Front Immunol 9:2009

Damián MR, Cortes-Perez NG, Quintana ET et al (2022) Functional foods, nutraceuticals and probiotics: a focus on human health. Microorganisms 10:1065. https://doi.org/10.3390/microorganisms10051065

Deal CL, Moskowitz RW (1999) Nutraceuticals as therapeutic agents in osteoarthritis: the role of glucosamine, chondroitin sulfate, and collagen hydrolysate. Rheum Dis Clin North Am 25:379–395. https://doi.org/10.1016/S0889-857X(05)70074-0

Dhara O, Chakrabarti P (2020). Oils and fats-largely unexplored source of nutraceuticals. Nov Tech Nutri Food Sci 5:394–406. https://doi.org/10.31031/NTNF.2020.05.000603

Dhingra AK, Rathi V, Chopra B (2021) Chapter 3.1.1—Resveratrol. In: Belwal T, Nabavi SM, Nabavi SF et al (eds) Naturally occurring chemicals against Alzheimer's disease. Academic Press, London, pp 33–47

Dosoky NS, Setzer WN (2018) Chemical composition and biological activities of essential oils of Curcuma species. Nutrients 10:1196. https://doi.org/10.3390/nu10091196

Dosoky NS, Setzer WN (2019) Chemical composition and biological activities of essential oils of curcuma species. Nutrients 10(9). https://doi.org/10.3390/nu10091196

Edgar S, Hopley B, Genovese L et al (2018) Effects of collagen-derived bioactive peptides and natural antioxidant compounds on proliferation and matrix protein synthesis by cultured normal human dermal fibroblasts. Sci Rep 8:10474. https://doi.org/10.1038/s41598-018-28492-w

Engels G, Brinckmann JA (2017) Nigella—Nigella sativa. HerbalGram 8–16

Fakhri S, Abbaszadeh F, Dargahi L, Jorjani M (2018) Astaxanthin: a mechanistic review on its biological activities and health benefits. Pharmacol Res 136:1–20

Fontanari D, Di Palma C, Giorgetti G et al (1994) Effects of S-adenosyl-L-methionine on cognitive and vigilance functions in the elderly. Curr Ther Res 55:682–689. https://doi.org/10.1016/S0011-393X(05)80803-3

Geng J, Dong J, Ni H, Lee MS, Wu T, Jiang K, Wang G, Zhou AL, Malouf R (2010) Ginseng for cognition. Cochrane Database Syst Rev. Issue 12. Art. No.: CD007769. https://doi.org/10.1002/14651858.CD007769.pub2. Accessed 9 Apr 2024

Gholamnezhad Z, Havakhah S, Boskabady MH (2016) Preclinical and clinical effects of Nigella sativa and its constituent, thymoquinone: a review. J Ethnopharmacol 190:372. https://doi.org/10.1016/j.jep.2016.06.061

Gigante A, Torcianti M, Boldrini E et al (2014) Vitamin K and D association stimulates in vitro osteoblast differentiation of fracture site derived human mesenchymal stem cells. J Biol Regul Homeost Agents 22:35–44

Giovannucci E (2002) A review of epidemiologic studies of tomatoes, lycopene, and prostate cancer. Exp Biol Med 227:852–859. https://doi.org/10.1177/153537020222701003

Goh YX, Jalil J, Lam KW et al (2022) Genistein: a review on its anti-inflammatory properties. Front Pharmacol 13:820969

Górska-Warsewicz H, Rejman K, Laskowski W, Czeczotko M (2019) Milk and dairy products and their nutritional contribution to the average polish diet. Nutrients 11:1771. https://doi.org/10.3390/nu11081771

Grandview Research (2023), Nutraceuticals Market Size, Share & Trends Analysis Report By Product (Dietary Supplements, Functional Foods, Functional Beverages), By Ingredient, By Application, By Region, And Segment Forecasts, 2023–2030. https://www.grandviewresearch.com/industry-analysis/nutraceuticals-market. Accessed on 9-Nov-2023.

Grandview Research (2022), Probiotics Market Size, Share & Trends Analysis Report By Product (Food & Beverages, Dietary Supplements), By Ingredient (Bacteria, Yeast), By Distribution Channel, By End-use, By Region, And Segment Forecasts, 2023 –2030. https://www.grandviewresearch.com/industry-analysis/probiotics-market. Accessed on 9-Nov-2023.

Grosso G, Pajak A, Marventano S et al (2014) Role of omega-3 fatty acids in the treatment of depressive disorders: a comprehensive meta-analysis of randomized clinical trials. PLoS One 9:e96905. https://doi.org/10.1371/journal.pone.0096905

Grzanna R, Lindmark L, Frondoza CG (2005) Ginger—an herbal medicinal product with broad anti-inflammatory actions. J Med Food 8:125–132. https://doi.org/10.1089/jmf.2005.8.125

Gulati OP, Ottaway PB (2008) Chapter 14—Botanical nutraceuticals (food supplements, fortified and functional foods) in the European Union with main focus on nutrition and health claims regulation. In: Bagchi D (ed) Nutraceutical and functional food regulations in the United States and around the world. Academic Press, San Diego, pp 199–219

Heard KJ (2008) Acetylcysteine for acetaminophen poisoning. N Engl J Med 359:285–292. https://doi.org/10.1056/nejmct0708278

Hemarajata P, Versalovic J (2013) Effects of probiotics on gut microbiota: mechanisms of intestinal immunomodulation and neuromodulation. Ther Adv Gastroenterol 6:39–51

Hewlings SJ, Kalman DS (2017). Curcumin: A review of its effects on human health. In Foods 6:92. https://doi.org/10.3390/foods6100092

Heymann T, Heinz P, Glomb M (2015) Lycopene inhibits the isomerization of β-carotene during quenching of singlet oxygen and free radicals. J Agric Food Chem 63. https://doi.org/10.1021/acs.jafc.5b00377

Hill CL, March LM, Aitken D et al (2016) Fish oil in knee osteoarthritis: a randomised clinical trial of low dose versus high dose. Ann Rheum Dis 75:23–29. https://doi.org/10.1136/annrheumdis-2014-207169

Honda M (2020) Nutraceutical and pharmaceutical applications of carotenoids. In: Eduardo J-L, Queiroz MI, Zepka LQ (eds) Pigments from microalgae handbook. Springer, Cham, pp 449–469

Hu ML, Rayner CK, Wu KL et al (2011) Effect of ginger on gastric motility and symptoms of functional dyspepsia. World J Gastroenterol 17:105–110. https://doi.org/10.3748/wjg.v17.i1.105

Hursel R, Viechtbauer W, Westerterp-Plantenga MS (2009) Effects of green tea on weight loss and weight maintenance. A meta-analysis. Int J Obes (Lond) 33:956–961. https://doi.org/10.1038/ijo.2009.135

Jackson MI, Jewell DE (2020) Docosahexaenoate-enriched fish oil and medium chain triglycerides shape the feline plasma lipidome and synergistically decrease circulating gut microbiome-derived putrefactive postbiotics. PLoS ONE 15(3). https://doi.org/10.1371/journal.pone.0229868

Jaganathan S, Vellayappan MV, Narasimhan G et al (2014) Chemopreventive effect of apple and berry fruits against colon cancer. World J Gastroenterol 20:17029. https://doi.org/10.3748/wjg.v20.i45.17029

Joob B, Wiwanitkit V (2015) Krill oil: new nutraceuticals. J Coast Life Med 3:669–670. https://doi.org/10.12980/JCLM.3.2015JCLM-2014-0082

Kalra EK (2003) Nutraceutical—definition and introduction. AAPS PharmSci 5(3):E25. https://doi.org/10.1208/ps050325

Keles MS, Demirci N, Yildirim A et al (2008) Protective effects of N-acetylcysteine and Ginkgo biloba extract on ischaemia-reperfusion-induced hepatic DNA damage in rats. Clin Exp Med 8:193–198. https://doi.org/10.1007/s10238-008-0005-1

Kendrick A (2012) Natural food and beverage colourings. In: Natural food additives, ingredients and flavourings. Woodhead Publishing Limited, Cambridge, pp 25–40. https://doi.org/10.1533/9780857095725.1.25

Kris-Etherton PM, Harris WS, Appel LJ (2002) Fish consumption, fish oil, omega-3 fatty acids, and cardiovascular disease. Circulation 106:2747–2757. https://doi.org/10.1161/01.CIR.0000038493.65177.94

Křížová L, Dadáková K, Kašparovská J, Kašparovský T (2019) Isoflavones. Molecules 24:1076. https://doi.org/10.3390/molecules24061076

Kumar Jha S, Roy P, Chakrabarty S (2021). Nutraceuticals with pharmaceutics: its importance and their applications. Int J Drug Dev & Res 13:002

Kumar A, Patil S (2022) Concept of nutraceuticals and pathya kalpana; a review. Int J Pharm Res Appl 75:74–81

Lambert JD, Hong J, Kim DH et al (2004) Piperine enhances the bioavailability of the tea polyphenol (-)-epigallocatechin-3-gallate in mice. J Nutr 134:1948–1952. https://doi.org/10.1093/jn/134.8.1948

Lim Y-S, Ok Y-J, Hwang S-Y et al (2019) Marine collagen as a promising biomaterial for biomedical applications. Mar Drugs 17:467. https://doi.org/10.3390/md17080467

Liu J, Wang L, Zhan S-Y, Xia Y (2011) Coenzyme Q10 for Parkinson's disease. Cochrane Database Syst Rev. https://doi.org/10.1002/14651858.CD008150.pub2

Luthuli S, Wu S, Cheng Y et al (2019) Therapeutic effects of fucoidan: a review on recent studies. Mar Drugs 17:487. https://doi.org/10.3390/md17090487

Mahluji S, Attari V, Mobasseri M et al (2013) Effects of ginger (Zingiber officinale) on plasma glucose level, HbA1c and insulin sensitivity in type 2 diabetic patients. Int J Food Sci Nutr 64:682. https://doi.org/10.3109/09637486.2013.775223

Majchrzak D, Mitter S, Elmadfa I (2004) The effect of ascorbic acid on total activity of black and green teas. Food Chem 88:447–451. https://doi.org/10.1016/j.foodchem.2004.01.058

Mancuso C, Santangelo R (2017) Panax ginseng and Panax quinquefolius: from pharmacology to toxicology. Food Chem Toxicol 107:362–372. https://doi.org/10.1016/j.fct.2017.07.019

Martinez-Banaclocha MA (2023) Targeting the cysteine redox proteome in Parkinson's disease: the role of glutathione precursors and beyond. Antioxidants 12:1373. https://doi.org/10.3390/antiox12071373

Mashhadi NS, Ghiasvand R, Askari G et al (2012) Anti-oxidative and anti-inflammatory effects of ginger in health and physical activity: review of current evidence. Int J Prev Med 4(Suppl 1):S36–S42

Matera R, Lucchi E, Valgimigli L (2023) Plant essential oils as healthy functional ingredients of nutraceuticals and diet supplements: a review. Molecules 28:901. https://doi.org/10.3390/molecules28020901

Markets and Markets (2022). Nutraceuticals Market By Type (Vitamins, Dietary Supplements, Herbs), Ingredient (Omega-3 Fatty Acids, Probiotics, Carotenoids), Application (Cardiovascular Health, Digestive Health, Bone Health), Distribution Channel (Retail Pharmacies, Online/E-commerce, Direct Selling), Region (North America, Europe, Asia-Pacific, Latin America, Middle East & Africa) - Forecast To 2027. https://www.marketsandmarkets.com/Market-Reports/nutraceutical-ingredient-market-1319.html. Accessed on 8-Nov-2023.

McMorris T, Mielcarz G, Harris RC, Swain JP, Howard A (2007) Creatine supplementation and cognitive performance in elderly individuals. Aging Neuropsychol Cognit 14:517–528. https://doi.org/10.1080/13825580600788100

Meghwal M, Goswami TK (2013) Piper nigrum and piperine: an update. Phytother Res 27:1121–1130. https://doi.org/10.1002/ptr.4972

Morcos NC (1997) Modulation of lipid profile by fish oil and garlic combination. J Natl Med Assoc 89(10):673–678

Mozaffarian D, Wu JHY (2011) Omega-3 fatty acids and cardiovascular disease: effects on risk factors, molecular pathways, and clinical events. J Am Coll Cardiol 58:2047–2067. https://doi.org/10.1016/j.jacc.2011.06.063

Nair KP (2019) Nutraceutical properties of turmeric. In: Turmeric (Curcuma longa L.) and ginger (Zingiber officinale Rosc.)—world's invaluable medicinal spices: the agronomy and economy of turmeric and ginger. Springer, Cham, pp 195–222

Nan Aftab KL, Vieira A (2010) Comparative antioxidant activities and synergism of resveratrol and oxyresveratrol. Nat Prod Res 24:1726–1733. https://doi.org/10.1080/14786410902990797

Nantz MP, Rowe CA, Muller CE et al (2012) Supplementation with aged garlic extract improves both NK and γδ-T cell function and reduces the severity of cold and flu symptoms: a randomized, double-blind, placebo-controlled nutrition intervention. Clin Nutr 31:337–344. https://doi.org/10.1016/j.clnu.2011.11.019

Nasir M, Hussain S, Jabbar S (2014) Tomato processing, lycopene and health benefits: a review. Sci Lett 3:1–5

Omar SH (2013) Garlic and cardiovascular diseases. In: Ramawat KG, Mérillon J-M (eds) Natural products: phytochemistry, botany and metabolism of alkaloids, phenolics and terpenes. Springer, Berlin, pp 3661–3696

Pandey MM, Rastogi S, Rawat AKS (2013) Indian traditional ayurvedic system of medicine and nutritional supplementation. Evid Based Complement Alternat Med 2013:376327

Pandi-Perumal SR, Trakht I, Srinivasan V et al (2008) Physiological effects of melatonin: role of melatonin receptors and signal transduction pathways. Prog Neurobiol 85:335–353. https://doi.org/10.1016/j.pneurobio.2008.04.001

Patel S (2015) Emerging trends in nutraceutical applications of whey protein and its derivatives. J Food Sci Technol 52:6847–6858. https://doi.org/10.1007/s13197-015-1894-0

Playford RJ, Weiser MJ (2021) Bovine colostrum: its constituents and uses. Nutrients 13:265. https://doi.org/10.3390/nu13010265

Priyanka S, Varsha R, Verma R, Babu AS (2023) Spirulina: a spotlight on its nutraceutical properties and food processing applications. J Microbiol Biotechnol Food Sci 12:4785. https://doi.org/10.55251/jmbfs.4785

PRNewswire (2023). Emerging Trends in the Nutraceuticals Industry Future Market Insights, Inc. https://www.prnewswire.com/news-releases/emerging-trends-in-the-nutraceuticals-uture-market-insights-inc.301948235.html#:~:text=The%20nutraceutical%20ingredients%20market%20is,US%24%201%2C233.4%20million%20in%202023. Accessed on 10-Nov-2023.

Rahman K, Lowe GM (2006) Garlic and cardiovascular disease: a critical review12. J Nutr 136:736S–740S. https://doi.org/10.1093/jn/136.3.736S

Rathi V, Dhingra AK, Chopra B (2021) Chapter 3.2.18—Withania somnifera. In: Belwal T, Nabavi SM, Nabavi SF et al (eds) Naturally occurring chemicals against Alzheimer's disease. Academic Press, London, pp 401–407

Reiter RJ, Mayo JC, Tan D-X et al (2016) Melatonin as an antioxidant: under promises but over delivers. J Pineal Res 61:253–278. https://doi.org/10.1111/jpi.12360

Rizzo G, Baroni L (2018) Soy, soy foods and their role in vegetarian diets. Nutrients 10:43

Sadowska AM, Verbraecken J et al (2006). Role of N-acetylcysteine in the management of COPD. In Therapeutics and Clinical Risk Management 1: 425–434.

Saw CLL, Huang Y, Kong AN (2010) Synergistic anti-inflammatory effects of low doses of curcumin in combination with polyunsaturated fatty acids: docosahexaenoic acid or eicosapentaenoic acid. Biochem Pharmacol 79:421–430. https://doi.org/10.1016/J.BCP.2009.08.030

Semwal RB, Semwal DK, Combrinck S, Viljoen AM (2015) Gingerols and shogaols: important nutraceutical principles from ginger. Phytochemistry 117:554–568. https://doi.org/10.1016/j.phytochem.2015.07.012

Sethi V, Garg M, Herve M, Mobasheri A (2022) Potential complementary and/or synergistic effects of curcumin and boswellic acids for management of osteoarthritis. Ther Adv Musculoskelet Dis 14:1759720X2211245

Shah B, Prasad Patel BP, Shrestha S, Prasad SM, Shah RK (2021) Current status and prospects of Ayurveda nutraceuticals in Kathmandu, Nepal. Healer J 2:8–16

Shoba G, Joy D, Joseph T, Majeed M, Rajendran R, Srinivas PS (1998) Influence of piperine on the pharmacokinetics of curcumin in animals and human volunteers. Planta Med 64:353–356. https://doi.org/10.1055/s-2006-957450

Statista (2018), Value of the nutraceutical market across India in 2012 and 2017 with a forecast for 2022. https://www.statista.com/statistics/920054/india-nutraceutical-market-value/#:~:text=In%202017%2C%20India's%20nutraceutical%20market.growth%20potential%20in%20the%20sector. Accessed on 9-Nov-2023.

Singh AK, Gupta AK, Manish, Singh PK (2014) Rasayana therapy: a magic contribution of ayurveda for healthy long life. Int J Res Ayurveda Pharm 5:41–47

Sinha R (2011) Charak Samhita. Chaukhamba Surbharti Prakashan Varanasi, Sutrasthana adhya 5th, 1st edn, reprint. p78

Srinivasan K (2007) Black pepper and its pungent principle-piperine: a review of diverse physiological effects. Crit Rev Food Sci Nutr 47:735–748. https://doi.org/10.1080/10408390601062054

Story EN, Kopec RE, Schwartz SJ, Harris GK (2010) An update on the health effects of tomato lycopene. Annu Rev Food Sci Technol 1:189–210. https://doi.org/10.1146/annurev.food.102308.124120

Surjushe A, Vasani R, Saple DG (2008) Aloe vera: a short review. Indian J Dermatol 53:163–166

Taku K, Melby M, Kronenberg F et al (2012) Extracted or synthesized soybean isoflavones reduce menopausal hot flash frequency and severity: systematic review and meta-analysis of randomized controlled trials. Menopause 19:776–790. https://doi.org/10.1097/gme.0b013e3182410159

Tang GY, Meng X, Gan RY et al (2019) Health functions and related molecular mechanisms of tea components: an update review. Int J Mol Sci 20:6196

Tenório M, Graciliano N, Moura F et al (2021) N-acetylcysteine (NAC): impacts on human health. Antioxidants 10:967. https://doi.org/10.3390/antiox10060967

Tewari S, Gupta V, Bhattacharya S (2000) Comparative study of antioxidant potential of tea with and without additives. Indian J Physiol Pharmacol 44:215–219

Tsen SY, Ameri F, Smith JS (2006) Effects of rosemary extracts on the reduction of heterocyclic amines in beef patties. J Food Sci 71:C469–C473. https://doi.org/10.1111/j.1750-3841.2006.00149.x

Ujjaliya N, Dash S, Kumar Jain S (2018) A review on nutraceuticals in ayurveda. World J Pharm Pharm Sci 7:278. https://doi.org/10.20959/wjpps20185-11447

van Ballegooijen AJ, Pilz S, Tomaschitz A et al (2017) The synergistic interplay between vitamins D and K for bone and cardiovascular health: a narrative review. Int J Endocrinol 2017:7454376. https://doi.org/10.1155/2017/7454376

Venkatramanujam S, Warren DS, Pendi-perumal SR, Zakharia R, Bhatnagar PK, Brzezinski A (2009) Melatonin and human reproduction: Shedding light on the darkness hormone. Gynecol Endocrinol 25(12):779–785. https://doi.org/10.3109/09513590903159649

Villani F, Galimberti M, Zunino F (1990) Effect of glutathione and N-acetylcysteine on in vitro and in VIVO cardiac toxicity of doxorubicin. Free Radic Res Commun 11:145–151. https://doi.org/10.3109/10715769009109677

Violi F (2010) Introduction to nutraceuticals special issue. Cardiovasc Ther 28:185–186. https://doi.org/10.1111/j.1755-5922.2010.00204.x

Walker LA, Koturbash I, Kingston R, ElSohly MA, Yates CR, Gurley BJ, Khan I (2020) Cannabidiol (CBD) in dietary supplements: perspectives on science, safety, and potential regulatory approaches. J Diet Suppl 17:493–502. https://doi.org/10.1080/19390211.2020.1777244

White CM (2019) A review of human studies assessing cannabidiol's (CBD) therapeutic actions and potential. J Clin Pharmacol 59:923

Williamson EM, Liu X, Izzo AA (2019) Trends in use, pharmacology, and clinical applications of emerging herbal nutraceuticals. Br J Pharmacol 177(6):1227–1240. https://doi.org/10.1111/bph.v177.6/issuetoc

Wu G (2013) Functional amino acids in nutrition and health. Amino Acids 45:407–411. https://doi.org/10.1007/s00726-013-1500-6

Zangeneh AR, Takhshid MA, Ranjbaran R et al (2021) Diverse effect of vitamin C and N-acetylcysteine on aluminum-induced eryptosis. Biochem Res Int 2021:6670656. https://doi.org/10.1155/2021/6670656

Zhao CF, Lei DJ, Song GH et al (2015) Characterisation of water-soluble proanthocyanidins of Pyracantha fortuneana fruit and their improvement in cell bioavailable antioxidant activity of quercetin. Food Chem 169:484–491. https://doi.org/10.1016/J.FOODCHEM.2014.07.091

Zhao JV, Burgess S, Fan B, Schooling CM (2022) l-Carnitine, a friend or foe for cardiovascular disease? A Mendelian randomization study. BMC Med 20:272. https://doi.org/10.1186/s12916-022-02477-z

Zuchi C, Ambrosio G, Lüscher TF, Landmesser U (2010) Nutraceuticals in cardiovascular prevention: lessons from studies on endothelial function. Cardiovasc Ther 28:187–201. https://doi.org/10.1111/j.1755-5922.2010.00165.x

The Importance of Natural Products in Drug Discovery: Opportunities and Success

Rupshikha Nath, Rajat Nath, Abhijit Mitra, Anupam Das Talukdar, Deepa Nath, Jayanta Kumar Patra, Manabendra Dutta Choudhury, and Sibashish Kityania

Contents

R. Nath · R. Nath · A. Das Talukdar · M. D. Choudhury · S. Kityania (✉)
Department of Life Science and Bioinformatics, Assam University, Silchar, Assam, India

A. Mitra
Prescientia Group of Institutions, Hailakandi, Assam, India

D. Nath
Department of Botany, Gurucharan College, Silchar, Assam, India

J. K. Patra
Dongguk University, Goyang-si, Republic of Korea

© The Author(s), under exclusive license to Springer Nature Singapore Pte Ltd. 2024

A. Das Talukdar et al. (eds.), *Traditional Resources and Tools for Modern Drug Discovery*, Interdisciplinary Biotechnological Advances,
https://doi.org/10.1007/978-981-97-4600-2_13

Abstract Natural products are biological chemicals created by living creatures such as plants, animals, microorganisms, and sea critters. Various endogenous and exogenous factors have compelled living organisms to produce such products over time having great ecological significance. Before realizing the diverse potential of various natural products, they were believed to benefit only specific organisms, aiding in defense, metabolism, attraction, and survival strategies. Inspired by nature, the earliest civilization used natural products for various purposes like medicine, cosmetics, dyes, and perfumes. In pharmaceutical research, natural products are essential because they provide prospective pathways for the discovery of novel medicinal molecules and have yielded excellent outcomes. Natural products possess diverse pharmacophores, stereochemistry, and metabolite-like properties. Their chemical diversity provides a potential for discovering novel compounds with varied properties. Unique biological activities, synergy, and complexity make them valuable for addressing disease targets, combating drug resistance, and inhibiting protein interactions. They inspire new compound designs, enhance drug potency, and align with sustainability goals. Natural products validate traditional medicine and hold promise for personalized medicine and neurological disorders. The bioactivity of natural compounds may now be predicted by researchers, thanks to the use of contemporary technology techniques like bioinformatics and data mining. This chapter covers the relevance of natural products in the medication development process.

Keywords Natural compounds · Pharmacokinetics · Drug discovery · Bioinformatics · Medicine · Bioactivity

1 Introduction

Throughout the history of medicine, natural resources such as plants, animals, and minerals have played an important role in the treatment of human illnesses. The roots of modern medicine are deeply embedded in traditional practices that have persisted globally for thousands of years. Indigenous knowledge of plant-based medicines, transmitted through generations, has significantly shaped various medical traditions (Brahmachari 2012). Because of their significance throughout history, natural goods keep generating curiosity, their role as experimental tools, and their current resurgence in public use through complementary and alternative medicine (Müller et al. 2004; Patwardhan et al. 2004). The World Health Organization (WHO) estimates that 80% of people worldwide use natural goods, herbal cures, and traditional medicine; this has a significant economic effect (Farnsworth et al. 1985; Arvigo and Balick 1993; Barnes 2003). The historical significance is evident in the ancient medical systems of nations such as Mesopotamia, China, and India, with early documentation discovered in books about the use of therapeutic plants such as the *Rig Veda*, *Sushruta Samhita* in India, the Chinese *Materia Medica* (Cragg and Newman 2001), the *Ebers Papyrus* in Egypt (Nakanishi 1999), and the work of

Pedanius Dioscorides in *De Materia Medica* in ancient Greece (Cragg and Newman 2001). The enduring reliance on crude plant-based preparations for primary healthcare in developing countries underscores the continued importance of natural products (Chang et al. 1986; Dev 1999; Schultes and Raffauf 1990). Since natural products have been used in contemporary medicine, active components have been isolated, marking significant milestones in the history of medicine. Morphine (1826) (Newman et al. 2000) and aspirin (1899) are some significant examples of natural product-derived medications introduced in the nineteenth century (Brahmachari 2012).

Natural products' concealed structural diversity makes them essential for medication development, exceeding that of synthetic compounds. In medicinal chemistry, 40% of the chemical scaffolding inherent in natural products is still absent, despite efforts to use diversity-oriented synthesis (Burke et al. 2003; Lee and Schneider 2001). Their uniqueness, diverse biological activities, and historic role in discoveries across chemistry and biology underline their therapeutic potential. Only 5–15% of land plants have been chemically analyzed, leaving vast potential for discovering novel chemicals with diverse applications (Brahmachari 2012). Natural compounds, often distinct from synthetics, interact with biomolecules and have contributed to about 50% of drugs introduced in the last two decades (Vuorela et al. 2004). They are essential for understanding cellular processes and are valuable lead structures for pharmacological development (Hung et al. 1996; Hinterding et al. 1998). Their distinct chemical space makes them appealing in therapeutic areas lacking lead compounds, and they serve as templates for building biologically relevant libraries, reinforcing their importance in balanced drug discovery programs (Henkel et al. 1999; Lee and Schneider 2001; Horton et al. 2003; Brahmachari 2012).

The challenge of addressing global health issues, including communicable and noncommunicable diseases, persists despite advances in drug development (Thomford et al. 2018). Pharmaceutical research and development (R&D) methodologies that rely on old methods must change to find viable medication candidates with minimum adverse effects (Thomford et al. 2018). Natural resources, historically significant for drug discovery, particularly in the development of anticancer and antimicrobial agents, are seen as a promising avenue. The use of plant-based medicines, rooted in historical practices and passed down through generations, has garnered renewed interest. A significant percentage of FDA (Food and Drug Administration) authorized drugs come from plants, showcasing the potential of natural resources (Newman and Cragg 2012; Carter 2011). However, challenges such as standardization, isolation of compounds, and the lack of rigorous clinical trials hinder wider acceptance (Li and Weng 2017; Leonti and Verpoorte 2017; Thomford et al. 2018). Overcoming these challenges involves leveraging advanced technologies, including quantum computing, computational biology, and artificial intelligence, to explore the combinatorial effects of plant-based compounds (Özdemir and Hekim 2018; Özdemir 2015; Thomford et al. 2018). Innovative extraction methods and technological advancements in analytical techniques contribute to the ongoing initiatives aimed at discovering new drugs from plants. The exploration of synergistic effects of compounds from whole plant extracts, along with the adoption of Good Manufacturing Practices, is gaining momentum in clinical trials, offering a holistic approach to therapeutic benefits (Wen et al. 2005a; Chan et al. 2012). The future of drug development is based on extensive, combinatorial evaluations of plant extracts and the integration of "-omics" technologies, automation, and big data in the synthesis and testing of pharmacologically relevant molecules from natural sources (Kim et al. 2015; Medema and Fischbach 2015).

2 Natural Product

2.1 *Different Types of Sources*

The field of natural product chemistry has seen explosive growth, becoming recognized as a separate field. Numerous unique chemotypes may be found in nature, which have produced chemicals with a variety of structural configurations and biological activities, contributing significantly to drug discovery (Brahmachari 2009). While around 200,000 natural compounds are currently known, only a small fraction of plant and microbial sources have been explored, presenting immense untapped potential (Brahmachari 2012; Jachak and Saklani 2007). Natural products, critical to pharmaceuticals for centuries, remain a viable technique for developing new medications, with the world's biodiversity offering a vast reservoir of novel drugs (Brahmachari 2009, 2012); some of the biological sources of natural products are mentioned in Table 1. The renewed interest in natural compounds, aided by modern chemical and biological tools, holds promise for discovering new therapies against various diseases.

2.2 *Significance of Natural Products*

Evaluating natural ingredients in the quest for new drugs presents several significant advantages that outweigh their limitations. Firstly, they offer unparalleled chemical diversity, especially from unexplored marine resources, possibly resulting in the identification of brand-new bioactive substances, with projections indicating a 100-fold increase in hit rate over synthetic substances (Clardy and Walsh 2004; Berdy 2005). Secondly, natural products occupy distinct chemical spaces, containing unique scaffolds not found in synthetic compound databases, presenting promising avenues for new drug starting points (Lam 2007; Bajorath 2000; Feher and Schmidt 2003). Thirdly, they act as models for combinatorial chemistry, producing libraries of analogs that may have improved drug-like characteristics (Lam 2007). Additionally, advancements in understanding genetics and biosynthesis enable the optimization and alteration of natural product biosynthesis resulting in the identification of fresh derivatives (Bibb 2005; Cundliffe 2006). Lastly, natural products not

Table 1 Some examples of natural product-based medicine

Sl. no	Compound extracted from nature	Source	Properties	References
1	Artemisinin and its derivatives (e.g., dihydroartemisinin, artemether, arteether, and artesunate)	*Artemisia annua L.*	Antimalarial treatments	Ziffer et al. (1997), Hsu (2006) and Gaudillière et al. (2001)
2	Caspofungin acetate	*Glarea lozoyensis*	Interferes with the formation of the fungal cell wall, causing the fungal cell to die	Graul (2002) and McCormack and Perry (2005)
3	Ertapenem	*Streptomyces cattleya*	Antimicrobial action with a wide range that targets both Gram-positive and Gram-negative bacteria	Sader and Gales (2001) Livermore et al. (2005) and Parakh et al. (2009)
4	Cefditoren pivoxil	*Cephalosporium* spp.	Broad-spectrum antibacterial drug	Wellington and Curran (2005) and Fuchs (2007)
5	Daptomycin	*Streptomyces roseosporus*	Use in the management of complex infections of the skin and its structures	Dhawan et al. (2009), Straus and Hancock (2006) and Fowler et al. (2006)
6	Fumagillin	*Aspergillus fumigatus*	Use to treat intestinal amoebiasis	Hanson and Eble (1949), Nor El-Din (1956) and Brahmachari (2012)
7	Amrubicin hydrochloride	*Streptomyces peucetius* var. *caesius*	Antitumor activity	Graul (2002) and Frantz and Smith (2003)
8	Trabectedin	*Ecteinascidia turbinata*	Use in the management of advanced soft tissue sarcoma	Carter and Keam (2007) and Grosso et al. (2007)
9	Ixabepilone	*Sorangium cellulosum*	An anticancer drug	Cardoso et al. (2008) and Pronzato (2008)
10	Romidepsin	*Chromobacterium violaceum*	Use against cutaneous and peripheral T-cell lymphoma	Piekarz et al. (2009) and Woo et al. (2009)
11	Galantamine hydrobromide	*Galanthus nivalis*	Used in the management of mild to severe Alzheimer's illness	Heinrich and Lee Teoh (2004) and Raskind (2003)

(continued)

Table 1 (continued)

Sl. no	Compound extracted from nature	Source	Properties	References
12	Capsaicin	*Capsicum* spp.	Apply as an ointment to alleviate peripheral neuropathy pain	Remadevi and Szallisi (2008), Simpson et al. (2008) and Knotkova et al. (2008)
13	Exenatide	*Heloderma suspectum*	Antidiabetic	Malhotra et al. (1992) and Wolfson (2007)
14	Pimecrolimus	*Streptomyces hygroscopicus* subsp. *ascomyceticus*	Use to treat eczema	Nghiem et al. (2002), Allen et al. (2003) and Boone et al. (2007)
15	Dronabinol	*Cannabis sativa*	As pain killer	Wade et al. (2003) and Russo et al. (2007)

only function as drugs but also aid in understanding targets and pathways (Gullo et al. 2006). Examples like erythromycin, penicillin, and the extraordinary capacity of microbes to create compounds that resemble drugs are demonstrated by statins without requiring chemical modifications, demonstrating how natural ingredients may be used to produce new drugs (Lam 2007).

2.3 The Process Involved

Technology advancements and cheminformatic techniques for screening library creation are transforming the conventional bioassay-guided isolation of natural compounds (Harvey et al. 2015). The conventional approach involved screening complex extract samples, but challenges arose due to the presence of various compounds. To address this, pre-fractionation strategies are employed, removing interfering compounds and enhancing screening efficiency. Natural product-based screening collections are now more successful because of this technique, which has demonstrated greater hit rates in high-throughput screening campaigns (Harvey et al. 2015; Appleton et al. 2007).

The investigation procedure for natural product hits involves extraction, fractionation, and purification to yield biologically active compounds (Koehn and Carter 2005). Technologies like liquid chromatography-mass spectrometry (LC-MS) assist in dereplication, avoiding duplication of effort by identifying known compounds (Strege 1999; Nielsen and Smedsgaard 2003). NMR advancements, including multidimensional methods and sensitivity improvements, revolutionize the structure determination of new molecules (Freeman and Morris 1979). High-resolution mass spectrometry (HRMS) helps determine molecular formulas accurately (Bax et al. 1986). SAR studies are facilitated by biosynthetic congeners and shotgun derivatization (Koehn and Carter 2005). Challenges persist in isolating active compounds from complex matrices, but innovations like online bioassays and frontal affinity chromatography combined with mass spectrometry (FAC-MS) simplify this (Chan et al. 2003; Zhu et al. 2003). Advanced technologies like liquid chromatography-nuclear magnetic resonance-mass spectrometry (LC-NMR-MS), aided by cryogenic probes and micro-coils, enable direct analysis of complex mixtures (Lommen et al. 2000; Exarchou et al. 2003). Despite hurdles in isolation, biosynthetic technologies show promise for natural product drug discovery (Khosla and Keasling 2003). These innovations have paved the way for more efficient and streamlined methods of discovering and characterizing bioactive compounds.

Moreover, in natural product research, metabolomics emerged later but became essential in understanding secondary metabolites and their biosynthetic pathways. Advanced analytical methods like high-resolution Fourier-transform mass spectrometry (HRFTMS) (Ohta et al. 2010; Hufsky et al. 2014) and NMR spectroscopy (Grkovic et al. 2014) aid in identifying and quantifying metabolites, allowing dereplication (Ebada et al. 2008), especially in intricate combinations (Yilmaz et al. 2011; Harvey et al. 2015). Metabolomics assists in predicting bioactive structures, prioritizing fractions for purification, and utilizing information technologies for efficient data analysis (Yuliana et al. 2011). This method detects and increases the synthesis of desired chemicals while also optimizing fermentation (Harvey et al. 2015). In traditional herbal medicine like TCM (traditional Chinese medicine; mixture of plant products), metabolomics contributes to studying complex mixtures' effects on biological systems (Youns et al. 2010; Yu et al. 2013) and quality control (Pelkonen et al. 2012). Metabolomics data aids in predicting biosynthetic precursors, enhancing the production of functional natural products like tanshinones, promising therapeutic agents for various health disorders (Cheng et al. 2013).

3 Drug Discovery

The journey from discovering a natural product (NP)-based drug to its market launch involves multiple stages: initial discovery, drug development, and clinical trials, Showing schematic diagram in Fig. 1 (A&B). This entire process can span around 10 years and cost over $800 million, primarily due to the extensive number of leads discarded along the way (Dickson and Gagnon 2004). Approximately 1 in 5000 lead molecules complete clinical trials, putting them on the market and approving their use (Brahmachari 2012). Approval requires submission of applications like the Investigational New Drug (IND) to regulatory bodies like the FDA or EMEA, followed by a Marketing Authorization Application (MAA) and New Drug Application (NDA), leading to responses such as "approval," "nonapproval," or "approvable," determining the drug's fate in the market (Brahmachari 2012).

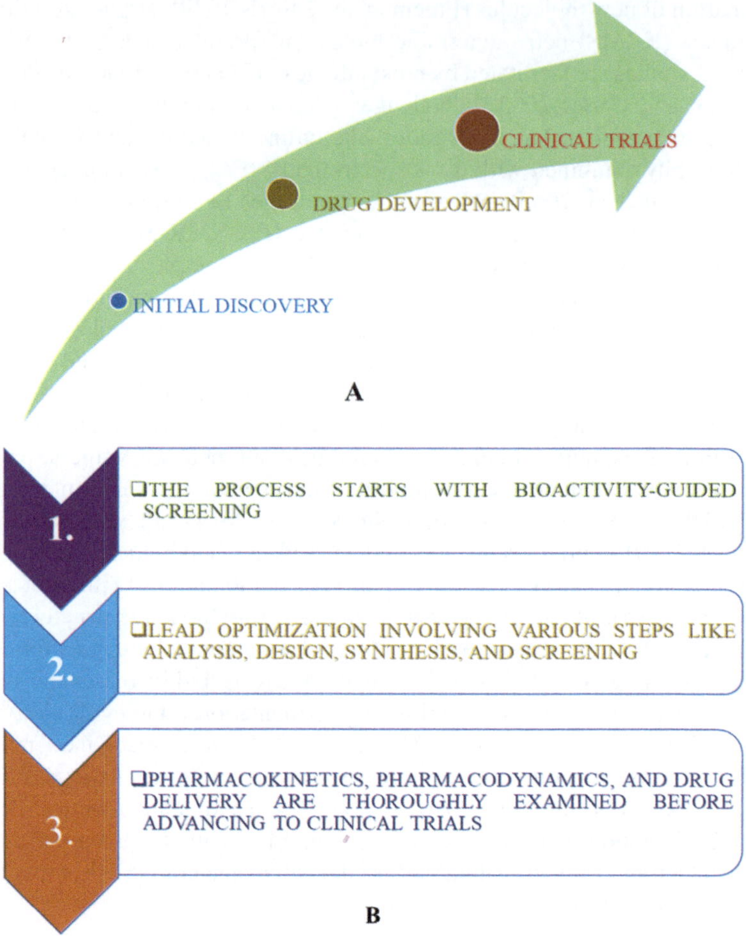

Fig. 1 (**A** and **B**) The journey from discovering a natural product (NP)-based drug to its market launch involves three stages

3.1 *Natural Product-Based Drugs*

Natural products, including bacteria, plants, and animals, are used as therapeutics and are called natural product-based pharmaceuticals. These medications have been used therapeutically for a very long time, and many conventional cures come from natural sources. They are crucial in modern medicine, offering a wide array of pharmacological activities and serving as the basis for numerous pharmaceuticals. The process of drug discovery, traditionally costly and time-intensive, has seen advancements in recent years, particularly with drug repositioning strategies that explore new uses for existing medications A few medications derived from natural products are included in Table 1 (The chemical structures of natural compounds are shown in Fig. 2).

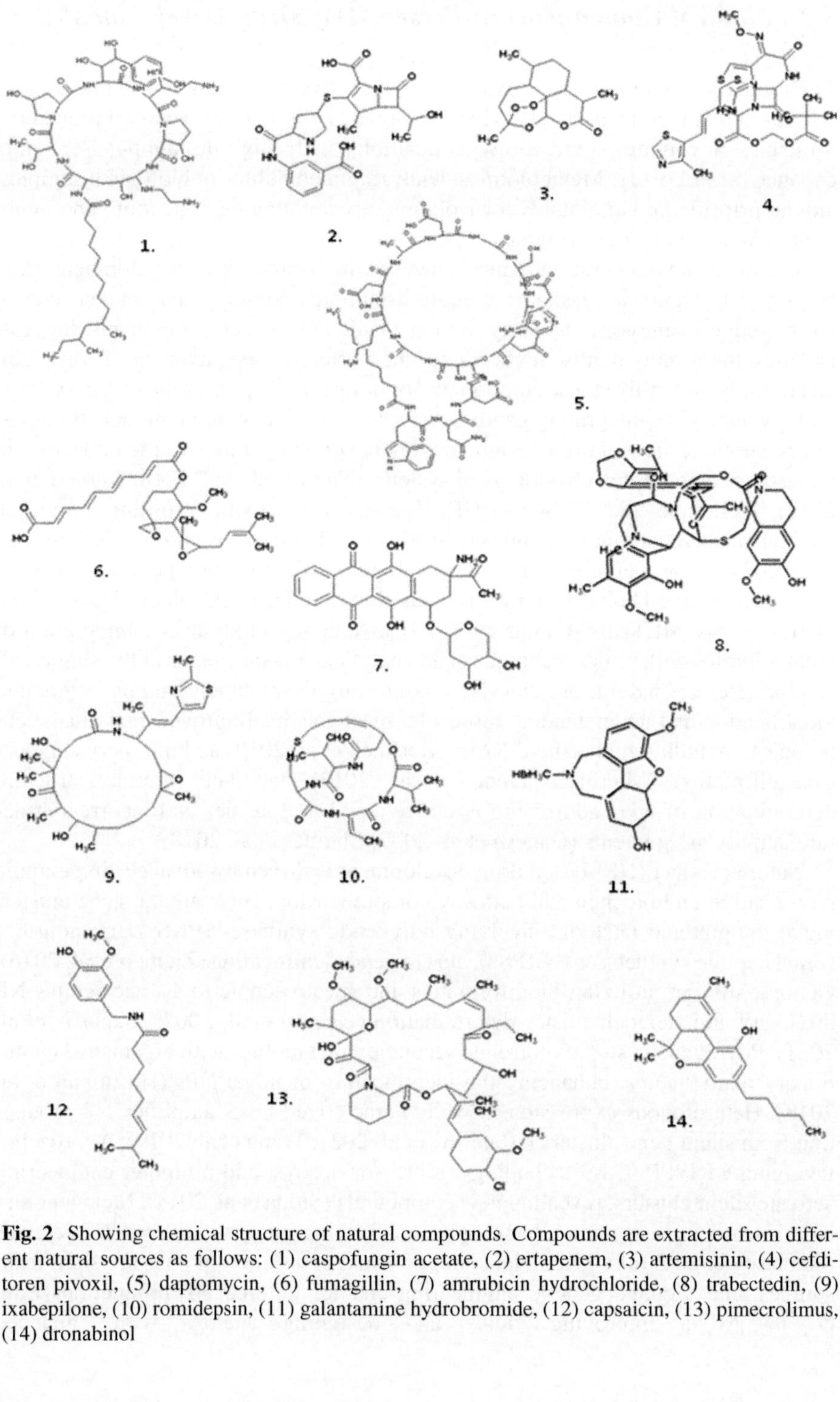

Fig. 2 Showing chemical structure of natural compounds. Compounds are extracted from different natural sources as follows: (1) caspofungin acetate, (2) ertapenem, (3) artemisinin, (4) cefditoren pivoxil, (5) daptomycin, (6) fumagillin, (7) amrubicin hydrochloride, (8) trabectedin, (9) ixabepilone, (10) romidepsin, (11) galantamine hydrobromide, (12) capsaicin, (13) pimecrolimus, (14) dronabinol

3.2 Natural Components in Present-Day Drug Development

Traditionally, screening crude extracts for bioactive chemicals is the first step in NP-based therapeutic discovery. This is followed by a rigorous isolation procedure. Strategies involving subfractions compatible with high-throughput screening enhance hit discovery. Metabolomics, analyzing metabolites in biological samples, aids in prioritizing compounds for isolation, accelerating dereplication, and annotating unknown analogs (Wagenaar 2008).

To create next-generation drugs, however, innovative drug development from natural ingredients necessitates a multidisciplinary strategy that makes use of cutting-edge technology. Synergy among medicinal extract components suggests isolating them individually might hinder their effectiveness, requiring fresh methods to study and utilize these compounds for innovative drugs. Embracing a systems biology approach, integrating genomics, proteomics, and computational strategies, offers a holistic view of drug design, moving beyond targeting specific molecules to understanding complex physiological systems (Thomford et al. 2018; Buriani et al. 2012). Techniques like NMR and HRMS assist in metabolite profiling, providing qualitative and quantitative insights (Eugster et al. 2011; Stavrianidi 2020; Atanasov et al. 2021). However, challenges in unambiguous identification persist, addressed by databases like Global Natural Products Social (GNPS) (Doak et al. 2014) and platforms like METLIN (Guijas et al. 2018; Atanasov et al. 2021). Integration of metabolomics with omics techniques and analytical advancements at the single-cell level accelerates natural product (NP)-based drug development, aiding compound identification and understanding molecular mechanisms. Improved analytical technologies, including higher-field NMR (Berlinck et al. 2019) and microcrystal electron diffraction (MicroED) (Jones et al. 2018), facilitate rigorous structure determination of NPs, addressing issues related to impurities and incorrect structure/activity assignments (Ganesh et al. 2007; Choules et al. 2018).

Nature product (NP)-based drug development is driven by advances in genomic modification and biochemical pathway comprehension. Biosynthetic gene clustering in fungus and bacteria, involving polyketide synthases (PKSs) and nonribosomal peptide synthetases (NRPSs), aids in gene identification (Ziemert et al. 2016). Genome mining, utilizing bioinformatics and spectroscopic tools, accelerates NP discovery and stereochemistry determination (Ganesh et al. 2007; Atanasov et al. 2021). Phylogenetic studies correlate secondary metabolites with organisms' evolutionary relationships, enhancing the identification of novel NPs (Hoffmann et al. 2018). Heterologous expression in well-characterized hosts amplifies NP production from silent gene clusters (Atanasov et al. 2021; Wang et al. 2019). Approaches involving CRISPR–Cas9 technology, repressor decoys, and promoter engineering activate silent clusters, revealing new compounds (Sidda et al. 2014). Metagenome-based searches and animal/plant genome mining unearth diverse NP classes (Atanasov et al. 2021). Microbial symbionts and plant microbiomes are significant sources, and targeted genetic engineering enables tailored NP production (Rath et al. 2011), improving yields and designing analogs with enhanced pharmacological properties (Atanasov et al. 2021). Biosynthetic engineering combined with chemical modification offers a viable strategy for NP drug candidate improvement, complementing total synthesis's complexity and cost (Preobrazhenskaya et al. 2010; Tevyashova et al. 2013).

Technological advancements are reinvigorating natural product (NP)-based drug development, particularly in combating infectious diseases and cancer. Strategies involving the human microbiome and engineering techniques optimize known NP classes and explore NP-induced immunogenic cell death for cancer treatment (Galon and Bruni 2019; Galluzzi et al. 2017; Diederich 2019). Challenges persist in botanical therapies due to variable NP compositions, but advancements in metabolomics and regulatory guidance aid in their development (Atanasov et al. 2015). Gut microbiota's impact on health opens new avenues, though still in the early stages (Scherlach and Hertweck 2018; Valencia et al. 2017). Computational tools like databases and machine learning (Dias et al. 2018) are pivotal, aiding genetic analysis and activity prediction (Schneider et al. 2016). Chemical modification and sustainable supply strategies are essential for clinical development, acknowledging the limitations of unaltered NPs (Atanasov et al. 2021).

4 Natural Products in Pharmaceutical Industries

Omics analysis generates vast datasets from genomics, transcriptomics, proteomics, metabolomics, and more, requiring computational tools for interpretation. Computational platforms like bioinformatics enable an understanding of pathophysiological effects, molecular interactions, and toxicological profiles of natural products. Machine learning aids in the virtual screening of compounds, while connectivity mapping associates disease-related gene signatures with potential drug responses (Korotcov et al. 2017; Oprea and Matter 2004). However, handling big data presents challenges; summarizing it into actionable insights is essential. Connectivity map (CMap) connects compounds, genes, and diseases, aiding in understanding molecular-level disease mechanisms (Brum et al. 2015; Roos 2001; Lamb et al. 2006). For clinical medicine to have relevant applications, integrating different omics data with clinical phenotypes is essential. Challenges persist, including tumor heterogeneity affecting cancer treatments and the need for more efficient models to evaluate candidate compounds' efficacy and safety during drug development (Kim et al. 2016; Chen et al. 2012). Several established methods used in the manufacturing of pharmaceuticals made from natural sources and were widely used in the pharmaceutical industry are discussed below.

4.1 Genomic Approach

The accurate identification of plant species for natural products is crucial for successful drug discovery, relying on genomic methods like DNA barcoding for precise species identification (Buriani et al. 2012). This approach surpasses traditional methods, enhancing accuracy in biodiversity inventories and herbal product authentication (Ghorbani et al. 2017; Thompson and Newmaster 2014; Cao et al. 2014). Bio-farming ensures consistency in pharmacological molecules after DNA barcoding authenticates the species (Pulice et al. 2016). Genomic techniques, including transcriptomic analysis, unveil molecular mechanisms behind natural product actions, aiding in understanding biological networks and drug targeting (Jones et al. 2015; Barbosa et al. 2015). Whole genome sequencing and transcriptomic analyses have revolutionized drug target exploration, identifying new targets for diseases like cancer (Vogelsang et al. 2012; Eckel-Passow et al. 2015). Extensive chemical databases and these technologies accelerate drug development, from design to clinical trials (Fishilevich et al. 2017).

4.2 Proteomic Approach

Proteomic platforms complement genomic and transcriptomic methods in ensuring natural product quality and understanding their mechanism of action (Bumpus et al. 2009; Martínez-Esteso et al. 2015). Techniques like mass spectrometry (MS) and two-dimensional electrophoresis (2D electrophoresis) offer insights into protein expression and function, aiding in profiling and ensuring product consistency (Kim et al. 2017). Proteomics helps in identifying and characterizing compounds, revealing multi-target effects, and can serve as a platform for molecular diagnostics (Lao et al. 2014). Identification of the target protein without altering the natural product has resulted in higher activity, employing methods like thermal proteome profiling (TPP) and cellular thermal shift assay (CETSA). Affinity chromatography is widely used to identify target proteins, but modifications can reduce the product's efficacy, emphasizing the need for nonmodifying approaches for successful target identification (Novick and Rubinstein 2012; Wang et al. 2015). Recent label-free methods measure natural product-target protein responses, allowing proteomic analysis identify several protein targets for a single natural substance (Schirle et al. 2012).

4.3 Identification of Proteins or Targets

Stability of proteins from rates of oxidation (SPROX) and drug affinity responsive target stability (DARTS) are methods that use natural products to discover target proteins and measure changes in protein stability (Mcfedries et al. 2013; West et al.

2010). CETSA, another technique, stabilizes target proteins with candidate compounds, useful for identifying anticancer therapeutic targets (Martinez Molina et al. 2013). For cellular-level investigation, a more sophisticated form called TPP combines mass spectrometry with high temperatures brought on by ligand binding (Huber et al. 2015). These methods reveal potential off-target proteins, aiding in understanding drug effects. RNA interference (RNAi) techniques like small interfering RNA (siRNA) and CRISPR–Cas9 offer tools for functional validation, assessing off-target effects, and understanding natural compound interactions at the genetic level, crucial for drug development studies, particularly in anticancer therapies (Ho et al. 2009; Bassik et al. 2013; Kasap et al. 2014; Thomford et al. 2018).

4.4 Metabolomic Approach

Metabolomics and metabonomics are powerful approaches to uncovering therapeutic compounds from natural products for global health. Metabolomics analyzes characteristic metabolites of natural products (Liu and Locasale 2017), while metabonomics investigates the more comprehensive metabolic reaction of live organisms to external stimuli (Nicholson and Lindon 2008). Recent approaches emphasize a systems biology perspective, revealing complete biological mechanisms and pharmacological effects. Utilizing technologies like UPLC–MS and NMR, metabolomic profiling has identified therapeutic compounds in various herbs. It's also been crucial in verifying species consistency, such as processed *Panax* species, employing NMR-based metabolomics and statistical analysis (Ekow Thomford et al. 2018; Park et al. 2014; Yan et al. 2015). The pharmacodynamic, pharmacokinetic, and toxicological features of natural products are revealed by this extensive method, providing important information for drug development (Thomford et al. 2018).

4.5 Automation of Drug Discovery

Even if there was first doubt about the impact of robots and the loss of jobs, the automation of drug research has revolutionized the industry. Pharmaceutical companies leverage high-throughput assays to accelerate drug discovery (Chapman 2003). Computer-aided design tools like ADAM and EVE streamline compound synthesis and hit finding (King et al. 2009; Sparkes et al. 2010). Advanced microfluidic systems handle compound screening and synthesis, expediting hypothesis testing within days and offering a clearer reflection of disease biology than conventional assays (Macconnell et al. 2017). Artificial intelligence (AI) and "organ-on-chip" technologies improve disease biology by mimicking and reducing human error, material consumption, testing time, and other aspects of drug research (Meanwell 2016; Gupta et al. 2018; Duch et al. 2007). Compound design is made easier by computational methods, which highlight characteristics like toxicity, excretion, metabolism, distribution, and absorption (ADMET) (Meanwell 2016). Automation paves the way for diverse compound libraries, aided by deep learning models predicting biological activity and ADMET properties. Compound manufacture is revolutionized by 3D printing, in-line purification, and microfluidics-based synthesis, while organoid-based systems imitate human biology to enable more precise drug identification (Godfrey et al. 2013; Thomford et al. 2018). Yet, challenges persist, such as fluidic surface instability and channel clogging in continuous flow microfluidic systems, demanding ongoing innovation for sustainable drug discovery automation (Thomford et al. 2018).

4.6 Computer-Aided Drug Design

Certainly, the development of synthetic compounds inspired by natural products is crucial for addressing global health issues, often uncovering new leads dismissed by stringent design criteria. Guidelines in drug design, like the "rule of three" and "rule of five," may limit innovation and show human bias, particularly in natural product applications (Lipinski 2000; Congreve et al. 2003; Van Molle et al. 2012; Ntie-Kang et al. 2014). Computational tools aid in simplifying complex natural compounds into smaller, potentially active molecules. Software like Scaffold Hunter (Wetzel et al. 2009; Lagunin et al. 2000) and PASS predict biological activities, successfully identifying anticancer agents and cytochrome P450 modulators (Rodrigues et al. 2016). While these methods show promise, simple derivatives may exhibit weaker activities than the original compounds. Computational software and approaches predict compound-target associations, employing similarity-based docking or feature comparison to known drugs (Reker et al. 2014). Additionally, computer-based toxicity assessments and in silico methods offer early detection, reducing time and costs in drug development (Dimasi et al. 2010; Chavan et al. 2014). Combination treatments are becoming more popular in drug design as a result of the recognition of tumor heterogeneity. They emphasize a combinatorial approach to successful treatment, especially when combined with targeted drugs like inhibitors of kinase (Baselga et al. 2012; Kawajiri et al. 2015; Swain et al. 2015; Thomford et al. 2018). Overall, computer-based drug design shows potential for improving lead compound optimization and advancing combinatorial treatment strategies.

5 Success Stories of Natural Products (NP)-Based Drugs

Natural products have a remarkable and diverse heritage in medication development. India's Ayurveda is one example of how natural goods, particularly those derived from plants, have been in use for a while in medical procedures all around the world (Kapoor 2018) and traditional Chinese medicine (Chang et al. 1986). Promising lead compounds for a range of ailments are still being produced by plants, such as curcuminoids for inflammation, psoralens for vitiligo, and *Rauwolfia* alkaloids for hypertension (Patwardhan and Vaidya 2010; Patwardhan 2000). Additionally, the resurgence of old molecules, like forskolin and coleonol, for conditions like obesity and atherosclerosis highlights the evolving applications of traditional knowledge (Das et al. 2009). Following World War II, exploration for resources of nature experienced a surge. Pharmaceutical firms concentrated on finding antibiotics, which resulted in the identification of important antibacterial drugs such as gentamicin and streptomycin from microbes employing fermentation-based technology (Davies 1999; Berdy 2005; Demain and Sanchez 2009), driven extensive research in natural product discovery.

Marine sources are found to produce natural products like discodermolide, bryostatins, dolostatins, and cephalostatins (Haefner 2003; Hay 1996); in addition, venoms and toxins from snakes and spiders also have contributed substantially to drug development (Chang 1999) (α-bungarotoxin), unveiling compounds with promising biological activities. The creation of drugs has been greatly influenced by natural commodities; in the pre-high-throughput screening period, more than 80% of drug compounds came from or were inspired by natural products (Ortholand and Ganesan 2004; Grabley and Thiericke 1998). Remarkably, since 1994, approximately 50% of approved pharmaceuticals are made of natural components, with the majority treating infectious and cancerous disorders (Newman et al. 2003). The years between 2001 and 2007 witnessed the introduction of numerous drugs derived from natural products, treating a spectrum of disorders from bacterial and fungal infections to genetic diseases and cancer (Butler 2008; Harvey 2008). These natural product-based drugs have not only achieved clinical success but also commercial viability, with plant-derived anticancer drugs like paclitaxel generating billions in sales of $1.6 billion in 2000 (Thayer 2003; Oberlies and Kroll 2004).

In the 1990s and early 2000s, despite earlier successes, the advent of alternative techniques like sensible pharmaceutical formulation and high-throughput screening in conjunction with chemistry in combinatorial form caused the pharmaceutical industry, especially major businesses, to move away from natural product discovery (Brahmachari 2012; Salvatella and Giralt 2003; Jain 2004). The belief in rapid and cost-effective drug development through synthetic compound libraries led to the downsizing or elimination of natural product operations (Rouhi 2003; Cordell 2002; Okuda 2002). Challenges like limited access, complex natural product chemistry, and concerns over intellectual property contributed to this shift. Despite efforts to speed up drug discovery, the slow nature of natural product research clashed with the industry's demand for rapid results (Lam 2007). While synthetic compounds offered promise, they couldn't fill the gap in certain areas like oncology and metabolic diseases where natural products historically excelled (Marris 2006). Drugs derived from natural sources, notably statins and antibiotics, continue to yield substantial revenues, underlining the enduring impact and value of natural products in pharmaceuticals (Brahmachari 2012).

Certainly, recent advancements in technology offer a promising revival for natural products as key sources of drug leads. Innovations like streamlined screening processes, improved sourcing, enhanced synthetic methodologies, and genomic insights into microbial and marine environments are reshaping natural product discovery (Lam 2007). Techniques like combinatorial biosynthesis and genome mining enable the creation of new derivatives with improved qualities (Bibb 2005; Cundliffe 2006). Natural products not only serve as drugs but also aid in understanding disease mechanisms, identifying new drug targets, and validating pathways, such as in anti-inflammatory and anticancer treatments. Microbial natural products, known for their therapeutic potential and favorable pharmacokinetic properties, have produced numerous drugs like antibiotics, cholesterol-lowering agents, anticancer drugs, and immunosuppressants directly from their natural forms without extensive modification, highlighting their immense value in drug discovery (Brahmachari 2012; Lam 2007). Overall, the diverse molecular reservoir of unstudied organisms presents a goldmine of opportunities for future discoveries and innovation in drug development.

6 Future Prospective

Novel methods are necessary to maintain the success of natural product medication development. Utilizing techniques like microbial genome mining and exploring untapped sources can expand the identification of novel, potent drug leads. With a large, uncharted microbial kingdom and expanding genetic databases, the prospects for finding naturally occurring compounds with biological activity are bright. Using organic synthesis, biotransformation, and combinatorial biosynthesis after lead identification yields a variety of analogs for enhanced characteristics or novel activities. Combinatorial chemistry may now legitimately use natural product leads as templates, and the time it takes to find new drugs can be shortened by combining these strategies with improvements in purification and screening methods (Lam 2007).

Moreover, omics approaches, like metabolomics and genomics, offer solutions to supply issues of minor yet potent natural products. Overexpression of biosynthetic gene clusters in cultivable organisms and transgenic plant strategies, employing overexpression and antisense suppression, increase compound production (Wang et al. 2011; Lange and Ahkami 2013). For instance, tobacco plants engineered with genes for artemisinin biosynthesis yielded double the compound compared to its natural source (Van Herpen et al. 2010). Microorganisms like *E. coli* and *S. cerevisiae* function as platforms for producing flavonoids, overcoming limitations in precursor availability for commercial-scale production (Malla et al. 2012). Metagenomic tools aid in identifying and exploiting these compounds' true producers, revolutionizing drug development from marine sources and expanding the scope of biosynthetic systems (Harvey et al. 2015).

In addition, emerging technologies like quantum computing, microfluidics, and AI aid in understanding plant compounds and their molecular effects, enhancing drug discovery. Innovations in extraction methods, including supercritical fluid extraction and molecular distillation, promise higher yields of bioactive compounds

(Özdemir and Hekim 2018; Özdemir 2015; Thomford et al. 2018). Synergistic effects of plant extracts, as opposed to single compounds, have shown therapeutic benefits, driving the need for comprehensive research into whole plant extracts and their molecular basis (Wen et al. 2005b).

7 Conclusion

The historical and ongoing significance of natural products in medicine spans thousands of years, deeply ingrained in traditional healthcare practices worldwide. These substances have contributed significantly to human health. Even in modern times, natural products remain crucial due to their diverse structures and unique biological activities, aiding in drug development and understanding cellular mechanisms. However, challenges like standardization and limited clinical trials pose hurdles to their wider acceptance. To overcome these obstacles, integrating advanced technologies like AI and "-omics" approaches becomes crucial, hinting at a future where these natural compounds will continue to shape drug discovery by providing holistic solutions and innovative therapeutic avenues.

References

Allen BR, Lakhanpaul M, Morris A, Lateo S, Davies T, Scott G, Cardno M, Ebelin ME, Burtin P, Stephenson TJ (2003) Systemic exposure, tolerability, and efficacy of pimecrolimus cream 1% in atopic dermatitis patients. Arch Dis Child 88:969–973

Appleton DR, Buss AD, Butler MS (2007) A simple method for high-throughput extract prefractionation for biological screening. Chimia 61:327–327

Arvigo R, Balick MJ (1993) Rainforest remedies: one hundred healing herbs of Belize. Lotus Press, New Delhi

Atanasov AG, Waltenberger B, Pferschy-Wenzig E-M, Linder T, Wawrosch C, Uhrin P, Temml V, Wang L, Schwaiger S, Heiss EH (2015) Discovery and resupply of pharmacologically active plant-derived natural products: a review. Biotechnol Adv 33:1582–1614

Atanasov AG, Zotchev SB, Dirsch VM, Supuran CT (2021) Natural products in drug discovery: advances and opportunities. Nat Rev Drug Discov 20:200–216

Bajorath J (2000) Chemoinformatics methods for systematic comparison of molecules from natural and synthetic sources and design of hybrid libraries. Mol Divers 5:305–313

Barbosa S, Carreira S, Bailey D, Abaitua F, O'hare, P. (2015) Phosphorylation and SCF-mediated degradation regulate CREB-H transcription of metabolic targets. Mol Biol Cell 26:2939–2954

Barnes J (2003) Pharmacovigilance of herbal medicines: a UK perspective. Drug Saf 26:829–851

Baselga J, Cortés J, Kim SB, Im SA, Hegg R, Im YH, Roman L, Pedrini JL, Pienkowski T, Knott A, Clark E, Benyunes MC, Ross G, Swain SM (2012) Pertuzumab plus trastuzumab plus docetaxel for metastatic breast cancer. N Engl J Med 366:109–119

Bassik MC, Kampmann M, Lebbink RJ, Wang S, Hein MY, Poser I, Weibezahn J, Horlbeck MA, Chen S, Mann M, Hyman AA, Leproust EM, Mcmanus MT, Weissman JS (2013) A systematic mammalian genetic interaction map reveals pathways underlying ricin susceptibility. Cell 152:909–922

Bax A, Aszalos A, Dinya Z, Sudo K (1986) Structure elucidation of the antibiotic desertomycin through the use of new two-dimensional NMR techniques. J Am Chem Soc 108:8056–8063

Berdy J (2005) Bioactive microbial metabolites. J Antibiot 58:1–26

Berlinck RG, Monteiro AF, Bertonha AF, Bernardi DI, Gubiani JR, Slivinski J, Michaliski LF, Tonon LA, Venancio VA, Freire VF (2019) Approaches for the isolation and identification of hydrophilic, light-sensitive, volatile and minor natural products. Nat Prod Rep 36:981–1004

Bibb MJ (2005) Regulation of secondary metabolism in streptomycetes. Curr Opin Microbiol 8:208–215

Boone B, Ongenae K, Van Geel N, Vernijns S, De Keyser S, Naeyaert JM (2007) Topical pimecrolimus in the treatment of vitiligo. Eur J Dermatol 17:55–61

Brahmachari G (2009) Mother nature: an inexhaustible source of drugs and lead molecules. In: Chemistry, biochemistry and pharmacology, 1st edn. Narosa Publishing House, New Delhi, pp 1–20

Brahmachari G (2012) Natural products in drug discovery: impacts and opportunities—an assessment. In: Bioactive natural products: opportunities and challenges in medicinal chemistry. World Scientific, Singapore

Brum AM, Van De Peppel J, Van Der Leije CS, Schreuders-Koedam M, Eijken M, Van Der Eerden BC, Van Leeuwen JP (2015) Connectivity map-based discovery of parbendazole reveals targetable human osteogenic pathway. Proc Natl Acad Sci U S A 112:12711–12716

Bumpus SB, Evans BS, Thomas PM, Ntai I, Kelleher NL (2009) A proteomics approach to discovering natural products and their biosynthetic pathways. Nat Biotechnol 27:951–956

Buriani A, Garcia-Bermejo ML, Bosisio E, Xu Q, Li H, Dong X, Simmonds MS, Carrara M, Tejedor N, Lucio-Cazana J, Hylands PJ (2012) Omic techniques in systems biology approaches to traditional Chinese medicine research: present and future. J Ethnopharmacol 140:535–544

Burke MD, Berger EM, Schreiber SL (2003) Generating diverse skeletons of small molecules combinatorially. Science 302:613–618

Butler MS (2008) Natural products to drugs: natural product-derived compounds in clinical trials. Nat Prod Rep 25:475–516

Cao M, Wang J, Yao L, Xie S, Du J, Zhao X (2014) Authentication of animal signatures in traditional Chinese medicine of Lingyang Qingfei wan using routine molecular diagnostic assays. Mol Biol Rep 41:2485–2491

Cardoso F, De Azambuja E, Lago LD (2008) Current perspectives of epothilones in breast cancer. Eur J Cancer 44:341–352

Carter GT (2011) Natural products and pharma 2011: strategic changes spur new opportunities. Nat Prod Rep 28:1783–1789

Carter NJ, Keam SJ (2007) Trabectedin: a review of its use in the management of soft tissue sarcoma and ovarian cancer. Drugs 67:2257–2276

Chan NW, Lewis DF, Rosner PJ, Kelly MA, Schriemer DC (2003) Frontal affinity chromatography–mass spectrometry assay technology for multiple stages of drug discovery: applications of a chromatographic biosensor. Anal Biochem 319:1–12

Chan K, Shaw D, Simmonds MS, Leon CJ, Xu Q, Lu A, Sutherland I, Ignatova S, Zhu Y-P, Verpoorte R (2012) Good practice in reviewing and publishing studies on herbal medicine, with special emphasis on traditional Chinese medicine and Chinese materia medica. J Ethnopharmacol 140:469–475

Chang C (1999) Looking back on the discovery of α-bungarotoxin. J Biomed Sci 6:368–375

Chang H-M, But PP, Yao S-C (1986) Pharmacology and applications of Chinese materia medica. World Scientific, Singapore

Chapman T (2003) Lab automation and robotics: automation on the move. Nature 421:665–666

Chavan S, Nicholls IA, Karlsson BC, Rosengren AM, Ballabio D, Consonni V, Todeschini R (2014) Towards global QSAR model building for acute toxicity: Munro database case study. Int J Mol Sci 15:18162–18174

Chen R, Mias GI, Li-Pook-Than J, Jiang L, Lam HY, Chen R, Miriami E, Karczewski KJ, Hariharan M, Dewey FE, Cheng Y, Clark MJ, Im H, Habegger L, Balasubramanian S, O'Huallachain M,

Dudley JT, Hillenmeyer S, Haraksingh R, Sharon D, Euskirchen G, Lacroute P, Bettinger K, Boyle AP, Kasowski M, Grubert F, Seki S, Garcia M, Whirl-Carrillo M, Gallardo M, Blasco MA, Greenberg PL, Snyder P, Klein TE, Altman RB, Butte AJ, Ashley EA, Gerstein M, Nadeau KC, Tang H, Snyder M (2012) Personal omics profiling reveals dynamic molecular and medical phenotypes. Cell 148:1293–1307

Cheng Q, He Y, Li G, Liu Y, Gao W, Huang L (2013) Effects of combined elicitors on tanshinone metabolic profiling and SmCPS expression in salvia miltiorrhiza hairy root cultures. Molecules 18:7473–7485

Choules MP, Klein LL, Lankin DC, Mcalpine JB, Cho S-H, Cheng J, Lee H, Suh J-W, Jaki BU, Franzblau SG (2018) Residual complexity does impact organic chemistry and drug discovery: the case of rufomyazine and rufomycin. J Org Chem 83:6664–6672

Clardy J, Walsh C (2004) Lessons from natural molecules. Nature 432:829–837

Congreve M, Carr R, Murray C, Jhoti H (2003) A 'rule of three' for fragment-based lead discovery? Drug Discov Today 8:876–877

Cordell GA (2002) Natural products in drug discovery–creating a new vision. Phytochem Rev 1:261–273

Cragg GM, Newman DJ (2001) Natural product drug discovery in the next millennium. Pharm Biol 39:8–17

Cundliffe E (2006) Antibiotic production by actinomycetes: the Janus faces of regulation. J Ind Microbiol Biotechnol 33:500

Das B, Tandon V, Lyndem LM, Gray AI, Ferro VA (2009) Phytochemicals from Flemingia vestita (Fabaceae) and Stephania glabra (Menispermeaceae) alter cGMP concentration in the cestode Raillietina echinobothrida. Comp Biochem Physiol Pt C Toxicol Pharmacol 149:397–403

Davies J (1999) Section 3. The advent of modern microbiology. In praise of antibiotics. ASM News 65:304–310

Demain AL, Sanchez S (2009) Microbial drug discovery: 80 years of progress. J Antibiot 62:5–16

Dev S (1999) Ancient-modern concordance in ayurvedic plants: some examples. Environ Health Perspect 107:783–789

Dhawan B, Gadepalli R, Kapil A (2009) In vitro activity of daptomycin against Staphylococcus aureus and vancomycin-resistant enterococcus faecium isolates associated with skin and soft tissue infections: first results from India. Diagn Microbiol Infect Dis 65:196–198

Dias T, Gaudêncio SP, Pereira F (2018) A computer-driven approach to discover natural product leads for methicillin-resistant *Staphylococcus aureus* infection therapy. Mar Drugs 17:16

Dickson M, Gagnon JP (2004) Key factors in the rising cost of new drug discovery and development. Nat Rev Drug Discov 3:417–429

Diederich M (2019) Natural compound inducers of immunogenic cell death. Arch Pharm Res 42:629–645

Dimasi JA, Feldman L, Seckler A, Wilson A (2010) Trends in risks associated with new drug development: success rates for investigational drugs. Clin Pharmacol Ther 87:272–277

Doak BC, Over B, Giordanetto F, Kihlberg J (2014) Oral druggable space beyond the rule of 5: insights from drugs and clinical candidates. Chem Biol 21:1115–1142

Duch W, Swaminathan K, Meller J (2007) Artificial intelligence approaches for rational drug design and discovery. Curr Pharm Des 13:1497–1508

Ebada SS, Edrada RA, Lin W, Proksch P (2008) Methods for isolation, purification and structural elucidation of bioactive secondary metabolites from marine invertebrates. Nat Protoc 3:1820–1831

Eckel-Passow JE, Lachance DH, Molinaro AM, Walsh KM, Decker PA, Sicotte H, Pekmezci M, Rice T, Kosel ML, Smirnov IV, Sarkar G, Caron AA, Kollmeyer TM, Praska CE, Chada AR, Halder C, Hansen HM, Mccoy LS, Bracci PM, Marshall R, Zheng S, Reis GF, Pico AR, O'Neill BP, Buckner JC, Giannini C, Huse JT, Perry A, Tihan T, Berger MS, Chang SM, Prados MD, Wiemels J, Wiencke JK, Wrensch MR, Jenkins RB (2015) Glioma groups based on 1p/19q, IDH, and TERT promoter mutations in tumors. N Engl J Med 372:2499–2508

Ekow Thomford N, Dzobo K, Adu F, Chirikure S, Wonkam A, Dandara C (2018) Bush mint (Hyptis suaveolens) and spreading hogweed (Boerhavia diffusa) medicinal plant extracts differentially affect activities of CYP1A2, CYP2D6 and CYP3A4 enzymes. J Ethnopharmacol 211:58–69

Eugster PJ, Guillarme D, Rudaz S, Veuthey J-L, Carrupt P-A, Wolfender J-L (2011) Ultra high pressure liquid chromatography for crude plant extract profiling. J AOAC Int 94:51–70

Exarchou V, Godejohann M, Van Beek TA, Gerothanassis IP, Vervoort J (2003) LC-UV-solid-phase extraction-NMR-MS combined with a cryogenic flow probe and its application to the identification of compounds present in Greek oregano. Anal Chem 75:6288–6294

Farnsworth NR, Akerele O, Bingel AS, Soejarto DD, Guo Z (1985) Medicinal plants in therapy. Bull World Health Organ 63:965

Feher M, Schmidt JM (2003) Property distributions: differences between drugs, natural products, and molecules from combinatorial chemistry. J Chem Inf Comput Sci 43:218–227

Fishilevich S, Nudel R, Rappaport N, Hadar R, Plaschkes I, Iny Stein T, Rosen N, Kohn A, Twik M, Safran M, Lancet D, Cohen D (2017) GeneHancer: genome-wide integration of enhancers and target genes in GeneCards. Database (Oxford) 2017:bax028

Fowler VG Jr, Boucher HW, Corey GR, Abrutyn E, Karchmer AW, Rupp ME, Levine DP, Chambers HF, Tally FP, Vigliani GA, Cabell CH, Link AS, Demeyer I, Filler SG, Zervos M, Cook P, Parsonnet J, Bernstein JM, Price CS, Forrest GN, Fätkenheuer G, Gareca M, Rehm SJ, Brodt HR, Tice A, Cosgrove SE (2006) Daptomycin versus standard therapy for bacteremia and endocarditis caused by *Staphylococcus aureus*. N Engl J Med 355:653–665

Frantz S, Smith A (2003) New drug approvals for 2002. Nat Rev Drug Discov 2:95–96

Freeman R, Morris GA (1979) Two-dimensional Fourier transformation in NMR. Bull Magn Res 1:1–26

Fuchs T (2007) Case study: Cefditoren Pivoxil: an oral prodrug of Cefditoren. In: Stella VJ, Borchardt RT, Hageman MJ, Oliyai R, Maag H, Tilley JW (eds) Prodrugs: challenges and rewards. Part 1. Springer, New York, NY

Galluzzi L, Buqué A, Kepp O, Zitvogel L, Kroemer G (2017) Immunogenic cell death in cancer and infectious disease. Nat Rev Immunol 17:97–111

Galon J, Bruni D (2019) Approaches to treat immune hot, altered and cold tumours with combination immunotherapies. Nat Rev Drug Discov 18:197–218

Ganesh T, Yang C, Norris A, Glass T, Bane S, Ravindra R, Banerjee A, Metaferia B, Thomas SL, Giannakakou P (2007) Evaluation of the tubulin-bound paclitaxel conformation: synthesis, biology, and SAR studies of C-4 to C-3′ bridged paclitaxel analogues. J Med Chem 50:713–725

Gaudillière B, Bernardelli P, Berna PP (2001) Chapter 28. To market, to market—2000. Annu Rep Med Chem 36:293–318

Ghorbani A, Saeedi Y, De Boer HJ (2017) Unidentifiable by morphology: DNA barcoding of plant material in local markets in Iran. PLoS One 12:e0175722

Godfrey AG, Masquelin T, Hemmerle H (2013) A remote-controlled adaptive medchem lab: an innovative approach to enable drug discovery in the 21st century. Drug Discov Today 18:795–802

Grabley S, Thiericke R (1998) Drug discovery from nature. Springer Science & Business Media, Berlin

Graul AI (2002) The year's new drugs. Drug News Perspect 15:29–43

Grkovic T, Pouwer RH, Vial ML, Gambini L, Noël A, Hooper JN, Wood SA, Mellick GD, Quinn RJ (2014) NMR fingerprints of the drug-like natural-product space identify Iotrochotazine A: a chemical probe to study Parkinson's disease. Angew Chem Int Ed 53:6070–6074

Grosso F, Jones RL, Demetri GD, Judson IR, Blay JY, Le Cesne A, Sanfilippo R, Casieri P, Collini P, Dileo P, Spreafico C, Stacchiotti S, Tamborini E, Tercero JC, Jimeno J, Di'ncalci M, Gronchi A, Fletcher JA, Pilotti S, Casali PG (2007) Efficacy of trabectedin (ecteinascidin-743) in advanced pretreated myxoid liposarcomas: a retrospective study. Lancet Oncol 8:595–602

Guijas C, Montenegro-Burke JR, Domingo-Almenara X, Palermo A, Warth B, Hermann G, Koellensperger G, Huan T, Uritboonthai W, Aisporna AE (2018) METLIN: a technology platform for identifying knowns and unknowns. Anal Chem 90:3156–3164

Gullo VP, Mcalpine J, Lam KS, Baker D, Petersen F (2006) Drug discovery from natural products. J Ind Microbiol Biotechnol 33:523–531

Gupta A, Müller AT, Huisman BJH, Fuchs JA, Schneider P (2018) Generative recurrent networks for de novo drug design. Mol Inform 37:1700111

Haefner B (2003) Drugs from the deep: marine natural products as drug candidates. Drug Discov Today 8:536–544

Hanson FR, Eble TE (1949) An antiphage agent isolated from Aspergillus Sp. J Bacteriol 58:527–529

Harvey AL (2008) Natural products in drug discovery. Drug Discov Today 13:894–901

Harvey AL, Edrada-Ebel R, Quinn RJ (2015) The re-emergence of natural products for drug discovery in the genomics era. Nat Rev Drug Discov 14:111–129

Hay ME (1996) Marine chemical ecology: what's known and what's next? J Exp Mar Biol Ecol 200:103–134

Heinrich M, Lee Teoh H (2004) Galanthamine from snowdrop—the development of a modern drug against Alzheimer's disease from local Caucasian knowledge. J Ethnopharmacol 92:147–162

Henkel T, Brunne RM, Müller H, Reichel F (1999) Statistical investigation into the structural complementarity of natural products and synthetic compounds. Angew Chem Int Ed 38:643–647

Hinterding K, Alonso-Díaz D, Waldmann H (1998) Organic synthesis and biological signal transduction. Angew Chem Int Ed 37:688–749

Ho CH, Magtanong L, Barker SL, Gresham D, Nishimura S, Natarajan P, Koh JLY, Porter J, Gray CA, Andersen RJ, Giaever G, Nislow C, Andrews B, Botstein D, Graham TR, Yoshida M, Boone C (2009) A molecular barcoded yeast ORF library enables mode-of-action analysis of bioactive compounds. Nat Biotechnol 27:369–377

Hoffmann T, Krug D, Bozkurt N, Duddela S, Jansen R, Garcia R, Gerth K, Steinmetz H, Müller R (2018) Correlating chemical diversity with taxonomic distance for discovery of natural products in myxobacteria. Nat Commun 9:803

Horton DA, Bourne GT, Smythe ML (2003) The combinatorial synthesis of bicyclic privileged structures or privileged substructures. Chem Rev 103:893–930

Hsu E (2006) The history of qing Hao in the Chinese materia medica. Trans R Soc Trop Med Hyg 100:505–508

Huber KV, Olek KM, Müller AC, Tan CS, Bennett KL, Colinge J, Superti-Furga G (2015) Proteome-wide drug and metabolite interaction mapping by thermal-stability profiling. Nat Methods 12:1055–1057

Hufsky F, Scheubert K, Böcker S (2014) New kids on the block: novel informatics methods for natural product discovery. Nat Prod Rep 31:807–817

Hung DT, Jamison TF, Schreiber SL (1996) Understanding and controlling the cell cycle with natural products. Chem Biol 3:623–639

Jachak SM, Saklani A (2007) Challenges and opportunities in drug discovery from plants. Curr Sci 92:1251–1257

Jain AN (2004) Virtual screening in lead discovery and optimization. Curr Opin Drug Discov Devel 7:396–403

Jones MJ, Goodman SJ, Kobor MS (2015) DNA methylation and healthy human aging. Aging Cell 14:924–932

Jones CG, Martynowycz MW, Hattne J, Fulton TJ, Stoltz BM, Rodriguez JA, Nelson HM, Gonen T (2018) The CryoEM method MicroED as a powerful tool for small molecule structure determination. ACS Cent Sci 4:1587–1592

Kapoor L (2018) CRC handbook of Ayurvedic medicinal plants. CRC, Boca Raton, FL

Kasap C, Elemento O, Kapoor TM (2014) DrugTargetSeqR: a genomics- and CRISPR-Cas9-based method to analyze drug targets. Nat Chem Biol 10:626–628

Kawajiri H, Takashima T, Kashiwagi S, Noda S, Onoda N, Hirakawa K (2015) Pertuzumab in combination with trastuzumab and docetaxel for HER2-positive metastatic breast cancer. Expert Rev Anticancer Ther 15:17–26

Khosla C, Keasling JD (2003) Metabolic engineering for drug discovery and development. Nat Rev Drug Discov 2:1019–1025

Kim E, Moore BS, Yoon YJ (2015) Reinvigorating natural product combinatorial biosynthesis with synthetic biology. Nat Chem Biol 11:649–659

Kim RS, Goossens N, Hoshida Y (2016) Use of big data in drug development for precision medicine. Expert Rev Precis Med Drug Dev 1:245–253

Kim SW, Lee SH, Min CW, Jo IH, Bang KH, Hyun D-Y, Agrawal GK, Rakwal R, Zargar SM, Gupta R, Kim ST (2017) Ginseng (Panax sp.) proteomics: an update. Appl Biol Chem 60:311–320

King RD, Rowland J, Oliver SG, Young M, Aubrey W, Byrne E, Liakata M, Markham M, Pir P, Soldatova LN, Sparkes A, Whelan KE, Clare A (2009) The automation of science. Science 324:85–89

Knotkova H, Pappagallo M, Szallasi A (2008) Capsaicin (TRPV1 agonist) therapy for pain relief: farewell or revival? Clin J Pain 24:142–154

Koehn FE, Carter GT (2005) The evolving role of natural products in drug discovery. Nat Rev Drug Discov 4:206–220

Korotcov A, Tkachenko V, Russo DP, Ekins S (2017) Comparison of deep learning with multiple machine learning methods and metrics using diverse drug discovery data sets. Mol Pharm 14:4462–4475

Lagunin A, Stepanchikova A, Filimonov D, Poroikov V (2000) PASS: prediction of activity spectra for biologically active substances. Bioinformatics 16:747–748

Lam KS (2007) New aspects of natural products in drug discovery. Trends Microbiol 15:279–289

Lamb J, Crawford ED, Peck D, Modell JW, Blat IC, Wrobel MJ, Lerner J, Brunet JP, Subramanian A, Ross KN, Reich M, Hieronymus H, Wei G, Armstrong SA, Haggarty SJ, Clemons PA, Wei R, Carr SA, Lander ES, Golub TR (2006) The connectivity map: using gene-expression signatures to connect small molecules, genes, and disease. Science 313:1929–1935

Lange BM, Ahkami A (2013) Metabolic engineering of plant monoterpenes, sesquiterpenes and diterpenes—current status and future opportunities. Plant Biotechnol J 11:169–196

Lao Y, Wang X, Xu N, Zhang H, Xu H (2014) Application of proteomics to determine the mechanism of action of traditional Chinese medicine remedies. J Ethnopharmacol 155:1–8

Lee M-L, Schneider G (2001) Scaffold architecture and pharmacophoric properties of natural products and trade drugs: application in the design of natural product-based combinatorial libraries. J Comb Chem 3:284–289

Leonti M, Verpoorte R (2017) Traditional Mediterranean and European herbal medicines. J Ethnopharmacol 199:161–167

Li F-S, Weng J-K (2017) Demystifying traditional herbal medicine with modern approach. Nat Plants 3:1–7

Lipinski CA (2000) Drug-like properties and the causes of poor solubility and poor permeability. J Pharmacol Toxicol Methods 44:235–249

Liu X, Locasale JW (2017) Metabolomics: a primer. Trends Biochem Sci 42:274–284

Livermore DM, Mushtaq S, Warner M (2005) Selectivity of ertapenem for Pseudomonas aeruginosa mutants cross-resistant to other carbapenems. J Antimicrob Chemother 55:306–311

Lommen A, Godejohann M, Venema D, Hollman P, Spraul M (2000) Application of directly coupled HPLC– NMR– MS to the identification and confirmation of quercetin glycosides and phloretin glycosides in apple peel. Anal Chem 72:1793–1797

Macconnell AB, Price AK, Paegel BM (2017) An integrated microfluidic processor for DNA-encoded combinatorial library functional screening. ACS Comb Sci 19:181–192

Malhotra R, Singh L, Eng J, Raufman JP (1992) Exendin-4, a new peptide from Heloderma suspectum venom, potentiates cholecystokinin-induced amylase release from rat pancreatic acini. Regul Pept 41:149–156

Malla S, Koffas MA, Kazlauskas RJ, Kim B-G (2012) Production of 7-O-methyl aromadendrin, a medicinally valuable flavonoid, in Escherichia coli. Appl Environ Microbiol 78:684–694

Marris E (2006) Marine natural products: drugs from the deep. Nature 443:904–906

Martinez Molina D, Jafari R, Ignatushchenko M, Seki T, Larsson EA, Dan C, Sreekumar L, Cao Y, Nordlund P (2013) Monitoring drug target engagement in cells and tissues using the cellular thermal shift assay. Science 341:84–87

Martínez-Esteso MJ, Martínez-Márquez A, Sellés-Marchart S, Morante-Carriel JA, Bru-Martínez R (2015) The role of proteomics in progressing insights into plant secondary metabolism. Front Plant Sci 6:504

Mccormack PL, Perry CM (2005) Caspofungin: a review of its use in the treatment of fungal infections. Drugs 65:2049–2068

Mcfedries A, Schwaid A, Saghatelian A (2013) Methods for the elucidation of protein-small molecule interactions. Chem Biol 20:667–673

Meanwell NA (2016) Improving drug design: an update on recent applications of efficiency metrics, strategies for replacing problematic elements, and compounds in nontraditional drug space. Chem Res Toxicol 29:564–616

Medema MH, Fischbach MA (2015) Computational approaches to natural product discovery. Nat Chem Biol 11:639–648

Müller WE, Schröder HC, Wiens M, Perovic-Ottstadt S, Batel R, Müller IM (2004) Traditional and modern biomedical prospecting: part II—the benefits. Evid Based Complement Alternat Med 1:133–144

Nakanishi K (1999) A brief history of natural products chemistry. In: Comprehensive natural products chemistry, vol 1, p 8

Newman DJ, Cragg GM (2012) Natural products as sources of new drugs over the 30 years from 1981 to 2010. J Nat Prod 75:311–335

Newman DJ, Cragg GM, Snader KM (2000) The influence of natural products upon drug discovery. Nat Prod Rep 17:215–234

Newman DJ, Cragg GM, Snader KM (2003) Natural products as sources of new drugs over the period 1981–2002. J Nat Prod 66:1022–1037

Nghiem P, Pearson G, Langley RG (2002) Tacrolimus and pimecrolimus: from clever prokaryotes to inhibiting calcineurin and treating atopic dermatitis. J Am Acad Dermatol 46:228–241

Nicholson JK, Lindon JC (2008) Systems biology: metabonomics. Nature 455:1054–1056

Nielsen KF, Smedsgaard J (2003) Fungal metabolite screening: database of 474 mycotoxins and fungal metabolites for dereplication by standardised liquid chromatography–UV–mass spectrometry methodology. J Chromatogr A 1002:111–136

Nor El-Din G (1956) The use of fumagillin in the treatment of amebiasis. Am J Trop Med Hyg 5:67–71

Novick D, Rubinstein M (2012) Ligand affinity chromatography, an indispensable method for the purification of soluble cytokine receptors and binding proteins. Methods Mol Biol 820:195–214

Ntie-Kang F, Lifongo LL, Judson PN, Sippl W, Efange SM (2014) How "drug-like" are naturally occurring anti-cancer compounds? J Mol Model 20:2069

Oberlies NH, Kroll DJ (2004) Camptothecin and taxol: historic achievements in natural products research. J Nat Prod 67:129–135

Ohta D, Kanaya S, Suzuki H (2010) Application of Fourier-transform ion cyclotron resonance mass spectrometry to metabolic profiling and metabolite identification. Curr Opin Biotechnol 21:35–44

Okuda T (2002) Trends in natural product-based drug discovery, and roles of biotech ventures and biological resource centers. In: World Federation for Culture Collections Newsletter, vol 35, pp 21–26

Oprea TI, Matter H (2004) Integrating virtual screening in lead discovery. Curr Opin Chem Biol 8:349–358

Ortholand J-Y, Ganesan A (2004) Natural products and combinatorial chemistry: back to the future. Curr Opin Chem Biol 8:271–280

Özdemir V (2015) Omics 2.0: an accelerator for global science, systems medicine and responsible innovation. OMICS 19:579–580

Özdemir V, Hekim N (2018) Birth of industry 5.0: Making sense of big data with artificial intelligence,"the internet of things" and next-generation technology policy. OMICS 22:65–76
Parakh A, Krishnamurthy S, Bhattacharya M (2009) Ertapenem. Kathmandu Univ Med J 7:454–460
Park HW, In G, Kim JH, Cho BG, Han GH, Chang IM (2014) Metabolomic approach for discrimination of processed ginseng genus (Panax ginseng and Panax quinquefolius) using UPLC-QTOF MS. J Ginseng Res 38:59–65
Patwardhan B (2000) Ayurveda: the designer medicine. Indian Drugs 37:213–227
Patwardhan B, Vaidya AD (2010) Natural products drug discovery: accelerating the clinical candidate development using reverse pharmacology approaches. Indian J Exp Biol 48:220
Patwardhan B, Vaidya AD, Chorghade M (2004) Ayurveda and natural products drug discovery. Curr Sci 86:789–799
Pelkonen O, Pasanen M, Lindon JC, Chan K, Zhao L, Deal G, Xu Q, Fan T-P (2012) Omics and its potential impact on R&D and regulation of complex herbal products. J Ethnopharmacol 140:587–593
Piekarz RL, Frye R, Turner M, Wright JJ, Allen SL, Kirschbaum MH, Zain J, Prince HM, Leonard JP, Geskin LJ, Reeder C, Joske D, Figg WD, Gardner ER, Steinberg SM, Jaffe ES, Stetler-Stevenson M, Lade S, Fojo AT, Bates SE (2009) Phase II multi-institutional trial of the histone deacetylase inhibitor romidepsin as monotherapy for patients with cutaneous T-cell lymphoma. J Clin Oncol 27:5410–5417
Preobrazhenskaya MN, Olsufyeva EN, Tevyashova AN, Printsevskaya SS, Solovieva SE, Reznikova MI, Trenin AS, Galatenko OA, Treshalin ID, Pereverzeva ER (2010) Synthesis and study of the antifungal activity of new mono-and disubstituted derivatives of a genetically engineered polyene antibiotic 28, 29-didehydronystatin A1 (S44HP). J Antibiot 63:55–64
Pronzato P (2008) New therapeutic options for chemotherapy-resistant metastatic breast cancer: the epothilones. Drugs 68:139–146
Pulice G, Pelaz S, Matías-Hernández L (2016) Molecular farming in Artemisia annua, a promising approach to improve anti-malarial drug production. Front Plant Sci 7:329
Raskind MA (2003) Update on Alzheimer drugs (galantamine). Neurologist 9:235–240
Rath CM, Janto B, Earl J, Ahmed A, Hu FZ, Hiller L, Dahlgren M, Kreft R, Yu F, Wolff JJ (2011) Meta-omic characterization of the marine invertebrate microbial consortium that produces the chemotherapeutic natural product ET-743. ACS Chem Biol 6:1244–1256
Reker D, Rodrigues T, Schneider P, Schneider G (2014) Identifying the macromolecular targets of de novo-designed chemical entities through self-organizing map consensus. Proc Natl Acad Sci USA 111:4067–4072
Remadevi R, Szallisi A (2008) Adlea (ALGRX-4975), an injectable capsaicin (TRPV1 receptor agonist) formulation for longlasting pain relief. IDrugs 11:120–132
Rodrigues T, Reker D, Schneider P, Schneider G (2016) Counting on natural products for drug design. Nat Chem 8:531–541
Roos DS (2001) Computational biology. Bioinformatics—trying to swim in a sea of data. Science 291:1260–1261
Rouhi AM (2003) Betting on natural products for cures. Chem Eng News 81:93–93
Russo EB, Guy GW, Robson PJ (2007) Cannabis, pain, and sleep: lessons from therapeutic clinical trials of Sativex, a cannabis-based medicine. Chem Biodivers 4:1729–1743
Sader HS, Gales AC (2001) Emerging strategies in infectious diseases: new carbapenem and trinem antibacterial agents. Drugs 61:553–564
Salvatella X, Giralt E (2003) NMR-based methods and strategies for drug discovery. Chem Soc Rev 32:365–372
Scherlach K, Hertweck C (2018) Mediators of mutualistic microbe–microbe interactions. Nat Prod Rep 35:303–308
Schirle M, Bantscheff M, Kuster B (2012) Mass spectrometry-based proteomics in preclinical drug discovery. Chem Biol 19:72–84

Schneider G, Reker D, Chen T, Hauenstein K, Schneider P, Altmann KH (2016) Deorphaning the macromolecular targets of the natural anticancer compound doliculide. Angew Chem Int Ed 55:12408–12411

Schultes RE, Raffauf RF (1990) The healing forest: medicinal and toxic plants of the Northwest Amazonia. Dioscorides Press, Portland, OR

Sidda JD, Song L, Poon V, Al-Bassam M, Lazos O, Buttner MJ, Challis GL, Corre C (2014) Discovery of a family of γ-aminobutyrate ureas via rational derepression of a silent bacterial gene cluster. Chem Sci 5:86–89

Simpson DM, Estanislao L, Brown SJ, Sampson J (2008) An open-label pilot study of high-concentration capsaicin patch in painful HIV neuropathy. J Pain Symptom Manag 35:299–306

Sparkes A, Aubrey W, Byrne E, Clare A, Khan MN, Liakata M, Markham M, Rowland J, Soldatova LN, Whelan KE, Young M, King RD (2010) Towards robot scientists for autonomous scientific discovery. Autom Exp 2:1

Stavrianidi A (2020) A classification of liquid chromatography mass spectrometry techniques for evaluation of chemical composition and quality control of traditional medicines. J Chromatogr A 1609:460501

Straus SK, Hancock RE (2006) Mode of action of the new antibiotic for gram-positive pathogens daptomycin: comparison with cationic antimicrobial peptides and lipopeptides. Biochim Biophys Acta 1758:1215–1223

Strege MA (1999) High-performance liquid chromatographic–electrospray ionization mass spectrometric analyses for the integration of natural products with modern high-throughput screening. J Chromatogr B Biomed Sci Appl 725:67–78

Swain SM, Baselga J, Kim SB, Ro J, Semiglazov V, Campone M, Ciruelos E, Ferrero JM, Schneeweiss A, Heeson S, Clark E, Ross G, Benyunes MC, Cortés J (2015) Pertuzumab, trastuzumab, and docetaxel in HER2-positive metastatic breast cancer. N Engl J Med 372:724–734

Tevyashova AN, Olsufyeva EN, Solovieva SE, Printsevskaya SS, Reznikova MI, Trenin AS, Galatenko OA, Treshalin ID, Pereverzeva ER, Mirchink EP (2013) Structure-antifungal activity relationships of polyene antibiotics of the amphotericin B group. Antimicrob Agents Chemother 57:3815–3822

Thayer A (2003) Bristol-Myers to settle suits. Chem Eng News 81:6–6

Thomford NE, Senthebane DA, Rowe A, Munro D, Seele P, Maroyi A, Dzobo K (2018) Natural products for drug discovery in the 21st century: innovations for novel drug discovery. Int J Mol Sci 19:1578

Thompson KA, Newmaster SG (2014) Retracted article: molecular taxonomic tools provide more accurate estimates of species richness at less cost than traditional morphology-based taxonomic practices in a vegetation survey. Biodivers Conserv 23:1411–1424

Valencia PM, Richard M, Brock J, Boglioli E (2017) The human microbiome: opportunity or hype. Nat Rev Drug Discov 16:823–824

Van Herpen TW, Cankar K, Nogueira M, Bosch D, Bouwmeester HJ, Beekwilder J (2010) Nicotiana benthamiana as a production platform for artemisinin precursors. PLoS One 5:e14222

Van Molle I, Thomann A, Buckley DL, So EC, Lang S, Crews CM, Ciulli A (2012) Dissecting fragment-based lead discovery at the von Hippel-Lindau protein:hypoxia inducible factor 1α protein-protein interface. Chem Biol 19:1300–1312

Vogelsang M, Wang Y, Veber N, Mwapagha LM, Parker MI (2012) The cumulative effects of polymorphisms in the DNA mismatch repair genes and tobacco smoking in oesophageal cancer risk. PLoS One 7:e36962

Vuorela P, Leinonen M, Saikku P, Tammela P, Rauha J-P, Wennberg T, Vuorela H (2004) Natural products in the process of finding new drug candidates. Curr Med Chem 11:1375–1389

Wade DT, Robson P, House H, Makela P, Aram J (2003) A preliminary controlled study to determine whether whole-plant cannabis extracts can improve intractable neurogenic symptoms. Clin Rehabil 17:21–29

Wagenaar MM (2008) Pre-fractionated microbial samples–the second generation natural products library at Wyeth. Molecules 13:1406–1426

Wang Y, Chen S, Yu O (2011) Metabolic engineering of flavonoids in plants and microorganisms. Appl Microbiol Biotechnol 91:949–956

Wang HZ, Chu ZZ, Chen CC, Cao AC, Tong X, Ouyang CB, Yuan QH, Wang MN, Wu ZK, Wang HH, Wang SB (2015) Recombinant passenger proteins can be conveniently purified by one-step affinity chromatography. PLoS One 10:e0143598

Wang B, Guo F, Dong S-H, Zhao H (2019) Activation of silent biosynthetic gene clusters using transcription factor decoys. Nat Chem Biol 15:111–114

Wellington K, Curran MP (2005) Spotlight on cefditoren pivoxil in bacterial infections. Treat Respir Med 4:149–152

Wen M-C, Wei C-H, Hu Z-Q, Srivastava K, Ko J, Xi S-T, Mu D-Z, Du J-B, Li G-H, Wallenstein S (2005a) Efficacy and tolerability of antiasthma herbal medicine intervention in adult patients with moderate-severe allergic asthma. J Allergy Clin Immunol 116:517–524

Wen MC, Wei CH, Hu ZQ, Srivastava K, Ko J, Xi ST, Mu DZ, Du JB, Li GH, Wallenstein S, Sampson H, Kattan M, Li XM (2005b) Efficacy and tolerability of anti-asthma herbal medicine intervention in adult patients with moderate-severe allergic asthma. J Allergy Clin Immunol 116:517–524

West GM, Tucker CL, Xu T, Park SK, Han X, Yates JR 3rd, Fitzgerald MC (2010) Quantitative proteomics approach for identifying protein-drug interactions in complex mixtures using protein stability measurements. Proc Natl Acad Sci USA 107:9078–9082

Wetzel S, Klein K, Renner S, Rauh D, Oprea TI, Mutzel P, Waldmann H (2009) Interactive exploration of chemical space with scaffold hunter. Nat Chem Biol 5:581–583

Wolfson W (2007) Leapin' lizards: amylin targets diabetes and obesity via incretins. Chem Biol 14:235–236

Woo S, Gardner ER, Chen X, Ockers SB, Baum CE, Sissung TM, Price DK, Frye R, Piekarz RL, Bates SE, Figg WD (2009) Population pharmacokinetics of romidepsin in patients with cutaneous T-cell lymphoma and relapsed peripheral T-cell lymphoma. Clin Cancer Res 15:1496–1503

Yan T, Fu Q, Wang J, Ma S (2015) UPLC-MS/MS determination of ephedrine, methylephedrine, amygdalin and glycyrrhizic acid in beagle plasma and its application to a pharmacokinetic study after oral administration of Ma Huang Tang. Drug Test Anal 7:158–163

Yilmaz A, Nyberg NT, Jaroszewski JW (2011) Metabolic profiling based on two-dimensional J-resolved 1H NMR data and parallel factor analysis. Anal Chem 83:8278–8285

Youns M, Hoheisel JD, Efferth T (2010) Toxicogenomics for the prediction of toxicity related to herbs from traditional Chinese medicine. Planta Med 76:2019–2025

Yu J, Nag SA, Zhang R (2013) Advances in translational pharmacological investigations in identifying and validating molecular targets of natural product anticancer agents. Curr Cancer Drug Targets 13:596–609

Yuliana ND, Khatib A, Choi YH, Verpoorte R (2011) Metabolomics for bioactivity assessment of natural products. Phytother Res 25:157–169

Zhu L, Chen L, Luo H, Xu X (2003) Frontal affinity chromatography combined on-line with mass spectrometry: a tool for the binding study of different epidermal growth factor receptor inhibitors. Anal Chem 75:6388–6393

Ziemert N, Alanjary M, Weber T (2016) The evolution of genome mining in microbes—a review. Nat Prod Rep 33:988–1005

Ziffer H, Highet RJ, Klayman DL (1997) Artemisinin: an endoperoxidic antimalarial from Artemisia annua L. Fortschr Chem Org Naturst 72:121–214

Traditional Medicinal Plants and Their Ethnomedicinal Values

Indrani Gogoi, Mregangka Dowara, and Pankaj Chetia

Contents

Abstract The use of traditional medicinal herbs relies on the indigenous knowledge, skills, customs, and beliefs of different cultures to uphold general wellness and prevent and treat ailments. Despite the development of modern medical facilities and the expansion of the healthcare system, a sizeable portion of the population from developing countries seeks out traditional medicines for various physical and mental health problems. Over the last few decades, a growing belief in traditional medicine, which has fewer side effects than allopathic medicine, has led to a resur-

I. Gogoi · M. Dowara · P. Chetia (✉)
Molecular Plant Taxonomy and Bioinformatics Research Laboratory, Department of Life Sciences, Dibrugarh University, Dibrugarh, Assam, India
e-mail: chetiapankaj@dibru.ac.in

© The Author(s), under exclusive license to Springer Nature Singapore Pte Ltd. 2024

A. Das Talukdar et al. (eds.), *Traditional Resources and Tools for Modern Drug Discovery*, Interdisciplinary Biotechnological Advances,
https://doi.org/10.1007/978-981-97-4600-2_14

gence of interest in medicinal plants. The WHO (World Health Organization) has also acknowledged the values and utility of medicinal plants from an ethnomedical perspective. The phytoconstituents in plants make them viable candidates for commercial manufacturing, and thus the majority of medicines prescribed globally are derived from plants. Surprisingly, for affordability and availability, herbs are still frequently chosen over pharmaceuticals in the healthcare system of most of the Asian countries including India. Herbal treatments have a long history in India, where they are important aspects of the regional indigenous traditions. Scientific evidence, however, has validated the traditional uses of several herbs as medicines for treating serious health problems including cancer and neurodegenerative disorders, ensuring their patient-friendly attributes and therapeutic potential. Since ethnomedicinal research is crucial for the discovery of novel medications from indigenous medicinal plants, it serves as the cornerstone of the pharmaceutical industry. But as a result of several factors related to urbanization and globalization, ethnomedicinal expertise has been slowly vanishing. To restore the virtues of ethnomedicine for the welfare of human society, it is therefore necessary to investigate, identify, and record traditional medicinal herbs. The current chapter intends to emphasize the ethnomedicinal advantages of medicinal plant resources and to grasp the prospects and challenges for the development of a promising base on plant-based therapies.

Keywords Indigenous knowledge · Herbal treatment · Ethnomedicine · Plant-based therapeutics · Local tribes

1 Introduction

Human civilization and plants have shared a very close relationship from time immemorial. The traditional knowledge of using medicinal plants to treat various diseases affirmed that plant-based medications have a rich history spanning ancient culture and scientific research. Many of these natural items have been brought to us by scientific research on traditional treatments used by different civilizations. The progressive advancement of pharmacology, phytochemistry, and other related fields, along with the optimism for plant-based scientific solutions, have created a trend regarding the potential benefits of natural goods. A lot of studies have documented the medicinal values of plants from various regions across the world, but there is still a need for proper exploration and identification of plants to find out the phytoconstituents crucial for treating disorders. The discoveries of modern plant-based therapeutics are solely based on ethnomedicine. In treating and recuperating from certain medical issues, such uses of folk medicine have been proven to be revolutionary. Folklore medicine includes a wide array of practices—Ayurveda, Unani, and Siddha from India; traditional medicine of China, Korea, Africa, and Islamic countries; acupuncture; Irani; and so on (Grover et al. 2002). All of these conventional methods may serve as valuable founts for contemporary drug research. The development of drug-derived medicinal plants is continuously creating spotlights of various pharmacological targets including cancer, respiratory tract disorders, and neurological diseases. Despite knowing the underlying mechanisms of action, the usage of different plant parts in treating illness is quite astonishing. However, the lack of quality control, standard methods, and sometimes usage of high doses could be the primary lacunae in promoting traditional healers. However, due to their efficaciousness, accessibility, and customary beliefs, a considerable segment of the population continues to rely on medicinal plant usage and herbal products to address a variety of health conditions.

The primary goal of modern drug research is to identify active molecules on specific targets through high-throughput screening techniques. The alkaloids, phenols, and terpenes present in plants are important pharmacological compounds for drug discovery. Bioactive compounds such as polyphenols exhibit antimicrobial, antidiabetic, antioxidant, and inflammatory properties. The flavone and flavopiridol isolated from *Dysoxylum binectariferum* have been reported to prevent life-threatening cancers by inhibiting protein kinases (Bhardwaj et al. 2023). A good number of novel drugs derived from plants have either received approval from the United States Food and Drug Administration (US FDA) or passed clinical trials. But regretfully, the expensive strategies of plant-based therapeutics development and isolation of active compounds in large quantities will make it more and more challenging to provide a novel, economically competitive treatment for any serious illness.

There are more than 45,000 plant species in India, most of which are found in the Andaman and Nicobar Islands, the Western Ghats, and the eastern Himalayas. Being a large repository of documented medicinal plants, India is also known as the botanical garden of the world. However, the lack of establishment of experimental data and sometimes their efficacy have become the major hindrance to the amalgamation of indigenous medicinal knowledge into modern health practices (Bode 2006). Moreover, proper botanical verification is mandatory for the characterization of potential therapeutic compounds available in the plants. Reports have been available regarding contamination and side effects of misidentified plants in herbal treatment. Cases of serious cardiac arrhythmias were reported from the USA when plantanin was inadvertently replaced with *Digitalis lantana* in an herbal supplement product. Similarly, borage is a commonly cultivated oilseed in most of the European regions. The toxic *Mandragora autumnalis*, which causes anticholinergic symptoms, can occasionally be mistaken for young borage leaves (Cornara et al. 2020). Micromorphological analysis, DNA barcoding, and phytochemical analysis could be important to confirm the identity of medicinal plants. Thus, the development of modern medicinal drugs from traditional sources is an integrated approach which includes the procurement of raw compounds on a mass scale and the implementation of products by clinical trials using advanced scientific research.

Understanding ethnomedicinal values not only enriches our perception of plant-based remedies but also underscores the cultural significance and relevance of these practices in various parts of the world. The ethnomedicinal wisdom of diverse ethnic groups acts as the foundational cornerstone for today's scientific research and

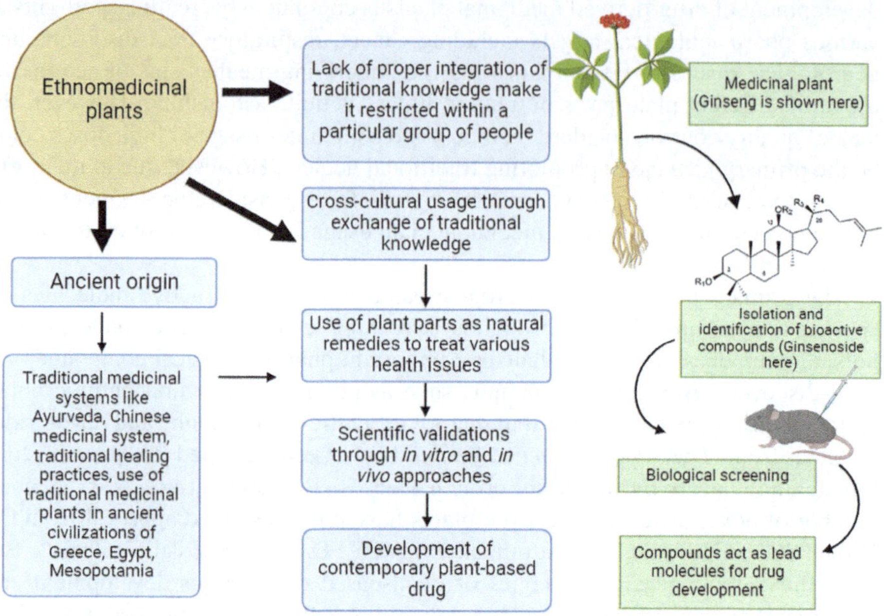

Fig. 1 Flow chart summarizing the key components of the chapter

methodologies in developing contemporary drugs. Therefore, it is imperative to highlight the historical background of medicinal plant usage in past. This chapter aims to showcase the rich heritage of ethnomedicinal wisdom, focusing on the historical journey and importance of ethnomedicine in harnessing the natural healing properties of plants. Moreover, by delving into ethnomedicinal values, this chapter sheds light on the treasure trove of scientific explanations of natural bioactive compounds that are reported to be derived from traditional medicinal plants with the efficacy of treating some serious health burdens. The flow chart (Fig. 1) has been constructed highlighting the key components of the chapter for comprehending the whole concept.

2 What Does History Say About Traditional Medicine?

Nature consistently provides a clear indication of the notable occurrences of coexistence. The concept of using medicinal plants is expanding with time, and their use is currently in demand. Medicinal herbs in particular have long served as a general indicator of the health of the ecosystem (Jamshidi-Kia et al. 2017). Studies showed that the most ancient kinds of drugs are derived from medicinal plants in various forms, either in their natural state or as pure compounds isolated from them. Archaeological studies also provide credible evidence that prehistoric humans were aware of the medicinal properties of plants (Halberstein 2005; Ahmad et al. 2014). The usage of plants for therapeutic purposes predates human civilization, and some of the earliest allusions can be found in the *Atharvaveda*, which is the basis of ayurveda medicine, a traditional form of Indian medicine that dates back to 2000 B.C.E. (Khan 2014).

With the passage of time and the rise of civilizations, humans acquired the ability to recognize and categorize natural resources that might be used to fulfill basic needs. Among these, the usage of plants and plant extracts for their medicinal purposes may be attributed to the most ancient narratives, customs, and literature that define their capabilities for relieving pain and treating diseases. Based on such properties of plants in the vicinity, herbal medicine systems have emerged and resulted in India's Ayurvedic and Unani systems, medicinal systems of parts of Asia including China and Tibet, and medicinal systems of America and African countries (Mamedov 2012).

It can be very challenging to pinpoint the precise moment to use plants as medications. Existing evidence says that plants were first cultivated for their herbal properties almost 60,000 years ago (Solecki 1975). Documents pertaining to medicinal plants date back at least 2500 years in Greece and Central Asia, and perhaps 5000 years in Egypt and China (Ang-Lee et al. 2001). In India, the first record of the therapeutic use of plants for drugs was discovered in Nagpur, which had 12 recipes for making drugs using more than 250 varieties of plants, including some alkaloid-containing plants (Kelly 2010). Around 2500 B.C., Chinese Emperor Shen Nung wrote a book containing 365 drugs or dried parts of medicinal plants. Of these plants, many are still in use today, including *Ephedra*, ginseng, *Podophyllum*, *Rhei rhisoma*, camphor, and *Theae folium* (Petrovska 2012). It was noteworthy that the people of Egypt and China were aware of the medicinal uses of plants since 27 centuries B.C. which was claimed by some earlier inscriptions (Jamshidi-Kia et al. 2017). Hippocrates (460–377 B.C.) is considered the father of allopathic medicine, and he documented 300 medicinal plants, by categorizing them according to their physiological action. Garlic was used for intestinal parasites; wormwood and *Centaurium umbellatum* were used to alleviate fevers; opium, henbane, and mandrake were used as narcotics; sea onion, parsley, asparagus, and garlic were used as diuretics; oak and pomegranate were used as astringents (Gorunović and Lukić 2001). Distinguished medical writer Celsus (25 B.C.–50 A.D.) incorporated quotations from about 250 plants having medicinal properties in his work *De Re Medica*. The most well-known author on medicinal plants in ancient times was Dioscorides, who traveled with the Roman Army and studied medicinal plants wherever he proceeded (Petrovska 2012). Interestingly, a large number of plants utilized in pharmacotherapy were brought in by the Arabs from India. These plants are largely of true medicinal value and are included throughout all global pharmacopoeias. The renowned Arabian works *Canon Medicinae* and *Liber Magnae Collectionis Simplicum Alimentorum Et Medicamentorum* of Avicenna (980–1037) and Ibn Baitar (1197–1248), respectively, which comprised descriptions of over a thousand medicinal plants, had been reviewed by European physicians during the Middle Ages. Across Europe, botanical gardens sprang up, and efforts were made to cultivate both native and imported therapeutic plants. *Cortex chinae*, which was derived from *Cinchona succirubra* Pavon's quinine bark, was introduced to European medicine in the seventeenth century and was known as "countess' powder" (Petrovska 2012).

During the early nineteenth century, there was a significant transformation in the understanding and utilization of medicinal plants. The exploration, separation, and confirmation of alkaloids from pomegranate, strychnos, ipecacuanha, quinine, and others marked the beginning of scientific pharmacy in parts of the world. With the improvement of chemical methods, more active phytocomponents such as tannins, etheric oils, saponins, hormones, vitamins, and so forth, were found (Srivastava 2018). In the early twentieth century, stabilization techniques for medicinal plants having components with medicinal values were brought out. After a series of chemical, physiological, and clinical research, numerous forgotten plants including *Aconitum, Colchicum, Filix mas, Hyosciamus, Opium, Punica granatum, Ricinius, Stramonium, Secale cornutum, Styrax*, and others have been introduced back into use in pharmacies, and drugs have been derived from them (Mazid et al. 2012).

3 An Overview of the Indian Medicinal Plant Usage

One of the most significant sources of medicine, whose use stretches back to the prehistoric era, is plants. Even though modern medicine and pharmaceutical research have advanced over the years, the use of plants for therapeutic purposes has become a basic part of daily life. According to the WHO, approximately 4.3 billion people, or above 80% of the world's population, rely on these conventional plant-based medical systems to meet their basic healthcare needs, and about 65% of Indian people are reliant on this system (Uniyal and Shiva 2005).

Indian medicinal system stretches back at least 5000 B.C., overlapping with the advent of the Indus Valley civilization around 4500 B.C. Even during the British era, when the Western healthcare system was promoted as a response to epidemic diseases such as malaria, smallpox, and cholera, this medical method was still in use in India. At least 7500 of the approximately 15,000 plant species that are found in India have been utilized for healthcare purposes, according to the Botanical Survey of India. A reasonable knowledge base has been established as a result of the approximately 1700 species whose biological characteristics and drug action have been documented and data available for about 1200 species, particularly those most commonly employed in the traditional medicinal system of India (Mazid et al. 2012).

Folk traditional medicine describes those medical practices that are typically bequeathed orally from one generation to the next and restricted to a specific geographic location or population, and tribe or community. The forests' enormous diversity of flora and fauna provides the tribes in isolated regions with the means of subsistence and healthcare they require (Rout et al. 2009). They have been using herbal remedies for medical conditions since ancient times (Xavier et al. 2014). Over 4635 tribes are dispersed across India, the majority of which are reliant on the resources of nearby forests to meet their basic requirements, according to the Anthropological Survey of India. Furthermore, the vast majority of India's population—roughly one billion people—continue to live in rural and village areas, often cut off from the conventional channels for disseminating medical knowledge. Millions of households continue to practice self-medication using locally sourced drugs based on medicinal plants, and South Asia offers evidence of a sizable practitioner population with varying degrees of specialization. Numerous species, many of which are native to South Asia, are utilized in these concoctions, which are frequently exclusive to India (Mazid et al. 2012). The study of indigenous remedies is currently receiving attention due to the immense amount of knowledge possessed by the nation's tribal groups (Muthu et al. 2006; Pei et al. 2009; Ragupathy and Newmaster 2009; Ramaya et al. 2009; Xavier et al. 2014). Research findings indicated that Arunachal Pradesh explored the traditional use of Mishitita, a bitter root that grows between 2000 and 3000 m in the districts of Dibang and Lohit and is used by tribal people to treat fever, backaches, and diarrhea (Mazid et al. 2012). Tripathi and Srivastava (2010), mentioned about 11 medicinal plants which include *Acalypha indica*, *Phyllanthus emblica*, *Jatropha curcas*, *Euphorbia thymifolia*, etc., and their usage as home remedies to treat various skin diseases in Uttar Pradesh (Anand et al. 2022). A list of some important Indian traditional medicinal plants and their uses against various diseases has been provided in Table 1.

3.1 Ethnomedicinal Evidence of Medicinal Plants from Assam

The use of plants for varied human necessities since ancient times reveals a lengthy history of human interactions with the environment. Indigenous civilizations all around the world have developed centuries-long abilities in the use of phytomedicines to prevent disease and sustain health. Despite significant advancements in the twentieth century in the fields of synthetic pharmaceuticals and antibiotics, plants remain a major source of drugs for both conventional and modern medical systems worldwide. Today, the majority of people in the entire globe (approximately 80%) acquire medical treatment primarily from plants and plant extracts (Ripunjoy 2016). Approximately 53.8 million people of different tribal communities live in villages that are dominated by forests on the Indian subcontinent, which makes up 15% of all Indian landmasses and is home to one of the largest emporia of ethnobotanical richness (Sajem and Gosai 2006).

Northeast India has been graced with a diverse range of biodiversity in plant genetic resources and is a vital part of the Indo-Myanmar biodiversity hotspot, one of the 25 worldwide biodiversity hotspots currently identified. It has a diverse range of flora, many of which are thought to have therapeutic significance. A huge number of tribal and ethnic tribes have settled for home in Northeast India over the ages, having come from different directions at different points in history. These communities rely on indigenous plant resources since they are vital to both their material and religious customs (Ripunjoy 2016). Therefore, Northeast India is a veritable

Table 1 List of plants along with their ethnomedicinal properties used by different tribes and communities of India

Botanical name	Used by	Parts used	Ethnomedicinal uses	References
Abrus precatorius L.	Jaunsari Tribe of Uttarakhand	Leaves, seeds, and roots	Decoction of leaves, seeds, and roots is used to cure fever, asthma, chest pain, and tuberculosis	Prakash (2015)
Acacia leucophloea (Roxb.) Willd.	Bhil, Garasia, Damor, and Kathodia tribes of Rajasthan	Leave	Leave decoction is taken orally during urinary tract bleeding	Katewa et al. (2004)
Achyranthes aspera L.	Chenchus tribes of Andhra Pradesh	Seeds and plant ash	Seed decoction is consumed for hydrophobia; plant ash is used to treat asthma	Reddy et al. (1988)
Acorus calamus L.	Guddies and Gujjar tribes of Himachal Pradesh	Roots	Root paste is applied topically on the chest of the patient suffering from bronchitis and chest congestion	Guleria and Vasishth (2009)
Adhatoda zeylanica Medik	Asur and Munda tribes of Bihar	Leaves	A leaf decoction given before and after delivery is claimed to have stimulating and healing properties	Jain et al. (1994)
Aegle marmelos (L.) Correa	Madiya, Muriya, Gond, and Bhatra tribes of Chhattisgarh	Leaves and fruits	Leaves and fruit pulps are used to treat diabetes, diarrhea, dysentery, and piles. Fruit pulp is also used to prepare tonic	Sahu et al. (2014)
Ageratum conyzoides L.	Bhil, Bhilala, and Patliya tribes of Madhya Pradesh	Leaf	Leaf juice is given in leukoderma daily during the morning for about 1 month	Thakur et al. (2014)
Asparagus racemosus Willd.	Jaintia tribes of Assam	Leaves	Powder of dried leaves is taken orally to cure stomach problems and urinary tract disorders	Sajem and Gosai (2006)
Argemone mexicana L.	Nath community of Haryana	Whole plant	The whole plant is used in eye disorders, mental retardation, skin problems, wound healing, and female sex disorders	Panghal et al. (2010)
Averrhoa carambola L.	Ao Naga tribe of Nagaland	Fruits and leaves	Fresh fruits or powders of dried fruits and dried leaves are good for high blood pressure, bladder, and intestinal problems	Kichu et al. (2015)
Callicarpa arborea Roxb.	Mizo tribe of Mizoram	Leaves and bark	Decoction of leaves and bark is used to treat stomach aches, diabetes, and convulsions	Laldingliani et al. (2022)

(continued)

Table 1 (continued)

Botanical name	Used by	Parts used	Ethnomedicinal uses	References
Carex nubigena D. Don	Adi tribe of Arunachal Pradesh	Stem	Oral stem extract from fresh plant is taken to treat jaundice	Ghosh et al. (2014)
Cassia tora L.	Kunabis tribe of Karnataka	Leaves	The external application of crushed leaves is applied to cure boils and hair follicle infections	Harsha et al. (2002)
Cassia occidentalis L.	Tharu tribe of Uttar Pradesh	Leaves	A paste of fresh leaves is applied topically over the leprosy wound	Kumar et al. (2013)
Catharanthus roseus (L.) G. Don	Santal tribe of West Bengal	Leave	Decoction of leaves is used to cure diabetes	Mandal et al. (2020)
Cynodon dactylon (L.) Pers.	Lepcha, Bhotia, and Nepali tribes of Sikkim	Whole plant	The decoction of the plant is used in dysentery, diarrhea, and piles	Tamang et al. (2023)
Eclipta prostrata (L.) L.	Korku tribe of Maharashtra	Whole plant	To treat rhinitis, 2–3 drops of whole plant juice are applied	Jagtap et al. (2006)
Elephantopus scaber L.	Garo and Khasi tribes of Meghalaya	Roots	The roots are crushed and given to patients with heart and liver trouble	Rao (1981)
Ficus religiosa L.	Koyas tribe of Telangana	Stem and bark	Aqueous extract of stem and bark are used to treat hepatitis, antidiabetic, and STDs	Gurrapu and Mamidala (2016)
Glycyrrhiza glabra L.	Punjab	Oil, gums, and root	Used for the treatment of peptic ulcer, hepatitis C, and pulmonary and skin diseases	Singh et al. (2018)
Gmelina arborea Roxb.	Zeliangrong tribe of Manipur	Bark	Ground paste of the bark is applied during bone fracture	Panmei et al. (2019)
Jasminum auriculatum Vahl	Velip, Kunabi, and Dhangars tribes of Goa	Flowers	Flowers are used to heal migraine	Ananth Nag (2016)
Leptadenia pyrotechnica (Forssk.) Decne	Niama, Damor, Baranda, Bhagora, and Angari tribes of Gujarat	Stem	The juice of boiled young stems is given orally to cure tuberculosis	Patel et al. (2014)
Ocimum sanctum L.	Kani tribes of Kerala	Leaves	It is used to treat variety of conditions like cold and flu, indigestion, stress, anxiety, earache, bronchitis, malaria, diabetes, heart disease, snakebite, asthma, hepatitis, tuberculosis, genitourinary disorders, etc.	Xavier et al. (2014)

(continued)

Table 1 (continued)

Botanical name	Used by	Parts used	Ethnomedicinal uses	References
Paederia scandens (Lour.) Merr.	Santhal tribes of Jharkhand	Leaves	Leaves are made into a paste and applied on skin infection	Mondal and Rahaman (2012)
Syzygium cumini (L.) Skeels	Irula tribe of Tamil Nadu	Seed	The mixture of seed powder with hot water or milk is taken to cure diabetes	Revathi and Parimelazhagan (2010)
Vitex heterophylla Blume ex Miq.	Reang tribe of Tripura	Leaves	Leaf paste is applied externally during bone fracture	Shil et al. (2014)

emporium and a key location for the study of indigenous knowledge systems due to its incredibly diverse tribal and ethnic communities and abundant biological resources.

Assam, known as the gateway to Northeast India, captivates both tourists and adventurers with its spectacular aesthetic and exquisite beauty. According to 2001 census, the region is home to 12.8% of India's total tribal population. For all of recorded history, the state's tremendous diversity in both population and flora has enabled its inhabitants an early advantage in observing and evaluating the abundant flora and wildlife to create their customary knowledge. The majority of tribal economies have historically been based on hunting and gathering or agriculture for sustenance (Sajem and Gosai 2006).

Assam is the home territory of numerous indigenous or tribal groups, including the Karbi, Zeme Naga, Kuki, Hmar, Hajong, Rengma Naga, Garo, Khasi, Mech, Jaintia, and Karbi, Mishing, Deuri, Rabha, Dimasa, Hajong, and Tiwa . As is well known, there is an immense variety of plants in Northeast India, and this is especially true in Assam, where the majority of the tribes use indigenous plants as folk remedies. The history of Assam's ethnobotanical research can be traced back by the adherence to individuals; this includes Kanjilal (1936), Hajra and Boissya (1980), Sharma Thakur (1999), Das et al (2006), Das and Tag (2006), Sajem and Gosai (2006), Buragohain and Konwar (2007), Kalita and Bora (2008), Sikdar and Dutta (2008), Sarma and Sarma (2010), Barukial and Sarma (2011), Sonowal and Baruah (2011), Abujam and Shah (2012), Gam (2013), Nath (2014), Talukdar (2014), etc. (Kalita et al. 2015; Bailung and Puzari 2016).

People in Assam live in remoteness and adhere to traditional beliefs because the majority of the state's districts are vulnerable to flooding and are historically backward; modernization has mostly bypassed these areas. Thus, from prehistoric times to the present, plants have been essential to human existence. Most people utilize a variety of locally grown plants as medication to treat common ailments. Even if some people have access to healthcare services, many still favor the more established practices of employing medicinal herbs to treat a variety of common ailments as quick remedies. It has additionally been found that the entire plant, as well as other parts like leaves, bark, stem, fruit, fruit sepal, root, tuber, rhizome, and bulb, is used to treat various illnesses. The majority of plants have multiple uses, including treating various ailments like blood purification, asthma, fever, piles, menstruation problems, diarrhea, jaundice, tonsillitis, and cough (Talukdar and Gupta 2020).

4 Phytoconstituents as Novel Compounds for Alternative Therapeutics

Investigating the potent components of plants is key to the development of bioactive compounds for use in traditional medicine. A large number of chemicals are effectively extracted from prospective plants with the advent of new technologies and improved knowledge of the isolated plant constituents, characterization, and analytical techniques. These compounds have greatly aided in the quest for new drugs derived from ethnomedicinal plants. The extraction of active chemicals from plants has been a key step in the development of pharmaceuticals and therapies, and this started with the early nineteenth-century extraction of morphine from opium (Ali et al. 2021). The discovery of drugs from medicinal plants produced early medications that are still in use today, which include quinine, digitoxin, cocaine, codeine, etc. Plant-based therapeutics have excellent contributions to cure some serious health disorders. For example, serpentine, an alkaloid, present in *Rouwolfia serpentina*, can be considered a breakthrough in treating high blood pressure and hypertension. It is almost established that plant-based compounds are the greatest source of high chemical variety, and therefore they can serve as the baseline for drug design.

4.1 Cancer Treatment

Cancer is one of the leading causes of mortality and morbidity across the globe. Surgery, radiation, and chemotherapy are commonly suggested by doctors, but these have serious adverse side effects and shorten the lifetime of the patients. Alongside such treatments, promising research on natural therapeutics for treating one of the scariest diseases, cancer, is still ongoing though half of the research has achieved its goals in this respective field. According to a report, 49% of medications used in cancer therapy were either natural products or derived from plant-based compounds (Cragg and Newman 2013). Some natural drugs such as Herceptin are available in the market for breast cancer treatment; however, these drugs are very expensive and have limited effectiveness in patients. In most cases, the tumor becomes resistant to a particular drug on regular use, and thereby patients have to shift to new drugs (Fridlender et al. 2015). The taxane dipertene (Fig. 2) available in the bark of *Taxus brevifolia* is well known for its cytotoxic activity from where the anticancer drug paclitaxel is formulated. Later, research revealed that other species of *Taxus* are also capable of production of this molecule. Another ethnomedicinal

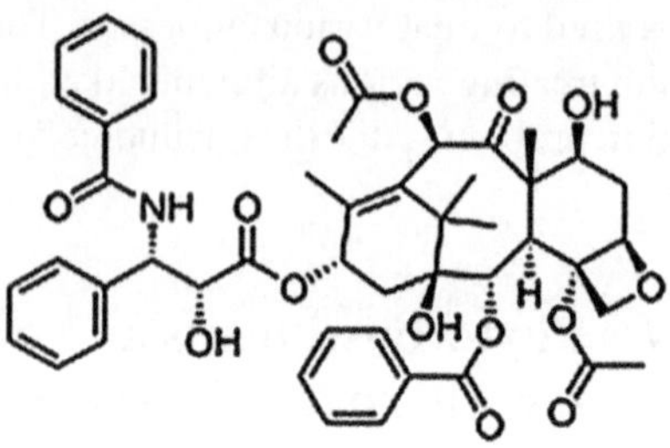

Fig. 2 Molecular structure of taxane from *Taxus brevifolia*

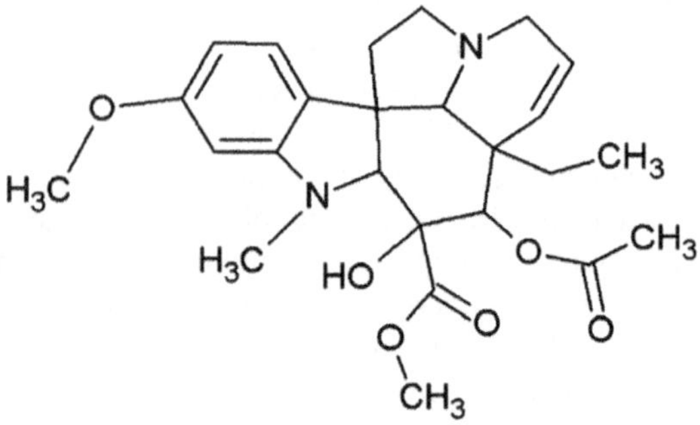

Fig. 3 Molecular structure of vindoline from *Vinca rosea*

plant *Vinca rosea* is famous among common people for its antitumor and antidiabetic properties. Vinca alkaloids, vindesine, and vindoline (Fig. 3) produced by this plant have preventive activity against cell division and proliferation (Koel et al. 2020). *Gymnema sylvestre* is another significant medicinal plant whose anticancer properties have been extensively studied. Important bioactive compounds reported from the plant include triterpene saponins (gymnemic acids) (Fig. 4), gymnemanol, gumarin, and gymmestrogenin which have potential anticancer activity against HeLa cells, human breast carcinoma cell lines, and melanoma cells (Khan et al. 2019). Curcumin is a polyphenol present in *Curcuma longa* rhizome that has been proven to be one of the potent chemopreventive agents. Curcumin can regulate cytokines, protein kinases, and transcription factors and act as an antioxidant, antiaging, anti-allergic, anti-inflammatory, and antitumor agent. It is noteworthy that most of the plants screened for their pharmacological properties in today's era have an extensive history of traditional use in various disease treatments. In a recent study, the lead compound Rautandiol B was isolated from medicinal plants *Neorautanenia mitis*, *Hydnora abyssinica*, and *Senna surattensis* and uncovered its potent activity against *Plasmodium falciparum*. Furthermore, the roots of these plants have been widely used for diarrhea, bacterial infection treatment, wound treatment, and piles and throat treatment in some African countries (Dawurung et al. 2021). Novel compounds (asiatic acid and loliolide) isolated from *Bacopa monnieri*, a very ancient traditional medicinal herb, also decipher their pharmacological actions against several neurological disorders in studies (Jeyasri et al. 2020). The peel of *Citrus aurantium* L. (Rutaceae) fruit contains hesperetine (Fig. 5), a naturally occurring phytoestrogen that has garnered popularity recently for its antitumor potential. Due

Fig. 4 Molecular structure of gymnemic acid from *Gymnema sylvestre*

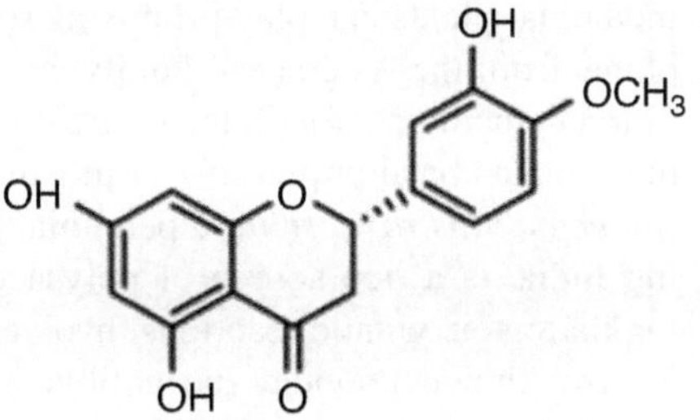

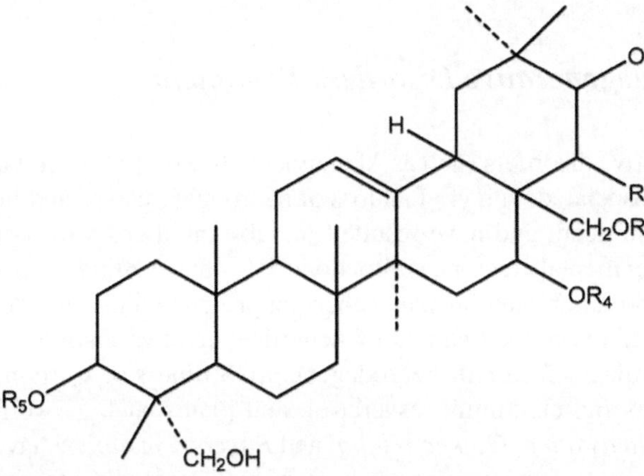

Fig. 5 Molecular structure of hesperetin from *Citrus aurantium*

to their structural diversity, it might be one of the potent drug candidates for alternative cancer treatment (Sohel et al. 2022). Structures of some novel compounds from medicinal plants having antitumor properties are shown as follows.

4.2 *Diabetes Treatment*

Due to the adverse side effects of antidiabetic medications, people are now switching to plant-based products. Traditional medicines are always the first choice for treating such severe health issues in developing countries. However, research has shown that some medicinal plants can treat diabetes more effectively than contemporary medications and are used as antidiabetic and antihyperlipidemic remedies (Fig. 6). Bioactive compounds found in *Momordica charantia*, *Allium cepa*, *Aloe vera*, *Pterocarpus marsupian*, *Tinospora crispa*, and many more plants have been investigated for their insulin-like behavior in diabetic patients and established as potent alternative therapeutics for treating diabetes (Tran et al. 2020). It has been demonstrated through recent findings that several phytochemicals found in medicinal plants can play promising roles in managing hyperglycemia (Fig. 5). The plants from the Asteraceae family are considered the most accepted Moroccan traditional medicine for diabetes treatment. The bioactive compounds available in these plants have impressive results in controlling diabetes (Bouyahya et al. 2021). *Anoectochilus roxburghii*, a perennial herb from Southeast Asian countries including India, is a rich source of polysaccharides that lower blood glucose levels by regulating enzymatic reactions involved in glucose metabolisms. This plant species is also known to yield the antihyperglycemic compound kinsenoside (Alam et al. 2022).

4.3 Neurodegenerative Disorders Treatment

Neurodegenerative disorders such as Alzheimer's disease (AD) and Parkinson's disease (PD) are associated with gradual loss of neuronal structure and function, which can culminate in death and have resulted in substantial costs for both people and society. Environmental factors, mitochondrial abnormalities, oxidative stress, inflammation, protein breakdown, and aberrant protein buildup in neurons are some of the frequently researched causes of neurodegenerative disorders. Nonetheless, aging is a significant factor in neurodegenerative illnesses. Certain natural compounds like ginseng, curcumin, resveratrol, and plants such as wolfberry, ashwagandha, *Centella asiatica*, *Ginkgo biloba*, and *Baccopa monnieri* have been used in clinical trials or in vivo settings to treat neurological diseases and symptoms (Ratheesh et al. 2017). Traditional Persian medicine has been shown to have positive effects on neurodegenerative disorders by upregulating antiapoptotic gene expression and downregulating proapoptotic protein functions (Farzaei et al. 2018). In the latest ethnopharmacological study, Sharma et al. (2022) documented a total of 73 ethnomedicinal plants of the Western Himalayas that are used to treat different neurological disorders including Huntington's disease, epilepsy, Alzheimer's, depression, Parkinson's disease, dementia, spinocerebellar ataxia, etc. Galantamine, a novel plant-based medication for Alzheimer's disease, is an alkaloid belonging to the Amaryllidaceae family that was extracted from *Narcissus tazetta* and evaluated for its neuroprotective activity. Biologically active phytoconstituents of *Oxalis corniculata* L. possess inhibitory activity against carbonic anhydrase and cholinesterase which may have uses in the treatment of Alzheimer's disease (Imran et al. 2020). Molecular modeling studies revealed that desmethoxyangonin isolated from dichloromethane extract of *Renealmia alpinia* demonstrated potent inhibitory activity toward MAO-B isoform (monoamine oxidase-B) which causes oxidative deamination of neurotransmitters in Parkinson's disease (Chaurasiya et al. 2017; Leuci et al. 2021).

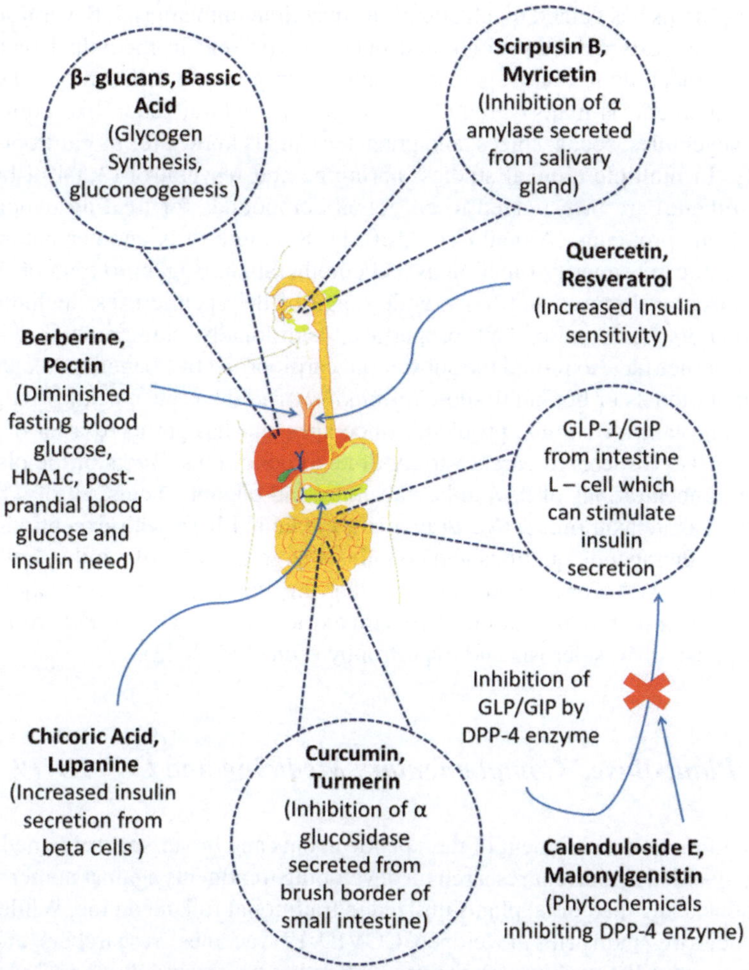

Fig. 6 Diagrammatic representation of different phytocomponents and their roles in controlling diabetes (as described by Alam et al. 2022 in their work)

4.4 *Cardiovascular Disease Treatment*

Cardiovascular diseases are the most common health issue in recent dates which primarily affect the blood circulatory system and heart. The WHO states that cardiovascular diseases are the major cause of rising mortality rates worldwide. The use of herbal products in treating cardiovascular problems is on the rise in the modern era due to the serious side effects and lack of economic viability of many synthetic drugs that act as antiplatelets, cholesterol-lowering agents, angiotensin-converting enzyme inhibitors, channel blockers, and vasodilators (Michel et al. 2020). In traditional medicinal systems, the usage of plants for lowering high blood pressure and heart problems has gained much attention since time immemorial. Several scientific research suggests that flavonoids and phenolics present in medicinal plants may reduce intracellular oxidative stress, which is connected to cell damage. The polyphenol quercetin is majorly found in a variety of natural foods like coffee, nuts, fruits, vegetables, cocoa, spices, and green teas and is known for its cardioprotective activity. In multiple clinical studies, naringine and resveratrol (3,5,4′-trihydroxy-trans-stilbene) are investigated to be potent compounds for treating patients with major heart problems (Alotaibi et al. 2021). Kaempferol is another natural compound that can primarily function as a chemotherapeutic agent to treat obesity and cardiovascular diseases and has a wide range of therapeutic uses, including anti-inflammatory and antioxidant properties. Additionally, numerous in vitro and in vivo studies have explored the substantial cardioprotective benefits of kaempferol in animal models or human tissues. *Terminalia arjuna* has been acknowledged as a medicinal plant for cardiac problems since ages and has garnered a lot of interest lately for its prospects to cure a variety of heart conditions. Barks of the plant have greater concentrations of flavonols, flavanole, and phenolic acids, all of which can enhance vasorelaxation. *Urtica dioica* is a medicinal herb with exceptional nutritional and therapeutic qualities and contains a large quantity of xanthophyll, leucoanthocyanidins, flavones, and flavanols that are thought to have a major role in lowering systolic blood pressure. Crocetin extracted from *Crocus sativus* also protects against atherosclerosis and hypertrophy (Din et al. 2023).

4.5 Plant-Based Complementary Medicine and COVID-19

The continuous development of therapeutic agents and broad-spectrum medications is greatly aided by current research on developing treatments against numerous viral infections using medicinal plants utilized in traditional folk medicine. While several manufacturing industries developed COVID-19 vaccines, researchers and many people were still searching for the most effective treatment, which included traditional medicines. Several medicinal plants have been utilized for cough and cold relief and immunity building since ancient times. Therefore, such plants have received attention for developing and designing new anti-infectives to prevent infectious agents. Couples of research have demonstrated that the herbal extract of medicinal plants can interact with important viral proteins that are associated with virulence, even though the general mechanism of herbal extracts is thought to be the prevention of viral reproduction. Many fruit and herbal extracts from different geographic locations have been touted as treatments for viral infections. According to certain research, *Andrographis paniculata* can lessen coronavirus symptoms and strengthen the immune system. A Sri Lankan herbal drink is utilized to prevent viral infections like SARS-CoV-2, which reduces the symptoms associated with fever. Although the WHO stated that there is no proof to support this assertion, people used to believe that consuming lemon juice with warm water can increase vitamin C levels and prevent COVID-19 disease (Mirzaie et al. 2020).

However, many clinical studies have displayed that natural compounds from plants could be the best drug candidates for treating viral infections. An important phytocomponent of *Canavalia ensiformis* is concanavalin A, and it can bind with glycosylated membrane protein, thereby preventing target cell recognition. Another study revealed that an aqueous extract of *Houttuynia cordata* exhibits anti-SARS activities by preventing 3CLpro (chymotrypsin-like protease) and RdRp proteins (Lau et al. 2008). According to an evaluation, the consumption of Chinese traditional medicines alone or in combination with marketed drugs reduced the clinical symptoms in COVID patients, which describes this effectiveness against the virus. Indian traditional medicines have also played significant roles in treating the SARS-CoV-2 virus. Studies signified that medicinal plants such as *Vitex trifolia, Gymnema sylvestre, Leucas aspera, Evolvulus alsinoides, Cassia alata*, and *Abutilon indicum* have well-known antiviral properties with the ability to decrease NF-kB inflammatory cytokines (Mirzaie et al. 2020). An in silico study of plant-derived active components demonstrated the SARS-CoV-2 inhibitory activity of 36 phytochemicals from Indian medicinal plants and concluded that Withaferin A and Withanolide D from *Withania somnifera* showed greater binding affinity with antigenic spike protein of the virus (Kashyap et al. 2022). It is hypothesized that some triterpenoid structures, for example, coagulins, withanolides, kamalachalcone, and pseudojervine, could serve as powerful drug candidates for inhibiting spike proteins which interact with hACE2, hence preventing SARS-CoV-2 from recognizing the host cell (Puttaswamy et al. 2020). Developing medications from phytopharmaceuticals is a popular strategy for finding environmentally acceptable medicinal compounds with negligible or nonexistent negative effects. Given the time constraints, it is imperative to devise a quick and practical approach to creating medications that block the molecular machinery of viruses.

5 Limitations of Therapeutics Development from Ethnomedicinal Plants

From a practical and scientific standpoint, ethnomedicinal knowledge and plants are receiving more attention in the field of alternative therapeutics. Its effectiveness has been demonstrated over many years, but the mechanism of folk remedies and natural products is much more intricate than the automatous explanation of a single bioactive component due to the use of partly fractionated extracts in most of the experimental studies (Pandey et al. 2011). The contemporary perspective on alternative medications is founded on the usage of traditional medicinal plant and natural products, their ecology, and traditional beliefs in human communities. Hence, currently conducted ethnopharmacological research prioritizes sharing and documenting indigenous medicinal knowledge as well as the scientific study of indigenous medications for better development. The amount of knowledge saved in electronic media has significantly expanded in the modern era with the emergence of medicinal plant databases (Dighe et al. 2010). These databases gathered a wealth of data on the traits of plants, their therapeutic uses, and their bioactive substances, together with pictures of the plants. But most of the time, data accuracy and maintenance are becoming the major limitations for the continuation of these databases. Data on medicinal plants are typically shared with consideration for the open accessibility of traditional knowledge. Some argue that if traditional knowledge is freely accessible to the public, it can be used as a modern commodity in the same way as any other kind of intellectual property (Brush 1993), which could be another challenge for such databases (Ningthoujam et al. 2012). Besides these, the future of ethnomedicinal knowledge is thought to be in jeopardy because of some of its shortcomings, which include the treatment and healing of only minor illnesses, the lack of written form, the position of local culture not being universal, and the restriction of medical knowledge to a small number of members of society. This also sheds light on the importance of ethnomedicinal knowledge and medicinal plant conservation in parallel to their usage in the modern era to find potent bioactive compounds for drug development.

6 Conclusion

The research interest in the field of ethnobotany and ethnomedicine is not new, and now it is rapidly evolving with the advent of advanced molecular screening techniques. Unquestionably, the analysis of active compounds from plants faced a troubled past, but it has grown over time to emerge as a mainstay in various domains like medicine, pharmacology, synthetic chemistry, and mainly nanotechnology. Understanding the value of plants as potential sources of therapeutics, all of these scientific fields have started intensive endeavors aimed at producing alternative therapeutics or isolating novel lead chemicals for treating life-threatening diseases. However, uncovering novel plant-based leads would be exceedingly challenging if concerns like biodiversity, indigenous knowledge, and the preservation of ethnomedical data were disregarded. Therefore, given their importance to the healthcare system, medicinal plants with therapeutic qualities ought to be cultivated and promoted. Likewise, the involvement of taxonomists and plant specialists is crucial for the accurate identification of therapeutic plants. The worldwide interest in natural products by the general public, researchers, and multinational corporations, along with the integration of cutting-edge technology and the ethnomedicinal benefits of plants, may help establish a golden avenue for future significant contributions to healthcare.

References

Abujam SS, Shah RK (2012) Study on the ethnomedicinal system of local people of Dibrugarh, Assam. Int J Pharm Innov 2:17–28

Ahmad M, Sultana S, Fazl-i-Hadi S, Ben Hadda T, Rashid S, Zafar M, Khan MA et al (2014) An ethnobotanical study of medicinal plants in high mountainous region of Chail valley (District Swat-Pakistan). J Ethnobiol Ethnomed 10:1–18

Alam S et al (2022) Antidiabetic phytochemicals from medicinal plants: prospective candidates for new drug discovery and development. Front Endocrinol 13:800714. https://doi.org/10.3389/fendo.2022.800714

Ali SA, Singh G, Datusalia AK (2021) Potential therapeutic applications of phytoconstituents as immunomodulators: pre-clinical and clinical evidences. Phytother Res 35(7):3702–3731

Alotaibi BS, Ijaz M, Buabeid M, Kharaba ZJ, Yaseen HS, Murtaza G (2021) Therapeutic effects and safe uses of plant-derived polyphenolic compounds in cardiovascular diseases: a review. Drug Des Devel Ther 15:4713–4732. https://doi.org/10.2147/DDDT.S327238

Anand U, Tudu CK, Nandy S, Sunita K, Tripathi V, Loake GJ et al (2022) Ethnodermatological use of medicinal plants in India: from ayurvedic formulations to clinical perspectives—a review. J Ethnopharmacol 284:114744. https://doi.org/10.1016/j.jep.2021.114744

Ananth Nag B (2016) A case study on medicinal plants-an integral part of Goan rituals of Goa, India. Int J Environ Sci 5(1):32–38

Ang-Lee MK, Moss J, Yuan CS (2001) Herbal medicines and perioperative care. JAMA 286(2):208–216

Bailung B, Puzari M (2016) Traditional use of plants by the Ahoms in human health management in upper Assam, India. J Med Plants Stud 4:48–51

Bhardwaj N, Sharma A, Tripathi N, Goel B, Ravikanth G, Guru SK, Jain SK (2023) New cycloartane triterpenoids from Dysoxylum malabaricum and their cytotoxic evaluation. Steroids 200:109315

Barukial J, Sarma JN (2011) Ethnomedicinal plants used by the people of Golaghat district, Assam. Indian J Med Aromat Plants 1:203–211

Bode M (2006) Taking traditional knowledge to the market: the commoditization of Indian medicine. Anthropol Med 13(3):225–236

Bouyahya A, El Omari N, Elmenyiy N, Guaouguaou FE, Balahbib A, Belmehdi O et al (2021) Moroccan antidiabetic medicinal plants: ethnobotanical studies, phytochemical bioactive compounds, preclinical investigations, toxicological validations and clinical evidences; challenges, guidance and perspectives for future management of diabetes worldwide. Trends Food Sci Technol 115:147–254. https://doi.org/10.1016/j.tifs.2021.03.032

Brush SB (1993) Indigenous knowledge of biological resources and intellectual property rights: the role of anthropology. Am Anthropol 95(3):653–671

Buragohain J, Konwar BK (2007) Ethnomedicinal plants used in skin diseases by some Indo-Mongoloid communities of Assam. Asian J Exp Sci 21:281–288

Chaurasiya ND, León F, Ding Y, Gómez-Betancur I, Benjumea D, Walker LA et al (2017) Interactions of Desmethoxyyangonin, a secondary metabolite from *Renealmia alpinia*, with human monoamine oxidase-A and oxidase-B. Evid Based Complement Alternat Med 2017:1. https://doi.org/10.1155/2017/4018724

Cornara L, Ambu G, Trombetta D, Denaro M, Alloisio S, Frigerio J et al (2020) Comparative and functional screening of three species traditionally used as antidepressants: *Valeriana officinalis* L., *Valeriana jatamansi* jones ex roxb. And *Nardostachys jatamansi* (D.Don) DC. Plan Theory 9:1–27. https://doi.org/10.3390/plants9080994

Cragg GM, Newman DJ (2013) Natural products: a continuing source of novel drug leads. Biochim Biophys Acta Gen Subj 6:3670–3695

Dawurung CJ, Nguyen MTH, Pengon J, Dokladda K, Bunyong R, Rattanajak R et al (2021) Isolation of bioactive compounds from medicinal plants used in traditional medicine:

Rautandiol B, a potential lead compound against *Plasmodium falciparum*. BMC Complement Med Ther 21:1–12. https://doi.org/10.1186/s12906-021-03406-y
Das NJ, Saikia SP, Sarkar S, Devi K (2006) Medicinal plants of North Kamrup district of Assam used in primary healthcare system. Indian J Trad Knowl 5(4):489–493
Das AK, Tag H (2006) Ethnomedicinal studies of the Khamti tribe of Arunachal Pradesh. Indian J Trad Knowl 5(3):317–322
Dighe NS, Nirmal SA, Musmade DS, Dhasade VV (2010) Herbal database management. Syst Rev Pharm 1(2):152–157
Din SRU, Saeed S, Khan SU, Kiani FA, Alsuhaibani AM, Zhong M (2023) Bioactive compounds (BACs): a novel approach to treat and prevent cardiovascular diseases. Curr Probl Cardiol 48(7):101664
Farzaei MH, Shahpiri Z, Mehri MR, Bahramsoltani R, Rezaei M, Raeesdana A, Rahimi R (2018) Medicinal plants in neurodegenerative diseases: perspective of traditional Persian medicine. Curr Drug Metab 19(5):429–442
Fridlender M, Kapulnik Y, Koltai H (2015) Plant derived substances with anti-cancer activity: from folklore to practice. Front Plant Sci 6:799
Gam NK (2013) Ethno Medicinal claims existing among Mising Tribes of Assam. Int J Sci Invent Today 2:284–291
Ghosh G, Ghosh DC, Melkania U, Majumdar U (2014) Traditional medicinal plants used by the Adi, Idu and Khamba tribes of Dehang-Debang biosphere Reserve in Arunachal Pradesh. Int J Agric Environ Biotech 7(1):165
Gorunović MS, Lukić PB (2001) Pharmakognosy. University of Belgrade–Faculty of Pharmacy, Belgrade
Grover JK, Yadav S, Vats V (2002) Medicinal plants of India with anti-diabetic potential. J Ethnopharmacol 81(1):81–100
Guleria V, Vasishth A (2009) Ethnobotanical uses of wild medicinal plants by Guddi and Gujjar tribes of Himachal Pradesh. Ethnobot Leafl 9:8
Gurrapu S, Mamidala E (2016) Medicinal plants used by traditional medicine practitioners in the management of HIV/AIDS-related diseases in tribal areas of Adilabad district, Telangana region. Am J Sci Med Res 2(1):239–245
Halberstein RA (2005) Medicinal plants: historical and cross-cultural usage patterns. Ann Epidemiol 15(9):686–699
Harsha VH, Hebbar SS, Hegde GR, Shripathi V (2002) Ethnomedical knowledge of plants used by Kunabi tribe of Karnataka in India. Fitoterapia 73(4):281–287
Hajra PK, Boissya AK (1980) Ethnobotanical notes on Miris (Missings) of Assam plain; Jain S K: Glimpses of Indian ethno botany. Oxford and IBH Publishers, New Delhi, pp 161–169
Imran M, Irfan A, Ibrahim M, Assiri MA, Khalid N, Ullah S et al (2020) Carbonic anhydrase and cholinesterase inhibitory activities of isolated flavonoids from *Oxalis corniculata* L. and their first-principles investigations. Ind Crop Prod 148:112285. https://doi.org/10.1016/j.indcrop.2020.112285
Jagtap SD, Deokule SS, Bhosle SV (2006) Some unique ethnomedicinal uses of plants used by the Korku tribe of Amravati district of Maharashtra, India. J Ethnopharmacol 107(3):463–469
Jain SP, Singh SC, Puri HS (1994) Medicinal plants of neterhat, Bihar, India. Int J Pharm 32(1):44–50
Jamshidi-Kia F, Lorigooini Z, Amini-Khoei H (2017) Medicinal plants: past history and future perspective. J Herbmed Pharmacol 7(1):1–7
Jeyasri R, Muthuramalingam P, Suba V, Ramesh M, Chen JT (2020) Bacopa monnieri and their bioactive compounds inferred multi-target treatment strategy for neurological diseases: a cheminformatics and system pharmacology approach. Biomol Ther 10:1–19. https://doi.org/10.3390/biom10040536
Kalita G, Rout S, Kumar Mishra R, Sarma P (2015) Traditionally used medicinal plants of Bajali sub-division, Barpeta District, Assam. J Med Plants Stud 3:8–17

Kalita D, Bora R (2008) Some folk medicines from Lakhimpur district, Assam. Indian J Trad Knowl 7:414–416

Kanjilal UN, Kanjilal PC, Das A, Purkayastha C (1936) Flora of Assam, vol II. Assam Govt. Press, Shillong, India

Kashyap D, Jakhmola S, Tiwari D, Kumar R, Moorthy NSHN, Elangovan M, Brás NF, Jha HC (2022) Plant derived active compounds as potential anti SARS-CoV-2 agents: an in-silico study. J Biomol Struct Dyn 40:10629–10650. https://doi.org/10.1080/07391102.2021.1947384

Katewa SS, Chaudhary BL, Jain A (2004) Folk herbal medicines from tribal area of Rajasthan, India. J Ethnopharmacol 92(1):41–46

Kelly K (2010) Old World and new: early medical care, 1700–1840. Infobase Publishing, New York

Khan H (2014) Medicinal plants in light of history: recognized therapeutic modality. J Evid Based Complement Altern Med 19:216–219. https://doi.org/10.1177/2156587214533346

Khan F, Sarker MMR, Ming LC, Mohamed IN, Zhao C, Sheikh BY, Tsong HF, Rashid MA (2019) Comprehensive review on phytochemicals, pharmacological and clinical potentials of *Gymnema sylvestre*. Front Pharmacol 10:1223. https://doi.org/10.3389/fphar.2019.01223

Kichu M, Malewska T, Akter K, Imchen I, Harrington D, Kohen J, Vemulpad SR, Jamie JF (2015) An ethnobotanical study of medicinal plants of Chungtia village, Nagaland, India. J Ethnopharmacol 166:5–17

Koel M, Kuhtinskaja M, Vaher M (2020) Extraction of bioactive compounds from *Catharanthus roseus* and *Vinca minor*. Sep Purif Technol 252:117438. https://doi.org/10.1016/j.seppur.2020.117438

Kumar A, Pandey VC, Singh AG, Tewari DD (2013) Traditional uses of medicinal plants for dermatological healthcare management practices by the Tharu tribal community of Uttar Pradesh, India. Genet Resour Crop Evol 60:203–224

Laldingliani TBC, Thangjam NM, Zomuanawma R, Bawitlung L, Pal A, Kumar A (2022) Ethnomedicinal study of medicinal plants used by Mizo tribes in Champhai district of Mizoram, India. J Ethnobiol Ethnomed 18(1):1–29

Lau KM, Lee KM, Koon CM, Cheung CSF, Lau CP, Ho HM et al (2008) Immunomodulatory and anti-SARS activities of Houttuynia cordata. J Ethnopharmacol 118:79–85. https://doi.org/10.1016/j.jep.2008.03.018

Leuci R, Brunetti L, Poliseno V, Tortorella P, Piemontese L, Laghezza A et al (2021) Cardiovascular and neurodegenerative diseases. Foods 10(1):29. https://doi.org/10.3390/foods10010029

Mamedov N (2012) Medicinal plants studies: history, challenges and prospective. Med Aromat Plants 01(8):1000e133. https://doi.org/10.4172/2167-0412.1000e133

Mandal A, Adhikary T, Chakraborty D, Ro P, Saha J, Barman A, Saha P (2020) Ethnomedicinal uses of plants by Santal tribe of Alipurduar district, West Bengal, India. Indian J Sci Technol 13(20):2021–2029

Mazid M, Khan TA, Mohammad F (2012) Medicinal plants of rural India: a review of use by Indian folks. Indo Global J Pharma Sci 2(3):286–304

Michel J, Abd Rani NZ, Husain K (2020) A review on the potential use of medicinal plants from Asteraceae and Lamiaceae plant family in cardiovascular diseases. Front Pharmacol 11:852

Mirzaie A, Halaji M, Dehkordi FS, Ranjbar R, Noorbazargan H (2020) A narrative literature review on traditional medicine options for treatment of corona virus disease 2019 (COVID-19). Complement Ther Clin Pract 40:101214

Mondal S, Rahaman CH (2012) Medicinal plants used by the tribal people of Birbhum district of West Bengal and Dumka district of Jharkhand in India. Indian J Tradit Knowl 11(4):674–679

Muthu C, Ayyanar M, Raja N, Ignacimuthu S (2006) Medicinal plants used by traditional healers in Kancheepuram District of Tamil Nadu, India. J Ethnobiol Ethnomed 2:1–10. https://doi.org/10.1186/1746-4269-2-43

Nath DW (2014) Ethno-medicinal Significance of Few Naturally Grown Plants Practices for Hepatitis in Sivasagar District of Assam, India. Eur Acad Res 2(4):5371–5385

Ningthoujam SS, Das TA, Potsangbam KS, Choudhury MD (2012) Challenges in developing medicinal plant databases for sharing ethnopharmacological knowledge. J Ethnopharmacol 141:9–32. https://doi.org/10.1016/j.jep.2012.02.042
Pandey M, Debnath M, Gupta S, Chikara SK (2011) Phytomedicine: an ancient approach turning into future potential source of therapeutics. J Pharmacogn Phyther 3:27–37
Panghal M, Arya V, Yadav S, Kumar S, Yadav JP (2010) Indigenous knowledge of medicinal plants used by Saperas community of Khetawas, Jhajjar District, Haryana, India. J Ethnobiol Ethnomed 6:1–11
Panmei R, Gajurel PR, Singh B (2019) Ethnobotany of medicinal plants used by the Zeliangrong ethnic group of Manipur, Northeast India. J Ethnopharmacol 235:164–182
Patel SK, Desai PR, Pandey VB (2014) Ethnomedicinal plants used by the tribals in Bhiloda taluka of Sabarkantha district, Gujarat. Indian J Adv Plant Res 1:33–36
Pei S, Zhang G, Huai H (2009) Application of traditional knowledge in forest management: ethnobotanical indicators of sustainable forest use. For Ecol Manage 257:2017–2021. https://doi.org/10.1016/j.foreco.2009.01.003
Petrovska BB (2012) Historical review of medicinal plants' usage. Pharmacogn Rev 6:1–5. https://doi.org/10.4103/0973-7847.95849
Prakash R (2015) Medicinal plants used by tribal communities: a study of Uttarakhand Himalayan region. Int J Human Soc Sci Invent 4(2):55–61
Puttaswamy H, Gowtham HG, Ojha MD, Yadav A, Choudhir G, Raguraman V, Kongkham B, Selvaraju K, Shareef S, Gehlot P, Ahamed F, Chauhan L (2020) *In silico* studies evidenced the role of structurally diverse plant secondary metabolites in reducing SARS-CoV-2 pathogenesis. Sci Rep 10:1–24. https://doi.org/10.1038/s41598-020-77602-0
Ragupathy S, Newmaster SG (2009) Valorizing the "Irulas" traditional knowledge of medicinal plants in the Kodiakkarai reserve Forest, India. J Ethnobiol Ethnomed 5:1–13. https://doi.org/10.1186/1746-4269-5-10
Ramaya S, Alaguchamy N, Maruthappan VM, Sivaperumal R, Sivalingam M, Krishnan A, Govindaraji V, Kannan K, Jayakumararaj R (2009) Wound healing ethnomedicinal plants popular among the Malayali tribes in Vattal Hills, Dharmapuri, TN, India. Ethnobot Leafl 13:1257–1271
Rao RR (1981) Ethnobotany of Meghalaya: medicinal plants used by Khasi and Garo tribes. Econ Bot 35(1):4–9
Ratheesh G, Tian L, Venugopal JR, Ezhilarasu H, Sadiq A, Fan TP, Ramakrishna S (2017) Role of medicinal plants in neurodegenerative diseases. Biomanuf Rev 2:1–16
Reddy MB, Reddy KR, Reddy MN (1988) A survey of medicinal plants of Chenchu tribes of Andhra Pradesh, India. Int J Crude Drug Res 26(4):189–196
Revathi P, Parimelazhagan T (2010) Traditional knowledge on medicinal plants used by the Irula tribe of Hasanur Hills, Erode District, Tamil Nadu, India. Ethnobot Leafl 2010(2):4
Ripunjoy S (2016) Indigenous knowledge on the exploitation and utilization of medicinal plants by the Thengal Kachari tribe of Jorhat district, Assam, north-East India. Curr Life Sci 2:92–101
Rout SD, Panda T, Mishra N (2009) Ethno-medicinal plants used to cure different diseases by tribals of Mayurbhanj district of North Orissa. Stud Ethno-Med 3(1):27–32
Sahu PK, Masih V, Gupta S, Sen DL, Tiwari A (2014) Ethnomedicinal plants used in the healthcare systems of tribes of Dantewada, Chhattisgarh, India. Am J Plant Sci 05:1632
Sajem AL, Gosai K (2006) Traditional use of medicinal plants by the Jaintia tribes in north Cachar Hills district of Assam, Northeast India. J Ethnobiol Ethnomed 2:1–7. https://doi.org/10.1186/1746-4269-2-33
Sarma R, Sarma HK (2010) Ethnomedicines of Sonapur, Kamrup district, Assam. Indian J Trad Knowl 9:163–165
Sharma K, Verma R, Kumar D, Nepovimova E, Kuča K, Kumar A, Raghuvanshi D, Dhalaria R, Puri S (2022) Ethnomedicinal plants used for the treatment of neurodegenerative diseases in Himachal Pradesh, India in Western Himalaya. J Ethnopharmacol 293:115318. https://doi.org/10.1016/j.jep.2022.115318

Sharma Thakur GC (1999) Indigenous health practices and system of cure among the tribes of Assam plains. In: Health, healers and healing: studies in medical anthropology. NL Publications, Dibrugarh, pp 239–249
Shil S, Choudhury MD, Das S (2014) Indigenous knowledge of medicinal plants used by the Reang tribe of Tripura state of India. J Ethnopharmacol 152(1):135–141
Singh A, Rani R, Sharma M (2018) Medicinal herbs of Punjab (India). Biol Forum 10:10–27
Sikdar M, Dutta U (2008) Traditional phytotherapy among the Nath people of Assam. Stud Ethno-Med 2(1):39–45
Sohel M, Sultana H, Sultana T, Al Amin M, Aktar S, Ali MC et al (2022) Chemotherapeutic potential of hesperetin for cancer treatment, with mechanistic insights: a comprehensive review. Heliyon 8:e08815. https://doi.org/10.1016/j.heliyon.2022.e08815
Sonowal R, Barua I (2011) Ethnomedical practices among the Tai-Khamyangs of Assam, India. Stud Ethno-Med 5(1):41–50
Solecki RS (1975) Shanidar IV, a Neanderthal flower burial in northern Iraq. Science 190(4217):880–881
Srivastava AK (2018) Significance of medicinal plants in human life. In: Synthesis of medicinal agents from plants. Elsevier, Amsterdam, pp 1–24
Talukdar S, Gupta A (2020) Ethnomedicinal knowledge of the Garo community of two villages in western Assam, India. J Herbal Med 20:100229
Talukdar M (2014) Ethnomedicinal Plants used by the Sonowal Kacharis of Bhekulajan Village in Dibrugarh District, Assam, NE India. Int Res J Environ Sci 3:54–57
Tamang S, Singh A, Bussmann RW, Shukla V, Nautiyal MC (2023) Ethno-medicinal plants of tribal people: a case study in Pakyong subdivision of East Sikkim, India. Acta Ecol Sin 43(1):34–46
Thakur A, Naqvi SMA, Aske DK, Sainkhediya J (2014) Study of some ethno medicinal plants used by tribals of Alirajpur, Madhya Pradesh, India. Indian Res J Agric For Sci 2320:6063
Tran N, Pham B, Le L (2020) Bioactive compounds in anti-diabetic plants: from herbal medicine to modern drug discovery. Biology (Basel) 9:1–31. https://doi.org/10.3390/biology9090252
Tripathi SC, Srivastava M (2010) Ethnomedicinal flora of Euphorbiaceae used in dermatological problems. Indian Trad Knowl 9(2):318–320
Uniyal B, Shiva V (2005) Traditional knowledge on medicinal plants among rural women of the Garhwal Himalaya, Uttaranchal. Indian J Trad Knowl 4(3):259
Xavier TF, Kannan M, Lija L, Auxillia A, Rose AKF, Kumar SS (2014) Ethnobotanical study of Kani tribes in Thoduhills of Kerala, South India. J Ethnopharmacol 152:78–90. https://doi.org/10.1016/j.jep.2013.12.016

Importance of Traditional Resources in Pharmaceutical Industries

Deepshikha Buragohain, Pritam Kundu, Rajat Nath, Sibashish Kityania, Anirudha Giri, and Anupam Das Talukdar

Contents

D. Buragohain · R. Nath · S. Kityania · A. Giri · A. Das Talukdar (✉)
Department of Life Science and Bioinformatics, Assam University, Silchar, Assam, India
e-mail: anupam@bioinfoaus.ac.in

P. Kundu
Pharmacy Discipline, Khulna University, Khulna, Bangladesh

© The Author(s), under exclusive license to Springer Nature Singapore Pte Ltd. 2024

A. Das Talukdar et al. (eds.), *Traditional Resources and Tools for Modern Drug Discovery*, Interdisciplinary Biotechnological Advances,
https://doi.org/10.1007/978-981-97-4600-2_15

1 Introduction

The pharmaceutical sector is the largest growing sector in the context of medical science worldwide. Since the introduction of human civilization to today, huge improvements and inventions are visible to improve the healthcare system through the development of the pharmaceutical sector. Scientists from all over the world have been conducting more and more researches incessantly to improve this sector. Although most of the threatening diseases have been cured by inventing many life-saving drugs, still there are many fatal diseases by which many people have been suffering and also have died every year. That's why more and more researches have been necessary always. In modern allopathic medicine system, numerous types of drugs have been used. Among these drugs, most of them were originated from natural resources. And this allopathic medicine system was developed from traditional medicine system, which is still very much popular, and well-accepted techniques in many parts of the world. As per the World Health Organization's definition, traditional medicine encompasses a comprehensive array of knowledge, skills, and practices aimed at maintaining health, diagnosing, treating, and preventing physical and mental illnesses. Rooted in the theories, beliefs, and experiences of various cultures, these practices often defy scientific explanation. Traditional medicine systems, including traditional Chinese medicine, Ayurveda, Siddha, Hekimi, Kampo, traditional Korean medicine, and Unani, have been practiced globally for centuries, evolving into well-structured and regulated medical systems. Serving as the world's oldest medical system, traditional medicine offers treatments for both physical and mental ailments. Despite some limitations and shortcomings, these systems remain invaluable repositories of human wisdom. They utilize a wide range of natural resources, including medicinal plants, microorganisms, marine-derived algae, fungi, and fishes, among others. All of these natural resources are main fundamentals of traditional medicine. Utilization of these resources paved the way of pure drug substances and their derivatives. Subsequently, through the use of computer-aided drug designing system and synthetic chemistry, numerous compounds have been invented having various types of pharmacological benefits.

"Traditional medicine" encompasses the healing practices of diverse cultures and communities worldwide, including Indian Ayurveda, Chinese, Japanese, African, and Arabic Unani medicine. It encompasses a broad spectrum of traditional medical approaches and techniques. Plants are used extensively in the majority of these traditional medical practices, and they may be paired with complementary therapies like aromatherapy, acupuncture, yoga, curanderismo, or acupuncture in addition to pharmaceuticals (Subbarayappa 2001).While some traditional medical systems are passed down orally from generation to generation, others are backed by extensive libraries of written works that document both theoretical ideas and practical techniques. In several regions of the world, the majority of people still get their main medical treatment from their own traditional medicine practices today. Traditional medicine is sometimes called "complementary and alternative medicine" when it is used outside of its original culture. Among the most widely used traditional medical systems nowadays are those from Africa, India, and China.

Allopathic medicine is a system in which physicians, nurses, pharmacists, and therapists treat illnesses and their symptoms with medication, radiation therapy, or surgery. It is used in most of the countries of the world. And most of the allopathic drugs are mainly originated from different types of natural resources, especially medicinal plants. Our world is a mysterious source of numerous types of natural products having medicinal moieties. These bioactive compounds have been serving the main basis of traditional medicine systems all over the world. Afterward, by using many separation techniques and computer-aided drug designing system or by synthetic chemistry, pure isolated compounds serve as lead molecules in allopathic system. But, still, natural products serve the most important role in providing drug molecules. Pharmaceutical companies from all over the world use these compounds to prepare the ideal dosage forms. There are some important reasons to choose natural products over synthetic products, for instance, (1) well tolerated, (2) better efficacy, (3) less side or adverse effects, (4) comparatively low price, (5) impart mental and spiritual satisfaction, and (6) aiding body's own immune system (Karimi et al. 2015). For those above reasons, in many times people prefer natural compounds over the synthetic ones. As a result, numbers of pharmaceutical companies are built up based on producing natural product-based medicines. And this number is continuously increasing. The decision to use plant-derived drug molecules and other alternative treatments is driven by a number of factors in industrialized nations in Europe and North America, including a dislike of "chemical" drugs and their associated negative or adverse effects, a hope for "natural" alternatives that are purported to have fewer or no side effects, misgivings about the philosophies of allopathic medicine, etc. (Marinac et al. 2007). Nevertheless, these kinds of care are also employed to enhance traditional treatments for persistent, crippling ailments like mental illnesses, cancer, diabetes, and cardiovascular disease. Furthermore, certain social, cultural, and religious perspectives regarding health and illness are frequently strongly associated with traditional medicine (Mans 2013).

1.1 Historical Use of Traditional Resources in Pharmaceuticals

1.1.1 An Overview of the Historical Utilization of Traditional Resources in Pharmaceuticals Industries

Natural products remain a highly fruitful avenue for identifying potential drug candidates. They offer a distinctive structural diversity unlike that found in conventional combinatorial chemistry, opening doors to the discovery of novel lead compounds with low molecular weights. Despite less than 10% of the world's biodiversity having been explored for its impact on living organisms, numerous significant natural lead compounds are yet to be unearthed. The key challenge lies in harnessing this abundant but untapped chemical diversity. It also suggests that there is still a great deal of valuable natural lead compounds to be found (Mishra and Tiwari 2011; Cragg and Newman 2005). The Chinese emperor Shen Nung described herbal medicines around 2800 B.C., which is the first known use of therapeutic herbs (Mans 2013). Around 2600 B.C., Mesopotamia revealed a significant discovery: cuneiform clay tablets recording the medicinal uses of oils derived from *Commiphora* species (myrrh) and Cupressus sempervirens (cypress). These ancient remedies, detailed on the tablets, continue to be utilized today for treating ailments such as colds, coughs, and inflammation (Dias et al. 2012). Dating back to ancient times, the Chinese Materia Medica (1100 B.C.), Shennong Herbal (~100 B.C., comprising 365 drugs), and the Tang Herbal (659 A.D., with over 850 drugs) stand as some of the earliest records documenting the uses of various bioactive natural products. Theophrastus, a pupil of Plato and the successor to Aristotle, penned the fourth-century B.C. Greek masterpiece *Historia Plantarum*, laying the groundwork for botanical science. Dioscorides, a distinguished physician, penned the manuscript *De Materia Medica* around 75 B.C., establishing it as a reliable guide to medicinal plants for more than 1500 years. Furthermore, Dioscorides, a Greek physician born circa 100 A.D., meticulously documented the collection, conservation, and utilization of medicinal herbs (Cragg and Newman 2005; Mans 2013). Galen's manuscript, written between A.D. 129 and 200 in Greece, adds to the body of evidence supporting the significance of plants in ancient medicine. Tang Shen-wei's extensive contributions encompass a diverse range of texts, notably delving into the realm of medicinal herbs. Among these, he meticulously explored the writings of eighth-century Arab scholars in Cordova, Moorish Spain, a renowned hub of scientific inquiry during that era. Additionally, his legacy includes 31 comprehensive volumes dedicated to Chinese medicinal plants, which were published in A.D. 1108. Within these works, Tang Shen-wei meticulously detailed the appearance, attributes, and therapeutic applications of numerous plants contemporary to his time, establishing a foundational pharmacopoeia for generations to come. Most of these works retained elaborative descriptions of medicinal plants. He discussed the therapeutic applications of various plants, including rye, mandrake, castor oil, opium, garlic, and thyme (Abourashed et al. 2003).

1.1.2 Key Historical Examples of Traditional Remedies in Drug Development

The beneficiary effects or roles of traditional medicine as well as traditional resources are too widespread to be ignored. These are the main fundamentals of modern research and development of finding newer and better drug molecules. Although many points of it are still a matter of debate, these offer a variety of activities in which it might be used to non-Western medical technologies or procedures. For instance, within any given herb or formulation, there exists a multitude of phytochemical groups, including but not limited to alkaloids, glycosides, tannins, terpenoids, flavonoids, and polyphenols. These phytochemicals exhibit biological effects either independently or in combination (Parasuraman et al. 2014). At present, more than 80% drugs of allopathic medicine are directly related to medicinal effects in traditional medicine, and most of them were found from various traditional resources (Manoharachary and Nagaraju 2016). Various traditional medicinal systems include Ayurvedic, Unani, homeopathic, traditional Chinese medicine, Kampo, traditional Korean medicine, traditional Aboriginal medicine, Russian herbal medicine, Siddha, and yoga and are mostly popular (Table 1). From the very beginning of their origin, these systems provide revolutionary welfare to modern allopathic system. Even, in many areas of the world, people still believe and depend on these systems, especially people from third-world countries as they can't cope with the increasing cost of treatment.

Table 1 Characteristics of various well-known traditional medicine systems (Castro 1999; Thas 2008; Yuan et al. 2016)

Name (origin)	Principles of theory	Current use in clinical symptoms or in research arena
Ayurveda (India, 400–1500 B.C.)	• Using natural components to restore balance and eliminate the disease's primary cause • Primarily focused on utilizing medicinal herbs and their extracts in a specific ratio to minimize toxicity and produce the best therapeutic result	• More people are becoming aware of the pharmacologically active ingredients in Ayurvedic medicine and how well they work as treatments
Unani medicine (Greece, 300 B.C.) (Ahmad 2003)	• It addresses a person's entire being, including their body, mind, and soul • An individual's physical attributes and innate personality are reflected in their temperament • The human body is prone to a wide range of ailments due to temperamental differences	• The WHO has recognized Unani as a complementary and alternative medical system • Mangrove plants have a variety of bioactive components that have been extracted and utilized in Unani therapy

(continued)

Table 1 (continued)

Name (origin)	Principles of theory	Current use in clinical symptoms or in research arena
Homeopathy (Germany, eighteenth century)	Stimulates healing responses to diseases by administering substances that mimic the symptoms of those diseases in healthy people	• This system is still popular to people in treating asthma, ear infections, allergies, arthritis, high blood pressure, dermatitis
Siddha (India, 10,000 B.C.)	• Emphasizes the holistic approach, physical, mental, social, and emotional well-being • Uses five main elements of nature, like soil, water, fire, air, and space	• Siddha medicine, renowned for its safety and efficacy, employs a combination of herbal and herbo-mineral remedies to address a spectrum of dermatological conditions including psoriasis, alopecia, diabetic ulcer, warts, pemphigus, pompholyx, and leprosy
Kampo (Japan, China, fifth–sixth century)	• Diseases are believed to be the result of psychological illnesses, and herbal remedies and soma are believed to have an equal impact on the body and the soul • The focus of Kampo therapy is the patient as a whole rather than the disease	• Examining the effectiveness of Kampo formulations and investigating associated mechanisms have also received attention. Kampo is thought to be extremely secure • To guarantee the safety and quality of Kampo formulations, both pharmaceutical companies and the government are heavily involved in process surveillance
Sasang constitutional medicine (SCM) (Korea, nineteenth century)	• It offers customized, constitution-specific therapies for a range of issues	• Despite Korea's excellent traditional healthcare system, 86% of the population still uses SCM • In both public and private hospitals, physicians trained in traditional medicine can provide Korean SCM
Traditional medicines in Africa	Generally speaking, they aim to reassemble the patients' social and mental equilibrium in accordance with norms and social ties	• Eighty percent of Africans either combine traditional medicine with conventional medicine or use it alone • Later the traditional medical system of Ghana was integrated into their national healthcare system
Russian herbal medicine (tenth century)	Traditional medical practices from Europe and Asia were included into its systems	• According to a recent survey, 14% of Russians regularly utilize herbal treatments, while 44% do so occasionally • There is a lot of promise in the gathering of plants with expectorant properties

1.1.3 Evolution of the Pharmaceutical Industry's Approach to Traditional Resources

The first pharmacies were privately owned by Arabs in the eighth century. Persian pharmacist, physician, philosopher, and poet Avicenna is known for his writings, which include *The Canon Medicine*, which have had a profound impact on the fields of pharmacy and medicine (Dias et al. 2012).

During the "Age of Herbals" in Europe, spanning from the fifteenth to the seventeenth centuries A.D., herbal medicine emerged as the predominant therapeutic approach for human ailments. Notably, many early herbalists were skilled in both medicine and botany. Physicians, especially those from the Mediterranean, commonly followed Dioscorides' approaches, while doctors from Germany and the Netherlands sought wisdom from "wise women," sometimes labeled as witches, to expand their understanding of medicinal plants. Notable publications from this era include the anonymous *Grete Herball* (1526), Gerard's *The Herbal or General History of Plants* (1597), and Culpeper's *The English Physician Enlarged* (1653), which were highly influential herbal guides during that period. The "doctrine of signatures" was a "scientific" technique that was primarily employed at the time to identify plants with therapeutic properties. A renowned professor from the medicine department at the University of Basel, Switzerland, named Paracelsus developed this method in the first five decades of the sixteenth century. This is a religious or spiritual belief system that is predicated on the idea that God has left plants in the universe for human happiness and has given them a "signature" that serves as a means of identification (Bennett 2007). Even so, the common names given to several plants during that time period still represent the doctrine. For instance, liverworts, characterized by their three-lobed leaves and belonging to the *Hepatica* genus within the Ranunculaceae family, were traditionally believed to possess properties beneficial for liver health. Similarly, wormwood (*Artemisia absinthium* L.) from the Asteraceae family was historically used to combat intestinal parasites, while certain species of snakeroots (genus *Ageratina*, Asteraceae) were thought to counteract the effects of snake venom. Lungworts, found in the *Pulmonaria* genus of the Boraginaceae family, were believed to alleviate lung-related ailments, and walnuts, belonging to the *Juglans* genus within the Juglandaceae family, were reputed to be effective in treating head ailments, as they were thought to bear a resemblance to the brain in their appearance (Bennett 2007).

The active element, morphine, was first isolated by German pharmacist Friedrich Sertürner from the opium poppy, *Papaver somniferum* L. (Papaveraceae), in 1803, setting the groundwork for the current pharmaceutical industry. After the Greek god of dreams, Morpheus, morphine was given its name. It was initially used as a painkiller and a "cure" for addiction of opium and alcohol. Later, it was found to be even more severely addictive compared with alcohol or opium. At present, it is used as a precursor for many bioactive opioids such as the antitussive codeine, thebaine. At present, morphine can only be used to relieve extremely severe chronic pain in patients with terminal cancer (Bennett 2007).

It is important to recognize knowledge from a wide range of nations and cultures served as the primary foundation for these discoveries, as well as many others made in the area of medicine. In fact, traditional medicine and therapies serve as the foundation for allopathic medicine as it exists today. Notably, local medicine has provided clues for the discovery of over 75% of the prescription medications derived from plants used globally (Wakdikar 2004). It should come as no surprise that using traditional medicine and old wisdom to find and create new medications to treat human illnesses continues to be a crucial approach.

2 Traditional Medicine Systems as Sources of Pharmaceuticals

2.1 Exploration of Traditional Medicine Systems as Reservoirs of Pharmaceutical Compounds

2.1.1 Ayurveda, Traditional Medicine, Unani Medicine, and Other Systems

Over 80% of individuals in poor nations lack the financial means to pay for even the most basic medical treatments, medications, and immunizations. Even though there is little evidence to support their safety or efficacy, complementary and alternative therapies are common among affluent people in both developed and developing nations. Historically, the application of natural goods as medicines has been documented through the usage of many of these bioactive natural ingredients in traditional medicines, cures, potions, and oils. The primary root of information on the applications of natural products got derived from medicinal plants is human experimentation over hundreds of years, including palatability tests and premature deaths, in an effort to find foods that could potentially treat illnesses (Kinghorn et al. 2011; Farnsworth et al. 1985). Worldwide recognized various traditional medicine system are thus led the basements of modern allopathic medicine system. The relentless researchers from all over the world are still trying to discover newer and better lead compounds to protect the mankind from the severe diseases. Ayurvedic, Unani, Siddha, Kampo, and Homeopathic systems have still tremendous accessibility toward many nations and races to combat diseases. In Table 2, we have put some important examples from these treatment systems that are still used to combat diseases.

Table 2 Medicinal uses of different natural products in different traditional medicine systems

Name of traditional medicine system	Name of medicinal source (common name)/compound	Clinical indications
Ayurvedic (Kumar et al. 2017; Pandey et al. 2013)	*Asparagus racemosus* (shatamuli)	1. Supplies many female hormones 2. Helps to keep the urinary tract healthy, boosts immunity, and purifies blood
	Commiphora mukul (guggul)	1. Increases white blood cell count to possess strong immunomodulating properties 2. Keeps the HDL to LDL ratio constant while lowering cholesterol and triglycerides and providing protection against the common cold
	Garcinia cambogia (garcinia)	1. Increases the synthesis of glycogen, which suppresses hunger. In this manner, the brain receives signals of happiness and fullness earlier 2. Highly enriched with vitamin C and used as a potential heart tonic
	Glycyrrhiza glabra (yashtimadhu)	1. A gentle laxative and it also relieves muscle spasms and tones and calms mucous membranes 2. As an antioxidant, a cancer preventive agent, a botanical enhancer, and for certain immunological processes such the synthesis of interferon
	Melia azadirachta (neem)	1. Has potent health-improving properties and is used as an astringent and tonic to aid in healing 2. Well known for its detoxifying properties 3. Has demonstrated the greatest advantages for the digestive, respiratory, urinary, and circulatory systems
	Piper longum (long pepper)	1. Has a revitalizing impact on the lungs and is a potent stimulant for the respiratory and digestive systems 2. Plays a crucial role in the release of metabolic heat energy
	Terminalia chebula (haritaki)	1. A tonic, expectorant, and purgative that is both safe and efficient 2. Possesses potent antimutagenic properties due to its high vitamin C concentration
	Zingiber officinale (ada)	1. Reduces the risk of gastrointestinal adverse effects and improves absorption 2. Improves digestion and to prevent nausea
	Curcuma longa (haldi)	1. Have strong anti-inflammatory and anticarcinogenic actions 2. Contains many antioxidants and nutrients
	Aloe barbadensis (Aloe vera)	1. Used in various inflammatory diseases, as well as in skin and liver disease 2. Used to treat flatulence, stomach pain 3. Used to cool body temperature in summer
	Ocimum sanctum (tulsi)	1. Leaves juice is used to treat cold, cough 2. Used in treating helminthiasis

(continued)

Table 2 (continued)

Name of traditional medicine system	Name of medicinal source (common name)/compound	Clinical indications
Unani (Hongal et al. 2014)	*Withania somnifera* (ashwagandha)	1. Has been demonstrated to be successful in defending against a range of biological, physical, and chemical stresses 2.1.1.1. 2. These characteristics may help patients with dental anxiety more effectively
	Acacia arabica (babul)	1. Good in treating dental plaque and gingival conditions 2. Showed very good antibacterial effect in oral infections
	Acacia catechu (Katha)	1. Good remedy in wound healing, dental caries, gingivitis, halitosis, and tonsillitis 2. Contains many important antioxidant molecules
	Allium sativum (lehsun)	Possesses very good antibacterial effect
	Salvadora Persica (meswak)	1. Shows analgesic property in toothache 2. Slightly antimicrobial 3. Contains very important essential oils
	Azadirachta indica (neem)	Has well-established analgesic, antioxidant, antiviral, antifungal, antibacterial, anti-inflammatory, and immunostimulant properties
Siddha (Ebenezer et al. 2021)	*Andrographis paniculata* (nilavembu)	1. Used to treat almost all types of pyrexia associated with muscle ache 2. Antiviral for arboviruses
	Amukkara choornam	A herbal mixture formula used to treat post-pyretic phase viruses that cause severe arthritic pain that can last for months or years
	Withania somnifera (Ashwagandha)	Having anti-inflammatory, neuroprotective, and immunomodulatory effects
	Kabasura kudineer	A herbal mixture formula used to treat respiratory diseases associated with fever
Kampo (Ravichandiran et al. 2017)	Juzen-taiho-to	Anemia, chronic disease, general fatigue, loss of appet ite, chemotherapy or surgery, colds
	Keishi-bushi-to	Rheumatoid arthritis
	Sho-saiko-to	Chronic gastroenteritis, orthostatic hypotension, chronic hepatitis, bronchial asthma, chronic tonsillitis, psychological problems
	Toki-shakuyaku-san	General fatigue, headache, menopausal syndrome, anemia, chronic nephritis, edema, sterility, cold limb
	Unsei-in	Allergic dermatitis, rhinitis, skin disease, gastritis, Behcet's disease, urticaria

Homeopathic (www.askdrshah.com/homeopathic-medicines.aspx)	Arnica	Arnica is commonly used for treating injuries, bruises, and muscle soreness
	Chamomilla	Chamomile is known for its calming properties and is used in homeopathy for conditions involving irritability, teething in infants, and digestive issues
	Nux vomica	*Nux vomica* is used for digestive issues, hangovers, and irritability
	Calcium carbonate	Commonly used in teething problems and issues related to the bones and joints
	Phosphorus	Used in homeopathy for conditions involving the nervous system, respiratory system, and digestive system
	Sulfur	Used in homeopathy for skin conditions, itching, and certain respiratory issues
	Cantharis	Used in homeopathy for conditions involving burning pain, particularly in the urinary and reproductive systems
	Sepia	Used in homeopathy for conditions related to hormonal imbalance, fatigue, and reproductive issues
	Thyroidinum	Used in homeopathy for conditions related to thyroid dysfunction. It may be recommended for symptoms such as fatigue, weight issues, and emotional disturbances
	Adrenalinum	Used in homeopathy for conditions related to stress and adrenal fatigue. It may be considered for symptoms such as fatigue, weakness, and stress-induced conditions
Others	*Salvia*	Aids in childbirth
	Alhagi maurorum	Effective in the treatment of epistaxis, dermatosis, pyrexia, leprosy, anorexia
	Ligusticum scoticum	Has antimicrobial capabilities
	Piptoporus betulinus	Antiseptic, disinfectant
	Agaricus campestris	Has antitumor properties
	Usnea dillenius	Cures scalp diseases
	Parmelia omphalodes	Cures inflammation on feet, cures burns and wounds
	Chondrus crispus, Mastocarpus stellatus	Cures colds, sore throats, chest infections including tuberculosis
	Porphyra umbilicalis	Acts as antitumor, laxative

2.1.2 Identification and Extraction of Active Compounds from Traditional Remedies

In order to get pure or isolated compounds from a medicinal source, we have to go through multiple extraction processes. This initial phase is pivotal for isolating targeted bioactive components from source materials for subsequent processing (such as separation and characterization). The essential procedure encompasses several steps: pre-washing the materials, freeze- or air-drying them, grinding to achieve homogeneity, and maximizing surface contact with the solvent. Throughout extraction, utmost care is necessary to preserve the integrity of potentially active ingredients, avoiding damage, distortion, or loss (Fabricant and Farnsworth 2001). The unique characteristics of the bioactive compound being targeted play a major role in the solvent system selection. The bioactive component can be extracted from natural products using a variety of solvent systems. Ethyl acetate, methanol, and other polar solvents are used in the extraction of polar compounds. More lipophilic substances can be extracted using dichloromethane or a 1:1 combination of dichloromethane and methanol. Because the desired compounds may be thermally labile and nonpolar to polar, the extraction methods' suitability must be taken into account. Sonification, heating under reflux, soxhlet extraction, and other techniques are frequently employed. Other contemporary extraction methods with specific benefits include surfactant-mediated extraction, solid-phase micro-extraction, microwave-assisted extraction, pressurized liquid extraction, supercritical fluid extraction, and solid-phase extraction (Huie 2002).

Separating plant extracts is important to find and understand the different helpful parts because they usually come mixed with various kinds of helpful components, or phytochemicals, that have different characteristics. To get pure compounds from these bioactive components, scientists usually use different types of chromatographic techniques like thin layer, column, flash, and high-performance liquid chromatography (HPLC). After getting pure compounds, they study their structure and how they work in the body. Besides chromatography, scientists can also find bioactive chemicals using methods like immunoassays, phytochemical screening assays, and Fourier-transform infrared spectroscopy (FTIR) (Sasidharan et al. 2011).

2.1.3 Case Studies Showcasing Successful Integration of Traditional Resources into Pharmaceutical Products

Type of species	Name	Uses in traditional medicine	Isolated pure compound used in modern allopathic medicine
Plants	*Schisandra chinensis*	Hepatoprotective	Schisandrin C, bifendate, bicyclol (Yang et al. 1994)
	Taxus brevifolia	Cytotoxic, anticancer	Paclitaxel (Yuan et al. 2016)
	Camptotheca acuminata	Antitumor	Camptothecin, irinotecan, topotecan (Yuan et al. 2016)
	Ginkgo biloba	Cerebral infarction	Ginkgolide B (Yuan et al. 2016)
	Polygonum multiflorum	Vascular dementia	Stilbene glycoside (Yuan et al. 2016)
	Ranunculus ternatus	Antituberculosis	Ternatolide (Yuan et al. 2016)
	Curcuma longa	Hyperlipoidemia	Cucurmin (Yuan et al. 2016)
	Catharanthus roseus	Antitumor	Vincristine, vinblastine (Yuan et al. 2016)
	Artemisia annua	Antimalarial	Artemisinin (Yuan et al. 2016)
	Corydalis yanhusuo	Analgesic	Tetrahydropalmatine (Yuan et al. 2016)
	Ligusticum chuanxiong	Heart arrhythmia	Tetramethylpyrazine
	Paeonia lactiflora	Analgesic	Paeoniflorin (Yuan et al. 2016)
	Epimedium brevicornum	Osteoporosis	Icariin (Yuan et al. 2016)
	Pueraria lobata	Diabetes	Puerarin (Yuan et al. 2016)
	Salvia miltiorrhiza	Cardiovascular and cerebrovascular diseases	Salvianolic acid (Yuan et al. 2016)
	Uncaria rhynchophylla	Antihypertensive	Rhynchophylline (Yuan et al. 2016)
	Saussurea lappa	Anti-gastric ulcer, antispasmodic	Costunolide (Yuan et al. 2016)
	Salix alba	Analgesic	Aspirin (Dias et al. 2012)
	Papaver somniferum	Analgesic	Opium alkaloids (Dias et al. 2012)
	Digitalis purpurea	Cardiotonic	Digitoxin (Dias et al. 2012)
	Cinchona succirubra	Antimalarial	Quinine (Dias et al. 2012)
	Pilocarpus jaborandi	Against glaucoma	Pilocarpine (Dias et al. 2012)
	Elaeocarpus grandis	Analgesic	Grandisines A, B (Dias et al. 2012)
	Galanthus nivalis	Treats dementia	Galantamine (Dias et al. 2012)
	Chondrodendron tomentosum	Muscle relaxant	Tubocurarine (Dias et al. 2012)
Fungi	*Aspergillus terreus*	Hyperlipoidemia	Lovastatin (Yuan et al. 2016)
	Penicillium notatum	Antibiotic	Penicillin (Dias et al. 2012)
	Amycolatopsis orientalis	Antibiotic	Vancomycin (Dias et al. 2012)
	Saccharopolyspora erythraea	Antibiotic	Erythromycin (Dias et al. 2012)
	Streptomyces peucetius	Antibiotic	Doxorubicin (Dias et al. 2012)

Type of species	Name	Uses in traditional medicine	Isolated pure compound used in modern allopathic medicine
Bacteria	*Chromobacterium violaceum*	Antitumor	Romidepsin (Yuan et al. 2016)
Marine product	*Aplidium albicans*	Antitumor	Plitidepsin (Dias et al. 2012)
	Spisula polynyma	Antitumor	Spisulosine (Dias et al. 2012)
	Dilophus ligatus	Antimicrobial	Crenuladial (Dias et al. 2012)
	Dilophus ligatus	Antimalarial	Aurepinnacin (Dias et al. 2012)
	Cryptotheca crypta	Antiviral	Spongouridine, spongothymidine (Dias et al. 2012)
	Bugula neritina	Antitumor	Bryostatin (Dias et al. 2012)

3 Biodiversity and Traditional Resources

3.1 The Relationship Between Biodiversity and Pharmaceutical Development

The variety among living things from various environments, such as desert, marine, and terrestrial ecosystems, as well as the ecological complexes they are a part of, is known as biodiversity. Biodiversity includes a wide range of species (plants, animals, fungi, and microorganisms), as well as a wide range of genes within those species and a wide range of ecosystems that those species live in. It is our planet's most intricate and significant characteristic. Life could not persist without biodiversity (Aerts et al. 2018; Young 1999).

The enormous genetic diversity found in plants, animals, and microorganisms presents a plethora of opportunities for improving humankind's ability to produce food, materials, and medicines. In contemporary pharmaceutical research, chemical variety and biodiversity are equivalent. The most important field in the history of medicine, disease prevention, and overall well-being has been and will remain drug discovery from wild species. Secondary metabolites are often created for specialized objectives and often serve crucial biological roles in plants, animals, and microbes. Therefore, tiny molecules generated in this way are "programmed" by nature to interact with large macromolecules (DNA, proteins) and thereby modify their functions. Lead structures and ready-made medications can both be obtained from natural materials. The wealth of information regarding the ecology, taxonomy, and applications of biologically significant organisms for medicine is too great to be wasted, including both traditional and contemporary scientific knowledge. As such, sustainability must be incorporated into all drug discovery programs' research models, whether they are natural or synthetic (Scannell and Bosley 2016; Young 1999) (Fig. 1).

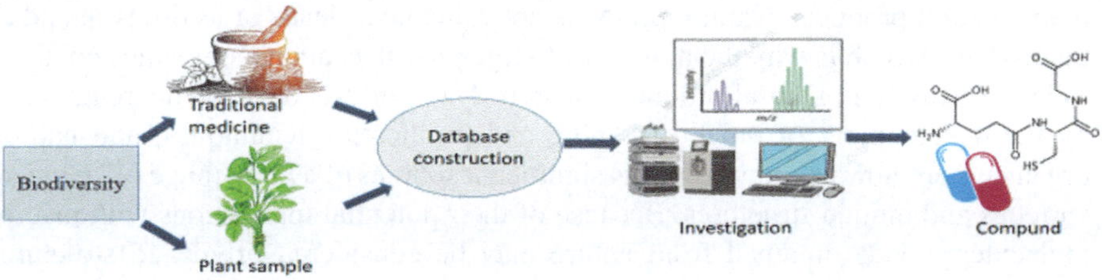

Fig. 1 Relationships between biodiversity and drug development

3.2 Historical Perspective and Present State of Drug Discovery

The drug discovery process is as old as human civilization. Without the development of drugs, human existence on Earth would not have been possible. From that time to today's modern world, there have been dramatic advancements in medical science. Over the past 200 years, there have been significant advancements in chemistry that have increased the availability of both synthetic analogs and pure molecules. Potent and effective extracts from biological sources were purified to derive active compound(s) having better properties than the virgin extracts. Advancements in cell biology and biochemistry during the twentieth century enabled the establishment of in vitro testing and screening, utilizing animal and, at times, human cell and tissue models. Currently, pharmaceuticals can be optimized without undergoing more difficult and confusing testing on humans and animals. This resulted in an increased focus on lead optimization and the development of more targeted medications with reduced side effects. The past 20 years have been defined by a phenomenon that can only be referred to as the "molecular biology revolution." We have been able to identify significant new biochemical targets for the treatment of disease through the identification of macromolecule targets for new drug designing as well as their role in diseases. Thus, there are string chances to discovered newer and better drug molecules from the previously treated sources.

The pharmaceutical industry has a great chance to find new and targeted treatments for the benefit of humanity, thanks to these potent techniques that are revealing a plethora of possible drug targets. High throughput screening and combinatorial chemistry can be used to quickly test a large range of naturally occurring materials, whether they are pure or impure. This has led to an increased appetite of the pharmaceutical companies to test a huge number of compounds . The importance of rescreening all available compounds—even those that haven't previously displayed intriguing activity—must be underlined as these new targets are determined. For older compounds, new targets and assays present fresh possibilities.

Approximately 35% of newly approved drugs in the decade between 1983 and 1994 were either synthetics based on natural products or compounds derived directly from natural products, with only around 5% of newly approved drug molecules being natural products. Natural products are valuable as leads or as drugs in and of themselves, which is why the majority of large pharmaceutical companies continue to invest heavily in natural product research. These efforts have been strengthened by the development of quick screening and purification techniques, and natural organisms are now acknowledged as significant sources of a large range of chemical varieties and unique structures. Because of their potential interactions with macromolecules, goods obtained from nature may be considered privileged structures having a considerably better likelihood of demonstrating biological activity as compared to random synthetic compounds.

3.3 *Conservation and Sustainable Use of Biodiversity-Rich Traditional Resources in Pharmaceuticals*

Although biodiversity becomes essential for the development of new drugs, biodiversity is alarmingly disappearing. The extinction rates of today are between 100 and 1000 times higher than those of previous eras (Pimm et al. 2014).

The well-being of humans may be indirectly impacted by biodiversity loss. Ecosystems become less vibrant and more vulnerable to shocks and disturbances as a result. It also fails to provide humans with essential needs. The natural protection provided by wetland habitats, such as the regulation of water run-off, is compromised, and as a result, the damage caused by floods and storms to coastal communities increases significantly. Natural disasters that have recently occurred in North America and Asia highlight this fact (Pushpangadan and Behl 2005). The loss of biodiversity affects the human diseases expansion, threatens production of food and water quality, reduces the availability of raw materials for biotechnology and drug discovery, and results in the loss of medical models (Chivian 2002). Not only does biodiversity aid in the healing process, but it also plays a major role in averting disease and illness because healthy ecosystems have the potential to safeguard human health. One of ecosystems' most important roles is to keep the balance between predators and prey, as well as between hosts, vectors, and parasites in plants, animals, and humans, therefore avoiding the emergence and expansion of infectious illnesses (Daszak et al. 2000; Ostfeld and Keesing 2000).

Malaria and leishmaniasis are two infectious diseases in humans that can be impacted by disturbing these equilibria through deforestation. Cholera is spread by increased algal blooms, which are caused by warming waters, fertilizer use, and sewage discharge, among other reasons. Variations in the quantity of acorns as well as in the populations of black-legged ticks, white-tailed deer, and white-footed mice can spread the Lyme disease (Walsh et al. 1993; Ostfeld et al. 1998). Plant medicine preparation and procurement costs have been impacted by notable changes in savannas, forests, and many other vegetation types. The degradation of spiritual locations, hallowed areas, and crevices has a tendency to diminish their worth and promote misuse of these lands (Anyinam 1995). For example, the Brazilian Amazon's deforestation during the past 30 years has reduced the availability of certain commonly used species of medicinal plants. The destruction of Amazonian forests could mean the end of many Brazilians' only access to healthcare, in addition to the loss of possible pharmaceutical medications for the developed nations. Many of these people live in rural and urban areas. This type of scenario is more or less common over the world (Shanley and Luz 2003). Even in rural areas, the area of forest has been reduced. As a result, the availability of medicinal plants has become scarce which also acts as precursor of high treatment cost. For instance, the freshly made herbal remedies are gradually being replaced by various mixtures, tinctures, and powdered forms so that they can be kept longer without going bad or losing their potency (Anyinam 1988). This continuous loss of forests directly hampers many animals especially lives on these areas. Animals like tigers, rhinos, bears, monkeys, spotted deer, and pangolins are overfished as a result of the growing use of medicinal animals. For these reasons, there is also scarcity of animal-derived drug products. So, it is high time to raise awareness to protect forests and animals, as well as biodiversity. If we neglect the conservation of nature and biodiversity, we will suffer in the long run (Alves and Rosa 2007).

4 Scientific Validation and Modern Pharmaceutical Research

4.1 The Importance of Scientific Validation in Utilizing Traditional Resources for Drug Development

Scientific validation plays an important role in modern pharmaceutical research. This is a process that uses rigorous scientific methods to confirm the effectiveness and safety of new drugs. Verification is essential to ensure the reliability, effectiveness, and safety of pharmaceuticals for use in human (Kapoor et al. 2018). Indigenous communities have long relied on traditional resources like medicinal plants for their therapeutic benefits, spanning centuries of healing practices. These plants harbor a wealth of compounds that can be harnessed through extraction to create potent medicines. One such illustration is the cinchona tree, whose bark has historically been employed in malaria treatment. Its active component, quinine, remains integral to contemporary antimalarial medications (Ku 2016).

Traditional resources have been used for centuries in various cultures around the world for the treatment of illness and disease. These resources, including plants, herbs, and minerals, have provided comfort and healing to countless individuals (Jaiswal et al. 2020).

The potential of these traditional resources in drug development and scientific validation is of great importance. Scientific validation involves rigorous testing and experimentation to determine the safety and effectiveness of a substance. This process ensures that potential medicines from traditional sources meet the standards required for human use. Without scientific verification, there is a risk that ineffective or harmful medicines will enter the market. By subjecting these resources to rigorous scientific scrutiny, their effectiveness can be objectively assessed. This not only builds trust among medical professionals but also encourages further research and investment in traditional medicine (Ashby et al. 2004).

Scientific validation allows for standardization of dosage and prescription. The chemical composition of traditional resources often varies depending on factors such as soil conditions and harvesting methods. Through scientific validation, researchers can identify the active ingredients responsible for therapeutic effects and develop standardized formulations that ensure consistent quality across different pharmaceutical batches. Standardize dosages and prescriptions to make these medicines more reliable and available to patients around the world. Numerous preclinical and clinical investigations have explored the cytoprotective, immunomodulatory, and immunoadjuvant capabilities of Ayurvedic treatments. Leveraging ethnopharmacological insights, alongside its holistic and systematic approach grounded in experiential knowledge, Ayurvedic medicine stands poised as a potent and innovative catalyst for discovering newer, safer, and more affordable therapeutic solutions (Patel et al. 2022).

4.2 Methods and Techniques Used to Validate Traditional Remedies for Pharmaceutical Use

There are various methods and techniques used to validate traditional treatments as medicines. One of the most commonly used methods is ethno botanical research, where researchers study indigenous communities and their traditional medical practices. By observing the plants used, preparation methods, and dosage forms, scientists can identify active ingredients that may have therapeutic effects. Another technique is phytochemical analysis, which involves separating and identifying the chemical components of plant extracts. This helps confirm the presence of bioactive compounds such as alkaloids, flavonoids, and terpenes that contribute to the plant's medicinal properties (Rahayu et al. 2021). Pharmacological studies are conducted to evaluate the biological activities of traditional remedies. These studies involve testing the extracts or isolated compounds on animals or cell cultures to assess their efficacy in treating specific diseases or conditions. In addition to these laboratory-based techniques, clinical trials are also essential for validating traditional remedies. These trials involve administering the remedy to human subjects under controlled conditions while monitoring its safety and effectiveness (Chow et al. 2011). Moreover, modern analytical techniques such as HPLC and mass spectrometry (MS) are employed to quantify active compounds present in traditional remedies accurately. Overall, validating traditional remedies for pharmaceutical use requires a multidisciplinary approach combining ethnobotany surveys, phytochemical analysis, pharmacological studies, clinical trials, and advanced analytical techniques (Li and Weng 2017).

T hese methods and techniques can ensure that traditional treatments meet rigorous standards of safety and effectiveness before being integrated into modern medicine.

4.3 Research Advancements and Breakthroughs in Incorporating Traditional Resources into Modern Pharmaceuticals

Advances in research and breakthroughs that incorporate traditional resources into modern medicine have opened new possibilities for treatment. Traditional resources such as plants and herbs have been used for centuries in different cultures around the world to treat various diseases (Reid et al. 2018). However, with the advent of modern medicine, these traditional treatments are often ignored or rejected. There is renewed interest in traditional resources because of their potential therapeutic benefits. Development of new medicines and treatments based on traditional sources is growing in modern medicine system. One example is the use of plants in cancer treatment. Many plant compounds have shown promising anticancer properties, leading to the development of drugs such as taxol, which is derived from the Pacific yew tree. Another example is the use of traditional Chinese medicine in the treatment of various diseases such as cardiovascular and respiratory diseases (Suranjan and Deekshit 2021). There are also several benefits to incorporating traditional resources into modern medicine. First, these natural substances are often more readily available and cheaper than synthetic drugs. Furthermore, they may have fewer side effects and be better tolerated by patients (Reid et al. 2018). Integrating traditional resources into modern medicine presents challenges. A major challenge is standardization, ensuring consistent quality and dosage between different batches of natural products. Another challenge is intellectual property rights. This is because many traditional treatments are part of cultural heritage and may not be easily patentable (Obeid et al. 2017).

Traditional resources for modern medicine have the potential to improve health outcomes. The potential therapeutic benefits of these natural resources may lead to safer and more effective treatments for various diseases.

5 Challenges and Controversies in Pharmaceutical Use of Traditional Resources

5.1 Ethical Considerations Surrounding the Integration of Traditional Resources into Pharmaceuticals

The pharmaceutical industry has long relied on traditional resources to develop new drugs and treatments. Many traditional plants and herbs are being overharvested, leading to a lack of natural habitat. This not only threatens ecosystem biodiversity but also limits opportunities for future drug development (Li and Weng 2017). Controversy also arises when pharmaceutical companies patent traditional resources or indigenous knowledge without proper compensation or recognition to local communities. This exploitation can lead to cultural appropriation and economic inequality. Given these challenges and controversies, it is important to promote sustainable harvesting practices and efforts to protect traditional resources (Girsberger 2000). Ethical concerns include the potential for exploitation of indigenous communities and their knowledge. Many traditional resources come from these communities, which have preserved medical practices for generations. When pharmaceutical companies profit from these resources without adequately compensating or including these communities, it can be considered a form of cultural appropriation and exploitation (Bubela and Gold 2012). Ethical considerations relate to the sustainability of traditional resources. As demand for natural ingredients increases, there is a risk that certain plant species may become overharvested or become extinct, disrupting the delicate balance within ecosystems. Integrating traditional resources into medicines can lead to promising medical advances, but must be done with careful consideration of ethical concerns. Pharmaceutical companies must engage in fair trade practices with indigenous communities and prioritize sustainability to avoid exploitation and environmental damage (Segger and Phillips 2015).

5.2 Intellectual Property Rights and Issues Related to Biopiracy

Intellectual property rights (IPR) play a vital role in safeguarding human creativity, encompassing inventions, literary and artistic works, and trademarks. Biopiracy occurs when individuals or entities exploit biological resources and traditional knowledge without permission, aiming for commercial benefit without adequately compensating the communities or nations that initially sourced these resources. Biopiracy is the exploitation of indigenous knowledge. Indigenous communities possess a wealth of traditional knowledge about medicinal plants, agricultural practices, and other natural resources that have been passed down through generations. Biopirates often exploit this knowledge without obtaining informed consent or compensating these communities appropriately (Suranjan and Deekshit 2021). Many developing countries are rich in biodiversity and possess unique genetic materials that can be used for various purposes such as pharmaceutical research or crop improvement. However, multinational corporations often extract these resources without proper authorization or benefit-sharing agreements with local communities. The consequences of biopiracy are the rights and livelihoods of indigenous peoples, and this hampers scientific research and development in developing countries (Singh and Singh 2015). Additionally, raising awareness about biopiracy among policymakers, researchers, and consumers is crucial in promoting ethical practices regarding intellectual property rights. Biopiracy poses significant challenges to intellectual property rights by exploiting indigenous knowledge and misappropriating genetic resources. Governments worldwide strengthen legal frameworks to protect traditional knowledge holders' rights while ensuring equitable benefit-sharing arrangements between resource-rich nations and corporations (Hamilton 2008).

5.3 Standardization, Quality Control Challenges, and Safety Concerns in Traditional Resource-Based Pharmaceuticals

For centuries, traditional resource-based medicines have been relied upon to address diverse health conditions. Sourced from natural reservoirs like plants, animals, and minerals, these remedies hold a rich history of efficacy. Nevertheless, ensuring standardized quality across these products presents notable hurdles, raising pertinent issues surrounding safety and quality control (Dhami 2013).

Modern drugs that are synthesized in laboratories under controlled conditions, traditional medicines are often prepared using ancient recipes passed down through generations. This lack of standardization makes it difficult to ensure consistent quality and potency in each batch produced (Nirmalatha and Sulekha 2016). Quality control is another major challenge faced by traditional resource-based pharmaceuticals. The variability in raw materials and preparation methods can lead to variations in the final product's composition. Additionally, the lack of regulatory oversight further complicates quality control efforts (Prakash et al. 2020).

Safety concerns also arise due to the use of natural ingredients that may contain toxic substances or interact with other medications. Traditional medicines often involve complex mixtures of plant extracts or animal parts, making it challenging to identify potential risks accurately (Rietjens et al. 2008). The regulatory bodies should establish guidelines for quality control and safety testing to ensure that these medicines meet acceptable standards. Traditional resource-based pharmaceuticals offer unique therapeutic potential, but their standardization poses significant challenges in terms of quality control and safety concerns (Patel et al. 2022).

6 Regulatory Landscape

6.1 Overview of Regulatory Frameworks Governing the Use of Traditional Resources in Pharmaceutical Industries

The regulatory landscape encompasses the intricate network of laws, regulations, and policies that govern industries and sectors within a nation. Crafted to uphold principles of fair competition, consumer protection, public safety, and economic equilibrium, it serves as a cornerstone shaping the business environment and guiding organizational conduct (Ogus 2002). The regulatory landscape is its ability to strike a balance between promoting innovation and protecting public interest. Regulations are necessary to prevent monopolistic practices, safeguard against fraud and deception, and ensure product safety. However, excessive regulation can stifle entrepreneurship and hinder economic growth. Therefore, it is essential for regulators to adopt a flexible approach that encourages innovation while addressing potential risks (Maresova 2023). Regulatory frameworks governing the use of traditional resources in pharmaceutical industries vary from country to country. In some countries, there are strict regulations in place that require companies to obtain permits or licenses before using traditional resources for commercial purposes. These regulations aim to prevent overexploitation and ensure that indigenous communities are fairly compensated for their knowledge and resources (Garattini and Padula 2017). Additionally, international agreements such as the Convention on Biological Diversity (CBD) provide a framework for countries to regulate access to genetic resources and share benefits arising from their utilization. The CBD recognizes the importance of traditional knowledge in pharmaceutical industries and encourages countries to develop national legislation that protects indigenous rights (Lakshmanan 2018). However, despite these regulatory frameworks, challenges still exist. Enforcement of regulations can be difficult due to limited resources or corruption. Furthermore, there is often a lack of awareness among pharmaceutical companies about the importance of respecting traditional knowledge and obtaining informed consent from indigenous communities (Garattini and Padula 2017).

The regulatory frameworks governing the use of traditional resources in pharmaceutical industries play a crucial role in ensuring sustainability and protecting indigenous rights. However, more efforts are needed to strengthen enforcement mechanisms and raise awareness among all stakeholders involved. Only through effective regulation can we strike a balance between modern medicine's advancements and preserving traditional knowledge for future generations (Forda et al. 2013).

6.2 Regulatory Challenges and Opportunities in Bringing Traditional Resource-Based Drugs to Market

For centuries, traditional resource-based medicines have served diverse cultures globally, harnessing the power of nature's bounty from plants, animals, and minerals to effectively address a myriad of health issues. Yet, navigating the path to modern market integration for these time-tested remedies presents a spectrum of regulatory hurdles and promising opportunities. One of the main challenges is the lack of scientific evidence to support the safety and efficacy of these traditional drugs. Unlike modern pharmaceuticals, which undergo rigorous testing and clinical trials before being approved for use, traditional medicines often lack this level of scientific scrutiny. This makes it difficult for regulatory authorities to assess their safety and effectiveness accurately (Vikas et al. 2017). Another challenge is the standardization of traditional resource-based drugs. These medicines are often prepared using ancient recipes or techniques that vary from one culture to another. The standardization makes it challenging to ensure consistent quality control across different batches or manufacturers (Sudha et al. 2009). However, there are also opportunities in bringing these traditional medicines to market. Many people today are seeking alternative forms of treatment that are more natural and holistic than conventional pharmaceuticals. Traditional resource-based drugs can offer a viable solution by providing effective treatments with fewer side effects (Patwardhan et al. 2008). The traditional resource-based drugs to market presents both regulatory challenges and opportunities. While there may be difficulties in terms of scientific evidence and standardization, embracing these natural remedies can provide alternative treatment options for individuals seeking holistic healthcare solutions while preserving cultural heritage and promoting sustainability (Mordeniz 2019).

6.3 Case Studies Illustrating Successful Compliance with Regulatory Requirement

Case studies illustrating successful compliance with regulatory requirements serve as valuable for businesses and organizations striving to meet legal obligations. One such case study is the pharmaceutical industry's adherence to Good Manufacturing Practices (GMP). GMP ensures that drugs are consistently produced and controlled according to quality standards (Tabersky et al. 2018). A prominent example of successful compliance is Johnson & Johnson's response to the Tylenol poisoning incident in 1982. Following the tampering of their product, the company swiftly recalled all Tylenol bottles from store shelves, redesigned packaging with tamper-evident seals, and implemented strict manufacturing processes. This proactive approach not only demonstrated their commitment to consumer safety but also set a benchmark for the entire industry (Kapoor et al. 2018). The pharmaceutical industry's response following the Tylenol poisoning incident demonstrates how swift action can restore consumer confidence and set new industry standards. Similarly, Google's adaptation of privacy settings showcases how companies can comply with data protection regulations while prioritizing user trust (Arnold and Oakley 2013).These cases emphasize the importance of proactivity, commitment to consumer safety or privacy, and adaptability in achieving successful compliance. By studying these examples, businesses can learn from best practices and develop strategies that enable them to meet regulatory requirements while fostering trust among consumers (Kabanov 2016).

7 Future Directions and Opportunities

7.1 Future Directions and Opportunities on the Importance of Traditional Resources in Pharmaceutical Industries

The pharmaceutical industry has historically drawn upon traditional sources in its quest for new drugs, tapping into the wealth of knowledge passed down through generations in indigenous communities. Plants, animals, and minerals long employed by these communities to address diverse health issues have served as valuable reservoirs of medicinal compounds. Nevertheless, the rise of modern medicine has ushered in an era increasingly dominated by synthetic compounds and laboratory-driven research methodologies. Consequently, there has been a waning interest in harnessing the potential of traditional resources for drug discovery, marking a notable departure from past practices (Suranjan and Deekshit 2021). Exploratory research is delving deeply into traditional medicine systems like Ayurveda, traditional Chinese medicine (TCM), and Indigenous healing practices, aiming to uncover potent compounds for integration into pharmaceutical development. One major advantage of traditional resources is their long history of use by different cultures around the world. These remedies have been tried and tested over generations, providing valuable insights into their safety and efficacy (Jaiswal et al. 2020). Moreover, traditional resources offer a vast array of chemical diversity that cannot be replicated synthetically. Plants alone contain thousands of bioactive compounds with medicinal properties waiting to be discovered. H arnessing this biodiversity through sustainable practices such as bioprospecting and ethnobotanical surveys can unlock new opportunities for drug discovery (Laraia et al. 2018). The future of pharmaceutical industries lies in the exploration and utilization of traditional resources. It is crucial to continue to invest in research and collaborations that prioritize the importance of traditional resources in drug discovery (Roy et al. 2011). The traditional medicine relies on natural remedies and ancient practices; modern pharmaceutical approaches utilize cutting-edge technology and synthetic compounds. Although these two approaches may seem distinct, there is immense potential for synergistic collaboration between them. Traditional medicine has been practiced for centuries and offers a wealth of knowledge about the healing properties of plants, herbs, and other natural substances. Many modern drugs have their roots in traditional remedies, as scientists have isolated active compounds from plants to create effective medications (Dhami 2013). Collaboration between traditional and modern pharmaceutical approaches can also lead to increased accessibility to healthcare worldwide. Traditional medicines are often more affordable than their synthetic counterparts, making them an attractive option for populations with limited resources. Traditional remedies should undergo rigorous testing to ensure their safety and efficacy before being integrated into mainstream healthcare practices. Additionally, efforts must be made to bridge any gaps in understanding between practitioners of traditional medicine and those trained in Western medical systems (Ghanemi 2017).

7.2 Emerging Trends and Innovations in Utilizing Traditional Resources for Drug Discovery

Traditional resources, such as plants and natural products, have been used for centuries by indigenous cultures for medicinal purposes. However, with the advent of modern technology and a deeper understanding of bioactive compounds, researchers are now able to harness the potential of these resources more effectively (Segneanu et al. 2017). Ethnobotany involves studying the relationship between people and plants, particularly how different cultures use plants for medicinal purpose. There has been a growing emphasis on sustainable sourcing and conservation efforts when it comes to utilizing traditional resources for drug discovery. With many plant species at risk due to habitat destruction or overharvesting, scientists are now working closely with local communities and governments to ensure responsible collection methods that protect biodiversity (Schuhmacher et al. 2016). Emerging trends and innovations in utilizing traditional resources for drug discovery offer immense potential for developing new treatments. The integration of ethnobotanical knowledge with scientific methods, along with the application of AI technologies, allows researchers to unlock nature's secrets more effectively (Patwardhan et al. 2008).

7.3 Research and Development Prospects in the Pharmaceutical Industry Related to Traditional Resources

Research and development (R&D) prospects in the pharmaceutical industry related to traditional resources hold immense potential for the future. Traditional resources, such as plants and herbs, have been used for centuries in various cultures to treat ailments. With advancements in technology and scientific understanding, there is now an opportunity to explore these resources further. Many plant species contain unique compounds that can be harnessed for medicinal purposes. For example, the bark of the Pacific yew tree led to the discovery of Taxol, a powerful anticancer drug. By conducting extensive research on traditional resources, scientists can uncover new therapeutic agents that may revolutionize healthcare (Schuhmacher et al. 2016). Moreover, R&D prospects in this field have significant economic implications. Developing countries often possess a wealth of traditional knowledge about medicinal plants. By investing in research and development related to these resources, pharmaceutical companies can tap into this knowledge base and create partnerships with local communities. However, there are challenges associated with researching traditional resources. Firstly, there is a need for stringent regulations to protect biodiversity and ensure ethical practices when sourcing materials from nature. The pharmaceutical industry related to traditional resources offers promising avenues for innovation and economic growth. H arnessing the power of nature's bounty responsibly and ethically can unlock new treatments that may improve global health outcomes while preserving our planet's biodiversity (Kingston 2011).

References

Abourashed EA, El-Alfy AT, Khan IA, Walker L (2003) Ephedra in perspective—a current review. Phytother Res 17:703–712

Aerts R, Honnay O, Van Nieuwenhuyse A (2018) Biodiversity and human health: mechanisms and evidence of the positive health effects of diversity in nature and green spaces. Br Med Bull 127:5–22

Ahmad J (2003) An introduction and development of Unani system of medicine and its potential herbal drugs used for liver diseases. Toyama Med Pharmaceut Univ 29:3–9

Alves RR, Rosa IM (2007) Biodiversity, traditional medicine and public health: where do they meet? J Ethnobiol Ethnomed 3:1–9

Anyinam CA (1988) Persistence with change: a rural-URBAN study of ethno-medical practices in contemporary Ghana 2700–2700

Anyinam C (1995) Ecology and ethnomedicine: exploring links between current environmental crisis and indigenous medical practices. Soc Sci Med 40:321–329

Arnold DG, Oakley JL (2013) The politics and strategy of industry self-regulation: the pharmaceutical industry's principles for ethical direct-to-consumer advertising as a deceptive blocking strategy. J Health Polit Policy Law 38:505–544

Ashby S, Chuah SH, Hoffmann R (2004) Industry self-regulation: a game-theoretic typology of strategic voluntary compliance. Int J Econ Bus 11:91–106

Bennett BC (2007) Doctrine of signatures: an explanation of medicinal plant discovery or dissemination of knowledge? Econ Bot 61:246–255

Bubela T, Gold ER (2012) Introduction: indigenous rights and traditional knowledge. In: Genetic Resources and Traditional Knowledge. Edward Elgar Publishing, Cheltenham

Castro M (1999) Homeopathy: a theoretical framework and clinical application. J Nurse Midwifery 44:280–290

Chivian E (2002) Biodiversity: its importance to human health. Center for Health and the Global Environment, Harvard Medical School, Cambridge, MA, p 23

Chow S-C, Yang L-Y, Lu Y (2011) Some controversial issues in clinical trials. Drug Inf J 45:163–174

Cragg GM, Newman DJ (2005) Biodiversity: a continuing source of novel drug leads. Pure Appl Chem 77:7–24

Daszak P, Cunningham AA, Hyatt AD (2000) Emerging infectious diseases of wildlife—threats to biodiversity and human health. Science 287:443–449

Dhami N (2013) Trends in pharmacognosy: a modern science of natural medicines. J Herbal Med 3:123–131

Dias DA, Urban S, Roessner U (2012) A historical overview of natural products in drug discovery. Metabolites 2:303–336

Ebenezer KS, Chidambaram T, Manivanan R, Babu V, Punniyamoorthy A, Palanisamy M (2021). Potential Non-toxic Indian Medicinal Plants may Use as a Traditional Medicine to Treat Covid-19 and Opportunistic Fungal Infections. Acta Scientific Pharmaceutical Sciences (ISSN: 2581–5423), 5(8)

Fabricant DS, Farnsworth NR (2001) The value of plants used in traditional medicine for drug discovery. Environ Health Perspect 109:69–75

Farnsworth NR, Akerele O, Bingel AS, Soejarto DD, Guo Z (1985) Medicinal plants in therapy. Bull World Health Organ 63:965

Forda SR, Bergström R, Chlebus M, Barker R, Andersen PH (2013) Priorities for improving drug research, development and regulation. Nat Rev Drug Discov 12:247–248

Garattini L, Padula A (2017) Between pharmaceutical patents and European patients: is a compromise still possible? Expert Opin Ther Pat 27:1073–1076

Ghanemi A (2017) International drugs markets database to improve global drugs accessibility. Res Soc Admin Pharm 13:880–881

Girsberger MA (2000) Intellectual property rights and traditional knowledge: background, terminology and issues arising. In: Methods for risk assessment of transgenic plants: IV. Biodiversity and biotechnology. Springer, Berlin

Hamilton C (2008) Intellectual property rights, the bioeconomy and the challenge of biopiracy. Genom Soc Policy 4:26

Hongal S, Torwane NA, Pankaj G, Chandrashekhar B, Gouraha A (2014) Role of Unani system of medicine in management of orofacial diseases: a review. J Clin Diagn Res 8:ZE12

Huie CW (2002) A review of modern sample-preparation techniques for the extraction and analysis of medicinal plants. Anal Bioanal Chem 373:23–30

Jaiswal M, Wanjari A, Noman O (2020) Pharmaceutical-analytical standardization of palasha kshara prepared by two different combinations and evaluation of their diuretic action in wister rats. Int J Curr Res Rev 12:146–149

Kabanov I (2016) Effective frameworks for delivering compliance with personal data privacy regulatory requirements. In: 2016 14th Annual Conference on Privacy, Security and Trust (PST). IEEE, Piscataway, NJ, pp 551–554

Kapoor D, Vyas RB, Dadrwal D (2018) An overview of analytical instrument qualification with reference of pharmaceutical industry. J Drug Deliv Therapeut 8:99–103

Karimi A, Majlesi M, Rafieian-Kopaei M (2015) Herbal versus synthetic drugs; beliefs and facts. J Nephropharmacol 4:27

Kinghorn AD, Pan L, Fletcher JN, Chai H (2011) The relevance of higher plants in lead compound discovery programs. J Nat Prod 74:1539–1555

Kingston DG (2011) Modern natural products drug discovery and its relevance to biodiversity conservation. J Nat Prod 74:496–511

Ku Y-W (2016) The development of cinchona cultivation and 'kina Gaku'in the Japanese empire, 1912–45. In: Environment, modernization and development in East Asia: perspectives from environmental history. Springer, Berlin

Kumar S, Dobos GJ, Rampp T (2017) The significance of ayurvedic medicinal plants. J Evid Based Complement Altern Med 22:494–501

Lakshmanan PK (2018) Implementation of the convention on biological diversity and its protocols in India. In: Locating India in the Contemporary International Legal Order, pp 275–306
Laraia L, Robke L, Waldmann H (2018) Bioactive compound collections: from design to target identification. Chem 4:705–730
Li F-S, Weng J-K (2017) Demystifying traditional herbal medicine with modern approach. Nat Plants 3:1–7
Manoharachary C, Nagaraju D (2016) Medicinal plants for human health and welfare. Ann Phytomed 5:24–34
Mans DR (2013) From forest to pharmacy: plant-based traditional medicines as sources for novel therapeutic compounds. Acad J Med Plants 1:101–110
Maresova P (2023) Impact of regulatory changes on innovations in the medical device industry: comment on "clinical decision support and new regulatory frameworks for medical devices: are we ready for it?—A viewpoint paper". Int J Health Policy Manag 12:1–4
Marinac JS, Buchinger CL, Godfrey LA, Wooten JM, Sun C, Willsie SK (2007) Herbal products and dietary supplements: a survey of use, attitudes, and knowledge among older adults. J Osteop Med 107:13–23
Mishra BB, Tiwari VK (2011) Natural products: an evolving role in future drug discovery. Eur J Med Chem 46:4769–4807
Mordeniz C (2019) Traditional and complementary medicine. BoD–Books on Demand, Norderstedt
Nirmalatha J, Sulekha G (2016) Quality standards in medicinal plants. Asian J Hortic 11:233–237
Obeid MA, Al Qaraghuli MM, Alsaadi M, Alzahrani AR, Niwasabutra K, Ferro VA (2017) Delivering natural products and biotherapeutics to improve drug efficacy. Ther Deliv 8:947–956
Ogus A (2002) Regulatory institutions and structures. Ann Publ Cooper Econ 73:627–648
Ostfeld RS, Keesing F (2000) Biodiversity and disease risk: the case of Lyme disease. Conserv Biol 14:722–728
Ostfeld RS, Keesing F, Jones CG, Canham CD, Lovett GM (1998) Integrative ecology and the dynamics of species in oak forests. Integr Biol 1:178–186
Pandey M, Rastogi S, Rawat A (2013) Indian traditional ayurvedic system of medicine and nutritional supplementation. Evid Based Complement Alternat Med 2013:376327
Parasuraman S, Thing GS, Dhanaraj SA (2014) Polyherbal formulation: concept of ayurveda. Pharmacogn Rev 8:73
Patel Y, Patel AA, Patel RJ (2022) A comparative study of biologic regulation in US, Canada, Australia, Europe and Singapore. Int J Drug Regul Affairs 10:28–34
Patwardhan B, Vaidya AD, Chorghade M, Joshi SP (2008) Reverse pharmacology and systems approaches for drug discovery and development. Curr Bioact Comp 4:201–212
Pimm SL, Jenkins CN, Abell R, Brooks TM, Gittleman JL, Joppa LN, Raven PH, Roberts CM, Sexton JO (2014) The biodiversity of species and their rates of extinction, distribution, and protection. Science 344:1246752
Prakash G, Chandra SA, Sandhya P, Bidur C, Samir D (2020) Pharmacopoeial comparison of in-process and finished product quality control test for pharmaceutical tablets. GSC Biol Pharmaceut Sci 11:155–165
Pushpangadan P, Behl H (2005) Environment & biodiversity: agenda for future. In: Third International Conference on plants & environmental pollution (ICPEP-3)
Rahayu SE, Purba R, Matondang I (2021) Ethnobotanical study of medicinal plants Inurug indigenous village, Bogor district, Indonesia. Plant Archiv 21:116
Ravichandiran V, Veerapandian M, Kumar KM (2017) Kampo medicine for renal inflammatory conditions. In: Significance of Japanese Kampo medicine in supportive care of heart failure/function, p 111
Reid A-M, Oosthuizen CB, Fibrich BD, Twilley D, Lambrechts IA, de Canha MN, Rademan S, Lall N (2018) Traditional medicine: the ancient roots of modern practice. In: Medicinal plants for holistic health and well-being. Elsevier, Amsterdam
Rietjens IM, Slob W, Galli C, Silano V (2008) Risk assessment of botanicals and botanical preparations intended for use in food and food supplements: emerging issues. Toxicol Lett 180:131–136

Roy A, McDonald R, Price A, Chaguturu R (2011) Recent trends in collaborative, open source drug discovery. Open Conf Proc J 2:130
Sasidharan S, Chen Y, Saravanan D, Sundram K, Latha LY (2011) Extraction, isolation and characterization of bioactive compounds from plants' extracts. Afr J Tradit Complement Altern Med 8:1–10
Scannell JW, Bosley J (2016) When quality beats quantity: decision theory, drug discovery, and the reproducibility crisis. PLoS One 11:e0147215
Schuhmacher A, Gassmann O, Hinder M (2016) Changing R&D models in research-based pharmaceutical companies. J Transl Med 14:1–11
Segger MCC, Phillips F-K (2015) Indigenous traditional knowledge for sustainable development: the biodiversity convention and plant treaty regimes. J For Res 20:430–437
Segneanu AE, Velciov SM, Olariu S, Cziple F, Damian D, Grozescu I (2017) Bioactive molecules profile from natural compounds. In: Asao T (ed) Amino acid—new insights and roles in plant and animal, pp 209–228
Shanley P, Luz L (2003) The impacts of forest degradation on medicinal plant use and implications for health care in eastern Amazonia. BioScience 53:573–584
Singh SP, Singh AM (2015) Bioprospecting: enhancing the value of biodiversity and intellectual property. In: Plant genetic resources and traditional knowledge for food security, pp 231–236
Subbarayappa B (2001) The roots of ancient medicine: an historical outline. J Biosci 26:135–143
Sudha A, Murty V, Chandra T (2009) Standardization of metal based herbal drugs. Am J Infect Dis 5:200–206
Suranjan B, Deekshit A (2021) Evaluating aged consumers insight regarding modern medicine and traditional medicine. J Drug Deliv Therapeut 11:100–104
Tabersky D, Woelfle M, Ruess J-A, Brem S, Brombacher S (2018) Recent regulatory trends in pharmaceutical manufacturing and their impact on the industry. Chimia 72:146–146
Thas JJ (2008) Siddha medicine—background and principles and the application for skin diseases. Clin Dermatol 26:62–78
Vikas B, Akhil B, Suja S, Sujathan K (2017) An exploration of phytochemicals from Simaroubaceae. Asian Pac J Cancer Prev 18:1765
Wakdikar S (2004) Global health care challenge: Indian experiences and new prescriptions. Electron J Biotechnol 7:02–03
Walsh J, Molyneux D, Birley M (1993) Deforestation: effects on vector-borne disease. Parasitology 106:S55–S75
Yang Y, Yang B, Jin L (1994) Retrospection, stratagem, and practice on innovative drug research and development of Chinese materia medica. Chin Tradit Herb Drug 40(10):1513–1519
Young RN (1999) Importance of biodiversity to the modern pharmaceutical industry. Pure Appl Chem 71:1655–1661
Yuan H, Ma Q, Ye L, Piao G (2016) The traditional medicine and modern medicine from natural products. Molecules 21:559

Documentation of Apoptosis-Promoting and Antimicrobial Compounds from the Apocynaceae Family

Mamta Goswami, Rajat Nath, Deepa Nath, Jayanta Kumar Patra, and Anupam Das Talukdar

Contents

Abstract Apocynaceae is the large flowering plant family that consists of various genera like *Alstonia*, *Calotropis*, *Catharanthus*, *Cerbera*, *Holarrhena*, *Kopsia*, *Nerium*, etc., that contain some very important phytochemicals that were reported to promote apoptosis and inhibit the microbes and attenuate the multidrug resistance. Apoptosis is programmed cell death that can interfere with various disease morphology, like apoptosis in cancerous tissue, and is useful for cancer treatments. Also blocked apoptosis could be one of the reasons for the cancer multidrug resistance. Several compounds like quercetin, oleandrin, kaempferol, isoquercetin, apigenin, and cardenolides were reported as apoptosis-promoting agent that increases the apoptotic rate, by mainly increasing the caspase activity and altering some other

M. Goswami · R. Nath · A. Das Talukdar (✉)
Department of Life Science and Bioinformatics, Assam University, Silchar, Assam, India
e-mail: anupam@bioinfoaus.ac.in

D. Nath
Department of Botany, Gurucharan College, Silchar, Assam, India

J. K. Patra
Dongguk University, Goyang-si, Republic of Korea

© The Author(s), under exclusive license to Springer Nature Singapore Pte Ltd. 2024

A. Das Talukdar et al. (eds.), *Traditional Resources and Tools for Modern Drug Discovery*, Interdisciplinary Biotechnological Advances,
https://doi.org/10.1007/978-981-97-4600-2_16

factors. Antimicrobial drug resistance was another global threat because these microbes cause some very fatal diseases like gonorrhea and tuberculosis. Bacteria are one of those that contribute the maximum percentage to the drug resistance due to the misuse and overuse of antibiotics; also fungi show multidrug resistance. Some phytoactive compounds from the Apocynaceae family were reported to attenuate the multidrug resistance by inhibiting the efflux pump that effluxes out the antimicrobial drug; also they increased the susceptibility of the antibiotics toward the resistance strains of the microbes. Some phytochemical was even found to inhibit the virus by blocking their replication and entry into the host. In this chapter, some important phytochemicals that promote apoptosis, show antimicrobial activity, and attenuate multidrug resistance are documented. These phytochemicals are being studied for their potential to inhibit antimicrobial drug resistance and promote apoptosis, which may help in treating various types of cancer.

Keywords Apocynaceae · Apoptosis · Cancer · Antimicrobial · Multidrug resistance

1 Introduction

Over the decades, some plant metabolites that are not directly associated with the primary functioning of the plant like photosynthesis, respiration, transpiration, etc., which we generally termed as a waste or secondary products of the plant, have the greatest role in the plant development; also, these compounds are active against various diseases like cancer, diabetes, rheumatism, Alzheimer's disease, urinary tract infection, diarrhea, fever, viral diseases, and many other (Provenza and Villalba 2010). Ancient people used plants to treat various diseases when there were no developed medical systems; these are the traditional knowledge that makes use of natural products from plant species and are of great importance in drug discovery research (Yuan et al. 2016). Various phytoactive compounds reported from different families, especially members of the Apocynaceae family, constitute a wide range of important phytoactive compounds that are active against various chronic and fatal diseases.

Apoptosis, a programmed cell death, identifiable by distinct morphological alterations such as cell blebbing, detachment, cytoplasmic shrinkage, and DNA fragmentation. It holds pivotal importance in numerous physiological functions essential for typical development and tissue homeostasis (Nguyen et al. 2015). Apoptosis generally involves two types of signal pathways, one that involves the death receptors called as extrinsic pathway and the other that involves the mitochondria known as the intrinsic pathway (Zimmermann and Green 2001). Both the pathways involve many proteins that mediate the programmed cell death. Caspase is one of the important proteins that have proteolytic activity and are mainly responsible for the pre-apoptotic changes; caspase-3 is the main or executioner caspase that does the main work and is common for both the pathways (Elmore 2007). Certain caspases function as initiator caspases, triggering the activation of downstream executioner caspases. Specifically in the extrinsic pathway, caspase-8 is activated, while in the intrinsic pathway, caspase-9 is activated (Rai et al. 2005; Elmore 2007). Several other important biochemical markers like Bax protein, Bcl-2 family protein, FasL, and Fas protein regulate the apoptosis process (Elmore 2007). Demotion or promotion of apoptosis could help in treating many diseases; one of those diseases is cancer where apoptosis promotion in cancerous tissues leads to cancer treatment (Dhandayuthapani et al. 2013). The resistance of cancer drugs can be attributed, in part, to the inhibition of apoptosis (Saraswathy and Gong 2013). Some apoptosis-promoting compounds were documented from the Apocynaceae family.

Not only for the cancer drug but for the microbial diseases, this multidrug resistance was a big problem and global threat also. Due to the misuse and overuse of antibiotics by physicians and human society, some microbial species developed resistance strains that are very less susceptible to any antibiotic or drug (Anand et al. 2020). Bacterial species like *Staphylococcus aureus*, *Pseudomonas aeruginosa*, *Klebsiella pneumonia*, etc., show resistance against maximum antibiotics (Figueiredo et al. 2017). Some important phytoactive compounds like conessine from the *Holarrhena antidysenterica* were reported to inhibit the efflux pump, thus combating multidrug resistance (Siriyong et al. 2017). Not only are they reported to be present and active in bacterial drug resistance, but also against fungal drug resistance. For example, vinblastine and vincristine were reported to inhibit *Trypanosoma cruzi* by inhibiting cytokinesis (Grellier et al. 1999). Compounds from this family were also reported to be active against the viral diseases and protozoal diseases like malaria and dysentery (Wang et al. 2012; Gu and Hao 2016). Structure of some important compounds from the different members of the apocynaceae family was shown in figure 3.

2 Apocynaceae Family

Apocynaceae comes under the order Gentianales and is a large flowering plant family that includes herbs, shrubs, trees, vines, and succulents and is commonly known as the dogbane family. It includes 215 genera and 1900 species and includes 5 subfamilies, Apocynoideae, Asclepiadoideae, Periplocoideae, Rauvolfioideae, and Secamonoideae (El-Fiki et al. 2019; Islam and Lucky 2019). The characteristic features of this family are its members often produce some milky, poisonous juice and have leaves with soft margins and clustered flowers. Leaves are simple, opposite with pinnate venation. Inflorescence is cymose which is in the terminal position or axillary with the bracteoles. Flowers are bisexual, pentamerous, or sometimes tetramerous and actinomorphic with quincuncial, quintuple, or rarely 4-partite calyx. The corolla is salverform, funnelform with five or fewer lobes that overlap to the right or left and sometimes rarely valvate. The anthers are typically sagittate, or connivent into a cone adhering to the pistil head, dehiscing longitudinal dehiscence with base rounded, cordate, sagittate, or extended into an empty spur connected with filaments that are short to make up a stamen that are five or rarely four in number; granular pollen, ring or cup-shaped, with two to five lobes or absent. Ovaries are superior, infrequently half-inferior, separate, or connate with one or two locules present, inside that one to two ovules present. Style is one in number with capitate, conical, or lamb-shaped pistil head, stigma absent. The plant species under the family bears fruit that can be follicle, drupe, berry, or capsule. Seeds bear a thick, hairy, scanty endosperm; some species bear seeds with no endosperm (Rahman and Akter 2015).

Several significant genera within this family include *Allamanda*, *Kopsia*, *Nerium*, *Alstonia*, *Tabernaemontana*, *Calotropis*, *Catharanthus*, *Cerbera*, *Holarrhena*, *Thevetia*, *Rauvolfia*, and *Wrightia*. Found across a wide geographical range from tropical to temperate zones, members of this family are employed in traditional medicine worldwide due to their diverse array of phytoactive compounds, including alkaloids, cardenolides, triterpenoids, phenols, saponins, tannins, flavonoids, iridoids, and others (Naidoo et al. 2021). *Alstonia scholaris* (R. Br.) is primarily found in the tropical rainforests of West and Central Africa, where it is employed in treating diseases such as diarrhea, malaria, dysentery, skin conditions, and catarrhal fever. *Carissa carandas* is mainly found in India, Sri Lanka, Java, Malaysia, Myanmar, and Pakistan, the decoction of the leaves is very useful for remittent fever, and the unripe fruit is astringent, appetizer, and antipyretic. *Catharanthus roseus* (L.), originally from Madagascar, has been employed in cancer therapy. Two alkaloids, vinblastine and vincristine, derived from this plant, serve as medications for treating cancer (Rahman and Akter 2015; Dey et al. 2017; Bhadane et al. 2018). Likewise, many species like *Nerium oleander*, *Tabernaemontana divaricata*, *Rauwolfia serpentina*, *Thevetia peruviana*, and *Wrightia tinctoria* were reported to produce some very important metabolites that are active against various diseases like cancer, diabetes, malaria, and many others, not only in metabolic disorders but also active against the disease caused by the bacteria, fungi, virus, and other microbes that sometimes become very fatal due to the resistance against the multiple drugs. Regarding multiple drug resistance, numerous metabolites that show promise for treating a range of illnesses have been identified in this family (Rahman and Akter 2015; Bhadane et al. 2018; Islam and Lucky 2019).

3 Ethnobotanical Importance of the Family Apocynaceae

In earlier days, there were no developed medical systems, so the people at that time relied upon the plants that were present around them for treatment; they employed various plant parts to address a range of persistent conditions such as cancer, diabetes, diarrhea, fever, and numerous other health issues. These plants are the food and medicine of those ethnic communities. Even nowadays, in areas where there are no hospitals, this type of traditional knowledge is used in emergencies which they acquired from their earlier generations. The use of herbal medicine has been documented in ancient manuscripts like vedas (Ayurveda, Siddha, Unani, Homeopathy).

This traditional knowledge nowadays is used in the area of drug discovery that is surveyed and documented by some researchers from ethnic groups (Chauhan 2020; Singha et al. 2022).

Apocynaceae family bears many ethnomedicinally important species that belong to the different genera. Some important genera of the Apocynaceae family like *Allamanda*, *Alstonia*, *Calotropis*, *Catharanthus*, *Cerbera*, *Geissospermum*, *Holarrhena*, *Nerium*, *Tabernaemontana*, *Thevetia*, *Rauvolfia*, *Wrightia*, and *Voacanga* were recorded in the ancient texts of Ayurveda, Siddha, Unani, Homeopathy, and Traditional Herbal Medicine. The local traditional healers of the Kancheepuram district of Tamil Nadu have revealed the use of seed juice of *Wrightia tinctoria* for treating stomachaches; also traditional healers from the Meru district of Kenya revealed the use of stem and root bark decoction of *Carissa edulis* for treating the same (Muthu et al. 2006; Gakuya et al. 2013; Bhadane et al. 2018). The Balinese communities in Indonesia have unveiled their utilization of *Alstonia scholaris* bark to treat diarrhea (Sujarwo et al. 2015). *Wrightia tinctoria* was also used for treating fever by the Indigenous healers from the Tamil Nadu's Kancheepuram region (Muthu et al. 2006; Bhadane et al. 2018); *Thevetia neriifolia* and *Voacanga africana* were used by the West Nigerian communities for treating the same with the leaves preparation (Ayoola et al. 2008; Bhadane et al. 2018). For treating skin diseases, various ethnic groups revealed the use of various plant parts like the traditional healers from Kancheepuram who revealed the use of *Wrightia tinctoria* oil as an anti-dandruff agent and *Rauvolfia tetraphylla* whole plant paste and castor oil applied in the skin (Muthu et al. 2006); the Brazilian community used stem bark preparation of *Geissospermum sericeum* (Steele et al. 2002; Bhadane et al. 2018); the Santal tribe of Thakurgaon, Bangladesh, used *Nerium indicum* leaves paste on the affected area (Rahmatullah et al. 2009; Bhadane et al. 2018); the inhabitants of rural areas in the Churu district of the Thar Desert in Rajasthan utilize a mixture of *Nerium indicum* leaves and root oil for similar medicinal purposes (Upadhyay et al. 2007). Rural communities of the Meru district of Kenya used root and stem bark preparation of *Carissa edulis* for treating malaria (Gakuya et al. 2013; Bhadane et al. 2018). The Santal tribe of Thakurgaon, Bangladesh, used crushed leaves of *Holarrhena antidysenterica* boiled in water for juice taken orally for treating dysentery (Rahmatullah et al. 2009; Bhadane et al. 2018). The Gujarat local communities used stem bark plaster of *Holarrhena antidysenterica* for treating rheumatism (Kaul and Atal 1983); also the rural communities of Rajasthan used *Carissa carandas* for treating the same (Upadhyay et al. 2007; Bhadane et al. 2018). The rural communities of India used *Alstonia scholaris* flowers for the treatment of asthma (Guha Bakshi et al. 1999; Bhadane et al. 2018). The local communities of Gujarat used to give seed preparation of *Holarrhena antidysenterica* orally to the child as a pediatric care (Kaul and Atal 1983). The residents of NR Pura Taluk in Chikmagalur district, Karnataka, employed *Rauwolfia serpentina* roots to treat snake bites, as documented by (Prakasha et al. 2010; Bhadane et al. 2018). Similarly, in the rural areas of Barisal district, Bangladesh, *Hemidesmus indicus* leaves were utilized to address urinary tract infections (Bhadane et al. 2018; Hossain and Rahman 2018). The whole plant mixture of *Catharanthus roseus* with the cow milk was taken by the communities of Kancheepuram of Tamil Nadu for the treatment of diabetes (Muthu et al. 2006; Bhadane et al. 2018).

4 Bioactivity of Phytoactive Compound from Apocynaceae Family

4.1 Apoptosis-Promoting Activity

Apoptosis refers to a regulated cell death distinguished by distinct cellular alterations like blebbing, detachment from neighboring cells, cytoplasmic shrinkage, fragmentation of DNA, and biochemical shifts such as the initiation of caspases through either intrinsic or extrinsic pathways involving mitochondria. The increase in the induction of apoptosis in cancerous tissues contributes to cancer treatment (Dhandayuthapani et al. 2013). Multidrug resistance in cancer treatment was due to many reasons like overexpression of ABC efflux transporters, blocked apoptosis, cell cycle alterations, etc. (Saraswathy and Gong 2013). Various phytoactive compounds from the plant species of the Apocynaceae family were reported to promote apoptosis, which could contribute toward the decrease of multidrug resistance of cancer drugs due to the impaired apoptotic pathway is listed in Table 1, the mode of action of those phytochemicals in apoptosis promotion was shown in Fig. 1. From the *Calotropis procera* (Aiton) Dryand, nine phytoactive compounds, namely, isorhamnetin-3-O-hexose-deoxyhexoside, kaempferol-3-O-hexose-deoxyhexoside, kaempferol-3-O-hexoside, coroglaucigenin, calotropagenin, Reseoside I e II, adynerin, quercetin-3-O-hexoside, roseoside, calotropin was reported to increase the caspase-3 activity (Rabelo et al. 2021). Thevebioside, extracted from *Thevetia peruviana*, was noted to increase the expression of Bcl-2-associated death promoter (Bad) and the presence of cleaved caspase-3 and cleaved poly (ADP-ribose) polymerase (PARP), while reducing the level of B-cell lymphoma-extra large (Bcl-xl) (Yao et al. 2020). From the seeds of *Cerbera manghas* L., neriifolin compound was extracted that was reported to increase DNA fragmentation with an increase in early apoptosis rate and increased caspase-3, caspase-8, and caspase-9 activity with increased Fas and Fas-L level in hepatic cancer (Zhao et al. 2011). From the tea extract of *Apocynum venetum* leaves, 10 different phytoactive compounds, namely, neochlorogenic acid, isoquercetin, chlorogenic acid, rutin, isochlorogenic acid C, isochlorogenic acid B, astragalin, rosmarinic acid, trans-cinnamic acid, and quercetin, were reported to increase caspase-3 and caspase-9 activity with an increase in mRNA expression of tumor suppressor gene p21 and p53 in hepatic cancer (Li et al. 2019). Oleandrin from the leaves of *Nerium oleander* was reported to increase apoptotic rate; increase caspase activity and PARP cleavage with decreased phosphorylated protein kinase (pAkt) expression by increasing the Jun N-terminal kinase (JNK) and mitogen-activated protein kinase (MAPK) activity; and also increase the FasL expression and increase apoptosis rate indicated by the increased level of the

Table 1 Apoptosis-promoting compounds from the Apocynaceae family

Phytoactive compound	Source plant	Isolated from the plant part	Therapeutic activity	Mode of action	Reference
Isorhamnetin-3-O-hexose-deoxyhexoside, quercetin-3-O-hexoside, kaempferol-3-O-hexoside, kaempferol-3-O-hexose-deoxyhexoside, coroglaucigenin, calotropagenin, adynerin, Reseoside I e II, roseoside, calotropin	*Calotropis procera* (Aiton) Dryand	Leaves	Anticancer, apoptosis-promoting	Increased caspase-3 activity	Rabelo et al. (2021)
Thevebioside	*Thevetia peruviana* (Pers) K. Schum	–	Anticancer, apoptosis-promoting	Increased fragmented nuclei; upregulated bad, cleaved caspase-3, cleaved PARP; downregulated Bcl-xl level	Yao et al. (2020)
Gallic acid, catechol, ellagic acid, caffeic acid, ferulic acid, *0*-coumaric acid, chlorogenic, syringic acid, cinnamic acid, salicylic acid, protocatechuic acid	*Calotropis procera*	Leaves	Anti-lung and colon cancer, apoptosis-promoting	Increased DNA fragmentation with increased early and late apoptosis rate	Bou Malhab et al. (2023)
Neriifolin	*Cerbera manghas* L.	Seeds	Anti-hepatic cancer, apoptosis-promoting	Increased DNA fragmentation with an increase in early apoptosis rate, increased caspase-3,-8,-9 activity, increased Fas and Fas-L level	Zhao et al. (2011)
Neochlorogenic acid, rutin, isoquercetin, isochlorogenic acid B, isochlorogenic acid C, chlorogenic acid, astragalin, trans-cinnamic acid, rosmarinic acid, quercetin	*Apocynum venetum*	Leaves as tea extracts	Anti-hepatic cancer	Increased caspase −3, −9 activity; increased mRNA expression of p21, p53	Li et al. (2019)

(continued)

Table 1 (continued)

Phytoactive compound	Source plant	Isolated from the plant part	Therapeutic activity	Mode of action	Reference
Oleandrin	*Nerium oleander*	Leaves	–	Increased apoptotic rate; increased caspase activity and PARP cleavage; decreased pAkt, increased MAPK and JNK activity, FasL expression	Raghavendra et al. (2007) and Wen et al. (2016)
4-Hydrobenzaldehyde, 2-hydroxy-4-methoxybenzaldehyde, 3-hydroxy-4-methoxybenzaldehyde, 2-hydroxy-4-methoxybenzoic acid, vanillic acid	*Hemidesmus indicus* (L.) R.Br.	Root	Anticancer (lymphoblastic leukemia)	Decreased mitochondrial potential; increased caspase-3,-8 activity and cleaved PARP	Turrini et al. (2018)
Caffeoylquinic acid, stearic acid, ursolic acid, isoquercitrin	*Vallaris glabra* Kuntz	Leaves	Apoptotic-promoting	Activated caspase-8,-9, and – 3 activity	Wong et al. (2013)
Alkaloids and triterpenes	*Alstonia scholaris* (L.) R. Br.	Leaves	Anti-lung cancer	Increased apoptosis rate; decreased Bcl-2, pro-caspase-8; increased cleaved caspase-8	Dey et al. (2017)
Antofine *N*-oxide	*Cynanchum vincetoxicum* (L.) pers.	–	Anticancer (leukemia)	Increased apoptosis rate	Dey et al. (2017)
β-Carboline	*Geissospermum vellosii* Allemão	Bark	Anticancer (prostate and ovarian)	Increased apoptotic cell	Dey et al. (2017)
Jerantinine A, tabercarpamines, cononusine, ervaluteine, vincamajicine, tacamonidine, 6-oxoibogaine, ervatensine	*Tabernaemontana corymbosa* Roxb. ex Wall	Leaves and stem bark	Anticancer	Increased apoptosis rate with decreased cell growth and increased necrosis	Dey et al. (2017)
β-Carboline	*Rauvolfia vomitoria* Afzel.	Leaves	Anticancer (prostate)	Increased apoptosis rate with decreased cell growth and increased DNA damage	Dey et al. (2017)

Tabernaemontanine, dregamine, 16-epidregamine, taberelegantine C and B, voacangine, vobasine	*Tabernaemontana elegans* Stapf.	Roots	Anticancer (colon cancer)	Increased apoptotic rate and caspase-3 activity	Dey et al. (2017)
Alkaloids, tanin, flavanoids, phenolics, saponins	*Rhazya stricta* Decne	Stem, leaves, and flower	Anticancer	Increased caspase-3, -7,-8,-9 activity, cleaved PARP, Bax; decreased Bid, Bcl-2 via PI3K/Akt/mTOR pathway inhibition; increased DNA fragmentation	Ullah et al. (2023)
Apigenin	*Himatanthus sucuuba* (Spruce) Woodson	Latex	Anticancer (colon cancer)	Enhances UVB-induced apoptosis	Herrera-Calderón et al. (2021)
Oleandrin	*Nerium oleander* L.	Leaves	Antitumorous	Increased caspase cleaved form of cytokeratin-18, increased apoptosis	El-Menshawi et al. (2010)
Rhazyaminine	*Rhazya stricta*	Leaves	Anti-breast cancer	Increased apoptosis	Iqbal et al. (2019)
Calactin, calatropagenin, usharin, afroside, calatoxin, ghamphoside	*Calotropis gigantea*	Leaves and stem	Anti-breast cancer	Increased apoptotic cell population, upregulated Bak1, Bax gene, downregulated Bcl2 gene, increased DNA fragmentation	Kharat and Kharat (2019)
Holamine, funtumine	*Holarrhena floribunda* G. Don	Leaves	Anticancer	Externalized phosphatidylserine, increased caspase activity	Badmus et al. (2019) and Badmus et al. (2020)
Peruvosidic acid, thevefoline, solanoside, neriifoside, neriifolin	*Thevetia peruviana* L.	Roots, leaves, and aerial parts	Anticancer (prostate)	Increased apoptotic rate with increased DNA fragmentation	Ramos-Silva et al. (2017) and Thomas-Charles and Fennell (2019)

(continued)

Table 1 (continued)

Phytoactive compound	Source plant	Isolated from the plant part	Therapeutic activity	Mode of action	Reference
12β,6′-dihydroxycalotrpin, 16α-hydroxycalotropin, 12β-hydroxycalactin, calotropin, calactin	*Pergularia tomentosa*	Leaves	Anticancer	Increased caspase-3 activity	Martucciello et al. (2022)
Vobtusine	*Voacanga foetida*	Bark	Anticancer (leukemia)	Activated caspase-3 and caspase-9	Hirasawa et al. (2020)
Calotropin	*Calotropis gigantea*	Stem bark	Anticancer (colon)	Increased apoptosis induction, cleaved caspase-3; decreased Bcl-2; reduced ATP level and increased ROS level	Winitchaikul et al. (2021)
Isovincathicine	*Catharanthus roseus*	Whole dry plant	Anti-lung cancer	Increased apoptosis induction, inhibited Mcl-1, Bcl-xL expression; decreased Bim_{EL} expression	Hiroki et al. (2024)

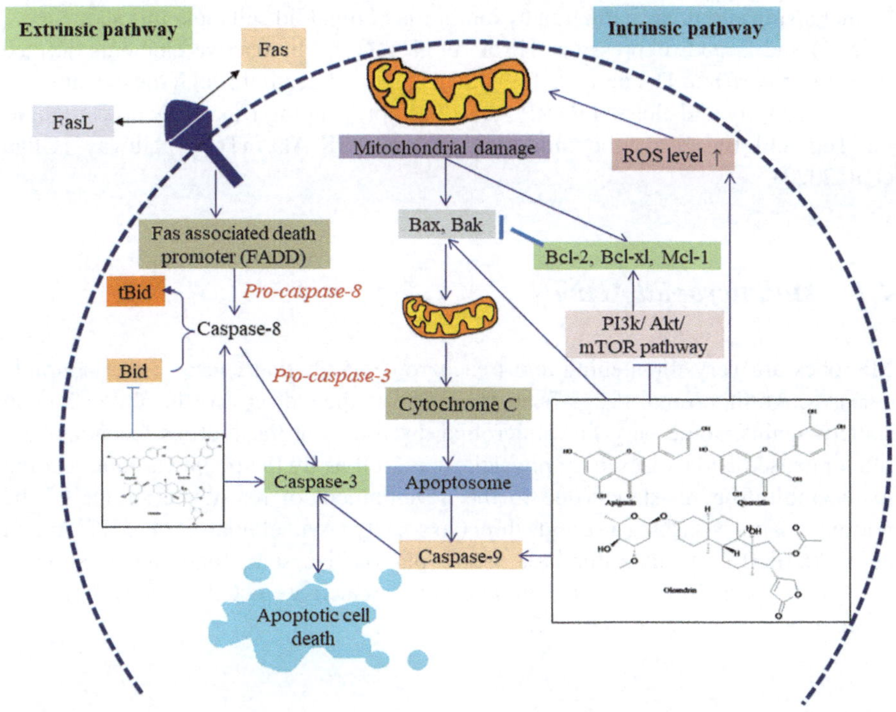

Fig. 1 Mechanism of the different phytochemicals in apoptosis promotion

cleaved form of cytokeratin-18 (Raghavendra et al. 2007; Saraswathy and Gong 2013; Wen et al. 2016). Several members of the Apocynaceae family like *Hemidesmus indicus* (L.) R.Br., *Vallaris glabra* Kuntz, *Alstonia scholaris* (L.) R. Br., *Tabernaemontana elegans* Stapf., *Pergularia tomentosa*, and *Voacanga foetida* were reported to increase caspase activity that increases the apoptosis induction because of the presence of many phytoactive compounds like caffeoylquinic acid, stearic acid, ursolic acid, isoquercitrin, tabernaemontanine, dregamine, 16-epidregamine, tabernaelegantine C and tabernaelegantinine B, voacangine, vobasine, vobtusine, and many others (Hirasawa et al. 2020; Wong et al. 2013; Dey et al. 2017; Turrini et al. 2018; Martucciello et al. 2022). β-carboline, isolated from the bark of *Geissospermum vellosii* Allemão and *Rauvolfia vomitoria* Afzel was reported to increase the apoptosis rate as observed by using different apoptosis assay of different cancer cell lines, similar type of result was observed by the Antofine *N*-oxide from *Cynanchum vincetoxcicum* (L.) pers., different phytoactive compound from leaves and stem bark of *Tabernaemontana corymbosa* Roxb. ex Wall. (Dey et al. 2017). Calotropin that was isolated from the stem bark of *Calotropis gigantea* was reported to increase apoptosis by an increase in cleaved caspase-3, decreased Bcl-2 aided by the reduced adenosine triphosphate (ATP) level, and increased reactive oxygen species (ROS) level (Winitchaikul et al. 2021). Isovincathicine from the *Catharanthus roseus* was found active against lung cancer by increasing apoptosis induction by inhibition of myeloid cell leukemia sequence-1 (Mcl-1) and Bcl-xL expression (Hiroki et al. 2024). Phytoactive elements derived from *Rhazya stricta* Decne were found to trigger cell death through the elevation of caspase activity and cleaved PARP, as well as upregulating Bax while downregulating Bid and Bcl-2 via the inhibition of the PI3K/Akt/mTOR pathway (Ullah et al. 2023).

4.2 Antimicrobial Activity

Microbes are very threatening and treacherous agents that cause some uncurable diseases. Antimicrobial drugs from the point of their discovery benefited human society against some very fatal microbial diseases, but these drugs nowadays are often misused and overused by physicians as well as by the people around leading to the multidrug resistance due to the development of resistance strains of the microbes. This possesses a serious threat to society (Ayukekbong et al. 2017; Anand et al. 2020). To attenuate this issue, some phytoactive compounds were extracted that were reported to decrease drug resistance, which are listed in Table 2 and the mode of action is shown in Fig. 2.

4.2.1 Antibacterial Activity

Bacteria are one of the microbes that show maximum drug resistance against various antibiotics. After the penicillin discovery in 1928 several other antibiotics discovered led to the over of them by the society and the bacteria selected some resistance strains (Nikaido 2009). Antibiotic drug resistance was a very threatening situation in the global scenario (Anand et al. 2020), so to attenuate this problem, some new compounds have been reported from the Apocynaceae family that were found active against various multidrug resistance strains of bacteria. *Himatanthus drasticus* leaf extract was found active against *Klebsiella pneumoniae, Staphylococcus aureus*, *Pseudomonas aeruginosa*, and *Listeria monocytogenes* because of the presence of quercetin, plumieride, rutin, plumericin, chlorogenic acid, and isoquercetin. Quercetin-3-lathyroside was observed to suppress the formation of their biofilm and disturb the integrity of bacterial membranes, and increase in the inhibitory effect of ciprofloxacin was also reported (Figueiredo et al. 2017). Conessine from *Holarrhena antidysenterica* inhibited efflux pump system MexAB-OprM through competitive inhibition which was responsible for the multidrug resistance in *Pseudomonas aeruginosa* (Siriyong et al. 2017). *Tabernaemontana stapfiana* Britten extract was found to inhibit MDR strains of *Klebsiella pneumonia, and Staphylococcus aureus* also increased the antibiotic susceptibility toward those bacterial strains (Tadeg et al. 2005). The stem bark of *Voacanga africana* Stapf was reported to decrease the *Salmonella enterica* cell proliferation (Awah et al. 2024). The *Decalepis hamiltonii* root extract displayed effectiveness against *Helicobacter*

Table 2 Antimicrobial compounds from the Apocynaceae family

Phytoactive compound	Source plant	Plant part	Therapeutic activity against microbial species	Mode of action or result observed	Reference
Antibacterial					
3-methylmannoside, tetradecanoic acid, 5,9,13-trimethylester, squalene, pentatriacontane, hexatriacontane, 2,4,4-trimethyl-3-hydroxymethyl-5a-(3-methyl-but-2-enyl)cyclohexane, 9,19-cycloergost24(28)-en-3-ol, 4,14-dimethyl-,acetate(3,beta.,4.Alpha., 5.Alpha.)	*Catharanthus roseus* (L.) G. Don	Leaves	*Pseudomonas aeruginosa, Staphylococcus aureus, Streptococcus pyogenes, Salmonella typhimurium, Escherichia coli*	Increased zone of inhibition	Arulvendhan et al. (2024)
Oxalic acid, proline, propanoic acid, 2-[(trimethylsilyl)oxy]-, trimethylsilyl ester, benzoic acid, 4-methyl-2-trimethyl-,trimethylsilyl ester, phenol4-(3,4-dihydro-2,2,4-trimethyl-2H-1-benzopyran-4-yl), 2′-methoxy-butyric acid, *p*-coumaryl alcohol, coumaran-3-one, 2-phenylthiophene, elymoclavin, phenylbenzofuran	*Caralluma europaea* (Guss.) N.E.Br.	Aerial parts	*Staphylococcus aureus, Escherichia coli, Pseudomonas aeruginosa, Klebsiella pneumoniae*	Increased zone of inhibition	Amrati et al. (2021)
Quercetin, plumieride, rutin, plumericin, chlorogenic acid, isoquercitrin, quercetin-3-lathyroside	*Himatanthus drasticus*	Leaves	*Staphylococcus aureus, Pseudomonas aeruginosa, Listeria monocytogenes, Klebsiella pneumoniae*	Decreased minimal inhibitory concentration and minimal bactericidal concentration; inhibited biofilm formation; disrupted bacterial membrane; increased the inhibitory effect of ciprofloxacin	Figueiredo et al. (2017)

(continued)

Table 2 (continued)

Phytoactive compound	Source plant	Plant part	Therapeutic activity against microbial species	Mode of action or result observed	Reference
Conessine	*Holarrhena antidysenterica*	–	*Pseudomonas aeruginosa*	Decreased minimal inhibitory concentration of antibiotics; inhibited efflux pump system MexAB-OprM through competitive inhibition	Siriyong et al. (2017)
Saponins, tannins, alkaloids, phenols, flavanoids, phytosterols, triterpenes	*Pergularia daemia*	Dried leaves	*Pseudomonas aeruginosa, Staphylococcus aureus, Bacillus subtilis, Escherichia coli*	Minimal inhibitory concentration (MIC) was very low in the case of gram-positive bacteria	Sarkodie et al. (2016)
Triterpenes, tannins, alkaloids	*Catharanthus roseus*	Flower	*Pseudomonas aeruginosa, Staphylococcus aureus*	Reduced bacterial colonies as indicated by light turbidity	Nayak and Pinto Pereira (2006) and Almagro et al. (2015)
Yohimbine, vincamine, scolpolamine, atropine, colchicine, allantoin, trigonelline, octopamine, synephrine, capsaicin, quercetin, apigenin, genistein, naringin, silymarin, silibinin, gallic acid, caffeic acid, chlorogenic acid, quinic acid	*Catharanthus roseus*	–	*Escherichia coli, Pseudomonas aeruginosa, Klebsiella pneumonia, Staphylococcus aureus, Bacillus subtilis, Enterococcus faecalis, Acinetobacter baumannii, Proteus mirabilis*	Decreased MIC values	Özçelik et al. (2011) and Almagro et al. (2015)
2-hydroxy-4-methoxy benzaldehyde	*Decalepis hamiltonii* (Wight & Arn)	Roots	*Helicobacter pylori*	Increased inhibition zone; ruptured cell wall and cavity formation in the bacterial cell	Srikanta et al. (2011)

–	*Voacanga africana* Stapf	Stem bark	*Salmonella enterica*	Reduced bacterial cell proliferation with decreased colonies	Awah et al. (2024)
Flavonoids, coumarins, alkaloids, tannins, saponins	*Tabernaemontana stapfiana* Britten	Stem and root bark, leaves	MDR strains of *Klebsiella pneumonia, Staphylococcus aureus*	Increased zone of inhibition (17 mm) observed in MDR strains of bacteria and increased microbial susceptibility to antibiotic	Tadeg et al. (2005)
Anti-viral					
Yohimbine, vincamine, scolpolamine, atropine, colchicine, allantoin, trigonelline, octopamine, synephrine, capsaicin, quercetin, apigenin, genistein, naringin, silymarin, silibinin, gallic acid, caffeic acid, chlorogenic acid, quinic acid	*Catharanthus roseus*	–	Herpes simplex virus 1, Parainfluenza-3	Inhibits viral growth as calculated by cytopathogenic effect	Özçelik et al. (2011) and Almagro et al. (2015)
Paniculatumosides C, D, and E	*Cynanchum paniculatum*	–	*Tobacco mosaic virus*, animal-infecting *Sindbis virus*, *Eastern equine encephalitis virus*, *Getah virus*	Suppresses the expression of viral subgenomic RNA	Gu and Hao (2016)
7-demethoxytylophorine, 7-demethoxytylophorine N-oxide	*Cynanchum komarovii*	Roots	*Tobacco mosaic virus*	Partial inhibition of viral agents	An et al. (2001) and Han et al. (2018)
Cynanoside A, cynanoside G, cynanoside M, glaucogenin C3-*O*-*β*-D-cymaropyranosyl-(1 → 4)-α-L-diginopyranosyl-(1 → 4)-*β*-D-cymaropyranoside	*Cynanchum atratum*	Aerial parts	*Tobacco mosaic virus*	Inhibited viral replication	Yan et al. (2014) and Han et al. (2018)

(continued)

Table 2 (continued)

Phytoactive compound	Source plant	Plant part	Therapeutic activity against microbial species	Mode of action or result observed	Reference
–	*Cynanchum paniculatum* (Bunge) Kitag	Roots	Bovine viral diarrhea (BVD) virus	Shows high toxicity against the virus at high concentration	Kim et al. (2017) and Han et al. (2018)
Alstotides	*Alstonia scholaris*	Leaves	Infectious bronchitis virus, dengue type 2 virus	Interferes with the S-protein that plays a role in virus entry	Nguyen et al. (2015) and Julsrigival et al. (2021)
Lupeol acetate, voacangine	*Tabernaemontana cymosa*	Seeds	Dengue type 2 virus, *Chikungunya virus*	Inhibited the viral entry and replication	Gómez-Calderón et al. (2017)
Antifungal					
3-methylmannoside, tetradecanoic acid, 5,9,13-trimethylester, squalene, pentatriacontane, hexatriacontane, 2,4,4-trimethyl-3-hydroxymethyl-5a-(3-methyl-but-2-enyl)cyclohexane, 9,19-cycloergost24(28)-en-3-ol, 4,14-dimethyl-, acetate(3,beta.,4.alpha., 5.alpha.)	*Catharanthus roseus* (L.) G. Don	Leaves	*Fusarium* sp., *Aspergillus tamari Aspergillus niger, Aspergillus flavus, Candida albicans*	Increased zone of inhibition	Arulvendhan et al. (2024)
Coumaran-3-one, 2-phenylthiophene, oxalic acid, proline, propanoic acid, 2-[(trimethylsilyl) oxy]-, trimethylsilyl ester, benzoic acid, 4-methyl-2-trimethyl-,trimethylsilyl ester, phenol4-(3,4-dihydro-2,2,4-trimethyl-2H-1-benzopyran-4-yl), 2′-methoxy-butyric acid, *p*-coumaryl alcohol, emyloclavin, phenylbenzofuran	*Caralluma europaea* (Guss.) N.E.Br.	Aerial parts	*Candida albicans, Saccharomyces cerevisiae*	Increased zone of inhibition	Amrati et al. (2021)

Yohimbine, vincamine, scolpolamine, atropine, colchicine, allantoin, trigonelline, octopamine, synephrine, capsaicin, quercetin, apigenin, genistein, naringin, silymarin, silibinin, gallic acid, caffeic acid, chlorogenic acid, quinic acid	*Catharanthus roseus*	–	*Candida albicans, Candida parapsilosis*	Decreased MIC values	Özçelik et al. (2011) and Almagro et al. (2015)
Vincristine, vinblastine	*Catharanthus roseus*	–	*Trypanosoma cruzi*	Inhibited nuclear division and cytokinesis without affecting microtubule assembly	Almagro et al. (2015) and Grellier et al. (1999))
Caudatin, qingyangshengenin	*Cynanchum wilfordii*	Roots	*Sclerotinia sclerotiorum, Magnaporthe oryzae*	Significant inhibitory activity on the mycelial growth	Huang et al. (2016) and Wang et al. (2021)
Kidjoranine, caudatin	*Cynanchum wilfordii*	Roots	*Blumeria graminis, Sphaerotheca humuli*	Inhibits the fungal growth	Yoon et al. (2011) and Wang et al. (2021)
Cynatratoside-C	*Cynanchum atratum*	Roots	*Ichthyophthirius multifiliis*	Completely inhibited the infective stage tomonts reproduction	Fu et al. (2014)
Alstovenine	*Alstonia venenata*	Bark	*Helminthosporium maydis, Helminthosporium sativum, Erysiphe polygoni*	Inhibited the spore germination	Singh et al. (1999) and Zhao et al. (2023)
Tricin-4′-O-β-L-arabinoside	*Alstonia macrophylla* Wall ex A. DC	Leaves	*Salmonella typhimurium, Candida albicans*	Significant increase in zone of inhibition	Parveen et al. (2010) and Zhao et al. (2023)

(continued)

Table 2 (continued)

Phytoactive compound	Source plant	Plant part	Therapeutic activity against microbial species	Mode of action or result observed	Reference
Scholarisin I, scholarisin II, scholarisin III, scholarisine F	*Alstonia scholaris*	Leaves and bark	*Gibberella pulicaris, Colletotrichum nicotianae*	Increased zone of inhibition with lowest MIC values	Wang et al. (2013)
Gluconic acid, quinic acid, vanillic acid, chlorogenic acid, catechin, procyanidin B dimer, procyanidin B trimer, epicatechin, cinchonain IIb, lariciresinol hexoside, Phloretin, cinchonain Ib	*Hancornia speciosa*	Stem bark	*Candida krusei, Candida albicans, Candida tropicalis*	Inhibited the formation of virulent filaments and partially inhibited the formation of hyphae	Silva et al. (2024)
Antiprotozoal					
Catharoseumine	*Catharanthus roseus*	Whole plant	*Plasmodium falciparum*	Inhibited falcipain-2	Wang et al. (2012)
Marcocarpamine, villastonine, alstonerine, alstophylline, 11-methoxyakuammicine, norfluorocurarine, pleiocarpamine, vincamajine	*Alstonia angustifolia* Wall. ex A. Dc.	Root	*Entamoeba histolytica, Plasmodium falciparum*	Active against both the protozoans	Wright et al. (1992) and Anand et al. (2020)
Echitamine, 6,7-seco-angustilobine B, β-amyrin	*Alstonia angustifolia*	Root bark	*Plasmodium falciparum*	Exhibited strong inhibition against *Plasmodium* (IC_{50} < 10 μg/mL)	Cimanga et al. (2019) and Anand et al. (2020)
Flavopereirine	*Geissospermum vellosii* Allemão	Stem bark	*Leishmania amazonensis*	Inhibited the promastigote form of the parasite; inhibited oligopeptidase B	da Silva e Silva et al. (2019) and Anand et al. (2020)
Bauerenol acetate	*Tabernaemontana longipes* Donn. Sm.	Leaves	*Trypanosoma brucei*	Inhibited the protozoan	Carothers et al. (2018) and Anand et al. (2020)

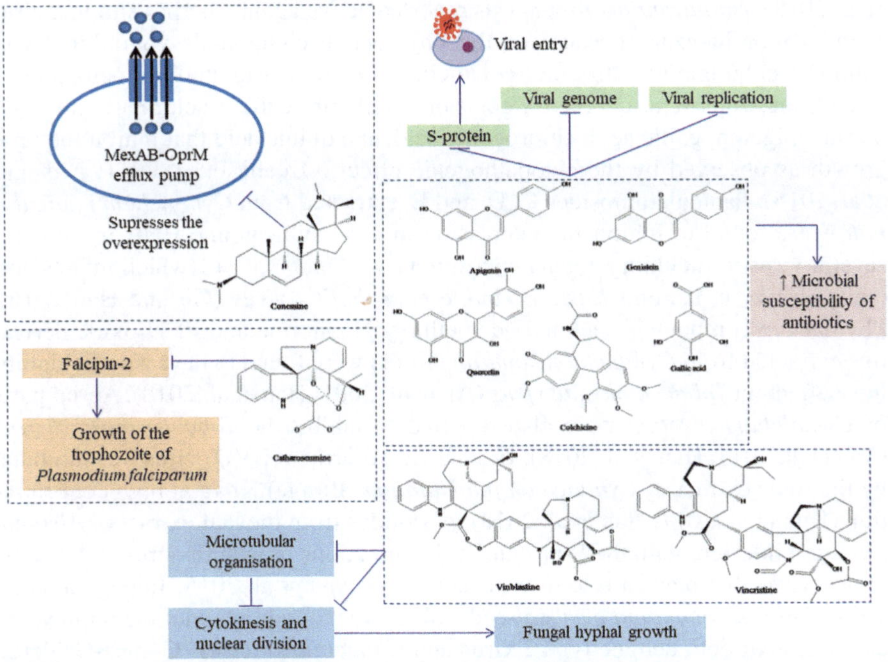

Fig. 2 Mechanism of the antimicrobial phytoactive compound from Apocynaceae family helping in combating multidrug resistance

pylori and induces a cavity formation in the bacterial membrane attributed to the existence of 2-hydroxy-4-methoxy benzaldehyde (Srikanta et al. 2011). *Catharanthus roseus* was also reported to be found active against *Escherichia coli*, *Staphylococcus aureus*, *Bacillus subtilis*, *Enterococcus faecalis*, *Pseudomonas aeruginosa*, *Acinetobacter baumannii*, *Streptococcus pyogenes*, *Klebsiella pneumonia*, *Salmonella typhimurium*, and *Proteus mirabilis* and reported the presence of 3-methylmannoside, squalene, tetradecanoic acid, 5,9,13-trimethylester, pentatriacontane, hexatriacontane, yohimbine, vincamine, scolpolamine, atropine, colchicine, allantoin, trigonelline, octopamine, synephrine, capsaicin, genistein, quercetin, apigenin, silibinin, naringin, silymarin, gallic acid, caffeic acid, chlorogenic acid, and quinic acid (Nayak and Pinto Pereira 2006; Özçelik et al. 2011; Almagro et al. 2015; Arulvendhan et al. 2024).

4.2.2 Antiviral Activity

Members of the Apocynaceae family were also active against very fatal viral diseases and reported to inhibit some very dangerous animal viruses like herpes simplex virus 1 that causes herpes, encephalitis virus that causes encephalitis, etc., and also affect some plant viruses like *Tobacco mosaic virus* (Gu and Hao 2016; Han et al. 2018). *Catharanthus roseus* was reported active against herpes simplex virus 1 and Parainfluenza-3 because of the phytoactive compounds yohimbine, vincamine, scolpolamine, atropine, colchicine, allantoin, trigonelline, octopamine, synephrine, capsaicin, quercetin, genistein, silibinin, caffeic acid, naringin, silymarin, apigenin, gallic acid, chlorogenic acid, and quinic acid that inhibit the viral growth as observed by the cytopathogenic effect (Özçelik et al. 2011; Almagro et al. 2015). Paniculatumosides C, D, and E, extracted from *Cynanchum paniculatum*, were found to inhibit the expression of viral subgenomic RNA in vitro for several viruses, including *Tobacco mosaic virus*, *Sindbis virus* (which infects animals), *Getah virus*, and *Eastern equine encephalitis virus* (Gu and Hao 2016). Phytoactive compounds such as 7-demethoxytylophorine and its N-oxide derivative extracted from *Cynanchum komarovii* roots were found to have a partial inhibitory effect on *Tobacco mosaic virus* (An et al. 2001; Han et al. 2018). Aerial parts of *Cynanchum atratum* were also reported to inhibit the *Tobacco mosaic virus* (Yan et al. 2014; Han et al. 2018). Bovine viral diarrhea (BVD) virus was inhibited by the roots extract of *Cynanchum paniculatum* (Bunge) Kitag at high concentration (Kim et al. 2017; Han et al. 2018). Alstotides from the leaf extract of *Alstonia scholaris* interfere with the S-protein of the infectious bronchitis virus and dengue type 2 virus that plays a role in virus entry (Nguyen et al. 2015; Julsrigival et al. 2021). *Tabernaemontana cymosa* seed extract was found to hinder the duplication and access of both dengue type 2 virus and *Chikungunya virus* (Gómez-Calderón et al. 2017).

4.2.3 Antifungal Activity

Some pathogenic fungi also contribute to multidrug resistance or antibiotic drug resistance; this drug resistance like bacterial drug resistance was reported to be attenuated by the phytoactive compounds present in the Apocynaceae family. *Catharanthus roseus* was reported as active against *Fusarium* sp., *Aspergillus niger*, *Candida parapsilosis*, *Aspergillus flavus*, *Aspergillus tamari*, *Candida albicans*, and *Trypanosoma cruzi* and reported the presence of 3-methylmannoside 5,9,13-trimethylester, squalene, tetradecanoic acid, pentatriacontane, hexatriacontane, yohimbine, vincamine, scolpolamine, atropine, colchicine, allantoin, trigonelline, octopamine, synephrine, capsaicin, quercetin, apigenin, genistein, naringin, silymarin, silibinin, gallic acid, vincristine, and vinblastine. Vincristine and vinblastine were found to inhibit cytokinesis by disorganizing microtubular organization (Nayak and Pinto Pereira 2006; Özçelik et al. 2011; Almagro et al. 2015;

Arulvendhan et al. 2024). Stem bark extract of *Hancornia speciosa* was found active against *Candida krusei*, *Candida albicans*, and *Candida tropicalis* by inhibiting the formation of virulent filament and partially inhibiting the formation of hyphae (Silva et al. 2024). Leaf extract of *Alstonia scholaris* was found to inhibit *Gibberella pulicaris* and *Colletotrichum nicotianae* and attenuate multidrug resistance which was observed by lowering the MIC value of antibiotics (Nguyen et al. 2015). Alstovenine from the bark extract of *Alstonia venenata* inhibits the spore germination of *Helminthosporium maydis*, *Helminthosporium sativum*, and *Erysiphe polygoni* (Nguyen et al. 2015). Cynatratoside-C from the *Cynanchum atratum* roots completely inhibited the infective stage tomonts reproduction of *Ichthyophthirius multifiliis* (Nguyen et al. 2015).

4.2.4 Antiprotozoal Activity

Many protozoans cause very fatal diseases like *Plasmodium* causes malaria, *Entamoeba histolytica* is the reason for amoebic dysentery, and *Leishmania* causes leishmaniasis; likewise, there are several other protozoans causing very fatal diseases. Phytoactive compound from the Apocynaceae family was reported to be against some of the protozoa. Catharoseumine from the *Catharanthus roseus* was reported to inhibit falcipain-2 which was responsible for the continuation of the lifecycle of *Plasmodium falciparum* (Wang et al. 2012). Various compounds extracted from *Alstonia angustifolia* Wall. ex A.DC. have been identified, including marcocarpamine, villastonine, norfluorocurarine, alstophylline, 11-methoxyakuammicine, pleiocarpamine, alstonerine, and vincamajine. These substances have shown inhibitory effects against *Plasmodium falciparum* and *Entamoeba histolytica* (Wright et al. 1992; Anand et al. 2020). The root bark of *Alstonia angustifolia* was found to be active against *Plasmodium falciparum* because of compounds like echitamine, 6,7-seco-angustilobine B, and β-amyrin (Cimanga et al. 2019; Anand et al. 2020). Flavopereirine from the stem bark of *Geissospermum vellosii* Allemão reported to inhibit *Leishmania amazonensis* by inhibiting the virulence factor and thus helps in the treatment of leishmaniasis (da Silva e Silva et al. 2019; Anand et al. 2020) Structure of some important compounds from the different members of the apocynaceae family was shown in (Fig. 3).

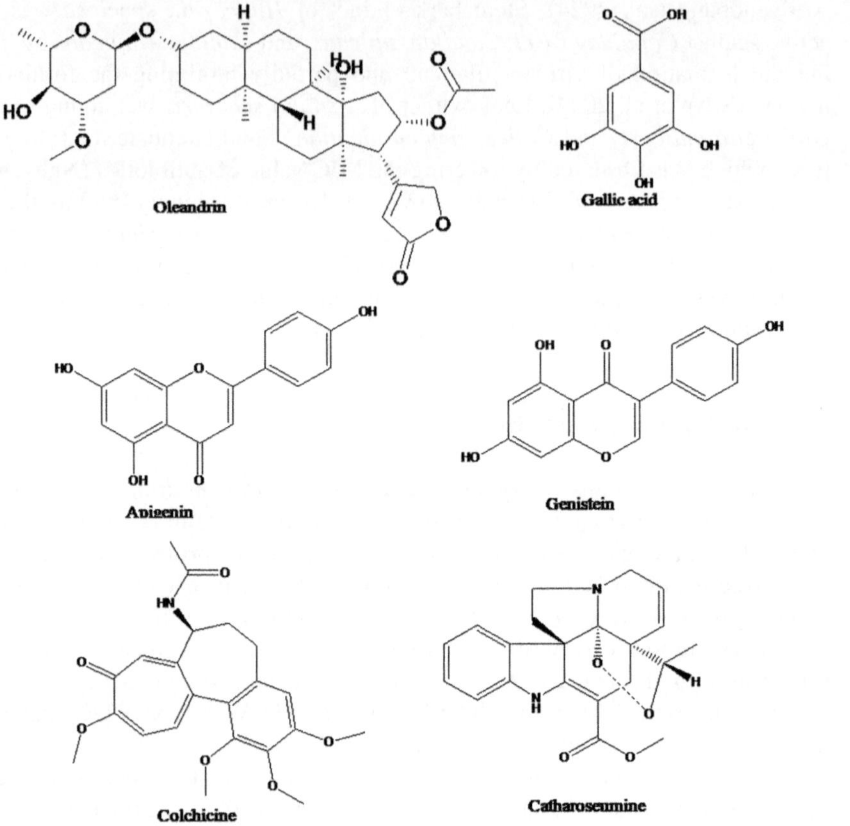

Fig. 3 Some important structures of the phytochemical compound from the Apocynaceae family. (Drawn using ChemDraw software)

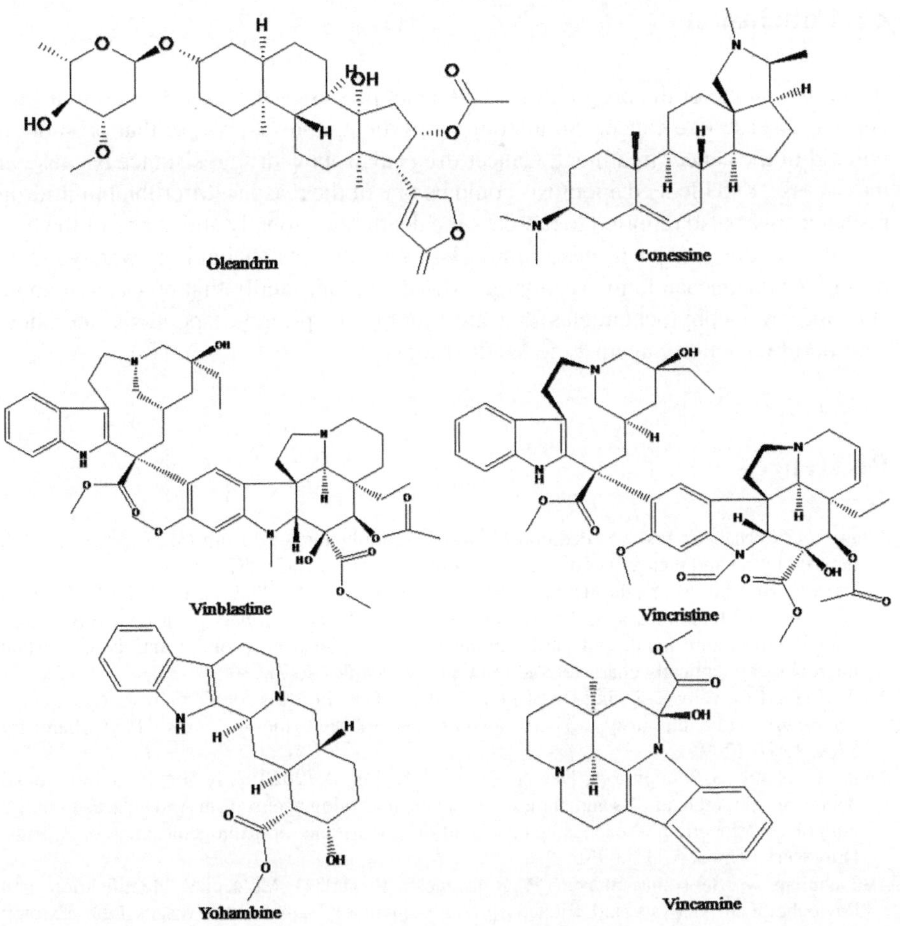

Fig. 3 (continued)

5 Future Prospects

A detailed study on the phytochemicals of the Apocynaceae family was missing, and it needs more attention. So far the multidrug resistance is concerned, a proper mode of action for the phytochemical of the Apocynaceae family is missing, so future work could be done revealing those behind actions. Also, the Apocynaceae family was present abundantly around the different ethnic tribes residing in different places, and they are using them as a source of medicine due to the lack of a developed medical system, but due to the lack of interaction with those ethnic groups, a proper ethnomedical documentation was missing. So more attentive documentation of such knowledge is required so that one can't misuse them.

6 Conclusion

Apoptosis is one of the programmed cell death processes that could result in a very chronic disease like cancer. So alterations in the apoptosis process that promote it will aid in the cancer treatment. Cancer drugs also show drug resistance because of the overuse, and blocked apoptosis could be one of the reasons. Microbial multidrug resistance was also reported to increase the death rate globally and is one of the biggest threats due to which maximum possible treatments available nowadays were failing. Apocynaceae family is a large flowering plant family that consists of some very important phytochemicals that are reported to promote apoptosis and show antimicrobial activity against the MDR strains.

References

Almagro L, Fernández-Pérez F, Pedreño MA (2015) Indole alkaloids from *Catharanthus roseus*: bioproduction and their effect on human health. Molecules 20(2):2973–3000

Amrati FE, Bourhia M, Saghrouchni H, Slighoua M, Grafov A, Ullah R, Ezzeldin E, Mostafa GA, Bari A, Ibenmoussa S, Bousta D (2021) *Caralluma europaea* (Guss.) n.e.Br.: anti-inflammatory, antifungal, and antibacterial activities against nosocomial antibiotic-resistant microbes of chemically characterized fractions. Molecules 26(3):636

An T, Huang RQ, Yang Z, Zhang DK, Li GR, Yao YC, Gao J (2001) Alkaloids from *Cynanchum komarovii* with inhibitory activity against the tobacco mosaic virus. Phytochemistry 58(8):1267–1269

Anand U, Nandy S, Mundhra A, Das N, Pandey DK, Dey A (2020) A review on antimicrobial botanicals, phytochemicals and natural resistance modifying agents from Apocynaceae family: possible therapeutic approaches against multidrug resistance in pathogenic microorganisms. Drug Resist Updat 51:100695

Arulvendhan V, Saravana Bhavan P, Rajaganesh R (2024) Molecular Identification and Phytochemical Analysis and Bioactivity Assessment of Catharanthus roseus Leaf Extract: Exploring Antioxidant Potential and Antimicrobial Activities. Applied biochemistry and biotechnology. 10.1007/s12010-024-04902-w. Advance online publication. https://doi.org/10.1007/s12010-024-04902-w

Awah LA, Taïwe GS, Babiaka SB, Cho-Ngwa F, Ngemenya MN (2024) In vivo antibacterial activity of the methanol extract of *Voacanga africana* Stapf (Apocynaceae) stem bark against clinical multidrug-resistant *Salmonella typhimurium* in Wistar rats. Sci Afr 23:e02118

Ayoola G, Coker H, Adesegun S, Adepoju-Bello A, Obaweya K, Ezennia EC, Atangbayila T (2008) Phytochemical screening and antioxidant activities of some selected medicinal plants used for malaria therapy in southwestern Nigeria. Trop J Pharm Res 7(3):1019–1024

Ayukekbong JA, Ntemgwa M, Atabe AN (2017) The threat of antimicrobial resistance in developing countries: causes and control strategies. Antimicrob Resist Infect Control 6:1–8

Badmus JA, Ekpo OE, Hussein AA, Meyer M, Hiss D (2019) Cytotoxic and cell cycle arrest properties of two steroidal alkaloids isolated from *Holarrhena floribunda* (G. Don) T. Durand & Schinz leaves. BMC Complement Altern Med 19:1–9

Badmus JA, Ekpo OE, Sharma JR, Sibuyi NRS, Meyer M, Hussein AA, Hiss DC (2020) An insight into the mechanism of holamine-and funtumine-induced cell death in cancer cells. Molecules 25(23):5716

Bhadane BS, Patil MP, Maheshwari VL, Patil RH (2018) Ethnopharmacology, phytochemistry, and biotechnological advances of family Apocynaceae: a review. Phytother Res 32(7):1181–1210

Bou Malhab LJ, Bajbouj K, Shehab NG, Elayoty SM, Sinoj J, Adra S, Taneera J, Saleh MA, Abdel-Rahman WM, Semreen MH, Alzoubi KH, Bustanji Y, El-Huneidi W, Abu-Gharbieh E (2023) Potential anticancer properties of calotropis procera: an investigation on breast and colon cancer cells. Heliyon 9(6):e16706

Carothers S, Nyamwihura R, Collins J, Zhang H, Park H, Setzer WN, Ogungbe IV (2018) Bauerenol acetate, the pentacyclic triterpenoid from *Tabernaemontana longipes*, is an antitrypanosomal agent. Molecules 23(2):355

Chauhan K (2020) Role of ethnobotany on Indian society: a review. J Arts Cult Philos Religion Lang Literature 4(2):109–111

Cimanga RK, Nsaka SL, Tshodi ME, Mbamu BM, Kikweta CM, Makila FB-M, Cos P, Maes L, Vlietinck AJ, Exarchou V, Tuenter E, Pieters L (2019) In vitro and in vivo antiplasmodial activity of extracts and isolated constituents of Alstonia congensis root bark. J Ethnopharmacol 242:111736

da Silva e Silva JV, Brigido HPC, de Albuquerque KCO, Carvalho JM, Reis JF, Faria LV, Coelho-Ferreira MR, Silveira FT, da Silva Carneiro A, Percário S (2019) Flavopereirine—an alkaloid derived from *Geissospermum vellosii*—presents leishmanicidal activity in vitro. Molecules 24(4):785

Dey A, Mukherjee A, Chaudhury M (2017) Chapter 10. Alkaloids from Apocynaceae: origin, pharmacotherapeutic properties, and structure-activity studies. In: Atta Ur R (ed) Studies in natural products chemistry, vol 52. Elsevier, Amsterdam, pp 373–488

Dhandayuthapani S, Marimuthu P, Hörmann V, Kumi-Diaka J, Rathinavelu A (2013) Induction of apoptosis in HeLa cells via caspase activation by resveratrol and genistein. J Med Food 16(2):139–146

El-Fiki M, El-Taher A, El-Gendy A, Lila M (2019) Morphological and anatomical studies on some taxa of family Apocynaceae. Al-Azhar J Agric Res 44(1):136–147

El-Menshawi BS, Fayad W, Mahmoud K, El-Hallouty SM, El-Manawaty M (2010) Screening of natural products for therapeutic activity against solid tumors. Indian J Exp Biol 48:258

Elmore S (2007) Apoptosis: a review of programmed cell death. Toxicol Pathol 35(4):495–516

Figueiredo C, Branco Santos JC, Castro Junior JAA, Wakui VG, Rodrigues JFS, Arruda MO, Monteiro AS, Monteiro-Neto V, Bomfim MRQ, Kato L, Nascimento da Silva LC, Grisotto MAG (2017) Himatanthus drasticus leaves: chemical characterization and evaluation of their antimicrobial, antibiofilm, antiproliferative activities. Molecules 22(6):910

Fu Y-W, Zhang Q-Z, Xu D-H, Liang J-H, Wang B (2014) Antiparasitic effect of cynatratoside-C from *Cynanchum atratum* against *Ichthyophthirius multifiliis* on grass carp. J Agric Food Chem 62(29):7183–7189

Gakuya D, Itonga S, Mbaria J, Muthee J, Musau J (2013) Ethnobotanical survey of biopesticides and other medicinal plants traditionally used in Meru central district of Kenya. J Ethnopharmacol 145(2):547–553

Gómez-Calderón C, Mesa-Castro C, Robledo S, Gómez S, Bolivar-Avila S, Diaz-Castillo F, Martínez-Gutierrez M (2017) Antiviral effect of compounds derived from the seeds of *Mammea americana* and *Tabernaemontana cymosa* on dengue and chikungunya virus infections. BMC Complement Altern Med 17:1–12

Grellier P, Sinou V, Garreau-de Loubresse N, Bylen E, Boulard Y, Schrével J (1999) Selective and reversible effects of vinca alkaloids on *Trypanosoma cruzi* epimastigote forms: blockage of cytokinesis without inhibition of the organelle duplication. Cell Motil Cytoskeleton 42(1):36–47

Gu XJ, Hao DC (2016) Recent advances in phytochemistry and pharmacology of C21 steroid constituents from Cynanchum plants. Chin J Nat Med 14(5):321–334

Guha Bakshi DN, Sensarma P, Pal DC (1999) A lexicon of medicinal plants in India. Naya Prokash, Calcutta

Han L, Zhou X, Yang M, Zhou L, Deng X, Wei S, Wang W, Wang Z, Qiao X, Bai C (2018) Ethnobotany, phytochemistry and pharmacological effects of plants in genus Cynanchum Linn. (Asclepiadaceae). Molecules 23(5):1194

Herrera-Calderón O, Calero-Armijos LL, Cardona-G W, Herrera-R A, Moreno G, Algarni MA, Alqarni M, El-Saber Batiha G (2021) Phytochemical screening of himatanthus sucuuba (spruce) Woodson (apocynaceae) latex, in vitro cytotoxicity and incision wound repair in mice. Plan Theory 10(10):2197

Hirasawa Y, Kaneda T, Morita H (2020) Vobtusine from Voacanga foetida (Blume) Rolfe induces apoptosis via activation of caspase pathway in human HL-60 leukemia cell line. Int J Adv Sci Eng Inform Technol 10(6):2536

Hiroki M, Abulikemu A, Totsuka C, Hirasawa Y, Kaneda T, Morita H (2024) Isovincathicine from *Catharanthus roseus* induces apoptosis in A549 cells. J Nat Med 78(1):216–225

Hossain U, Rahman MO (2018) Ethnobotanical uses and informant consensus factor of medicinal plants in Barisal district, Bangladesh. Bangladesh J Plant Taxon 25(2):241

Huang L-J, Wang B, Zhang J-X, Yuan C-M, Gu W, Mu SZ, Hao X-J (2016) Synthesis and evaluation of antifungal activity of C21-steroidal derivatives. Bioorg Med Chem Lett 26(8):2040–2043

Iqbal W, Alkarim S, Kamal T, Choudhry H, Sabir J, Bora RS, Saini KS (2019) Rhazyaminine from Rhazya stricta inhibits metastasis and induces apoptosis by downregulating Bcl-2 gene in MCF7 cell line. Integr Cancer Ther 18:1534735418809901

Islam MS, Lucky RA (2019) A study on different plants of Apocynaceae family and their medicinal uses. J Pharm Res 4(1):40–44

Julsrigival J, Sirisa-ard P, Julsrigival S, Akarchariya N (2021) Antiviral medicinal plants found in Lanna traditional medicine. Chin Herbal Med 13(4):494–501

Kaul M, Atal C (1983) Studies on *Holarrhena antidysenterica* wall. 1. Botany, medico-ethnobotany and distribution. J Ethnopharmacol 8(3):349–356

Kharat KR, Kharat AS (2019) The Calotropis Gigantea Methanolic extract induces apoptosis in human breast carcinoma cells. Iran J Med Sci 44(6):483–492

Kim W, Oh T-S, Park Y-J (2017) Anti-viral effect of herbal medicine Korean traditional *Cynanchum paniculatum* (BGE.) kitag extracts. Afr J Tradit Complement Altern Med 14(3):194–198

Li C, Huang G, Tan F, Zhou X, Mu J, Zhao X (2019) In vitro analysis of antioxidant, anticancer, and bioactive components of *Apocynum venetum* tea extracts. J Food Qual 2019:1–13

Martucciello S, Paolella G, Romanelli AM, Sposito S, Meola L, Cerulli A, Masullo M, Piacente S, Caputo I (2022) Pro-apoptotic and pro-autophagic properties of cardenolides from aerial parts of *Pergularia tomentosa*. Molecules 27(15):4874

Muthu C, Ayyanar M, Raja N, Ignacimuthu S (2006) Medicinal plants used by traditional healers in Kancheepuram District of Tamil Nadu, India. J Ethnobiol Ethnomed 2:1–10

Naidoo CM, Naidoo Y, Dewir YH, Murthy HN, El-Hendawy S, Al-Suhaibani N (2021) Major bioactive alkaloids and biological activities of Tabernaemontana species (Apocynaceae). Plan Theory 10:313. https://doi.org/10.3390/plants10020313

Nayak B, Pinto Pereira LM (2006) *Catharanthus roseus* flower extract has wound-healing activity in Sprague Dawley rats. BMC Complement Altern Med 6:1–6

Nguyen PQT, Ooi JSG, Nguyen NTK, Wang S, Huang M, Liu DX, Tam JP (2015) Antiviral cystine knot α-amylase inhibitors from *Alstonia scholaris*. J Biol Chem 290(52):31138–31150

Nikaido H (2009) Multidrug resistance in bacteria. Annu Rev Biochem 78:119–146

Özçelik B, Kartal M, Orhan I (2011) Cytotoxicity, antiviral and antimicrobial activities of alkaloids, flavonoids, and phenolic acids. Pharm Biol 49(4):396–402

Parveen M, Khanam Z, Ali A, Ahmad SM (2010) A novel antimicrobial flavonoidic glycoside from the leaves of *Alstonia macrophylla* wall ex a. DC (Apocynaceae). Chin Chem Lett 21(5):593–595

Prakasha H, Krishnappa M, Krishnamurthy Y, Poornima S (2010) Folk medicine of NR Pura taluk in Chikmagalur district of Karnataka. Indian J Tradit Knowl 9(1):55–60

Provenza FD, Villalba JJ (2010) The role of natural plant products in modulating the immune system: an adaptable approach for combating disease in grazing animals. Small Rumin Res 89(2–3):131–139

Rabelo AC, Borghesi J, Carreira ACO, Hayashi RG, Bessa F, Barreto RDSN, da Costa RP, Cantanhede Filho AJ, Carneiro FJC, Miglino MA (2021) *Calotropis procera* (Aiton) Dryand

(Apocynaceae) as an anti-cancer agent against canine mammary tumor and osteosarcoma cells. Res Vet Sci 138:79–89

Raghavendra PB, Sreenivasan Y, Manna SK (2007) Oleandrin induces apoptosis in human, but not in murine cells: dephosphorylation of Akt, expression of FasL, and alteration of membrane fluidity. Mol Immunol 44(9):2292–2302

Rahman A, Akter M (2015) Taxonomy and traditional medicinal uses of apocynaceae (dogbane) family of Rajshahi District, Bangladesh. Res Rev J Bot Sci 4(4):1–12

Rahmatullah M, Mollik MAH, Azam A, Islam MR, Chowdhury MAM, Jahan R, Chowdhury MH, Rahman T (2009) Ethnobotanical survey of the Santal tribe residing in Thakurgaon District, Bangladesh. Am Eur J Sustain Agric 3(4):889–898

Rai NK, Tripathi K, Sharma D, Shukla VK (2005) Apoptosis: a basic physiologic process in wound healing. Int J Low Extrem Wounds 4(3):138–144

Ramos-Silva A, Tavares-Carreón F, Figueroa M, De la Torre-Zavala S, Gastelum-Arellanez A, Rodríguez-García A, Galán-Wong LJ, Avilés-Arnaut H (2017) Anticancer potential of Thevetia peruviana fruit methanolic extract. BMC Complement Altern Med 17:1–11

Saraswathy M, Gong S (2013) Different strategies to overcome multidrug resistance in cancer. Biotechnol Adv 31(8):1397–1407

Sarkodie JA, Squire SA, Oppong Bekoe E, Fosu Domozoro CY, Kretchy IA, Ahiagbe MK, Frimpong-Manso S, Oboba Kwakyi NA, Edoh DA, Sakyiama M, Lamptey VK, Affedzi-Obresi S, Duncan JL, Debrah P, N'guessa BB, Asiedu-Gyekye JI, Kwadwo Nyarko A (2016) Antioxidant and antimicrobial capacities of ethanolic extract of Pergularia daemia leaves: a possible substitute in diabetic management. J Complement Integr Med 13(3):239–245

Silva VBD, Almeida-Bezerra JW, Novais MHG, Farias NS, Coelho JJ, Ribeiro PRV, Canuto KM, Coutinho HDM, Morais-Braga MFB, Oliveira AFM (2024) Chemical composition, antifungal, and anti-virulence action of the stem bark of *Hancornia speciosa* Gomes (Apocynaceae) against *Candida* spp. J Ethnopharmacol 321:117506

Singh S, Sarma B, Srivastava J, Singh U, Ray A (1999) Antifungal activity of Δ 3-alstovenine, a plant alkaloid isolated from Alstonia venenata. Folia Microbiol 44:510–512

Singha S, Nath R, Das S, Kityania S, Talukdar AD, Nath D (2022) Phytochemicals and their bioactivity from plants of Dryopteridaceae family. In: Bioactive compounds in bryophytes and pteridophytes. Springer, Berlin, pp 1–18

Siriyong T, Srimanote P, Chusri S, Yingyongnarongkul BE, Suaisom C, Tipmanee V, Voravuthikunchai SP (2017) Conessine as a novel inhibitor of multidrug efflux pump systems in *Pseudomonas aeruginosa*. BMC Complement Altern Med 17(1):405

Srikanta BM, Nayaka MA, Dharmesh SM (2011) Inhibition of *Helicobacter pylori* growth and its cytotoxicity by 2-hydroxy 4-methoxy benzaldehyde of *Decalepis hamiltonii* (Wight & Arn); a new functional attribute. Biochimie 93(4):678–688

Steele JC, Veitch NC, Kite GC, Simmonds MS, Warhurst DC (2002) Indole and β-Carboline alkaloids from *Geissospermum sericeum*. J Nat Prod 65(1):85–88

Sujarwo W, Keim AP, Savo V, Guarrera PM, Caneva G (2015) Ethnobotanical study of Loloh: traditional herbal drinks from Bali (Indonesia). J Ethnopharmacol 169:34–48

Tadeg H, Mohammed E, Asres K, Gebre-Mariam T (2005) Antimicrobial activities of some selected traditional Ethiopian medicinal plants used in the treatment of skin disorders. J Ethnopharmacol 100(1):168–175

Thomas-Charles C, Fennell H (2019) Anti-prostate cancer activity of plant-derived bioactive compounds: a review. Curr Mol Biol Rep 5:140–151

Turrini E, Calcabrini C, Tacchini M, Efferth T, Sacchetti G, Guerrini A, Paganetto G, Catanzaro E, Greco G, Fimognari C (2018) In vitro study of the cytotoxic, cytostatic, and antigenotoxic profile of *Hemidesmus indicus* (L.) R. Br.(Apocynaceae) crude drug extract on T lymphoblastic cells. Toxins 10(2):70

Ullah MF, Abuduhier FM, Bhat SH, Ahmad A, Ajmal MR, Mustafa SK (2023) Cytotoxic and anti-metastatic action mediates the anti-proliferative activity of Rhazya stricta Decne inducing apoptotic cell death in human cancer cells: implication in chemopreventive mechanism. Emirates J Food Agric 8:683–692

Upadhyay B, Roy S, Kumar A (2007) Traditional uses of medicinal plants among the rural communities of Churu district in the Thar Desert, India. J Ethnopharmacol 113(3):387–399

Wang L, He H-P, Di Y-T, Zhang Y, Hao X-J (2012) Catharoseumine, a new monoterpenoid indole alkaloid possessing a peroxy bridge from *Catharanthus roseus*. Tetrahedron Lett 53(13):1576–1578

Wang W, Cheng M-H, Wang X-H (2013) Monoterpenoid indole alkaloids from Alstonia rupestris with cytotoxic, anti-inflammatory and antifungal activities. Molecules 18(6):7309–7322

Wang L, Cai F, Zhao W, Tian J, Kong D, Sun X, Liu Q, Chen Y, An Y, Wang F, Liu X, Wu Y, Zhou H (2021) *Cynanchum auriculatum* Royle ex Wight., *Cynanchum bungei* Decne. And *Cynanchum wilfordii* (maxim.) Hemsl.: current research and prospects. Molecules 26(23):7065

Wen S, Chen Y, Lu Y, Wang Y, Ding L, Jiang M (2016) Cardenolides from the Apocynaceae family and their anticancer activity. Fitoterapia 112:74–84

Winitchaikul T, Sawong S, Surangkul D, Srikummool M, Somran J, Pekthong D, Kamonlakorn K, Nangngam P, Parhira S, Srisawang P (2021) Calotropis gigantea stem bark extract induced apoptosis related to ROS and ATP production in colon cancer cells. PLoS One 16(8):e0254392

Wong SK, Lim YY, Chan EW (2013) Botany, uses, phytochemistry and pharmacology of selected Apocynaceae species: a review. Pharmacogn Commun 3(3):2–11

Wright C, Allen D, Cai Y, Phillipson J, Said I, Kirby G, Warhurst D (1992) In vitro antiamoebic and antiplasmodial activities of alkaloids isolated from *Alstonia angustifolia* roots. Phytother Res 6(3):121–124

Yan Y, Zhang JX, Liu KX, Huang T, Yan C, Huang LJ, Liu S, Mu SZ, Hao XJ (2014) Seco-pregnane steroidal glycosides from the roots of *Cynanchum atratum* and their anti-TMV activity. Fitoterapia 97:50–63

Yao C, Su L, Zhang F, Zhu X, Zhu Y, Wei L, Jiao X, Hou Y, Chen X, Wang W, Wang J, Zhu X, Zou C, Zhu S, Xu Z (2020) Thevebioside, the active ingredient of traditional Chinese medicine, promotes ubiquitin-mediated SRC-3 degradation to induce NSCLC cells apoptosis. Cancer Lett 493:167–177

Yoon M-Y, Choi NH, Min BS, Choi GJ, Choi YH, Jang KS, Han S-S, Cha B, Kim J-C (2011) Potent in vivo antifungal activity against powdery mildews of pregnane glycosides from the roots of *Cynanchum wilfordii*. J Agric Food Chem 59(22):12210–12216

Yuan H, Ma Q, Ye L, Piao G (2016) The traditional medicine and modern medicine from natural products. Molecules 21(5):559

Zhao Q, Guo Y, Feng B, Li L, Huang C, Jiao B (2011) Neriifolin from seeds of *Cerbera manghas* L. induces cell cycle arrest and apoptosis in human hepatocellular carcinoma HepG2 cells. Fitoterapia 82(5):735–741

Zhao MX, Cai J, Yang Y, Xu J, Liu WY, Akihisa T, Li W, Kikuchi T, Feng F, Zhang J (2023) Traditional uses, chemical composition and pharmacological activities of Alstonia R. Br. (Apocynaceae): a review. Arab J Chem 16(8):104857

Zimmermann KC, Green DR (2001) How cells die: apoptosis pathways. J Allergy Clin Immunol 108(4, Supplement):S99–S103

Modern Drug Research and Its Impact on Pharmaceutical Industries

Rajat Nath, Sibashish Kityania, Subrata Das, Deepa Nath, Jayanta Kumar Patra, and Anupam Das Talukdar

Contents

R. Nath · S. Kityania · A. Das Talukdar (✉)
Department of Life Science and Bioinformatics, Assam University, Silchar, Assam, India
e-mail: anupam@bioinfoaus.ac.in

S. Das
Department of Botany, Karimganj College, Karimganj, Assam, India

D. Nath
Department of Botany, Gurucharan College, Silchar, Assam, India

J. K. Patra
Dongguk University, Goyang-si, Republic of Korea

© The Author(s), under exclusive license to Springer Nature Singapore Pte Ltd. 2024

A. Das Talukdar et al. (eds.), *Traditional Resources and Tools for Modern Drug Discovery*, Interdisciplinary Biotechnological Advances,
https://doi.org/10.1007/978-981-97-4600-2_17

Abstract Modern drug research has revolutionized the pharmaceutical industry, reshaping healthcare and patient well-being. The constant demand for innovative drugs across therapeutic domains drives pharmaceutical companies to navigate intricate drug discovery and development processes. The discovery phase spans target identification, hit exploration, lead optimization, and candidate selection, while development entails chemical synthesis refinement, toxicity assessments, clinical trials, and regulatory clearance. These multifaceted stages, though time-intensive and costly, face added challenges due to stringent regulations, environmental concerns, and patent expirations, collectively constraining R&D productivity. To address these challenges, the industry seeks innovative and collaborative strategies among academia, industry, and government research entities. By converging efforts, stakeholders can sustainably deliver high-quality medicines. This discourse extensively examines the preclinical discovery stage and development processes. It also underscores the pharmaceutical sector's predicaments and novel approaches that promise industry longevity. Contemporary drug research marks a new epoch of innovation, fundamentally reshaping pharmaceutical landscapes. Innovations in research methodologies, technologies, and collaborative networks have ushered transformative changes. These advancements expedite drug discovery, enabling precision-engineered therapeutics with heightened efficacy. From target identification through high-throughput screening to molecular modeling, cutting-edge tools expedite potential drug identification. Moreover, modern drug research unravels intricate molecular pathways and disease mechanisms, fostering personalized medicine and improved therapeutic outcomes. Tailored treatments based on individual profiles mitigate adverse effects, revolutionizing patient care. Biologics, gene therapies, and precision medicine exemplify this paradigm shift toward targeted interventions. However, challenges persist along the journey from laboratory to patient bedside. The formidable costs and risks inherent in drug research and clinical trials mandate strategic collaborations between academia, industry, and regulatory bodies. Ethical considerations regarding intellectual property, patient privacy, and equitable treatment access necessitate thoughtful assessment. As drug research continues to reshape pharmaceutical prospects, cohesive efforts remain crucial to overcoming obstacles and steering the industry toward sustained innovation.

Keywords Modern drug research · Pharmaceutical industry · Drug discovery and development · Precision medicine · Personalized treatment

1 Introduction

The constant pillar of innovation is what has been at the heart of the pharma industry that saved so many lives throughout the years. In this particular context, the breakthroughs that have arisen as a result of such contributions have opened up new ways of handling curable and chronic illnesses. New drugs are being discovered in areas like cardiovascular, metabolic, arthritic pain, depression–anxiety, oncology, women health, gastroenteric health, and infections. This has improved health conditions leading to better quality of life and increased lifespan (Khanna 2012).

As a leading pillar in healthcare advancement, diseases constantly present new and unique challenges for which innovative solutions are sought out within the pharmaceutical industry (Milanesi et al. 2020). Technological advances and a better comprehension of organism functioning have led to considerable reformations in drug research through the years (Trenfield et al. 2019; Taiwo and Ezugwu 2020; Tandon et al. 2020). The chapter "Modern Drug Research and Its Impact on Pharmaceutical Industries" looks into history, primary research methods, obstacles as well as possibilities, and broad ramifications of today's approach in modern-day drug research and production. We begin by acknowledging that drug research forms a vital part in determining how the history of pharma will develop as we move to this study. Innovation in therapeutics has moved from the past conventional time taking approach to modern technological environment using advanced sophisticated resources and teamwork. The second shift brings about a new phase in personalized medication that entails using an individual's genes and disease features (Doestzada et al. 2018; Pucci et al. 2019; Von Krogh et al. 2020; Huptych et al. 2021; Jian et al. 2023).

This trek down the modern-day pharmaceutical research lanes will take us back all way along the steps taken toward current approaches. We shall review the elements of modern drug discovery such as high-throughput screening (HTS), bioinformatics, targeted drugs, and how they have shaped into how the pharmaceuticals industry is operating today. On the other hand, change has some difficulties linked to it. Though diseases have become very complex, drugs are no longer effective, while new regulations make it very hard to overcome these obstacles. As we evaluate the impact of modern drug research on the pharmaceutical industry, we will also explore the economic implications, moral questions, as well as the fluctuating drug approval paradigm (Alanis 2005).

We shall witness the triumphs and tribulations of drug researches through the use of case studies so as to learn from victories and understand failure valuable lessons. In addition to this, we shall sneak a little into the future, seeing upcoming developments in conjunction with artificial intelligence as well as machine learning that are destined to become pioneering avenues of medicine in the future.

2 Historical Perspective

Drug discovery and development have created a storyline filled with memorable events that will always be an important part of medical history. In civilizations, like Mesopotamia, the foundation of pharmacology was established through their use of willow bark, which contained salicylates known for providing relief from pain (Sabatowski et al. 2004). This practice, documented in the ancient Sumerian text, has been suggested as an early precursor to aspirin use (Montinari et al. 2019). The synthesis of aspirin by Felix Hoffmann in 1897 at Bayer, Germany, was a significant milestone in the modern era. That revelation did not only present one of the most widely utilized analgesics, but it also brought out the systematic development of medicines (Montinari et al. 2019). Following this, Alexander Fleming's discovery of penicillin in 1928 led to the beginning of the era of antibiotics which ultimately changed how we view and treat bacterial infection (Olsen 2013; Haider 2023).

The development of pharmaceuticals went through a major shift from "happy accidents" to more logical and systematic discovery approach in the middle of the twentieth century (Bengtson and Settersten 2016). The revelation of the DNA double helix shape by James Watson and Francis Crick in 1953 led to the establishment of the basis for molecular biology, tremendously changing understanding of genomic data and its role in disease formation (Markel 2021). This groundbreaking discovery established the basis for creating medications with specific target and paved the way for the advancements of biotechnology, within drug research and development.

The study of molecular structures underwent a revolution in the twentieth century with the advancement of X-ray crystallography and the introduction of polymerase chain reaction (PCR). X-ray crystallography allowed for detailed examination of molecular structures, while PCR enabled the amplification of specific DNA sequences. These technologies have become crucial tools in drug research, enabling a more profound comprehension of molecular interactions and helping the identification of possible targets for drugs (Salon et al. 2011; Wiggin et al. 2014; Blass 2015).

Technological advancements toward the end of the twentieth century revolutionized the landscape for drug discovery. Combinatorial chemistry introduced in the 1970s by Mario J. Molina and F. Sherwood Rowland made it easy for the scientists to facilitate the swift production of various chemicals in short time leading to the quick screening of many probable candidate drugs. The use of high-throughput screening greatly sped up the drug discovery process by enabling scientists to quickly explore vast quantities of compounds simultaneously (Laszlo 2007).

One major event in the 1990s was the successful completion of the International Human Genome Project (IHGP), considered an important milestone. As an end result, this ambitious endeavor created an authentic portrait of the human genome representing the basic structure of human genetic data (Collins and McKusick 2001; Raggio 2002; Gannett 2008). It was this genome revolution that laid the groundwork for personalized medicines that allowed treatment to be tailored to each given patient which thus introduced a new era of precision medications (Brittain et al. 2017; Carrasco-Ramiro et al. 2017). Gaining a sense of these historical developments will help understand how today's medicine studies came about. Serendipity, investigative science, and technologic innovation continue playing a deterministic role on the direction of pharmaceutical research and development (Ashley 2016; Ahmed et al. 2020).

3 Key Components of Modern Drug Research

3.1 High-Throughput Screening and Computational Methods

The first pivotal component in modern drug research involves the utilization of high-throughput screening (Liu et al. 2004). HTS has revolutionized the process of drug development as it enables the evaluation of chemical libraries. By automating the process, HTS expedites the identification of drug candidates thereby enhancing productivity during the phases of research (Hajare et al. 2013; Aldewachi et al. 2021). Computational methods, alongside HTS, play a crucial role in contemporary drug research. The utilization of molecular modeling and simulation tools has become imperative in anticipating molecular behaviors and interactions. This mechanized approach significantly eradicates the need for extensive trial-and-error experimentation, amplifying the accuracy and triumph rates of drug design (White 2000; Latour 2008; Durrant and McCammon 2011; Frenkel and Smit 2023).

3.2 Omics Technologies (Genomics, Proteomics, Metabolomics)

Omics technologies, such as genomics, proteomics, and metabolomics, provide an understanding of the molecular aspects of diseases and the possible outcomes of medical treatments. They represent a transformation, in drug research (Russell et al. 2013; Singh et al. 2023). The field of genomics has been greatly influenced by throughput sequencing technologies leading to a shift in our understanding of diseases at the genetic level. Through genome-wide association studies (GWAS) and other scale genomic investigations, we are able to identify genetic variants that are connected to specific diseases. Identifying genomes plays a role in research as it allows scientists to pinpoint genes associated with disease pathways. Moreover, genomic information is used to categorize patients and personalize treatments based on each individual's makeup optimizing effectiveness while minimizing adverse effects (Witte 2010; Tam et al. 2019; Uffelmann et al. 2021).

Proteomics involves an examination of proteins, which play a role in biological processes. Detection of protein patterns associated with diseases is made possible through techniques like high-throughput protein profiling and sophisticated mass spectrometry. Proteomics helps validate targets in drug development by shedding light on the functions of proteins within disease pathways. Moreover, it aids in identifying biomarkers that are vital for selecting patients in trials and monitoring treatment responses. The integration of data improves our understanding of diseases and optimizes approaches to drug development (Auffray et al. 2009; Hasin et al. 2017; Kan et al. 2017; Huizar et al. 2020).

Through the systematic analysis of small molecules, metabolism provides valuable insights into chronic metabolic changes associated with pathological states and drug responses (Tyagi et al. 2021). Metabolic profiling of metabolites in biological samples facilitates the identification of abnormal metabolic pathways associated with disease states. Metabolomics plays an important role in target validation within the realm of clinical research by providing a comprehensive view of biochemical ramifications of drug interventions. Besides metabolomic profiling facilitates the identification of biomarkers, which in turn affects clinical trial monitoring and the design of diagnostic tools (Kaddurah-Daouk et al. 2008; Patel and Ahmed 2015; Johnson et al. 2016; Blasco et al. 2018).

Integrating the fields of genomics, proteomics, and metabolomics in the drug discovery process enables a better understanding of disease mechanisms and potential therapeutic targets. It enables scientists to decode complex interactions between proteins, genes, and metabolites, resulting in comprehensive systems for formulating precise and efficient therapies. As these cutting-edge technologies continue to develop, they undeniably propel the medical research advancement, simplifying the process of drug discovery and development.

3.3 Genomics and Personalized Medicine

The integration of genomics in medical research marks a paradigm shift toward personalized medicine. Genomic data provides valuable insight into individual genetic variation, providing a deeper understanding of patient-specific responses to drugs. This personalized approach allows the tailoring of treatments, optimizing therapeutic results while minimizing adverse effects. The genomic data extends beyond individualized treatment planning, impacting drug discovery strategies. By unraveling the genetic basis of diseases, researchers can identify new targets for drug intervention. The intersection of genomics and drug research opens avenues for more effective and targeted therapeutic interventions (Ginsburg and Willard 2009; Katsios and Roukos 2010; Sadee 2011; Hassan et al. 2022).

3.4 Biotechnology and the Rise of Biopharmaceuticals

Biotechnology has emerged as a transformative force in modern medical research (Green 2023). The development of biopharmaceuticals including monoclonal antibodies and recombinant proteins represents an important advance. This biological complexity provides new avenues for therapeutic intervention, especially for traditionally challenging targets. Although the development of biopharmaceuticals presents challenges such as complexity and cost considerations, the ability to meet unmet medical needs makes it the breed of novel drugs cornerstone. Biopharmaceuticals contribute significantly to the differentiation of therapeutic modalities, expanding the range of therapies available. Major aspects of modern pharmaceutical research, including quantitative analytics, computational methods, genomics, and biotechnology, collectively reflect broad changes in pharmaceuticals; they find out not only that these advances make the medical system more efficient but also pave the way for precision, personalized medicine interventions and shape the future of innovation (Ayers and Day 2015; Kesik-Brodacka 2018; Romano and Tatonetti 2019; Rastogi et al. 2022; Zhu et al. 2022).

4 Impact on Pharmaceutical Industries of Modern Drug Research

4.1 *Accelerating Drug Development Timelines*

The implementation of contemporary drug exploration method has brought about a fresh epoch of effectiveness in the pharmaceutical sector, notably in hastening the progression of drug development. Through rapid identification of potential drug prospects from extensive compound libraries, high-throughput screening and combinatorial chemistry efficiently condense the preclinical stage. Employing automated technologies, miniaturization, and parallel processing within HTS, scientists are capable of simultaneously evaluating numerous compounds, thereby diminishing the duration traditionally dedicated to hit identification. The amalgamation of advanced computational tools further expedites the enhancement of molecular structures, augmenting the efficiency of lead optimization. This acceleration in early-stage investigation ultimately culminates in a decreased time-to-market for groundbreaking therapies (White 2000; Latour 2008; Durrant and McCammon 2011; Hajare et al. 2013; Aldewachi et al. 2021; Frenkel and Smit 2023).

4.2 *Improving Success Rates in Clinical Trials*

Throughout history, the success rates of clinical trials have posed a significant obstacle in the process of drug development. These rates have often been hindered by high attrition rates. However, in the realm of modern drug research, there is a shift toward precision medicine and personalized therapeutics. This approach leads to a more informed and stratified approach to clinical trials. One of the key factors contributing to this shift is the utilization of biomarkers that have been discovered through the exploration of genomes, proteomics, and metabolomics. These biomarkers allow for the identification of specific patient subgroups who are more likely to respond positively to a particular therapy. By focusing on these subgroups, there is an increased likelihood of successful trial outcomes. Moreover, this focused approach also streamlines the process of patient recruitment, minimizing variability within the populations under study. As a result, pharmaceutical companies are experiencing a substantial reduction in the financial constraints associated with lengthy and unsuccessful clinical trials (Abramovitz et al. 2019; Kiriiri et al. 2020; Fountzilas et al. 2022; Sun et al. 2022).

4.3 Personalized Medicine and Its Market Implications

The emergence of personalized medicine as a direct consequence of modern drug research has extensive implications for pharmaceutical industries, both in terms of research methodologies and market dynamics. The ability to tailor treatments based on genetic and molecular characteristics to individual patient profiles represents a paradigm shift away from the one-size-fits-all approach. Although this provides improved patient outcomes and satisfaction, it necessitates a reconsideration of business models. The traditional breakthrough drug paradigm, characterized by broad commercial appeal, is giving way to a more targeted approach where treatments are customized for specific genetic or molecular subpopulations. This transformation not only aligns with evolving patient expectations but also presents new challenges in market segmentation, pricing strategies, and industry competition dynamics (Behl et al. 2022; Camela et al. 2022; Sharma et al. 2022; Yamamoto et al. 2022).

The influence of contemporary medicine exploration on the pharmaceutical sectors is revolutionary and diverse. Beyond the lab, these progresses reshape exploration strategies, redefine market dynamics, and bring in fresh challenges and chances. While venturing through this intricate territory, the quest for originality intertwines with the urgency of attaining economic sustainability and providing palpable enhancements in patient consequences.

5 Case Studies

5.1 Successful Examples of Modern Drug Research

5.1.1 Imatinib (Gleevec)

Imatinib, a fundamental drug in precision medicine, showcases the transformative potential of focused treatments. Imatinib, engineered through a logical design methodology, specifically aims at the fusion protein BCR-ABL, the root cause behind chronic myeloid leukemia (CML). Its triumph stems from its capacity to modify the aberrant signaling pathway, leading to remarkable outcomes in individuals with CML. The narrative of imatinib not only underscores the feasibility of targeted therapeutics but also sets a precedent for tailoring treatments according to a deep comprehension of the molecular aspects of diseases (Capdeville et al. 2002; La Rosee et al. 2002; Lydon and Druker 2004).

5.1.2 Herceptin (Trastuzumab)

An illustration of the successful transformation of the molecule knowledge into therapeutic practice is the personalized medicine paradigm called herceptin. The drug herceptin, used for HER2-positve breast cancer, binds selectively on the surplus HER2 receptors. Clinical impact is significant and has resulted in better outcomes in cases of HER2+ tumor patients. In addition, herceptin does not only prove to be precision medicine but also becomes an inspiration for developing medicines based on patients' molecular profiles (Pegram et al. 2000; Sakamoto and Mitsuyama 2000; Harries and Smith 2002).

5.1.3 Pembrolizumab (Keytruda)

Immunotherapy represents progress and includes pembrolizumab—an immune checkpoint inhibitor. Understanding how one's immune system deals with cancer informs the development of pembrolizumab as a treatment that blocks the PD–1 pathway and thus releases one's own attack against cancerous cells. Immunotherapy is a rapidly developing trend in contemporary pharmaceutical science. It proved its effectiveness in many tumors, in particular melanoma and pulmonary neoplasm (Emens et al. 2016; Alsaab et al. 2017; Prasad et al. 2018).

5.2 Instances of Failure and Lessons Learned

5.2.1 Thalidomide

Thalidomide is a tale, from history that serves as a reminder of the crucial need for thorough safety assessments. Back in the 1960s, thalidomide was originally marketed as a sedative and nausea medication and tragically led to severe birth defects. This heartbreaking incident underscores the significance of preclinical testing especially when it comes to drugs meant for use during pregnancy (Kim and Scialli 2011; Vargesson 2015; Woolf 2022; Aggarwal et al. 2023).

5.2.2 Vioxx (Rofecoxib)

The withdrawal of Vioxx because of its heart risks demonstrates the challenges in evaluating the long-term safety of medications. Rofecoxib, a pain medication belonging to the COX 2 inhibitor class, unexpectedly caused side effects. This case emphasizes the need for monitoring drugs after they are approved as well as the importance of continuously assessing the balance between risks and benefits throughout a medication's lifespan (Berman 2020; Rogers et al. 2020).

5.2.3 Flibanserin (Addyi)

Flibanserin a medication intended for the treatment of sexual dysfunction faced difficulties, in terms of effectiveness and safety. The initial rejection by authorities followed by approval, with strict safety protocols, highlights the complexities involved in developing drugs for conditions that have subjective objectives. This case underscores the importance of conducting evaluations and understanding the specific needs of patients (Bueter and Jukola 2020).

5.3 Case Studies on Collaboration Between Academia and Industry

5.3.1 Statins

A successful collaboration between academia and industry would imply effective application of statins, also known as cholesterol-lowering drugs. Statins are made possible by the studies that established the connection between HMG-CoA reductase and cholesterol synthesis in academic research. Afterward, inter-industry correlation brought about statin formulations like lovastatin and revolutionized cardiac therapy with evidence on the relationship between scientific research and pharmaceutical creation (Feng et al. 2006; Cheng et al. 2020).

5.3.2 Remdesivir

A more recent example of such a collaboration is the antiviral drug remdesivir used in the treatment of COVID-19. Remdesivir, initially intended to treat Ebola, was repurposed to combat COVID-19 jointly using academic institutions, government bodies, and pharmaceutical industries. It is a perfect example of flexibility that comes with joint actions in addressing new health concerns (Cleary et al. 2020; Lexchin 2021; Garibaldi et al. 2022).

5.3.3 Monoclonal Antibodies for Cancer Treatment

Cancer treatment using monoclonal antibodies, such as rituximab and trastuzumab, was also collaborative. Cancer cell surface antigen research helped to understand the antigen, and industry collaboration helped translated that understanding to therapeutic monoclonal antibody. Such collaborations have overhauled the cancer treatment paradigms showing why the necessity of collaborative efforts (Dalrymple et al. 2006; Buck et al. 2010; Cope et al. 2018; Castelli et al. 2019).

6 Future Trends

6.1 *Advanced Technologies Shaping Drug Research*

6.1.1 Artificial Intelligence (AI) and Machine Learning

In its current state, the growing interconnection between AI and machine learning cannot be divorced from the future direction of drug discovery. The future is promising for these tools that can accelerate target identification, predict compound interaction, and boost lead optimization in the process of drug discovery. Artificial intelligence (AI)-driven algorithms can scan through huge databases, discover latent correlations, and direct researchers toward the direction of new treatment avenues. Combining high-throughput screening, aided by artificial intelligence as well as computational drug design, is expected to increase efficiency and accuracy during early-stage drug development (Gupta et al. 2021; Paul et al. 2021; Vatansever et al. 2021; Patel and Shah 2022).

6.1.2 Quantum Computing

The development of quantum computing is expected to enable computation that even traditional computers are unlikely to cope with. As such, this would enable quicker molecular interaction simulations during research into drugs, thereby providing more accurate conclusions about how these are going to behave. It allows one to solve complex biological processes which will lead to discovery of new drugs faster than the conventional computing systems (Cao et al. 2018; Blunt et al. 2022; Zinner et al. 2022; Wang et al. 2023).

6.2 *Omics Integration and Personalized Therapeutics*

6.2.1 Multi-Omics Approaches

Omics will be integrated including genomics, proteomics, and metabolomics among others. Researchers also identify detailed molecular markers and potential therapeutic targets using multi-omics techniques which provide insight into disease causes. The period of "big data" in medicinal services joined with enhanced analytics will permit analysts to uncover complexes of illness on a framework level, setting the way toward progressively particular and explicit therapeutic intercessions (Higdon et al. 2015; Turanli et al. 2018; John et al. 2020).

6.2.2 Single-Cell Omics

The development of new technologies has led to advanced single-cell omics techniques. Single-cell genomics, proteomics, and metabolomics allow for studying individual cells and exploring within-cell heterogeneity and responses to treatment. This highly granular level of analysis has huge utility in finding rare cells, appreciating molecular pluralism in health conditions and customizing therapeutics as per tailoring therapies to individual cellular profiles (Buriani et al. 2019; Li et al. 2021; Liu et al. 2021).

7 Conclusion

The landscape of drug research is going through an unprecedented shift owing to a convergence of technology advancements, novel therapeutic modalities, and growing ethical considerations. From imatinib to Casp9, mRNA vaccines reflect adaptation in pharmaceutical innovation. Looking ahead, some issues become apparent, affecting the course of drug research and development.

Utilizing state-of-the-art technology such as artificial intelligence and quantum computing will change the scope of drug research. Instead, it helps in better identification of the targets and compound optimization efficiency and new ways to replicating all biological processes. However, the trend of AI and high-throughput screening convergence coupled with the promise of quantitative computing in molecular simulations promises decoding drug-target interactions at unprecedented level of accuracy in the future.

Multi-omics and single-cell technologies will facilitate delivery of holistic comprehension of molecular basis of diseases under omics integration. Personalized medicine promises treatment interventions that have been adapted to each patient on the basis of their genomes, proteomics, and metabolomics. However, it is important to note that such a shift toward precision medicine does not only benefit therapy effectiveness but also creates an era of personalized healthcare.

Paradigmatic shift in treatment entails investigation of advanced therapeutic modalities like gene therapies, RNA-based therapeutics, and cell therapies. With the advancement of medical technology, genetic diseases and tumors that were considered incurable before are now becoming targets of intervention. These breakthroughs in genetic editing technologies and CAR-T therapy in oncology underline what it can be for the future.

Despite the tremendous advancements, a multitude of ethical dilemmas arises. Genetic editing and the discovery of new drugs necessitate robust frameworks to safeguard patient well-being, protect data privacy, and ensure equitable access. Regulatory bodies play a crucial role in addressing these challenges, striking a balance between fostering innovation and upholding public health.

Global collaboration and open science initiatives act as vital pillars for the future of pharmaceutical research. By connecting research communities worldwide through collaborative platforms and shared data repositories, the pace of discovery is accelerated. Embracing open science principles enhances not only the reproducibility of research but also democratizes knowledge access, fostering inclusivity in scientific advancements.

As we conclude our examination of contemporary drug research and forthcoming trends, it becomes apparent that the path ahead is marked by both excitement and accountability. The convergence of cutting-edge technologies, novel therapeutic approaches, and ethical quandaries propels the pharmaceutical industry into an era where medical possibilities are incessantly pushed. The future of drug research embodies a societal commitment to enhancing healthcare outcomes, addressing global health issues, and ushering in an era where personalized and precise medications are universally accessible.

References

Abramovitz M, Williams C, De PK, Dey N, Willis S, Young B, Andreopoulou E, Symmans WF, Sicklick JK, Schilsky RL (2019) Precision medicine clinical trials: successes and disappointments, challenges and opportunities–lessons learnt. In: Predictive biomarkers in oncology: applications in precision medicine, pp 593–603

Aggarwal R, Nautiyal A, Joshi Y (2023) Thalidomide-the disastrous drug, p 241

Ahmed Z, Zeeshan S, Mendhe D, Dong X (2020) Human gene and disease associations for clinical-genomics and precision medicine research. Clin Transl Med 10(1):297–318

Alanis AJ (2005) Resistance to antibiotics: are we in the post-antibiotic era? Arch Med Res 36(6):697–705

Aldewachi H, Al-Zidan RN, Conner MT, Salman MM (2021) High-throughput screening platforms in the discovery of novel drugs for neurodegenerative diseases. Bioengineering 8(2):30

Alsaab HO, Sau S, Alzhrani R, Tatiparti K, Bhise K, Kashaw SK, Iyer AK (2017) PD-1 and PD-L1 checkpoint signaling inhibition for cancer immunotherapy: mechanism, combinations, and clinical outcome. Front Pharmacol 8:561

Ashley EA (2016) Towards precision medicine. Nat Rev Genet 17(9):507–522

Auffray C, Chen Z, Hood L (2009) Systems medicine: the future of medical genomics and healthcare. Genome Med 1:1–11

Ayers D, Day PJ (2015) Systems medicine: the application of systems biology approaches for modern medical research and drug development. Mol Biol Int 2015:698169

Behl T, Kaur I, Sehgal A, Singh S, Albarrati A, Albratty M, Najmi A, Meraya AM, Bungau S (2022) The road to precision medicine: eliminating the "one size fits all" approach in Alzheimer's disease. Biomed Pharmacother 153:113337

Bengtson VL, Settersten R Jr (2016) Handbook of theories of aging. Springer, Berlin

Berman H (2020) How Vioxx exposed conflicts of interest at the Food and Drug Administration and the New England Journal of Medicine. Food Drug Law J 75(4):552–573

Blasco H, Patin F, Descat A, Garçon G, Corcia P, Gelé P, Lenglet T, Bede P, Meininger V, Devos D (2018) A pharmaco-metabolomics approach in a clinical trial of ALS: identification of predictive markers of progression. PLoS One 13(6):e0198116

Blass BE (2015) Basic principles of drug discovery and development. Elsevier, Amsterdam

Blunt NS, Camps J, Crawford O, Izsák R, Leontica S, Mirani A, Moylett AE, Scivier SA, Sunderhauf C, Schopf P (2022) Perspective on the current state-of-the-art of quantum computing for drug discovery applications. J Chem Theory Comput 18(12):7001–7023

Brittain HK, Scott R, Thomas E (2017) The rise of the genome and personalised medicine. Clin Med 17(6):545

Buck E, Mulvihill M, Iwata KK (2010) Pursuit of personalized anticancer therapy: leveraging collaboration between academia and the biotech/pharmaceutical industry. Mount Sinai J Med 77(4):358–365

Bueter A, Jukola S (2020) Sex, drugs, and how to deal with criticism: the case of Flibanserin. In: Uncertainty in pharmacology: epistemology, methods, and decisions, pp 451–470

Buriani A, Fortinguerra S, Sorrenti V, Gabbia D, Carrara M (2019) Single-cell omics in personalized medicine. In: Single-cell omics. Elsevier, Amsterdam, pp 221–236

Camela E, Potestio L, Fabbrocini G, Ruggiero A, Megna M (2022) New frontiers in personalized medicine in psoriasis. Expert Opin Biol Ther 22(12):1431–1433

Cao Y, Romero J, Aspuru-Guzik A (2018) Potential of quantum computing for drug discovery. IBM J Res Dev 62(6):20

Capdeville R, Buchdunger E, Zimmermann J, Matter A (2002) Glivec (STI571, imatinib), a rationally developed, targeted anticancer drug. Nat Rev Drug Discov 1(7):493–502

Carrasco-Ramiro F, Peiró-Pastor R, Aguado B (2017) Human genomics projects and precision medicine. Gene Ther 24(9):551–561

Castelli MS, McGonigle P, Hornby PJ (2019) The pharmacology and therapeutic applications of monoclonal antibodies. Pharmacol Res Perspect 7(6):e00535

Cheng F, Ma Y, Uzzi B, Loscalzo J (2020) Importance of scientific collaboration in contemporary drug discovery and development: a detailed network analysis. BMC Biol 18(1):1–9

Cleary EG, Jackson MJ, Folchman-Wagner Z, Ledley FD (2020) Foundational research and NIH funding enabling emergency use authorization of remdesivir for COVID-19. MedRxiv.2020.2007. 2001.20144576

Collins FS, McKusick VA (2001) Implications of the human genome project for medical science. JAMA 285(5):540–544

Cope AP, Barnes MR, Belson A, Binks M, Brockbank S, Bonachela-Capdevila F, Carini C, Fisher BA, Goodyear CS, Emery P (2018) The RA-MAP consortium: a working model for academia–industry collaboration. Nat Rev Rheumatol 14(1):53–60

Dalrymple M, Taylor D, Kettleborough C, Bryans J, Solari R (2006) Academia–industry partnerships in drug discovery. Expert Opin Drug Discovery 1(1):1–6

Doestzada M, Vila AV, Zhernakova A, Koonen DP, Weersma RK, Touw DJ, Kuipers F, Wijmenga C, Fu J (2018) Pharmacomicrobiomics: a novel route towards personalized medicine? Protein Cell 9(5):432–445

Durrant JD, McCammon JA (2011) Molecular dynamics simulations and drug discovery. BMC Biol 9(1):1–9

Emens LA, Butterfield LH, Hodi FS, Marincola FM, Kaufman HL (2016) Cancer immunotherapy trials: leading a paradigm shift in drug development. J Immunother Cancer 4:1–8

Feng C, Han A, Ye C, Xu R, Li M (2006) The HMG-CoA reductase pathway, statins and angioprevention. Semin Ophthalmol 21:29–35

Fountzilas E, Tsimberidou AM, Vo HH, Kurzrock R (2022) Clinical trial design in the era of precision medicine. Genome Med 14(1):1–27

Frenkel D, Smit B (2023) Understanding molecular simulation: from algorithms to applications. Elsevier, Amsterdam

Gannett, L. (2008). The Human Genome Project. The Stanford Encyclopedia of Philosophy (Fall 2008 Edition)

Garibaldi BT, Wang K, Robinson ML, Betz J, Caleb Alexander G, Andersen KM, Joseph CS, Mehta HB, Korwek K, Sands KE (2022) Real-world effectiveness of remdesivir in adults hospitalized with coronavirus disease 2019 (COVID-19): a retrospective, multicenter comparative effectiveness study. Clin Infect Dis 75(1):e516–e524

Ginsburg GS, Willard HF (2009) Genomic and personalized medicine: foundations and applications. Transl Res 154(6):277–287

Green E (2023) Innovations in biotechnology: transforming science, technology, and business. Int Multidiscip J Sci Technol Bus 2(02):5–8

Gupta R, Srivastava D, Sahu M, Tiwari S, Ambasta RK, Kumar P (2021) Artificial intelligence to deep learning: machine intelligence approach for drug discovery. Mol Divers 25:1315–1360

Haider R (2023) Penicillin and the antibiotics revolution global history. Asian J Pharmaceut Res 13(1):55

Hajare AA, Salunkhe SS, Mali SS, Gorde SS, Nadaf SJ, Pishawikar SA (2013) Review on: high-throughput screening is an approach to drug discovery. Am J Pharm Tech Res 4:112–129

Harries M, Smith I (2002) The development and clinical use of trastuzumab (Herceptin). Endocr Relat Cancer 9(2):75–85

Hasin Y, Seldin M, Lusis A (2017) Multi-omics approaches to disease. Genome Biol 18(1):1–15

Hassan M, Awan FM, Naz A, deAndrés-Galiana EJ, Alvarez O, Cernea A, Fernández-Brillet L, Fernández-Martínez JL, Kloczkowski A (2022) Innovations in genomics and big data analytics for personalized medicine and health care: a review. Int J Mol Sci 23(9):4645

Higdon R, Earl RK, Stanberry L, Hudac CM, Montague E, Stewart E, Janko I, Choiniere J, Broomall W, Kolker N (2015) The promise of multi-omics and clinical data integration to identify and target personalized healthcare approaches in autism spectrum disorders. Omics 19(4):197–208

Huizar CC, Raphael I, Forsthuber TG (2020) Genomic, proteomic, and systems biology approaches in biomarker discovery for multiple sclerosis. Cell Immunol 358:104219

Huptych M, Potucek J, Lhotská L (2021) Towards personalized medication. pHealth 2021. IOS Press, Amsterdam, pp 49–57

Jian J, He D, Gao S, Tao X, Dong X (2023) Pharmacokinetics in pharmacometabolomics: towards personalized medication. Pharmaceuticals 16(11):1568

John A, Qin B, Kalari KR, Wang L, Yu J (2020) Patient-specific multi-omics models and the application in personalized combination therapy. Future Oncol 16(23):1737–1750

Johnson CH, Ivanisevic J, Siuzdak G (2016) Metabolomics: beyond biomarkers and towards mechanisms. Nat Rev Mol Cell Biol 17(7):451–459

Kaddurah-Daouk R, Kristal BS, Weinshilboum RM (2008) Metabolomics: a global biochemical approach to drug response and disease. Annu Rev Pharmacol Toxicol 48:653–683

Kan M, Shumyatcher M, Himes BE (2017) Using omics approaches to understand pulmonary diseases. Respir Res 18(1):1–20

Katsios C, Roukos DH (2010) Individual genomes and personalized medicine: life diversity and complexity. Pers Med 7(4):347–350

Kesik-Brodacka M (2018) Progress in biopharmaceutical development. Biotechnol Appl Biochem 65(3):306–322

Khanna I (2012) Drug discovery in pharmaceutical industry: productivity challenges and trends. Drug Discov Today 17(19–20):1088–1102

Kim JH, Scialli AR (2011) Thalidomide: the tragedy of birth defects and the effective treatment of disease. Toxicol Sci 122(1):1–6

Kiriiri GK, Njogu PM, Mwangi AN (2020) Exploring different approaches to improve the success of drug discovery and development projects: a review. Fut J Pharmaceut Sci 6(1):1–12

La Rosee P, O'Dwyer M, Druker B (2002) Insights from pre-clinical studies for new combination treatment regimens with the Bcr-Abl kinase inhibitor imatinib mesylate (Gleevec/Glivec) in chronic myelogenous leukemia: a translational perspective. Leukemia 16(7):1213–1219

Laszlo P (2007) On the self-image of chemists, 1950–2000. In: The public image of chemistry. World Scientific, Singapore, pp 329–367

Latour RA (2008) Molecular simulation of protein-surface interactions: benefits, problems, solutions, and future directions. Biointerphases 3(3):FC2–FC12

Lexchin J (2021) Are academia–pharma partnerships essential for novel drug discovery in the time of the COVID-19 pandemic? Expert Opin Drug Discovery 16(5):475–479

Li Y, Ma L, Wu D, Chen G (2021) Advances in bulk and single-cell multi-omics approaches for systems biology and precision medicine. Brief Bioinform 22(5):bbab024

Liu B, Li S, Hu J (2004) Technological advances in high-throughput screening. Am J Pharmacogenomics 4:263–276

Liu J, Qu S, Zhang T, Gao Y, Shi H, Song K, Chen W, Yin W (2021) Applications of single-cell omics in tumor immunology. Front Immunol 12:697412

Lydon NB, Druker BJ (2004) Lessons learned from the development of imatinib. Leuk Res 28:29–38

Markel H (2021) The secret of life: Rosalind Franklin, James Watson, Francis crick, and the discovery of DNA's double helix. WW Norton, New York

Milanesi M, Runfola A, Guercini S (2020) Pharmaceutical industry riding the wave of sustainability: review and opportunities for future research. J Clean Prod 261:121204

Montinari MR, Minelli S, De Caterina R (2019) The first 3500 years of aspirin history from its roots–a concise summary. Vascul Pharmacol 113:1–8

Olsen A (2013) The antibiotic revolution. Spring 2013. Kuhn's Philosophy of Science, p 6

Patel S, Ahmed S (2015) Emerging field of metabolomics: big promise for cancer biomarker identification and drug discovery. J Pharm Biomed Anal 107:63–74

Patel V, Shah M (2022) Artificial intelligence and machine learning in drug discovery and development. Intell Med 2(3):134–140

Paul D, Sanap G, Shenoy S, Kalyane D, Kalia K, Tekade RK (2021) Artificial intelligence in drug discovery and development. Drug Discov Today 26(1):80

Pegram MD, Konecny G, Slamon DJ (2000) The molecular and cellular biology of HER2/neu gene amplification/overexpression and the clinical development of herceptin (trastuzumab) therapy for breast cancer. Adv Breast Cancer Manag:57–75

Prasad V, Kaestner V, Mailankody S (2018) Cancer drugs approved based on biomarkers and not tumor type—FDA approval of pembrolizumab for mismatch repair-deficient solid cancers. JAMA Oncol 4(2):157–158

Pucci C, Martinelli C, Ciofani G (2019) Innovative approaches for cancer treatment: Current perspectives and new challenges. Ecancermedicalscience 13:961

Raggio N (2002) Cracking the human genome: a look into how competitive forces create and reshape collaborations. SWITCH 17(1):10

Rastogi S, Rastogi P, Mendiratta N (2022) Bioinformatics: methods and applications-genomics, proteomics and drug discovery. PHI Learning, New Delhi

Rogers A, Grieve K, MacDonald TM, Nishioka S (2020) COX-2 inhibitors: communication of accumulating risk evidence and a product withdrawal. In: Communicating about risks and safe use of medicines: real life and applied research. Springer, Berlin, pp 131–146

Romano JD, Tatonetti NP (2019) Informatics and computational methods in natural product drug discovery: a review and perspectives. Front Genet 10:368

Russell C, Rahman A, Mohammed AR (2013) Application of genomics, proteomics and metabolomics in drug discovery, development and clinic. Ther Deliv 4(3):395–413

Sabatowski R, Schafer D, Kasper S, Brunsch H, Radbruch L (2004) Pain treatment: a historical overview. Curr Pharm Des 10(7):701–716

Sadee W (2011) Genomics and personalized medicine. Elsevier, Amsterdam, pp 2–4

Sakamoto G, Mitsuyama S (2000) New molecule-targeting therapy with herceptin (trastuzumab), an anti-HER2 (c-erB-2) monoclonal antibody. Breast Cancer 7:350–357

Salon JA, Lodowski DT, Palczewski K (2011) The significance of G protein-coupled receptor crystallography for drug discovery. Pharmacol Rev 63(4):901–937

Sharma M, Sehrawat R, Luthra S, Daim T, Bakry D (2022) Moving towards industry 5.0 in the pharmaceutical manufacturing sector: challenges and solutions for Germany. IEEE Trans Eng Manag:1–18

Singh S, Sarma DK, Verma V, Nagpal R, Kumar M (2023) Unveiling the future of metabolic medicine: omics technologies driving personalized solutions for precision treatment of metabolic disorders. Biochem Biophys Res Commun 682:1

Sun D, Gao W, Hu H, Zhou S (2022) Why 90% of clinical drug development fails and how to improve it? Acta Pharm Sin B 12(7):3049–3062

Taiwo O, Ezugwu AE (2020) Smart healthcare support for remote patient monitoring during covid-19 quarantine. Inform Med Unlocked 20:100428
Tam V, Patel N, Turcotte M, Bossé Y, Paré G, Meyre D (2019) Benefits and limitations of genome-wide association studies. Nat Rev Genet 20(8):467–484
Tandon A, Dhir A, Islam AN, Mäntymäki M (2020) Blockchain in healthcare: a systematic literature review, synthesizing framework and future research agenda. Comput Ind 122:103290
Trenfield SJ, Awad A, Madla CM, Hatton GB, Firth J, Goyanes A, Gaisford S, Basit AW (2019) Shaping the future: recent advances of 3D printing in drug delivery and healthcare. Expert Opin Drug Deliv 16(10):1081–1094
Turanli B, Karagoz K, Gulfidan G, Sinha R, Mardinoglu A, Arga KY (2018) A network-based cancer drug discovery: from integrated multi-omics approaches to precision medicine. Curr Pharm Des 24(32):3778–3790
Tyagi R, Kumar P, Sharma U (2021) Metabolomics techniques: a brief update. In: Epigenetics and metabolomics. Elsevier, Amsterdam, pp 1–29
Uffelmann E, Huang QQ, Munung NS, De Vries J, Okada Y, Martin AR, Martin HC, Lappalainen T, Posthuma D (2021) Genome-wide association studies. Nat Rev Methods Primers 1(1):59
Vargesson N (2015) Thalidomide-induced teratogenesis: history and mechanisms. Birth Defects Res C Embryo Today 105(2):140–156
Vatansever S, Schlessinger A, Wacker D, Kaniskan HÜ, Jin J, Zhou MM, Zhang B (2021) Artificial intelligence and machine learning-aided drug discovery in central nervous system diseases: state-of-the-arts and future directions. Med Res Rev 41(3):1427–1473
Von Krogh G, Kucukkeles B, Ben-Menahem SM (2020) Lessons in rapid innovation from the COVID-19 pandemic. MIT Sloan Manag Rev 61(4):8–10
Wang P-H, Chen J-H, Yang Y-Y, Lee C, Tseng YJ (2023) Recent advances in quantum computing for drug discovery and development. IEEE Nanotechnol Mag 17:26
White RE (2000) High-throughput screening in drug metabolism and pharmacokinetic support of drug discovery. Annu Rev Pharmacol Toxicol 40(1):133–157
Wiggin GR, Patel JC, Marshall FH, Jazayeri A (2014) Molecular biology for medicinal chemists. In: The handbook of medicinal chemistry, pp 239–265
Witte JS (2010) Genome-wide association studies and beyond. Annu Rev Public Health 31:9–20
Woolf AD (2022) Thalidomide tragedy, 1950s. In: History of modern clinical toxicology. Elsevier, Amsterdam, pp 165–175
Yamamoto Y, Kanayama N, Nakayama Y, Matsushima N (2022) Current status, issues and future prospects of personalized medicine for each disease. J Pers Med 12(3):444
Zhu Y, Ouyang Z, Du H, Wang M, Wang J, Sun H, Kong L, Xu Q, Ma H, Sun Y (2022) New opportunities and challenges of natural products research: when target identification meets single-cell multiomics. Acta Pharm Sin B 12:4011
Zinner M, Dahlhausen F, Boehme P, Ehlers J, Bieske L, Fehring L (2022) Toward the institutionalization of quantum computing in pharmaceutical research. Drug Discov Today 27(2):378–383

The Market Value of Traditionally Used Medicinal Plants

Pedro Antonio Martin-Cervantes, Parisa Ziarati, and Barbara Sawicka

Contents

Abstract Recently, the consumption of medicinal plants and dietary supplements derived from natural resources has become widespread in many developed societ-

P. A. Martin-Cervantes (✉)
Department of Financial Economics and Accounting, University of Valladolid, Valladolid, Spain
e-mail: pedroantonio.martin@uva.es

P. Ziarati
Halal Research Center of IRI, Food and Drug Administration, Ministry of Health and Medical Education, Tehran, Iran
e-mail: parziarati@gmail.com

B. Sawicka
Department of Plant Production Technology and Commodities Science, University of Life Sciences, Lublin, Poland
e-mail: barbara.sawicka@up.lublin.pl

© The Author(s), under exclusive license to Springer Nature Singapore Pte Ltd. 2024

A. Das Talukdar et al. (eds.), *Traditional Resources and Tools for Modern Drug Discovery*, Interdisciplinary Biotechnological Advances,
https://doi.org/10.1007/978-981-97-4600-2_18

ies. This chapter discusses the growing popularity of the consumption of medicinal plants, their products, and dietary supplements in developed societies. With the belief in the lack of side effects, low cost, and antioxidant properties in preventing and treating many diseases, such as cancer, diabetes, overweight, osteoporosis, and hypercholesterolemia, people are increasingly turning to natural products. In fact, the use of medicinal plants is safer and their employment can be carried out in much lower doses than medicine purely based on components of chemical origin. Manufacturers take advantage of this trend on the market and develop marketing strategies, focusing on health-promoting aspects, preventing diseases, and healing properties of herbal plants. It is recommended to use medicinal plants to maintain health and prevent side effects associated with chemical drugs. Also, high healthcare costs contribute to the increased demand for natural products with health benefits. Therefore, the marketing of herbal products focuses mainly on some health-promoting, disease-preventing, or healing properties, which often refer to natural bioactive compounds. However, you should be careful and know about potential side effects or interactions with other drugs, because medicinal plants can also cause adverse reactions. Valuable and reliable information on the properties of medicinal plants is extremely important for consumers to make informed choices.

Keywords Medicinal plants market · Regulatory agencies · Alternative medicines · Ethnobotany · Phytotherapies

1 Introduction

Traditional medicine continues to play a major role among people worldwide, slightly to meet *primary healthcare needs*. In recent years, the aromatic, herbal, and medicinal plants marketed for fresh/ dried consumption have been flourishing, especially in the form of eco-friendly certified packaged herbs (Martínez de Arano et al. 2021; Zaccardelli et al. 2021). In light of many investigations, detecting that dissatisfaction with conventional medicine was the most main objective for a preferred maneuver on herbal medicine (Ghod Elahi et al. 2022; Sawicka et al. 2022; Welz et al. 2018). For instance, more than three billion people use traditional herbs and medicines coming mainly from forest areas (Taghouti et al. 2022; WHO 2013). The huge market of herbal/ medicinal and aromatic plants is not limited to food preparation, based on their flavor, but also as many technical innovations and investments, due to the special consideration for those products for the treatment of various health challenges in different national healthcare settings (Bourgou et al. 2021; Cheminal et al. 2020; Ikinyom et al. 2023). In recent years, drug discovery programs in dominant therapeutic areas in cancer, neuropharmacological, inflammation, metabolic and infectious, cardiovascular, and gastrointestinal are based upon plant-derived compounds (Dehelean et al. 2021; Jiménez-González et al. 2023; Paşayeva et al. 2021).

Many attractions of such eco-friendly attitudes exist. The plants used for various therapies are reachable by comparatively long shelf life, and rich diversity. The most important advantage of herbal medicine is the minimal side effects and comparatively low price in comparison to synthetic medicines. The repercussion of the medicinal plants sector as a major resource base for the traditional medicine and herbal *industry* mainly depends on the awareness and interest of several stakeholders involved in their supply chain: the farmers, supportive government policies, availability of guaranteed markets, fruitful price levels, and determine to simple and opportune agro-techniques (George 2011).

In many countries, whether developed or developing, the use of herbal medicines is achieved on an occasional basis, as opposed to the almost predestined use of pharmaceutical products (Chapman and Chomchalow 2005; Munasinghe 2010), namely in France, medicinal and aromatic plants (MAP) or herbal, medicinal and aromatic plants (HMAP) industries typically apply to flavored herbs, medicinal plants, and plants cultivated on the based on the fragrance and perfume industries (Bergoënd and Piasentin 2022), while at least 70% of the population of the developing world directly depended on traditional medicine for primary medical care (Abdallah et al. 2023; Astutik et al. 2019; Jeelani et al. 2018).

The global medicine market is more than one trillion US dollars (Calixto 2019). About 35% of medicines have originated from natural products. T he last decade plainly observed an enormous stream of approval and endorsement of public interest in natural therapies both in developing and developed countries, which results in those herbal remedies being accessible not only in drug stores but also in food stores and supermarkets (Cragg et al. 1999; Fatima and Siddique 2021; Suh Nchang et al. 2023)

As reported by WHO (2013), nearly 80% of the total world population trusts in herbal remedies for the administration of their essential health system as a result of the scarcity of advanced health services and insufficiency of finances and income (Birjees et al. 2022). However, increasing major concerns concerning the safety of herbal medicines subsequently of researches and studies revealed that high levels of heavy metals were present in some herbal medicines, in addition to potential side effects or interactions with other drugs, because medicinal plants can also cause adverse reactions. Valuable and reliable information on the properties of medicinal plants is extremely important for consumers to make informed choices.

For a brief dissection of the main world markets of medicinal plants, this chapter is structured as follows: First, a brief characterization of the sociocultural significance and scope of medicinal plants is presented. This is followed by a classification of medicinal plants according to their origin, preliminary to two sections that respectively outline the traditional markets for medicinal plants and discuss several concrete cases of successful implementation of phytotherapies in the present. Subsequently, the main international regulatory agencies are listed. The chapter then describes the main handicap of economic and financial research on medicinal plants. Finally, it concludes by highlighting the negative aspects that can be associated with the application of traditional medicines and explores new lines of research that can lead to a flourishing analysis of the market of medicinal plants.

2 The World of Medicinal Plants: A Brief Characterization

There is no area more deeply rooted in human development than that of medicinal plants. Since the dawn of time, cultures such as the Romans (Hornblower et al. 2012) or the Celts (Wagner et al. 2020), if we refer to the European case, have made use of them in search of medicinal remedies, magic, alchemy, etc. Such is the case of the laurel in the Greco-Roman tradition (Paulson 2007) or even hemlock, which, although it was the means by which characters such as Polemarchus, Photius, Socrates, and Theramenes were forced to commit suicide (Barceloux 2008), it also had its own medicinal use. The echo of these ancient civilizations continues to be reflected in the four corners of the universe through ethnobotany in countries as distant as Bolivia (Macía et al. 2005), Ethiopia (Chekole et al. 2015), South Africa (Botha et al. 2004), Sudan (Issa et al. 2018), Thailand (Maneenoon et al. 2015), or Nepal (Kunwar et al. 2013), just to name a few latitudes where its use is ancestral. In fact, modern medicine can be considered as an evolution of traditional medicine (Pérez-Lloréns et al. 2023) in an evolution that went "from the Garden to the Pill Box" (Athni and Athni 2019). An example of this is the use since time immemorial of willow bark in the relief of pain and fever, a remote ancestor of acetylsalicylic acid, a medicament that would give rise to a veritable "aspirin war" in terms of its mass marketing (Mann and Plummer 1991).

3 Classification of Medicinal Plants According to the Natural Environment They Come From

According to Fig. 1, we can classify the diversity of medicinal plants found in different natural environments, each with its own properties and therapeutic uses.

Among the medicinal plants of vegetable origin, we have chamomile, an annual herbaceous plant of the Asteraceae family (Franke and Schilcher 2007). It has small daisy-like flowers with an apple aroma. The dried flowers are used to prepare chamomile infusions, which historically have been used to treat digestive problems, relieve stress, and promote sleep. Regarding medicinal plants of arboreal origin, it is common to highlight several types of willow, such as *Salix alba* (white willow) (Maistro et al. 2019) and *Salix purpurea* (red willow) (Aliferis et al. 2015), trees that contain salicin, a compound that in the body is converted to salicylic acid, a precursor of aspirin. Willow bark has been used for centuries as a natural analgesic and anti-inflammatory to relieve pain and fever (Miner and Hoffhines 2007).

Finally, the sacred lotus is a representative of medicinal plants of aquatic origin. Perennial in nature, it is known for its beautiful pink flowers and floating leaves. Various parts of the plant, including seeds, leaves, and roots, are used in traditional Asian medicine (Mukherjee 2014), as its components are rich in antioxidants and are used in infusions and extracts to improve heart health and reduce constipation.

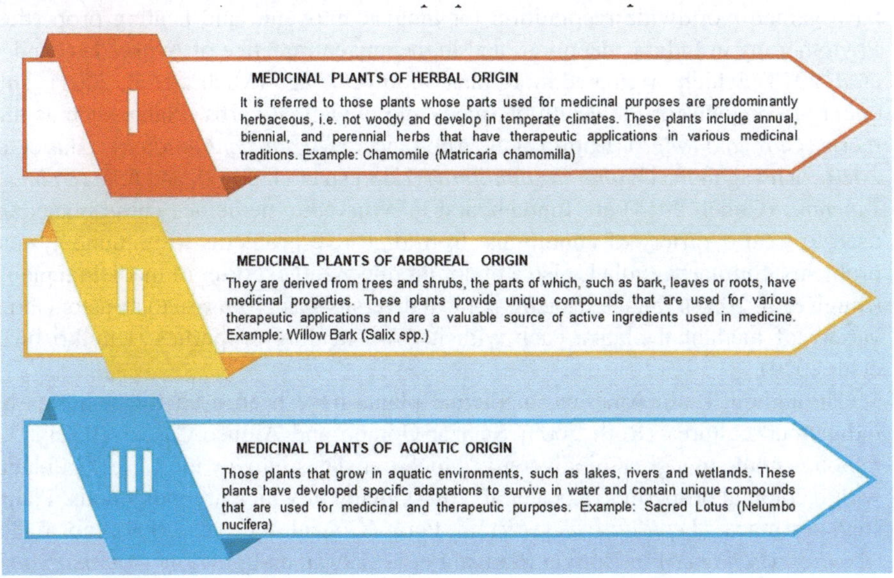

Fig. 1 Preliminary classification of medicinal plants. (Source: Own elaboration)

4 Traditional Medicinal Plant Markets

Basically, we can distinguish the following markets dedicated to medicinal plants: European Union, China, India, Latin America, and Africa. In the European Union, medicinal plants have a long tradition dating back to the ancient cultures of Greece and Rome, where they were valued for their healing properties. Phytotherapy in Europe has historically been linked to alchemical and herbal practices (Karpenko 1995). Today, the European Union strictly regulates medicinal plants to ensure their safety and efficacy. The EU Herbal Medicinal Products Directive sets quality and safety standards for herbal products. In the European Union, the most widely produced medicinal plants are chamomile (*Matricaria chamomilla*) (Chauhan et al. 2021), yarrow (Achillea millefolium) (Saeidnia et al. 2011), valerian (*Valeriana officinalis*) (Patočka and Jakl 2010), and echinacea (*Echinacea purpurea*) (Hostettmann 2003). These plants are used in a variety of products, from herbal teas to herbal supplements and ointments. In addition, herbs such as ginkgo (*Ginkgo biloba*) (Czigle et al. 2018) and garlic (*Allium sativum*) (Pittler and Ernst 2007) are also popular for their medicinal properties. About medicinal plant markets in China, they enjoy a rich tradition in herbal medicine dating back more than 2000 years.

Traditional Chinese medicine (TCM) is based on the use of plants, animals, and minerals to restore balance to the body (Marshall 2020). Plants such as ginseng (*Panax ginseng*) (Potenza et al. 2023), astragalus (*Astragalus membranaceus*) (Auyeung et al. 2016), St. John's wort (Ng et al. 2017), and licorice root (*Glycyrrhiza uralensis*) (Wang et al. 2013) are widely used in traditional Chinese medicine and have gained worldwide recognition for their adaptogenic and healing properties. Phytotherapy in India is deeply rooted in the ancient practice of Ayurveda (Pandey et al. 2013), which originated more than 5000 years ago (Mishra et al. 2001). This country is known for its wealth of spices and medicinal herbs. Plants such as turmeric (*Curcuma longa)* (Luthra et al. 2001), neem (*Azadirachta indica*) (Islas et al. 2020), ashwagandha (*Withania somnifera*) (Lopresti et al. 2019), and tulsi (*Ocimum sanctum*) (Cohen 2014) are fundamental in Ayurvedic medicine. These plants are used to treat a variety of conditions, from digestive problems to immune system problems. Similarly, India is also a major producer and exporter of medicinal plants (Singh et al. 2022). In fact, its geographical diversity has given rise to an astonishing variety of medicinal plants, each with its own healing properties (Unnikrishnan et al. 2020).

Throughout Latin America, medicinal plants have been used for centuries by indigenous cultures (Roth 2008; Salazar-Gómez and Alonso-Castro 2022). The region is home to a great biodiversity, which translates into a wide variety of plants with medicinal properties (Roca et al. 2004), many of which are indigenous. Plants such as maca (*Lepidium meyenii*) in Peru (Gonzales et al. 2009), coca leaf (*Erythroxylum coca*) in Bolivia (Sauvain et al. 1997), and graviola (*Annona muricata*) (Rady et al. 2018) in several Latin American countries are highly valued for their medicinal properties. In addition, herbs such as rue (*Ruta graveolens*) (Aronson 2016) and chamomile (*Matricaria chamomilla*) (Cabada-Aguirre et al. 2023) are common in traditional medicine in the region. Phytotherapy in Latin America reflects the fusion of indigenous practices with European and African influences (Castro Braga 2021; Papenfuss 2022). It should be noted that Africa has a rich tradition of herbal medicine that has been transmitted from generation to generation. African communities have used plants such as moringa (*Moringa oleifera*) (Pareek et al. 2023), baobab (*Adansonia* sp.) (Chapotin et al. 2006), aloe vera (Grace et al. 2008), and yohimbine bark (*Pausinystalia yohimbe*) (Raman et al. 2012) to treat various conditions, from fevers to skin problems. African phytotherapy is often based on traditional knowledge transmitted orally (Devine et al. 2022), whereas the exploitation and commercialization of these plants often face challenges related to sustainability and biodiversity conservation (Dzoyem et al. 2013; Schippmann et al. 2002).

In this continent, medicinal plants are fundamental in health care, especially in rural areas (Mahomoodally 2013). For illustrative purposes, Table 1 shows several prototypical plants from certain countries or regions.

Table 1 Enumeration of the distinctive medicinal plants of each country or geographic area, determining their main uses and applications

Medicinal plant	Continent	Country of origin	Main uses
Ginseng (*Panax ginseng*)	Asia	China, South Korea	Improves endurance, increases energy, reduces stress
Aloe vera (*Aloe barbadensis* Miller)	Africa	South Africa	Treatment of burns, irritated skin, digestion
Turmeric (*Curcuma longa*)	Asia	India	Anti-inflammatory properties, digestive support, antioxidant
Chamomile (*Matricaria chamomilla*)	Europe	Germany, Hungary	Calming for the nervous system, relieves insomnia, digestive problems
Ginkgo biloba (*Ginkgo biloba*)	Asia	China	Improves cerebral circulation, prevents memory loss, antioxidant
Ginger (*Zingiber officinale*)	Asia	India, China	Anti-inflammatory properties, pain relief, nausea
Echinacea (*Echinacea purpurea*)	North America	United States	Strengthens the immune system, helps fight colds and flu
Licorice (*Glycyrrhiza glabra*)	Europe, Asia	Greece, Turkey	Anti-inflammatory, cough relief, gastrointestinal problems
Garlic (*Allium sativum*)	Europe, Asia	China, India	Reduces blood pressure, strengthens the immune system, antibacterial properties
Rosemary (*Rosmarinus officinalis*)	Europe, Asia	Spain, Italy	Stimulates memory, antioxidant properties, improves digestion
Calendula (*Calendula officinalis*)	Europe	Spain	Healing, anti-inflammatory, skin care
Sage (*Salvia officinalis*)	Europe, Asia	Greece, Italy	Improves digestion, reduces perspiration, relieves sore throat
Stevia (*Stevia rebaudiana*)	South America	Paraguay, Brazil	Natural sweetener, controls blood sugar, helps in weight loss
Hibiscus (*Hibiscus sabdariffa*)	Africa	Sudan, Senegal	Reduces blood pressure, improves heart health, diuretic
Cinnamon (*Cinnamomum verum*)	Asia	Sri Lanka	Stabilizes blood sugar levels, antioxidant properties, improves digestion
Lavender (*Lavandula angustifolia*)	Europe, Asia, Africa	France, Bulgaria	Calming for the nervous system, helps in sleep, relieves stress
Basil (*Ocimum basilicum*)	Asia, Africa	India, Egypt	Anti-inflammatory properties, improves digestion, relieves anxiety
Boldo (*Peumus boldus*)	South America	Chile	Stimulates digestion, protects the liver, relieves gastric problems
Valerian (*Valeriana officinalis*)	Europe, Asia	Germany, Russia	Natural calming, relieves anxiety, improves sleep quality
Dandelion (*Taraxacum officinale*)	Europe, Asia, North America	United States, China	Detoxifying, diuretic, improves liver function

Source: Own elaboration

5 Specific Cases in Which the Use of Medicinal Plants Has Been Beneficial

The specific applications of medicinal plants are practically innumerable, but the following five cases of more recent studies showing the versatility of this area of human health should be highlighted:

- Prevention of malaria in Africa with artemisinin: Artemisinin, a compound extracted from the plant *Artemisia annua* (sweet wormwood), has been used successfully to treat malaria in many parts of Africa (Bigira et al. 2014; Ndeffo Mbah et al. 2015). In malaria-endemic regions, artemisinin-based treatments have proven to be effective and more affordable than some traditional pharmaceutical treatments. Organizations such as the World Health Organization (WHO) have supported and promoted the use of this herbal therapy to combat malaria (WHO 2022).
- Glycemic control with cinnamon in type 2 diabetes: Scientific studies have shown that cinnamon, a spice derived from the bark of the *Cinnamomum verum* tree, can help control blood sugar levels in people with type 2 diabetes (Deyno et al. 2019; Silva et al. 2022; Zarezadeh et al. 2023). Some people have chosen to incorporate cinnamon into their daily diet to help regulate their glucose levels, thus complementing their medical treatment.
- Chronic pain relief with kratom in the United States: Despite the controversy surrounding kratom (*Mitragyna speciosa*), a plant native to Southeast Asia, many people in the United States have used this plant to relieve chronic pain (Chin and Mark-Lee 2018; Halpenny 2017) and reduce opioid withdrawal symptoms. Some research suggests that the alkaloids present in kratom have analgesic properties and may help overcome opioid addiction (Boyer et al. 2008).
- Treatment of respiratory diseases with herbs in China during the COVID-19 pandemic: During the COVID-19 pandemic, China incorporated herbal treatments into its approach to disease treatment (Lyu et al. 2021; Zhu et al. 2021). The formula "Qingfei Paidu decoction," which includes a combination of Chinese herbal medicines, was used as part of the treatment to alleviate symptoms.
- Respiratory infections in COVID-19 patients: This integrative approach has been considered part of China's success in treating the pandemic in its early phases (Wang et al. 2022).
- Use of Medicinal Plants in Primary Care Programs in Brazil: In Brazil, the Unified Health System (SUS) has incorporated medicinal plants into primary care programs (Villas Bôas Gde and Gadelha 2007). Through programs such as "Farmacias Vivas," communities can grow and use local medicinal plants for the treatment of common diseases. These programs have improved access to health care in rural areas and have strengthened traditional knowledge about the medicinal properties of plants (de Moura Monteiro and Pessoa 2022).

Fig. 2 List of the most representative national and transnational regulatory agencies. (Source: Own elaboration from Fan et al. (2012))

6 Regulatory Agencies for Medicinal Plants

In the absence of univocal regulation at the global level, there is no homogeneous legislation for all medicinal plants worldwide, as evidenced in Fig. 2, which lists the main agencies and regulatory bodies.

Likewise, Table 2 compiles a selection of medicinal plants that are commonly marketed in the European Union, following the precepts of the EMA (EMA 2023), its regulatory agency.

7 The Relative Opacity of the Economic-Financial Data Coming from the Markets of Medicinal Plants

Although it is obvious that medicinal plants are becoming an increasingly powerful transnational market, there are several problems that hinder their analysis. First, the lack of regulation at the international level, as well as the establishment of a single market free of customs barriers. Similarly, it is very difficult for researchers to access up-to-date open data, with the clear exception of seminal works on the North American market such as Blumenthal (1999a, b) or Brevoort (1995), who detail the

Table 2 Selection of medicinal plants commonly used within the European Union

No	*Name*	No	Name
1	*Achillea millefolium* L., yarrow (*Achillea millefolium* L.)	42	*Harpagophyti* radix (*Harpagophytum procumbens* D.C. i *H. zeyheri* Decne)
2	*Agni casti* fructus (*Vitex agnus-castus* L.)	43	*Hederae helicis* folium (*Hedera helix* L.)
3	*Agrimonia eupatoria* L., agrimony (*Agrimonia eupatoria* L.)	44	*Helichrysi* flos (*Helichrysum arenarium* (L.) Moench)
4	*Agropyri repentis* rhizoma; syn. *Graminis rhizoma* (*Agropyron repens* (L.) P. Beauv.)	45	Herbal tea combinations for use as sedatives (*Humulus lupulus* L., flos (hop strobile; *Lavandula angustifolia* Mill., flos (lavender), *Melissa officinalis* L., folium (melissa leaf), *Passiflora incarnata* L., herba (passion flower), *Valeriana officinalis* L., radix (valerian root))
5	*Allium sativus* bulbus (*Allium sativum* L.)	46	*Herniaria glabra* L.; *Herniaria hirsuta* L.; *Herniaria incana* Lam., Rupturewort
6	*Aloes* folii succus siccatus (*Aloe barbadensis* Mill. and *Aloe* (various species Th.)	47	*Hieracium pilosella* L., mouse-ear hawkweed
7	*Aloysia citrodora* Paláu (syn. *Aloysia triphylla* (L'Hér.) Kuntze; *Verbena triphylla, Lemon verbena* leaf	48	*Ilex paraguariensis* A.St.-Hil, maté leaf
8	Anisi aetheroleum (*Pimpinella anisum* L.)	49	*Juglandis* folium (*Juglans regia* L.)
9	*Arnicae* flos (*Arnica montana* L.)	50	*Juniperi* aetheroleum (*Juniperus communis* L.)
10	*Artemisia absinthium* L., wormwood	51	*Lecithinum* ex soya (*Glycine max* (L.) Merr.)
11	*Avena sativa*, oat fruit	52	*Leonuri cardiacae* herba (*Leonurus cardiaca* L.)
12	*Avenae* fructus (*Avena sativa* L.)	53	*Leonurus cardiaca* L., motherwort (*Leonurus cardiaca* L.)
13	*Betula pendula* Roth / *Betula pubescens* Ehrh., Birch Leaf	54	Lini semen (*Linum usitatissimum* L.)
14	*Betulae* folium (*Betula pubescens* Ehrh.)	55	*Liquiritiae* radix (syn.: *Glycyrrhizae radix*) (*Glycyrrhiza glabra* L., *G. inflata Batali; G. uralensis* Fisch.)
15	*Calendula officinalis* L., *Calendula* flower	56	*Lupuli* flos (*Humulus lupulus* L.)
16	*Camellia sinensis* (L.) Kuntze, green tea	57	*Malva sylvestris* L., mallow flower
17	*Capsella bursa-pastoris* (L.) Medik., shepherd's purse	58	*Marrubii* herba (*Marrubium vulgare* L.)
18	*Capsicum annuum* L. var. *minimum* (Miller) Heiser, *Capsicum*	59	*Meliloti* herba (*Melilotus officinalis* L.)
19	*Carum carvi* L., caraway fruit	60	*Mentha x piperita* L., *peppermint* leaf
20	*Carvi* fructus (*Carum carvi* L.)	61	*Menthae piperitae* aetheroleum

(continued)

Table 2 (continued)

No	*Name*	No	Name
21	Caryophylli flos aetheroleum (*Syzygium aromaticum* L.)	62	Myrrha (*Commiphora myrrha* Nees.)
22	*Cassia senna* L.; *Cassia angustifolia* Vahl, senna leaf	63	*Ononidis* radix (*Ononis spinosa* L.)
23	*Centella asiatica* (L.) Urb., Centella	64	*Origani dictamni* herba
24	*Cetraria islandica* (L.) Acharius s.l., Iceland moss	65	*Potentilla erecta* (L.) Raeusch., tormentil
25	*Chamaemelum nobile* (L.) All. (*Anthemis nobilis* L.), Roman chamomile flower	66	*Primula veris* L.; *Primula elatior* (L.) Hill, *Primula* flower
26	*Chamomillae romanae* flos	67	*Rhamni purshianae* cortex (*Rhamnus purshianus* D.C.)
27	*Cichorium intybus* radix	68	*Rhamnus frangula* L., frangula bark
28	*Cola nitida* (Vent.) Schott et Endl. (*C. vera* K. Schum.); *Cola acuminata* (P. Beauv.) Schott et Endl. (*Sterculia acuminata* P. Beauv.), Cola	69	*Ribes nigrum* L., blackcurrant leaf
29	*Cucurbita pepo* L., pumpkin seed	70	*Rubus idaeus* L., raspberry leaf
30	*Eleutherococci* radix (*Eleutherococcus senticosus* Maxim.)	71	*Rusci* rhizoma (*Ruscus aculeatus* L.)
31	*Equiseti* herba (*Equisetum arvense* L.)	72	*Salix* (various species including *S. purpurea* L.; *S. daphnoides* Vill.; *S. fragilis* L.), willow bark
32	*Eschscholzia californica* Cham., California poppy	73	*Salvia officinalis* L., sage leaf
33	*Eucalypti aetheroleum* (*Eucalyptus globulus* Labill.; *Eucalyptus polybractea* R.T. Baker; *Eucalyptus smithii* R.T. Baker)	74	*Sennae* folium (*Senna alexandrina, Senna angustifolia, Senna obovata* Mill.).
34	*Foeniculi amari* fructus (*Foeniculum vulgare* Mill.)	75	Silybi mariani fructus (*Silybum marianum* (L.) Gaertn.; *Carduus marianus* L.)
35	*Frangulae* cortex (*Frangula alnus* Mill.)	76	*Solanum dulcamara* L., woody nightshade stem
36	*Fraxini folium* (*Fraxinus excelsior* L.)	77	*Thymi* aetheroleum (*Thymus vulgaris* L.; *Thymus zygis* L.)
37	*Fucus vesiculosus* L., bladderwrack	78	*Tilia cordata* Miller, *Tilia platyphyllos* Scop., Tilia × vulgaris Heyne, lime flower
38	*Gentiana lutea* L., gentian root	79	*Urticae* folium (*Urtica dioica* L.)
39	*Ginkgo biloba* L., ginkgo leaf	80	*Vaccinii macrocarpi* fructus (*Vaccinium macrocarpon* Aiton)
40	*Ginseng* radix (*Panax ginseng* L.)	81	*Verbasci* flos (*Verbascum thapsus* L.)

(continued)

Table 2 (continued)

No	*Name*	No	Name
41	*Grindelia robusta* Nutt., *Grindelia squarrosa* (Pursh) Dunal, *Grindelia humilis* Hook. et Arn., *Grindelia camporum* Greene, gumweed herb	82	*Vitis vinifera* L., (grapevine leaf)

Source: Own elaboration from EMA (2023)

size of this market during the 1990s, as well as the total consumer spending in this pharmaceutical subsector, specifying the policies of mergers and acquisitions, and the entry onto the scene of the giants of the pharmaceutical industry. Note that to this day, it continues to be a highly atomized industry, as can be seen in Table 3.

8 The Dangers of Traditional Medicine Related to Medicinal Plants

One of the main dangers of traditional medicine from medicinal plants is the lack of standardization and regulation of their production and consumption (Ekor 2014). Traditional practices often lack strict regulations, which can lead to variability in the quality and safety of the herbal products used (Heinrich 2015). Similarly, soil and water contamination can lead to the accumulation of heavy metals and other toxic substances in medicinal plants (Alengebawy et al. 2021). In addition, some plants contain natural chemical constituents that can be toxic in large quantities or when improperly combined with other substances, which can lead to side effects and toxicity (Jaishankar et al. 2014; Poppenga 2010). It should be noted that medicinal plants may interact with conventional drugs and other treatments, which may alter their efficacy or cause serious side effects (Déciga-Campos et al. 2022). People often do not inform their physicians about the use of traditional medicines, which can lead to unwanted interactions and health problems (Ameade et al. 2018).

Therefore, accurate identification of medicinal plant species is crucial to ensure their safety and efficacy (Techen et al. 2004). However, in many traditional practices, plants are identified visually, which can lead to errors. Species confusion can lead to accidental ingestion of toxic plants (Cornara et al. 2018). Some medicinal plants come from endangered species or are unsustainably extracted from the wild (Chen et al. 2016). Overfishing and overexploitation can lead to biodiversity loss and the disappearance of important medicinal plants, negatively affecting the communities that depend on them (Morris 2010).

Another factor that is always present when implementing traditional medicine is, in some cases, the lack of education on safe practices for collection, preparation, and administration of medicinal plants which can lead to inappropriate use and health problems (Fokunang et al. 2011). Inadequate transmission of traditional knowledge and lack of access to reliable information can increase the risks

Table 3 World's leading companies involved in the production of medicinal plants

Company	Country	Specialization
A. Vogel	Switzerland	Plant-based health products, including tinctures and tablets
Bio-Botanica Inc.	United States	Botanical extracts and natural essential oils derived from medicinal plants
Bioforce AG	Switzerland	Natural products and plant-based supplements, including herbal products and medicinal plant extracts
Bioforce USA	United States	Supplements and plant-based health products, including tinctures and tablets
Bionorica SE	Germany	Phytotherapeutic drugs, including medicinal plant-based products
Chengdu Hawk Bio-Engineering Co., Ltd	China	Production and export of herbal extracts and medicinal plants
Euromed Pharma	Spain	Production and commercialization of standardized extracts of medicinal plants
Gaia Herbs	United States	Plant-based supplements, including capsules and tinctures of medicinal plants
Indena	Italy	Extracts of medicinal plants for the pharmaceutical and supplement industry
Martin Bauer Group	Germany	Production of herbal infusions, extracts, and plant-based products
Naturex	France	Extracts of medicinal plants and natural products for various industries
PhytoLab	Germany	Production of extracts of medicinal plants and essential oils
Phytomed Herbal Solutions	South Africa	Extracts of medicinal plants and herbal products
Pierre Fabre Médicament	France	Herbal medicines for various medical conditions
Pukka Herbs	United Kingdom	Organic teas and supplements based on medicinal plants
Sido Muncul	Indonesia	Production of supplements and medicines based on traditional Indonesian plants
Traditional Medicinals	United States	Production of medicinal herbal teas and plant-based wellness products

Source: Own elaboration

associated with traditional medicine (Ssenku et al. 2022). Such practices can discourage the use of evidence-based medical treatments, which can lead to delays in the diagnosis and treatment of serious diseases. Resistance to adopting medical practices supported by scientific studies can have serious consequences for people's health (Ventola 2010). It should not be forgotten that overexploitation and unsustainable collection of medicinal plants can threaten biodiversity and natural ecosystems. The loss of plant species can negatively impact the food chain and the health of the ecosystem in general (Alves and Ierecê 2007). In summary, although traditional herbal medicine holds an important place in many cultures, it is essential to address the dangers associated with its use. Adequate regulation, public education, conservation of the biodiversity, and collaboration between traditional and modern medicine systems are essential to minimize the risks and take advantage of the benefits of medicinal plants in a safe and effective way.

The similarities between these markets lie in the rich cultural tradition of herbal medicine, which has been passed down from generation to generation. In all these places, medicinal plants are deeply rooted in the beliefs and practices of local communities (Aziz et al. 2018). Moreover, all these markets face similar challenges, such as regulation, species conservation, and integration of herbal medicine into modern health systems. Differences lie in the plant varieties used, historical and cultural influences (Eshete and Molla 2021), as well as region-specific regulations. For example, traditional Chinese medicine is based on principles of balance and energy, whereas Indian Ayurveda has a different conception of the doshas (body elements) (Patwardhan et al. 2005). Phytotherapy practices in Latin America and Africa vary greatly due to the cultural and geographic diversity of these regions (Ameh et al. 2010).

9 Summary

The new lines of research related to the commercialization of medicinal plants should be directed to the search for plants yet to be exploited (Harvey 2000) and to the analysis of global regulation, both of medicinal plants of traditional Chinese medicine (Chen et al. 2016), and of any type of practice related to phytotherapy, regardless of the region it comes from. In this sense, it would be very interesting if initiatives such as the International Regulatory Cooperation for Herbal Medicines (IRCH) (WHO 2023), whose implementation is still too recent to be able to predict the results of this global regulation plan, were to bear fruit. Finally, we should stress the importance of the proliferation of databases and updated studies, in both cases open to the public; otherwise, it would be impossible to carry out an accurate global analysis of the commercialization of medicinal plants, the main handicap we have to face today.

References

Abdallah EM, Alhatlani BY, de Menezes R, Martins CHG (2023) Back to nature: medicinal plants as promising sources for antibacterial drugs in the post-antibiotic era. Plan Theory 12(17):3077. https://doi.org/10.3390/plants12173077

Alengebawy A, Abdelkhalek ST, Qureshi SR, Wang M-Q (2021) Heavy metals and pesticides toxicity in agricultural soil and plants: ecological risks and human health implications. Toxics 9(3):42. https://doi.org/10.3390/toxics9030042

Aliferis KA, Chamoun R, Jabaji S (2015) Metabolic responses of willow (Salix purpurea L.) leaves to mycorrhization as revealed by mass spectrometry and 1H NMR spectroscopy metabolite profiling. Front Plant Sci 6:344. https://doi.org/10.3389/fpls.2015.00344

Alves RRN, Ierecê RML (2007) Biodiversity, traditional medicine and public health: where do they meet? J Ethnobiol Ethnomed 3(1):14. https://doi.org/10.1186/1746-4269-3-14

Ameade EPK, Ibrahim M, Ibrahim H-S, Habib RH, Gbedema SY (2018) Concurrent use of herbal and orthodox medicines among residents of Tamale, Northern Ghana, who patronize hospitals and herbal clinics. Evid Based Complement Alternat Med 2018:1–8. https://doi.org/10.1155/2018/1289125

Ameh SJ, Obodozie OO, Abubakar MS, Garba MM (2010) Current phytotherapy an inter-regional perspective on policy, research and development of herbal medicine. J Med Plant Res 4:1508–1516. https://doi.org/10.5897/JMPR10.282

Aronson JK (2016) Rutaceae. In: Aronson JK (ed) Meyler's side effects of drugs, vol 7, 16th edn. Elsevier, Amsterdam, pp 265–279. https://doi.org/10.1016/b978-0-444-53717-1.01425-6

Astutik S, Pretzsch J, Ndzifon Kimengsi J (2019) Asian medicinal plants' production and utilization potentials: a review. Sustainability 11(19):5483. https://doi.org/10.3390/su11195483

Athni TS, Athni SS (2019) The evolution of modern medicine: garden to pill box. In: Joshee N, Dhekney SA, Parajuli P (eds) Medicinal plants: from farm to pharmacy. Springer Nature, Cham, pp 1–16. https://doi.org/10.1007/978-3-030-31269-5_1

Auyeung KK, Han Q-B, Ko JK (2016) Astragalus membranaceus: a review of its protection against inflammation and gastrointestinal cancers. Am J Chin Med 44(01):1–22. https://doi.org/10.1142/s0192415x16500014

Aziz MA, Adnan M, Khan AH, Shahat AA, Al-Said MS, Ullah R (2018) Traditional uses of medicinal plants practiced by the indigenous communities at Mohmand Agency, FATA. Pak J Ethnobiol Ethnomed 14(1):2

Barceloux DG (2008) Poison Hemlock (Conium maculatum L.). In: Barceloux DG (ed) Medical toxicology of natural substances: foods, fungi, medicinal herbs, plants, and venomous animals. Wiley, Hoboken, NJ, pp 796–799. https://doi.org/10.1002/9780470330319.ch131

Bergoënd A, Piasentin J (2022) Medicinal plants, aromatic herbs and fragrance plants in France: a small but thriving sector with a strong traditional base and a dynamic research network. Chronica Hortic 60(4):24–27

Bigira V, Kapisi J, Clark TD, Kinara S, Mwangwa F, Muhindo MK, Osterbauer B, Aweeka FT, Huang L, Achan J, Havlir DV, Rosenthal PJ, Kamya MR, Dorsey G (2014) Protective efficacy and safety of three antimalarial regimens for the prevention of malaria in young Ugandan children: a randomized controlled trial. PLoS Med 11(8):e1001689. https://doi.org/10.1371/journal.pmed.1001689

Birjees M, Ahmad M, Zafar M, Nawaz S, Jehanzeb S, Ullah F, Zaman W (2022) Traditional knowledge of wild medicinal plants used by the inhabitants of Garam Chashma valley, district Chitral, Pakistan. Acta Ecol Sin 42(2):19–33. https://doi.org/10.1016/j.chnaes.2020.12.006

Blumenthal M (1999a) Harvard study estimates consumers spend 5.1 billion on herbal products? HerbalGram 45:68

Blumenthal M (1999b) Herb industry sees mergers, acquisitions, and entry by pharmaceutical giants in 1998. HerbalGram 45:68–97

Botha J, Witkowski ETF, Sackleton CM (2004) Market profiles and trade in medicinal plants in the Lowveld, South Africa. Environ Conserv 31(1):38–46. https://doi.org/10.1017/s0376892904001067

Bourgou S, Ben Haj Jilani I, Karous O, Megdiche-Ksouri W, Ghrabi-Gammar Z, Mohamed L, Khabbach A, El Haissoufi M, Lamchouri F, Greveniotis V, Avramakis M, Hatzilazarou S, Anestis I, Tsoktouridis G, Krigas N (2021) Medicinal-cosmetic potential of the local endemic plants of crete (Greece), Northern Morocco and Tunisia: priorities for conservation and sustainable exploitation of neglected and underutilized phytogenetic resources. Biology 10(12):1344. https://doi.org/10.3390/biology10121344

Boyer EW, Babu KM, Adkins JE, McCurdy CR, Halpern JH (2008) Self-treatment of opioid withdrawal using kratom (Mitragynia speciosa korth). Addiction 103(6):1048–1050. https://doi.org/10.1111/j.1360-0443.2008.02209.x

Brevoort P (1995) The U.S. botanical market. An overview. HerbalGram 36:49–59

Cabada-Aguirre P, López López AM, Mendoza KCO, Garay Buenrostro KD, Luna-Vital DA, Mahady GB (2023) Mexican traditional medicines for women's reproductive health. Sci Rep 13(1):2807. https://doi.org/10.1038/s41598-023-29921-1

Calixto JB (2019) The role of natural products in modern drug discovery. An Acad Bras Cienc 91(3):e20190105. https://doi.org/10.1590/0001-3765201920190105

Castro Braga F (2021) Brazilian traditional medicine: historical basis, features and potentialities for pharmaceutical development. J Trad Chin Med Sci 8:S44–S50. https://doi.org/10.1016/j.jtcms.2020.06.005

Chapman K, Chomchalow N (2005) Production of medicinal plants in Asia. Acta Hortic 679:45–59. https://doi.org/10.17660/ActaHortic.2005.679.6

Chapotin SM, Razanameharizaka JH, Holbrook NM (2006) Water relations of baobab trees (Adansonia spp. L.) during the rainy season: does stem water buffer daily water deficits? Plant Cell Environ 29:1021–1032. https://doi.org/10.1111/j.1365-3040.2005.01456.x

Chauhan R, Singh S, Kumar V, Kumar A, Kumari A, Rathore S, Kumar R, Singh S (2021) A comprehensive review on biology, genetic improvement, agro and process technology of German Chamomile (Matricaria chamomilla L.). Plan Theory 11(1):29. https://doi.org/10.3390/plants11010029

Chekole G, Asfaw Z, Kelbessa E (2015) Ethnobotanical study of medicinal plants in the environs of Tara-gedam and Amba remnant forests of Libo Kemkem District, northwest Ethiopia. J Ethnobiol Ethnomed 11(1). https://doi.org/10.1186/1746-4269-11-4

Cheminal A, Kokkoris IP, Strid A, Dimopoulos P (2020) Medicinal and aromatic Lamiaceae plants in Greece: linking diversity and distribution patterns with ecosystem services. Forests 11(6):661. https://doi.org/10.3390/f11060661

Chen S-L, Yu H, Luo H-M, Wu Q, Li C-F, Steinmetz A (2016) Conservation and sustainable use of medicinal plants: problems, progress, and prospects. Chinas Med 11(1):37. https://doi.org/10.1186/s13020-016-0108-7

Chin K-Y, Mark-Lee WF (2018) A review on the antinociceptive effects of Mitragyna speciosa and its derivatives on animal model. Curr Drug Targets 19(12):1359–1365. https://doi.org/10.2174/1389450118666170925154025

Cohen MM (2014) Tulsi-Ocimum sanctum: a herb for all reasons. J Ayurveda Integr Med 5(4):251. https://doi.org/10.4103/0975-9476.146554

Cornara L, Smeriglio A, Frigerio J, Labra M, Di Gristina E, Denaro M, Mora E, Trombetta D (2018) The problem of misidentification between edible and poisonous wild plants: reports from the Mediterranean area. Food Chem Toxicol 119:112–121. https://doi.org/10.1016/j.fct.2018.04.066

Cragg GM, Boyd MR, Khanna R, Newman DJ, Sausville EA (1999) Natural product drug discovery and development. In: Romeo JT (ed) Phytochemicals in human health protection, nutrition, and plant defense, vol 33. Springer, Berlin, pp 1–29. https://doi.org/10.1007/978-1-4615-4689-4_1

Czigle S, Tóth J, Jedlinszki N, Háznagy-Radnai E, Csupor D, Tekeľová D (2018) Ginkgo biloba food supplements on the European Market—adulteration patterns revealed by quality control of selected samples. Planta Med 84(06/07):475–482. https://doi.org/10.1055/a-0581-5203

de Moura Monteiro GB, Pessoa BHS (2022) Semeando saúde: o cultivo de um jardim medicinal na Atenção Básica em Saúde, um relato de experiência. Revista Brasileira de Medicina de Família e Comunidade 17(44):3046. https://doi.org/10.5712/rbmfc17(44)3046

Déciga-Campos M, Ventura-Martínez R, González-Trujano ME, Silveira D (2022) Editorial: pharmacological interaction between drugs and medicinal plants. Front Pharmacol 13:1081090. https://doi.org/10.3389/fphar.2022.1081090

Dehelean CA, Marcovici I, Soica C, Mioc M, Coricovac D, Iurciuc S, Cretu OM, Pinzaru I (2021) Plant-derived anticancer compounds as new perspectives in drug discovery and alternative therapy. Molecules 26(4):1109. https://doi.org/10.3390/molecules26041109

Devine SNO, Kolog EA, Atinga R (2022) Toward a knowledge-based system for African traditional herbal medicine: a design science research approach. Front Artif Intell 5:856705. https://doi.org/10.3389/frai.2022.856705

Deyno S, Eneyew K, Seyfe S, Tuyiringire N, Peter EL, Muluye RA, Tolo CU, Ogwang PE (2019) Efficacy and safety of cinnamon in type 2 diabetes mellitus and pre-diabetes patients: a meta-analysis and meta-regression. Diabetes Res Clin Pract 156:107815. https://doi.org/10.1016/j.diabres.2019.107815

Dzoyem JP, Tshikalange E, Kuete V (2013) Medicinal plants market and industry in Africa. In: Kuete V (ed) Medicinal plant research in Africa. Elsevier, Amsterdam, pp 859–890. https://doi.org/10.1016/b978-0-12-405927-6.00024-2

Ekor M (2014) The growing use of herbal medicines: issues relating to adverse reactions and challenges in monitoring safety. Front Pharmacol 4:177. https://doi.org/10.3389/fphar.2013.00177

EMA (2023) European Medicines Agency. European Union herbal monograph. https://www.ema.europa.eu/en/medicines/field_ema_web_categories%253Aname_field/Herbal/field_ema_herb_outcome/european-union-herbal-monograph-254. Accessed 15 Aug 2023

Eshete MA, Molla EL (2021) Cultural significance of medicinal plants in healing human ailments among Guji semi-pastoralist people, Suro Barguda District, Ethiopia. J Ethnobiol Ethnomed 17(1):61. https://doi.org/10.1186/s13002-021-00487-4

Fan T-P, Deal G, Koo H-L, Rees D, Sun H, Chen S, Dou J-H, Makarov VG, Pozharitskaya ON, Shikov AN, Kim YS, Huang Y-T, Chang YS, Jia W, Dias A, Wong VC, Chan K (2012) Future development of global regulations of Chinese herbal products. J Ethnopharmacol 140(3):568–586. https://doi.org/10.1016/j.jep.2012.02.029

Fatima A, Siddique YH (2021) Role of flavonoids against COVID-19. Open Biol J 9(1):47–55. https://doi.org/10.2174/1874196702109010047

Fokunang CN, Ndikum V, Tabi OY, Jiofack RB, Ngameni B, Guedje NM, Tembe-Fokunang EA, Tomkins P, Barkwan S, Kechia F, Asongalem E, Ngoupayou J, Torimiro NJ, Gonsu KH, Sielinou V, Ngadjui BT, Angwafor III, Nkongmeneck A, Abena OM et al (2011) Traditional medicine: past, present and future research and development prospects and integration in the National Health System of Cameroon. Afr J Tradit Complement Altern Med 8(3):284. https://doi.org/10.4314/ajtcam.v8i3.65276

Franke R, Schilcher H (2007) Relevance and use of chamomile (*Matricaria recutita* L.). Acta Hortic 749:29–43. https://doi.org/10.17660/actahortic.2007.749.2

George P (2011) Concerns regarding the safety and toxicity of medicinal plants—an overview. J Appl Pharmaceut Sci 1(6):40–44

Ghod Elahi M, Hekmati M, Esmaeili D, Ziarati P, Yousefi M (2022) Antibacterial activity of modified Carvacrol against *Staphylococcus epidermidis* and *Pseudomonas aeruginosa*. Novelty Clin Med 1(4):1051. https://doi.org/10.22034/ncm.2022.350997.1051

Gonzales GF, Gonzales C, Gonzales-Castañeda C (2009) *Lepidium meyenii* (Maca): a plant from the Highlands of Peru—from tradition to science. Forsch Komplementarmed Klass Naturheilkd 16(6):373–380. https://doi.org/10.1159/000264618

Grace OM, Simmonds MSJ, Smith GF, Van Wyk AE (2008) Therapeutic uses of Aloe L. (*Asphodelaceae)* in southern Africa. J Ethnopharmacol 119(3):604–614. https://doi.org/10.1016/j.jep.2008.07.002

Halpenny GM (2017) Mitragyna speciosa: balancing potential medical benefits and abuse. ACS Med Chem Lett 8(9):897–899. https://doi.org/10.1021/acsmedchemlett.7b00298

Harvey A (2000) Strategies for discovering drugs from previously unexplored natural products. Drug Discov Today 5:294–300. https://doi.org/10.1016/s1359-6446(00)01511-7

Heinrich M (2015) Quality and safety of herbal medical products: regulation and the need for quality assurance along the value chains. Br J Clin Pharmacol 80(1):62–66. https://doi.org/10.1111/bcp.12586

Hornblower S, Spawforth A, Eidinow E (2012) The Oxford Classical Dictionary, 4th edn. Oxford University Press, Oxford. https://doi.org/10.1093/acref/9780199545568.001.0001

Hostettmann K (2003) Geschichte einer Pflanze am Beispiel von Echinacea [History of a plant: the example of Echinacea]. *Forschende Komplementärmedizin Und Klassische*. Naturheilkunde 10:9–12. https://doi.org/10.1159/000071678

Ikinyom N, Lamwaka AV, Malagala AT, Ndyomugyenyi EK (2023) Ethnobotanical study of nutraceutical plants used to manage opportunistic infections associated with HIV/AIDS in Acholi sub-region, Northern Uganda. Trop Med Health 51(1):50. https://doi.org/10.1186/s41182-023-00540-w

Islas JF, Acosta E, Buentello Z, Delgado-Gallegos JL, Moreno-Treviño MG, Escalante B, Moreno-Cuevas JE (2020) An overview of Neem (Azadirachta indica) and its potential impact on health. J Funct Foods 74:104171. https://doi.org/10.1016/j.jff.2020.104171

Issa TO, Mohamed YS, Yagi S, Ahmed RH, Najeeb TM, Makhawi AM, Khider TO (2018) Ethnobotanical investigation on medicinal plants in Algoz area (South Kordofan), Sudan. J Ethnobiol Ethnomed 14(1):31. https://doi.org/10.1186/s13002-018-0230-y

Jaishankar M, Tseten T, Anbalagan N, Mathew BB, Beeregowda KN (2014) Toxicity, mechanism and health effects of some heavy metals. Interdiscip Toxicol 7(2):60–72. https://doi.org/10.2478/intox-2014-0009

Jeelani SM, Rather GA, Sharma A, Lattoo SK (2018) In perspective: Potential medicinal plant resources of Kashmir Himalayas, their domestication and cultivation for commercial exploitation. J Appl Res Med Aromat Plants 8:10–25. https://doi.org/10.1016/j.jarmap.2017.11.001

Jiménez-González V, Benítez G, Pastor JE, López-Lázaro M, Calderón-Montaño JM (2023) Evaluation of anticancer activity of 76 plant species collected in Andalusia (Spain) against lung cancer cells. Plan Theory 12(18):3275. https://doi.org/10.3390/plants12183275

Karpenko V (1995) Medical aspects of the late European alchemy. Bull Indian Inst Hist Med 25(1–2):202–215

Kunwar RM, Mahat L, Acharya RP, Bussmann RW (2013) Medicinal plants, traditional medicine, markets and management in far-west Nepal. J Ethnobiol Ethnomed 9(1):24. https://doi.org/10.1186/1746-4269-9-24

Lopresti AL, Smith SJ, Malvi H, Kodgule R (2019) An investigation into the stress-relieving and pharmacological actions of an ashwagandha (Withania somnifera) extract: a randomized, double-blind, placebo-controlled study. Medicine 98(37):e17186. https://doi.org/10.1097/md.0000000000017186

Luthra PM, Singh R, Chandra R (2001) Therapeutic uses of Curcuma longa (turmeric). Indian J Clin Biochem 16(2):153–160. https://doi.org/10.1007/bf02864854

Lyu M, Fan G, Xiao G, Wang T, Xu D, Gao J, Ge S, Li Q, Ma Y, Zhang H, Wang J, Cui Y, Zhang J, Zhu Y, Zhang B (2021) Traditional Chinese medicine in COVID-19. Acta Pharm Sin B 11(11):3337–3363. https://doi.org/10.1016/j.apsb.2021.09.008

Macía MJ, García E, Vidaurre PJ (2005) An ethnobotanical survey of medicinal plants commercialized in the markets of La Paz and El Alto, Bolivia. J Ethnopharmacol 97(2):337–350. https://doi.org/10.1016/j.jep.2004.11.022

Mahomoodally MF (2013) Traditional medicines in Africa: an appraisal of ten potent African medicinal plants. Evid Based Complement Alternat Med 2013:1–14. https://doi.org/10.1155/2013/617459

Maistro EL, Terrazzas PM, Perazzo FF, Gaivão IODM, Sawaya ACHF, Rosa PCP (2019) Salix alba (white willow) medicinal plant presents genotoxic effects in human cultured leukocytes. J Toxic Environ Health A 82(23–24):1223–1234. https://doi.org/10.1080/15287394.2019.1711476

Maneenoon K, Khuniad C, Teanuan Y, Saedan N, Prom-in S, Rukleng N, Kongpool W, Pinsook P, Wongwiwat W (2015) Ethnomedicinal plants used by traditional healers in Phatthalung Province, Peninsular Thailand. J Ethnobiol Ethnomed 11(1):43. https://doi.org/10.1186/s13002-015-0031-5

Mann CC, Plummer ML (1991) The Aspirin Wars: money, medicine, and 100 years of rampant competition. Alfred A. Knopf, New York

Marshall AC (2020) Traditional Chinese medicine and clinical pharmacology. In: Hock F, Gralinski M (eds) Drug discovery and evaluation: methods in clinical pharmacology. Springer, Berlin, pp 455–482. https://doi.org/10.1007/978-3-319-68864-0_60

Martínez de Arano I, Maltoni S, Picardo A, Mutke S (2021) Non-wood forest products for people, nature and the green economy. In: Recommendations for policy priorities in Europe A white paper based on lessons learned from around the Mediterranean. European Forest Institute and FAO. https://doi.org/10.36333/k2a05

Miner J, Hoffhines A (2007) The discovery of aspirin's antithrombotic effects. Tex Heart Inst J 34(2):179–186

Mishra L, Singh BB, Dagenais S (2001) Ayurveda: a historical perspective and principles of the traditional healthcare system in India. Altern Ther Health Med 7(2):36–42

Morris RJ (2010) Anthropogenic impacts on tropical forest biodiversity: a network structure and ecosystem functioning perspective. Philos Trans R Soc B Biol Sci 365(1558):3709–3718. https://doi.org/10.1098/rstb.2010.0273

Mukherjee S (2014) Status of female education among Santal, Kheria Sabar and Birhor tribal communities of Puruliya district, West Bengal, India. Trans Inst Indian Geographers 36(2):271–278. https://www.scopus.com/inward/record.uri?eid=2-s2.0-84924415843&partnerID=40&md5=8542b4a58d4ea1fa07faaa8631a5c452

Munasinghe M (2010) Making development more sustainable: sustainomics framework and practical applications, 2nd edn. MIND Press, Colombo

Ndeffo Mbah ML, Parikh S, Galvani AP (2015) Comparing the impact of Artemisinin-based combination therapies on Malaria transmission in Sub-Saharan Africa. Am J Trop Med Hyg 92(3):555–560. https://doi.org/10.4269/ajtmh.14-0490

Ng QX, Venkatanarayanan N, Ho CYX (2017) Clinical use of Hypericum perforatum (St John's wort) in depression: a meta-analysis. J Affect Disord 210:211–221. https://doi.org/10.1016/j.jad.2016.12.048

Pandey MM, Rastogi S, Rawat AKS (2013) Indian traditional ayurvedic system of medicine and nutritional supplementation. Evid Based Complement Alternat Med 2013:1–12. https://doi.org/10.1155/2013/376327

Papenfuss M (2022) Santería in Catemaco, Mexico: hybrid (re)configurations of religious meaning and practice. Interdiscip J Religion Transform Contemp Soc 9(2):375–394. https://doi.org/10.30965/23642807-bja10044

Pareek A, Pant M, Gupta MM, Kashania P, Ratan Y, Jain V, Pareek A, Chuturgoon AA (2023) Moringa oleifera: an updated comprehensive review of its pharmacological activities, ethnomedicinal, phytopharmaceutical formulation, clinical, phytochemical, and toxicological aspects. Int J Mol Sci 24(3):2098. https://doi.org/10.3390/ijms24032098

Paşayeva L, Demirpolat E, Fatullayev H, Tugay O (2021) Phytochemical composition and cytotoxic activity of four Cousinia Species against human hepatocellular carcinoma (HepG2) cell line. Phytomed Plus 1(4):100068. https://doi.org/10.1016/j.phyplu.2021.100068

Patočka J, Jakl J (2010) Biomedically relevant chemical constituents of Valeriana officinalis. J Appl Biomed 8(1):11–18. https://doi.org/10.2478/v10136-009-0002-z

Patwardhan B, Warude D, Pushpangadan P, Bhatt N (2005) Ayurveda and traditional Chinese medicine: a comparative overview. Evid Based Complement Alternat Med 2(4):465–473. https://doi.org/10.1093/ecam/neh140

Paulson R (2007) Sin and evil: moral values in literature. Yale University Press, London

Pérez-Lloréns JL, Critchley AT, Cornish ML, Mouritsen OG (2023) Saved by seaweeds (II): traditional knowledge, home remedies, medicine, surgery, and pharmacopoeia. J Appl Phycol 35(5):2049–2068. https://doi.org/10.1007/s10811-023-02965-6

Pittler MH, Ernst E (2007) Clinical effectiveness of garlic (Allium sativum). Mol Nutr Food Res 51(11):1382–1385. https://doi.org/10.1002/mnfr.200700073

Poppenga RH (2010) Poisonous plants. In: Luch A (ed) molecular, clinical and environmental toxicology: clinical toxicology, vol 2. Birkhäuser, Basel, pp 123–175. https://doi.org/10.1007/978-3-7643-8338-1_4

Potenza MA, Montagnani M, Santacroce L, Charitos IA, Bottalico L (2023) Ancient herbal therapy: a brief history of Panax ginseng. J Ginseng Res 47(3):359–365. https://doi.org/10.1016/j.jgr.2022.03.004

Rady I, Bloch MB, Chamcheu R-CN, Banang Mbeumi S, Anwar MR, Mohamed H, Babatunde AS, Kuiate J-R, Noubissi FK, El Sayed KA, Whitfield GK, Chamcheu JC (2018) Anticancer properties of Graviola (Annona muricata): a comprehensive mechanistic review. Oxid Med Cell Longev 2018:1–39. https://doi.org/10.1155/2018/1826170

Raman V, Avula B, Galal A, Wang YH, Khan IA (2012) Identification of Yohimbe Bark (Pausinystalia johimbe) using microscopy and UPLC-MS techniques. Planta Med 78(05). https://doi.org/10.1055/s-0032-1307529

Roca W, Espinoza C, Panta AL (2004) Agricultural applications of biotechnology and the potential for biodiversity valorization in Latin America and the Caribbean. AgBioforum 7(1–2):13–22

Roth I (2008) Ethnobotany in South America. In: Selin H (ed) Encyclopaedia of the history of science, technology, and medicine in non-western cultures. Springer Nature, Cham, pp 898–899. https://doi.org/10.1007/978-1-4020-4425-0_9432

Saeidnia S, Gohari A, Mokhber-Dezfuli N, Kiuchi F (2011) A review on phytochemistry and medicinal properties of the genus Achillea. Daru 19(3):173–186

Salazar-Gómez A, Alonso-Castro AJ (2022) Medicinal plants from Latin America with wound healing activity: ethnomedicine, phytochemistry, preclinical and clinical studies—a review. Pharmaceuticals 15(9):1095. https://doi.org/10.3390/ph15091095

Sauvain M, Rerat C, Moretti C, Saravia E, Arrazola S, Gutierrez E, Lema A-M, Muñoz V (1997) A study of the chemical composition of *Erythroxylum coca* var. *coca* leaves collected in two ecological regions of Bolivia. J Ethnopharmacol 56(3):179–191. https://doi.org/10.1016/s0378-8741(97)00030-5

Sawicka B, Ziarati P, Messaoudi M, Agarpanah J, Skiba D, Bienia B, Barbaś P, Rebiai A, Krochmal-Marczak B, Yeganehpoor F (2022) Role of herbal bioactive compounds as a potential bioavailability enhancer for active pharmaceutical ingredients. In: Handbook of research on advanced phytochemicals and plant-based drug discovery. IGI Global, Hershey, PA, pp 450–495. https://doi.org/10.4018/978-1-6684-5129-8.ch023

Schippmann U, Leaman DJ, Cunningham AB (2002) Impact of cultivation and gathering of medicinal plants on biodiversity: global trends and issues. In: Biodiversity and the ecosystem approach in agriculture, forestry and fisheries: Satellite Event on the Occasion of the Ninth Regular Session of the Commission on Genetic Resources for Food and Agriculture

Silva ML, Bernardo MA, Singh J, de Mesquita MF (2022) Cinnamon as a complementary therapeutic approach for dysglycemia and dyslipidemia control in type 2 diabetes mellitus and its molecular mechanism of action: a review. Nutrients 14(13):2773. https://doi.org/10.3390/nu14132773

Singh PA, Bajwa N, Chinnam S, Chandan A, Baldi A (2022) An overview of some important deliberations to promote medicinal plants cultivation. J Appl Res Med Aromat Plants 31:100400. https://doi.org/10.1016/j.jarmap.2022.100400

Ssenku JE, Okurut SA, Namuli A, Kudamba A, Tugume P, Matovu P, Wasige G, Kafeero HM, Walusansa A (2022) Medicinal plant use, conservation, and the associated traditional knowledge in rural communities in Eastern Uganda. Trop Med Health 50(1):39. https://doi.org/10.1186/s41182-022-00428-1

Suh Nchang A, Shinyuy LM, Noukimi SF, Njong S, Bambara S, Kalimba EM, Kamga J, Ghogomu SM, Frederich M, Talom JLL, Souopgui J, Robert A (2023) Knowledge about asymptomatic malaria and acceptability of using Artemisia afra Tea among Health Care Workers (HCWs) in Yaounde, Cameroon: a cross-sectional survey. Int J Environ Res Public Health 20(13):6309. https://doi.org/10.3390/ijerph20136309

Taghouti I, Cristobal R, Brenko A, Stara K, Markos N, Chapelet B, Hamrouni L, Buršić D, Bonet J-A (2022) The market evolution of medicinal and aromatic plants: a global supply chain analysis and an application of the Delphi method in the Mediterranean Area. Forests 13(5):808. https://doi.org/10.3390/f13050808

Techen N, Crockett S, Khan I, Scheffler B (2004) Authentication of medicinal plants using molecular biology techniques to compliment conventional methods. Curr Med Chem 11(11):1391–1401. https://doi.org/10.2174/0929867043365206

Unnikrishnan R, Dev SA, Jayaraj R (2020) Pitfalls and promises of raw drug identification techniques in the ayurvedic industry: an overview. 3 Biotech 10(11):497. https://doi.org/10.1007/s13205-020-02482-0

Ventola CL (2010) Current issues regarding complementary and alternative medicine (CAM) in the United States: Part 1: the widespread use of cam and the need for better-informed health care professionals to provide patient counseling. P T 35(8):461–468

Villas Bôas Gde K, Gadelha CA (2007) Oportunidades na indústria de medicamentos e a lógica do desenvolvimento local baseado nos biomas brasileiros: bases para a discussão de uma política nacional. Cad Saude Publica 23(6):1463–1471. https://doi.org/10.1590/s0102-311x2007000600021

Wagner C, De Gezelle J, Komarnytsky S (2020) Celtic provenance in traditional herbal medicine of medieval Wales and classical antiquity. Front Pharmacol 11. https://doi.org/10.3389/fphar.2020.00105

Wang X, Zhang H, Chen L, Shan L, Fan G, Gao X (2013) Liquorice, a unique "guide drug" of traditional Chinese medicine: a review of its role in drug interactions. J Ethnopharmacol 150(3):781–790. https://doi.org/10.1016/j.jep.2013.09.055

Wang X, Ma T, Zhang W, Chu Q (2022) Effectiveness and safety research of Qingfei Paidu (QFPD) in treatment of COVID-19: an up-to-date systematic review and meta-analysis. Chinas Med 17(1):122. https://doi.org/10.1186/s13020-022-00675-8

Welz AN, Emberger-Klein A, Menrad K (2018) Why people use herbal medicine: insights from a focus-group study in Germany. BMC Complement Altern Med 18(1):92. https://doi.org/10.1186/s12906-018-2160-6

WHO (2013) WHO traditional medicine strategy: 2014–2023. World Health Organization, Geneva

WHO (2022) Departmental news: tackling emerging antimalarial drug resistance in Africa. https://www.who.int/news/item/18-11-2022-tackling-emerging-antimalarial-drug-resistance-in-africa. Accessed 15 Aug 2023

WHO (2023) International Regulatory Cooperation for Herbal Medicines (IRCH). WHO, Geneva

Zaccardelli M, Roscigno G, Pane C, Celano G, Di Matteo M, Mainente M, Mencherini AV, Esposito T, Vitti A, De Falco E (2021) Essential oils and quality composts sourced by recycling vegetable residues from the aromatic plant supply chain. Ind Crop Prod 162:113255. https://doi.org/10.1016/j.indcrop.2021.113255

Zarezadeh M, Musazadeh V, Foroumandi E, Keramati M, Ostadrahimi A, Mekary RA (2023) The effect of cinnamon supplementation on glycemic control in patients with type 2 diabetes or with polycystic ovary syndrome: an umbrella meta-analysis on interventional meta-analyses. Diabetol Metab Syndr 15(1):127. https://doi.org/10.1186/s13098-023-01057-2

Zhu X-B, Guo M, Zhang Z-H, Sun L-H, Liu L, Zhou L-J, Shan C-L, Yang Y, Kan L-D, Li L-C (2021) Chinese herbal injections for coronavirus disease 2019 (COVID-19): a narrative review. Integr Med Res 10(4):100778. https://doi.org/10.1016/j.imr.2021.100778

Endophytes as Potential Contenders of Their Host Plants for Bioactive Drug Discovery: A Review

Tsegu Kiros, Yadessa Melaku, Chuleui Jung, Xuebo Hu, and Aman Dekebo

Contents

T. Kiros
Department of Chemistry, Haramaya University, Dire Dawa, Ethiopia

Department of Applied Chemistry, Adama Science and Technology University, Adama, Ethiopia

Y. Melaku (✉)
Department of Applied Chemistry, Adama Science and Technology University, Adama, Ethiopia

C. Jung
Department of Plant medicals, Andong National University, Andong, Republic of Korea
e-mail: cjung@andong.ac.kr

X. Hu
Lab of Drug Discovery and Molecular Engineering, College of Plant Science and Technology, Huazhong (Central China) Agricultural University, Shizishan, Nanhu, Wuhan, China

A. Dekebo (✉)
Department of Applied Chemistry, Adama Science and Technology University, Adama, Ethiopia

Department of Plant medicals, Andong National University, Andong, Republic of Korea

Institute of Pharmaceutical Sciences, Adama Science and Technology University, Adama, Ethiopia
e-mail: amandekeb@gmail.com

© The Author(s), under exclusive license to Springer Nature Singapore Pte Ltd. 2024

A. Das Talukdar et al. (eds.), *Traditional Resources and Tools for Modern Drug Discovery*, Interdisciplinary Biotechnological Advances,
https://doi.org/10.1007/978-981-97-4600-2_19

Abstract Endophytes, the plant-associated microbes, have currently gained significant interest due to their massive biodiversity for the discovery of novel and alternative bioactive compounds. Review articles previously published were not systematically reviewed in a way suitable for biochemical studies of endophytes. This has posed a great challenge to researchers from developing countries, in which study on endophytes is unfamiliar. The present review is intended to give an insight into the method of exploration of endophytes and associated bioactive agents and shed light on the studies of endophytes in the context of some countries, like Ethiopia. External surface disinfection and reducing into small segments of the selected plant, followed by plating onto an appropriate agar medium, are the main steps routinely applied to isolate endophytes. The active endophytic microbe is then fermented, and culture filtrate is extracted with ethyl acetate and subjected to different chromatographic techniques to fractionate and isolate bioactive compounds. The toxicity, moderate biological potency, and low yield of medicinal plant-based natural products are still some of the challenges in world health medicine. In the near future, therefore, there is a great opportunity to discover new, effective, and safe bioactive natural molecules from endophytes.

Keywords Bioactive compounds · Endophytes · Host plants · Symbiotic interaction

1 Introduction

Botanical sources, beyond their direct therapeutic usage, have been known as reservoirs of untold numbers of internal microbes, the so-called endophytes. Endophytes are plant-associated microbes (mostly filamentous fungi and bacteria) living within the living tissues of their hosts for at least one period of their life phase (Yu et al. 2010). Their presence is unobtrusive and asymptomatic to the host plant, rather for securing their residency. Medicinal plants are the dominant source of endophytes, followed by crops (Yu et al. 2010). Large numbers of endophytic microbes, mainly fungi and bacteria, have a symbiotic interaction with their host plant, in which both sides are mutually beneficial. That is, the host plant serves a source of food and home to the endophytes, whereas the associated microbes protect their host from external enemies. And such a mutual relationship helps the endophytes produce secondary metabolites (Abdou et al. 2010; Alvin et al. 2014). In other words, endophytes are the guardians or elicitors of their host plants from external attackers via the release of antibiotics or hydrolytic enzymes. The protection mechanism of the endophytes can either be via direct exudation of chemicals or strengthening of the host plant's self-defense capacity. Besides, plant hormone secretion, siderophores production, nitrogen fixation, mineral solubilization (like phosphorus), and enzymatic activities (such as ethylene suppression by 1-aminocyclopropane-1-carboxylate deaminase) are other mechanisms by which endophytic microbes have played an essential part in the development of their hosts (Abdou et al. 2010; Alvin et al. 2014). On the contrary, there may be a probability for endophytes to become pathogens when there is a scarcity of foods and other stressful circumstances, and the opposite is true. And this scenario makes the endophyte-host plant interactions mutualistic, commensal, or parasitic (Izumi et al. 2008; Prasad and Dagar 2014; Souza et al. 2016). According to some expectations, the number of endophytic microbes found in different parts (i.e., seed, tuber, root, stem, leaf, and fruit) of a plant is close to one million, though only a few of them have been described. Today, many researchers have attempted to isolate and document endophytic microbes, especially filamentous fungi from various medicinally important plants, and examine their antimicrobial activities. And, since then, numerous bioactive natural products have been investigated and reported from those endophyte isolates (Izumi et al. 2008). However, the endophytes that have been studied in detail so far are still very few in number with respect to their huge biodiversity, unlike epiphytic microbes, which are intensively studied. Moreover, the review articles published so far were not systematically reviewed in a way suitable for biochemical studies of endophytes. This has posed a great challenge to researchers from developing countries such as Ethiopia, where studying endophytes is unfamiliar. Taking this and the biodiversity of endophytes and their potential in producing bioactive metabolites into consideration, a compiled review that contains the most important parts of the endophyte study, including the exploration methods, is therefore much needed to carry out an extensive and continuous investigation on endophytes, thereby increasing the chance of finding novel bioactive natural products. The purpose of the present review is, therefore, to give an insight into the exploration methods of potential endophytes and associated bioactive agents that can be considered contenders for their host medicinal plants and shed light on the studies of endophytes in the context of some countries, such as Ethiopia.

1.1 Endophyte Mode of Transmission

Endophytes inhabit an entirely secure and protected environment, in contrast to epiphytes that reside outside the plants. There are two types of transmission modes in which endophytes follow during their invasion of the host plant: one is via horizontal and another is via vertical (Izumi et al. 2008). The horizontal mode of transmission is carried out through airborne spores and/or airborne inoculum. In contrast, the vertical way of transmission occurs via seeds. Endophytes, which are transmitted via the vertical mode, can never appear on the external part of the host plant and are the real ones (Izumi et al. 2008). The possible mechanism of transmission may be either via passive diffusion, in which those originated from the one next to the soil ecosystem are the chief colonizers of the plants, or via active selection, in which only specific microbial species win the competition to inhabit the interior part of the hosts (Izumi et al. 2008). Endophytes can be found in the anthosphere (flowers), spermosphere (seeds), caulosphere (stems), rhizosphere (roots), carposphere (fruits), and phylloplane (leaves) (Souza et al. 2016). During their distribution, endophytes like phytopathogens colonize the intercellular spaces of plants as well as vascular tissues by gaining entry into the plant via penetration of root hair cells or secretion of cell-wall-degrading enzymes like cellulases. More briefly, endophytes penetrate the plant tissue through natural openings like stomata, lenticels, hydathodes, or wounds (Souza et al. 2016).

1.2 Endophyte-Host Plant Symbiotic Interaction

Endophytes can have different types of interactions with their host, starting from symbiotic via mutualism up to parasitism, while they grow in the interior part of the plant. Such relationships depend on various factors, such as genetic history, environmental influences, infection patterns, and mechanisms of transmission. And the complications of the relationship endophytes make with their host plants can vary from plant to plant and endophyte to endophyte (Patil et al. 2016). Endophytes may also interact with other living species, such as mycorrhizal fungi. The endophyte-host plant interaction occurs via some stages, including attaching to the surface of the host, followed by infecting and inhabiting the interior part of the plant. Multiplication, different phenotypic character expression, and the expansion of territories to other plant individuals are the final stages of the endophytes' interaction with their hosts. As mentioned above, the endophytes of one plant may also be an infection to other plant species under stressful conditions or in a changed environment (Delaye et al. 2013). Of course, the probability of becoming pathogens to other plants is rare for endophytes since their existence from the beginning is dependent on friendly interactions. But the big question is: how do the endophytes manage and secure their existence within their host plant without any apparent harm? Asymptomatic colonization is an indication of an antagonistic balance between the host and the endophyte. Endophytes possess the same virulence factors as pathogens, such as producing exoenzymes, which enable them to penetrate and invade the internal part of the host plant. Many of the endophytes can also produce phytotoxic metabolites. The host plant can therefore respond with preformed and induced defense agents like those of the pathogens. Therefore, provided that the microbial attack and the plant's self-protection are balanced in this way, the relationship becomes asymptomatic. Endophytic microbe-host plant positive interaction is so crucial in determining and influencing the capability of endophytes to biosynthesis secondary metabolites, which also affect the host plant in synthesizing phytoalexins to protect against phytopathogens under stressed conditions. From this fact, it is possible to say that environmental factors have played a decisive role in the biosynthetic ability of host plants and their associated microbes to produce secondary metabolites with different chemical profiles and pharmacological activity (Kumara et al. 2012). Whether it is balanced or not, the endophyte-host interaction in general is up to the nature of the associates, the pathogenicity of the microbes, and the self-protective ability of the host.

1.3 Endophytes from Medicinal Plants

Botanicals are well-known sources of plenty of chemical agents with special characters and wide diversity. And beyond that, they also harbor many of the hitherto unknown microbial species. Following the vital uses of phytochemicals and their impact on the biochemical nature of plant-associated microbes, researchers have been interested once again in botanical sources for exploration of their associated microorganisms. The ability of endophytes to mimic the biosynthesis chemistry of their host plant to produce the same or alternative biocompounds is another reason to attract researchers. Moreover, endophytes have an extra advantage economically and ecologically in producing interesting chemical agents with great novelty and diverse biological activities, thereby helping to conserve biodiversity. Hence, exploring such precious biological sources from their hosts, which leads to the isolation of associated microchemicals, is so crucial (Padhi et al. 2013). It is believed that close to 300,000 botanical species inhabit our planet, and every single plant is expected to host more than a million internal microbes. Unfortunately, the plant species that have been investigated so far for their endophyte sources are few in number relative to their endophytic biology. It is, therefore, very apparent that the probability of obtaining new and diversified endophytic microbes from various plants inhabiting different ecosystems is high (Prasad and Dagar 2014). Many researchers are engaged in isolation of endophytic microbes particularly filamentous fungi and evaluating their antimicrobial activities. And such efforts have led to the isolation of numerous bioactive agents from the already-explored microbes. But, very sadly, the number of endophytes previously studied for their associated chemical entities is few in comparison to their wide biodiversity. Hence, investing a huge amount of time and great effort into these hidden microbes may be profitable to find interesting drugs with better efficacy and safety (Patil et al. 2016). In the present review, some of the endophytic microbes (fungi and bacteria) isolated and reported so far from various species of medicinal plants, together with their antimicrobial activities, are listed in Table 1.

Table 1 Examples of fungal and bacterial endophytes discovered from medicinal plants

Plant source	Endophyte name	Biological activity	References
Aloe vera	*Bacillus tequilensis*, *Chryseobacterium indologenes*, *Bacillus aerophilus*, and *Pseudomonas entomophila*	Antibacterial and antifungal	Martinez-Klimova et al. (2017)
Polygonum cuspidatum	*Streptomyces* sp.	Antibacterial and antifungal	Martinez-Klimova et al. (2017)
Annona crassiflora	*Rhizoctonia* sp.	Antibacterial and antifungal	de Mendonca et al. (2015)
Curcuma xanthorrhiza	*Xylaria* sp., *xanthorrhiza*	Cytotoxic	Hammerschmidt et al. (2015)
Azadirachta indica	*Chaetomium colletotrichum*, *Curvularia trichoderma*, *Streptomyces*, *Streptosporangium*, and *Nocardia*	Antioxidant, antibacterial, and antifungal	Kumaresan et al. (2015)
Rauwolfia serpentine	*Aspergillus awamori* and *Penicillium* sp.	Antibacterial and antifungal	Nath et al. (2015)
Aegle marmelos	*Aspergillus flavus*	Antioxidant and antimicrobial	Patil et al. (2015)
Tabebuia argentea	*Penicillium* sp., *Aspegerillus niger*, *Alternaria* sp., and *Alternaria alternata*	Anticancer	Chen et al. (2014)
Garcinia nobilis	*Penicillium* sp.	Antibacterial	Jouda et al. (2014)
Xanthium sibiricum	*Eupenicillium* sp.	Antibacterial	Li et al. (2014a)
Ginkgo biloba	*Chaetomium globosum*	Cytotoxic	Li et al. (2014b)
Pittosporum mannii	*Diaporthe* cf. *ueckeri*	–	Kemkuignou et al. (2023) and Mahdi et al. (2014)
Prosopis chilensis	*Emericella*, *Aspergillus*, and *Chaetomium*	Antimicrobial and antioxidant	
Hypericum mysorense	*Bionectria ochroleuca*, *Chaetomium globosum*, and *Aspergillus* sp.	Antibacterial and antioxidant	Samaga and Rai (2016)
Curcuma longa	*Penicillium* sp.	Antibacterial	Singh et al. (2014)
Crotalaria pallida	*Alternaria* sp., *Penicillium* sp., and *Aspergillus flavus*	Antioxidant and antimicrobial	Singh et al. (2014)
Ocimum tenuiflorum	*Penicillium citrinum*	Antibacterial	Lai et al. (2013)
Terminalia brownii	*Rhizopus oryzae*, *Aspergillus niger*, and *Aspergillus flavus*	Antibacterial and antifungal	Basha et al. (2012)

Generally, it is so challenging to differentiate whether or not a particular endophytic microbe really originated from the internal part of uninfected plant tissue or was mistakenly isolated from other sources. To overcome such difficulty, some methods, such as surface sterilization of the host plant and targeting the appropriate microbe growing on appropriate media, histological observation in combination with the molecular method, followed by identification using selective chemical analysis (such as the immunological method), and finally amplification of the microbial biological material directly from the host plant, have routinely been applied (Schulz et al. 2006).

2 Bioactive Compounds from Endophytes

Today, endophytic microbes are ascertained for their ability to produce the same or similar chemical agents as their hosts. The ability of endophytic microbes to biosynthesize new bioactive compounds may possibly be attributed to the endophyte-host plant metabolic interaction. The production and storage of chemical entities may therefore be influenced by the possible factors originating from the two associates, plants and endophytes. There are many assumptions regarding the synthesis of bioactive molecules by endophytes. On one hand, the host plant and associated microbes can biosynthesize secondary compounds independently, like in the history of gibberellin. On the other hand, the endophytes can incorporate the host plant's genetic material into their own biological material via genetic recombination during the course of evolution with the host plants, which at the end allows them to replicate the synthetic ability of their host plant for the same chemical(s). For instance, *Fusarium solani*, an endophytic fungus, could produce a starting chemical agent used for the later synthesis of the anticancer drug camptothecin. However, to synthesize camptothecin, *Fusarium solani* uses a host plant enzyme called strictosidine synthase (Kusari et al. 2012). This indicates how much the association between endophytes and their host contributes to the biosynthesis of the same molecules by the associates simultaneously (Heinig et al. 2013). One of the most successful stories of endophyte studies for their associated chemicals is the history of the famous anticancer drug, taxol. As it is known, taxol first originated from *Taxus brevifolia*, the Pacific yew tree. Then, later, the compound was isolated from other plants of the same genus. Unfortunately, the economic and ecological infeasibility of these slow-growing plants has forced researchers to find other alternatives for the production of taxol. Following many efforts, the endophytic fungus, the so-called *Taxomyces andreanae*, which originated from the same plant, was also found to be the producer of the same paclitaxel. At the beginning, the assumption was that the endophytic fungi, including *Taxomyces andreanae*, could not biosynthesize paclitaxel on their own independently; rather, they mimicked the host and later accumulated it in their cell walls (Heinig et al. 2013). However, such an assumption was ruled out by the recently reported finding, which revealed that the sequences with significant homology in the endophytes' genetic materials did not match the genes of *Taxus* spp. used to biosynthesis taxol (Heinig et al. 2013). This is an indication that *Taxus* spp. and associated endophytic fungi can biosynthesize taxol independently. Moreover, the above fact was supported by another finding, which stated that a collection of potential genes that participated in the biosynthesis of paclitaxel are also found in the *Corylus avellana* originated endophytic fungus, *Penicillium aurantiogriseum.* All the above findings have therefore disqualified the hypothesis, which states that there may be a possibility of gene transfer horizontally between the host plant and their endophytic fungus. Currently, the discovery of *Taxomyces andreanae* has solved the problems associated with overharvesting of slow-growing yew trees, ecological imbalance, and the high price of the anticancer drug, taxol. This is because taxol nowadays could be produced via fermentation from the aforementioned endophytic fungus (like that of penicillin), and this makes the drug available to all cancer patients (Abdou et al. 2010; Schulz et al. 2006). Searching and producing natural products from endophytic microbes via fermentation is, therefore, cost-effective, continuous, less toxic, very effective, and environmentally friendly. And this may help to minimize or avoid the overharvesting of slow-growing and rare plant species like yew trees and conserve the world's ever-dwindling biodiversity. Because of these rationales, scientists and researchers have refocused on the world of microbes. In most review articles, endophytic filamentous fungi have been mentioned as sources of many bioactive compounds. In this review, some of the antibacterial, antifungal, and anticancer molecules explored from fungal endophytes so far are also mentioned below.

2.1 Antimicrobial Compounds

So far, plenty of antimicrobial metabolites have been exploited as bioactive natural products from endophytes. According to many review articles, the classes of fungal endophytes, namely, *Phoma*, *Xylaria*, *Colletotrichum*, *Fusarium*, *Penicillium*, and *Aspergillus*, have extensively been reported as sources of numerous antimicrobial compounds. In this review also, some antifungal and antibacterial compounds originated from the aforementioned, and other fungal endophytic classes have been mentioned below (Fig. 1).

The fungal endophytes and plant sources of the compounds in Fig. 1, together with their antimicrobial activity, are shown below (Table 2).

2.2 Anticancer Compounds

Though the role of plant-based natural products in the research and development of anticancer drugs is indispensable, the search for substitute chemicals with greater originality and diversity is still in great need. Active metabolites isolated

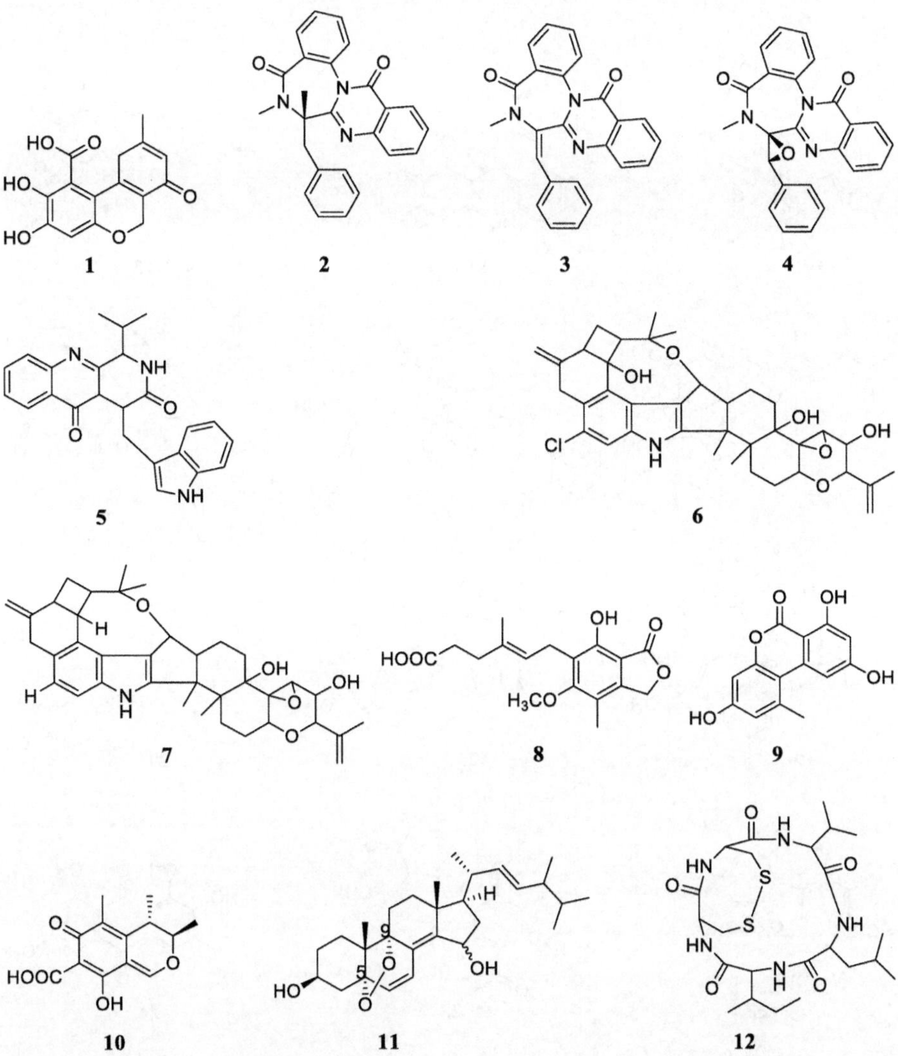

Fig. 1 Antimicrobial compounds originated from fungal endophytes of medicinal plants

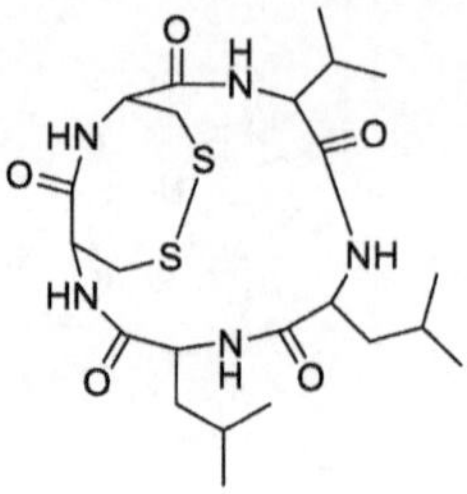

13

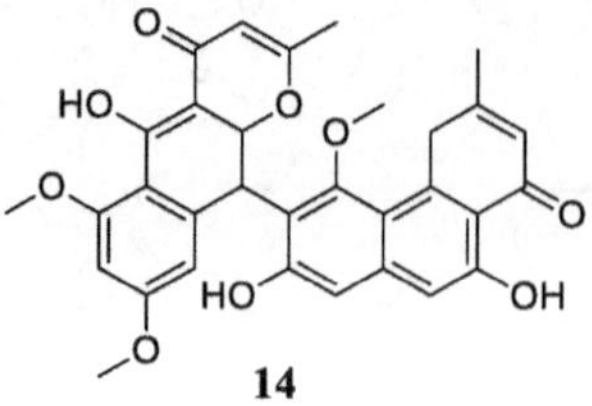

14

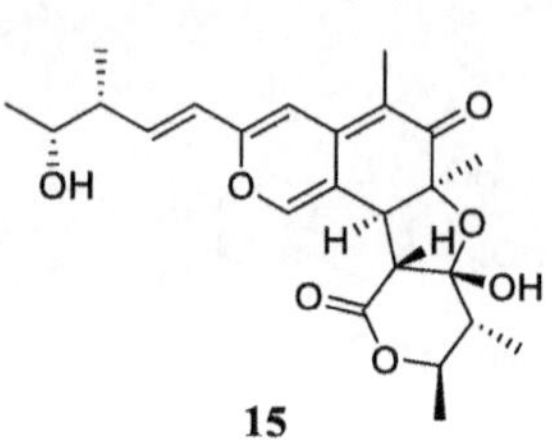

15

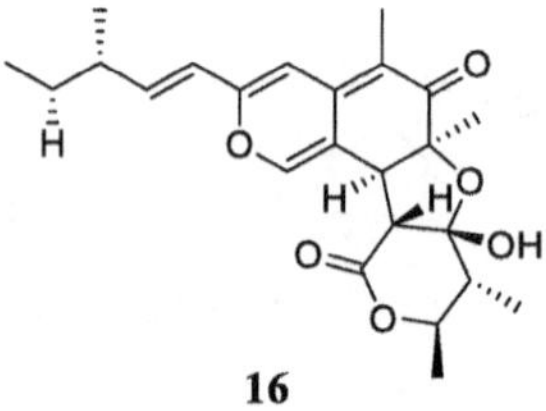

16

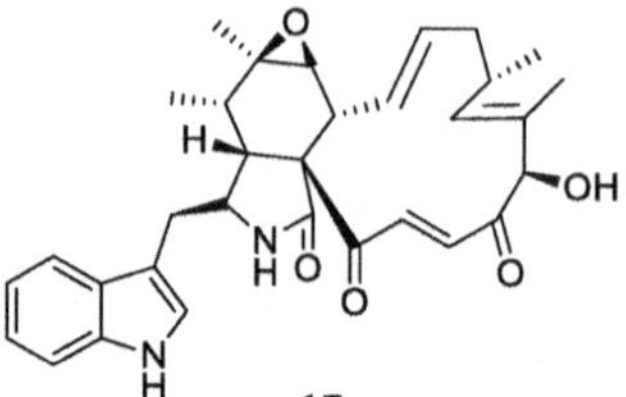

17

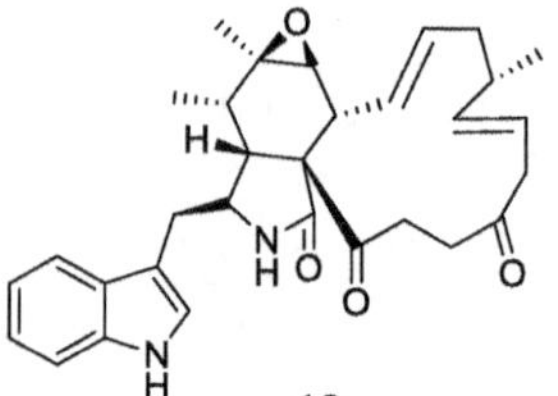

18

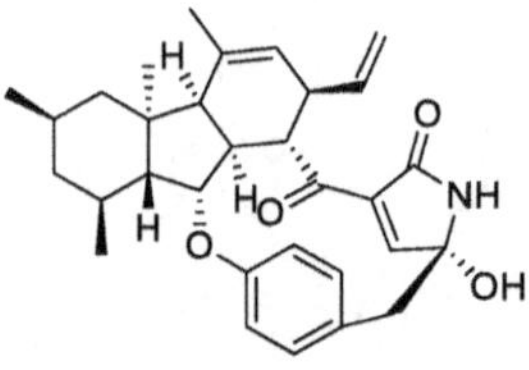

19

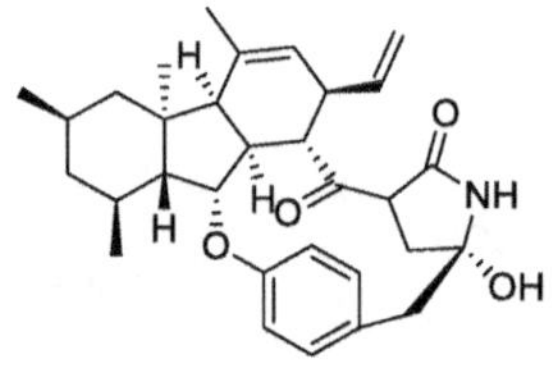

20

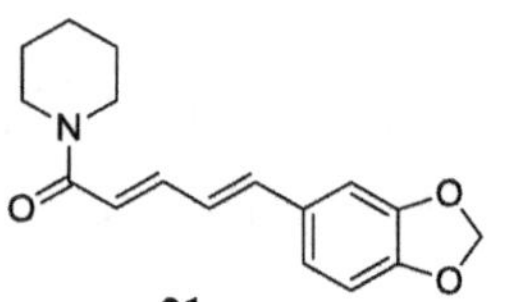

21

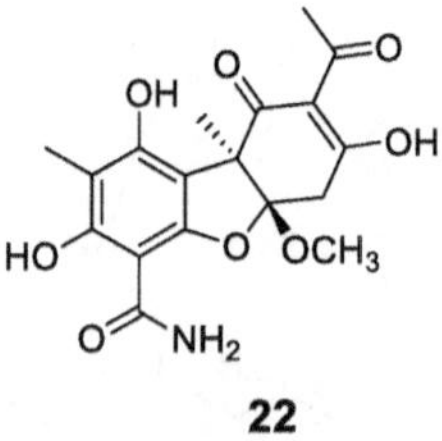

22

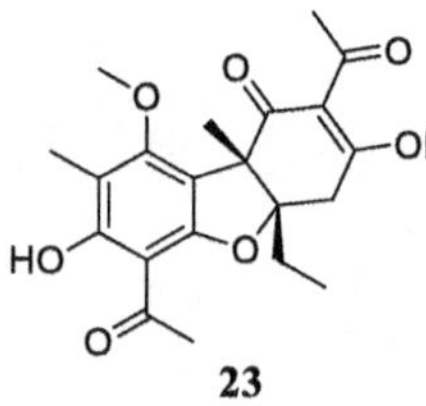

23

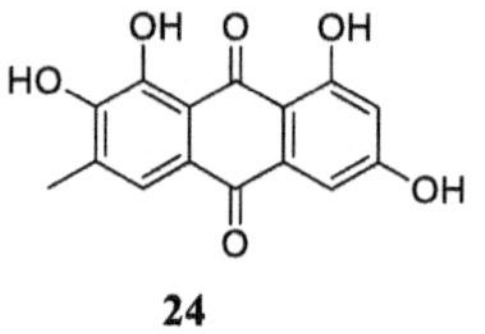

24

Fig. 1 (continued)

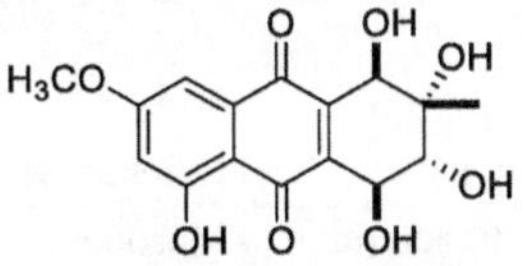
25

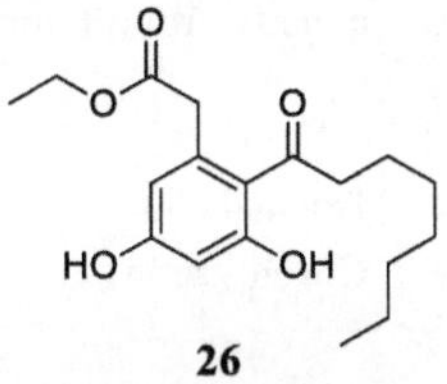
26

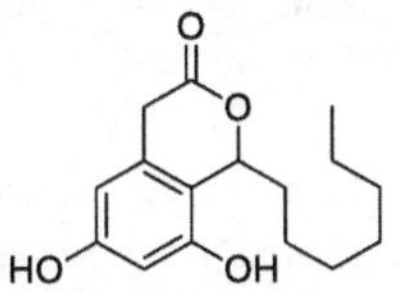
27

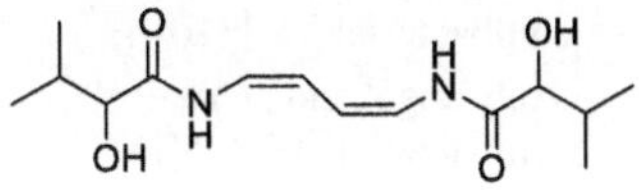
28

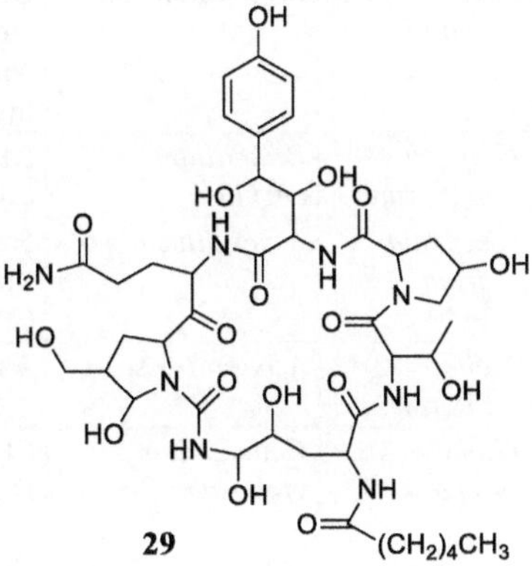
29

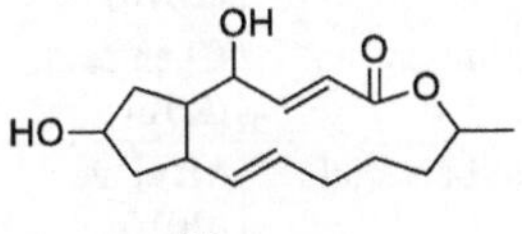
30

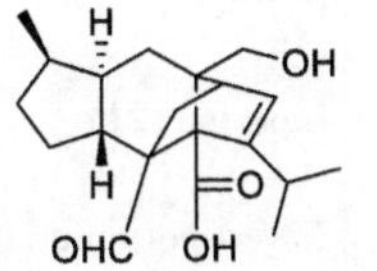
31

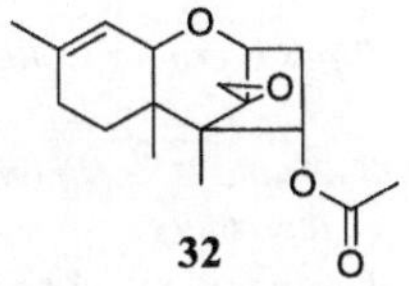
32

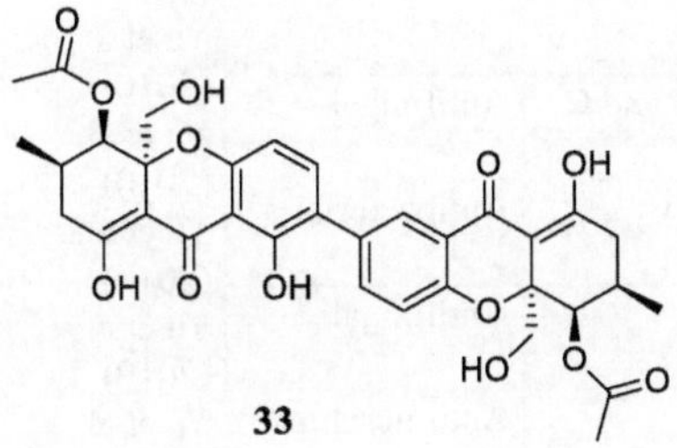
33

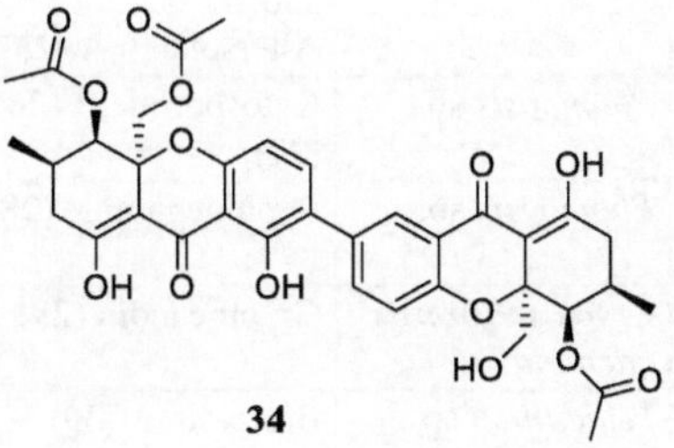
34

Fig. 1 (continued)

from endophytes could be another method for finding new anticancer medications with minimal side effects. Lots of fungal endophyte-based anticancer compounds have been reported so far. To save time and space and for simplicity, some representative anticancer agents obtained from fungal endophytes have been listed in Fig. 2.

The names of the abovementioned anticancer compounds (Fig. 2), combined with their plant and fungal endophyte sources, are tabulated in Table 3.

Table 2 Some antimicrobial compounds from fungal endophytes originated from various plant sources

Plant source	Endophyte name	Isolated compound	Biological activity	References
Garcinia nobilis	*Penicillium* sp.	Citromycetin (**1**)	Antibacterial	Martinez-Klimova et al. (2017)
Taxus brevifolia	*Penicillium* sp.	Benzomalvin A (**2**), B (**3**) and C (**4**), fiscalin B (**5**), penitrems A (**6**) and B (**7**), mycophenolic acid (**8**)	Antibacterial	Stierle and Stierle (2000)
Ocimum tenuiflorum	*Penicillium citrinum*	Alternariol (**9**) and citrinin (**10**)	Antibacterial	Lai et al. (2013)
Avicennia marina	*Aspergillus niger*	5α,9α-epidioxy-sterol (**11**), Malformins A (**12**) and C(**13**)	Antibacterial	Lai et al. (2013)
Melia azedarach	*Aspergillus* sp.	Asperpyrone A(**14**)	Antifungal and antibacterial	Xiao et al. (2014)
Ginkgo biloba	*Chaetomium globosum*	Chaetomugilin A (**15**) and D (**16**), Chaetoglobosins A (**17**) and C (**18**)	Antifungal and antibacterial	Yu et al. (2010)
Maize	*Acremonium zeae*	Pyrrocidines A (**19**) and B (**20**)	Antifungal	Yu et al. (2010)
Piper nigrum	*Colletotrichum gloeosporioides*	Piperine (**21**)	Antimicrobial	Chen et al. (2014)
Saurauia scaberrinae	*Phoma* sp.	Cercosporamide (**22**) and Phomodione (**23**)	Antibacterial	Yu et al. (2010)
Urospermum picroides	*Ampelomyces* sp.	3-O-methylalaternin (**24**) and Altersolanol A (**25**)	Antibacterial	Yu et al. (2010)
Mangrove plant	*Phomopsis* sp.	Cytosporone B (**26**) and C (**27**)	Antifungal	Yu et al. (2010)
Garcinia dulcis	*Phomopsis* sp.	Phomoenamide (**28**)	Antibacterial	Yu et al. (2010)
Tripterygium wilfordii	*Cryptosporiopsis quercina*	Cryptocandin (**29**)	Antifungal	Yu et al. (2010)
Panax ginseng	*Penicillium* sp.	Brefeldin A (**30**)	Antimicrobial	Yu et al. (2010)
Garcinia dulcis	*Xylaria* sp.	Sordaricin (**31**)	Antifungal	Yu et al. (2010)
Llex cornuta lindl *Pittosporum mannii*	*Trichoderma harzianum* *Diaporthe* cf. *ueckeri*	Trichodermin (**32**) Dicerandrols A (**33**) and B (**34**)	Antifungal Antibacterial (Kemkuignou et al. 2023)	Yu et al. (2010)

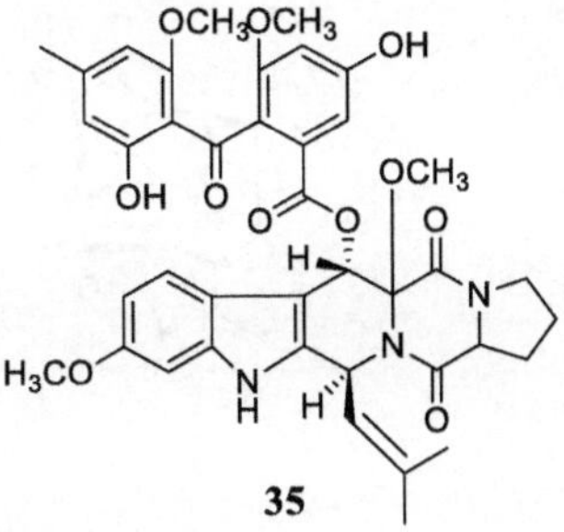

35

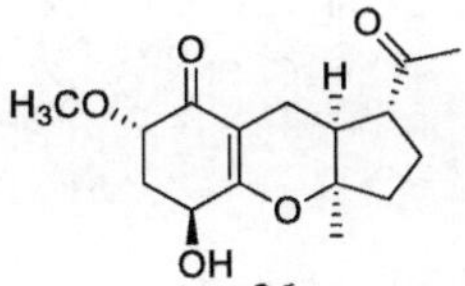

36

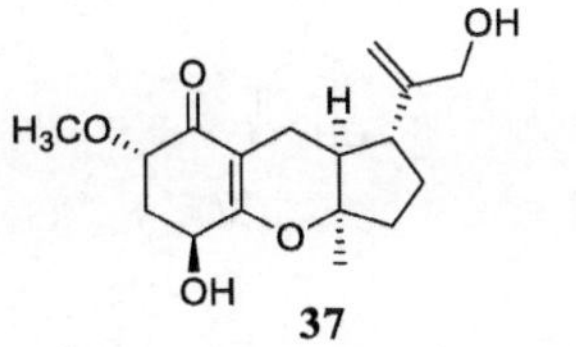

37

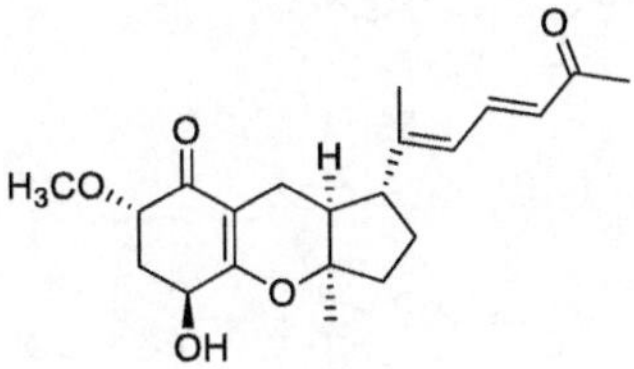

38

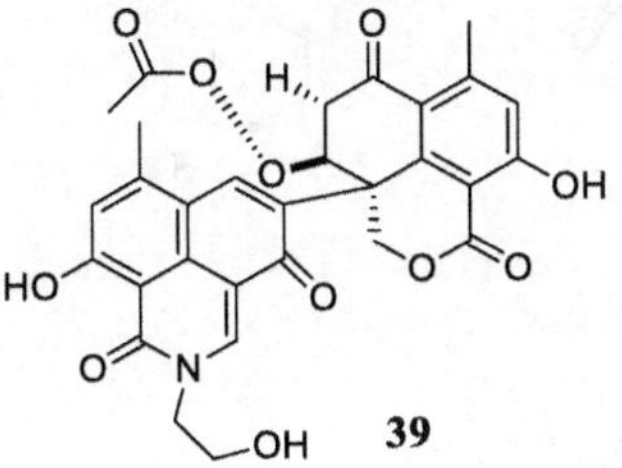

39

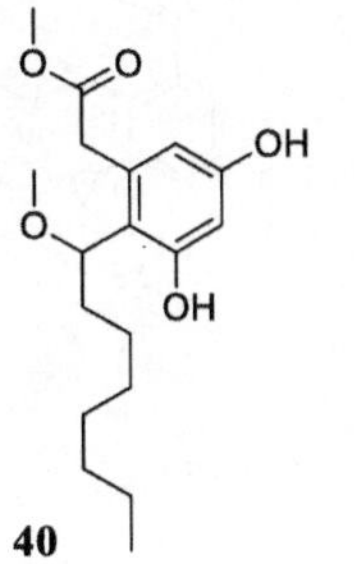

40

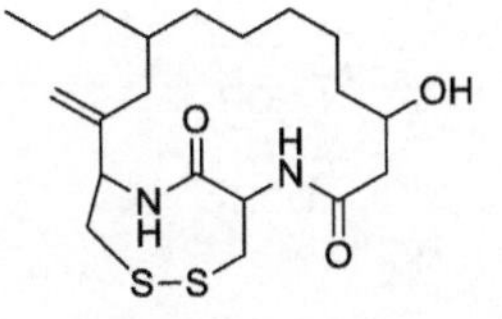

41

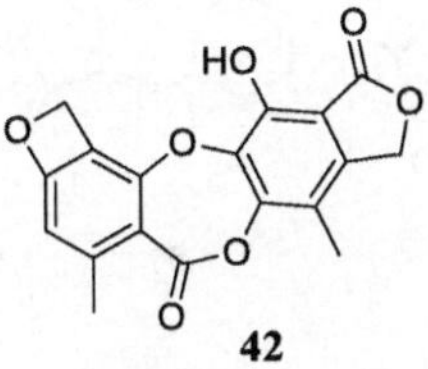

42

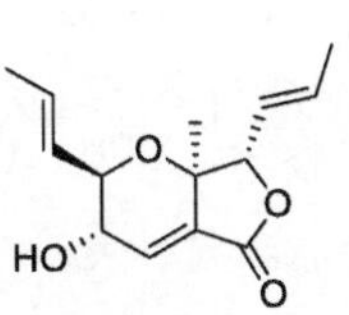

43

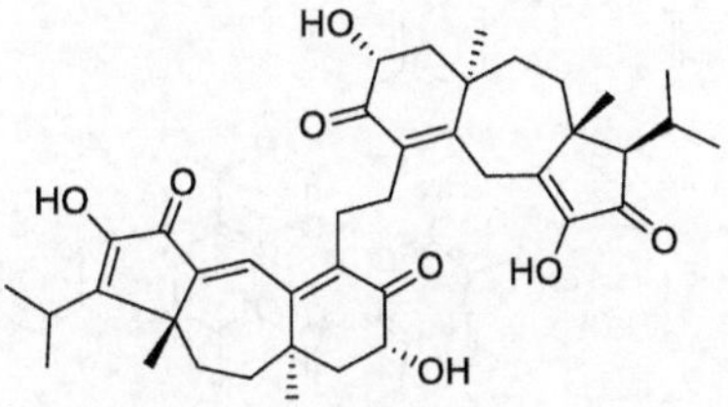

44

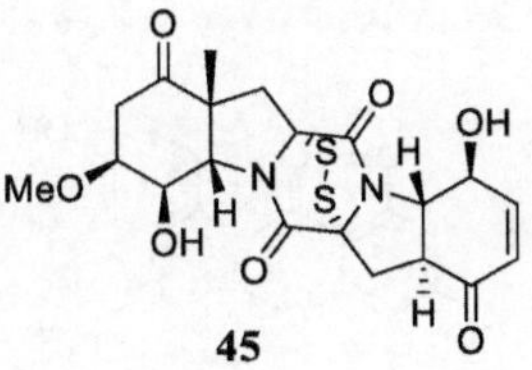

45

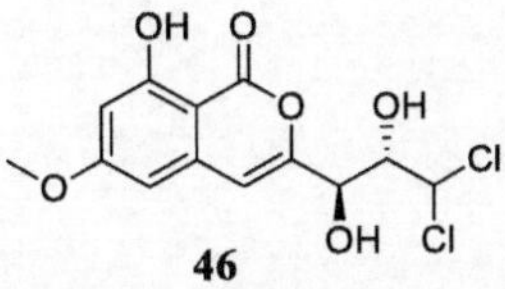

46

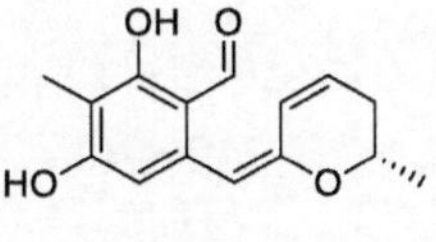

47

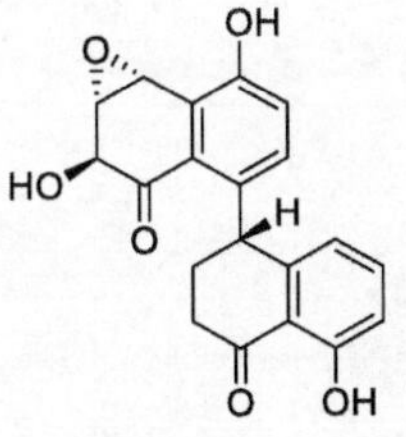

48

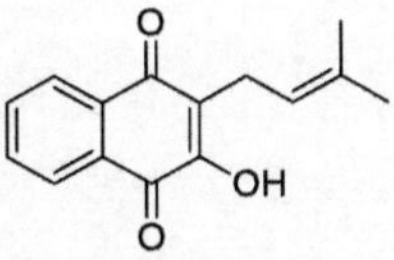

49

Fig. 2 Anticancer molecules produced by fungal endophytes originated from medicinal plants

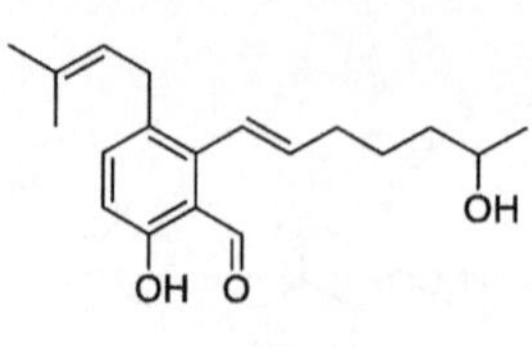

50

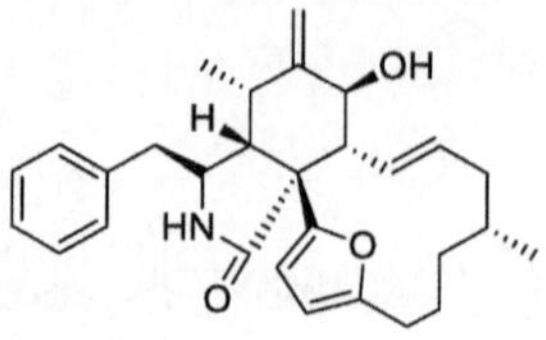

51

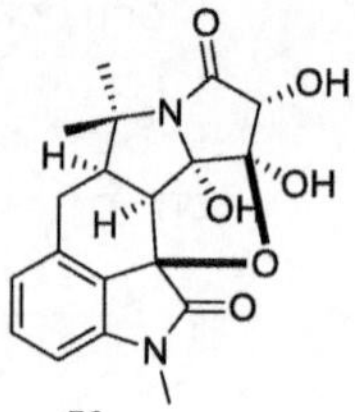

52

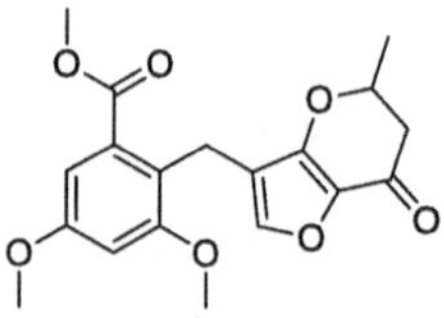

53

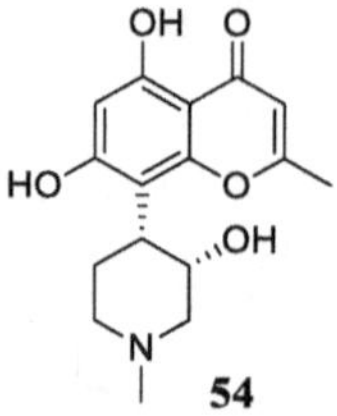

54

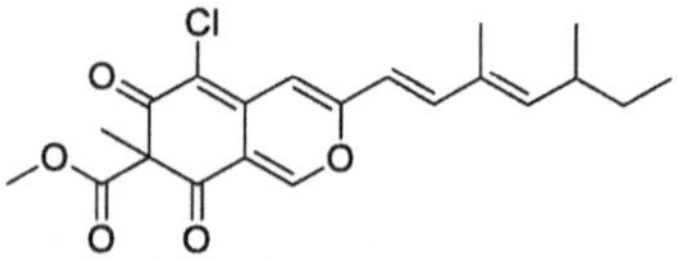

55

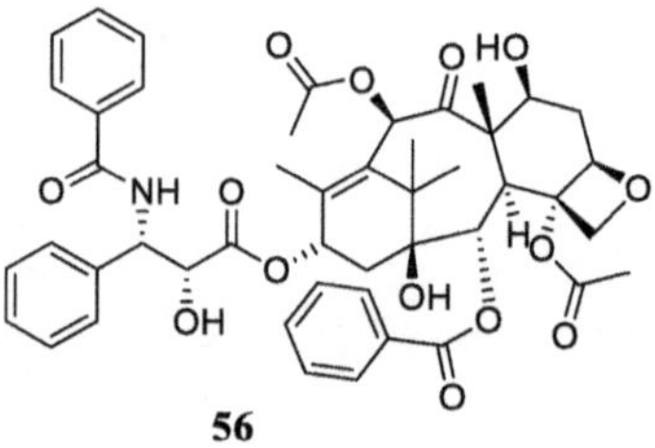

56

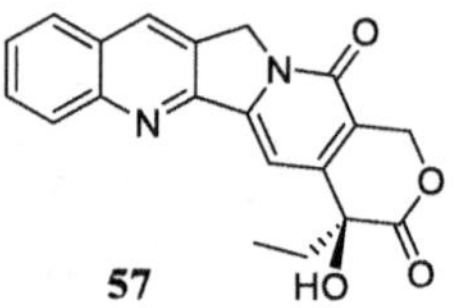

57

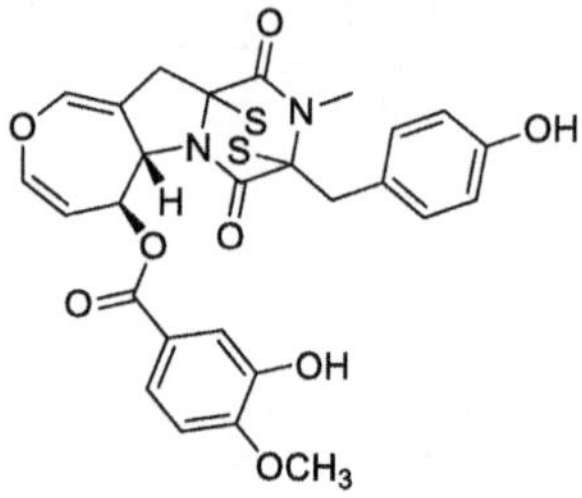

58

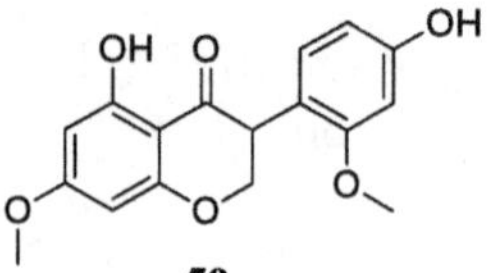

59

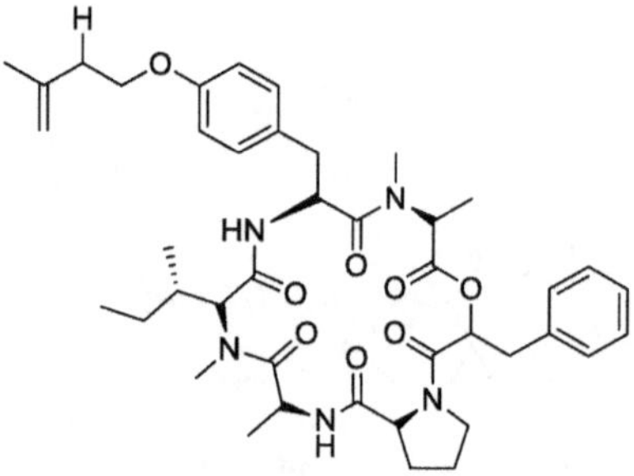

60

Fig. 2 (continued)

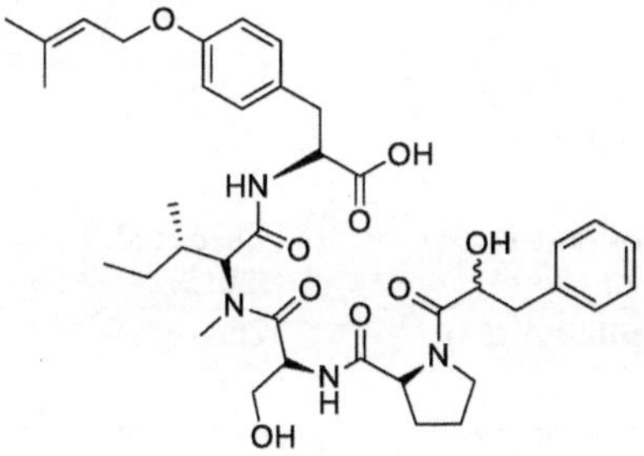

61

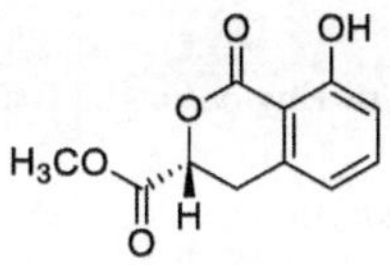

62

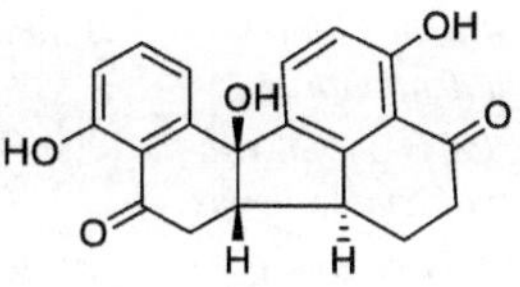

63

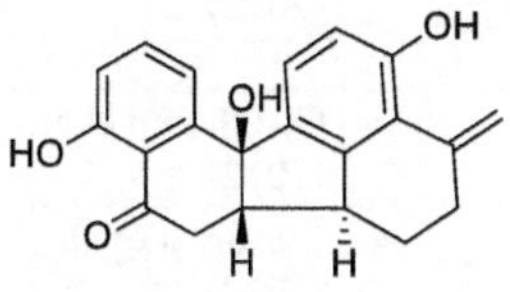

64

Fig. 2 (continued)

Table 3 Some representative anticancer compounds derived from fungal endophytes associated with medicinal plants

Medicinal plants	Fungal endophytes	Anticancer compounds	References
Diphylleia sinensis	*Aspergillus fumigatus*	Fumitremorgin D (Du and Su 2014)	Liang et al. (2015)
Acanthus ilicifolius	*Aspergillus flavipes*	Meroterpenoids (**36–38**)	Bai et al. (2015)
Panax notoginseng	*Penicillium manginii*	Duclauxamide A1(**39**)	Cao et al. (2015)
Aegiceras corniculatum	*Dothiorella* sp.	Dothiorelone F (**40**)	Du and Su (2014)
Pongamia pinnata	*Phomopsis glabrae*	PM181110 (**41**)	Verekar et al. (2014)
Kandelia candel	*Phomopsis* sp.	Phomopsidone A (**42**)	Zhang et al. (2014)
Olyralati folia	*Libertella blepharis*	3-epi-Waol A (**43**)	Adames et al. (2015)
Fallopia japonica	*Cercospora* sp.	Cercosporene F (**44**)	Feng et al. (2014)
Avicennia marina	*Penicillium brocae*	Penicibrocazine A (**45**)	Meng et al. (2014)
Myoporum bontioides	*Trichoderma* sp.	Dichlorodiaportinol A (**46**)	Li et al. (2014b)
Panax ginseng	*Penicillium melinii* and *Penicillium janthinellum*	Ginsenocin (**47**)	Zheng et al. (2013)
Kandelia candel	*Cladosporium* sp.	Cladosporone A (**48**)	Ai et al. (2015)
Tabebuia argentea	*Aspergillus niger*	Lapachol (**49**)	Chen et al. (2014)

(continued)

Table 3 (continued)

Medicinal plants	Fungal endophytes	Anticancer compounds	References
Aegiceras corniculatum	*Pestalotiopsis* sp.	Pestalol B (**50**)	Sun et al. (2014)
Eupatorium adenophorum	*Phoma multirostrata*	Multirostratin A (**51**)	Chen et al. (2015)
Paris polyphylla var. yunnanensis	*Aspergillus versicolor*	Aspergilline A (**52**)	Zhou et al. (2014)
Tripterygium wilfordii	*Penicillium* sp.	Penifupyrone (**53**)	Chen et al. (2014)
Dysoxylum binectariferum Hook.f	*Fusarium proliferatum*	Rohitukine (**54**)	Kumara et al. (2012)
Eugenia jambolana	*Cephalotheca faveolata*	Sclerotiorin(**55**)	Giridharan et al. (2012)
Rhizophora annamalayana	*Fusarium oxysporum*	Taxol (**56**)	Elavarasi et al. (2012)
Miquelia dentata	*Fomitopsis* sp.	Camptothecine (**57**)	Shweta et al. (2013)
Astragalus lentiginosus	*Emericella* sp.	Secoemestrin D (**58**)	Xu et al. (2013)
Cajanus cajan	*Hypocrea lixii*	Cajanol (**59**)	Zhao et al. (2013)
Sonneratia caseolaris	*Bionectria ochroleuca*	Pullularins E (**60**) and F (**61**)	Ebrahim et al. (2012)
Alternaria tenuissima	*Erythrophleum fordii*	Tenuissimasatin (**62**)	Fang et al. (2012)
Artemisia annua	*Hypoxylon truncatum*	Daldinone C(**63**) and D(**64**)	Gu et al. (2007)
Pittosporum mannii	*Diaporthe* cf. *ueckeri*	Dicerandrols A (**33**) and B (**34**)	Kemkuignou et al. (2023)

3 Classification and Application of Endophytes

Endophyte research that has been done thus far has focused mainly on fungal endophytes. Accordingly, the word endophyte, most of the time, refers to endophytic fungi. Endophytic fungi are generally classified as those that colonize some grasses (known as clavicipitaceous endophytes or C-endophytes) and those that can be extracted from asymptomatic tissues of angiosperms, conifers, ferns, and nonvascular plants (also known as non-clavicipitaceous endophytes or NC-endophytes) (Rodriguez et al. 2009; Wani et al. 2015). Such classification is based on variations in taxonomy, plant hosts, evolutionary relatedness, and ecological functions. Non-clavicipitaceous, based on mode of transmission between host generations and ecological function, *in planta* biodiversity levels, *in planta* colonization, tissue(s) colonized, and host range, are further divided into functional classes 2, 3, and 4. Following their biodiversity, importance to ecology, huge potential applications, and capability of switching between endophytic and other life styles, NC-endophytes have now gained a great deal of attention (Rodriguez et al. 2009; Wani et al. 2015).

Additionally, endophytes can be classified as temporary, systemic, or genuine endophytes. Systemic or real endophytes enter into their hosts vertically via seeds and form a mutual relationship with their host plant without exhibiting any signs of illness at any phase of their life cycle. On the contrary, transient endophytes enter the host plant in a horizontal way via airborne spores, and under normal circumstances, they do not cause any visible signs of disease. But, under a scarcity of resources and stressed conditions, transient endophytes may become pathogenic (Wani et al. 2015).

Besides their pharmaceutical applications, endophytes have also played a great role in biotechnology (such as in agriculture, biotransformation, biodegradation, and bioremediation processes), nanotechnology, and as a source of biofuels and extracellular enzymes. Some of the most important applications of endophytes are listed and discussed below.

3.1 Application in Agriculture

Endophytes contribute a lot to the enhancement of plant development and production, directly or indirectly. For instance, endophytic microbes improve the development potential of host plants by providing adequate nutrient supply (like sulfate, phosphorus, nitrates, and siderophores) and plant growth hormones (such as indole acetic acid, gibberellic acid, ethylene, and auxins), thereby leading to the production of new and economically important crops. Hence, endophytes nowadays are considered as green fertilizers applied for effective and sustainable agricultural practices without any harm to the ecosystem (Kaul et al. 2014). Endophytes can also enhance the biomass and chlorophyll contents of the host plants, either by directly providing growth hormones or by letting the host plant generate hormones. Moreover, endophytes are very helpful to increase the production of crops by enhancing the crop's immune system against biotic and abiotic stresses through the production of effective biochemicals and modification of the genetic expression of the crops (Wani et al. 2015). Endophytes can also play a crucial role in agriculture and horticulture as alternative and/or promising candidates for biocontrol of pests, insects, weeds, etc. For instance, endophytic bacteria that were isolated from the root of Ethiopian coffee were found to be a potential biocontrol agent to antagonize *Meloidogyne incognita*, the tomato root-knot nematode, using both *in vitro* and *in vivo* analysis (Hallmann et al. 2009).

3.2 Application in Biotransformation

Biotransformation is the mechanism of converting exogenous substances with the aid of biological agents. Endophytes have the ability to break down large molecules with great complexity into smaller, simpler ones. In addition, endophytic microbes remove toxic materials generated by the host plant during the process of defending itself from external attackers (Kaul et al. 2014). For instance, when the host plant is attacked by pests, it starts to transform hydroxamine acids into compounds having a toxic nature, such as benzoxazolinon-2-benzoxazolinone (BOA) or 2-hydroxy-1,4-benzoxazin-3-(2H)-one (HBOA). Endophytes such as *Fusarium sambucinum* have therefore a great ability to convert these toxins to the nontoxic molecule N-(2-hydroxyphenyl) malonamic acid (Zikmundova et al. 2002). Endophytes can also biotransform existing compounds into completely new ones. The *Artemisia annua* originated endophytic fungus, *Trichoderma roseum*, for example, converts artemisinic acid into two new metabolites with better activity (Gaur et al. 2014).

3.3 Application in Biodegradation/Bioremediation Process

Biodegradation or bioremediation is a technique for getting rid of environmental contaminants and pollutants using microorganisms as treatment agents. Currently, endophytes have been confirmed for their ability to degrade environmental pollutants, including polycyclic aromatic hydrocarbons, using extracellular enzymes such as lignin catabolic enzymes, lipoidases, proteinases, cellulases, pectinases, and phenol oxidases. Also, endophytes play key roles in the bioremediation process, a technique used to remove toxic substances from the environment using a green strategy. To mention a few, endophytes engulf toxic heavy metals, degrade toxic secondary metabolites, remove greenhouse gases from the air, and inhibit the emergence of pests on host plants (Yadav and Yadav 2017).

3.4 As a Source of Biofuels

Surprisingly, endophytes are found to be a promising source of biofuel, according to recent research. This is because endophytes (especially fungi) produce volatile hydrocarbons that have the potential to be used as fuels and were named mycodiesels, which have shared properties similar to those of normal diesel fuel (Strobel 2014). This is due to the fact that endophytes isolated from their host botanicals have been confirmed to contain large amounts of methyl esters and are known as efficient precursors for biofuels. For instance, the *Jatropha curcas*-based *Colletotrichum truncatum* endophyte has been found as a source of oil from its hexane extract, and its fatty acid composition was discovered to be comparable to that of *Jatropha* oil, according to Kumar and Kaushik (2013) report.

3.5 Application in Nanotechnology

One of the most studied nanoparticles is made from silver metal. Silver-based nanoparticles have been confirmed for their wide and multiple biological activities, such as anticancer and antimicrobial activities. Silver nanoparticles are synthesized most of the time via chemical approaches (Birla et al. 2009). This synthesis approach causes the presence of trace toxic chemicals, which is undesirable in medical applications. This limitation has forced researchers to search for another strategy for safe nanomaterial synthesis. In the nanotechnology discipline, designing green methods for the synthesis of nanoscale materials is a very important issue. To date, using endophytic microbes in the production of nanoparticles has been found to be an environmentally friendly and green strategy and a better alternative for chemical synthesis. So far, different fungal endophytes have been applied for the synthesis of nanoparticles, of which silver nanoparticles (Ag-Nps) are the dominant ones (Verma et al. 2010).

3.6 Sources of Extracellular Enzymes

Filamentous endophytic fungi are the most preferred microorganisms for the production of useful enzymes since they are easily cultivable, thereby providing a sufficient amount of extracellular enzymes. Fungal enzymes are more advantageous than enzymes originated from other biological sources, such as plants and animals, in various applications due to their stability (at high temperatures and extreme pH) (Bhagobaty and Joshi 2012). Cellulases, chitinases, amylases, lipases, proteases, pectinases, and lactase are examples of extracellular enzymes made by endophytes. These enzymes have been widely applied in numerous industrial processes, including textile, leather, baking, brewing, paper, pulp, bioremediation, and biosensing. The enzymes made by diverse endophytic microbes often depend on the host plants and their ecological conditions (Bhagobaty and Joshi 2012).

4 Exploration Methods on Endophytes and Their Bioactive Agents

In the study of endophytes, identifying a plant material suspected of being reservoir of potential endophytes and selecting the appropriate isolation method are the two critical steps. In other words, focusing on the procedures and strategies that improve the chances of finding endophytes with the probability of producing novel active metabolites is very vital.

4.1 Plant Selection Method

Having a plant material that has been suspected to harbor novel endophytic microbes is the key step in an endophytic study. According to several reviewed articles, plant species (Yu et al. 2010) inhabiting areas rich in biodiversity, such as tropical rainforests and mangrove forests, and under the influence of biotic and abiotic strains;

(Abdou et al. 2010) with great resistance to phytopathogens; (Alvin et al. 2014) occupying unique environmental settings like ancient land masses; and (Izumi et al. 2008) known for their traditional medicines with an old age and endemism nature, are expected to be reservoirs of many potential endophytes, thereby contributing to the discovery of bioactive drugs. Hence, it is so important to give great attention to these selection strategies and take them into consideration during the attempt to explore potential endophytes, which increases the probability of finding new and effective chemical agents (Yu et al. 2010; Patil et al. 2016).

4.2 *Endophyte Isolation Method*

In the exploration of endophytes, the isolation procedure is another important step. The procedure to be followed is expected to have a selective ability to target the potential endophytic microbes by eliminating the epiphytes from the plant surface. Washing and disinfecting the outer surface of selected plant part(s) using different disinfectants is the first procedure that should be conducted. Surface sterilization is most frequently utilized to eradicate attached contaminants and epiphytes so as to detect and quantify endophytes found only inside the host plant's tissues. Then, the sterilized plant part is either powdered or reduced in size into small segments, which are then placed directly onto appropriate culture plates or broth agar aseptically (Prasad and Dagar 2014; Martinez-Klimova et al. 2017). Briefly, the plant samples are cleaned using autoclaved distilled water, sterilized with hydrogen peroxide, and then rinsed with 70% ethanol followed by 3% hypochlorite with tween 20 (surfactant). Lastly, they are rinsed with autoclaved distilled water. The aliquots of the last rinsed water should be cultured on nutrient media for sterility checking. That is, if there is no microbial growth observed on the medium, it indicates the completion of surface sterilization. It is crucial to remember that the endophytic isolation procedures should be performed within 24 h of harvesting the fresh plant material; if not, it should be preserved in the refrigerator (4 °C) for the survival of the microbes (Yu et al. 2010). And each procedure should be carried out aseptically.

4.3 *Isolation of Fungal Endophyte*

The surface disinfected plant sample is cut aseptically into small segments (since colonization may be very limited) and then plated onto an appropriate nutrient medium. Sabouraud Dextrose Agar (SDA) and Potato Dextrose Agar (PDA) are the most commonly used nutrient media for endophytic fungi isolation, supplemented with antibacterial agents (mostly chloramphenicol) to inhibit the growth of any bacteria. Plates are then incubated at 27 °C for some days to promote the outgrowth of endophytic fungi and monitor any hyphal tip growth. After observation of some hyphal growth, subculturing of the individual fungal colony is followed. Lastly, each cultured endophytic fungus is checked for purity and transferred to a freshly prepared PDA plate (Izumi et al. 2008; Souza et al. 2016; Schulz et al. 2006).

4.4 *Isolation of Bacterial Endophyte*

Two possible methods can be employed to isolate endophytic bacteria from different plant parts. The first method is somewhat similar to that of the fungal endophyte isolation, i.e., surface disinfected plant material is reduced in size into small segments (close to 2 mm), impregnated on appropriate nutrient agar dishes, and incubated at 37 °C for some days until visible growth is observed. This method is mainly selected for fast-growing microorganisms, like endophytic fungi and actinobacteria, whereas it is seldom applied for endophytic bacteria (Schulz et al. 2006). The second method involves macerating the surface disinfected plant material to a fine powder using a sterile pestle and mortar, followed by suspending and serially diluting with sterile water by agitating in an orbital shaker. The suspension is then streaked onto Petri plates containing nutrient culture media such as Tryptic Soy Agar (TSA) supplemented with antifungal agents to block the growth of any fungal species. Finally, the plates are incubated at 37 °C for some days (depending on the nature of the bacterial strain and culture medium). Sometimes autoclaved water or phosphate buffer solution (PBS) is added to the surface disinfected plant tissue prior to maceration (in proportion to 1.00 g of leaves or root and 0.50 g of stem for 3 mL of buffer) to enhance the maceration and make the suspension homogenous. Maceration is the best technique used for isolating endophytic bacteria from surface-sterilized plant tissue, which is also helpful in identifying the wide range of cultivable bacterial and slow-growing fungal endophytes (Izumi et al. 2008; Souza et al. 2016; Schulz et al. 2006). After incubation, the isolated bacterial and fungal colonies can be identified via morphological and physiological characterization (by observing cell shape, colony color, and texture), fatty methyl ester analysis, and molecular methods based on particular gene sequencing of 16S rDNA.

4.5 *Preliminary In Vitro Antimicrobial Activity Test of Endophyte Isolates*

Endophytic microbes have currently gained more interest due to their enormous potential for producing novel bioactive molecules. Lots of endophytes, especially endophytic fungi, have produced many compounds biologically active against bacterial, fungal, and herbal species. The agar disc diffusion method is the preferred and frequently used one for preliminary screening assays of endophytic fungi and bacteria against various bacterial and fungal strains. For example, to assess the antibacterial activity of fungal endophytes, nutrient agar (NA) medium is dispensed into Petri dishes, inoculated with the bacterial suspension, and evenly dispersed with a sterile cotton bud. The endophytic fungal isolate cultured on PDA is then put on the surface of the NA medium that has already been seeded with the bacterial strain. Another PDA medium, empty of any fungal colony, can serve as a negative control. The plates are subjected to incubation at 37 °C for 24 h (depending on the bacterial growth nature). The antibacterial activity, after incubation, is proven by visualizing and measuring the inhibition zones in replicate (Izumi et al. 2008; Souza et al. 2016; Schulz et al. 2006).

4.6 Endophyte Isolates Fermentation and Extraction of Bioactive Compounds

The fungal and bacterial endophytes showing antimicrobial potential in primary screening are submitted to the fermentation process to scale up the yield. Briefly, each active endophytic fungus is picked from actively growing margins and fermented in an Erlenmeyer flask with PDA for some hours at room temperature (depending on the experiment protocol) in static conditions free of antibiotics. The culture broth, after incubation, is filtered through a double layer of sterile muslin cloth in order to harvest mycelium. The same procedure can also be applied to the endophytic bacterial isolate using Mueller-Hinton agar. The culture filtrates are then extracted with water and ethyl acetate, followed by solvent evaporation. The search for bioactive agents from fungal and bacterial endophytes is more advantageous and reproducible to scale up the products since the endophytic microbes can be supplied continuously via the fermentation process. Thus, endophytes can be considered as green and renewable sources of enough amounts of natural products, thereby helping to conserve the ecosystem of the world. In addition, the fermentation of endophytes reduces the overharvesting of plants by supplying a sufficient amount of secondary metabolites, thereby conserving the natural ecosystem (Izumi et al. 2008; Souza et al. 2016; Schulz et al. 2006).

4.7 Secondary Assay and Activity-Guided Identification of Pure Compounds

The endophytic fungal and bacterial isolates with a comparatively wider antimicrobial range displayed in the initial assay and fermented are designated for the secondary assay. The secondary antimicrobial test is performed using the same technique as the preliminary screening assay mentioned above. That is, the culture filtrate and ethyl acetate extracts are equidistantly placed on NA plates containing the inoculated bacterial and fungal strains. The plates can be incubated at 37 °C for 24 h and at 28 °C for 48 h for antibacterial and antifungal activities, respectively, and antimicrobial activity is determined by visualization and inhibition zone measurement. The active ethyl acetate extract is then subjected to different chromatographic techniques to fractionate and isolate the compounds responsible for the observed activity. Each collected fraction is also subjected to a biological assay using the approaches mentioned above. The active and pure fractions are then subjected to spectroscopic analysis (Izumi et al. 2008; Souza et al. 2016; Schulz et al. 2006).

5 Concluding Notes and Future Prospects

This day, the toxicity, moderate biological potency, and low yield of medicinal plant-based natural products are some of the challenges in world health medicine. Due to this, researchers have shifted their attention toward other alternatives, in which endophytes are at the frontier. Endophytes are one possible source of new bioactive molecules considered to be the counterparts of their host plants. Many studies have proved that endophytic microbes produce novel natural drugs as a means of resistance against pathogens. Till now, plenty of research has been done on endophytes, and a great number of bioactive natural products have been found. However, the bioactive compounds that have been isolated so far are limited to a few endophyte species. Besides, the uncultivable nature and difficulty of knowing the biosynthesis, regulation, and intermediates, as well as the unclear interaction with the host plant of the endophytes and their natural products, are among the challenges researchers have faced in the endophytic studies. In the near future, therefore, there may be a great probability of obtaining new and safe chemical agents with better activity and low toxicity from endophytes by targeting the active one using sensitive screening, isolation, purification, and identification methods. To do so, it is our hope that the present compiled review may give an insight to those who are interested in the isolation of endophytic microbes and associated bioactive secondary metabolites and may also shed light on the study of endophytes in the context of some developing countries, where research works in this field are almost nonexistent.

Competing Interests The authors declare no conflict of interests regarding the financial and content of this chapter.

Funding Statement We thank Adama Science and Technology University for the research fund with grant number: ASTU/AS-R/003/2020. CJ was partially supported from the National Research Foundation of Korea (NRF-2018R1A6A1A03024862) grant. AD is grateful to the World Academy of Sciences (TWAS) and the United Nations Educational, Scientific and Cultural Organization (UNESCO) for supporting this research with funds under the TWAS Research Grant Award_20–274 RG/CHE/AF/AC_G–FR3240314163).

References

Abdou R, Scherlach K, Dahse H-M, Sattler I, Hertweck C (2010) Botryorhodines A–D, antifungal and cytotoxic depsidones from Botryosphaeria rhodina, an endophyte of the medicinal plant Bidens pilosa. Phytochemistry 71(1):110–116

Adames I, Ortega HE, Asai Y, Kato M, Nagaoka K, TenDyke K et al (2015) 3-epi-Waol a and Waol C: polyketide-derived γ-lactones isolated from the endophytic fungus Libertella blepharis F2644. Tetrahedron Lett 56(1):252–255

Ai W, Lin X, Wang Z, Lu X, Mangaladoss F, Yang X et al (2015) Cladosporone a, a new dimeric tetralone from fungus Cladosporium sp. KcFL6'derived of mangrove plant Kandelia candel. J Antibiot 68(3):213–215

Alvin A, Miller KI, Neilan BA (2014) Exploring the potential of endophytes from medicinal plants as sources of antimycobacterial compounds. Microbiol Res 169(7–8):483–495

Bai Z-Q, Lin X, Wang J, Zhou X, Liu J, Yang B et al (2015) New meroterpenoids from the endophytic fungus aspergillus flavipes AIL8 derived from the mangrove plant Acanthus ilicifolius. Mar Drugs 13(1):237–248

Basha NS, Ogbaghebriel A, Yemane K, Zenebe M (2012) Isolation and screening of endophytic fungi from Eritrean traditional medicinal plant Terminalia brownii leaves for antimicrobial activity. Int J Green Pharmacy 6(1):40

Bhagobaty R, Joshi S (2012) Enzymatic activity of fungi endophytic on five medicinal plant species of the pristine sacred forests of Meghalaya, India. Biotechnol Bioprocess Eng 17(1):33–40

Birla S, Tiwari V, Gade A, Ingle A, Yadav A, Rai M (2009) Fabrication of silver nanoparticles by Phoma glomerata and its combined effect against *Escherichia coli, Pseudomonas aeruginosa and Staphylococcus aureus*. Lett Appl Microbiol 48(2):173–179

Cao P, Yang J, Miao C-P, Yan Y, Ma Y-T, Li X-N et al (2015) New duclauxamide from Penicillium manginii YIM PH30375 and structure revision of the duclauxin family. Org Lett 17(5):1146–1149

Chen M-J, Fu Y-W, Zhou Q-Y (2014) Penifupyrone, a new cytotoxic funicone derivative from the endophytic fungus Penicillium sp. HSZ-43. Nat Prod Res 28(19):1544–1548

Chen Z-M, Chen H-P, Li Y, Feng T, Liu J-K (2015) Cytochalasins from cultures of endophytic fungus Phoma multirostrata EA-12. J Antibiot 68(1):23–26

de Mendonca AN, e Feliphe BHMP, Lemes RML, Ruiz ALTG, de Carvalho JE, Ikegaki M (2015) Potential antimicrobial and antiproliferative activity of the crude extract of the endophytic fungus Rhizoctonia sp. from Annona crassiflora. Afr J Pharm Pharmacol 9(10):327–332

Delaye L, García-Guzmán G, Heil M (2013) Endophytes versus biotrophic and necrotrophic pathogens—are fungal lifestyles evolutionarily stable traits? Fungal Divers 60(1):125–135

Du X-P, Su W-J (2014) Two new polyketides from mangrove endophytic fungus Dothiorella sp. Chem Nat Compd 50(2):214–216

Ebrahim W, Kjer J, El Amrani M, Wray V, Lin W, Ebel R et al (2012) Pullularins E and F, two new peptides from the endophytic fungus Bionectria ochroleuca isolated from the mangrove plant Sonneratia caseolaris. Mar Drugs 10(5):1081–1091

Elavarasi A, Rathna GS, Kalaiselvam M (2012) Taxol producing mangrove endophytic fungi fusarium oxysporum from Rhizophora annamalayana. Asian Pac J Trop Biomed 2(2):S1081–S10S5

Fang ZF, Yu SS, Zhou WQ, Chen XG, Ma SG, Li Y et al (2012) A new isocoumarin from metabolites of the endophytic fungus Alternaria tenuissima (Nees & T. Nees: Fr.) Wiltshire. Chin Chem Lett 23(3):317–320

Feng Y, Ren F, Niu S, Wang L, Li L, Liu X et al (2014) Guanacastane diterpenoids from the plant endophytic fungus Cercospora sp. J Nat Prod 77(4):873–881

Gaur R, Tiwari S, Jakhmola A, Thakur JP, Verma RK, Pandey R et al (2014) Novel biotransformation processes of artemisinic acid to their hydroxylated derivatives 3β-hydroxyartemisinic acid and 3β, 15-dihydroxyartemisinic by fungus Trichothecium roseum CIMAPN1and their biological evaluation. J Mol Catal B: Enzym 106:46–55

Giridharan P, Verekar SA, Khanna A, Mishra P, Deshmukh SK (2012) Anticancer activity of sclerotiorin, isolated from an endophytic fungus Cephalotheca faveolata Yaguchi, Nishim & Udagawa. Indian J Exp Biol 50:464

Gu W, Ge H, Song Y, Ding H, Zhu H, Zhao X et al (2007) Cytotoxic benzo [j] fluoranthene metabolites from Hypoxylon truncatum IFB-18, an endophyte of Artemisia annua. J Nat Prod 70(1):114–117

Hallmann J, Mekete T, Sikora R, Kiewnick S (2009) Endophytic bacteria from Ethiopian coffee plants and their potential to antagonise Meloidogyne incognita. Nematology 11(1):117–127
Hammerschmidt L, Ola A, Mueller WE, Lin W, Mandi A, Kurtan T et al (2015) Two new metabolites from the endophytic fungus Xylaria sp. isolated from the medicinal plant Curcuma xanthorrhiza. Tetrahedron Lett 56(10):1193–1197
Heinig U, Scholz S, Jennewein S (2013) Getting to the bottom of Taxol biosynthesis by fungi. Fungal Divers 60(1):161–170
Izumi H, Anderson IC, Killham K, Moore ER (2008) Diversity of predominant endophytic bacteria in European deciduous and coniferous trees. Can J Microbiol 54(3):173–179
Jouda J-B, Kusari S, Lamshöft M, Talontsi FM, Meli CD, Wandji J et al (2014) Penialidins A–C with strong antibacterial activities from Penicillium sp., an endophytic fungus harboring leaves of Garcinia nobilis. Fitoterapia 98:209–214
Kaul A, Schuster E, Jennings BH (2014) The Groucho co-repressor is primarily recruited to local target sites in active chromatin to attenuate transcription. PLoS Genet 10(8):e1004595
Kemkuignou BM, Lambert C, Schmidt K, Schweizer L, Anoumedem EGM, Kouam SF et al (2023) Unreported cytochalasins from an acid-mediated transformation of cytochalasin J isolated from Diaporthe cf. ueckeri. Fitoterapia 166:105434
Kumar S, Kaushik N (2013) Endophytic fungi isolated from oil-seed crop Jatropha curcas produces oil and exhibit antifungal activity. PLoS One 8(2):e56202
Kumara PM, Zuehlke S, Priti V, Ramesha BT, Shweta S, Ravikanth G et al (2012) Fusarium proliferatum, an endophytic fungus from Dysoxylum binectariferum hook. F, produces rohitukine, a chromane alkaloid possessing anti-cancer activity. Antonie Van Leeuwenhoek 101(2):323–329
Kumaresan S, Karthi V, Senthilkumar V, Balakumar B, Stephen A (2015) Biochemical constituents and antioxidant potential of endophytic fungi isolated from the leaves of *Azadirachta indica* a. Juss (neem) from Chennai, India. J Acad Ind Res 3(8):355–361
Kusari S, Verma VC, Lamshoeft M, Spiteller M (2012) An endophytic fungus from Azadirachta indica a. Juss. That produces azadirachtin. World J Microbiol Biotechnol 28(3):1287–1294
Lai D, Brötz-Oesterhelt H, Müller WE, Wray V, Proksch P (2013) Bioactive polyketides and alkaloids from *Penicillium citrinum*, a fungal endophyte isolated from *Ocimum tenuiflorum*. Fitoterapia 91:100–106
Li G, Kusari S, Lamshöft M, Schüffler A, Laatsch H, Spiteller M (2014a) Antibacterial secondary metabolites from an endophytic fungus, Eupenicillium sp. LG41. J Nat Prod 77(11):2335–2341
Li C, Gong B, Cox DG, Li C, Wang J, Ding W (2014b) Dichlorodiaportinol A–A new chlorine-containing isocoumarin from an endophytic fungus Trichoderma sp. 09 from Myoporum bontioides A. Gray and its cytotoxic activity. Pharmacogn Mag 10(Suppl 1):S153
Liang Z, Zhang T, Zhang X, Zhang J, Zhao C (2015) An alkaloid and a steroid from the endophytic fungus Aspergillus fumigatus. Molecules 20(1):1424–1433
Mahdi T, Mohamed I, Yagi S (2014) Endophytic fungal communities associated with ethno-medicinal plants from Sudan and their antimicrobial and antioxidant prospective. Ecosystems 4:5
Martinez-Klimova E, Rodríguez-Peña K, Sánchez S (2017) Endophytes as sources of antibiotics. Biochem Pharmacol 134:1–17
Meng L-H, Li X-M, Lv C-T, Huang C-G, Wang B-G (2014) Brocazines A–F, cytotoxic bisthiodiketopiperazine derivatives from Penicillium brocae MA-231, an endophytic fungus derived from the marine mangrove plant Avicennia marina. J Nat Prod 77(8):1921–1927
Nath A, Chattopadhyay A, Joshi S (2015) Biological activity of endophytic fungi of Rauwolfia serpentina Benth: an ethnomedicinal plant used in folk medicines in Northeast India. Proc Natl Acad Sci Sect B Biol Sci 85(1):233–240
Padhi L, Mohanta YK, Panda SK (2013) Endophytic fungi with great promises: a review. J Adv Pharm Educ Res 3(3):152
Patil M, Patil R, Maheshwari V (2015) Biological activities and identification of bioactive metabolite from endophytic aspergillus flavus L7 isolated from Aegle marmelos. Curr Microbiol 71(1):39–48

Patil RH, Patil MP, Maheshwari VL (2016) Bioactive secondary metabolites from endophytic fungi: a review of biotechnological production and their potential applications. In: Studies in natural products chemistry, vol 49. Elsevier, Amsterdam, pp 189–205

Prasad M, Dagar S (2014) Identification and characterization of endophytic bacteria from fruits like avacado and black grapes. Int J Curr Microbiol App Sci 3(8):937–947

Rodriguez R, White J Jr, Arnold AE, Redman aRa. (2009) Fungal endophytes: diversity and functional roles. New Phytol 182(2):314–330

Samaga PV, Rai VR (2016) Diversity and bioactive potential of endophytic fungi from Nothapodytes foetida, Hypericum mysorense and Hypericum japonicum collected from Western Ghats of India. Ann Microbiol 66(1):229–244

Schulz BJE, Boyle CJC, Sieber TN (2006) In: Sieber TN (ed) Microbial root endophytes, 9th edn. Springer, Berlin

Shweta S, Gurumurthy BR, Ravikanth G, Ramanan US, Shivanna MB (2013) Endophytic fungi from Miquelia dentata Bedd., produce the anti-cancer alkaloid, camptothecine. Phytomedicine 20(3–4):337–342

Singh D, Rathod V, Ninganagouda S, Hiremath J, Singh AK, Mathew J (2014) Optimization and characterization of silver nanoparticle by endophytic fungi Penicillium sp. isolated from Curcuma longa (turmeric) and application studies against MDR E. coli and S aureus. Bioinorg Chem Appl 2014:1

Souza IF, Napoleão TH, de Sena KX, Paiva PM, de Araújo JM, Coelho LC (2016) Endophytic microorganisms in leaves of Moringa oleifera collected in three localities at Pernambuco state, northeastern Brazil. Microbiol Res J Int 13:1–7

Stierle AA, Stierle DB (2000) Bioactive compounds from four endophytic Penicillium sp. of a Northwest Pacific yew tree. In: Studies in natural products chemistry, vol 24. Elsevier, Amsterdam, pp 933–977

Strobel G (2014) The use of endophytic fungi for the conversion of agricultural wastes to hydrocarbons. Biofuels 5(4):447–455

Sun J-F, Lin X, Zhou X-F, Wan J, Zhang T, Yang B et al (2014) Pestalols A–E, new alkenyl phenol and benzaldehyde derivatives from endophytic fungus Pestalotiopsis sp. AcBC2 isolated from the Chinese mangrove plant Aegiceras corniculatum. J Antibiot 67(6):451–457

Verekar SA, Mishra PD, Sreekumar ES, Deshmukh SK, Fiebig H-H, Kelter G et al (2014) Anticancer activity of new depsipeptide compound isolated from an endophytic fungus. J Antibiot 67(10):697–701

Verma VC, Kharwar RN, Gange AC (2010) Biosynthesis of antimicrobial silver nanoparticles by the endophytic fungus *Aspergillus clavatus*. Nanomedicine 5(1):33–40

Wani ZA, Ashraf N, Mohiuddin T, Riyaz-Ul-Hassan S (2015) Plant-endophyte symbiosis, an ecological perspective. Appl Microbiol Biotechnol 99(7):2955–2965

Xiao J, Zhang Q, Gao Y-Q, Shi X-W, Gao J-M (2014) Antifungal and antibacterial metabolites from an endophytic aspergillus sp. associated with Melia azedarach. Nat Prod Res 28(17):1388–1392

Xu Y-m, Espinosa-Artiles P, Liu MX, Arnold AE, Gunatilaka AL (2013) Secoemestrin D, a Cytotoxic Epitetrathiodioxopiperizine, and Emericellenes A–E, Five Sesterterpenoids from Emericella sp. AST0036, a fungal endophyte of Astragalus lentiginosus 1. J Nat Prod 76(12):2330–2336

Yadav A, Yadav K (2017) Exploring the potential of endophytes in agriculture: a minireview. Adv Plants Agric Res 6(4):00221

Yu H, Zhang L, Li L, Zheng C, Guo L, Li W et al (2010) Recent developments and future prospects of antimicrobial metabolites produced by endophytes. Microbiol Res 165(6):437–449

Zhang W, Xu L, Yang L, Huang Y, Li S, Shen Y (2014) Phomopsidone a, a novel depsidone metabolite from the mangrove endophytic fungus Phomopsis sp. A123. Fitoterapia 96:146–151

Zhao J, Li C, Wang W, Zhao C, Luo M, Mu F et al (2013) Hypocrea lixii, novel endophytic fungi producing anticancer agent cajanol, isolated from pigeon pea (Cajanus cajan [L.] M illsp.). J Appl Microbiol 115(1):102–113

Zheng C-J, Xu L-L, Li Y-Y, Han T, Zhang Q-Y, Ming Q-L et al (2013) Cytotoxic metabolites from the cultures of endophytic fungi from Panax ginseng. Appl Microbiol Biotechnol 97(17):7617–7625

Zhou M, Miao M-M, Du G, Li X-N, Shang S-Z, Zhao W et al (2014) Aspergillines A–E, highly oxygenated hexacyclic indole–tetrahydrofuran–tetramic acid derivatives from Aspergillus versicolor. Org Lett 16(19):5016–5019

Zikmundova M, Drandarov K, Bigler L, Hesse M, Werner C (2002) Biotransformation of 2-benzoxazolinone and 2-hydroxy-1, 4-benzoxazin-3-one by endophytic fungi isolated from Aphelandra tetragona. Appl Environ Microbiol 68(10):4863–4870

Importance of Traditional Medicines in the Treatment of Neurological Diseases

Reena Haobam, Pooja Devi Nongthombam, and Yaiphabi Chanu Tongbram

Contents

Abstract Since ancient times, traditional medicines have been used for the treatment of varied neurological diseases and symptoms as evidenced in the relics of Indian Ayurvedic medicines and Chinese traditional medicines. Numerous studies have shown the efficacy of various compounds derived from traditionally used medicinal plants in improving the conditions of patients with neurological disorders and in preclinical investigations of neurological disorders. Many medicinal plants

R. Haobam (✉) · P. D. Nongthombam · Y. C. Tongbram
Department of Biotechnology, Manipur University, Canchipur, Imphal, Manipur, India

© The Author(s), under exclusive license to Springer Nature Singapore Pte Ltd. 2024

A. Das Talukdar et al. (eds.), *Traditional Resources and Tools for Modern Drug Discovery*, Interdisciplinary Biotechnological Advances,
https://doi.org/10.1007/978-981-97-4600-2_20

and their isolated compounds have demonstrated anti-inflammatory and antioxidant effects in various peripheral nervous systems. Since oxidative stress and inflammation are considered to play a significant role in neurodegeneration and various brain pathogenesis, they provide a certain extent of protection against neurological disorders. Treatment with synthetic drugs is cost-effective; however, clinical trials have failed because of various adverse side effects. Moreover, many drugs available for the treatment of neurodegenerative diseases are for symptomatic treatments rather than for curing the diseases. This highlights the critical need for alternate medicines to treat such diseases. The main objective of this chapter is to provide a detailed comprehensive review of the potential of traditionally used medicines against neurological disorders and to inspire the development of improved, efficient, and innovative therapeutic strategies for the treatment of neurological diseases.

Keywords Oxidative stress · Inflammation · Traditional medicine · Medicinal plants · Phytocompounds · Neurological disorders

Abbreviations

6-OHDA	6-Hydroxydopamine
βA	Beta amyloid
ΔΨm	Mitochondrial membrane potential
AChE	Acetylcholinesterase
AD	Alzheimer's disease
AKT/GSKβ	Protein kinase B/glycogen synthase kinase β
BACE1	β-Site APP cleavage enzyme 1
BBB	Blood-brain barrier
CHM	Chinese herbal medicine
CNS	Central nervous system
COX-2	Cyclooxygenase-2
CREB	cyclic-AMP response element binding protein
GPx	Glutathione peroxidase
GSH	Glutathione
GSDMD	Gasdermin D
GSK3β	Glycogen synthase kinase-3 beta
ICV-STZ	Intracerebroventricular streptozotocin
IL-1β	Interleukin-1 beta
IL-6	Interleukin 6
INF-γ	Interferon gamma
iNOS	Inducible nitric oxide synthase
iPSCs	Induced pluripotent stem cells
JNK/SAPK	Jun amino-terminal kinases/stress-activated protein kinases
KLF4	Kruppel-like factor 4
MDA	Malondialdehyde
MPP	1-Methyl-4-phenylpyridinium
MPTP	1-Methyl-4-phenyl-1,2,3,6-tetrahydropyridine

NGF	Nerve growth factors
NF-κB	Nuclear factor kappa B
ND	Neurological diseases
NLRP3	Nucleotide-binding domain, leucine-rich-containing family, pyrin domain-containing-3
NO	Nitric oxide
OGD	Oxygen-glucose deprivation
PD	Parkinson's disease
P13K	Phosphoinositide 3-kinases
ROS	Reactive oxygen species
SIRT1	Sirtuin 1 protein
SNpc	Substantia nigra pars compacta
SOD	Superoxide dismutase
TCM	Traditional Chinese medicine
TH	Tyrosine hydroxylase
TNF-α	Tumor necrosis factor-alpha

1 Introduction

Traditional medicine is an ancient health and healing practice, knowledge, and religious belief that has been used over generations (Bussmann and Sharon 2006). Before modern medicine existed, traditional medicine played a significant role in the treatment of various ailments (Yuan et al. 2016). Ancient civilizations such as the Egyptians, Greeks, Romans, Chinese, and Indians had a prevalence of traditional medicinal methods (Sen et al. 2011). Traditional medicines, especially the Indian traditional system (Ayurveda) and traditional Chinese medicine (TCM), are the most ancient medical practices continuously being used even today. The Chinese have developed a holistic system of therapy and medicine over many centuries of practice and study in the diagnosis, treatment, and prevention of various diseases including cardiovascular, cancer, and neurological diseases (NDs) (Wang et al. 2013). In India, over 5000 years ago, Ayurveda originated as one of the oldest holistic treatment systems in the world (Mukherjee et al. 2012). In recent years, there has been an increasing consideration of the significance of traditional medicine because of its positive impacts on human health. Many countries have implemented measures to include traditional medicine in their healthcare system (Ikram et al. 2015). Plants possessing medicinal properties have been utilized in traditional folklore for a long history, mainly for neuroprotection, an increase in lifespan, and enhancement of cognitive function (Brimson et al. 2021). The potential of medicinal plants and compounds derived from it in the treatment of NDs has garnered a lot of attention in recent years. However, further investigations need to be conducted to determine the effectiveness, safety, and tolerability of various herbal formulations.

NDs affect millions of people each year throughout the world. The most prevalent NDs are Alzheimer's disease (AD) and Parkinson's disease (PD) (Velmurugan et al. 2018). In AD, memory loss, cognitive impairment, disorientation, aggressiveness, and mood swings lead to impairment in behavior and functionality (Cummings 2000). The global population of individuals affected by AD currently exceeds 50 million (Pritam et al. 2022). It is projected that the number of people affected will double every 5 years, resulting in an estimated total of over 150 million AD patients by the year 2050 (Pritam et al. 2022). PD is a movement disorder caused by the selective death of SNpc dopaminergic neurons, decreasing nigrostriatal dopamine levels (Tamano et al. 2018). The current therapeutic approaches that focus on neurotransmitters, enzyme replacement, or regulation have shown limited efficacy in providing short-term symptomatic relief; however, it has not been successful in slowing down the progression of the diseases (Majumder and George 2023).

Many people are shifting their attention to traditional medicine as it is based on age-old practices, beliefs, and experiences with minimal risk of adverse effects, a vast range of potential therapeutic benefits, cost-effectiveness, and easy availability (Gyasi et al. 2016).

2 Pathophysiology of Neurological Disorders

In ND, misfolded protein aggregation is considered the most prevalent pathogenic mechanism (Soto 2003). AD is an age-related, progressive, and incurable ND that leads to memory loss, behavioral changes, and a decline in cognitive function (Zvěřová 2019). It is characterized by the aberrant accumulation of amyloid-beta protein and the development of neurofibrillary tangles composed of abnormal tau proteins within the brain (Zvěřová 2019). Currently, there is a lack of effective treatment that can reverse the adverse effects associated with AD. PD is considered the second most prevalent neurological disorder after AD (Chin-Chan et al. 2015). It is characterized by the gradual deterioration of dopaminergic neurons in the region of SNpc, resulting in impaired movement (Chin-Chan et al. 2015). Misfolding of alpha-synuclein proteins leads to the formation of Lewy bodies that help in dopaminergic neuron degeneration (Fahn 2003). Bradykinesia, hypokinesia, resting tremors, stiffness, and non-motor symptoms, including psychosis, are common symptoms observed in PD patients (Fahn 2003). Levodopa is currently the most effective drug for alleviating PD symptoms (Schapira et al. 2009). However, the exact etiology of numerous NDs remains unknown.

2.1 Oxidative Stress and Neuroinflammation in the Pathophysiology of Neurological Disorders

Oxidative stress and neuroinflammation have been hypothesized as major factors that cause neuronal degeneration in both AD and PD (Gao et al. 2014). Reactive oxygen species (ROS) are produced endogenously within the human body due to oxygen consumption and processes related to cellular signaling (Chen and Zhong 2014). During normal conditions, the body's natural antioxidant system controls the ROS balance (Chen and Zhong 2014). On the other hand, under pathological circumstances, a disruption in the equilibrium between the generation and reduction of ROS leads to an increase in ROS levels (Sayre et al. 2008). The process of oxidative phosphorylation produces oxidants and superoxide radicals, leading to mitochondria as the main source of ROS production in the cells (Cadenas and Davies 2000). The misfolding and aggregation of proteins, such as amyloid beta and alpha-synuclein, can be stimulated by ROS (Abramov et al. 2020). One of the first changes seen in the etiology of AD is oxidative stress in the brain, which may play a significant role in AD development (Kamat et al. 2016). Neurons in the SN can be damaged by oxidative stress, leading them to become susceptible to degeneration in PD (Trist et al. 2019). ROS are also capable of targeting and damaging the lipids found in cell membranes. Because of this, cell membranes may lose their fluidity and integrity, which may interfere with multiple signaling pathways and functions (Juan et al. 2021).

A neuroinflammatory response, characterized by microglial activation and increased in the production of pro-inflammatory cytokines, can be triggered by oxidative stress in the brain (Leszek et al. 2016). Neuroinflammation promotes AD development and accompanies plaque and tangle formation, impairing learning and memory (Wright et al. 2013). Activated microglia and astrocytes cause chronic neuroinflammation by releasing proinflammatory cytokines, such as IL-1β, TNF-α, and INF-γ, affecting the brain's tissues (Wang et al. 2015). These conditions were observed in AD patients (Wang et al. 2015). PD pathogenesis is also associated with chronic neuroinflammation. Since neuroinflammation affects dopaminergic pathways in the nigrostriatal region, it might have an important role in PD development (Leszek et al. 2016). In PD patients, activated microglia were determined in the striatum and substantia nigra, and their cerebrospinal fluid contains higher levels of pro-inflammatory mediators such as IL-1β and TNF-α (Illes 2020). So, there is considerable interest in investigating the neuroinflammatory signaling pathways within the CNS as potential targets for pharmacotherapy (Schwenkgrub et al. 2017). Hence, oxidative stress and neuroinflammation are gaining more attention as possible factors contributing to the pathophysiology of neurological disorders.

Medicinal plants were found to have beneficial effects on human health as they possess different classes of secondary metabolites, including phytochemical components like phenolics, flavonoids, alkaloids, terpenoids, etc. (Swallah et al. 2020). Antioxidant and anti-inflammatory activities are the most important and extensively studied aspects of phytoconstituents (Forni et al. 2019). This protective mechanism can halt the progression of various neurological disorders. Figure 1 illustrates the importance of medicinal plants and phytocompounds in the regulation of oxidative stress and neuroinflammation.

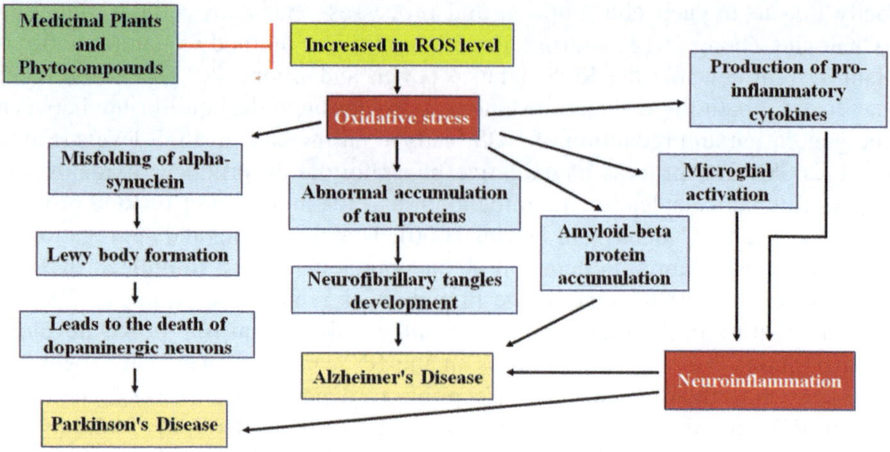

Fig. 1 The role of medicinal plants and phytocompounds in oxidative stress and neuroinflammation in the context of Parkinson's and Alzheimer's disease

3 Role of Medicinal Plants in Neuroprotection

Traditional medical systems have extensively used many plants over the centuries, including those with medicinal properties, to treat various diseases, including neurological disorders (Adams et al. 2007). Neuroprotection refers to various measures and agents that halt or reduce the progression of neurological disorders and protect the nervous system from degeneration (Mandel et al. 2003). Various medicinal plants possess antioxidants, including flavonoids, polyphenols, and vitamins, that play a major role in protecting nerve cells against oxidative stress and cellular damage induced by free radicals (Cui et al. 2020). The potential benefits of these compounds have gained more attention as they possess polyphenols, flavonoids, carotenoids, alkaloids, and others that help to improve brain health and overcome the risk of neurological disorders (John et al. 2020; Rahman et al. 2021). Inflammation has been linked to neurological conditions (Wright et al. 2013; Leszek et al. 2016). Certain medicinal plants possess anti-inflammatory activities, so this will help to slow down the disease progression of CNS disorders (Wright et al. 2013; Leszek et al. 2016). Compounds such as curcumin found in turmeric and resveratrol contained in grapes can inhibit the production of molecules that promote inflammation, thereby lowering the progression of neurological disorders (Velmurugan et al. 2018; John et al. 2020). The potential of certain phytomedicines to inhibit the accumulation of amyloid plaques and tau aggregates has been studied, considering that this could contribute to neuroprotection in AD. Damage to nerve cells and the development of NDs can result from prolonged exposure to environmental toxins and neurotoxic substances (Nabi and Tabassum 2022). Certain medicinal plants may be able to protect NDs from these toxins. For example, plants like

B. monnieri, contain compounds with neuroprotective effects against toxins like paraquat and aluminum, which might be effective in combating environmental neurotoxicity (Hosamani et al. 2016; Jyoti et al. 2007). Although many antioxidant-rich natural products have neuroprotective effects, very few of them have been observed to promote neuronal regeneration like bacosides of *B. monnieri* (Shalini et al. 2021), salvianolic acid B (Shu et al. 2018), berberine (Chen et al. 2020), and *Ginkgo biloba* (Tchantchou et al. 2009). Several studies have investigated the potential of certain phytomedicines to modulate the BBB, which might help with the specific delivery of therapeutic compounds to the brain and some compounds have the potential to cross the BBB such as withanamides (Vareed et al. 2014), huperzine A (De La Garza et al. 2016), baicalein (Liang et al. 2017), and berberine (Wang et al. 2005). Research has shown that some plants, including *Curcuma longa*, *B. monnieri*, and *Ginkgo biloba*, can stimulate neurogenesis (Shalini et al. 2021; Sasikumar et al. 2022; Ude et al. 2013).

4 Neuroprotective Activity of Medicinal Plants and Natural Compounds Used for NDs

In the past decades, numerous studies have emerged addressing the potential impact of traditionally used medicinal plants on various disorders. The origins of medicine in India may be traced back to ancient times, as evidenced by early references to the medicinal properties of plants in the Rig Veda, a renowned compilation of Sanskrit hymns, created around 4500 to 1600 B.C. (Iriti et al. 2010). Ayurvedic scriptures have highlighted various herbs and plant-based products for treating many diseases, including CNS disorders (Mythri and Srinivas Bharath 2012). Among other herbs, ashwagandha and Brahmi are considered to promote the proper functioning of the nerves (Dubey and Chinnathambi 2019). Ayurveda-based drug development seems promising and faster because the formulations/drugs are determined based on their extensive use in the population and verified in clinical trials (Dubey and Chinnathambi 2019). In order to ensure a holistic strategy to manage neurological disorders, it would be beneficial to get counsel from a certified Ayurvedic practitioner and a neurologist. For many thousands of years, TCM, which includes Chinese herbal medicine (CHM), has been used in therapeutic strategies for treating neurological diseases (Sreenivasmurthy et al. 2017). Chinese herbs and their active molecules might play a significant role in many signaling pathways involved in neurogenesis (Sreenivasmurthy et al. 2017). Based on these perspectives, TCM could be used in treating many neurological disorders. Here, we have outlined some of the Indian and Chinese medicinal plants and some active components used for AD and PD. The role of traditional Indian and Chinese medicinal plants in neurological disorders is detailed in Table 1. The role of some natural compounds in neurological disorders are also presented in Table 2.

Table 1 List of some well-characterized Indian and Chinese traditional medicinal plants with neuroprotective properties

Sl. no.	Plant name	Active components	Mechanism of action	Neurological disorders	References
1	*Curcuma longa*	Curcumin	Curcumin hinders the aggregation of beta-amyloid protein and enhances macrophage activity. It also decreases mitochondrial depolarization, cytochrome c release, ROS, and caspase 3/9 activation. Curcumin downregulates proinflammatory mediators	AD, PD, schizophrenia, epilepsy, etc.	Adami and Bottai (2022), Liu et al. (2011) and Selim et al. (2019)
2	*Bacopa monnieri*	Bacosaponins A, B, and C; bacopasides III–V; and bacoside A and B	Bacoside A boosts the antioxidant defense system and improves memory *B. monnieri* improves cognitive function and memory by inhibiting AChE activity *B. monnieri* reduces lipid peroxidation and aggregation of beta-amyloid protein *B. monnieri* also lowers oxidative stress and regenerates nerve cells	AD, PD, epilepsy, mental illness, asthma, rheumatism, and insomnia	Rao et al. (2012), Banerjee et al. (2021) and Aguiar and Borowski (2013)
3	*Withania somnifera*	Withanone, withanolides, isopelletierine, sitoindoside VII and VIII, and withanamides	Withanone suppressed the formation of beta-42 Root extracts of the plant alleviates PD symptoms and protects dopaminergic neurons. Withanamides possess the capability to cross the BBB	AD, PD	Du et al. (2021), Bharti et al. (2016), Prakash et al. (2013) and Vareed et al. (2014)
4	*Convolvulus pluricaulis* Choisy	Convolamine and convolvine, scopoletin, scopoline, flavonoids, glycosides, and coumarins	*C. pluricaulis* reduces tau protein expression and reduces tau protein-induced oxidative stress *C. pluricaulis* reduces the levels of glutathione, GPx, MDA, and protein carbonyl	AD, PD	Olakkaran and Antony (2019) and Bihaqi et al. (2012)

5	*Ginkgo biloba*	EGb761 (terpenoids (ginkgolides A, B, C, J, and bilobalide), flavonoid glycosides (kaempferol, isorhamnetin, quercetin, etc.)) and organic acids	EGb761 attenuates pro-inflammatory and enhances anti-inflammatory effects It reverses amyloid pathology in AD model It inhibits the loss of striated dopamine level and TH-positive cells in SNpc, lipid peroxidation, and superoxide radical production It also activates CREB in hippocampal neurons	AD, PD	Wang et al. (2016), Wan et al. (2016), Rojas et al. (2008) and Tchantchou et al. (2009)
6	*Huperzia serrata*	*Lycopodium* (alkaloids) constitutes huperzine A (Hup A), huperzine B (Hup B), N-methyl-Hup B, huperzine, and carinatumin A and carinatumin B	Hup A shows non-cholinergic effects. It inhibits oxidative stress, mitochondrial dysfunction, and apoptosis Hup A decreases the overexpression of TNF-α and 1 L-1β, and it inhibits the activation of NFκB pathway	AD, schizophrenia	Wu et al. (2011), Sreenivasmurthy et al. (2017), Qian and Ke (2014), Wang et al. (2005), Tao et al. (2016) and Zhang et al. (2008)
7	*Scutellaria baicalensis* Georgi	Baicalein	Baicalein attenuates mitochondrial dysfunction, apoptosis, and lipid peroxidation and also increases GSH level. It modulates neuroinflammatory NLRP3/caspase-1/GSDMD pathway. It also hinders the aggregation of amyloid protein	PD	Rong-Xiu et al. (2017), Li et al. (2012), Lee et al. (2005), Rui et al. (2020), Lu et al. (2011) and Song et al. (2021b)
8	*Salvia miltiorrhiza*	Salvianolic acid A and B, danshensu, dihydrotanshinone, cryptotanshinone, tanshinone I, tanshinone IIA, and tanshinone IIB	Sal B enhances the endogenous enzymes SOD and GPx activities and inhibits GSK3β activity. It reduces the expression level of iNOS, TNF-α, IL-1β mRNA, and COX-2 and attenuates lipid peroxidation. It also regulates AKT/GSKβ signaling pathway Danshensu shows antioxidation by activating PI3K/AKT/Nrf2 signaling pathway	PD	Sreenivasmurthy et al. (2017), Zhang et al. (2016), Tang et al. (2016), Wang et al. (2010), Lee et al. (2013), Shu et al. (2018) and Wang et al. (2020)

Table 2 List of some well-characterized phytocompounds that are active against neurological diseases

Sl. no.	Phyto-compounds	Plant source	Mechanism of action	Neurological disorders	References
1	Berberine	*Berberis aristata*, *Coptis japonica*, *Coptis chinensis*, *Xanthorhiza simplicissima*, and *Hydrastis canadensis*	Berberine activates class III P13K/Beclin-1 pathway to promote autophagic clearance of Aβ and also decreases the BACE1 protein level. It reduces tau phosphorylation by inhibiting NF-κB signaling, lowers the levels of TNF-α and IL-1β, improves GSH activity, and reduces lipid peroxidation. It also ameliorates cytotoxicity by activating the signaling pathway of PI3K/Akt and antioxidation	AD, PD	Cheng et al. (2022), Huang et al. (2017), He et al. (2017) and Deng et al. (2020)
2	Quercetin	*Platycladus orientalis* (L.) Franco, *Ginkgo biloba* L., *Alpinia officinarum* Hance, *Styphnolobium japonicum* (L.) Schott, etc. It is also found in fruits and vegetables	Quercetin attenuates the expressions of KLF4 to alleviates the apoptotic rate. It reduces lipid peroxidation and increases antioxidant enzyme activities. Quercetin has a non-cholinergic effect. It also alleviates iNOS over-expression	PD, AD	Song et al. (2021b), Li et al. (2016), Xi et al. (2012), Nabavi et al. (2012), Adedara et al. (2017) and Zhang et al. (2011)
3	Resveratrol	*Polygonum cuspidatum* and *Ventrum grandiflorum*	Resveratrol activates SIRT1 protein and reduces oxidative damage and cell death Resveratrol reduces deposition of Aβ1–42 and inhibits NF-κB signaling pathway Resveratrol inhibits α-synuclein aggregation and amyloid beta formation and reduces inflammation and oxidative stress	AD, PD, hypertension	Gambini et al. (2015), Cao et al. (2018), Zhao et al. (2018), Zhang et al. (2018) and Ma et al. (2014)

4.1 Curcuma longa

Curcuma longa, also referred to as turmeric, a perennial rhizomatous herb, has been widely used in Ayurveda for its therapeutic effects on various ailments (Kocaadam and Şanlier 2017). *C. longa* is one of the spices that is commonly used In India and other Asian countries. The main active component of *C. longa* is curcumin, and it is obtained from the rhizome of the plant (Kocaadam and Şanlier 2017). The herb provides a diverse array of medical properties among which anti-inflammatory, anticancer, antioxidant effects, and significant potential to enhance immune response (Witika et al. 2021). Also, it shows beneficial impacts against NDs such as AD, PD, schizophrenia, and epilepsy (Adami and Bottai 2022). Curcumin plays an important role by downregulating the expression level of transcription factors, enzymes, and cytokines (Patel et al. 2020). Curcumin possesses the capacity to effectively hinder the aggregation of beta amyloid protein (Mishra and Palanivelu 2008). In AD, curcumin demonstrates its effects by enhancing macrophage activity (Mishra and Palanivelu 2008). In PD, curcumin exhibits its protective mechanism by reducing mitochondrial depolarization, release of cytochrome c, ROS, and caspase-3/9 activation (Liu et al. 2011). Curcumin administration downregulates proinflammatory mediators such as TNF-α, IL-1β, and IL-6 (Selim et al. 2019). Various analogs of curcumin have been synthesized and assessed because of its wide range of pharmacological properties. The curcumin analogs demonstrate inhibitory properties on AChE, tau protein, βA1–42 aggregation, and BACE1 (Patil et al. 2020). Additionally, certain analogs possess the capacity to scavenge free radicals and chelate metal ions (Ak and Gülçin 2008).

4.2 Bacopa monnieri

Bacopa monnieri, referred to as Brahmi or water hyssop, is a perennial creeping plant that grows in India (Singh 2013). It has been extensively used in Ayurveda for thousands of years as a cognitive enhancer and treatment for neuropharmacological disorders. It is categorized as a "medhya rasayana" in Ayurveda (Singh 2013). It has been used for centuries as a therapeutic candidate for many diseases including epilepsy, mental illness, asthma, rheumatism, and insomnia (Rao et al. 2012). The active components of *B. monnieri* are bacosaponins A, B, and C; bacopasides III–V; and bacoside A and B (Rao et al. 2012). It is used for its medicinal properties such as antioxidant, anti-stress, anti-inflammatory, and antimicrobial nature (Ratheesh et al. 2017). Bacoside A has the potential to boost the antioxidant defense system and improve memory. Bacosides slow the progression of PD and protect the cholinergic neurons in patients (Banerjee et al. 2021). *B. monnieri* has been observed to enhance cognitive function and memory in a rat model by inhibiting AChE activity (Aguiar and Borowski 2013). In addition, the administration of *B. monnieri* extract has demonstrated significant lipoxygenase activity, lipid peroxidation, and aggregation of beta-amyloid protein in the brain of AD mice model (Aguiar and Borowski 2013). *B. monnieri* lowers the intracellular oxidative stress and the levels of ROS, regenerates nerve cells, and boosts memory (Mathew et al. 2010). *B. monnieri* prophylaxis significantly enhanced oxidative equilibrium by restoring paraquat-induced ROS, malondialdehyde, and hydroperoxides levels and inhibiting mitochondrial dysfunction (Hosamani et al. 2016).

4.3 Withania somnifera

Withania somnifera, commonly referred to as ashwagandha or Indian ginseng, is a medicinal plant with neuroprotective properties and has been widely used in the Indian traditional system for over a period of 4000 years (RajaSankar et al. 2009). They are widely distributed in the subtropical areas. The main active components of *W. somnifera* are alkaloids, Withanone, withanolides, isopelletierine, and sitoindoside VII and VIII (Bharti et al. 2016). They are found to possess anti-inflammatory activity, antioxidant activity, cholinergic activity, reactive oxygen scavenging activity, and body defense mechanism-enhancing properties (Zahiruddin et al. 2020). The root extract of ashwagandha has recently gained attention as a potential treatment for NDs. Recent studies have shown that the root extracts of ashwagandha and its compound have the potential to reduce oxidative stress and neuroinflammation (Bharti et al. 2016). Many researchers have also highlighted the impact of *W. somnifera* and its phytoconstituents in AD and PD. The compound Withanone exhibits significant potency by suppression of amyloid beta-42 formation. Withanone (WS-2) inhibits the amyloid-beta formation in the brain (Du et al. 2021). It also prevents cognitive deficits and reduces inflammation and oxidative stress in AD rats (Pandey et al. 2018). The root extract of ashwagandha protects dopaminergic neurons by inhibiting apoptotic pathways and oxidative stress in mice model of PD (Prakash et al. 2013). This root extract alleviates the Parkinson-like symptoms in PD mice (Prakash et al. 2013). The compound withanamides exhibits the capability to cross the BBB in AD mice model (Vareed et al. 2014). The abovementioned properties of *W. somnifera* and its compounds may contribute to its neuroprotective effects.

4.4 Convolvulus pluricaulis *Choisy*

Convolvulus pluricaulis, or Shankhapushpi, is a perennial herb characterized by its prostrate or suberect dispersing hairs (Kulkarni et al. 2012). This plant species is predominantly found in the plains of India (Kulkarni et al. 2012). Its active components are alkaloids such as convolamine and convolvine, scopoletin, scopoline, flavonoids, glycosides, and coumarins (Olakkaran and Antony 2019). Since ancient times, *C. pluricaulis* has been regarded as a brain tonic that may slow down aging and facilitate the regeneration of neural tissues (Olakkaran and Antony 2019). The plant has been categorized in "Medhya Rasayana" within the Ayurvedic tradition (Gupta and Fernandes 2019). In India, the aerial parts of *C. pluricaulis* have been widely used for centuries for various neurological disorders (Gupta and Fernandes 2019). In an in vivo AD model of *Drosophila*, *C. pluricaulis* reduces the expression of tau protein and is found to possess antioxidant properties and alleviate the oxidative stress generated by tau protein (Olakkaran and Antony 2019). The presence of alkaloids and flavonoids in C. pluricaulis extract acts as potent antioxidants, and it increases the survival rate and the overall lifespan of the animals (Olakkaran and Antony 2019). Scopolamine-induced amnesia was reversed by scopoletin and scopoline, and it also inhibits the activity of acetylcholinesterase (Malik et al. 2016). It is probable that the presence of scopoletin and scopoline in *C. pluricaulis* may be responsible for its ability to improve memory and might prevent amyloid-beta accumulation (Malik et al. 2016). *C. pluricaulis* shows neuroprotective activity and prevents aluminum-induced impairment of cholinergic and antioxidant activity in AD models (Bihaqi et al. 2009). *C. pluricaulis* also decreased the levels of glutathione, GPx, MDA, protein carbonyl, and GSH in rat brains of AD (Bihaqi et al. 2012). These characteristics of *C. pluricaulis* might play an important role in treating various NDs.

4.5 Ginkgo biloba

Ginkgo biloba has been used traditionally in Chinese herbal medicine. It is considered one of the oldest living trees. The standardized leaf extract from the *G. biloba* tree, labeled EGb761, has many active compounds such as terpenoids (ginkgolides A, B, C, J, and bilobalide), flavonoid glycosides (kaempferol, isorhamnetin, quercetin, etc.), and organic acids (Wang et al. 2016). In EGb761, the major active compounds are flavonoids and terpenoids. The medicinal properties of *G. biloba* have been used in treating diseases for a long time, including varied neurological symptoms similar to those of AD, clinical symptoms of aging, and peripheral vascular disease (Sreenivasmurthy et al. 2017). Studies have proven that *G. biloba* possesses neuroprotective properties (Wang et al. 2016). One of the studies reported that EGb761 attenuates pro-inflammatory and enhances anti-inflammatory effects, thereby improving cognitive impairment and reversing amyloid pathology in AD model of mice (Wan et al. 2016). EGb761 inhibited loss of striated dopamine level and TH-positive cells in SNpc, lipid peroxidation, and superoxide radical production leading to the improvement of motor dysfunction in MPTP-induced PD model of mice (Rojas et al. 2008). Moreover, EGb761 inhibited neurotoxicity and attenuated the behavioral deficit in the 6-OHDA-induced PD model of rats (Kim et al. 2004). Among its bioactive constituents, quercetin and bilobalide induce neurogenesis and synaptogenesis through the activation of CREB in hippocampal neurons (Tchantchou et al. 2009). For the therapy of cognitive deficit, flavonoids, and terpenoids are the responsible pharmacological effects and can penetrate the BBB (Ude et al. 2013). All of these findings show that *G. biloba* could have therapeutic potential and be effective in treating NDs.

4.6 Huperzia serrata

Huperzia serrata, a club moss (Sreenivasmurthy et al. 2017) also referred to as Qian Ceng Ta, is widely used traditionally in Chinese herbal medicine (CHM) for the treatment of strain, swelling, fever, and contusions; to alleviate pain; and as an antidote for poisoning and even schizophrenia (Wu et al. 2011; Sreenivasmurthy et al. 2017). *Lycopodium* (alkaloids) an active compound of *H. serrata* constitutes huperzine A (Hup A), huperzine B (Hup B), N-methyl-Hup B, huperzinine, carinatumin A, and carinatumin B (Wu et al. 2011; Sreenivasmurthy et al. 2017). Hup A is the major constituent among them and a potent competitive reversible inhibitor of AChE (Wu et al. 2011; Sreenivasmurthy et al. 2017). Hup A can easily penetrate the BBB (De La Garza et al. 2016). In addition, Hup A shows non-cholinergic effects on AD (Qian and Ke 2014). Hup A can remarkably enhance cognitive deficit, memory impairment, and daily activities of AD patients (Wang et al. 2016). Studies have found that Hup A has a neuroprotective effect by exerting its properties against oxidative stress, mitochondrial dysfunction, and apoptosis induced by Aβ and iron overload (Wang and Tang 2005; Tao et al. 2016). In an *in vivo* model, Hup A can effectively decrease the overexpression of pro-inflammatory cytokines like TNF-α and 1 L-1β, and it inhibits activation of NFκB, attenuated iNOS, COX-2, and overexpression of NO in C6 cells induced by OGD (Zhang et al. 2008). Besides, Hup A increases neurite outgrowth and enhances the secretion of NGFs (Tang et al. 2005). Hence, Hup A could be a suitable therapeutic agent in treating NDs.

4.7 Scutellaria baicalensis *Georgi*

Scutellaria baicalensis Georgi is a TCM, which is an important herbaceous plant. Baicalein is the main bioactive flavone constituent from the dried root of *S. baicalensis* Georgi (Rong-Xiu et al. 2017). Studies have proven that baicalein has pharmacological properties. Baicalein was found to reduce oxidative stress and anti-inflammatory activity; inhibits the aggregation of disease-specific amyloid proteins, anti-apoptosis activity, and excitotoxicity; and stimulates the neurogenesis and differentiation of neuronal cells (Li et al. 2017). It can effectively penetrate the BBB (Liang et al. 2017). Baicalein attenuates rotenone-induced apoptosis by ameliorating the mitochondrial dysfunction caused by oxidative damage in PC12 cells (Li et al. 2012). It also significantly attenuates apoptosis caused by the generation of ROS, JNK/SAPK activation, and reduced ΔΨm level induced by 6-OHDA in SH-SY5Y cells (Lee et al. 2005). Baicalein administration effectively modulates neuroinflammatory NLRP3/caspase-1/GSDMD pathway induced by MPTP in PD mice model (Rui et al. 2020). Moreover, aggregation of amyloid protein is hindered by Baicalein by inhibiting both oligomerization and fibrillation of α-synuclein and Aβ (Lu et al. 2011). Baicalein ameliorates oxidative stress and intracellular ROS by increasing GSH levels and decreasing lipid peroxidation in MPP + -treated SH-SY5Y cells (Song et al. 2021a).

4.8 Salvia miltiorrhiza

Salvia miltiorrhiza is a herb used in TCM for treating cardiovascular, cerebrovascular, and neurological disorders (Zhang et al. 2016). Its bioactive ingredients are salvianolic acid A and B; danshensu; which are water-soluble phenolic compounds; and dihydrotanshinone, cryptotanshinone, tanshinone I, tanshinone IIA, and tanshinone IIB, which are lipophilic in nature. Among them, salvianolic acid B (Sal B) is the major ingredient in the water extract (Sreenivasmurthy et al. 2017; Zhang et al. 2016). Sal B has antioxidant properties as it attenuates the free radical-induced stress by enhancing the endogenous enzymes SOD and GPx activities and inhibiting GSK3β activity in SH-SY5Y-APPsw cells (Tang et al. 2016). Studies in *in vivo* models have reported that Sal B has anti-inflammatory activity by reducing iNOS, TNF-α, IL-1β mRNA, and COX-2 expression levels, and it also attenuates lipid peroxidation (Wang et al. 2010; Lee et al. 2013). It is proven that Sal B potentially improves the proliferation and differentiation of iPSCs into neurons by regulating the AKT/GSKβ signaling pathway, which would be a potential treatment strategy for neural disease using iPSCs-derived neural stem cells (NSCs) (Shu et al. 2018). Besides danshensu, one of its bioactive ingredients significantly ameliorates motor impairment and exerts neuroprotective effects by antioxidation via activating PI3K/AKT/Nrf2 signaling pathway in PD mice model (Wang et al. 2020).

4.9 *Berberine*

Berberine, a TCM, is an isoquinoline quaternary alkaloid active compound found in different medicinal plants such as *Berberis aristata*, *Coptis japonica*, *Coptis chinensis*, *Xanthorhiza simplicissima*, and *Hydrastis canadensis* (Cheng et al. 2022). It has many medicinal properties like anticancer, antidiabetic, antibacterial, anti-inflammatory, antidepressant, antiprotozoal, anti-hyperlipidemia, antidepressant, anti-HIV, antinociceptive, and used in the NDs treatment (Hussien et al. 2018; Singh et al. 2019). Berberine ameliorates memory impairment and enhances autophagy through the class III P13K/Beclin-1 pathway to promote autophagic clearance of Aβ. It also decreases the BACE1 protein level, which results in the inhibition of the Aβ production in a mouse model of AD (Huang et al. 2017). Berberine reduces tau phosphorylation by inhibiting NF-κB signaling, inflammation (lowered the expression ofTNF-α and IL-1β), and oxidative stress (improved GSH activity and reduced lipid peroxidation) to reduce cognitive impairment in AD mice (He et al. 2017). Berberine alleviates cytotoxicity in SH-SY5Y cells by activating the signaling pathway of PI3K/Akt and antioxidation (Deng et al. 2020). Studies have highlighted that berberine attenuates spatial memory impairment by inhibiting AChE in the ICV-STZ-induced AD model of rats. Furthermore, berberine promotes cell proliferation and reduced apoptosis, as well as reduced cell death by regulating miR-188/NOS1 in BV2 and N2a cells (Chen et al. 2020). These studies proved that berberine is a potential neuroprotective molecule, which is further supported by its permeability through the BBB (Wang et al. 2005).

4.10 *Quercetin*

Quercetin is a flavonoid bioactive compound that is widely distributed in the plant kingdom and is used in TCM for its potent antioxidant activity (Li et al. 2016; Song et al. 2021b; Xu et al. 2019). The major sources of quercetin in Chinese herbal medicines are *Platycladus orientalis* (L.) Franco, *Ginkgo biloba* L., *Alpinia officinarum* Hance, *Styphnolobium japonicum* (L.) Schott, etc. (Song et al. 2021b). It is also found in fruits and vegetables (Li et al. 2016). It has many potential benefits to resist various diseases, such as anticarcinogenic, anti-inflammatory, antiviral, antioxidant, and psychostimulant activities (Li et al. 2016; Xu et al. 2019). Quercetin inhibits the expression of apoptotic target genes of KLF4 by increasing the ratio of bcl-2 to bax expression, thus alleviating the apoptotic rate and caspase-3 enzyme activity in H_2O_2-induced oxidative stress in SH-SY5Y cells (Xi et al. 2012). Quercetin also reduces lipid peroxidation, increases antioxidant enzyme activities such as catalase and SOD, and increases reduced GSH levels in sodium fluoride (NaF)-induced oxidative stress in rat brains (Nabavi et al. 2012). Quercetin has a non-cholinergic effect as it inhibits AChE activity in manganese-induced increase AChE activity in rat brains, which might escalate the neurotransmission and consequently increase motor function and locomotion (Adedara et al. 2017). Quercetin alleviates iNOS overexpression, resulting in the reduction of NO over-production in 6-OHDA-induced inflammation in PC12 cells (Zhang et al. 2011).

4.11 *Resveratrol*

Resveratrol, a natural phenolic compound, belongs to the group of stilbenoids. This compound was found to be present in traditionally used medicinal plants such as *Polygonum cuspidatum* and *Ventrum grandiflorum* to treat cardiovascular diseases and hypertension (Gambini et al. 2015). Grapes, berries, and red wines are the main sources of resveratrol and its derivatives (Zhang et al. 2018). Resveratrol exhibits antioxidant, anticancer, anti-inflammatory, and cardioprotective properties (Cao et al. 2018). Resveratrol reduces oxidative damage and cell death by activating SIRT1 protein, which delays the biological process of aging (Cao et al. 2018). The antioxidant and anti-inflammatory property of resveratrol shows neuroprotective effects in neurological disorders such as AD and PD. Piceatannol, the derivative of resveratrol, has been shown to inhibit amyloid beta formation (Choi et al. 2016). Resveratrol also exhibits antioxidant, anti-aggregation of amyloid beta, and radical scavenging activities (Ma et al. 2014). Resveratrol has been observed to reduce inflammation generated by Aβ1–42 in rat astrocytes and microglial cell lines by inhibiting the NF-κB signal pathway (Zhao et al. 2018). Resveratrol inhibits α-synuclein aggregation and cytotoxicity and reduces inflammation and oxidative stress in the mouse model of PD (Zhang et al. 2018).

5 Advantages of Phytoconstituents Over Synthetic Drugs

The traditional medicine system used medicinal plants safely for centuries. Plant-derived compounds have advantages over synthetic drugs for treating neurological disorders. Plants possess a diverse range of active constituents that have been shown to have potential therapeutic applications. Since medicinal plants have a long record of use, plant compounds have often been considered safer and more effective than synthetic compounds or drugs (Tohma et al. 2017; Hussain et al. 2018). Plant-derived compounds draw attention to those who prefer natural or holistic approaches to their healthcare. On the other hand, synthetic drugs exhibit adverse effects due to their newly developed chemical structures. Most currently used synthetic drugs have poor biological stability, ineffective targeting, and cause unwanted side effects (Sgarbossa et al. 2015). New interest in phytochemicals for ND treatment is mainly due to the adverse side effects of synthetic drugs. Although plant-based medicines may help some neurological problems, it is important to remember that not all of them are effective.

6 Conclusion

The prevalence of NDs is increasing, and there is currently no cure for it. Many synthetic drugs have adverse side effects; hence, people are searching for bioactive compounds made from plants with fewer adverse effects. More attention has been focused on Ayurveda and Chinese traditional medicines for treating complex diseases. Moreover, it has a beneficial synergic effect in the treatment and prevention of physiological and pathological conditions of NDs. Research has shown that their antioxidant and anti-inflammatory characteristics are primarily responsible for neuroprotection, making them beneficial for therapeutic use. Hence, the bioactive compounds isolated from these traditional medicines could be an important source in developing drugs to combat the NDs. However, further studies are required to elucidate the cellular and molecular mechanisms underlying their neuroprotective properties. This could be further extended with clinical trials to promote the development of safe and effective drugs for treating NDs.

References

Abramov AY, Potapova EV, Dremin VV, Dunaev AV (2020) Interaction of oxidative stress and misfolded proteins in the mechanism of neurodegeneration. Life 10(7):101. https://doi.org/10.3390/life10070101

Adami R, Bottai D (2022) Curcumin and neurological diseases. Nutr Neurosci 25(3):441–461. https://doi.org/10.1080/1028415X.2020.1760531

Adams M, Gmünder F, Hamburger M (2007) Plants traditionally used in age related brain disorders—a survey of ethnobotanical literature. J Ethnopharmacol 113(3):363–381. https://doi.org/10.1016/j.jep.2007.07.016

Adedara IA, Ego VC, Subair TI, Oyediran O, Farombi EO (2017) Quercetin improves neurobehavioral performance through restoration of brain antioxidant status and acetylcholinesterase activity in manganese-treated rats. Neurochem Res 42:1219–1229. https://doi.org/10.1007/s11064-016-2162-z

Aguiar S, Borowski T (2013) Neuropharmacological review of the nootropic herb Bacopa monnieri. Rejuvenation Res 16(4):313–326. https://doi.org/10.1089/rej.2013.1431

Ak T, Gülçin I (2008) Antioxidant and radical scavenging properties of curcumin. Chem Biol Interact 174(1):27–37. https://doi.org/10.1016/j.cbi.2008.05.003

Banerjee S, Anand U, Ghosh S, Ray D, Ray P, Nandy S, Deshmukh GD, Tripathi V, Dey A (2021) Bacosides from Bacopa monnieri extract: an overview of the effects on neurological disorders. Phytother Res 35(10):5668–5679. https://doi.org/10.1002/ptr.7203

Bharti VK, Malik JK, Gupta RC (2016) Ashwagandha: multiple health benefits. In: Nutraceuticals. Academic Press, New York, pp 717–733. https://doi.org/10.1016/B978-0-12-802147-7.00052-8

Bihaqi SW, Sharma M, Singh AP, Tiwari M (2009) Neuroprotective role of convolvulus pluricaulis on aluminium induced neurotoxicity in rat brain. J Ethnopharmacol 124(3):409–415. https://doi.org/10.1016/j.jep.2009.05.038

Bihaqi SW, Singh AP, Tiwari M (2012) Supplementation of convolvulus pluricaulis attenuates scopolamine-induced increased tau and amyloid precursor protein (AβPP) expression in rat brain. Indian J Pharm 44(5):593. https://doi.org/10.4103/0253-7613.100383

Brimson JM, Brimson S, Prasanth MI, Thitilertdecha P, Malar DS, Tencomnao T (2021) The effectiveness of Bacopa monnieri (Linn.) Wettst. As a nootropic, neuroprotective, or antidepressant supplement: analysis of the available clinical data. Sci Rep 11(1):596. https://doi.org/10.1038/s41598-020-80045-2

Bussmann RW, Sharon D (2006) Traditional medicinal plant use in northern Peru: tracking two thousand years of healing culture. J Ethnobiol Ethnomed 2(1):1–18. https://doi.org/10.1186/1746-4269-2-47

Cadenas E, Davies KJ (2000) Mitochondrial free radical generation, oxidative stress, and aging. Free Radic Biol Med 29(3–4):222–230. https://doi.org/10.1016/S0891-5849(00)00317-8

Cao W, Dou Y, Li A (2018) Resveratrol boosts cognitive function by targeting SIRT1. Neurochem Res 43:1705–1713. https://doi.org/10.1007/s11064-018-2586-8

Chen Z, Zhong C (2014) Oxidative stress in Alzheimer's disease. Neurosci Bull 30:271–281. https://doi.org/10.1007/s12264-013-1423-y

Chen M, Li L, Liu C, Song L (2020) Berberine attenuates Aβ-induced neuronal damage through regulating miR-188/NOS1 in Alzheimer's disease. Mol Cell Biochem 474:285–294. https://doi.org/10.1007/s11010-020-03852-1

Cheng Z, Kang C, Che S, Su J, Sun Q, Ge T, Guo Y, Lv J, Sun Z, Yang W, Li B (2022) Berberine: a promising treatment for neurodegenerative diseases. Front Pharmacol 13:845591. https://doi.org/10.3389/fphar.2022.845591

Chin-Chan M, Navarro-Yepes J, Quintanilla-Vega B (2015) Environmental pollutants as risk factors for neurodegenerative disorders: Alzheimer and Parkinson diseases. Front Cell Neurosci 9:124. https://doi.org/10.3389/fncel.2015.00124

Choi B, Kim S, Jang BG, Kim MJ (2016) Piceatannol, a natural analogue of resveratrol, effectively reduces beta-amyloid levels via activation of alpha-secretase and matrix metalloproteinase-9. J Funct Foods 23:124–134. https://doi.org/10.1016/j.jff.2016.02.024

Cui X, Lin Q, Liang Y (2020) Plant-derived antioxidants protect the nervous system from aging by inhibiting oxidative stress. Front Aging Neurosci 12:209. https://doi.org/10.3389/fnagi.2020.00209

Cummings JL (2000) Cognitive and behavioral heterogeneity in Alzheimer's disease: seeking the neurobiological basis. Neurobiol Aging 21(6):845–861. https://doi.org/10.1016/S0197-4580(00)00183-4

De La Garza R, Verrico CD, Newton TF, Mahoney JJ III, Thompson-Lake DG (2016) Safety and preliminary efficacy of the acetylcholinesterase inhibitor huperzine a as a treatment for cocaine use disorder. Int J Neuropsychopharmacol 19(3):pyv098. https://doi.org/10.1093/ijnp/pyv098

Deng H, Jia Y, Pan D, Ma Z (2020) Berberine alleviates rotenone-induced cytotoxicity by antioxidation and activation of PI3K/Akt signaling pathway in SH-SY5Y cells. Neuroreport 31(1):41–47. https://doi.org/10.1097/WNR.0000000000001365

Du Z, Li M, Ren J, Qu X (2021) Current strategies for modulating Aβ aggregation with multifunctional agents. Acc Chem Res 54(9):2172–2184. https://doi.org/10.1021/acs.accounts.1c00055

Dubey T, Chinnathambi S (2019) Brahmi (Bacopa monnieri): an ayurvedic herb against the Alzheimer's disease. Arch Biochem 676:108153. https://doi.org/10.1016/j.abb.2019.108153

Fahn S (2003) Description of Parkinson's disease as a clinical syndrome. Ann N Y Acad Sci 991(1):1–14. https://doi.org/10.1111/j.1749-6632.2003.tb07458.x

Forni C, Facchiano F, Bartoli M, Pieretti S, Facchiano A, D'Arcangelo D, Norelli S, Valle G, Nisini R, Beninati S, Tabolacci C (2019) Beneficial role of phytochemicals on oxidative stress and age-related diseases. Biomed Res Int 2019:1. https://doi.org/10.1155/2019/8748253

Gambini J, Inglés M, Olaso G, Lopez-Grueso R, Bonet-Costa V, Gimeno-Mallench L, Mas-Bargues C, Abdelaziz KM, Gomez-Cabrera MC, Vina J, Borras C (2015) Properties of resveratrol: in vitro and in vivo studies about metabolism, bioavailability, and biological effects in animal models and humans. Oxid Med Cell Longev 2015:837042. https://doi.org/10.1155/2015/837042

Gao HM, Zhou H, Hong JS (2014) Oxidative stress, neuroinflammation, and neurodegeneration. In: Neuroinflammation and neurodegeneration, pp 81–104. https://doi.org/10.1007/978-1-4939-1071-7_5

Gupta GL, Fernandes J (2019) Protective effect of convolvulus pluricaulis against neuroinflammation associated depressive behavior induced by chronic unpredictable mild stress in rat. Biomed Pharmacother 109:1698–1708. https://doi.org/10.1016/j.biopha.2018.11.046

Gyasi RM, Asante F, Abass K, Yeboah JY, Adu-Gyamfi S, Amoah PA (2016) Do health beliefs explain traditional medical therapies utilisation? Evidence from Ghana. Cogent Soc Sci 2(1):1209995. https://doi.org/10.1080/23311886.2016.1209995

He W, Wang C, Chen Y, He Y, Cai Z (2017) Berberine attenuates cognitive impairment and ameliorates tau hyperphosphorylation by limiting the self-perpetuating pathogenic cycle between NF-κB signaling, oxidative stress and neuroinflammation. Pharmacol Rep 69:1341–1348. https://doi.org/10.1016/j.pharep.2017.06.006

Hosamani R, Krishna G, Muralidhara. (2016) Standardized Bacopa monnieri extract ameliorates acute paraquat-induced oxidative stress, and neurotoxicity in prepubertal mice brain. Nutr Neurosci 19(10):434–446. https://doi.org/10.1179/1476830514Y.0000000149

Huang M, Jiang X, Liang Y, Liu Q, Chen S, Guo YI (2017) Berberine improves cognitive impairment by promoting autophagic clearance and inhibiting production of β-amyloid in APP/tau/PS1 mouse model of Alzheimer's disease. Exp Gerontol 91:25–33. https://doi.org/10.1016/j.exger.2017.02.004

Hussain G, Rasul A, Anwar H, Aziz N, Razzaq A, Wei W, Ali M, Li J, Li X (2018) Role of plant derived alkaloids and their mechanism in neurodegenerative disorders. Int J Biol Sci 14(3):341. https://doi.org/10.7150/ijbs.23247

Hussien HM, Abd-Elmegied A, Ghareeb DA, Hafez HS, Ahmed HE, Abd El-moneam N (2018) Neuroprotective effect of berberine against environmental heavy metals-induced neurotoxicity and Alzheimer's-like disease in rats. Food Chem 111:432–444. https://doi.org/10.1016/j.fct.2017.11.025

Ikram RRR, Abd Ghani MK, Abdullah N (2015) An analysis of application of health informatics in traditional medicine: a review of four traditional medicine SystemsInt. J Med Inform 84(11):988–996. https://doi.org/10.1016/j.ijmedinf.2015.05.007

Illes P (2020) P2X7 receptors amplify CNS damage in neurodegenerative diseases. Int J Mol Sci 21(17):5996. https://doi.org/10.3390/ijms21175996

Iriti M, Vitalini S, Fico G, Faoro F (2010) Neuroprotective herbs and foods from different traditional medicines and diets. Molecules 15(5):3517–3555. https://doi.org/10.3390/molecules15053517

John T, Samuel B, Abolaji O, Folashade O, Oyetooke A, Oluwatosin F (2020) Functional foods and bioactive compounds: roles in the prevention, treatment and management of neurodegenerative diseases. GSC Biol Pharm Sci 11(2):297–313. https://doi.org/10.30574/gscbps.2020.11.2.0143

Juan CA, Pérez de la Lastra JM, Plou FJ, Pérez-Lebeña E (2021) The chemistry of reactive oxygen species (ROS) revisited: outlining their role in biological macromolecules (DNA, lipids and proteins) and induced pathologies. Int J Mol Sci 22(9):4642. https://doi.org/10.3390/ijms22094642

Jyoti A, Sethi P, Sharma D (2007) Bacopa monniera prevents from aluminium neurotoxicity in the cerebral cortex of rat brain. J Ethnopharmacol 111(1):56–62. https://doi.org/10.1016/j.jep.2006.10.037

Kamat PK, Kalani A, Rai S, Swarnkar S, Tota S, Nath C, Tyagi N (2016) Mechanism of oxidative stress and synapse dysfunction in the pathogenesis of Alzheimer's disease: understanding the therapeutics strategies. Mol Neurobiol 53:648–661. https://doi.org/10.1007/s12035-014-9053-6

Kim MS, Lee JI, Lee WY, Kim SE (2004) Neuroprotective effect of Ginkgo biloba L. extract in a rat model of Parkinson's disease. Phytother Res 18(8):663–666. https://doi.org/10.1002/ptr.1486

Kocaadam B, Şanlier N (2017) Curcumin, an active component of turmeric (Curcuma longa), and its effects on health. Crit Rev Food Sci Nutr 57(13):2889–2895. https://doi.org/10.1080/10408398.2015.1077195

Kulkarni R, Girish KJ, Kumar A (2012) Nootropic herbs (Medhya Rasayana) in Ayurveda: an update. Pharmacogn Rev 6(12):147. https://doi.org/10.4103/0973-7847.99949

Lee HJ, Noh YH, Kim YS, Kim KY, Chung YH, Lee WB, Kim SS (2005) Baicalein attenuates 6-hydroxydopamine-induced neurotoxicity in SH-SY5Y cells. Eur J Cell Biol 84(11):897–905. https://doi.org/10.1016/j.ejcb.2005.07.003

Lee YW, Kim DH, Jeon SJ, Park SJ, Kim JM, Jung JM, Lee HE, Bae SG, Oh HK, Son KHH, Ryu JH (2013) Neuroprotective effects of salvianolic acid B on an Aβ25–35 peptide-induced mouse model of Alzheimer's disease. Eur J Pharmacol 704(1–3):70–77. https://doi.org/10.1016/j.ejphar.2013.02.015

Leszek J, Barreto GE, Gąsiorowski K, Koutsouraki E, Ávila-Rodrigues M, Aliev G (2016) Inflammatory mechanisms and oxidative stress as key factors responsible for progression of neurodegeneration: role of brain innate immune system. CNS Neurol Disord Drug Targets 15(3):329–336. https://doi.org/10.2174/1871527315666160202125914

Li XX, He GR, Mu X, Xu B, Tian S, Yu X, Meng FR, Xuan ZH, Du GH (2012) Protective effects of baicalein against rotenone-induced neurotoxicity in PC12 cells and isolated rat brain mitochondria. Eur J Pharmacol 674(2–3):227–233. https://doi.org/10.1016/j.ejphar.2011.09.181

Li Y, Yao J, Han C, Yang J, Chaudhry MT, Wang S, Liu H, Yin Y (2016) Quercetin, inflammation and immunity. Nutrients 8(3):167. https://doi.org/10.3390/nu8030167

Li Y, Zhao J, Hölscher C (2017) Therapeutic potential of baicalein in Alzheimer's disease and Parkinson's disease. CNS Drugs 31:639–652. https://doi.org/10.1007/s40263-017-0451-y

Liang W, Huang X, Chen W (2017) The effects of baicalin and baicalein on cerebral ischemia: a review. Aging Dis 8(6):850. https://doi.org/10.14336/AD.2017.0829

Liu Z, Yu Y, Li X, Ross CA, Smith WW (2011) Curcumin protects against A53T alpha-synuclein-induced toxicity in a PC12 inducible cell model for parkinsonism. Pharmacol Res 63(5):439–444. https://doi.org/10.1016/j.phrs.2011.01.004

Lu JH, Ardah MT, Durairajan SSK, Liu LF, Xie LX, Fong WFD, Hasan MY, Huang JD, El-Agnaf OM, Li M (2011) Baicalein inhibits formation of α-synuclein oligomers within living cells and prevents Aβ peptide fibrillation and oligomerisation. ChemBioChem 12(4):615–624. https://doi.org/10.1002/cbic.201000604

Ma T, Tan MS, Yu JT, Tan L (2014) Resveratrol as a therapeutic agent for Alzheimer's disease. Biomed Res Int 2014:350516. https://doi.org/10.1155/2014/350516

Majumder S, George S (2023) Nanotechnology-enabled phytochemicals from Ayurvedic herbs for the treatment of neurodegenerative disorders: an overview. In: Ayurvedic herbal preparations in neurological disorders, pp 611–633. https://doi.org/10.1016/B978-0-443-19084-1.00001-6

Malik J, Karan M, Vasisht K (2016) Attenuating effect of bioactive coumarins from convolvulus pluricaulis on scopolamine-induced amnesia in mice. Nat Prod Res 30(5):578–582. https://doi.org/10.1080/14786419.2015.1025398

Mandel S, Grünblatt E, Riederer P, Gerlach M, Levites Y, Youdim MB (2003) Neuroprotective strategies in Parkinson's disease: an update on progress. CNS Drugs 17:729–762. https://doi.org/10.2165/00023210-200317100-00004

Mathew J, Paul J, Nandhu MS, Paulose CS (2010) Bacopa monnieri and Bacoside-A for ameliorating epilepsy associated behavioral deficits. Fitoterapia 81(5):315–322. https://doi.org/10.1016/j.fitote.2009.11.005

Mishra S, Palanivelu K (2008) The effect of curcumin (turmeric) on Alzheimer's disease: an overview. Ann Indian Acad Neurol 11(1):13. https://doi.org/10.4103/0972-2327.40220

Mukherjee PK, Nema NK, Venkatesh P, Debnath PK (2012) Changing scenario for promotion and development of Ayurveda–way forward. J Ethnopharmacol 143(2):424–434. https://doi.org/10.1016/j.jep.2012.07.036

Mythri RB, Srinivas Bharath MM (2012) Curcumin: a potential neuroprotective agent in Parkinson's disease. Curr Pharm Des 18(1):91–99. https://doi.org/10.2174/138161212798918995

Nabavi SF, Nabavi SM, Latifi AM, Mirzaei M, Habtemariam S, Moghaddam AH (2012) Mitigating role of quercetin against sodium fluoride-induced oxidative stress in the rat brain. Pharm Biol 50(11):1380–1383. https://doi.org/10.3109/13880209.2012.675341

Nabi M, Tabassum N (2022) Role of environmental toxicants on neurodegenerative disorders. Front Toxicol 4:837579. https://doi.org/10.3389/ftox.2022.837579

Olakkaran S, Antony A (2019) Convolvulus pluricaulis (Shankhapushpi) ameliorates human microtubule-associated protein tau (hMAPτ) induced neurotoxicity in Alzheimer's disease drosophila model. J Chem Neuroanat 95:115–122. https://doi.org/10.1016/j.jchemneu.2017.10.002

Pandey A, Bani S, Dutt P, Satti NK, Suri KA, Qazi GN (2018) Multifunctional neuroprotective effect of Withanone, a compound from Withania somnifera roots in alleviating cognitive dysfunction. Cytokine 102:211–221. https://doi.org/10.1016/j.cyto.2017.10.019

Patel SS, Acharya A, Ray RS, Agrawal R, Raghuwanshi R, Jain P (2020) Cellular and molecular mechanisms of curcumin in prevention and treatment of disease. Crit Rev Food Sci Nutr 60:887. https://doi.org/10.1080/10408398.2018.1552244

Patil P, Thakur A, Sharma A, Flora SJS (2020) Natural products and their derivatives as multifunctional ligands against Alzheimer's disease. Drug Dev Res 81(2):165–183. https://doi.org/10.1002/ddr.21587

Prakash J, Yadav SK, Chouhan S, Singh SP (2013) Neuroprotective role of Withania somnifera root extract in Maneb–Paraquat induced mouse model of parkinsonism. Neurochem Res 38:972–980. https://doi.org/10.1007/s11064-013-1005-4

Pritam P, Deka R, Bhardwaj A, Srivastava R, Kumar D, Jha AK, Jha NK, Villa C, Jha SK (2022) Antioxidants in Alzheimer's disease: current therapeutic significance and future prospects. Biology 11(2):212. https://doi.org/10.3390/biology11020212

Qian ZM, Ke Y (2014) Huperzine a: is it an effective disease-modifying drug for Alzheimer's disease? Front Aging Neurosci 6:216. https://doi.org/10.3389/fnagi.2014.00216

Rahman MH, Bajgai J, Fadriquela A, Sharma S, Trinh TT, Akter R, Jeong YJ, Goh SH, Kim CS, Lee KJ (2021) Therapeutic potential of natural products in treating neurodegenerative disorders and their future prospects and challenges. Molecules 26(17):5327. https://doi.org/10.3390/molecules26175327

RajaSankar S, Manivasagam T, Sankar V, Prakash S, Muthusamy R, Krishnamurti A, Surendran S (2009) Withania somnifera root extract improves catecholamines and physiological abnormalities seen in a Parkinson's disease model mouse. J Ethnopharmacol 125(3):369–373. https://doi.org/10.1016/j.jep.2009.08.003

Rao RV, Descamps O, John V, Bredesen DE (2012) Ayurvedic medicinal plants for Alzheimer's disease: a review. Alzheimers Res Ther 4:1–9. https://doi.org/10.1186/alzrt125

Ratheesh G, Tian L, Venugopal JR, Ezhilarasu H, Sadiq A, Fan TP, Ramakrishna S (2017) Role of medicinal plants in neurodegenerative diseases. Biomanuf Rev 2:1–16. https://doi.org/10.1007/s40898-017-0004-7

Rojas P, Serrano-García N, Mares-Sámano JJ, Medina-Campos ON, Pedraza-Chaverri J, Ögren SO (2008) EGb761 protects against nigrostriatal dopaminergic neurotoxicity in 1-methyl-4-phenyl-1, 2, 3, 6-tetrahydropyridine-induced parkinsonism in mice: role of oxidative stress. Eur J Neurosci 28(1):41–50. https://doi.org/10.1111/j.1460-9568.2008.06314.x

Rong-Xiu LIU, Guo-Hu SONG, Pei-Gen WU, Zhang XW, Hui-Juan HU, Jia LIU, Xiao-Su MIAO, Zhi-Yan HOU, Wen-Quan WANG, Sheng-Li WEI (2017) Distribution patterns of the contents of five biologically activate ingredients in the root of Scutellaria baicalensis. Chin J Nat 15(2):152–160. https://doi.org/10.1016/S1875-5364(17)30030-4

Rui W, Li S, Xiao H, Xiao M, Shi J (2020) Baicalein attenuates neuroinflammation by inhibiting NLRP3/caspase-1/GSDMD pathway in MPTP-induced mice model of Parkinson's disease. Int J Neuropsychopharmacol 23(11):762–773. https://doi.org/10.1093/ijnp/pyaa060

Sasikumar P, Aswathy M, Prem PT, Radhakrishnan KV, Chakrapani PSB (2022) Plant derived bioactive compounds and their potential to enhance adult neurogenesis. Phytomed Plus 2(1):100191. https://doi.org/10.1016/j.phyplu.2021.100191

Sayre LM, Perry G, Smith MA (2008) Oxidative stress and neurotoxicity. Chem Res Toxicol 21(1):172–188. https://doi.org/10.1021/tx700210j

Schapira AHV, Emre M, Jenner P, Poewe WJEJON (2009) Levodopa in the treatment of Parkinson's disease. Eur Neurol 16(9):982–989. https://doi.org/10.1111/j.1468-1331.2009.02697.x

Schwenkgrub J, Zaremba M, Joniec-Maciejak I, Cudna A, Mirowska-Guzel D, Kurkowska-Jastrzębska I (2017) The phosphodiesterase inhibitor, ibudilast, attenuates neuroinflammation in the MPTP model of Parkinson's disease. PLoS One 12(7):e0182019. https://doi.org/10.1371/journal.pone.0182019

Selim NM, Elgazar AA, Abdel-Hamid NM, Abu El-Magd MR, Yasri A, El Hefnawy HM, Sobeh M (2019) Chrysophanol, physcion, hesperidin and curcumin modulate the gene expression of pro-inflammatory mediators induced by LPS in HepG2: in silico and molecular studies. Antioxidants 8(9):371. https://doi.org/10.3390/antiox8090371

Sen S, Chakraborty R, De B (2011) Challenges and opportunities in the advancement of herbal medicine: India's position and role in a global context. J Herb Med 1(3–4):67–75. https://doi.org/10.1016/j.hermed.2011.11.001

Sgarbossa A, Giacomazza D, Di Carlo M (2015) Ferulic acid: a hope for Alzheimer's disease therapy from plants. Nutrients 7(7):5764–5782. https://doi.org/10.3390/nu7075246

Shalini VT, Neelakanta SJ, Sriranjini JS (2021) Neuroprotection with Bacopa monnieri–a review of experimental evidence. Mol Biol Rep 48(3):2653–2668. https://doi.org/10.1007/s11033-021-06236-w

Shu T, Liu C, Pang M, He L, Yang B, Fan L, Zhang S, Wang X, Liu B, Rong L (2018) Salvianolic acid B promotes neural differentiation of induced pluripotent stem cells via PI3K/AKT/GSK3β/β-catenin pathway. Neurosci Lett 671:154–160. https://doi.org/10.1016/j.neulet.2018.02.007

Singh HK (2013) Brain enhancing ingredients from āyurvedic medicine: quintessential example of Bacopa monniera, a narrative review. Nutrients 5(2):478–497. https://doi.org/10.3390/nu5020478

Singh AK, Singh SK, Nandi MK, Mishra G, Maurya A, Rai A, Rai GK, Awasthi R, Sharma B, Kulkarni GT (2019) Berberine: a plant-derived alkaloid with therapeutic potential to combat Alzheimer's disease. Cent Nerv Syst Agents Med Chem 19(3):154–170. https://doi.org/10.2174/1871524919666190820160053

Song Q, Peng S, Zhu X (2021a) Baicalein protects against MPP+/MPTP-induced neurotoxicity by ameliorating oxidative stress in SH-SY5Y cells and mouse model of Parkinson's disease. Neurotoxicology 87:188–194. https://doi.org/10.1016/j.neuro.2021.10.003

Song L, Zhang J, Lai R, Li Q, Ju J, Xu H (2021b) Chinese herbal medicines and active metabolites: potential antioxidant treatments for atherosclerosis. Front Pharmacol 12:675999. https://doi.org/10.3389/fphar.2021.675999

Soto C (2003) Unfolding the role of protein misfolding in neurodegenerative diseases. Nat Rev Neurosci 4(1):49–60. https://doi.org/10.1038/nrn1007

Sreenivasmurthy SG, Liu JY, Song JX, Yang CB, Malampati S, Wang ZY, Huang YY, Li M (2017) Neurogenic traditional Chinese medicine as a promising strategy for the treatment of Alzheimer's disease. Int J Mol Sci 18(2):272. https://doi.org/10.3390/ijms18020272

Swallah MS, Sun H, Affoh R, Fu H, Yu H (2020) Antioxidant potential overviews of secondary metabolites (polyphenols) in fruits. Int J Food Sci 2020:1. https://doi.org/10.1155/2020/9081686

Tamano H, Morioka H, Nishio R, Takeuchi A, Takeda A (2018) AMPA-induced extracellular Zn2+ influx into nigral dopaminergic neurons causes movement disorder in rats. Neurotoxicology 69:23–28. https://doi.org/10.1016/j.neuro.2018.08.008

Tang LL, Wang R, Tang XC (2005) Effects of huperzine a on secretion of nerve growth factor in cultured rat cortical astrocytes and neurite outgrowth in rat PC12 cells. Acta Pharmacol Sin 26(6):673–678. https://doi.org/10.1111/j.1745-7254.2005.00130.x

Tang Y, Huang D, Zhang MH, Zhang WS, Tang YX, Shi ZX, Deng L, Zhou DH, Lu XY (2016) Salvianolic acid B inhibits Aβ generation by modulating BACE1 activity in SH-SY5Y-APPsw cells. Nutrients 8(6):333. https://doi.org/10.3390/nu8060333

Tao LX, Huang XT, Chen YT, Tang XC, Zhang HY (2016) Acetylcholinesterase-independent protective effects of huperzine a against iron overload-induced oxidative damage and aberrant iron metabolism signaling in rat cortical neurons. Acta Pharmacol Sin 37(11):1391–1400. https://doi.org/10.1038/aps.2016.78

Tchantchou F, Lacor PN, Cao Z, Lao L, Hou Y, Cui C, Klein WL, Luo Y (2009) Stimulation of neurogenesis and synaptogenesis by bilobalide and quercetin via common final pathway in hippocampal neurons. J Alzheimers Dis 18(4):787–798. https://doi.org/10.3233/JAD-2009-1189

Tohma H, Gülçin İ, Bursal E, Gören AC, Alwasel SH, Köksal E (2017) Antioxidant activity and phenolic compounds of ginger (Zingiber officinale Rosc.) determined by HPLC-MS/MS. J Food Meas Charact 11:556–566. https://doi.org/10.1007/s11694-016-9423-z

Trist BG, Hare DJ, Double KL (2019) Oxidative stress in the aging substantia nigra and the etiology of Parkinson's disease. Aging Cell 18(6):e13031. https://doi.org/10.1111/acel.13031

Ude C, Schubert-Zsilavecz M, Wurglics M (2013) Ginkgo biloba extracts: a review of the pharmacokinetics of the active ingredients. Clin Pharmacokinet 52:727–749. https://doi.org/10.1007/s40262-013-0074-5

Vareed SK, Bauer AK, Nair KM, Liu Y, Jayaprakasam B, Nair MG (2014) Blood–brain barrier permeability of bioactive withanamides present in Withania somnifera fruit extract. Phytother Res 28(8):1260–1264. https://doi.org/10.1002/ptr.5118

Velmurugan BK, Rathinasamy B, Lohanathan BP, Thiyagarajan V, Weng CF (2018) Neuroprotective role of phytochemicals. Molecules 23(10):2485. https://doi.org/10.3390/molecules23102485

Wan W, Zhang C, Danielsen M, Li Q, Chen W, Yuanjin CHAN, Li Y (2016) EGb761 improves cognitive function and regulates inflammatory responses in the APP/PS1 mouse. Exp Gerontol 81:92–100. https://doi.org/10.1016/j.exger.2016.05.007

Wang R, Tang XC (2005) Neuroprotective effects of huperzine a: a natural cholinesterase inhibitor for the treatment of Alzheimer's disease. Neurosignals 14(1–2):71–82. https://doi.org/10.1159/000085387

Wang X, Wang R, Xing D, Su H, Ma C, Ding Y, Du L (2005) Kinetic difference of berberine between hippocampus and plasma in rat after intravenous administration of Coptidis rhizoma extract. Life Sci 77(24):3058–3067. https://doi.org/10.1016/j.lfs.2005.02.033

Wang SX, Hu LM, Gao XM, Guo H, Fan GW (2010) Anti-inflammatory activity of salvianolic acid B in microglia contributes to its neuroprotective effect. Neurochem Res 35:1029–1037. https://doi.org/10.1007/s11064-010-0151-1

Wang Z, Wan H, Li J, Zhang H, Tian M (2013) Molecular imaging in traditional Chinese medicine therapy for neurological diseases. Biomed Res Int 2013:1. https://doi.org/10.1155/2013/608430

Wang WY, Tan MS, Yu JT, Tan L (2015) Role of pro-inflammatory cytokines released from microglia in Alzheimer's disease. Ann Transl Med 3(10):136. https://doi.org/10.3978/j.issn.2305-5839.2015.03.49

Wang ZY, Liu JG, Li H, Yang HM (2016) Pharmacological effects of active components of Chinese herbal medicine in the treatment of Alzheimer's disease: a review. Am J Chin Med 44(08):1525–1541. https://doi.org/10.1142/S0192415X16500853

Wang T, Li C, Han B, Wang Z, Meng X, Zhang L, He J, Fu F (2020) Neuroprotective effects of Danshensu on rotenone-induced Parkinson's disease models in vitro and in vivo. BMC Complement Med Ther 20(1):1–10. https://doi.org/10.1186/s12906-019-2738-7

Witika BA, Makoni PA, Matafwali SK, Mweetwa LL, Shandele GC, Walker RB (2021) Enhancement of biological and pharmacological properties of an encapsulated polyphenol: curcumin. Molecules 26(14):4244. https://doi.org/10.3390/molecules26144244

Wright AL, Zinn R, Hohensinn B, Konen LM, Beynon SB, Tan RP, Clark IA, Abdipranoto A, Vissel B (2013) Neuroinflammation and neuronal loss precede Aβ plaque deposition in the hAPP-J20 mouse model of Alzheimer's disease. PLoS One 8(4):e59586. https://doi.org/10.1371/journal.pone.0059586

Wu TY, Chen CP, Jinn TR (2011) Traditional Chinese medicines and Alzheimer's disease. Taiwan J Obstet Gynecol 50(2):131–135. https://doi.org/10.1016/j.tjog.2011.04.004

Xi J, Zhang B, Luo F, Liu J, Yang T (2012) Quercetin protects neuroblastoma SH-SY5Y cells against oxidative stress by inhibiting expression of Krüppel-like factor 4. Neurosci Lett 527(2):115–120. https://doi.org/10.1016/j.neulet.2012.08.082

Xu D, Hu MJ, Wang YQ, Cui YL (2019) Antioxidant activities of quercetin and its complexes for medicinal application. Molecules 24(6):1123. https://doi.org/10.3390/molecules24061123

Yuan H, Ma Q, Ye L, Piao G (2016) The traditional medicine and modern medicine from natural products. Molecules 21(5):559. https://doi.org/10.3390/molecules21050559

Zahiruddin S, Basist P, Parveen A, Parveen R, Khan W, Ahmad S (2020) Ashwagandha in brain disorders: a review of recent developments. J Ethnopharmacol 257:112876. https://doi.org/10.1016/j.jep.2020.112876

Zhang HY, Zheng CY, Yan H, Wang ZF, Tang LL, Gao X, Tang XC (2008) Potential therapeutic targets of huperzine a for Alzheimer's disease and vascular dementia. Chem Biol Interact 175(1–3):396–402. https://doi.org/10.1016/j.cbi.2008.04.049

Zhang ZJ, Cheang LCV, Wang MW, Lee SMY (2011) Quercetin exerts a neuroprotective effect through inhibition of the iNOS/NO system and pro-inflammation gene expression in PC12 cells and in zebrafish. Int J Mol Med 27(2):195–203. https://doi.org/10.3892/ijmm.2010.571

Zhang XZ, Qian SS, Zhang YJ, Wang RQ (2016) Salvia miltiorrhiza: a source for anti-Alzheimer's disease drugs. Pharm Biol 54(1):18–24. https://doi.org/10.3109/13880209.2015.1027408

Zhang LF, Yu XL, Ji M, Liu SY, Wu XL, Wang YJ, Liu RT (2018) Resveratrol alleviates motor and cognitive deficits and neuropathology in the A53T α-synuclein mouse model of Parkinson's disease. Food Funct 9(12):6414–6426. https://doi.org/10.1039/C8FO00964C

Zhao H, Wang Q, Cheng X, Li X, Li N, Liu T, Li J, Yang Q, Dong R, Zhang Y, Zhang L (2018) Inhibitive effect of resveratrol on the inflammation in cultured astrocytes and microglia induced by Aβ1–42. Neuroscience 379:390–404. https://doi.org/10.1016/j.neuroscience.2018.03.047

Zvěřová M (2019) Clinical aspects of Alzheimer's disease. Clin Biochem 72:3–6. https://doi.org/10.1016/j.clinbiochem.2019.04.015

Role of Ethnomedicinal Resources in Curing Metabolic Disease

Shabana Begum, Rajat Nath, Anupam Das Talukdar, Jayanta Kumar Patra, and Pranab Behari Mazumder

Contents

Abstract Ethnomedicinal resources refer to indigenous knowledge and techniques for utilizing medicinal plants for therapeutic purposes in particular contexts of culture. These resources usually involve the use of the medicinal plant as a whole or various plant sections such as roots, stems, leaves, fruits, seeds, and flowers for various medical advantages. Numerous traditional medicinal herbs and their bioactive

S. Begum · P. B. Mazumder (✉)
Natural Product and Biomedicine Research Laboratory, Department of Biotechnology, Assam University, Silchar, Assam, India

R. Nath · A. Das Talukdar
Department of Life Science and Bioinformatics, Assam University, Silchar, Assam, India

J. K. Patra
Dongguk University, Goyang-si, Republic of Korea

© The Author(s), under exclusive license to Springer Nature Singapore Pte Ltd. 2024

A. Das Talukdar et al. (eds.), *Traditional Resources and Tools for Modern Drug Discovery*, Interdisciplinary Biotechnological Advances,
https://doi.org/10.1007/978-981-97-4600-2_21

compounds such as polyphenols have shown promising results in treating metabolic diseases. Metabolic disease is a broad term referring to a group of diseases marked by abnormalities in the body's metabolism, in the uptake of food and energy, leading to serious complications like type 2 diabetes (T2D), nonalcoholic fatty liver disease (NAFLD), cardiovascular conditions, obesity, and hypercholesterolemia. These disorders frequently entail abnormalities with calorie management, regulation of lipids metabolism, glucose utilization, and insulin uptake. The development and advancement of metabolic conditions are significantly influenced by dietary habits, genetics, and exercise. Current treatment options include lifestyle interventions like diet modification, physical activity, and medical therapies like chemotherapy, surgery, and enzyme replacement therapy; however, these treatments can be challenging to maintain over time. Therefore, an alternative treatment method that may provide more secure and effective therapy for metabolic disorders includes the utilization of ethnomedicinal resources especially therapeutic plants and their natural bioactive constituents. Researchers have reported that several medicinal plants and their secondary metabolites such as resveratrol, curcumin, berberine, anthocyanin, etc. have promising medicinal benefits against metabolic diseases, by targeting multiple important proteins that are crucial for maintaining the normal function of the body, thus, showcasing the potential of ethnomedicinal resources in modern medicinal practices.

Keywords Ethnomedicinal resources · Medicinal plants · Bioactive compounds · Metabolic diseases

1 Introduction

Ethnomedicinal resources are the therapeutic plants, knowledge, and customs that regional tribes and indigenous populations have long utilized to treat illnesses and preserve health. Ethnomedicine resources also include the use of animal-derived products for therapeutic purposes (Harvey et al. 2015). However, medicinal plants have been the primary source of unprocessed pharmaceuticals utilized in various conventional healthcare systems across the globe. Approximately 200 million plant species have been estimated to exist worldwide, out of which the World Health Organization (WHO) has listed 21,000 different plant species with therapeutic qualities and traditional medicinal practitioners in India utilize over 2500 different plant species for various therapeutic purposes (Jayaprasad et al. 2012). The World Health Organization has also reported that 80% of the world's human population depends on medicinal plants regarding their healthcare (Gurib-Fakim 2006). Therefore, these resources are an invaluable repository of information that is carried from one generation to another, offering insightful details regarding the healing properties of the natural world against various human ailments (Bouafia et al. 2021; Siddique et al. 2021). Therefore, documenting cultural and indigenous knowledge is essential to maintain and preserve the knowledge, as the therapeutic potential of the medicinal plants are identified through extensive self-experiments that took years to develop (Chhetri et al. 2008). Further, documentation can also be used for relevant research and further pharmacological testing. In addition, the integration of indigenous treatments into regional healthcare systems has become a growing trend in emerging countries, and researchers are exploring the enormous therapeutic potential of ethnomedical knowledge for treating a wide range of disorders (Das 2021; Jain et al. 2010). These resources can also serve as an alternative medicinal system when contemporary options may be limited (Kumar et al. 2021). The potential of these resources is vast as 50% of all pharmaceutical drugs used in clinical trials worldwide are either extracted or derived from therapeutic plants (Gurib-Fakim 2006).

2 Metabolic Diseases

Metabolic diseases or metabolic disorders refer to a class of intricate medical conditions that affect the body's capacity to utilize energy and nutrients. These medical conditions have the potential to impair the body's normal processes, which can result in a variety of symptoms and consequences. These diseases are usually inherited and can also result from genetic mutations that alter particular biological processes, accumulating harmful materials or depleting vital substances (Sobrido et al. 2019). Inflammation, endoplasmic reticulum stress, and eating disorder also have a serious impact on the onset and course of metabolic diseases, particularly obesity, and diabetes (Erbay and Hotamıslıgil 2014; Fernandes-da-Silva et al. 2021; Yu and Muehleman 2023). Insulin resistance, hypertension, hypercholesterolemia, abdominal obesity, hyperglycemia, and dyslipidemia are the hallmarks of metabolic disease, which raises the possibility of cardiovascular conditions, type 2 diabetes, brain attack, chronic renal disease, nonalcoholic fatty liver disease (NAFLD), and cancer (Grundy et al. 2004). Symptoms of this disease include coma, nausea, vomiting, jaundice, muscle weakness, high blood sugar, edema of the brain, cutaneous conditions, and respiratory issues (Huang 2014). Because of the complex nature of the disease, there are few therapeutic options available for its treatment. Current treatment options for metabolic disease can be divided into three categories: pharmaceutical treatments which include chemotherapies, lifestyle interventions such as diet modification and physical activities, and traditional Chinese medicinal remedies such as the utilization of medicinal plants and acupuncture (Zhao et al. 2023). Other treatment options include enzyme replacement therapy and surgery for inherited metabolic disorders (Santamaria et al. 2013). Though these treatment options are effective, however, these therapies can be difficult to remain effective over time. Therefore, additional research is required to comprehend the underlying mechanisms of metabolic diseases to develop effective and diverse therapy options (Zhao et al. 2023). Ethnomedicinal resources, particularly medicinal plants, have recently drawn a lot of attention in treating metabolic diseases because of their therapeutic efficacy against the disease. Studies have suggested that the curative activity of medicinal plants is mainly because of the naturally occurring polyphenols such as flavonoids and non-flavonoids. These naturally occurring compounds can target multiple biochemical pathways, making them alternative and effective pharmaceutical medications in the treatment of various metabolic diseases (Nyakudya et al. 2020; Wang et al. 2022a).

3 Plants and Metabolic Disease

Medicinal plants have been extensively utilized to cure and prevent various diseases since ancient times, moreover, these plants have shown great therapeutic potential against metabolic diseases. Studies suggest that the plant extract and their polyphenols have potential antioxidant and anti-inflammatory properties, which helps to stabilize blood pressure, enhance the metabolism of lipids, and improve the regulation of blood sugar levels and also helps in maintaining a healthy weight, thereby preventing the development of metabolic diseases (Wang et al. 2022a). Several medicinal plants are known to act against metabolic diseases. Some of the best curative plants and the types of metabolic diseases they act upon are listed in Table 1.

4 Major Phytochemicals Used for the Treatment of Metabolic Diseases

Ethnomedicinal resources, particularly medicinal plants, act against metabolic disease through naturally occurring secondary metabolites. Some of the most common secondary metabolites that act against various metabolic disease conditions such as type 2 diabetes, cardiovascular diseases, obesity, and nonalcoholic fatty liver disease include resveratrol, curcumin, anthocyanin, quercetin, and berberine. Another compound such as aspalathin is also known to possess cardioprotective activity. Other active phytochemicals against obesity are capsaicin, kaempferol, fisetin, dioscin, and genistein. Plant polyphenols such as naringenin, hesperidin, and ellagic acid have been proclaimed to have beneficial effects against type 2 diabetes. In contrast, compounds such as myricitrin, rutin, silybin, and ginsenosides are known to act against nonalcoholic fatty liver disease. Furthermore, natural compounds that have shown potential effects against hypercholesterolemia include resveratrol, hesperidin, ellagic acid, berberine, naringenin, monascin, and ankaflavin. The structures of major phytochemicals are elucidated in Fig. 1.

Table 1 Medicinal plants used against metabolic diseases

Sl. no.	Plant	Disease	Reference
1	*Acacia Arabica*	Type 2 diabetes	Hegazy et al. (2013)
2	*Albizzia lebbeck* Benth	Type 2 diabetes	Ahmed et al. (2014)
3	*Aloe barbadensis*	Type 2 diabetes	Manjunath et al. (2016)
4	*Amberboa ramosa*	Type 2 diabetes	Paul et al. (2023)
5	*Annona muricata*	Type 2 diabetes	Olasehinde et al. (2022)
6	*Cistus laurifolius*	Type 2 diabetes	Orhan et al. (2013)
7	*Cyclocarya paliurus*	Type 2 diabetes	Wang et al. (2021)
8	*Mitragyna speciosa*	Type 2 diabetes	Zhang et al. (2023)
9	*Momordica charantia* var. *abbreviata* Ser.	Type 2 diabetes	Sun et al. (2023)
10	*Momordica charantia*	Type 2 diabetes	Ma et al. (2017)
11	*Oxalis corniculata*	Type 2 diabetes	Al-Qalhati et al. (2016)
12	*Pueraria lobate*	Type 2 diabetes	Chen et al. (2018)
13	*Pteropyrum scoparium*	Type 2 diabetes	Al-Qalhati et al. (2016)
14	*Trifolium alexandrinum*	Type 2 diabetes	Abdou et al. (2022)
15	Soybean leaves	Type 2 diabetes	Zang et al. (2015)
16	*Terminalia bellirica*	Type 2 diabetes	Latha and Daisy (2013)
17	*Vernonia anthelmintica*	Type 2 diabetes	Rajan et al. (2022)
18	*Zingiber capitatum*	Type 2 diabetes	Hasan et al. (2023)
19	*Zingiber officinale*	Type 2 diabetes	Akash et al. (2015)
20	*Aconitum heterophyllum*	Obesity	Kumar et al. (2019)
21	*Acorus calamus*	Obesity	Athesh and Jothi (2017)
22	*Agave sisalana*	Dyslipidemic, hypoglycemic, hyperglycemic, and obesity	Chege et al. (2023)
23	*Allium cepa*	Obesity	Momoh et al. (2022)
24	*Aloe barbadensis*	Obesity	Pothuraju et al. (2016)
25	*Aronia melanocarpa*	Obesity	Zhu et al. (2020)
26	*Camellia sinensis*	Obesity, metabolic syndrome, diabetes, cardiovascular diseases and cancer	Yang et al. (2018)
27	*Gymnema sylvestre*	Obesity	Pothuraju et al. (2016)
28	*Hibiscus sabdariffa*	Obesity	Diez-Echave et al. (2020)
29	*Moringa oleifera*	Obesity	Bais et al. (2014)

(continued)

Table 1 (continued)

Sl. no.	Plant	Disease	Reference
30	*Panax notoginseng*	Obesity	Zhang et al. (2021)
31	Olive	Obesity	Mikami et al. (2021)
32	Raspberry	Obesity and type 2 diabetes	Zhao et al. (2018)
33	*Zanthoxylum bungeanum*	Obesity	Wang et al. (2020)
34	*Allium fistulosum*	NAFLD	Hwang et al. (2018)
35	*Camellia oleifera*	NAFLD	Yeh et al. (2020)
36	*Crataegus oxyacantha*	NAFLD	Saeedi et al. (2018)
37	*Curcuma longa*	NAFLD	Lee et al. (2020)
38	*Cyclosorus terminans*	NAFLD	Songtrai et al. (2022)
39	*Cynara cardunculus*	NAFLD	Oppedisano et al. (2020)
40	*Gynostemma pentaphyllum*	NAFLD	Hwang et al. (2018)
41	*Dillenia indica*	NAFLD	Poornima et al. (2022)
42	*Gynostemma pentaphyllum*	NAFLD	Bae et al. (2018)
43	*Morinda citrifolia*	NAFLD	Wang et al. (2022b)
44	*Panax ginseng*	NAFLD	Charytoniuk et al. (2017)
45	*Phyllanthus emblica*	NAFLD	Luo et al. (2022)
46	*Phyllanthus niruri*	NAFLD	Al Zarzour et al. (2017)
47	*Portulaca oleracea*	NAFLD	Darvish Damavandi et al. (2021)
48	Sweet cherry	NAFLD	Charytoniuk et al. (2017)
49	*Tetrastigma hemsleyanum*	NAFLD	Xiao et al. (2023)
50	*Uncaria tomentosa*	NAFLD	Araujo et al. (2018)
51	*Amaranthus viridis*	Hypercholesterolemia	Salvamani et al. (2016)
52	*Curcuma longa*	Hypercholesterolemia	Ashraf et al. (2022)
	Ganoderma lucidum	Hypercholesterolemia	Meneses et al. (2016)
53	*Eleutherine americana*	Hypercholesterolemia	Saragih et al. (2014)
54	*Leontopodium alpinum*	Hypercholesterolemia	Wang et al. (2016)
55	*Leonurus sibiricus*	Hypercholesterolemia	Abedin et al. (2020)
56	*Panax quinquefolius*	Hypercholesterolemia	Liu et al. (2021)

(continued)

Table 1 (continued)

Sl. no.	Plant	Disease	Reference
57	*Solanum macrocarpon*	Hypercholesterolemia	Dougnon et al. (2014)
58	*Thymus vulgaris*	Hypercholesterolemia	Salehi et al. (2018)
59	*Alisma orientale*	Cardiovascular diseases	Wu et al. (2023)
60	*Callicarpa tomentosa*	Cardiovascular diseases	Noushida et al. (2024)
61	*Carthamus tinctorius*	Cardiovascular diseases	Yu et al. (2019)
62	*Crataegus oxyacantha*	Cardiovascular diseases	Cuevas-Durán et al. (2017)
63	*Garcinia kola*	Cardiovascular diseases	Olatoye et al. (2024)
64	*Ginkgo biloba*	Cardiovascular diseases	Yang et al. (2017)
65	*Gundelia tournefortii*	Cardiovascular diseases	Rafiee et al. (2017)
66	*Hibiscus sabdariffa*	Cardiovascular diseases	Sapian et al. (2023)
67	*Icacina trichantha*	Cardiovascular diseases	Abu et al. (2023)
68	*Melissa officinalis*	Cardiovascular diseases	Draginic et al. (2021)
69	*Pimpinella anisum*	Cardiovascular diseases	Pontes et al. (2019)
70	*Punica granatum*	Cardiovascular diseases	Razani et al. (2017)
71	*Ribes nigrum*	Cardiovascular diseases	Gubina-Vakulyck et al. (2018)
72	*Rosmarinus officinalis*	Cardiovascular diseases	Cuevas-Durán et al. (2017)
73	*Salvia miltiorrhiza*	Cardiovascular diseases	Ren et al. (2019)
74	*Theobroma cacao*	Cardiovascular diseases	Beshel et al. (2022)
75	*Urtica simensis*	Cardiovascular diseases	Tesfaye et al. (2021)

5 Mode of Action of Bioactive Compounds Against Metabolic Diseases

5.1 *Type 2 Diabetes (T2D)*

Type 2 diabetes (T2D), a metabolic disease, marked by the body's inability to metabolize a hormone, namely, insulin, which regulates glucose levels resulting in insulin resistance and alleviating insulin production. Studies suggested that natural phytocompounds such as anthocyanin (Li et al. 2015; Yan and Zheng 2017), berberine (Dong et al. 2017; Jiang et al. 2015), curcumin (Chuengsamarn et al. 2014; Mohiti-Ardekani et al. 2019), ellagic acids (Al Tamimi et al. 2022; Ghadimi et al. 2021), hesperidin, quercetin (Abdou et al. 2022; Homayouni et al. 2018), resveratrol (Movahed et al. 2013; Vlavcheski et al. 2020), and naringenin (Ahmed et al. 2017; Zygmunt et al. 2010) are effective against type 2 diabetes. These compounds

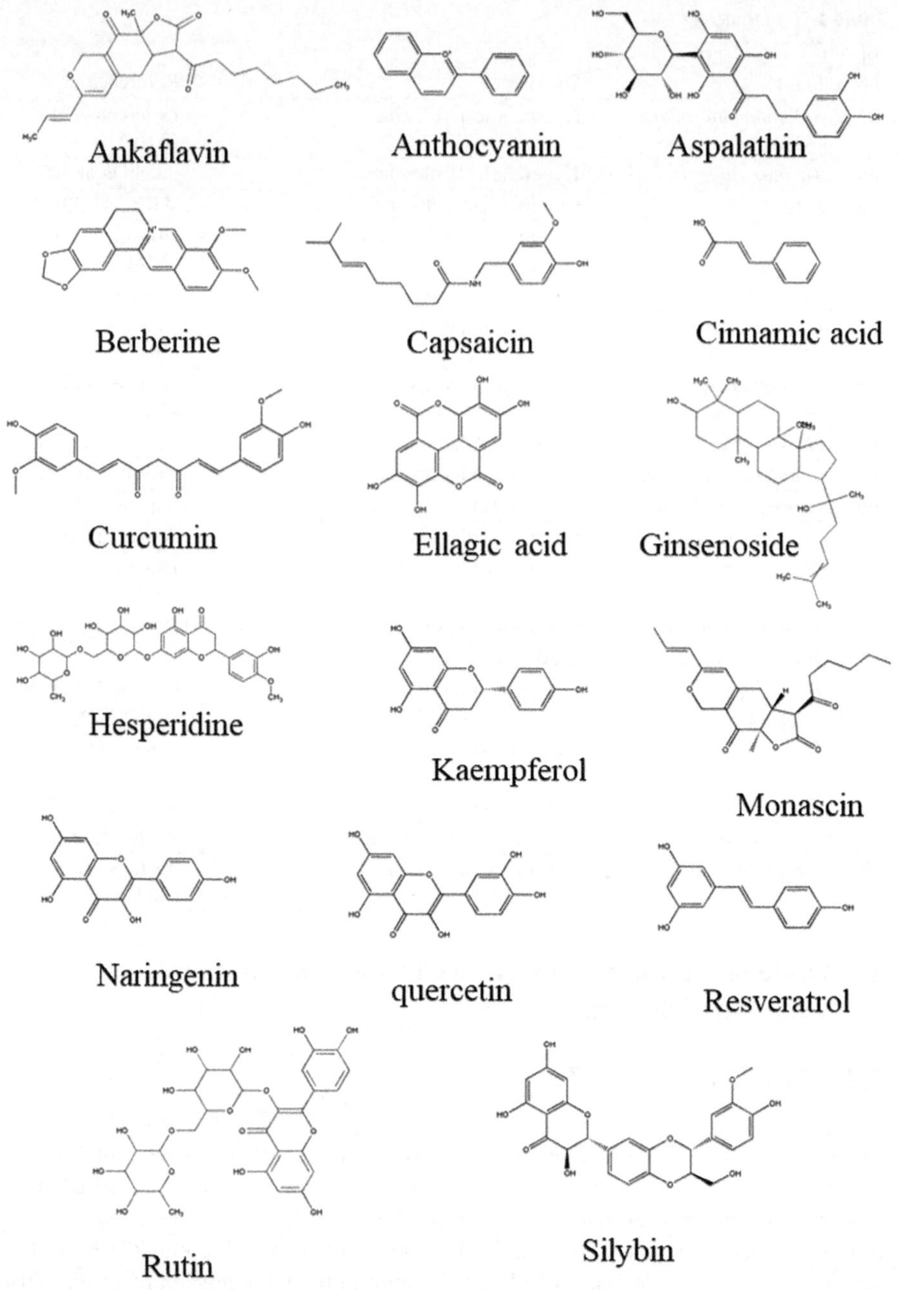

Fig. 1 Structure of some major phytochemicals against metabolic diseases

decrease blood glucose levels as well as regulate lipid profile by suppressing the expression of total cholesterol (TC), total glyceraldehyde (TG), low-density lipoprotein (LDL), and high-density lipoprotein (HDL). These bioactive compounds such as resveratrol, anthocyanin, curcumin, and naringenin also increase insulin sensitivity by upregulating biochemical processes that are vital for maintaining cellular energy homeostasis such as upregulation of glucose transporter type 4 (GLUT4) and berberine, ellagic acid, hesperidin, and quercetin stimulate the manifestation of pathways dealing with adenosine monophosphate-activated protein kinase (AMPK). These compounds also decrease insulin resistance by alleviating cytokines that are associated with inflammation such as interleukin-6 (IL-6) and tumor necrosis factor-alpha (TNF-α). The pharmacological activities of these compounds are projected in Fig. 2.

5.2 *Obesity*

Obesity is one of the primary risk factors for metabolic conditions, characterized by energetic imbalance mainly due to enhanced uptake of foods rich in calories. Many metabolic disorders and chronic illnesses like type 2 diabetes, muscle weakness, cardiovascular disease, and certain types of cancers are associated with it (de la Peña et al. 2023). Therapeutic plants and their bioactive chemical compounds such as quercetin, capsaicin, anthocyanin, berberine, kaempferol, resveratrol, fisetin, dioscin, and curcumin help to combat obesity by reducing appetite; lowering and regulating lipid and carbohydrate metabolism such as suppressing the expression of total cholesterol (TC), total glyceraldehyde (TG), high-density lipoprotein (HDL), blood glucose levels, and low-density lipoprotein (LDL); alleviating adipogenesis; improving the expenditure of energy by stimulating browning of adipocytes in white adipose tissue; modifying gut microbiome, decreasing the inflammation associated with obesity; and by suppressing biochemical pathways such as MAPK (Shang et al. 2021). Bioactive natural compounds such as curcumin and resveratrol not only suppress the expression of genes that lead to inflammation such as IL-6 (interleukin-6), and TNF-α (tumor necrosis factor-alpha) but also decrease the production of insulin and homeostasis model assessment insulin resistance (HOMA-IR) (Jazayeri-Tehrani et al. 2019; Shang et al. 2021; Timmers et al. 2011). Figure 3 illustrates the specific pharmacological activities of these compounds.

5.3 *Nonalcoholic Fatty Liver Disease*

Nonalcoholic fatty liver disease (NAFLD) is among the chief causes of chronic liver diseases, closely linked to metabolic diseases, characterized by the excess the buildup of fats in the liver leading to serious complications like cirrhosis of the liver, hepatocellular carcinoma, and metabolic-associated steatohepatitis (NASH) (Gujral

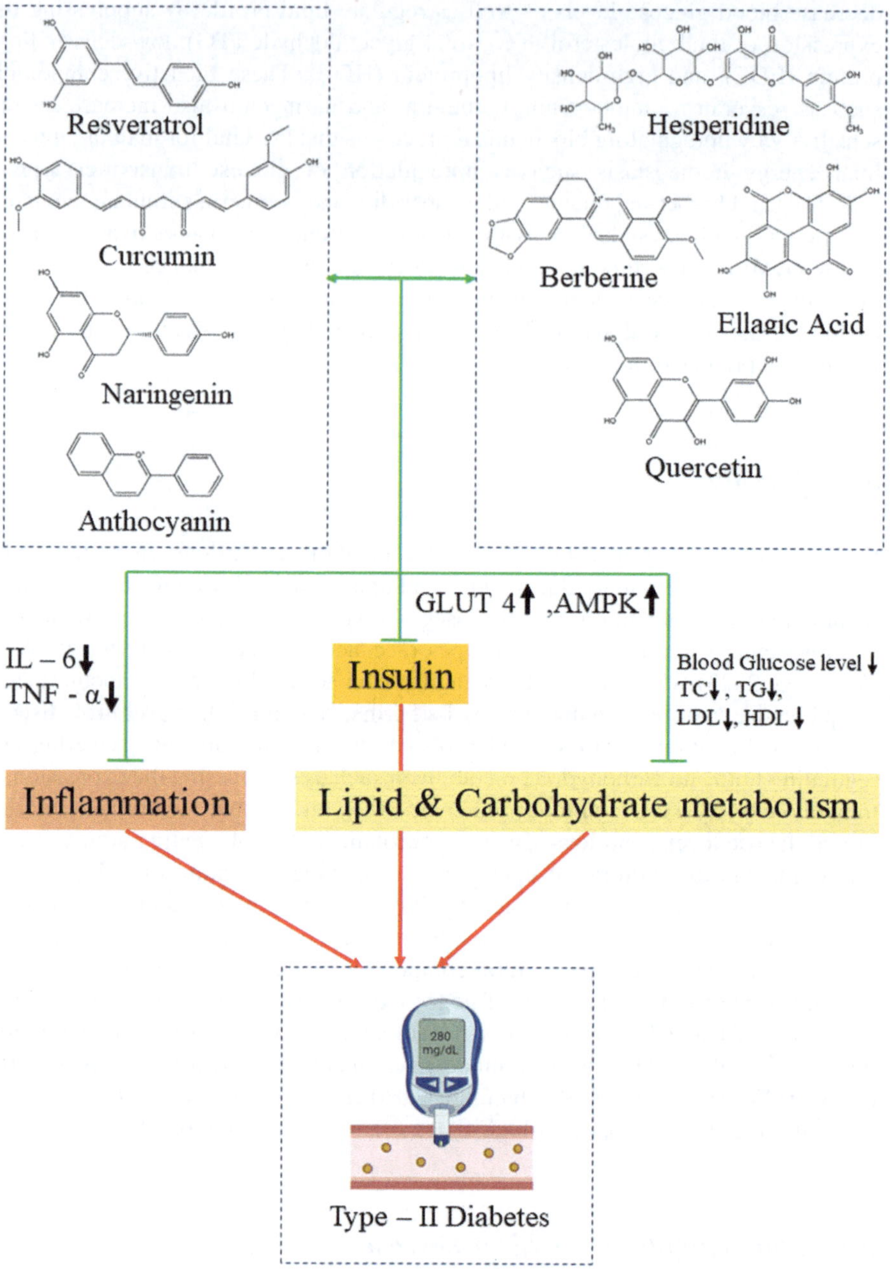

Fig. 2 Therapeutic activities of some major bioactive compounds against type 2 diabetes

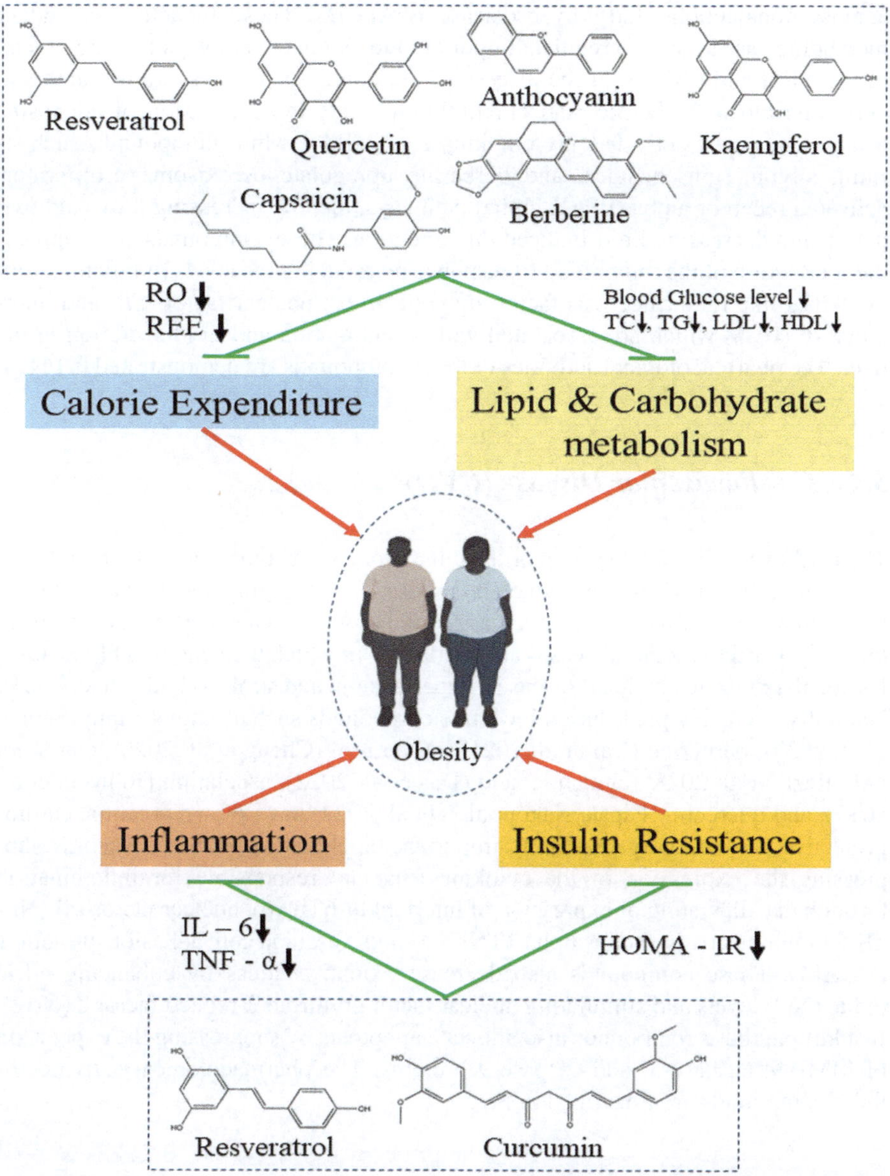

Fig. 3 Therapeutic activities of some major bioactive compounds against obesity

et al. 2022). Natural plant compounds such as resveratrol (Charytoniuk et al. 2017), anthocyanin (Charytoniuk et al. 2017; Fan et al. 2019), myricitrin (Gao et al. 2018; Kim et al. 2020), curcumin (Um et al. 2013; Wu et al. 2019), berberine (Zhang et al. 2019), rutin (Tung et al. 2021), silybin (Shuang et al. 2021), and ginsenoside (Hu et al. 2017; Liang et al. 2021) have been proclaimed to possess biological activity against nonalcoholic fatty liver disease (NAFLD). These bioactive secondary metabolites are known to regulate important biochemical pathways that are crucial to the onset and severity of the disease. Compounds such as resveratrol, anthocyanin, myricitrin, berberine, and curcumin stimulate the expression of adenosine 5-monophosphate-activated protein kinase (AMPK), while compounds such as rutin, silybin, ginsenosides, and berberine upregulate peroxisome proliferator-activated receptor alpha (PPAR-alpha) pathways, thereby increasing fatty acid oxidation and decreasing lipid-induced inflammation. These compounds also suppress the production of the inflammatory response by reducing the levels of inflammatory cytokines like tumor necrosis factor-alpha (TNF-α), nuclear factor- κB, and interleukin-6 (IL-6) which are associated with inflammation and lipid accretion in the liver. The pharmacological activities of these compounds are demonstrated in Fig. 4.

5.4 Cardiovascular Disease (CVD)

The cardiovascular (CV) system or the circulatory system maintains the blood flow throughout the body's tissues, which is essential for supplying nutrition and oxygen to different parts of the body as well as eliminating toxins and carbon dioxide from the body. Cardiovascular diseases are conditions in which the heart and blood arteries are affected causing heart diseases, hypertension, and stroke (Gaidai et al. 2023). Naturally occurring plant-based bioactive compounds such as anthocyanin (Sapian et al. 2022), berberine (Cai et al. 2021), resveratrol (Cheng et al. 2020), curcumin (Abolfazli et al. 2023), cinnamic acid (Das et al. 2022), aspalathin (Johnson et al. 2017), and quercetin (Papakyriakopoulou et al. 2022) have shown potential cardioprotective properties by regulating important biochemical pathways such as suppressing the expression of the cytokines that are responsible for inflammation through the alleviating of expression of interleukin-6 (IL-6), nuclear factor-κB (NF-κB), tumor necrosis factor-alpha (TNF-α), and vascular cell adhesion protein 1 (VCAM). These compounds also decrease oxidative stress by enhancing nitric oxide (NO) levels and stimulating nuclear factor erythroid 2-related factor 2 (Nrf2). In addition, these compounds also alleviate apoptosis by suppressing the expression of MMP-9, caspase-3, and caspase-9 proteins. The pharmacological activities of these compounds are illustrated in Fig. 5.

5.5 Hypercholesterolemia

Hypercholesterolemia is a medical condition marked by increased blood cholesterol, especially low-density lipoprotein (LDL). Hypercholesterolemia increases the chance of heart diseases such as atherosclerotic heart disease and stroke (Hervas and Ascaso 2019; Nordestgaard et al. 2013). Natural bioactive compounds such as resveratrol (Rašković et al. 2019), hesperidin (Bawazeer et al. 2016), and

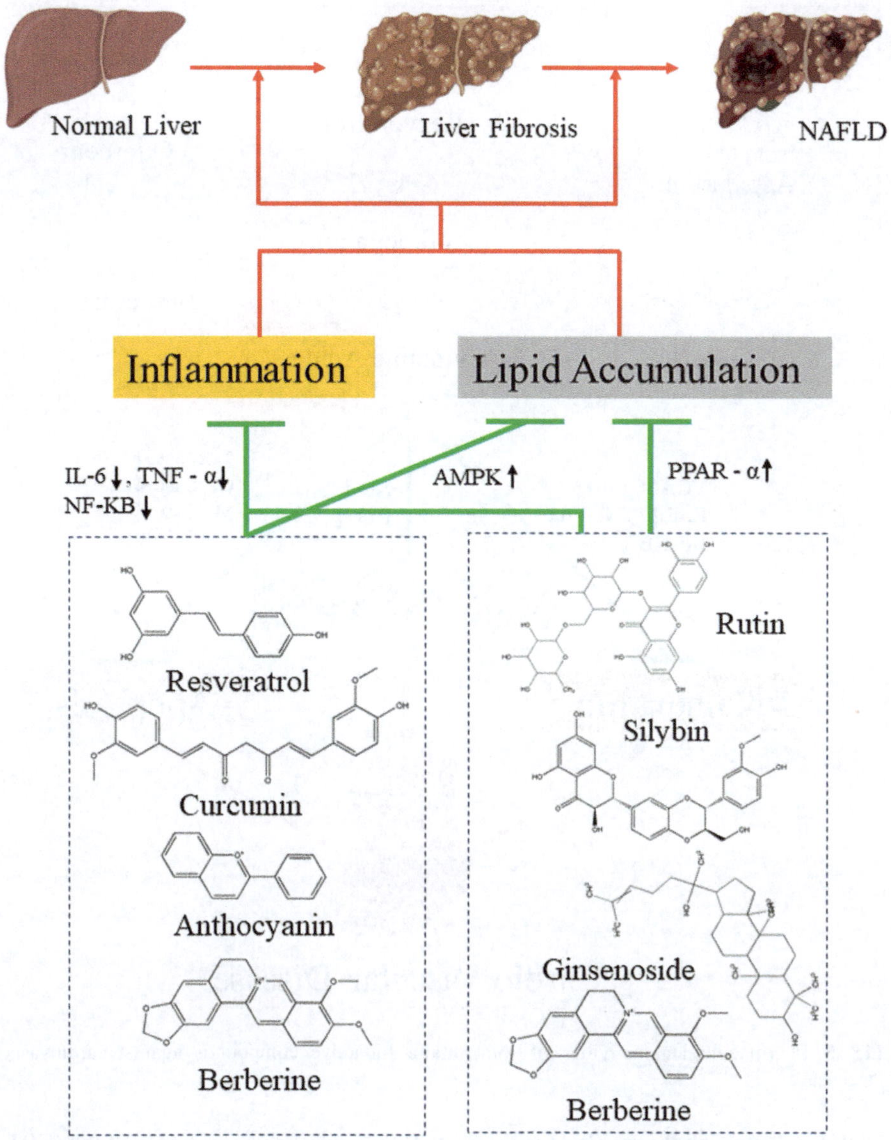

Fig. 4 Pharmacological activities of some major bioactive compounds against nonalcoholic fatty liver disease

naringenin (Bawazeer et al. 2017) decrease the expression of LDL by stimulating the upregulation of sterol regulatory element binding proteins (SREBPs) that upon activation, enhance the expression of low-density lipoprotein receptors (LDLR). In addition, compounds like ellagic acid (Kubota et al. 2019), berberine (Pang et al. 2015), monascin, and ankaflavin (Lee et al. 2018) also decrease the expression of

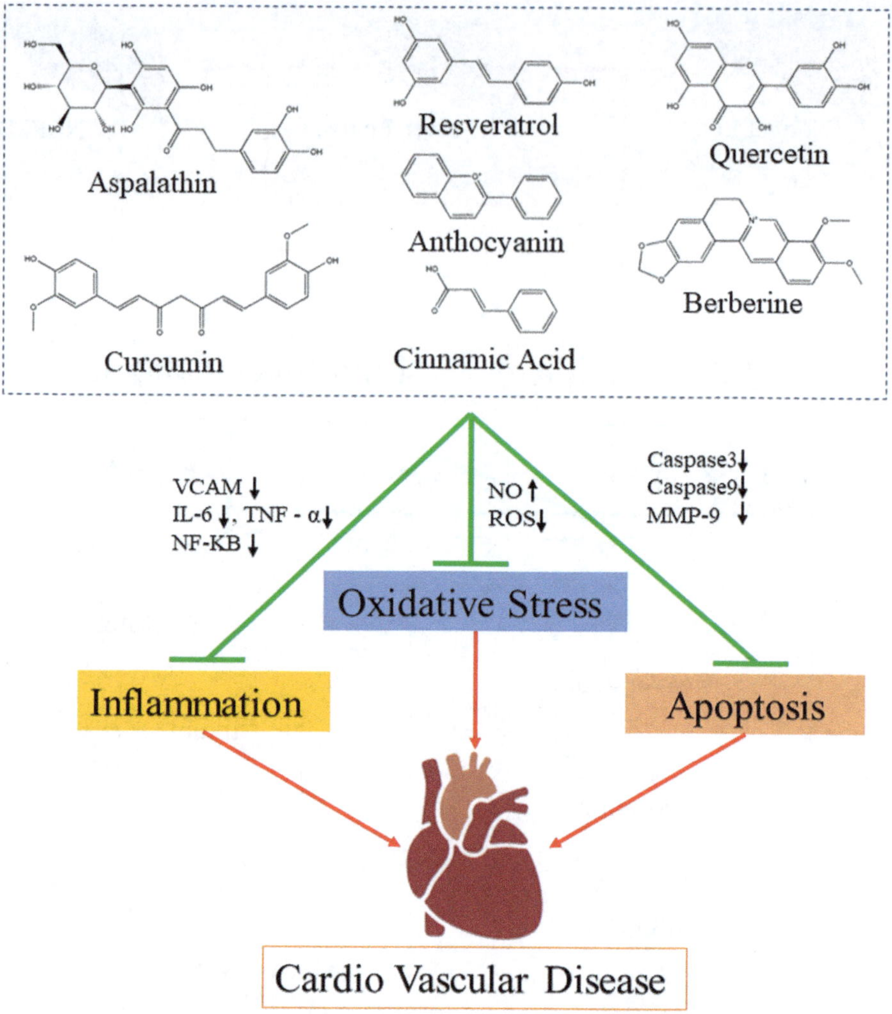

Fig. 5 Pharmacological activities of some major bioactive compounds against cardiovascular disease

LDL by enhancing the levels of active LDLR through the upregulation of the (ERK) extracellular signal-regulated kinase pathway. These compounds also simultaneously increase high-density lipoprotein (HDL) levels. Figure 6 demonstrates the distinct pharmacological activities of these compounds.

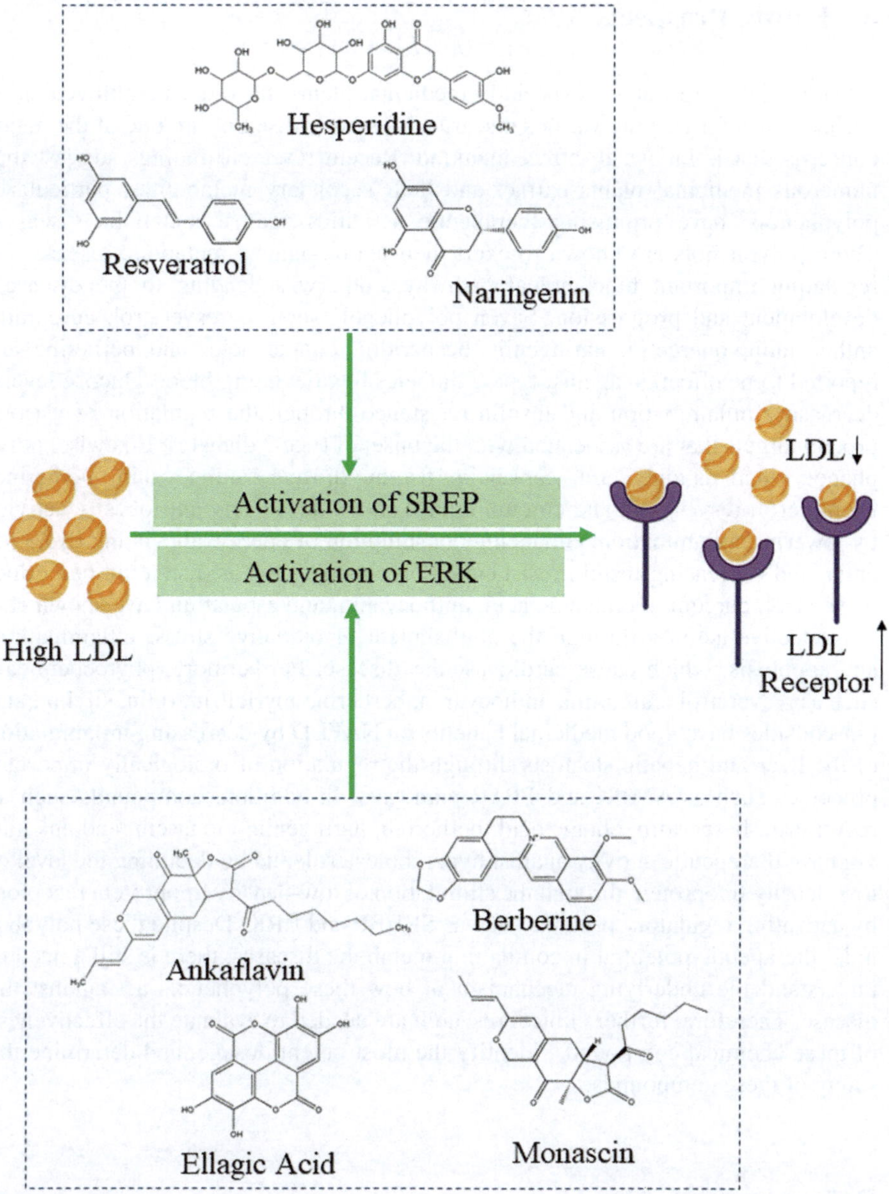

Fig. 6 Pharmacological activities of some major bioactive compounds against hypercholesterolemia

6 Future Prospects

Ethnomedicinal resources, especially medicinal plants, have been employed since ancient times for treating various diseases. Metabolic diseases are one of the major concerns that is known to affect mankind. Recent research findings suggest that numerous medicinal plants extract and their secondary metabolites, particularly polyphenols, have promising therapeutic activities against metabolic diseases. These polyphenols are known to exert their action against metabolic diseases by regulating important biochemical pathways, otherwise leading to the disease's development and progression. Seven polyphenols such as resveratrol, curcumin, anthocyanin, quercetin, naringenin, hesperidin, ellagic acid, and berberine are reported to be effective against type 2 diabetes by alleviating blood glucose levels, decreasing inflammation and insulin resistance through the regulation of various protein targets that are associated with the onset of type 2 diabetes. Likewise, polyphenols such as quercetin, capsaicin, fisetin, dioscin, anthocyanin, berberine, kaempferol, resveratrol, and curcumin have shown promising anti-obesity activity by lowering inflammation, enhancing consumption of energy, alleviating dyslipidemia, and decreasing insulin resistance. Polyphenols such as quercetin, berberine, resveratrol, curcumin, cinnamic acid, anthocyanin, and aspalathin have shown cardioprotective activity through the abolishment of oxidative stress, inflammation, and apoptosis, which cause cardiovascular disease. Furthermore, phytochemicals such as resveratrol, curcumin, anthocyanin, berberine myricitrin, rutin, silybin, and ginsenosides have good medicinal benefits on NAFLD by decreasing inflammation of the liver and hepatic steatosis through the regulation of biologically important processes such as AMPK and PPAR pathways. In addition, compounds such as resveratrol, hesperidin, ellagic acid, berberine, naringenin, monascin, and ankaflavin have therapeutic activity against hypercholesterolemia by declining the level of low-density lipoprotein through the stimulation of low-density lipoprotein receptors by activating regulatory proteins such as SREBP and ERK. Despite these polyphenols' therapeutic potential in combatting metabolic diseases, there is still a need to understand the underlying mechanism of how these polyphenols act against the disease. Therefore, further clinical research are needed to evaluate the effectiveness of these chemical compounds, identify the most potent dosage, and determine the safety of these compounds.

References

Abdou HM, Hamaad FA, Ali EY, Ghoneum MH (2022) Antidiabetic efficacy of Trifolium alexandrinum extracts hesperetin and quercetin in ameliorating carbohydrate metabolism and activating IR and AMPK signaling in the pancreatic tissues of diabetic rats. Biomed Pharmacother 149:112838

Abedin M, Rahman S, Islam M, Hayder S, Ashraf M, Pelc J, Sayed M (2020) Honeyweed (Leonurus sibiricus) and buckwheat (*Fagopyrum esculentum*) supplemented diet improve

growth performance, lipid profile and hematological values of broiler chicks. bioRxiv, 2020.2006. 2018.159764
Abolfazli S, Mortazavi P, Kheirandish A, Butler AE, Jamialahmadi T, Sahebkar A (2023) Regulatory effects of curcumin on nitric oxide signaling in the cardiovascular system. Nitric Oxide 143:16
Abu O, Umar A, Eiremiokhae C (2023) Investigation of the cardio protective capacity of aqueous extract of Icacina Trichantha leaves in rats exposed to CCl4. J Genet Cell Biol 6(1):322–328
Ahmed D, Kumar V, Verma A, Gupta PS, Kumar H, Dhingra V, Mishra V, Sharma M (2014) Antidiabetic, renal/hepatic/pancreas/cardiac protective and antioxidant potential of methanol/dichloromethane extract of Albizzia Lebbeck Benth. Stem bark (ALEx) on streptozotocin induced diabetic rats. BMC Complement Altern Med 14:1–17
Ahmed OM, Hassan MA, Abdel-Twab SM, Azeem MNA (2017) Navel orange peel hydroethanolic extract, naringin and naringenin have anti-diabetic potentials in type 2 diabetic rats. Biomed Pharmacother 94:197–205
Akash MSH, Rehman K, Tariq M, Chen S (2015) Zingiber officinale and type 2 diabetes mellitus: evidence from experimental studies. Crit Rev Eukaryot Gene Expr 25(2):91
Al Tamimi JZ, Alshammari GM, AlFaris NA, Alagal RI, Aljabryn DH, Albekairi NA, Alkhateeb MA, Yahya MA (2022) Ellagic acid protects against non-alcoholic fatty liver disease in streptozotocin-diabetic rats by activating AMPK. Pharm Biol 60(1):25–37
Al Zarzour RH, Ahmad M, Asmawi MZ, Kaur G, Saeed MAA, Al-Mansoub MA, Saghir SAM, Usman NS, Al-Dulaimi DW, Yam MF (2017) Phyllanthus niruri standardized extract alleviates the progression of non-alcoholic fatty liver disease and decreases atherosclerotic risk in Sprague–Dawley rats. Nutrients 9(7):766
Al-Qalhati IRS, Waly M, Al-Attabi Z, Al-Subhi LK (2016) Protective effect of Pteropyrum scoparium and Oxalis corniculata against streptozotocin-induced diabetes in rats. FASEB J 30:1176
Araujo LC, Feitosa KB, Murata GM, Furigo IC, Teixeira SA, Lucena CF, Ribeiro LM, Muscará MN, Costa SK, Donato J Jr (2018) Uncaria tomentosa improves insulin sensitivity and inflammation in experimental NAFLD. Sci Rep 8(1):11013
Ashraf H, Butt MS, Nadeem M, Aadil RM, Rusu AV, Trif M (2022) Microencapsulated curcumin from Curcuma longa modulates diet-induced hypercholesterolemia in Sprague Dawley rats. Front Nutr 9:1026890
Athesh K, Jothi G (2017) Pharmacological screening of anti-obesity potential of Acorus calamus Linn. In high fat cafeteria diet fed obese rats. Asian J Pharm Clin Res 10(4):384–390
Bae U-J, Park E-O, Park J, Jung S-J, Ham H, Yu K-W, Park Y-J, Chae S-W, Park B-H (2018) Gypenoside UL4-rich gynostemma pentaphyllum extract exerts a hepatoprotective effect on diet-induced nonalcoholic fatty liver disease. Am J Chin Med 46(06):1315–1332
Bais S, Singh GS, Sharma R (2014) Anti-obesity and hypolipidemic activity of Moringa oleifera leaves against high fat diet-induced obesity in rats. Adv Biol 2014:162914
Bawazeer NA, Choudhry H, Zamzami MA, Abdulaal WH, Middleton B, Moselhy SS (2016) Role of hesperetin in LDL-receptor expression in hepatoma HepG2 cells. BMC Complement Altern Med 16:1–7
Bawazeer NA, Choudary H, Zamzami MA, Abdulaal WH, Zeyadi M, Al Bukhari A, Middleton B, Moselhy SS (2017) Possible regulation of LDL-receptor by naringenin in HepG2 hepatoma cell line. Afr J Tradit Complement Altern Med 14(1):278–287
Beshel JA, Beshel FN, Nwangwa JN, Okon IA, Ejim CI, Owu DU (2022) Cardioprotective role of Theobroma cacao against isoproterenol-induced acute myocardial injury. Cardiovasc Hematol Agents Med Chem 20(1):75–80
Bouafia M, Amamou F, Gherib M, Benaissa M, Azzi R, Nemmiche S (2021) Ethnobotanical and ethnomedicinal analysis of wild medicinal plants traditionally used in Naâma, Southwest Algeria. Vegetos 34:654–662
Cai Y, Xin Q, Lu J, Miao Y, Lin Q, Cong W, Chen K (2021) A new therapeutic candidate for cardiovascular diseases: Berberine. Front Pharmacol 12:631100

Charytoniuk T, Drygalski K, Konstantynowicz-Nowicka K, Berk K, Chabowski A (2017) Alternative treatment methods attenuate the development of NAFLD: A review of resveratrol molecular mechanisms and clinical trials. Nutrition 34:108–117

Chege BM, Nyaga NM, Kaur PS, Misigo WO, Khan N, Wanyonyi WC, Mwangi PW (2023) The significant antidyslipidemic, hypoglycemic, antihyperglycemic, and anti-obesity activities of the aqueous extracts of agave Sisalana juice are partly mediated via modulation of calcium signaling pathways. Heliyon 9(2):e12400

Chen X, Yu J, Shi J (2018) Management of diabetes mellitus with puerarin, a natural isoflavone from Pueraria lobata. Am J Chin Med 46(08):1771–1789

Cheng CK, Luo JY, Lau CW, Chen ZY, Tian XY, Huang Y (2020) Pharmacological basis and new insights of resveratrol action in the cardiovascular system. Br J Pharmacol 177(6):1258–1277

Chhetri HP, Yogol NS, Sherchan J, Anupa K, Mansoor S, Thapa P (2008) Phytochemical and antimicrobial evaluations of some medicinal plants of Nepal. Kathmandu Univ J Sci Eng Technol 4(1):49–54

Chuengsamarn S, Rattanamongkolgul S, Phonrat B, Tungtrongchitr R, Jirawatnotai S (2014) Reduction of atherogenic risk in patients with type 2 diabetes by curcuminoid extract: a randomized controlled trial. J Nutr Biochem 25(2):144–150

Cuevas-Durán RE, Medrano-Rodríguez JC, Sánchez-Aguilar M, Soria-Castro E, Rubio-Ruíz ME, Del Valle-Mondragón L, Sánchez-Mendoza A, Torres-Narvaéz JC, Pastelín-Hernández G, Ibarra-Lara L (2017) Extracts of Crataegus oxyacantha and Rosmarinus officinalis attenuate ischemic myocardial damage by decreasing oxidative stress and regulating the production of cardiac vasoactive agents. Int J Mol Sci 18(11):2412

Darvish Damavandi R, Shidfar F, Najafi M, Janani L, Masoodi M, Akbari-Fakhrabadi M, Dehnad A (2021) Effect of Portulaca Oleracea (purslane) extract on liver enzymes, lipid profile, and glycemic status in nonalcoholic fatty liver disease: a randomized, double-blind clinical trial. Phytother Res 35(6):3145–3156

Das P (2021) Designing a framework of Ethnomedicinal plant knowledge integration using OSS. In: Research anthology on usage and development of open source software. IGI Global, Hershey, PA, pp 466–479

Das G, Gonçalves S, Heredia JB, Romano A, Jiménez-Ortega LA, Gutiérrez-Grijalva EP, Shin HS, Patra JK (2022) Cardiovascular protective effect of cinnamon and its major bioactive constituents: an update. J Funct Foods 97:105045

de la Peña I, Afable T, Dahilig-Talan VR, Cruz P (2023) Review of plant extracts and active components: mechanisms of action for the treatment of obesity-induced cognitive impairment. Brain Sci 13(6):929

Diez-Echave P, Vezza T, Rodriguez-Nogales A, Ruiz-Malagón AJ, Hidalgo-Garcia L, Garrido-Mesa J, Molina-Tijeras JA, Romero M, Robles-Vera I, Pimentel-Moral S (2020) The prebiotic properties of Hibiscus sabdariffa extract contribute to the beneficial effects in diet-induced obesity in mice. Food Res Int 127:108722

Dong K, Shang J, Tao L (2017) Effect of berberine combined with metformin on serum inflammatory factors and islet function in type 2 diabetes mellitus. J Clin Pathol Res 37:1418–1422

Dougnon TV, Bankolé HS, Klotoé JR, Sènou M, Fah L, Koudokpon H, Akpovi C, Dougnon TJ, Addo P, Loko F (2014) Treatment of hypercholesterolemia: screening of Solanum macrocarpon Linn (Solanaceae) as a medicinal plant in Benin. Avicenna J Phytomed 4(3):160

Draginic N, Jakovljevic V, Andjic M, Jeremic J, Srejovic I, Rankovic M, Tomovic M, Nikolic Turnic T, Svistunov A, Bolevich S (2021) Melissa officinalis L. as a nutritional strategy for Cardioprotection. Front Physiol 12:661778

Erbay E, Hotamıslıgil GS (2014) In: Bray GA, Bouchard C (eds) Inflammatory causes of obesity and metabolic diseases. CRC, Boca Raton, FL, p 573

Fan M, Choi Y-J, Tang Y, Bae SM, Yang HP, Kim E-K (2019) Efficacy and mechanism of polymerized anthocyanin from grape-skin extract on high-fat-diet-induced nonalcoholic fatty liver disease. Nutrients 11(11):2586

Fernandes-da-Silva A, Miranda CS, Santana-Oliveira DA, Oliveira-Cordeiro B, Rangel-Azevedo C, Silva-Veiga FM, Martins FF, Souza-Mello V (2021) Endoplasmic reticulum stress as the basis of obesity and metabolic diseases: focus on adipose tissue, liver, and pancreas. Eur J Nutr 60(6):2949–2960

Gaidai O, Cao Y, Loginov S (2023) Global cardiovascular diseases death rate prediction. Curr Probl Cardiol 48(5):101622

Gao J, Chen S, Qiu Z, Fang L, Zhang L, Guo C, Chen T, Qiu L (2018) Myricitrin ameliorates ethanol-induced steatosis in mouse AML12 liver cells by activating AMPK, and reducing oxidative stress and expression of inflammatory cytokines. Mol Med Rep 17(5):7381–7387

Ghadimi M, Foroughi F, Hashemipour S, Rashidi Nooshabadi M, Ahmadi MH, Ahadi Nezhad B, Khadem Haghighian H (2021) Randomized double-blind clinical trial examining the Ellagic acid effects on glycemic status, insulin resistance, antioxidant, and inflammatory factors in patients with type 2 diabetes. Phytother Res 35(2):1023–1032

Grundy SM, Brewer HB Jr, Cleeman JI, Smith SC Jr, Lenfant C (2004) Definition of metabolic syndrome: report of the National Heart, Lung, and Blood Institute/American Heart Association conference on scientific issues related to definition. Circulation 109(3):433–438

Gubina-Vakulyck GI, Gorbach TV, Tkachenko AS, Domarev AP, Martynova SM, Denysenko SA, Shcholok TT, Molchanova AV (2018) RIBES NIGRUM fruit extract shows CARDIOPROTECTIVE properties in experimental acute ischemic heart disease. Studia Universitatis Vasile Goldis Seria Stiintele Vietii (Life Sci Ser) 28(3):122–130

Gujral JK, Karunanand B, Sarin M, Gaurav A, Kashyap P (2022) Non-alcoholic fatty liver disease (NAFLD): s review of pathophysiology with clinical correlation grade 1 and grade 2 NAFLD. Indian J Basic Appl Med Res 12(1)

Gurib-Fakim A (2006) Medicinal plants: traditions of yesterday and drugs of tomorrow. Mol Aspects Med 27(1):1–93

Harvey AL, Edrada-Ebel R, Quinn RJ (2015) The re-emergence of natural products for drug discovery in the genomics era. Nat Rev Drug Discov 14(2):111–129

Hasan T, Islam A, Ara Khanom Riva R, Rahman MN, Ahmed S, Islam MA, Daula ASU (2023) Phytochemicals from Zingiber capitatum rhizome as potential α-glucosidase, α-amylase, and glycogen phosphorylase inhibitors for the management of type-II diabetes mellitus: inferences from in vitro, in vivo and in-silico investigations. Arab J Chem 16(10):105128

Hegazy GA, Alnoury AM, Gad HG (2013) The role of acacia Arabica extract as an antidiabetic, antihyperlipidemic, and antioxidant in streptozotocin-induced diabetic rats. Saudi Med J 34(7):727–733

Hervas SM, Ascaso FJ (2019) Hypercholesterolemia. In: Encyclopedia of endocrine diseases, vol 1, pp 320–326. https://doi.org/10.1016/B978-0-12-801238-3.65340-0

Homayouni F, Haidari F, Hedayati M, Zakerkish M, Ahmadi K (2018) Blood pressure lowering and anti-inflammatory effects of hesperidin in type 2 diabetes; a randomized double-blind controlled clinical trial. Phytother Res 32(6):1073–1079

Hu C, Sun L, Yang Q, Lu D, Luo S (2017) Ginsenosides from stems and leaves of ginseng prevent ethanol-induced lipid accumulation in human L02 hepatocytes. Chin J Integr Med 23:438–444

Huang J (2014) Pediatric intensive care unit and inborn errors of metabolism. In: Chinese Pediatric Emergency Medicine, pp 337–339

Hwang J-T, Shin EJ, Chung M-Y, Park JH, Chung S, Choi H-K (2018) Ethanol extract of Allium fistulosum inhibits development of non-alcoholic fatty liver disease. Nutr Res Pract 12(2):110

Jain D, Baheti A, Jain S, Khandelwal K (2010) Use of medicinal plants among tribes in Satpuda region of Dhule and Jalgaon districts of Maharashtra—an ethnobotanical survey. Indian J Tradit Knowl 9(1):152–157

Jayaprasad B, Thamayandhi D, Sharavanan P (2012) Traditionally using antidiabetic medicinal plants in Tamil Nadu. Int J Res Pharmaceut Biosci 2(1):1–8

Jazayeri-Tehrani SA, Rezayat SM, Mansouri S, Qorbani M, Alavian SM, Daneshi-Maskooni M, Hosseinzadeh-Attar M-J (2019) Nano-curcumin improves glucose indices, lipids, inflammation, and Nesfatin in overweight and obese patients with non-alcoholic fatty liver disease (NAFLD): a double-blind randomized placebo-controlled clinical trial. Nutr Metab 16:1–13

Jiang S-J, Dong H, Li J-B, Xu L-J, Zou X, Wang K-F, Lu F-E, Yi P (2015) Berberine inhibits hepatic gluconeogenesis via the LKB1-AMPK-TORC2 signaling pathway in streptozotocin-induced diabetic rats. World J Gastroenterol: WJG 21(25):7777

Johnson R, Dludla PV, Muller CJ, Huisamen B, Essop MF, Louw J (2017) The transcription profile unveils the cardio-protective effect of aspalathin against lipid toxicity in an in vitro H9c2 model. Molecules 22(2):219

Kim DY, Kim SR, Jung UJ (2020) Myricitrin ameliorates hyperglycemia, glucose intolerance, hepatic steatosis, and inflammation in high-fat diet/streptozotocin-induced diabetic mice. Int J Mol Sci 21(5):1870

Kubota S, Tanaka Y, Nagaoka S (2019) Ellagic acid affects mRNA expression levels of genes that regulate cholesterol metabolism in HepG2 cells. Biosci Biotechnol Biochem 83(5):952–959

Kumar JS, Ruchi J, Nilesh J (2019) Anti-obesity potential of Aconitum heterophyllum roots. Int J Pharm Bio Sci 10(3):65–74

Kumar M, Rawat S, Nagar B, Kumar A, Pala NA, Bhat JA, Bussmann RW, Cabral-Pinto M, Kunwar R (2021) Implementation of the use of ethnomedicinal plants for curing diseases in the Indian Himalayas and its role in sustainability of livelihoods and socioeconomic development. Int J Environ Res Public Health 18(4):1509

Latha RCR, Daisy P (2013) Therapeutic potential of octyl gallate isolated from fruits of Terminalia bellirica in streptozotocin-induced diabetic rats. Pharm Biol 51(6):798–805

Lee C-L, Wen J-Y, Hsu Y-W, Pan T-M (2018) The blood lipid regulation of Monascus-produced monascin and ankaflavin via the suppression of low-density lipoprotein cholesterol assembly and stimulation of apolipoprotein A1 expression in the liver. J Microbiol Immunol Infect 51(1):27–37

Lee YS, Lee DY, Kwon DY, Kang OH (2020) Improvement effect of non-alcoholic fatty liver disease by Curcuma longa L. extract. Kor J Med Crop Sci 28(4):276–286

Li D, Zhang Y, Liu Y, Sun R, Xia M (2015) Purified anthocyanin supplementation reduces dyslipidemia, enhances antioxidant capacity, and prevents insulin resistance in diabetic patients. J Nutr 145(4):742–748

Liang W, Zhou K, Jian P, Chang Z, Zhang Q, Liu Y, Xiao S, Zhang L (2021) Ginsenosides improve nonalcoholic fatty liver disease via integrated regulation of gut microbiota, inflammation and energy homeostasis. Front Pharmacol 12:622841

Liu Z, Qu C-Y, Li J-X, Wang Y-F, Li W, Wang C-Z, Wang D-S, Song J, Sun G-Z, Yuan C-S (2021) Hypoglycemic and hypolipidemic effects of malonyl ginsenosides from American ginseng (Panax quinquefolius L.) on type 2 diabetic mice. ACS Omega 6(49):33652–33664

Luo X, Zhang B, Pan Y, Gu J, Tan R, Gong P (2022) Phyllanthus emblica aqueous extract retards hepatic steatosis and fibrosis in NAFLD mice in association with the reshaping of intestinal microecology. Front Pharmacol 13:893561

Ma C, Yu H, Xiao Y, Wang H (2017) Momordica charantia extracts ameliorate insulin resistance by regulating the expression of SOCS-3 and JNK in type 2 diabetes mellitus rats. Pharm Biol 55(1):2170–2177

Manjunath K, Subash K, Tadvi N, Manikanta M, Rao U (2016) Effect of Aloe vera leaf extract on blood glucose levels in alloxan induced diabetic rats. Natl J Physiol Pharm Pharmacol 6(5):471

Meneses ME, Martínez-Carrera D, Torres N, Sánchez-Tapia M, Aguilar-López M, Morales P, Sobal M, Bernabé T, Escudero H, Granados-Portillo O (2016) Hypocholesterolemic properties and prebiotic effects of Mexican Ganoderma lucidum in C57BL/6 mice. PLoS One 11(7):e0159631

Mikami T, Kim J, Park J, Lee H, Yaicharoen P, Suidasari S, Yokozawa M, Yamauchi K (2021) Olive leaf extract prevents obesity, cognitive decline, and depression and improves exercise capacity in mice. Sci Rep 11(1):12495

Mohiti-Ardekani J, Asadi S, Ardakani AM, Rahimifard M, Baeeri M, Momtaz S (2019) Curcumin increases insulin sensitivity in C2C12 muscle cells via AKT and AMPK signaling pathways. Cogent Food Agric 5(1):1577532

Momoh BJ, Okere SO, Anyanwu GO (2022) The anti-obesity effect of Allium cepa L. leaves on high fat diet induced obesity in male Wistar rats. Clin Complement Med Pharmacol 2(3):100035

Movahed A, Nabipour I, Lieben Louis X, Thandapilly SJ, Yu L, Kalantarhormozi M, Rekabpour SJ, Netticadan T (2013) Antihyperglycemic effects of short term resveratrol supplementation in type 2 diabetic patients. Evid Based Complement Alternat Med 2013:851267

Nordestgaard BG, Chapman MJ, Humphries SE, Ginsberg HN, Masana L, Descamps OS, Wiklund O, Hegele RA, Raal FJ, Defesche JC (2013) Familial hypercholesterolaemia is underdiagnosed and undertreated in the general population: guidance for clinicians to prevent coronary heart disease: consensus statement of the European atherosclerosis society. Eur Heart J 34(45):3478–3490

Noushida N, Nayak RP, Sultana R, Salian TR, Alobid S, Almadani ME, Ahmad F, Gilkaramenthi R, Asdaq SMB, Almoteer AI (2024) Cardioprotective effects of Callicarpa tomentosa leaf extract in Wistar albino rats against isoproterenol-induced myocardial necrosis: phytochemical analysis and in vitro antioxidant study. J King Saud Univ Sci 36(3):103100

Nyakudya TT, Tshabalala T, Dangarembizi R, Erlwanger KH, Ndhlala AR (2020) The potential therapeutic value of medicinal plants in the management of metabolic disorders. Molecules 25(11):2669

Olasehinde OR, Afolabi OB, Owolabi OV, Akawa AB, Omiyale OB (2022) GC–MS analysis of phytochemical constituents of methanolic fraction of Annona muricata leaf and its inhibition against two key enzymes linked to type II diabetes. Sci Afr 16:e01178

Olatoye FJ, Akindele A, Awodele O (2024) The role of Kolaviron, a bioflavonoid from Garcinia kola, in the management of cardiovascular diseases: a systematic review. Heliyon 10:e27333

Oppedisano F, Muscoli C, Musolino V, Carresi C, Macrì R, Giancotta C, Bosco F, Maiuolo J, Scarano F, Paone S (2020) The protective effect of Cynara cardunculus extract in diet-induced NAFLD: involvement of OCTN1 and OCTN2 transporter subfamily. Nutrients 12(5):1435

Orhan N, Aslan M, Şüküroğlu M, Orhan DD (2013) In vivo and in vitro antidiabetic effect of Cistus laurifolius L. and detection of major phenolic compounds by UPLC–TOF-MS analysis. J Ethnopharmacol 146(3):859–865

Pang B, Zhao L-H, Zhou Q, Zhao T-Y, Wang H, Gu C-J, Tong X-L (2015) Application of berberine on treating type 2 diabetes mellitus. Int J Endocrinol 2015:905749

Papakyriakopoulou P, Velidakis N, Khattab E, Valsami G, Korakianitis I, Kadoglou NP (2022) Potential pharmaceutical applications of quercetin in cardiovascular diseases. Pharmaceuticals 15(8):1019

Paul RK, Ahmad I, Patel H, Kumar V, Raza K (2023) Phytochemicals from Amberboa ramosa as potential DPP-IV inhibitors for the management of type-II diabetes mellitus: inferences from in-silico investigations. J Mol Struct 1271:134045

Pontes VCB, Rodrigues DP, Caetano A, Gamberini MT (2019) Preclinical investigation of the cardiovascular actions induced by aqueous extract of Pimpinella anisum L. seeds in rats. J Ethnopharmacol 237:74–80

Poornima M, Sindhu G, Billu A, Sruthi C, Nisha P, Gogoi P, Baishya G, Raghu K (2022) Pretreatment of hydroethanolic extract of Dillenia indica L. attenuates oleic acid induced NAFLD in HepG2 cells via modulating SIRT-1/p-LKB-1/AMPK, HMGCR & PPAR-α signaling pathways. J Ethnopharmacol 292:115237

Pothuraju R, Sharma RK, Rather SA, Singh S (2016) Comparative evaluation of anti-obesity effect of Aloe vera and Gymnema sylvestre supplementation in high-fat diet fed C57BL/6J mice. J Intercult Ethnopharmacol 5(4):403

Rafiee L, Keshvari M, Atar AM, Hamidzadeh Z, Dashti GR, Rafieian-Kopaei M, Asgary S (2017) Effect of Gundelia tournefortii L. on some cardiovascular risk factors in an animal model. J Herbmed Pharmacol 6(4):191–195

Rajan M, Chandran V, Shahena S, Anie Y, Mathew L (2022) In vitro and in silico inhibition of α-amylase, α-glucosidase, and aldose reductase by the leaf and callus extracts of Vernonia anthelmintica (L.) Willd. Adv Tradit Med 22:1–15

Rašković A, Ćućuz V, Torović L, Tomas A, Gojković-Bukarica L, Ćebović T, Milijašević B, Stilinović N, Hogervorst JC (2019) Resveratrol supplementation improves metabolic control in rats with induced hyperlipidemia and type 2 diabetes. Saudi Pharmaceut J 27(7):1036–1043

Razani Z, Dastani M, Kazerani HR (2017) Cardioprotective effects of pomegranate (Punica granatum) juice in patients with ischemic heart disease. Phytother Res 31(11):1731–1738

Ren J, Fu L, Nile SH, Zhang J, Kai G (2019) Salvia miltiorrhiza in treating cardiovascular diseases: a review on its pharmacological and clinical applications. Front Pharmacol 10:463061

Saeedi G, Jeivad F, Goharbari M, Gheshlaghi GH, Sabzevari O (2018) Ethanol extract of Crataegus oxyacantha L. ameliorate dietary non-alcoholic fatty liver disease in rat. Drug Res 68(10):553–559

Salehi E, Afshar S, Mehrizi MZ, Chehrei A, Asadi M (2018) Direct reduction of blood serum cholesterol using Thymus vulgaris L.: preliminary biosorption study. Process Biochem 67:155–164

Salvamani S, Gunasekaran B, Shukor MY, Shaharuddin NA, Sabullah MK, Ahmad SA (2016) Anti-HMG-CoA reductase, antioxidant, and anti-inflammatory activities of Amaranthus viridis leaf extract as a potential treatment for hypercholesterolemia. Evid Based Complement Alternat Med 2016:8090841

Santamaria F, Montella S, Mirra V, De Stefano S, Andria G, Parenti G (2013) Respiratory manifestations in patients with inherited metabolic diseases. Eur Respir Rev 22(130):437–453

Sapian S, Taib IS, Katas H, Latip J, Zainalabidin S, Hamid ZA, Anuar NNM, Budin SB (2022) The role of anthocyanin in modulating diabetic cardiovascular disease and its potential to be developed as a nutraceutical. Pharmaceuticals 15(11):1344

Sapian S, Ibrahim Mze AA, Jubaidi FF, Mohd Nor NA, Taib IS, Abd Hamid Z, Zainalabidin S, Mohamad Anuar NN, Katas H, Latip J (2023) Therapeutic potential of Hibiscus sabdariffa Linn. In attenuating cardiovascular risk factors. Pharmaceuticals 16(6):807

Saragih B, Pasiakan M, Wahyudi D (2014) Effect of herbal drink plants Tiwai (Eleutherine Americana Merr) on lipid profile of hypercholesterolemia patients. Int Food Res J 21(3):1163

Shang A, Gan R-Y, Xu X-Y, Mao Q-Q, Zhang P-Z, Li H-B (2021) Effects and mechanisms of edible and medicinal plants on obesity: An updated review. Crit Rev Food Sci Nutr 61(12):2061–2077

Shuang C, Xiao-Jie P, Chao-Liang G, Yi-Tong G, Zhang P-F, Ting-Ting Y, Ji-Yu Z, Qing-Xian H, Cheng L-H, Guang-Ji W (2021) Silybin alleviates hepatic lipid accumulation in methionine-choline deficient diet-induced nonalcoholic fatty liver disease in mice via peroxisome proliferator-activated receptor α. Chin J Nat Med 19(6):401–411

Siddique Z, Shad N, Shah GM, Naeem A, Yali L, Hasnain M, Mahmood A, Sajid M, Idrees M, Khan I (2021) Exploration of ethnomedicinal plants and their practices in human and livestock healthcare in Haripur District, Khyber Pakhtunkhwa, Pakistan. J Ethnobiol Ethnomed 17:1–22

Sobrido M-J, Bauer P, De Koning T, Klopstock T, Nadjar Y, Patterson MC, Synofzik M, Hendriksz CJ (2019) Recommendations for patient screening in ultra-rare inherited metabolic diseases: what have we learned from Niemann-pick disease type C? Orphanet J Rare Dis 14:1–14

Songtrai S, Pratchayasakul W, Arunsak B, Chunchai T, Kongkaew A, Chattipakorn N, Chattipakorn SC, Kaewsuwan S (2022) Cyclosorus terminans extract ameliorates insulin resistance and non-alcoholic fatty liver disease (NAFLD) in high-fat diet (HFD)-induced obese rats. Nutrients 14(22):4895

Sun K, Ding M, Fu C, Li P, Li T, Fang L, Xu J, Zhao Y (2023) Effects of dietary wild bitter melon (Momordica charantia var. abbreviate Ser.) extract on glucose and lipid metabolism in HFD/STZ-induced type 2 diabetic rats. J Ethnopharmacol 306:116154

Tesfaye BA, Berhe AH, Wondafrash DZ, Berhe DF (2021) Cardioprotective effect of crude extract and solvent fractions of Urtica simensis leaves on cyclophosphamide-induced myocardial injury in rats. J Exp Pharmacol Volume 13:147–160

Timmers S, Konings E, Bilet L, Houtkooper RH, van de Weijer T, Goossens GH, Hoeks J, van der Krieken S, Ryu D, Kersten S (2011) Calorie restriction-like effects of 30 days of resveratrol

supplementation on energy metabolism and metabolic profile in obese humans. Cell Metab 14(5):612–622
Tung Y-T, Zeng J-L, Ho S-T, Xu J-W, Li S, Wu J-H (2021) Anti-NAFLD effect of Djulis hull and its major compound, rutin, in mice with high-fat diet (HFD)-induced obesity. Antioxidants 10(11):1694
Um MY, Hwang KH, Ahn J, Ha TY (2013) Curcumin attenuates diet-induced hepatic steatosis by activating AMP-activated protein kinase. Basic Clin Pharmacol Toxicol 113(3):152–157
Vlavcheski F, Den Hartogh DJ, Giacca A, Tsiani E (2020) Amelioration of high-insulin-induced skeletal muscle cell insulin resistance by resveratrol is linked to activation of AMPK and restoration of GLUT4 translocation. Nutrients 12(4):914
Wang L, Ladurner A, Latkolik S, Schwaiger S, Linder T, Hošek J, Palme V, Schilcher N, Polanský OE, Heiss EH (2016) Leoligin, the major lignan from edelweiss (Leontopodium nivale subsp. alpinum), promotes cholesterol efflux from THP-1 macrophages. J Nat Prod 79(6):1651–1657
Wang Y, Yang SH, Zhong K, Jiang T, Zhang M, Kwan HY, Su T (2020) Network pharmacology-based strategy for the investigation of the anti-obesity effects of an ethanolic extract of Zanthoxylum bungeanum maxim. Front Pharmacol 11:572387
Wang H, Tang C, Gao Z, Huang Y, Zhang B, Wei J, Zhao L, Tong X (2021) Potential role of natural plant medicine Cyclocarya paliurus in the treatment of type 2 diabetes mellitus. J Diabetes Res 2021:1
Wang S, Du Q, Meng X, Zhang Y (2022a) Natural polyphenols: a potential prevention and treatment strategy for metabolic syndrome. Food Funct 13(19):9734–9753
Wang R, Wang L, Wu H, Zhang L, Hu X, Li C, Liu S (2022b) Noni (Morinda citrifolia L.) fruit phenolic extract supplementation ameliorates NAFLD by modulating insulin resistance, oxidative stress, inflammation, liver metabolism and gut microbiota. Food Res Int 160:111732
Wu B, Xiao Z, Zhang W, Chen H, Liu H, Pan J, Cai X, Liang G, Zhou B, Shan X (2019) A novel resveratrol-curcumin hybrid, a19, attenuates high fat diet-induced nonalcoholic fatty liver disease. Biomed Pharmacother 110:951–960
Wu Y, Wang X, Yang L, Kang S, Yan G, Han Y, Fang H, Sun H (2023) Therapeutic effects of Alisma orientale and its active constituents on cardiovascular disease and obesity. Am J Chin Med 51(03):623–650
Xiao L, Xiong H, Deng Z, Peng X, Cheng K, Zhang H, Jiang L, Sun Y (2023) Tetrastigma hemsleyanum leaf extracts ameliorate NAFLD in mice with low-grade colitis via the gut–liver axis. Food Funct 14(1):500–515
Yan F, Zheng X (2017) Anthocyanin-rich mulberry fruit improves insulin resistance and protects hepatocytes against oxidative stress during hyperglycemia by regulating AMPK/ACC/mTOR pathway. J Funct Foods 30:270–281
Yang Y, Li Y, Wang J, Sun K, Tao W, Wang Z, Xiao W, Pan Y, Zhang S, Wang Y (2017) Systematic investigation of Ginkgo biloba leaves for treating cardio-cerebrovascular diseases in an animal model. ACS Chem Biol 12(5):1363–1372
Yang CS, Wang H, Sheridan ZP (2018) Studies on prevention of obesity, metabolic syndrome, diabetes, cardiovascular diseases and cancer by tea. J Food Drug Anal 26(1):1–13
Yeh W-J, Ko J, Huang W-C, Cheng W-Y, Yang H-Y (2020) Crude extract of Camellia oleifera pomace ameliorates the progression of non-alcoholic fatty liver disease via decreasing fat accumulation, insulin resistance and inflammation. Br J Nutr 123(5):508–515
Yu Z, Muehleman V (2023) Eating disorders and metabolic diseases. Int J Environ Res Public Health 20(3):2446
Yu G, Luo Z, Zhou Y, Zhang L, Wu Y, Ding L, Shi Y (2019) Uncovering the pharmacological mechanism of Carthamus tinctorius L. on cardiovascular disease by a systems pharmacology approach. Biomed Pharmacother 117:109094
Zang Y, Zhang L, Igarashi K, Yu C (2015) The anti-obesity and anti-diabetic effects of kaempferol glycosides from unripe soybean leaves in high-fat-diet mice. Food Funct 6(3):834–841

Zhang YP, Deng YJ, Tang KR, Chen RS, Liang S, Liang YJ, Han L, Jin L, Liang ZE, Chen YN (2019) Berberine ameliorates high-fat diet-induced non-alcoholic fatty liver disease in rats via activation of SIRT3/AMPK/ACC pathway. Curr Med Sci 39(1):37–43

Zhang X, Zhang B, Zhang C, Sun G, Sun X (2021) Effect of Panax notoginseng saponins and major anti-obesity components on weight loss. Front Pharmacol 11:601751

Zhang P, Wei W, Zhang X, Wen C, Ovatlarnporn C, Olatunji OJ (2023) Antidiabetic and anti-oxidant activities of Mitragyna speciosa (kratom) leaf extract in type 2 diabetic rats. Biomed Pharmacother 162:114689

Zhao L, Zou T, Gomez NA, Wang B, Zhu M-J, Du M (2018) Raspberry alleviates obesity-induced inflammation and insulin resistance in skeletal muscle through activation of AMP-activated protein kinase (AMPK) α1. Nutr Diabetes 8(1):39

Zhao X, An X, Yang C, Sun W, Ji H, Lian F (2023) The crucial role and mechanism of insulin resistance in metabolic disease. Front Endocrinol 14:1149239

Zhu Y, Zhang J, Wei Y, Hao J, Lei Y, Zhao W, Xiao Y, Sun A (2020) The polyphenol-rich extract from chokeberry (Aronia melanocarpa L.) modulates gut microbiota and improves lipid metabolism in diet-induced obese rats. Nutr Metab 17:1–15

Zygmunt K, Faubert B, MacNeil J, Tsiani E (2010) Naringenin, a citrus flavonoid, increases muscle cell glucose uptake via AMPK. Biochem Biophys Res Commun 398(2):178–183

Fermented Food of North East India and Its Potential to Serve as Pharmaceutical Agents

Sanjib Kumar Das, Pranjan Barman, Bhargab Jyoti Sahariah, and Manoj Kumar Deka

Contents

Abstract North East India is incredibly diverse, with over 300 ethnic communities and dialects, as well as a strong traditional knowledge system. These ethnic communities created fermented cuisine using locally accessible raw resources that includes green vegetables, different types of bamboo, milk, soybeans, pork, fish, cereal, and consumed it as part of their regular diet. According to study, these foods have a positive effect on health since they contain a high concentration of probiotics. These probiotics aid in digestion and nutrition absorption as well as also beneficial for immune system. Furthermore, fermented food has been claimed to heal and prevent a variety of ailments. They are reported to have the ability to induces physiological benefits by reducing the level of cholesterol, neutralising free radicals,

S. K. Das
Department of Biotechnology, NETES Institute of Technology and Science, Mirza, Assam, India

P. Barman
Department of Biotechnology, Gauhati University, Guwahati, Assam, India
e-mail: pranjan.barman@gauhati.ac.in

B. J. Sahariah · M. K. Deka (✉)
Department of Pharmaceutics, NETES Institute of Pharmaceutical Science, Mirza, Assam, India
e-mail: manojpharmacy86@gmail.com

© The Author(s), under exclusive license to Springer Nature Singapore Pte Ltd. 2024

A. Das Talukdar et al. (eds.), *Traditional Resources and Tools for Modern Drug Discovery*, Interdisciplinary Biotechnological Advances,
https://doi.org/10.1007/978-981-97-4600-2_22

acting as anticarcinogens, inducing a low-glycemic response, and lowering the risk of life-threatening diseases like cancer, cardiovascular diseases, etc. Intense metabolic processes carried out by a diverse range of microorganisms result in the formation of fermented meals by converting macromolecules in the food matrix into bioactive components such as bioactive peptides (BAPs). Therefore, in this chapter, we addressed different fermented food products created from various resources (vegetables, milk, meat, bamboo, and soybeans) by various ethnic communities of North East India, as well as their potential to function as pharmaceutical agents for health advantages.

Keywords Fermented food · Probiotic · Nutraceuticals · Pharmacological agent

1 Introduction

Ethnic fermented foods can be defined as the foods prepared by various tribal communities using their indigenous knowledge on food fermentation (Ibrahim et al. 2023). North East India is endowed with rich biodiversity and endemic plant wealth of ecological, economic, and cultural significance. The region is extremely diverse, with around more than 300 ethnic communities and an equal number of dialects with a rich traditional knowledge system (Borah et al. 2022; Pohsnem et al. 2023). Since the dawn of time, diverse ethnic populations living in this region have produced and consumed various ethnic fermented food products in their regular diet. They produced fermented meals using locally accessible resources like leafy vegetables, milk, bamboo, soybeans, pork, fish, and cereal (Das et al. 2016, 2022; Sarma and Gupta 2022). Initially, the advantages of fermented foods for health were unclear, but in recent decade many researchers reported the health benefit of fermented foods due to which scientists have been more interested in the fermentation process and the fermented materials it creates (Behera and Balaji 2021; Sharma and Yaiphathoi 2020; Kumari et al. 2022). It is reported probiotic microorganisms are primarily responsible for fermented foods' therapeutic and flavor-enhancing qualities. These probiotics help the body digest food and assimilate nutrients (Merenstein et al. 2023; Soemarie et al. 2021). The positive impact of these probiotic microorganisms on the immune system and general health are also well documented (Marx et al. 2020; Tamang and Lama 2022). Many of them develop nutraceuticals to make functional foods with enhanced nutrient bioavailability and antimicrobial bioactive compounds that make them efficient bio-preservatives (Lando et al. 2023; Grujović et al. 2022). Therefore, these microbes may have high possibilities to act as an antimicrobial agent against multidrug resistant pathogens. Additionally, it has been claimed that fermented foods can be used to cure and prevent a number of illnesses. They are reported to have the ability to induce physiological benefits by reducing the level of cholesterol, neutralising free radicals, acting as anticarcinogens, inducing a low-glycemic response, and lowering the risk of life threaded diseases like cancer, cardiovascular diseases, etc.(Pohsnem et al. 2023; Zommiti et al. 2020; Zacarchenco et al. 2022). Bioactive peptides are other important molecules which are also present in fermented food (Zaky et al. 2022). A variety of dietary proteins from plants and animals contain bioactive peptides, which lactic acid bacteria (LAB) can release during fermentation. These biological entities are typically buried inside the structure of parent proteins but become active when the proteins are broken (Jakubczyk et al. 2020; Jia et al. 2021). These molecules are also reported as antibacterial, anticancer, antihypertensive, antioxidative agent, etc. (Laya et al. 2023; Pohsnem et al. 2023; Şanlier et al. 2019). Intense metabolic processes carried out by a diverse range of microorganisms result in the formation of fermented meals by converting macromolecules in the food matrix into bioactive components such as bioactive peptides (BAPs) (Hayes 2021; Singh et al. 2022). BAPs formed from fermented foods have been demonstrated to have antihypertensive, antioxidant, and antibacterial properties; however, little is known about how BAPs are produced through fermented foods, and many of the foods' BAP resources have not been fully utilized (Guo et al. 2023; Dullius et al. 2020; Patil et al. 2022). Therefore, taking into account the significance and potential health benefits of fermented food in this chapter, we addressed several fermented food products made from various resources (vegetables, milk, meat, bamboo, and soybeans) by various ethnic groups of people, as well as their potential to serve as pharmaceutical agents. Some major health benefit of fermented foods are given in Fig. 1.

Fig. 1 Overall health benefit of fermented food

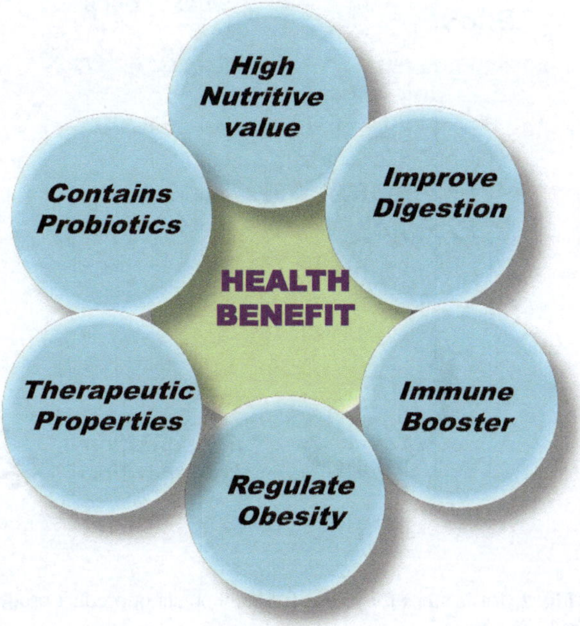

2 Major Fermented Food Products of North East India

2.1 Bamboo Shoot

A significant component of North East Indian traditional cuisine is bamboo shoots. The various ethnic communities of North East India can safely eat about 16 distinct varieties of bamboo plants (Banerjee et al. 2022; Basumatary et al. 2017). *Dendrocalamus hamiltonii*, *Bambusa nutans*, *Bambusa balcooa*, *Bambusa tulda*, *Bambusa nana*, *Bambusa arundinacea*, *Dendrocalamus giganteus*, *Melocanna bambusoides*, and *Teinostachyum wightii shoots* are frequently employed to produce fermented bamboo shoots (Banerjee et al. 2022; Deb and Jamir 2020; Yeptho et al. 2021). The production process of the fermented bamboo shoot varies from tribes to tribes as these tribes have their own traditional method to prepare these products. Below is a list of several traditional fermented bamboo shoot items that the indigenous North East India population frequently eat. Basic steps for bamboo fermentation are shown in Fig. 2, and bamboo shoot and its fermented products are shown in Fig. 3.

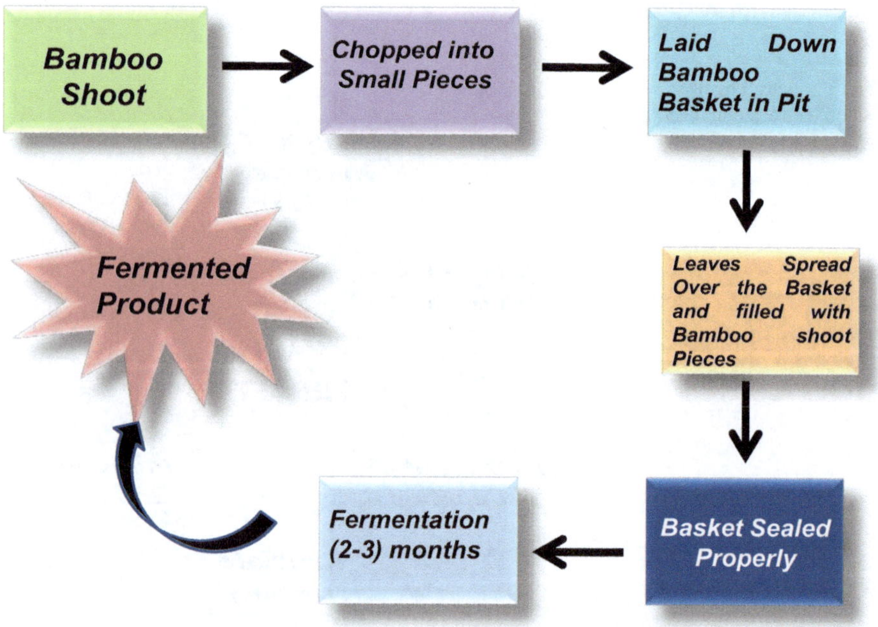

Fig. 2 Basic steps for fermented bamboo shoot product production. However, these steps varies tribe to tribe

Fig. 3 Bamboo shoot and its fermented product

2.1.1 Soibum

"Soibum" is among the most well-liked solid fermented foods of Manipur that is made using tender shoots of *Dendrocalamus giganteus*, *Bambusa tulda*, *Bambusa balcooa*, and *Bambusa pallida* (Behera and Balaji 2021).The "Meetei" community's distinctive delicacy, "Iromba," is made of "soibum." The Kwatha region of Manipur state is home to the traditional method of producing fermented bamboo shoots, which involves using colocasia leaves or polythene sheets (Jeyaram et al. 2009; Premlata et al. 2020). These materials are used to line a locally built chamber that is filled with thinly sliced bamboo shoots. By sufficiently compressing the bamboo shoots during the addition and covering of the upper surface, the densely packed chamber is rendered airtight. The chamber's bottom is perforated to make it easier to drain the juice that is generated during fermentation. "Soibum" production takes around 5 to 8 months, after which the fermented product is removed by opening the seal (Joshi et al. 2023). *Leuconostoc fallax*, *Lactobacillus lactis*, and *Lactobacillus curvatus* are the microorganisms that were isolated from soibum. "Soibum" is said to have anti-diarrheal properties, lower obesity and hyperlipidemia, strengthen the immune system, and stabilize the gastrointestinal tract (Behera and Balaji 2021).

2.1.2 Ekung

In Arunachal Pradesh, "Ekung" is a well-known traditional meal made with liquid bamboo. "Ekung" is made from the young shoots of *Bambusa balcooa*, *Bambusa tulda*, and *Dendrocalamus hamiltonii* (Sonar et al. 2015). When making "Ekung," the gathered bamboo species are cut into little pieces and put in a basket with a pit.

The entrance is completely covered with layers of leaves, and the basket is shut firmly. To provide weight and ensure efficient drainage of the liquid exudates generated during fermentation, stones and big bricks are used. The basket made of bamboo can be opened to harvest the fermented products—which are often hidden inside parent proteins' structures and become active when the proteins are broken—after the fermentation process, which lasts for 2 to 3 months (Nongdam 2015). The microorganism *Lactobacillus plantarum, Lactobacillus brevis, Lactobacillus casei, Limosilactobacillus fermentum*, and *Tetragenococcus halophilus* were isolated from "Ekung." It is reported that "Ekung" reduces cholesterol and irritable bowel syndrome (IBS), heal common cold and respiratory tract infections, improve digestive health, and also act as an anti-diarrheal agent (Behera and Balaji 2021). Some other fermented bamboo shoots products are listed in Table 1.

Table 1 Fermented bamboo products of North East India (Behera and Balaji 2021)

Sl. no.	Local name	Tribe	State	Microbial communities	Health benefit
1	Soidon	Meetei	Manipur	*Lactobacillus brevis, Lactobacillus lactis, Lactobacillus curvatus, Leuconostoc fallax*	Improve digestive health; immune booster; reduces obesity and hyperlipidemia; stabilize gastrointestinal tract; anti-diarrheal
2	Soijim	Meetei	Manipur	*Lactobacillus brevis, Leuconostoc lactis, Leptodactylus fallax, Leuconostoc mesenteroides*	Improve digestive health; immune booster; reduces obesity and hyperlipidemia; stabilize gastrointestinal tract; anti-diarrheal improved glucose metabolism and energy homeostasis
4	Eup	Nishi	Arunachal Pradesh	Lactobacillus *brevis, Lactobacillus casei, Limosilactobacillus fermentum, Leuconostoc mesenteroides, Leptodactylus fallax*	Lowers cholesterol, prevents respiratory tract infections, irritable bowel syndrome, ulcerative colitis, and other inflammatory bowel diseases, antimicrobial
5	Hirring	Nishi	Arunachal Pradesh	*Lactobacillus brevis, Lactobacillus, Leuconostoc lactis*	Lowers cholesterol, irritable bowel syndrome, a type of inflammatory bowel disease (ulcerative colitis)
6	Khorisa/ Bah gaj	Karbi	Assam	*Lactobacillus* spp.	Improve digestive health; immune booster; reduces obesity and hyperlipidemia; stabilize gastrointestinal tract

2.2 Soybean

Soybean (*Glycine max* (L.) Merrill) is a summer leguminous crop produced in rain-fed environments up to 1500-m elevation in the North East India. Because of the Mongolian population, the North East India has a long tradition of consuming diverse soybean cuisines. In North East India, soybeans are traditionally used to manufacture a range of fermented meals. Ethnic fermented soybean cuisine is the least expensive form of plant protein when compared to animal and milk products, and it is widely available to the North East's rural poor (Pohsnem et al. 2023; Lee et al. 2015). Kinema of Nepali, Tungrymbai of Khasi, Aakhone of Sema Naga, Hawaijar of Meitei, Bekang of Mizo, and Peruyaan of Apatani are some of the most popular ethnic non-salted, sticky fermented soybean cuisines in North East India (Kumari et al. 2023). Basic steps for the preparation of fermented soybean are shown in Fig. 4. Some fermented soybean product names and their health benefits are listed in Table 2.

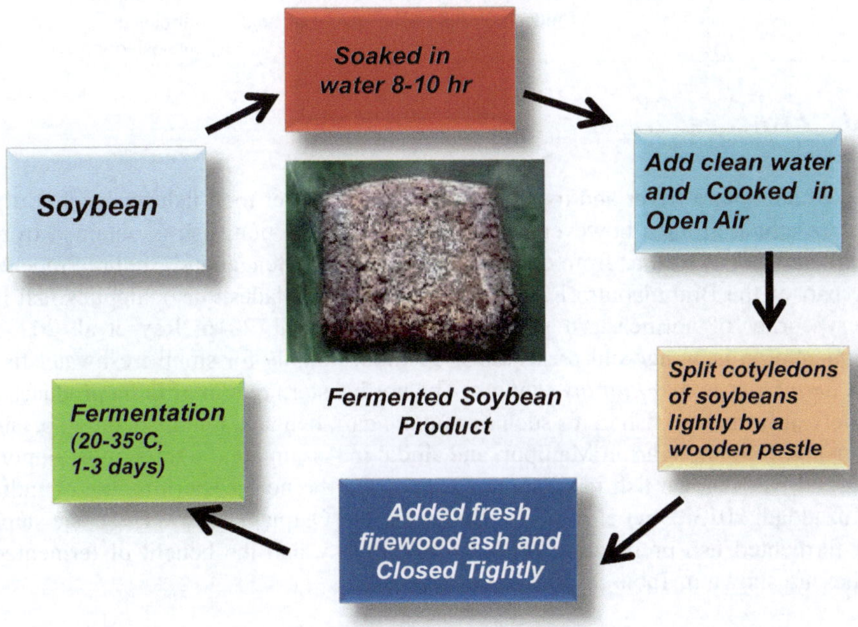

Fig. 4 Basic steps for fermented soybean product production. However, these steps varies tribe to tribe

Table 2 Fermented soybean products of the northeastern region (Tamang et al. 2009; Pohsnem et al. 2023)

Sl. no.	Product name	Tribe	State	Microorganism	Health benefit
1	Kinema	*Limboo/Rai*	Sikkim, Darjeeling hills	*Bacillus subtilis*, *Enterococcus faecium*, *Candida parapsilosis*, *Geotrichum candidum*	Antioxidant, digested protein
2	Hawaijar	*Meitei*	Manipur	*Bacillus subtilis*, *Bacillus licheniformis*, *Bacillus cereus*, *Staphylococcus aureus*, *Staphylococcus sciuri*, *Alkaligenes* sp., *Providencia rettgeri*	Antioxidant, anticancer, anti-osteoporosis, and hypocholesterolemic effects
3	Tungrymbai	*Khasi*	Meghalaya	*Bacillus subtilis* and *Enterococcus faecium*, *Saccharomyces bayanus*, *Saccharomycopsis fibuligera*, *Candida parapsilosis*, *Geotrichum candidum*	Antioxidant, antimicrobial, anti-diarrheal, low-cholesterol content
4	Aakhone	*Naga*	Nagaland	*Bacillus subtilis*, *Bacillus* sp.	Antioxidant, anticancer, anti-osteoporosis
5	Peruyyan	*Apatani*	Arunachal Pradesh	*Bacillus subtilis*, *Bacillus* sp., lactic acid bacteria	Antioxidant, anticancer, anti-osteoporosis

2.3 *Fish*

The Brahmaputra River and its lakes provide fresh water food fish for local northeastern ethnic groups; however, for commercial production, fish is obtained from local markets, as well as from distant markets in West Bengal and Andhra Pradesh; the part of the Brahmaputra River in Assam and Bangladesh also supplies fish in every corner of northeastern villages (Majumdar et al. 2016; Ray et al. 2023). Fermentation is an age-old preservation method, notably for small freshwater fish like *Setipinna* sp. and *Puntius sophore*. The northeastern region of India produces a variety of fermented fish items such as shidal, ngari, hentaak, lonailish, tungtap, and numsing. Hentak, ngari in Manipur, and shidal in Assam are the three most important and well-known fish in fermented form from the northeastern region of India (Muzaddadi 2013; Devi et al. 2015; Belleggia and Osimani 2023). The basic steps for fermented fish preparation are shown in Fig. 5, and the benefit of fermented fishes are shown in Table 3.

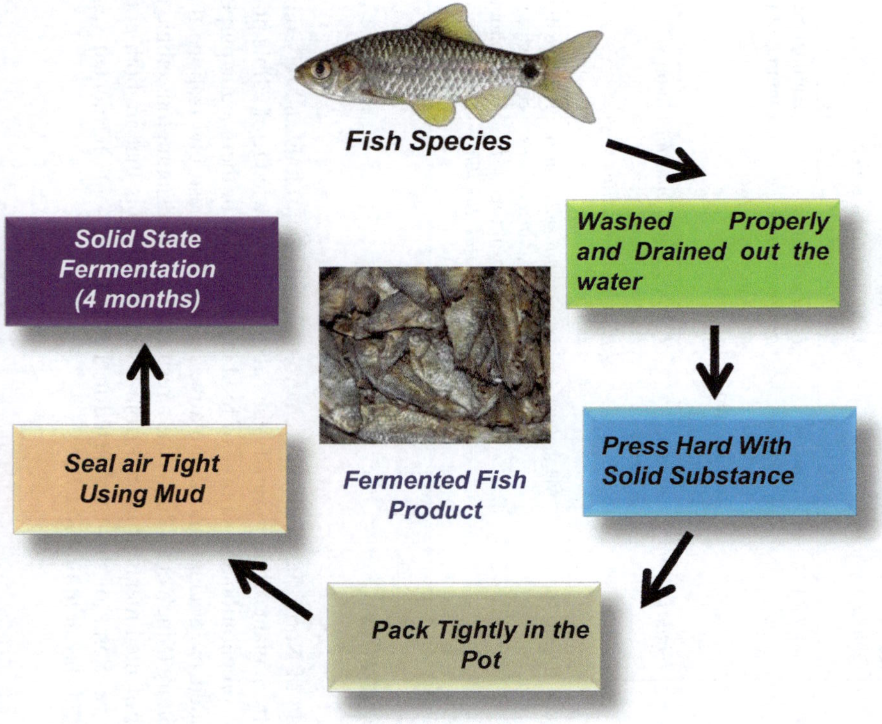

Fig. 5 Basic steps for fermented fish product production. However, these steps varies tribe to tribe

Table 3 Fermented fish products of the northeastern region of India (Muzaddadi 2013; Singh et al. 2018)

Sl. no.	Product name	Organism used	State	Microorganism	Health benefit
1	Ngari	*Puntius sophore*, *Puntisticto*, *Setipinna* sp.	Manipur	*Lactobacillus plantarum*, *Bacillus subtilis*, *B. pumilus and Miocrococcus* sp.	Antioxidant, enhance digestion, improve GI track
2	Hentak	*Esomus danricus*, *Puntius sophore*	Manipur	*Lb. fructosus*, *Lb. amylophilus*, *Enterococcus faecium*, *Bacillus subtilis*	Antioxidant
3	Shidal	*Puntius sophore*	Tripura, Arunachal Pradesh, Assam, and Nagaland	*L. plantarum*, *P. pentosaceus*, *P. acidilactici*, *E. faecium*, *E. faecalis*	Antimalarial, antimicrobial
4	Lona Illis	*Tenulosailisha*	Assam, Tripura	*Lactococcus plantarum*, *Lactobacillus plantarum*, *Bacillus subtilis*	Improve digestion
5	Tungtap	*Danio* spp.	Meghalaya	*Lb. coryniformis* subsp. *Torquens*, *Lactococcus lactis* subsp. *Cremoris*, *Lb. fructosus*, *Bacillus subtilis*	Improve digestion, cell surface Hydrophobicity, and lower cholesterol

2.4 Vegetables

The people of North East India are environmentally conscious and rely on plants and food from plant origin to satisfy the necessities. Knowledge of edible plants, as well as the preparation and storage of food material and for therapeutic purposes, is greatly aided by gradual and cumulative learning in these civilizations that live close to nature (Singh et al. 2014). As a result, this region's numerous ethnic groups created fermented meals produced from a variety of plants that are both nutritious and curative. Some of them are listed in Table 4, and some fermented vegetable products are shown in Fig. 6.

3 Conclusion

The northeastern states are primarily agricultural, with agriculture and allied activities employing roughly two-thirds of the workforce. The majority of the land possesses forests, abundant with natural ingredients. The technological advancement in the fermentation process utilized by ethnic groups reveals a significant relationship between them and nature, as well as an appreciation for microbial association with

Table 4 Fermented vegetable products of North East India (Das and Deka 2012)

Sl. no	Product name	Organism used	State	Microorganism	Health benefit
1	Gundruk	*Brasicca juncea*, *Puntisticto*, *Brassica oleracea*, *Raphanus sativus*	Sikkim	*Lactobacillus brevis*, *L. paracasei*, *Pediococcus pentosaceus*, and *Leuconostoc fallax*	Good appetizer, improve digestion
2	Sinki	*Raphanus sativus*	Sikkim	*Lactobacillus plantarum* and *L. fermentum*	Good appetizer, improve digestion
3	Ziang-sang	*Brassica* sp.	Manipur and Nagaland	*Lactobacillus plantarum*, *L. brevis*, and *Pediococcus acidilactici*	Condiment; protect cells from damage
4	Goyang	*Cardamine macrophylla*	Arunachal Pradesh and Sikkim	*Lactobacillus plantarum*, *L. brevis*, *Lactococcus lactis*, *Enterococcus faecium*, and *Pediococcus pentosaceus*	Improve digestion; rich in vitamins
5	Anishi	*Colocasia* sp.	Nagaland	*Lactobacillus plantarum*, *L. brevis*, and *Leuconostoc fallax*	Antimicrobial

Fig. 6 Some fermented vegetables products. (**a**) Gundruk. (**b**) Anishi. (**c**) Sinki

this. The prominent microbial population found in foods and beverages suggests that people in this area used indigenous microbiota for spontaneous fermentation. The kinds of foods that are fermented in this region are also determined by weather conditions. The selection of the potential microbial population and raw material improvement, the use of immobilized systems and/or enzymes, probiotic activity research, and the adoption of genetically modified organisms will aid in the industrialization of the region's fermented food items. Collaboration across several universities will result in the standardization of food products and an extension of shelf life strength. A retail establishment selling traditional fermented foods from the states in the North East India ought to be developed to aid in the effective marketing of packaged commodities. This would help to sustain the regional economy and enhance rural people's living conditions.

References

Banerjee S, Basak M, Dutta S, Chanda C, Dey S, Dey A, Somkuwar BG, Kharlyngdoh E, Das M (2022) Sustainable uses of bamboo by indigenous people with special emphasis on north-East India. In: Indigenous people and nature. Elsevier, Amsterdam, pp 543–576

Basumatary A, Middha SK, Usha T, Basumatary AK, Brahma BK, Goyal AK (2017) Bamboo shoots as a nutritive boon for Northeast India: an overview. 3 Biotech 7:1–6

Behera P, Balaji S (2021) Health benefits of fermented bamboo shoots: the twenty-first century green gold of Northeast India. Appl Biochem Biotechnol 193:1800–1812

Belleggia L, Osimani A (2023) Fermented fish and fermented fish-based products, an ever-growing source of microbial diversity: a literature review. Food Res Int 172:113112

Borah JR, Gilroy JJ, Evans KL, Edwards DP (2022) The value of shifting cultivation for biodiversity in Northeast India. Divers Distrib 28(9):1979–1992

Das AJ, Deka SC (2012) Fermented foods and beverages of the north-East India. Int Food Res J 19(2):377

Das G, Patra JK, Singdevsachan SK, Gouda S, Shin HS (2016) Diversity of traditional and fermented foods of the seven sister states of India and their nutritional and nutraceutical potential: a review. Front Life Sci 9(4):292–312

Das D, Sarkar S, Dihingia A, Afzal NU, Wann SB, Kalita J, Dewanjee S, Manna P (2022) A popular fermented soybean food of Northeast India exerted promising antihyperglycemic potential via stimulating PI3K/AKT/AMPK/GLUT4 signaling pathways and regulating muscle glucose metabolism in type 2 diabetes. J Food Biochem 46(12):e14385

Deb CR, Jamir B (2020) Ethnic fermented food products of Nagaland, India. J Food Chem Nanotechnol 6(1):18–27

Devi KR, Deka M, Jeyaram K (2015) Bacterial dynamics during yearlong spontaneous fermentation for production of ngari, a dry fermented fish product of Northeast India. Int J Food Microbiol 199:62–71

Dullius A, Fassina P, Giroldi M, Goettert MI, de Souza CFV (2020) A biotechnological approach for the production of branched chain amino acid containing bioactive peptides to improve human health: A review. Food Res Int 131:109002

Grujović MŽ, Mladenović KG, Semedo-Lemsaddek T, Laranjo M, Stefanović OD, Kocić-Tanackov SD (2022) Advantages and disadvantages of non-starter lactic acid bacteria from traditional fermented foods: potential use as starters or probiotics. Compr Rev Food Sci Food Saf 21(2):1537–1567

Guo Q, Chen P, Chen X (2023) Bioactive peptides derived from fermented foods: preparation and biological activities. J Funct Foods 101:105422

Hayes M (2021) Bioactive peptides in preventative healthcare: an overview of bioactivities and suggested methods to assess potential applications. Curr Pharm Des 27(11):1332–1341

Ibrahim SA, Yeboah PJ, Ayivi RD, Eddin AS, Wijemanna ND, Paidari S, Bakhshayesh RV (2023) A review and comparative perspective on health benefits of probiotic and fermented foods. Int J Food Sci Technol 58(10):4948–4964

Jakubczyk A, Karaś M, Rybczyńska-Tkaczyk K, Zielińska E, Zieliński D (2020) Current trends of bioactive peptides—new sources and therapeutic effect. Food Secur 9(7):846

Jeyaram K, Singh TA, Romi W, Devi AR, Singh WM (2009) Traditional fermented foods of Manipur. Indian J Tradit Knowl 8:115–121

Jia L, Wang L, Liu C, Liang Y, Lin Q (2021) Bioactive peptides from foods: production, function, and application. Food Funct 12(16):7108–7125

Joshi B, Indira A, Oinam S, Koul A, Chongtham N (2023) Fermented bamboo shoots: a potential source of nutritional and health supplements. In: Bamboo science and technology. Singapore, Springer Nature, pp 201–236

Kumari R, Sanjukta S, Sahoo D, Rai AK (2022) Functional peptides in Asian protein rich fermented foods: production and health benefits. Syst Microbiol Biomanuf 2(1):1–13

Kumari V, Thakur R, Kumari J, Kumari A, Khajuria D (2023) Nutritional improvement in soybean (Glycine max (L.) Merrill) through plant breeding and biotechnological interventions. Crop Pasture Sci 75(1)

Lando V, Valduga NZ, Moroni LS (2023) Functional characterization of lactobacilli strains with antimicrobial activity against salmonella spp. and cell viability in fermented dairy product. Biocatal Agric Biotechnol 47:102605

Laya A, Wangso H, Fernandes I, Djakba R, Oliveira J, Carvalho E (2023) Bioactive ingredients in traditional fermented food condiments: emerging products for prevention and treatment of obesity and type 2 diabetes. J Food Qual 2023

Lee C, Choi MS, Kim HT, Yun HT, Lee B, Chung YS et al (2015) Soybean [Glycine max (L.) Merrill]: importance as a crop and pedigree reconstruction of Korean varieties. Plant Breed Biotechnol 3:179–196

Majumdar RK, Roy D, Bejjanki S, Bhaskar N (2016) Chemical and microbial properties of shidal, a traditional fermented fish of Northeast India. J Food Sci Technol 53:401–410

Marx W, Scholey A, Firth J, D'Cunha NM, Lane M, Hockey M, Jacka F (2020) Prebiotics, probiotics, fermented foods and cognitive outcomes: a meta-analysis of randomized controlled trials. Neurosci Biobehav Rev 118:472–484

Merenstein D, Pot B, Leyer G, Ouwehand AC, Preidis GA, Elkins CA, Hill C, Lewis ZT, Shane AL, Zmora N, Petrova MI (2023) Emerging issues in probiotic safety: 2023 perspectives. Gut Microbes 15(1):2185034

Muzaddadi AU (2013) Naturally evolved fermented fish products of Northeast India (Seedal and Shidal)—a comparative study. Indian J Nat Prod Resour 4(2):170–177

Nongdam P (2015) Traditional fermented bamboo shoot foods of north-East India and their characteristic natural microbial flora. In: 10th World Bamboo Congress, Korea, pp 1–13

Patil PJ, Usman M, Zhang C, Mehmood A, Zhou M, Teng C, Li X (2022) An updated review on food-derived bioactive peptides: focus on the regulatory requirements, safety, and bioavailability. Compr Rev Food Sci Food Saf 21(2):1732–1776

Pohsnem JM, Ramakrisnan E, Parasar DP (2023) Fermented food products in the Himalayan belt (north East India) and their health benefits. Int J Gastronomy Food Sci 31:100676

Premlata T, Sharma V, Bisht MS, Nirmala C (2020) Edible bamboo resources of Manipur: consumption pattern of young shoots, processing techniques and their commercial status in the local market. Indian J Tradit Knowl 19:73–82

Ray B, Nath S, Murmu P, Das D (2023) Fermented fish products of north-East India. Chron Aquat Sci 1(1):23–29

Şanlier N, Gökcen BB, Sezgin AC (2019) Health benefits of fermented foods. Crit Rev Food Sci Nutr 59(3):506–527

Sarma U, Gupta S (2022) An overview on ethnic fermented food and beverages of India: interplay of microbes, immunity and nutrition. Nutr Health 28(3):331–339

Sharma I, Yaiphathoi S (2020) Role of microbial communities in traditionally fermented foods and beverages in north East India. In: Recent advancements in microbial diversity. Academic Press, New York, pp 445–470

Singh TA, Devi KR, Ahmed G, Jeyaram K (2014) Microbial and endogenous origin of fibrinolytic activity in traditional fermented foods of Northeast India. Food Res Int 55:356–362

Singh SS, De Mandal S, Lalnunmawii E, Senthil Kumar N (2018) Antimicrobial, antioxidant and probiotics characterization of dominant bacterial isolates from traditional fermented fish of Manipur, north-East India. J Food Sci Technol 55:1870–1879

Singh BP, Aluko RE, Hati S, Solanki D (2022) Bioactive peptides in the management of lifestyle-related diseases: current trends and future perspectives. Crit Rev Food Sci Nutr 62(17):4593–4606

Soemarie YB, Milanda T, Barliana MI (2021) Fermented foods as probiotics: A review. J Adv Pharm Technol Res 12(4):335–339

Sonar NR, Vijayendra SVN, Prakash M, Saikia M, Tamang JP, Halami PM (2015) Nutritional and functional profile of traditional fermented bamboo shoot based products from Arunachal Pradesh and Manipur states of India. Int Food Res J 22(2):788–797

Tamang JP, Lama S (2022) Probiotic properties of yeasts in traditional fermented foods and beverages. J Appl Microbiol 132(5):3533–3542

Tamang JP, Chettri R, Sharma RM (2009) Indigenous knowledge of northeast women on production of ethnic fermented soybean foods. Indian J Tradit Knowl 8(1)

Yeptho L, Ajungla T, Kuotsu K (2021) Ethnic study on bastenga, a fermented bamboo shoot product of Nagaland, India. Curr Sci 120:715–718

Zacarchenco PB, Pimentel TC, Alves ATS, Spadoti LM, Esmerino EA, Silva MC, da Cruz AG (2022) Health benefits of probiotics: an overview. In: Advances in dairy microbial products, pp 235–245

Zaky AA, Simal-Gandara J, Eun JB, Shim JH, Abd El-Aty AM (2022) Bioactivities, applications, safety, and health benefits of bioactive peptides from food and by-products: a review. Front Nutr 8:815640

Zommiti M, Feuilloley MG, Connil N (2020) Update of probiotics in human world: a nonstop source of benefactions till the end of time. Microorganisms 8(12):1907

Pharmaceutical and Nutraceutical Potential of Heritage Herbs

Tanjima Tarique Laskar, Monica Arora, and Anand Gaurav

Contents

T. T. Laskar
School of Pharmaceutical Sciences, University of Science and Technology Meghalaya, Baridua, Meghalaya, India

M. Arora (✉)
Faculty of Health Sciences, Villa College, QI Campus, Rahdhebai Hingun, Male', Republic of Maldives

A. Gaurav (✉)
Department of Pharmaceutical Sciences, School of Health Sciences and Technology, UPES, Dehradun, Uttarakhand, India

Faculty of Pharmaceutical Sciences, UCSI University, Taman Connaught, Cheras, Kuala Lumpur, Malaysia

© The Author(s), under exclusive license to Springer Nature Singapore Pte Ltd. 2024

A. Das Talukdar et al. (eds.), *Traditional Resources and Tools for Modern Drug Discovery*, Interdisciplinary Biotechnological Advances,
https://doi.org/10.1007/978-981-97-4600-2_23

Abstract Since antiquity, the medicinal herbs have been a vital source to heal a broad range of human ailments. In the past decades, the upturn of interest in natural drugs has originated due to the beliefs that the natural source of medicine is provisionally healthier than the synthetic products owing to its lesser side effects and availability at affordable prices. As per the WHO, about 4 billion people (75–85%) around the globe depend on folk medicines to cure various ailments. The benefit of medicinal herbs in human healthcare is definitely indisputable as even now numerous curative folk herbs are being utilized globally. It is of uttermost necessity to assess the wealthy legacy of traditional system of medicines. As the ethnomedicinal plants holds remarkable importance, they are widely used by the numerous heritages for their traditional curing properties. The study of ethnomedicinal plants enlightens the cultural practice, ecological knowledge, and the potential for the modern medicinal research. In this regard some ethnomedicinal plants such as *Withania somnifera*, *Ocimum sanctum*, *Cinnamomum cassia*, *Crataeva nurvala*, *Tinospora cordifolia*, *Uraria picta*, and *Solanum xanthocarpum* are investigated in this study to highlight their iatric importance in pharmaceutical and nutraceutical disciplines.

Keywords Botanical description · Phytochemical profile · Therapeutic potential · Pharmacological action

1 Introduction

Since times immemorial herbal plant has been serving as potent source of phytomedicines owing to the presence of wondrous assortment of precious phytochemical constituents. In various traditional system of medicines such as the Atharva Veda or Brahma Veda (a sacred Indian holy book), Ayurveda (ancient Indian book dealing with natural holistic medicines), Siddha (traditional medicinal system of Southern India), and Unani systems (Perso-Arabic traditional medicine), the ancient scholastic work on rich heritage of medicines of plant origin was included. From the last couple of decades, a notable progress for the revival of inquisitiveness regarding the investigation of folk medicinal plants has been observed to meet the basic healthcare needs of people. As per an estimation, around 13,000 species of herbal plants are reported globally to possess various remedial properties and are used as source of drugs (Gordon and David 2001). The phytomedicinal action of plants sources are typically due to the presence of combinations of secondary metabolic products of plants. This phytomedicinal effects are unique to each species of plant owing to the presence of their taxonomically distinct class of secondary products and also because of the synergistic influence of different active phytochemicals (Wink 2000). Systematic phytochemical screening of plant species routinely takes place in many laboratories in order to discover and unravel the new bioactive components of plant species. The research and exploration of plants for phytochemical identification and evaluation should be extended in order to boost up the process of drug discovery. Scientific methods should be used to examine the remedies which will ensure the safety profile of plants; as well as it will standardize and control the quality of the phytomedicinal products ultimately leading to the approval of the phytomedicines for use in healthcare facilities. An example of lead that has been synthesized from the analogues of naturally occurring phytoconstituents include the traditional anti-asthmatic substances sodium cromoglycate and sodium cromolyn which is found in furanochromone khellin (visammin) (Cor et al. 1970). The exploration of phytoconstituents as well as pharmacological evaluation of plants will provide a firm insight for the development of new life-saving drugs. Despite having extensive breakthroughs and impressive accomplishments made in the discipline of modern drug discovery and Western medical science, researchers are striving to comprehend the perplexed biological mechanism of various health implications by utilizing the accessible curative medicaments. A consequential holistic approach is emergence of traditional herbal medicines in the healthcare facilities which serves as a remedial alternative for treatment of the passive illness. The World Health Organization (WHO) has advised to integrate herbal medicines with modern medical research in order to properly address the different diseases that exist globally. Natural products-based therapeutics are particularly useful owing to the fact that there is wide availability and inexpensiveness by which the quality of life can be ameliorated. Therefore, identification and recognition of novel opportunities for natural products-based prevention and treatment may be remarkably propitious in battling the burden of numerous diseases (Palliyaguru et al. 2016). Plants with therapeutic qualities are used since decades throughout the world, giving contemporary researchers the chance to evaluate their effectiveness against a range of medical conditions. The therapeutic efficacy and the mechanism underlying the activity of bioactive compounds of herbal plants is still being investigated by researchers. However, some of these components have found their way into the Western pharmaceutical and nutraceutical markets as sources of dietary supplements and health supplements with the promising effects against a various health implications. By the application of modern scientific approaches and novel research concepts, the pharmaceutical and nutraceutical properties of the traditional herbal medicinal plants can be a cardinal source for the development of new medicines to treat intractable health disorders (Singh et al. 2010; Mirjalili et al. 2009). In this regard various ethnomedicinal plants such as *Withania somnifera*, *Ocimum sanctum*, *Cinnamomum cassia*, *Crataeva nurvala*, *Tinospora cordifolia*, *Uraria picta*, and *Solanum xanthocarpum* are explored for use in various health disorders.

2 *Withania somnifera*

2.1 Botanical Description

The medicinal plant *Withania somnifera*, also known as ashwagandha, is extensively used in Asia primarily in the Ayurvedic, Siddha, and Unani systems of medicine. The roots of *Withania somnifera* roots are used to treat a variety of illnesses, and it is included in the medicinal formulations of various traditional system of medicines (Pattanayak et al. 2010). It is a stellate-tomentose, upright, greyish shrub with long tuberous roots which attains a height of 30–75 cm. The plant has petiolate leaves which are sub-acute, oblong to broadly ovate, and placed alternately or sub-oppositely with a lamina measuring 5–10 cm in length and 2.5–7 cm in width. Flowers appear as green-colored, single, bisexual flowers, and the calyx of the fruit is enlarged, accrescent, and gamosepalous consisting of five lobes that measure 3 to 5 mm. The campanulate, greenish-yellow corolla has 5–8-mm-sized lobes. Fruit is an orange-red berry which is enclosed in the enlarged calyx, and it has milk-coagulating properties. Seeds are yellow, discoid, reniform, and arranged in large numbers and have a chromosome number of 48 (Schonbeck-Temesy 1972) (Fig. 1).

2.2 Phytochemical Profile

The plant contains around 12 types of alkaloids, 40 withanolides (steroidal lactones), and numerous sitoindosides. The primary chemical component of these plants is withanolides which is mostly found in the leaves and range in concentration from 0.001 to 0.5% dry weight (DW) (Kapoor 2001; Atta-ur-Rahman et al. 1993; Choudary et al. 1996; Rastogi and Mehrotra 1998; Bandyopadhyay et al. 2007; Atal et al. 1975; Anonymous 2004). The first compound which was identified and extracted from the plant is Withaferin A which is well-known due to its anticancerous properties, and it constitutes about 0.3% of dry weight of leaves (Glotter 1991; Leet et al. 1982; Tursunova et al. 1977; Abraham et al. 1968). Several other compounds belonging to withanolide group are also reported which includes withanolide D, G, H, I, J, K, N, O, Q, R, U, and withanolide y (Withaferin A and Withanolide D are the most potent pharmacological agents) (Krison and Glotter 1980). Roots of *Withania somnifera* possesses a dimeric thiowithanolide called ashwagandhanolide and fruits contain withanamides A–I (Subaraju et al. 2006; Jayaprakasam et al. 2004). Several other compounds were also identified which includes withasomniferin-A, withasomidienone, and withasomniferols A-C

Fig. 1 (**a**) Whole plant, (**b**) leaves, (**c**) seeds, (**d**) fruits, and (**e**) roots of *Withania somnifera*

(Elsakka et al. 1990; Gupta and Rana 2007). The leaves also consist of five unidentified alkaloids (0.09%) such as chlorogenic acid, withanone, condensed tannins, calystegines, and flavonoids. The root consists of alkaloids (0.13–0.31%) such as somniferine, somniferinine, withananine, withanine, pseudowithanine, nicotine, DL-isopelletierine, isopelletierine, choline, cuscohygrine, tropine, pseudotropine, 3-tropyltigloate, 3α-tigloyloxytropane, anaferine, mesoanaferine, anahygrine, hygrine, hentriacontane, visamine, withasomnine, and ashwagandhine (Ganzera et al. 2003). Seven new withanolide glycosides are identified which includes Withanosides I, Withanosides II, Withanosides III, Withanosides IV, Withanosides V, Withanosides VI, and Withanosides VII. The roots also contain several compounds such as withanicil, dulcitol, hentriacontane, and sugars, whereas the berries contain numerous amino acids (Johri et al. 2005).

2.3 *Therapeutic Potential of* Withania somnifera

Numerous therapeutic explorations have been carried out on the chemical constituents of *Withania somnifera* owing to its miraculous pharmaceutical and nutraceutical potentials. Currently it is sold in the US market as herbal supplements, and it is mostly recommended as an energy enhancer for improving athletic performance. This is due to the fact that the root contains very high amount of starch which is likely to have a positive impact on exercise performance since it acts as a carbohydrate supplement (Bucci 2000). Apart from nutraceutical properties, *Withania somnifera* also exhibits notable pharmacological actions. A brief overview of the pharmacological findings is explained below in Table 1.

3 *Ocimum sanctum*

3.1 Botanical Description

Ocimum sanctum L. (Tulsi) has been used since decades as a herbal remedy in the old Ayurvedic and Unani systems. Its many medicinal potentials have been well-documented in the conventional medical system. It is distinguished by the presence of an astringent taste and powerful scent. Tulsi is an erect, branched subshrub that attains 30 to 60 cm in length upon maturation. It contains greenish-purple leaves which has a pleasant aromatic odor. The plant is having a length of 5 cm, petiolate, ovate, acute or obtuse leaves, sub-serrate, whole, or dentate/toothed edges. The flowers are whorl-enclosed, elongated, violet racemes. (Pattanayak et al. 2010). The plant is acrid and bitter in taste. The fruits are tiny sized with presence of reddish-yellow seeds (Prajapati et al. 2003; Gupta et al. 2002) (Fig. 2).

3.2 Phytochemical Profile

The plant of *Ocimum sanctum* consists of a variety of phytoconstituents including saponins, glycosides, flavonoids, alkaloids triterpenoids, anthocyans, and tannins. The leaves contain 71% eugenol (volatile oil, mainly responsible for the therapeutic potentials), 20% eugenol methyl ether, 3% carvacrol, carvicol, linalool, euginal, caryophyllene, ursolic acid, ascorbate, tartarate, carotene, molludistin, and luteolin. Leaves and stem possess some phenolic compounds such as rosameric acid, circimaritin, isothymosin, and apigenin. Two flavonoids, namely, orientin and vicenin, are present. The leaves also contain traces of vitamins and minerals. The seeds contain sitosterol and five types of fatty acids such as linoleic (52.23%), linolenic (16.63%), palmitic (11.69%), oleic (13.82%), and stearic (3.19%) acid. Mucilage which is present in seeds contain some levels of sugars (xylose and polysaccharides) (Kelm et al. 2000; Gupta et al. 2002; Shishodia et al. 2003; Mukherji 1987; Raju et al. 1999; Ravid et al. 1997; Khanna and Bhatia 2003; Lawrence et al. 1972; Gupta 1996; Shah and Qadry 1998; Yanpallewar et al. 2004; Nair et al. 1982; IDMA 2002).

Table 1 Pharmacological actions of *Withania somnifera*

Sl no.	Pharmacological action	Description
1	Antivenom effect	Venom hyaluronidases is an enzyme which causes the toxins to spread rapidly by destructing the composition and rigidity of the extracellular matrix of tissues. External application of glycowithanolides inhibit the activity of the enzyme of snake venoms including cobra and viper (Machiah et al. 2006; Lizano et al. 2003)
2	Antiepileptic effect	The root extract exerts antiepileptic activity by various mechanism of action such as inhibiting the generation of ROS promoting the level of GABA in brain, suppressing the butyrylcholinesterase activity in the brain and by regeneration of neurites (Kulkarni and Dhir 2008)
3	Effect on the cardiopulmonary system	Alkaloids present in the roots has potent effect on the heart health and respiratory organs. The alkaloids act on the brain stem and stimulates the vasomotor and respiratory centers. The alkaloids also produce direct cardio-depressant and cardiotonic effect by blocking the ganglions (Malhotra et al. 1981)
4	Anti-stress effect	The seeds contain glycosides (sitoindosides VII and VIII) which is used for anti-stress effect (Singh et al. 1982)
5	Effect on the endocrine system	The root extract stimulates thyroidal activity by enhancing the levels of T3 and T4 hormones (Panda and Kar 1997)
6	Antioxidant activity	Sitoindosides and glycowithanolides promote antioxidant effect by increasing the levels of antioxidant enzymes particularly catalase, superoxide dismutase, and glutathione peroxidase (Bhattacharya et al. 1997)
7	Effect on the nervous system	An alkaloid found in roots called ashwagandholine has taming, mildly depressing (tranquillizing) relaxant, and antispasmodic effects on a variety of substances which triggers contractions of smooth muscles of trachea, intestine, uterine, and vascular systems. Sitoindosides and withaferin increase acetylcholinesterase activity in the brain. Additionally, it is said to have a memory-improving and cognitive-enhancing impact (Schliebs et al. 1997; Malhotra et al. 1965a, b)
8	Immunomodulatory action	Glycowithanolides in combination with sitoindosides induces phagocytosis, enhances the activity of lysosomal enzymes and hemolytic antibody formation, and activates peritoneal macrophages. It also lowers the production of TNF-α and IL-1 (Ghosal et al. 1989; Ziauddin et al. 1996; Dhuley 1997)
9	Nootropic effect	Withanoside IV significantly improves amnesia or memory deficits and also ameliorate neuronal dysfunction in Alzheimer's disease (Gupta and Rana 2007)
10	Hypoglycemic, diuretic, and hypocholesterolemic action	The plant is reported to raise the plasma HMG-CoA reductase activity and HDL cholesterol levels and lowers the level of triglycerides, LDL, and VLDL lipoproteins. The roots exert hypoglycemic, diuretic, and hypocholesterolemic properties. Urine salt and volume are markedly increased as a result of its diuretic activity (Visavadiya and Narasimhacharya 2006; Hemalatha et al. 2006; Andallu and Radhika 2000)

(continued)

Table 1 (continued)

Sl no.	Pharmacological action	Description
11	Antitumor effect	Withaferin A exhibits antitumor effect against lung adenoma. The root extract is useful for the treatment of skin cancer. Withanolides and Withaferin A effectively inhibits breast, lung, and colon cancer. Nuclear factor-kappaB (NF-kappaB) activation controls a number of genes that govern metastasis, cellular proliferation, carcinogenesis, and inflammation (Prakash et al. 2002; Jayaprakasam et al. 2003; Ichikawa et al. 2006)

Fig. 2 *Ocimum sanctum* L. (**a**) Plant, (**b**) leaves, (**c**) stems, and (**d**) flowers

3.3 Therapeutic Potential of Ocimum sanctum

Ocimum sanctum is enriched in vitamins, minerals, proteins, fats, and carbohydrates which are responsible for enhancing the activity of GIT by helping in utilization of nutrients which are obtained from edible sources. As a general vitalizer, it helps in enhancing physical endurance and brain function (Singh and Chaudhuri 2018). During the last few decades, *Ocimum sanctum* has exhibited notable activities, and some of those outstanding findings are presented in Table 2.

Table 2 Pharmacological actions of *Ocimum sanctum*

Sl no.	Pharmacological action	Description
1	Antidiabetic effect	*Ocimum sanctum* stimulates physiological pathways of insulin secretion and it also regulates corticosteroid- induced diabetic mellitus by lowering serum glucose and cortisol (Hannan et al. 2006; Gholap and Kar 2004)
2	Cardioprotective effect	*Ocimum sanctum* reduces superoxide dismutase and catalase activities of myocardium and also reduces rise in plasma cAMP level (caused due to chronic-resistant stress), glutathione (GSH) and LDH levels. Urosolic acid present in *Ocimum sanctum* causes protection against drug induced lipid peroxidation (Sood et al. 2006; Sharm et al. 2001; Balanehru and Nagarajan 1992)
3	Hypolipidemic effect	Leaves extract causes reduction of serum triglycerides, cholesterol, phospholipids, low density lipoproteins, and significantly increase the high-density lipoproteins (Sarkar et al. 1994)
4	Wound healing effect	*Ocimum sanctum is used* in the treatment of keloid scars (occurs due to deposition of collagen) (Shetty et al. 2006)
5	Radioprotective effect	Polysaccharides present in *Ocimum sanctum* prevent liposomal lipids and plasmid DNA from oxidative damage. Two flavonoids, namely, orientin and vicenin, provides protection against death caused due to radiation (Subramanian et al. 2005; Uma Devi et al. 2000)
6	Antioxidant activity	Linoleic acid which is widely found in fixed oil is responsible for exerting antibacterial activity (Singh et al. 2005)
7	Gastroprotective effect	The leaf extract suppresses pepsin secretion, inhibits lipid peroxidation, and increases the production of mucin secretion and cellular mucus as well as enhances the lifespan of mucosal cells (Goel et al. 2005)
8	Immunomodulatory action	Aqueous extract of the plant causes an increase in white blood cell counts and enhances the phagocytic activity (Mukherjee et al. 2005)
9	Central nervous system	Stem, leaf, and calli extract are reported to prevent tonic convulsions an act as anticonvulsant agent (Jaggi et al. 2003)
10	Anthelmintic activity	Eugenol present in leaves is found to cause muscle paralysis in worms (Asha et al. 2001)
11	Antitumor effect	Leaf extract modulates the activity of aryl hydrocarbon hydroxylase, cytochrome b5, and glutathione S-transferase (GST) which promotes detoxifying effect of carcinogens (Pandey and Madhuri 2010)

4 *Cinnamomum cassia*

4.1 *Botanical Description*

Cinnamomum cassia, an evergreen, aromatic bushy tree widely known as a spice. Cinnamon, the chief component responsible for spicy flavor, is present in the bark of young branches. *C. cassia* consists of hard greyish-brown bark with thickness of 0.5 in. which grows up to 33–49 ft. in length. The leaves are elliptic, narrow,

Fig. 3 *Cinnamomum cassia*: (**a**) leaves, (**b**) dried twig, (**c**) dried bark, and (**d**) stem

leathery, and elongated with a length of 6–10 in. and width of 2–3 in. Flowers are tiny, whitish which blooms the in axillary panicles. Fruits are ellipsoid, and upon maturation it turns from greenish to black-purple color (Senanayake and Wijesekera 2003) (Fig. 3).

4.2 Phytochemical Profile

There has been a lot extensive work carried out about the chemical compounds, and 160 compounds were reported to be present in the tree. Terpenoids represents the most abundant compound present in *Cinnamomum cassia*. Bark contains components such as cinncassiol (A–G), cinnacasside (A–F), cinnamoid (A–E), cinnacassoside (A–D), ethyl cinnamate, cinnacasol, and perseanol (Pham et al. 2019; Li et al. 2013; Liu et al. 2018; Zeng et al. 2017). The leaves contain 1-terpineol, *cis*-β-terpineol, α-terpineol, caryophyllene, *cis*-2-methoxycinnamic acid, and eugenol (Rao et al. 2018; Pannee et al. 2014). The twig contains cinnacassin (F–O), cinnacassin (A_1–A_7), cinnacassolide (A–C), evofolin B, and coumarin (Xiong et al. 2018; Zeng et al. 2014; Ngoc et al. 2012; He et al. 2016; Luo et al. 2013; Kim et al. 2017).

4.3 Therapeutic Potential of Cinnamomum cassia

Cinnamomum cassia has been widely used for fragrance ancient times. It contains a significant amount of minerals, vitamins, carbohydrates, proteins, fats, fibers, and moisture. The various mineral contents present in *Cinnamomum cassia* are chromium, manganese, iron, zinc, magnesium, phosphorus, calcium, potassium (highest amount), and sodium. The various types of vitamins include vitamins A, B12, B9, B6, B3, B2, B1, C, D, E, and K and β-carotene (highest content) (Shahzadi et al. 2021; Banu and Lunghar 2023; Kaur et al. 2018). The tree is also possessing various therapeutic activities which are summarized in Table 3:

Table 3 Pharmacological actions of *Cinnamomum cassia*

Sl no.	Pharmacological action	Description
1	Antitumor effect	Cinnamate, procyanidins, cinnamaldehyde, and cinnamyl alcohol are reported to exhibit action against cancer of the cervix, lungs, breast, and oral Neck and head cells (Patil et al. 2017; Ohnuma et al. 2011)
2	Anti-inflammatory and analgesic effect	Cinnamaldehyde inhibits the activity of cyclooxygenase and decreases generation of nitric oxide (Pannee et al. 2014)
3	Anti-tyrosinase activity	*Trans*-cinnamaldehyde reduces the melanin content of the cells and inhibits the expression of tyrosinase enzyme (Chou et al. 2013)
4	Immunomodulatory action	Cinncassiol G, Cinnacasol, and Cinnacasside F inhibits the proliferation of B and T cell (Zeng et al. 2017)
5	Antidiabetic and anti-obesity effect	Polyphenols lowers the activity of α-glucosidase, maltase, and sucrase. Further, it also boosts the storage of lipids in fat cells and enhances the oxidation of fatty acids (Uslu et al. 2018; Lee et al. 2016)
6	Antibacterial activity	Cinnamaldehyde is an effective antibacterial agent which exhibits action against gram-positive and gram-negative bacteria (Ooi et al. 2006)
7	Antifungal effect	Cinnamaldehyde exerts its antiproliferative activity on oral candidiasis caused by *Candida albicans*, and it is also effective *against Candida tropicalis*, *Candida glabrata*, and *Candida krusei* (Taguchi et al. 2010)
8	Antioxidant activity	Aqueous extract causes an increase WBC counts and enhances phagocytic activity (Jain et al. 2011)

Fig. 4 Different parts of *Crataeva nurvala*: flower (**a**), dried bark (**b**), and fruits (**c**)

5 *Crataeva nurvala*

5.1 Botanical Description

Crataeva nurvala is a branching plant growing in dry, hot climates and shady places which is distributed in various parts of the world mostly in tropical and subtropical regions. The flowers have a diameter of 5–8 cm and are fragrant, polygamous, whitish to milky white, and have trifoliate leaves. The plant's inflorescences resemble thick terminal corymbs with tiny bracts. The fruit is primarily berries with seeds embedded in the golden flesh and globes with a woody exterior. The bark is coated in many lenticels and is wrinkly and grey-white in color (Soosamma et al. 2010) (Fig. 4).

5.2 Phytochemical Profile

From the stem bark; phytoconstituents such as diosgenin, cadabicine diacetate, friedelin, and lupeol; and its acetate and betulinic acid can be extracted. *Crataeva nurvala* fruits include cetyl and ceryl alcohol, triacontanol, and glucocapparin. Leaf and root contents include stachydrine, diosgenin, β-sitosterol, varunol, and betulinic acid are also present in the bark, leaves, and roots (Md Enamul et al. 2008; Bopana and Saxena 2008; Parvin et al. 2011).

5.3 *Therapeutic Potential of* Crataeva nurvala

Crataeva nurvala is a renowned herbal medicine owing to its anti-inflammatory, lithotropic, contraceptive, and febrifuge properties. It also has a remedial effect in vomiting, kidney stone, gastric irritation, bladder stone, and rheumatic fever (Ghani 1998). *Crataeva nurvala* is also used for wound healing, breathing problems, joint lubrication, heart and lung weakness, skin moisture, and memory loss (Bhattacharjee and Shashidhara 2012). Various biological activities exhibited by the plant is summarized in Table 4.

6 *Tinospora cordifolia*

6.1 *Botanical Description*

Tinospora cordifolia is a giant-sized climbing plant with numerous coiled branches. The bark is greyish-white color; stem is odorous, bitter, fleshy, and filiform climber; and the powdered form of the stem has a creamish to dark brown appearance (Upadhyay et al. 2010; Tiwari et al. 2018). Partially and halfway twisted leaves are present which are simple, round, heart-shaped, pulvinate, alternating, and long-petioled (about 15 cm). Lamina is oval and membranous, with a length of 10–20 cm.

Table 4 Pharmacological actions of *Crataeva nurvala*

Sl no.	Pharmacological action	Description
1	Anti-urolithiatic effect	Lupeol reduces the level of oxalates and promotes diuretic and anti-urolithiatic activity (Shumaia et al. 2012; Iqbal et al. 2011)
2	Antifertility effect	Enhances estrogenic activity by expulsing the eggs from the fallopian tube (Bhaskar et al. 2009)
3	Antinociceptive activity	Inhibits the production and release of histamines, cytokines prostaglandins, and polypeptide kinins (Alam et al. 2006)
4	Antidiabetic effect	Increase insulin secretion from pancreas and also increase utilization of glucose by peripheral tissues (Prabhat et al. 2010)
5	Cardioprotective effect	Lupeol increases the activity of myocardium by suppressing the enzymatic activity of lactate dehydrogenase and creatine phosphokinase (Reeta et al. 2010)
6	Anticancer activity	Phytosterols and triterpenoids have anticancer effect. Lupeol induces the apoptotic death of cancerous cells of lungs (SMMC7721 cells) by downregulating the expression of DR3 in SMMC7721 cells (Zhang et al. 2009; Liby et al. 2007; Setzer and Setzer 2003; Bradford and Awad 2007)
7	Anti-HIV effect	β-Sitosterol causes inhibition of HIV disease (Fujioka et al. 1994)
8	Anti-diarrheal activity	Significantly retards enteropooling and intestinal transit by reducing the weight and volume of intestinal content (Inayathulla et al. 2010)

Fig. 5 Different parts of *Tinospora cordifolia*: (**a**) leaves, (**b**) fruits, (**c**) flowers, (**d**) dried stem, and (**e**) seeds

Flowers are greenish-yellow in color and unisexual and occupies axillary position. The male flower appears in groups, whereas female flowers occur in single/solitary form (Arul et al. 2005). The fruits contain single seeds which matures during winter, and flower blooms usually in the summer (Spandana et al. 2013). Roots are aerial, thread-like, squairshin, and sometimes lengthens continuously till touches ground, and the roots are ariel in nature with a penta arch primary structure (Sinha and Sharma 2015; Singh et al. 2003). Seeds have curve shape and endocarp is ornamented (Misra and Prakash 1969) (Fig. 5).

6.2 Phytochemical Profile

Many terpenoids, including furanosporide, ecdysterone, makisterone, etc., are found in the plant. Additionally, several glucosides have been isolated including cordifolioside A, B, C, D, and E and tinocordifolioside (Khuda et al. 1964; Hanuman et al. 1986a; b; Gagan et al. 1996). Various alkaloids such as palmatine, beberine, tembeterine, choline, jatrorrhizine, 1,2-substituted pyrrolidine, and tinosporine (Choudhary et al. 2013; Bisset and Nwaiwu 1983; Mahajan et al. 1985; Pathak et al. 1995a). Steroids including giloinsterol, ß-sitosterol, and 20 α-hydroxy ecdysone are present (Khaleque et al. 1970; Dixit and Khosa 1971; Pathak et al. 1995b).

6.3 *Therapeutic Potential of* Tinospora cordifolia

The plant is a common ingredient in Ayurvedic treatments from ancient times due to the presence of variety of nutrients including proteins (11.2%), carbohydrates, vitamins (A, B6, C), folic acid, and minerals (calcium, phosphorus, potassium, salt, magnesium and iron). It has a strong antioxidant content and strengthens the immune system. The various therapeutic utilities are summarized in Table 5.

7 *Uraria picta*

7.1 *Botanical Description*

Uraria picta is a very valuable herb widely used in different traditional medicine systems. It has a height of 0.5–2.0 m, and upon maturation the stem becomes woody and covered with short straight and hooked hairs. Leaves are dimorphic in nature, and at young stage they appear as simple, but upon maturation they appear as odd-pinnate. Flowers are purplish or blue colored, smaller in size, have raceme type

Table 5 Pharmacological actions of *Tinospora cordifolia*

Sl no.	Pharmacological action	Description
1	Antioxidant effect	Stem ethanol extract improves antioxidant activity by controlling the generation of reactive oxygen species. It also controls the glutathione level and lipid peroxidation process (Jayaprakash et al. 2015; Bafna and Balaraman 2005)
2	Antimicrobial effect	Methanolic extract of the plant is effective against gram-positive and gram-negative bacteria. The plant also contains antifungal properties against *Aspergillus fumigatus*, *Aspergillus flavus*, and *Aspergillus niger* (Shanthi and Nelson 2013; Singh et al. 2014; Allemailem et al. 2019)
3	Wound healing activity	The plant is attributed to promote the synthesis of collagen and speeds up the process of wound healing (Shanbhag et al. 2005)
4	Anti-osteoporotic effect	It has been observed that ecdysteroids specifically beta-ecdysone (Ecd) promotes the formation of bone cells. Moreover, it causes the thickness of joint cartilage to significantly grow (Abiramasundari et al. 2012; Gao and Cai 2008)
5	Immunomodulatory activity	An immunomodulatory effect has been reported for cordioside, syringin, cordiofolioside A, and cordiol isolate from *T. cordifolia* (Kapil and Sharma 1997)
6	Antidiabetic effect	Tannins, saponin, cardiac glycosides, flavonoids, and alkaloids such as magnoflorine, palmatine, and jatrorrhizine exerts potent hypoglycemic effect (Patel and Mishra 2016)
7	Antitoxic activity	Alkaloids present in the plant exerts protective effect against nephrotoxicity (Gupta and Sharma 2011)

Fig. 6 Different parts of *Uraria picta*: flower (**a**), leaves (**b**), and stem (**c**)

inflorescence which are bracteate, grow up to a length of 1.5 ft., and located in an axis. Corolla is papilionaceous, calyx have a length of 4 mm, and sepals grow up to a length of 4–5 mm. Pods have 2–6 seeds with smooth and polished surface containing 3–6 segments, and each segment is having a length of 5–9 mm and 2–3 mm in diameter (Waghire et al. 2011) (Fig. 6).

7.2 *Phytochemical Profile*

U. picta contains numerous important bioactive compounds including tannins, saponins, flavonoids, isoflavanones, phenolics, triterpenes cardiac glycosides, and steroids (Rahman and Parvin 2014). The aerial parts contain flavones and glycosides, while the seeds contain canavanine, amino acids, albumin protein, and different fatty acids such as oleic acid, linoleic acid, palmitic acid, and linolenic acid (Waghire et al. 2013).

7.3 *Therapeutic Potential of* Uraria picta

U. picta contains various amino acids, long-chain fatty acids, starch, sugar, fibers, and macro elements (Saxena et al. 2014). The therapeutic uses are listed in Table 6.

Table 6 Pharmacological actions of *Uraria picta*

Sl no.	Pharmacological action	Description
1	Antimicrobial effect	Isoflavanones extracted from bark and roots exerts antibacterial activity against various gram-positive and gram-negative bacteria (Khalili et al. 2012)
2	Alzheimer effect	The aqueous extract of plant is useful for inhibition of acetylcholinesterase and butyrylcholinesterase (Veronica et al. 2013)
3	Antioxidant activity	The aqueous extract of leaves, stems, and roots consists of phenolics and flavonoids exhibits promising antioxidant activity (Mohan et al. 2019).
4	Hypolipidemic effect	The levels of serum β lipoprotein fat elements and apoprotein levels and LDL are lowered (Khanna et al. 1991)
5	Anti-nephrotic activity	The plant lowers the elevated levels of urine urea, BUN levels, and conjointly elevated levels of urine and serum creatinine (Udvant et al. 2021)
6	Antihypertensive effect	Flavonoid and phenolic compound promotes antihypertensive activity by suppressing the level of angiotensin-converting enzymes (Singh et al. 2011)
7	Bone healing activity	Extract of leaves and stems are used to cure bone fractures or other bone-related problems (Sheth 2005)

8 *Solanum xanthocarpum*

8.1 *Botanical Description*

Solanum xanthocarpum is a weed located in different regions of the world. It is a bright greenish herb with zigzag pattern of stem containing numerous prickles. It consists of numerous branches and the younger branches which are covered densely with stellate tomentum. The prickles are erect, yellowish, shinning, compressed, and glabrous with a length of around 1.3 cm. Leaves are 5–10 in numbers with a length of 2.5–5.7 cm. Petioles are having a length of 1.3–2.5 m and 1.3–2 cm in diameter, berries have green-white strips at young stage but becomes yellow upon maturation and remains surrounded by the enlarged calyx of 1.3 cm length which is densely hairy with prickles. Seeds are glabrous with a diameter of 2.5 mm. Lobes are acute, linear-lanceolate, and hairy outside having a length of around 11 mm. The ovary is ovoid in shape and have glabrous style. Filaments are glabrous with small opening pores (Kirtikar and Basu 2005) (Fig. 7).

Fig. 7 Different parts of *Uraria picta*: whole plant (**a**), leaves (**b**), flower (**c**), and berry (**d**)

8.2 Phytochemical Profile

The berries of *Solanum xanthocarpum* consists of a wide variety of phytoconstituents which exhibits various therapeutical activities. The berries contain alkaloid saponins and glycoalkaloid such as solasonine and several steroidal alkaloids like solanacarpine and solamargine, solasonine, solasonine, solasodine, solasonine, solasodine, ß-sitosterol, and solanine (Saiyed and Kanga 1936; Gupta and Dutt 1938; Tupkari et al. 1972; Friedman and McDonald 1997; Gubarev et al. 1998; Cham 1994). The berries are enriched with solasodine, diosgenin, and carpesterol (McGehee et al. 2000; Heal et al. 2001). Fruit contains numerous phytoconstituents like caffeic acid, aesculetin, aesculin, diosgenin, campesterol, and daucosterol (Heble et al. 1968; Bector and Puri 1971).

8.3 *Therapeutic Potential of* **Solanum xanthocarpum**

The plant possesses phytochemicals as wells as carbohydrate, amino acids, vitamins, fatty acids, protein, fibers, and various minerals (calcium, potassium, phosphorus and sodium). The plant exerts various pharmacological action which are enlisted in Table 7.

Table 7 Pharmacological actions of *Solanum xanthocarpum*

Sl no.	Pharmacological action	Description
1	Anti-asthmatic effect	Extract of the plant has asthma relieving and antihistaminic and antiallergic property because of its bronchodilating effect (Davies 1990)
2	Cardiovascular effect	The plant exhibits cardiovascular protective effect by directly sensitizing the atrium by increasing the Ca^{2+} permeability and also by down regulating the beta adrenoceptors (Govindan et al. 1999)
3	Antibacterial activity	Methanolic extract of plant containing flavonoids and polyphenols has excellent antibacterial activity against various (Nithya et al. 2018)
4	Antifertility effect	Solasodine present in the berry has antifertility action (Singh and Singh 2014)
5	Anti-inflammatory activity	Stigmasterol, carpesterol, lupeol, and diosgenin have massive anti-inflammatory potential (Singh and Singh 2014)
6	Anticancer effect	Lupeol, apigenin, and solamargine have anticancer property (Chaturvedi et al. 2008)
7	Wound healing activity	Leaves extract showed reduced epithelization period and increased tensile strength of wounds (Dewangan et al. 2012)

9 Conclusion

In conclusion, the phytochemical profile, traditional usages, and various therapeutic potential of *Withania somnifera*, *Ocimum sanctum*, *Cinnamomum cassia*, *Crataeva nurvala*, *Tinospora cordifolia*, *Uraria picta*, and *Solanum xanthocarpum* have been explained in details in the study. Various scientific researches have validated that these traditional exerts numerous biological effects including anti-asthmatic effect, anticancer effect, cardiovascular effect, anti-inflammatory effect, antifertility effect, wound healing activity, Alzheimer effect, hypolipidemic effect, anti-nephrotic activity, antimicrobial effect, antihypertensive effect, bone healing, anti-HIV effect, anti-urolithiatic effect, anti-tyrosinase activity, radioprotective effect, gastroprotective effect, anthelmintic activity, antiepileptic effect, nootropic effect, antivenom effect, and antiviral effect. Because of the diverse therapeutic applicability, the plants are used in the clinic throughout various countries of the world. Nevertheless, there is still a need for sufficient research and more investigations to explore about the toxicity, side effects, alimentotherapy, and health products of these traditional plants. Thus, the study aims to provide a deep insight of the potential use of the herbs, plants, and trees which can be utilized for innovation of alternative medicines, and it is crucial to safeguard the efficacy of *Withania somnifera*, *Ocimum sanctum*, *Cinnamomum cassia*, *Crataeva nurvala*, *Tinospora cordifolia*, *Uraria picta*, and *Solanum xanthocarpum* to ensure and harbor conservation for further clinical development.

References

Abiramasundari G, Sumalatha KR, Sreepriya M (2012) Effects of Tinospora cordifolia (Menispermaceae) on the proliferation, osteogenic differentiation and mineralization of osteoblast model systems in-vitro. J Ethnopharmacol 141:474–480

Abraham A, Kirson I, Glotter E, Lavie D (1968) A chemotaxonomic study of Withania somnifera (L.) dun. Phytochemistry 7:957–962

Alam MA, Haque ME, Shilpi JA, Daulla KA (2006) Antinociceptive effect of the crude ethanolic extract of Crataeva nurvala. Buch on mice. Bangl J Vet Med 14(1):65–68

Allemailem KS, Almatroudi A, Alsahli MA, Khan A, Khan MA (2019) Tinospora cordifolia aqueous extract alleviates cyclophosphamide-induced immune suppression, toxicity and systemic candidiasis in immunosuppressed mice: in vivo study in comparison to antifungal drug fluconazole. Curr Pharm Biotechnol 20:1–5

Andallu B, Radhika B (2000) Hypoglycemic, diuretic and hypocholesterolemic effect of winter cherry (Withania somnifera, Dunal) root. Indian J Exp Biol 38(6):607–609

Anonymous (2004) Monograph: Withania somnifera. Altern Med Rev 9:211–214

Arul V, Miyazaki S, Dhananjayan R (2005) Studies on the anti-inflammatory, antipyretic and analgesic properties of the leaves of Aegle marmelos Corr. J Ethnopharmacol 96:159–163

Asha MK, Prashanth D, Murali B, Padmaja R, Amit A (2001) Anthelmintic activity of essential oil of *Ocimum sanctum* and eugenol. Fitoterapia 72:669–670

Atal CK, Gupta OP, Ranghunathan K, Dhar KL (1975) Pharmacognosy and phytochemistry of Withania somnifera (Linn) Dunal (Ashwagandha). Central Council for Research in Indian medicine and homeopathy, New Delhi

Atta-ur-Rahman, Abbas S, Dur e-Shawar, Jamal AS, Choudhary MI (1993) New withanolides from Withania spp. J Nat Prod 56:1000–1006

Bafna PA, Balaraman R (2005) Anti-ulcer and anti-oxidant activity of pepticare, a herbomineral formulation. Phytomedicine 12:264–270

Balanehru S, Nagarajan B (1992) Intervention of adriamycin induced free radical damage. Biochem Int 28:735–744

Bandyopadhyay M, Jha S, Tepfer D (2007) Changes in morphological phenotypes and withanolide composition of Ri-transformed roots of Withania somnifera. Plant Cell Rep 26:599–609

Banu AT, Lunghar J (2023) Cinnamon as a potential nutraceutical and functional food ingredient. In: Herbs, spices and their roles in nutraceuticals and functional foods. Academic Press, Cambridge, MA, pp 257–278

Bector NP, Puri AS (1971) J Assoc Physicians India 19:741–744

Bhaskar VH, Profulla KM, Balakrishnan BR, Balakrishnan N, Sangameswaran B (2009) Evaluation of the anti-fertility activity of stem bark of Crataeva nurvala buch-hum. Afr J Biotechnol 8(22):6453–6456

Bhattacharjee A, Shashidhara SC (2012) Aswathanarayana : phytochemical and ethnopharmacological profile of *Crataeva nurvala* Buch-hum (Varuna): a review. Asian Pac J Trop Biomed 2:S1162–S1168. https://doi.org/10.1016/S2221-1691(12)60379-7

Bhattacharya SK, Satyan KS, Chakrabarti A (1997) Effect of Trasina, an Ayurvedic herbal formulation, on pancreatic islet superoxide dismutase activity in hyperglycaemic rats. Indian J Exp Biol 35:297–299

Bisset NG, Nwaiwu J (1983) Quaternary alkaloids of Tinospora species. Planta Med 48:275–279

Bopana N, Saxena S (2008) Crataeva nurvala: a valuable medicinal plant. Int J Geogr Inf Syst 14(1):107–127

Bradford PG, Awad AB (2007) Phytosterols as anticancer compounds. Mol Nutr Food Res 51:161–170

Bucci LR (2000) Selected herbals and human exercise performance. Am J Clin Nutr 72:624S–636S

Cham BE (1994) Asia Pac J Pharmacol 9:113–118

Chaturvedi PK, Bhui K, Shukla Y (2008) Lupeol: connotations for chemoprevention. Cancer Lett 263(1):1–13

Chou ST, Chang WL, Chang CT, Hsu SL, Lin YC, Shih Y (2013) Cinnamomum cassia essential oil inhibits α-MSH-induced melanin production and oxidative stress in murine B16 melanoma cells. Int J Mol Sci 14:19186–19201

Choudary MI, Abbas S, Jamal AS, Atta-ur-Rahman (1996) Withania somnifera—a source of exotic withanolides. Heterocycles 42:555–563

Choudhary N, Siddiqui MB, Azmat S, Khatoon S (2013) Tinospora cordifolia: ethnobotany, phytopharmacology and phytochemistry aspects. IJPSR 4:891

Cor JS, Beach JF, Blair A, Clark AJ, King J, Lee TB et al (1970) Disodium chromoglycate. Adv Drug Res 5:190–196

Davies RJ (1990) Asthma in clinical medicine. In: Kumar PJ, Clark ML (eds) A text book for medical students and doctors, 2nd edn. Bailliere Tindall, London, pp 664–668

Dewangan H, Bais M, Jaiswal V, Verma VK (2012) Potential wound healing activity of the ethanolic extract of solanum xanthocarpum schrad and wendl leaves. Pak J Pharm Sci 25(1):189–195

Dhuley JN (1997) Effect of some Indian herbs on macrophage functions in ochratoxin A treated mice. J Ethnopharmacol 58:15–20

Dixit SN, Khosa RL (1971) Chemical investigations on Tinospora cordifolia (wild.) miers. Indian J Appl Chem 34:46–47

Elsakka M, Grigorescu E, Stanescu U, Stanescu U, Dorneanu V (1990) New data referring to chemistry of Withania somnifera species. Rev Med Chir Soc Med Nat Iasi 94(2):385–387

Friedman M, McDonald GM (1997) Crit Rev Plant Sci 16:55–132

Fujioka T, Kashiwada Y, Kilkuskie RE, Cosentino LM, Ballas LM, Jiang JB et al (1994) Anti-AIDS agents, 11. Betulinic acid and platanic acid as anti-HIV principles from Syzygium claviflorum, and the anti-HIV activity of structurally related triterpenoids. J Nat Prod 57(2):243–247

Gagan VD, Pradhan P, Sipahimalan AT, Banerji A, Palmatosides F (1996) Diterpene furan glucosides from Tinospora cordifolia-structural elucidation by 2D NMR spectroscopy. Indian J Chem 35B:630–634

Ganzera M, Choudhary MI, Khan IA (2003) Quantitative HPLC analysis of withanolides in Withania somnifera. Fitoterapia 74(1–2):68–76

Gao L, Cai G (2008) Beta-ecdysterone induces osteogenic differentiation in mouse mesenchymal stem cells and relieves osteoporosis. Biol Pharm Bull 31:2245–2249

Ghani A (1998) Medicinal plants of Bangladesh: chemical constituents and uses. Asiatic Society of Bangladesh, Dhaka

Gholap S, Kar A (2004) Hypoglycemic effects of some plant extracts are possibly mediated through inhibition in corticosteroid concentration. Pharmazie 59:876–878

Ghosal S, Lal J, Srivastava R et al (1989) Immunomodulatory and CNS effects of sitoindosides IX and X, two new glycowithanolides from Withania somnifera. Phytother Res 3:201–206

Glotter E (1991) Withanolides and related ergostane-type steroids. Nat Prod Rep 8:415–440

Goel RK, Sairam K, Dorababu M, Prabha T, Rao CV (2005) Effect of standardized extract of *Ocimum sanctum* Linn.on gastric mucosal offensive and defensive factors. Indian J Exp Biol 43:715–721

Gordon MC, David JN (2001) Naturan product drug discovery in the next millennium. Pharm Boil 39:8–17

Govindan S, Viswanathan S, Vijayasekaran V, Alagappan R (1999) A pilot study on the clinical efficacy of solanum xanthocarpum and Solanum trilobatum in bronchial asthma. J Ethnopharmacol 66(2):205–210

Gubarev MI, Enioutina EY, Taylor JL, Visic DM, Daynes RA (1998) Phytother Res 12:79–88

Gupta SC (1996) Validation of herbage yield, oil yield and major component of various Ocimum species/varieties (chemotypes) harvested at different stages of maturity. J Essent Oil Res 8:275–279

Gupta MP, Dutt S (1938) J Indian Chem Soc 15:95–100

Gupta GL, Rana AC (2007) Withania somnifera (Ashwagandha): A Review. Pharmacogn Rev 1:129–136

Gupta R, Sharma V (2011) A meliorative effects of Tinospora cordifolia root extract on histopathological and biochemical changes induced by aflatoxin-b in mice kidney. Toxicol Int 18:94–98
Gupta SK, Prakash J, Srivastava S (2002) Validation of traditional claim of Tulsi, Ocimum sanctum Linn. As a medicinal plant. Indian J Exp Biol 40:765–773
Hannan JM, Marenah L, Ali L, Rokeya B, Flatt PR, Abdel-Wahab YH (2006) *Ocimum sanctum* leaf extracts stimulate insulin secretion from perfused pancreas, isolated islets and clonal pancreatic beta-cells. J Endocrinol 189:127–136
Hanuman JB, Mishra AK, Sabata B (1986a) A natural phenolic lignan from Tinospora cordifolia Miers. J Chem Soc 1181–1185:1181
Hanuman JB, Bhatt RK, Sabata BK (1986b) A diterpenoid furanolactone from Tinospora cordifolia. Phytochemistry 25:1677–1680
He S, Zeng KW, Jiang Y, Tu PF (2016) Nitric oxide inhibitory constituents from the barks of Cinnamomum cassia. Fitoterapia 112:153–160
Heal KG, Sheikh NA, Hollingdale MR, Morrow WJ, Taylor MW (2001) Vaccine 19:4153–4161
Heble MR, Narayanaswami S, Chadha MS (1968) Diosgenin and β-Sitosterol: Isolation from Solanum Xanthocarpum Tissue Cultures. American Association for the Advancement of Science Stable 161(13):1145
Hemalatha S, Wahi AK, Singh PN, Chansouria JP (2006) Hypolipidemic activity of aqueous extract of Withania coagulans Dunal in albino rats. Phytother Res 20(7):614–617
Ichikawa H, Takada Y, Shishodia S, Jayaprakasam B, Nair MG, Aggarwal BB (2006) Withanolides potentiate apoptosis, inhibit invasion, and abolish osteoclastogenesis through suppression of nuclear factor-kappaB (NF-kappaB) activation and NFkappaB-regulated gene expression. Mol Cancer Ther 5(6):1434–1445
IDMA (2002) Indian herbal pharmacopoeia. IDMA, Mumbai, p 272
Inayathulla SWR, Asif K, Mukesh S (2010) Ethanolic extract of crataeva nurvala root bark in experimental animals. IJPPS 2:158–161
Iqbal S, Ishu B, Vivek A, Suchi B (2011) Prospective randomized clinical trial comparing phytotherapy with potassium citrate in management of minimal burden (≤8 mm) nephrolithiasis. Urol Ann 3(2):75–81
Jaggi RK, Madaan R, Singh B (2003) Anticonvulsant potential of holy basil, *Ocimum sanctum* Linn., and its cultures. Indian J Exp Biol 41:1329–1333
Jain DP, Pancholi SS, Patel R (2011) Synergistic antioxidant activity of green tea with some herbs. J Adv Pharm Technol Res 2(3):177–183
Jayaprakasam B, Zhang Y, Seeram N, Nair M (2003) Growth inhibition of tumor cell lines by withanolides from Withania somnifera leaves. Life Sci 74(1):125–132
Jayaprakasam B, Strasburg GA, Nair MG (2004) Potent lipid peroxidation inhibitors from Withania somnifera fruits. Tetrahedron 60:3109–3121
Jayaprakash R, Ramesh V, Sridhar MP, Sasikala C (2015) Antioxidant activity of ethanolic extract of Tinospora cordifolia on N-nitrosodiethylamine (diethyl nitrosamine) induced liver cancer in male wistar albino rats. J Pharm Bioallied Sci 7(S1):40–45
Johri S, Jamwal U, Rasool S, Kumar A, Verma V, Qazi GN (2005) Purification and characterization of peroxidases from Withania somnifera (AGB 002) and their ability to oxidize IAA. Plant Sci 169:1014–1021
Kapil A, Sharma S (1997) Immunopotentiating compounds from Tinospora cordifolia. J. Ethnopharmacol. 58:89–95
Kapoor LD (2001) Handbook of ayurvedic medicinal plants. CRC, London, pp 337–338
Kaur G, Invally M, Khan MK, Jadhav P (2018) A nutraceutical combination of Cinnamomum cassia & Nigella sativa for type 1 diabetes mellitus. J Ayurveda Integr Med 9(1):27–37
Kelm MA, Nair MG, Strasburg GM, DeWitt DL (2000) Antioxidant and cyclooxygenase inhibitory phenolic compounds from Ocimum sanctum Linn. Phytomedicine 7:7–13
Khaleque A, Maith MAW, Huq MS, Abul BK, Tinospora cordifolia IV. (1970) Isolation of heptacosanol, ß sitosterol and three other compounds tinosporine, cordifol and cordifolone. Pak J Sci Ind Res 14:481–483

Khalili MA, Norhayati AH, Emynur SS, Mohamad FG, Azlina M, Tajul ZM, Ahmad ZAL (2012) Screening of seven types Terengganu herbs for their potential antibacterial activity against selected food spencer microorganisms. Borneo Sci 2012:11–27

Khanna N, Bhatia J (2003) Antinociceptive action of Ocimum sanctum (Tulsi) in mice: possible mechanisms involved. J Ethnopharmacol 88:293–296

Khanna A, Chander R, Kapoor N (1991) Hypolipidemic activity of Abana in rats. Fitoterapia 62(3):271–275

Khuda MQI, Khaleque A, Ray N (1964) Tinospora cordifolia constituents of plants fresh from the field. Sci Res 1:177–183

Kim GJ, Lee JY, Choi HG, Kim SY, Kim E, Shim SH, Nam JW, Kim SH, Choi H (2017) Cinnamomum lactone, a new butyrolactone from the twigs of Cinnamomum cassia and its inhibitory activity of matrix metalloproteinases. Arch Pharm Res 40:304–310

Kirtikar KR, Basu BD (2005) Indian Medicinal Plants, vol 3. International Book Distributors, Dehradun, pp 1759–1761

Krison I, Glotter E (1980) 14α-hydroxy steroids form W. somnifera (L) Dunal. J Chem Res Synop 10:338–339

Kulkarni SK, Dhir A (2008) Withania somnifera: an Indian ginseng. Prog Neuropsychopharmacol Biol Psychiatry 32(5):1093–1105

Lawrence BM, Hogg JW, Terhune SJ, Pichitakul N (1972) Essential oils and their constituents. In: IX. The oils of Ocimum sanctum and Ocimum basilicum from Thailand. Flav Industry, pp 47–49

Lee SG, Siaw JA, Kang HW (2016) Stimulatory effects of cinnamon extract (*Cinnamomum cassia*) during the initiation stage of 3T3-L1 adipocyte differentiation. Food Secur 5:83

Leet JE, Hussain SF, Minard RD, Shamma M (1982) Heterocycles 19:2355–2360

Li YQ, Kong DX, Wu H (2013) Analysis and evaluation of essential oil components of cinnamon barks using GC-MS and FTIR spectroscopy. Ind Crop Prod 41:269–278

Liby KT, Yore MM, Sporn MB (2007) Triterpenoids and rexinoids as multifunctional agents for the prevention and treatment of cancer. Nat Rev Cancer 7:357–369

Liu X, Fu J, Yao XJ, Yang J, Liu L, Xie TG, Jiang PC, Jiang ZH, Zhu GY (2018) Phenolic constituents isolated from the twigs of Cinnamomum cassia and their potential neuroprotective effects. J Nat Prod 81:1333–1342

Lizano S, Domont G, Perales J (2003) Natural phospholipase A(2) myotoxin inhibitor proteins from snakes, mammals and plants. Toxicon 42(8):963–977

Luo Q, Wang SM, Lu Q, Luo J, Cheng YX (2013) Identification of Compounds from the WaterSoluble Extract of Cinnamomum cassia Barks and Their Inhibitory Effects against High-Glucose-Induced Mesangial Cells. Molecules 18:10930–10943

Machiah DK, Girish KS, Gowda TV (2006) A glycoprotein from a folk medicinal plant, Withania somnifera, inhibits hyaluronidase activity of snake venoms. Comp Biochem Physiol C Toxicol Pharmacol 143(2):158–161

Mahajan VR, Jolly CI, Kundnani KM (1985) A new hypoglycaemic agent from Tinospora cordifolia. Indian Drugs 23:119–120

Malhotra CL, Mehta VL, Das PK, Dhalla NS (1965a) Studies on Withania-ashwagandha, Kaul. V. The effect of total alkaloids (ashwagandholine) on the central nervous system. Indian J Physiol Pharmacol 9:127–136

Malhotra CL, Mehta VL, Prasad K, Das PK (1965b) Studies on Withania ashwagandha, Kaul. IV. The effect of total alkaloids on the smooth muscles. Indian J Physiol Pharmacol 9:9–15

Malhotra CL, Das PK, Dhalla NS, Prasad K (1981) Studies on Withania ashwagandha, Kaul. III. The effect of total alkaloids on the cardiovascular system and respiration. Indian J Med Res 49:448–460

McGehee DS, Krasowski MD, Fung DL, Wilson B, Groner GA, Moss JM (2000) Anesthesiology 93:510–519

Md. Enamul H, Md. Nahidul I, Dipankar DG, Mahbub H, Hossain US, Biazid AS (2008) Triterpenoids from the stem bark of Crataeva nurvala. Dhaka Univ. J Pharm Sci 17(1):71–74

Mirjalili MH, Moyano E, Bonfill M, Cusido RM, Palazón J (2009) Steroidal lactones from Withania somnifera, an ancient plant for novel medicine. Molecules 14(7):2373–2393
Misra B, Prakash B (1969) Study of medicinal plant and drug. Bhava Prakash Nighantu 1:26
Mohan B, Saxena HO, Kakkar A, Mishra MK (2019) Determination of antioxidant activity, total phenolic and flavonoid contents in leaves, stem and roots of Uraria picta Desv. Environ Conserv J 20(3):1–8
Mukherjee R, Dash PK, Ram GC (2005) Immunotherapeutic potential of *Ocimum sanctum* (L) in bovine subclinical mastitis. Res Vet Sci 79:37–43
Mukherji SP (1987) Ocimum—a cheap source of eugenol. Sci Rep 28:599–610
Nair AGR, Gunasegaran R, Joshi BS (1982) Chemical investigation of certain south Indian plants. Indian J Chem 21B:979
Ngoc TM, Khoi NM, Ha DT, Nhiem NX, Tai BH, Van Don D, Van Luong H, Son DC, Bae K (2012) Xanthine oxidase inhibitory activity of constituents of *Cinnamomum cassia* twigs. Bioorg Med Chem Lett 22:4625–4628
Nithya M, Ragavendran C, Natarajan D (2018) Antibacterial and free radical scavenging activity of a medicinal plant solanum xanthocarpum. Int J Food Prop 21(1):313–327
Ohnuma T, Matsumoto T, Itoi A, Kawana A, Nishiyama T, Ogura K, Hiratsuka A (2011) Enhanced sensitivity of A549 cells to the cytotoxic action of anticancer drugs via suppression of Nrf2 by procyanidins from Cinnamomi cortex extract. Biochem Biophys Res Commun 413:623–629
Ooi LS, Li Y, Kam SL, Wang H, Wong EY, Ooi VE (2006) Anti-microbial activities of cinnamon oil and cinnamaldehyde from the Chinese medicinal herb Cinnamomum cassia Blume. Am J Chin Med 34(3):511–522
Palliyaguru DL, Singh SV, Kensler TW (2016 Jun) Withania somnifera: from prevention to treatment of cancer. Mol Nutr Food Res 60(6):1342–1353
Panda S, Kar A (1997) Evidence for free radical scavenging activity of ashwagandha root powder in mice. Indian J Physiol Pharmacol 41:424–426
Pandey G, Madhuri S (2010 Nov) Pharmacological activities of Ocimum sanctum (tulsi): a review. Int J Pharm Sci Rev Res 5(1):61–66
Pannee C, Chandhanee I, Wacharee L (2014) Anti-inflammatory effects of essential oil from the leaves of *Cinnamomum cassia* and cinnamaldehyde on lipopolysaccharide-stimulated J774A.1 cells. J Adv Pharm Technol Res 5:164–170
Parvin S, Abdul KM, Abdul M, Ekramul H, Ashik MM, Mir Imam IW (2011) Triterpenoids and phytosteroids from stem bark of Crataeva nurvala buch ham. J Appl Pharmaceut Sci 1(9):47–50
Patel MB, Mishra S (2016) Hypoglycemic activity of alkaloidal fraction of Tinospora cordifolia. Pharma Innov 5:104
Pathak AK, Agarwal AK, Jain DC, Sharma RP, Howarth OW (1995a) NMR studies of 20 hydroxyecdysones, a steroid isolated from Tinospora cordifolia. Indian J Chem 34:674–676
Pathak AK, Jain DC, Sharma PR (1995b) Chemistry and biological activities of the genus Tinospora. Int J Pharm 33:277–287
Patil M, Choudhari AS, Pandita S, Islam MA, Raina P, Kaul-Ghanekar R (2017) Cinnamaldehyde, Cinnamic acid, and Cinnamyl alcohol, the bioactives of Cinnamomum cassia exhibit HDAC8 inhibitory activity: an in vitro and in silico study. Pharmacogn Mag 13:S645–S651
Pattanayak P, Behera P, Das D, Panda SK, Ocimum sanctum Linn. (2010) A reservoir plant for therapeutic applications: an overview. Pharmacogn Rev 4(7):95
Pham VC, Nguyen TTA, Vu TO, Gao TQ, Min BS, Kim JA (2019) Five new diterpenoids from the barks of *Cinnamomum cassia* (L.) J. Presl. Phytochem Lett 32:23–28
Prabhat D, Ranjan S, Mekap S, Pani S (2010) Phytochemical and pharmacological screening of the plant Crataeva magna against alloxan induced diabetes in rats. J Pharm Sci Res 2(4):257–263
Prajapati ND, Purohit SS, Sharma AK, Kumar T (2003) A hand book of medicinal plant, 1st edn. Agrobios, Jodhpur, p 367
Prakash J, Gupta SK, Dinda AK (2002) Withania somnifera root extract prevents DMBA-induced squamous cell carcinoma of skin in Swiss albino mice. Nutr Cancer 42(1):91–97

Rahman AHM, Parvin MIA (2014) Study of medicinal uses on Fabaceae family at Rajshahi, Bangladesh. Res Plant Sci 2:6–8

Raju PM, Ali M, Velasco-Negueruela A, Perez-Alonso MJ (1999) Volatile constituents of the leaves of Ocimum sanctum L. J Essent Oil Res 11:159–161

Rao ZL, Xu F, Wen TQ, Wang F, Sang WT, Zeng N (2018) Protective effects of essential oils from Rimulus cinnamon on endotoxin poisoning mice. Biomed Pharmacother 101:304–310

Rastogi RP, Mehrotra BN (1998) Compendium of Indian medicinal plants. New Delhi, Central Drug Research Institute

Ravid Z, Putievsky E, Katzir I, Lewinsohn E (1997) Enantiomeric composition of linalol in the essential oils of Ocimum species and in commercial basil oils. Flav Frag 12:393–396

Reeta J, Deepa S, Johri PK (2010) Study of hypolipidemic activity of selected medicinal plant extracts in normal and induced hyperlipidemic rats. Biochem Cell Arch 10(2):215–224

Saiyed IZ, Kanga DD (1936) Proc Indiana Acad Sci 4:255

Sarkar A, Lavania SC, Pandey DN, Pant C (1994) Changes in the blood lipid profile after administration of *Ocimum sanctum* (Tulsi) leaves in the normal albino rabbits. Indian J Physiol Pharmacol 38:311–312

Saxena HO, Soni A, Mohammad N, Choubey SK (2014) Phytochemical screening and elemental analysis in different plant parts of Uraria picta Desv.: a Dashmul species. J Chem Pharm Res 6:756–760

Schliebs R, Liebmann A, Bhattacharya SK et al (1997) Systemic administration of defined extracts from Withania somnifera (Indian ginseng) and Shilajit differentially affects cholinergic but not glutamatergic and GABAergic markers in rat brain. Neurochem Int 30:181–190

Schonbeck-Temesy E (1972) In: Rechinger KH (ed) Flora Iranica, vol 100. Akademische Druck-u. Verlagsanstalt, Graz, Austria, pp 29–26

Senanayake UM, Wijesekera RO (2003) Chemistry of cinnamon and cassia. In: Cinnamon and Cassia. CRC, London, pp 96–136

Setzer WN, Setzer MC (2003) Plant-derived triterpenoids as potential antineoplastic agents. Mini Rev Med Chem 3:540–556

Shah CS, Qadry JS (1998) A text book of pharmacognosy, p 216

Shahzadi M, Rizwan B, Tufail T, Basharat S, Shehzadi S (2021) Functional and nutraceutical characterization of cinnamon. Pak BioMed J 4(2):187–192

Shanbhag T, Shenoy S, Rao MC (2005) Wound healing profile of Tinospora cordifolia. Indian Drugs 42:217–221

Shanthi V, Nelson R (2013) Antibacterial activity of Tinospora cordifolia (Willd) Hook. F. Thoms on urinary tract pathogens. Int J Curr Microbiol App Sci 2:190–194

Sharm M, Kishore K, Gupta SK, Joshi S, Arya DS (2001) Cardiaprotective potential of *Ocimum sanctum* Linn in isoproterenol induced myocardial infraction in rats. Mol Cell Biochem 498:39–46

Sheth AK (2005) The herbs of Ayurveda, vol 4. AKSheth Publisher, p 1044

Shetty S, Udupa S, Udupa L, Somayaji N (2006) Wound healing activity of *Ocimum sanctum* Linn with supportive role of antioxidant enzymes. Indian J Physiol Pharmacol 50:163–168

Shishodia S, Majumdar S, Banerjee S, Aggarwal BB (2003) Urosolic acid inhibits nuclear factor-kappaB activation induced by carcinogenic agents through suppression of IkappaB alpha kinase and p65 phosphorylation: correlation with downregulation of cyclooxygenase 2, matrix metalloproteinase 9, and cyclin D1. Cancer Res 63:4375–4383

Shumaia P, Abdul KM, Ajijur RM, Mir IB, Ekramul HM (2012) Antibacterial activities and brine shrimp lethality bioassay of the chloroform extract of stem bark of Crataeva nurvala buch ham. Int J Pharm Sci Res 3(3):830–834

Singh D, Chaudhuri PK (2018 Aug) A review on phytochemical and pharmacological properties of holy basil (Ocimum sanctum L.). Ind Crop Prod 1(118):367–382

Singh OM, Singh TP (2014) Phytochemistry of solanum xanthocarpum: an amazing traditional healer. J Sci Ind Res 69:732–740

Singh N, Nath R, Lata A et al (1982) Withania somnifera (ashwagandha), a rejuvenating herbal drug which enhances survival during stress (an adaptogen). Int J Crude Drug Res 20:29–35

Singh SS, Pandey SC, Srivastava S, Gupta VS, Patro B, Ghosh AC (2003) Chemistry moreover, medicinal properties of Tinospora cordifolia (Guduchi). Indian J Pharm 35:83–91

Singh S, Malhotra M, Majumdar DK (2005) Antibacterial activity of *Ocimum sanctum* L.fixed oil. Indian J Exp Biol 43:835–837

Singh G, Sharma PK, Dudhe R, Singh S (2010) Biological activities of Withania somnifera. Ann Biol Res 1(3):56–63

Singh S, Singh HN, Gangwar AK, Devi KS, Niyogi D, Waghaye JY (2011) Effect of Uraria picta on bone healing in rabbits: a radiographic, angiographic and histopathological study. Indian J Vet Surg 32(2):107–110

Singh K, Panghal M, Kadyan S, Chaudhary U, Yadav JP (2014) Antibacterial activity of synthesized silver nanoparticles from Tinospora cordifolia against multi-drug resistant strains of pseudomonas aeruginosa isolated from burn patients. J Nanomed Nanotechnol 5:1–6

Sinha A, Sharma HP (2015) A medicinal plant: micropropagation and phytochemical screening of Tinospora cordifolia (Wild.) Miers ex Hook F & Thoms. IJAPBC 4:114–121

Sood S, Narang D, Thomas MK, Gupta YK, Maulik SK (2006) Effect of *Ocimum sanctum* Linn. on cardiac changes in rats subjected to chronic restraint stress. J Ethnopharmacol 108:423–427

Soosamma J, Madhavi T, Bincy R, Jincy S, Vinutha. (2010) Phytochemistry and pharmacology of an important Indian medicinal plant Crataeva nurvala Buch ham. Res J Pharmacogn Phytochem 2(4):275–280

Spandana U, Liakhat Ali SL, Nirmala T, Santhi M, Babu SD (2013) A review on Tinospora cordifolia. IJCPR 4:61–68

Subaraju GV, Vanisree M, Rao CV, Sivaramakrishna C, Sridhar P, Jayaprakasam B, Nair MG (2006) Ashwagandhanolide, a bioactive dimeric thiowithanolide isolated from the roots of Withania somnifera. J Nat Prod 69:1790–1792

Subramanian M, Chintalwar GJ, Chattopadhyay S (2005) Antioxidant and radioprotective properties of an *Ocimum sanctum* polysaccharide. Redox Rep 10:257–264

Taguchi Y, Takizawa T, Ishibashi H, Sagawa T, Arai R, Inoue S, Yamaguchi H, Abe S (2010) Therapeutic effects on murine oral candidiasis by oral administration of cassia (Cinnamomum cassia) preparation. Nippon Ishinkin Gakkai Zasshi 51(1):13–21

Tiwari P, Nayak P, Prusty SK, Sahu PK (2018) Phytochemistry and pharmacology of Tinospora cordifolia. Syst Rev Pharm 9:70–78

Tupkari SV, Saoji AN, Deshmukh VK (1972) Planta Med 22:184–187

Tursunova RN, Maslennikova VA, Abubakirov NK (1977) Withanolides in the vegetable kingdom. Chem Nat Comp 13:131–138

Udvant PB, Khairnar SJ, Jadhav GB, Sable RR, Arprit MK (2021) Effect of extract of Uria Picta Jacq. on experimentally induced hypertension in Wistar rats. Biosci Biotechnol Res Asia 18(3):523–532

Uma Devi P, Ganasoundari A, Vrinda B, Srinivasan KK, Unnikrishnan MK (2000) Radiation protection by the Ocimum flavonoids orientin and vicenin: mechanisms of action. Radiat Res 154:455–460

Upadhyay AK, Kumar K, Kumar A, Mishra HS (2010) Tinospora cordifolia (wild.) Hook. f. and Thoms. (Guduchi)–validation of the ayurvedic pharmacology through experimental and clinical studies. Int J Ayurveda Res 1:112–121

Uslu GA, Gelen V, Uslu H, Ozen H (2018) Effects of Cinnamomum cassia extract on oxidative stress, immunoreactivity of iNOS and impaired thoracic aortic reactivity induced by type II diabetes in rats. Braz J Pharm Sci 54(3)

Veronica OO, Ganiyu O, Emmanuel OI (2013) Antioxidant and anticholinesterase activities of aqueous extract of Uraria picta (Jacq.) DC. Afr J Pharm Pharmacol 7(41):2768–2773

Visavadiya NP, Narasimhacharya AV (2006) Hypocholesteremic and antioxidant effects of Withania somnifera (Dunal) in hypercholesteremic rats. Phytomedicine 14:136

Waghire HB, Survase SA, Pokle DS (2011) A preliminary study on the germination of Uraria picta (Jacq.) DC. J Ecobiotechnol 3:28–30

Waghire HB, Shaikh FK, Jaiwal BV, Pokle DS (2013) Polymorphism of albumin like proteins in three species of genus Uraria. Int J Res Pharmaceut Biomed Sci 4:939–942

Wink M (2000) Introduction biochemistry, role and biotechnology of secondary products. In: Wink M (ed) Biochemistry of secondary product metabolism. CRC, Boca Raton, FL, pp 1–16

Xiong YK, Liu XL, Mu ZJ, Yuan E, Sun MS, You QT, Liu ZY (2018) Differences of chemical constituents of volatile oil from *Cinnamomum* before and after drying and comparison of anti-drug resistant bacteria activities. Res Explor Lab 37:56–59

Yanpallewar SU, Rai S, Kumar M, Acharya SB (2004) Evaluation of antioxidant and neuroprotective effect of Ocimum sanctum on transient cerebral ischemia and long term cerebral hypoperfusion. Pharmacol Biochem Behav 79(1):155–164

Zeng JF, Xue YB, Shu PH, Qian HQ, Sa RJ, Xiang M, Li XN, Luo ZW, Yao GM, Zhang YH (2014) Diterpenoids with immunosuppressive activities from *Cinnamomum cassia*. J Nat Prod 77:1948–1954

Zeng JF, Zhu HC, Lu JW, Hu LZ, Song JC, Zhang YH (2017) Two new geranyl phenyl acetate glycosides from the barks of *Cinnamomum cassia*. Nat Prod Res 31:1812–1818

Zhang L, Zhang Y, Zhang L, Yang X, Lv Z (2009) Lupeol, a dietary triterpene inhibited growth, and induced apoptosis through down-regulation of DR3 in SMMC7721 cells. Cancer Invest 27:163–170

Ziauddin M, Phansalkar N, Patki P et al (1996) Studies on the immunomodulatory effects of ashwagandha. J Ethnopharmacol 50:69–76

Mushroom Bioactives: Traditional Resources with Nutraceutical Importance

Charu Agnihotri, Aarzoo, Shekhar Agnihotri, Shwet Kamal, and Bhim Pratap Singh

Contents

Abstract Mushrooms are widely recognized for their significant contributions to human well-being, earning their status as functional foods due to their rich nutritional and therapeutic attributes. These fungi encompass diverse natural constituents, including polysaccharides, glycoproteins, phenolic compounds, and proteins, all exhibiting immunomodulatory, anticarcinogenic, prebiotic, and antimicrobial

C. Agnihotri · Aarzoo · S. Agnihotri · B. P. Singh (✉)
Deparment of Agriculture and Environmental Sciences, NIFTEM-K, Sonepat, Haryana, India
e-mail: bpsingh@niftem.ac.in

S. Kamal
Directorate of Mushroom Research (DMR), Solan, Himachal Pradesh, India

© The Author(s), under exclusive license to Springer Nature Singapore Pte Ltd. 2024

A. Das Talukdar et al. (eds.), *Traditional Resources and Tools for Modern Drug Discovery*, Interdisciplinary Biotechnological Advances,
https://doi.org/10.1007/978-981-97-4600-2_24

properties. In recent years, particularly during the COVID-19 pandemic, mushroom and associated secondary metabolites have garnered increased recognition for their efficacy in nutraceuticals and pharmaceuticals. The pandemic has increased a shift in people's dietary preferences, aligning them more closely with healthier food options and nutritional supplements. Various mushroom species, including *Lentinus edodes*, *Ganoderma* species, *Pleurotus* species, and diverse wild mushrooms, known for their bioactive molecules, notably polysaccharides and proteins, have demonstrated immunomodulatory characteristics through their precise interactions with cellular receptors. Furthermore, ongoing research endeavors are focused on the targeted delivery of bioactive compounds derived from mushrooms, which are responsible for inhibiting cancerous cell growth and enhancing the phenomenon of apoptosis. This research direction is driven by increasing environmental and biological stress, which creates harmful molecules called reactive oxygen species (ROS). These ROS play a pivotal role in disrupting critical mitochondrial signaling pathways, including MAPKs, caspase 3, caspase 9, and others. Numerous components from specific mushroom varieties hold the potential to rectify these pathological conditions and reinstate cellular homeostasis. In summary, the book chapter provides facts about the potential beneficial effect of mushroom extracts and their active compounds and sheds light on the underlying targeted studies in the sector of pharmaceuticals and nutraceuticals.

Keywords Mushrooms · Immunomodulatory · Anticarcinogenic · Reactive oxygen species

1 Introduction

In recent years, infectious diseases have been on the rise globally, with several pathogens causing severe infections and posing significant threats to human health. To mitigate these health challenges, the extensive and indiscriminate use of antibiotics and drugs has become a common practice. However, this approach can have adverse consequences, such as antimicrobial resistance. Significantly, this misuse of antibiotics and drugs may facilitate the transmission of acquired resistance to subsequent generations. To address these challenges and ensure the sustainable management of infectious diseases, adopting a more judicious and evidence-based approach to the use of antibiotics and drugs is crucial. Researchers shifted more toward natural and nutraceutical compounds to achieve these goals. Mushrooms have constituted an integral component of the human diet since ancient times. However, over the last two decades, they have extended their utility into nutraceuticals, cosmeceuticals, and nanomedicines. Notably, mushrooms themselves and mushroom byproducts like mushroom stem have become subjects of critical research, owing to their rich assortment of nutraceutical components and their applications in both the cosmeceuticals and pharmaceutical industries. They are recognized as a sustainable source of antimicrobial agents. Several mushroom varieties, including *Pleurotus eryngii*, *Pleurotus ostreatus*, *Hericium erinaceus*, *Agaricus bisporus*, etc., have demonstrated substantial antimicrobial activity against various fungal and bacterial pathogens. Both these mushrooms' fruiting body and mycelium contain bioactive compounds responsible for potent antioxidant properties. In the fields of nanomedicines and drug discovery, approximately 200 wild mushroom species find application, further emphasizing their importance in scientific research exhibit specific arch and applications (Anusiya et al. 2021; Ferreira et al. 2016; Li et al. 2021; Martins 2016).

Specific nutraceutical compounds, such as polysaccharides, proteins, glycopeptides, and terpenes, confer therapeutic benefits of mushrooms. Among these components, polysaccharides, especially β-glucans with specific glycosidic linkages, have been successfully extracted from various mushroom varieties, showcasing notable bioactivity against pathogens. These particular bioactive constituents possess properties that augment the immune system while demonstrating antidiabetic, hepatoprotective, antimicrobial, and antitumor attributes. Apart from that, the utilization of β-glucans for improving damage induced by stressed environmental conditions that expedite skin aging or enhance chronic inflammation holds potential as a promising, natural, and cost-efficient strategy for upholding skin homeostasis equilibrium (Sousa et al. 2023). The mushroom polysaccharides are utilized significantly in food industries as dietary fibers, packaging, and edible coating materials. In pharmaceuticals, these polysaccharides are used as gut health products and are widely used for gastric ulcer injuries (Hao et al. 2023). Fungal polysaccharides such as lentinan, schizophyllan, and grifolin have clinical significance. Those polysaccharides possess antioxidant activity and regulate peroxidase enzymes such as superoxide dismutase, glutathione peroxidase, catalase, etc. Furthermore, in recent years, researchers have directed their focus toward ergosterol, a compound within mushrooms that serves as a precursor to vitamin D2 in ultraviolet (UV) rays (Kaur et al. 2023; Sen et al. 2023).

This chapter explores how mushroom bioactives can be good for our health. The chapter also covers how mushrooms can act like medicine and benefit our well-being. The goal is to understand how mushrooms could play a significant role in keeping us healthy in the future. We are exploring the potential of mushrooms and their associated bioactive compounds for our overall well-being and sustainability.

2 Therapeutic Attributes of Mushroom Bioactives

The *Agaricus bisporus*, *Flammulina velutipes*, *Lentinula edodes*, and *Pleurotus* species are the most widely grown mushrooms worldwide. Because of their medical benefits, including anticarcinogenic, antioxidant, immune enhancer, radical scavenging, cholesterol-lowering, cardiovascular, antimicrobial, anti-inflammatory, liver injury protective, detoxicating, antiobesity, antidiabetic, and analgesic qualities, mushrooms have been used for therapeutic purposes. From ancient times, people have been using edible mushrooms for food, and they are known for their taste, texture, and therapeutic qualities. Regarding nutrition, mushrooms are an essential source of fats, proteins, fibers, and minerals (Bach et al. 2017; Sande et al. 2019). The bioactive compounds found in mushrooms have a profoundly intriguing effect on consumers because they can improve health and lower the risk of disease. Due to their abundant bioactive compounds, mushrooms are a significant source of different polysaccharides and protein-polysaccharide complexes in modern medicine. In addition to polysaccharides, mushrooms also include a variety of physiologically significant substances, such as steroids, triterpenes, lectins, phenolics, alkaloids, lactones, and antibiotics (Badalyan et al. 2019; Chugh et al. 2022). Pantothenic acid, or vitamin B5, is found in mushrooms, which is critical in converting carbohydrates into glucose. Minerals like selenium and ergothioneine are naturally occurring antioxidants that protect human cells from oxidative damage. β-glucans are also found in mushrooms and are known to boost both innate and adaptive immune responses. Certain bioactive substances found in mushrooms have health advantages, including the promotion of immunological function as well as anti-inflammatory, antitumor, antioxidant, antiviral, and antibacterial properties (Kour et al. 2022).

2.1 Mushroom Bioactives as Immunomodulators

Anything that can change an immune response is considered an immunomodulator. Immunosuppressants suppress or deactivate the immune system, whereas immunostimulants stimulate or activate it. Plants are well known for containing bioactive immunomodulatory compounds (Reka et al. 2021). In the current health and wellness sectors, immunomodulators play a pivotal role, as the immune system is the primary barrier against illness. To keep the body's homeostasis, the immune system in any healthy organism produces a variety of immunomodulators (Enshasy et al. 2013). Several immunoregulatory substances, including polysaccharides, lectins, terpenes and terpenoids, and fungal immunomodulatory proteins (FIPs), are found in mushrooms (Acharya et al. 2016). Polysaccharides are the most often reported natural immunomodulators derived from mushrooms (Chakraborty et al. 2019). Medicinal mushrooms (MMs) are a valuable natural source of immunomodulators. Medicinal mushrooms, a subset of all mushrooms, are generally understood to be macroscopic fungi used for medicinal purposes in the form of extracts or whole powdered mushrooms. MMs have been used as herbal remedies for human health for thousands of years. These uses include the treatment of gastrointestinal issues, asthmatic conditions, and infectious diseases (Chaturvedi et al. 2018; Davis et al. 2020; Venturella et al. 2021).

Medicinal mushrooms have gained more recognition and are widely used as antimicrobial, antiviral, and anticancer agents. Mushrooms possess a wide range of clinical applications and great therapeutic potential due to their antimicrobial, prebiotic, anticancer, antioxidant, and anti-inflammatory properties (Kour et al. 2022; Singdevsachan et al. 2016).

2.2 *Mushroom Bioactives as Antioxidants*

Compared to other plants and vegetables, mushrooms contain a higher concentration of antioxidant compounds (Kour et al. 2022). Antioxidants found in mushrooms are essential because they fight against reactive oxygen species (ROS), which is a significant cause of diabetes, inflammation, cancer, and heart disease in human cells (Khursheed et al. 2020). Large concentrations of solid antioxidants like glutathione and ergothioneine are closely associated with the inhibitory power of fungal species found in mushrooms against premature mortality (Mwangi et al. 2022). Ascorbic acid, flavonoids, phenols, tocopherols, carotenoids, and other oxidatively resistant compounds are among the naturally occurring antioxidants found in mushrooms. Mushrooms also contain phenols, including p-hydroxybenzoic acid, tyrosine, catechol, tri-cinnamic acid, vanillic acid, and p-coumaric acid having significant impact on its medicinal potential (Hamza et al. 2023). These days, oxidative stress and poor nutrition are the two main causes of diseases. Many diseases have oxidative stress as their primary cause, brought on by an imbalance in metabolic processes. As a superior antioxidant agent, mushrooms can be used extensively in the nutraceutical and cosmeceutical industries. Mushrooms improve food flavor and are easily added to a regular diet either as a supplement or directly (Gupta and Kaplan 2018).

2.3 *Mushroom Bioactives as Anti-Inflammatory*

The body's natural immune response to conditions like autoimmune diseases, heart disease, memory loss, age-related illnesses, and chronic pain is inflammation. The critical factors in the development of tumors are inflammation and infections. Cancer treatment targeting inflammation and apoptosis may be very successful (Iman et al. 2016). Mushrooms are known to have both essential and dispensable amino acid contents, which are known to affect prostaglandin metabolism and contribute to their anti-inflammatory properties (Table 1) (Muszyńska et al. 2018). Additionally, ergosterol may have immunomodulatory and anti-inflammatory properties (Stastny et al. 2022). Long-term effects are common with many forms of inflammation, but with a healthy diet, we can reduce inflammation in the body. One potential tactic to reduce inflammation is to consume functional foods, such as fish, berries, vegetables, legumes, and mushrooms, which have anti-inflammatory qualities (Du et al. 2018). Various studies have promoted the anti-inflammatory characteristics of mushrooms. Mannosgalactan, fucomannogalactan, and fucogalactan are

Table 1 Anti-inflammatory properties of some species of mushrooms with most probable responsible compounds

Bioactive compound	Mushroom species	Mechanism of action	References
Polysaccharides, β-glucans	*A. bisporus* *L. edodes* *T. versicolor* *I. obliquus* *P. ostreatus* *G. lucidum*	Repression of pro-inflammatory proteins: cyclooxygenase-2, inducible nitric oxide synthase (iNOS), and interleukin-1 (IL-1) Inhibition of IκB-α subunit degeneration	Muszyńska et al. (2018)
Triterpene Lucidenic acid, ganoderic acid	*Ganoderma lucidum*	The secretion of TNF-α and IL-6 and inflammatory mediator NO and prostaglandin E 2 (PGE 2) from LPS-stimulated murine RAW264.7 cells was suppressed. Downregulated LPS-dependent expression of inducible nitric oxide synthase (iNOS) and COX-2 in RAW264.7 cells	Du et al. (2018)
Ethanol	*Macrolepiota procera*	Inhibited NO production	Taofiq et al. (2016c)
Glucans β-Glucans	*P. ostreatus*	Most active in inhibiting NF-κB/AP-1 activity	Stastny et al. (2022)
β-D-glucan (T-5-N)	*Phallus indusiate*	Anti-inflammatory properties	Assemie and Abaya (2022)
Polysaccharides, pyrogallol, hydroxybenzoic acid derivatives, flavonoids	*A. bisporus*	Anti-inflammatory	Elsayed et al. (2014)
A *β*-D-glucan called T-5-N	*Phallus indusiate*	Anti-inflammatory properties	Assemie and Abaya (2022)
β-D-glucan	Pholiota *nameko*	Anti-inflammatory properties	Abreu et al. (2019)
Diterpenes Steroids	*Wolfiporia cocos*	Anti-inflammatory properties	Baosong et al. (2020)
Extracts	*Hypholoma lateritium*	Inhibition of COX-2 and activation of Nrf2	Ványolós et al. (2020)

polysaccharides derived from *A. bisporus* that have been identified for their analgesic and anti-inflammatory properties. These compounds showed potential therapeutic applications in addressing pain and inflammation (Usman et al. 2021). It has been extensively reported that the traditional edible mushroom *Ganoderma resinaceum* can alleviate neurodegenerative diseases marked by inflammation and oxidative stress (Kou et al. 2022). *Cordyceps militaris* (L.) Fr. has inhibited inflammation, NO production, and IL-6 and TNF-α; reduced level of NO; and inhibited iNOS mRNA expressions (Taofiq et al. 2016c).

3 Mushroom as Medicines

The use of edible mushrooms for promoting longevity and treating ailments has been recognized by ancient and modern civilizations for thousands of years (Moumita and Das 2022). In Asian countries like China, Japan, and India, mushrooms like button mushrooms (*Agaricus* spp.), oyster mushrooms (*Pleurotus* spp.), and shiitake mushrooms (*Lentinus* spp.) are commonly consumed (Hyder and Dutta 2021). Species such as *Ganoderma* (Lingzhi), *Lentinus* (shiitake), *Auricularia*, *Pleurotus*, *Agaricus*, and *Cordyceps* are among the most well-known with potential medical benefits (Acharya et al. 2016).

Pleurotus ostreatus is frequently grown on a large scale for its medicinal and nutritional qualities. Its cultivation is incredibly economical and easily adaptable to the surroundings. It is also a bioremediation process and requires no special controlled conditions (Rai et al. 2021). The genus *Pleurotus* of mushrooms contains antiviral, antioxidant, anti-inflammatory, antihypertensive, antihypercholesterolemic, antiatherogenic, and antihyperlipidemic properties. These characteristics interfere with the risk factors for the severity of COVID-19, indicating the pharmacological potential of these mushrooms (Reis et al. 2022).

Considerable research has been done on shiitake mushrooms because they are among the most popular and widely used mushroom species. Its qualities and potential advantages that could be given to users and positively impact their health have been thoroughly researched. Shiitake mushrooms are recognized to contain vitamins, minerals, and proteins (Ahmad et al. 2023). Higher concentrations of provitamin D2; vitamins B1, B2, B6, and B12; and niacin can be found in shiitake mushrooms (Fukushima-Sakuno 2020).

Yarsagumba, also known as *Cordyceps sinensis* or *Ophiocordyceps sinensis*, belongs to the Ophiocordycipitaceae fungal family and is known for its parasitic relationship with the larvae of ghost moths. This unique fungus produces a fruiting body highly prized for its herbal medicinal properties (Sen et al. 2023). The anti-aging benefits are well known in yarsagumba, an excellent tonic for nourishing the body and the brain, and regular use is thought to improve biological and immune function (Sen et al. 2023).

The erinacines (A–I), a class of cyathin diterpenoids isolated from the mycelium of *Hericium erinaceus*, also known as lion's mane or yamabushitake, and the hericenones (C–H), benzyl alcohol derivatives extracted from the fruiting body, are other highly significant and well-researched bioactive metabolites. These two compounds exhibit both neuroprotective and neurotropic properties, and they can both easily cross the blood-brain barrier (Kawagishi et al. 2008). Both in vitro and in vivo, they stimulate the synthesis of nerve growth factor (NGF). Though it is most commonly used to treat neurodegenerative diseases and cognitive impairment, this medicinal mushroom also has antioxidative, anti-inflammatory nature, anticancer, immunostimulant, antidiabetic, antimicrobial, hypolipidemic, and antihyperglycemic properties (Chong et al. 2020).

Many pharmacological characteristics of *Ganoderma lucidum* include cytotoxic, antidiabetic, immunomodulatory, antihypertensive, anticancer, antimutagenic, anti-aging, antimicrobial, and hepatoprotective effects are investigated. These characteristics are primarily attributable to the two main metabolite groups found in *G. lucidum*: polysaccharides and triterpenes/triterpenoids. Triterpene compounds, such as lucidones, lucinedic acids, ganoderic acids, ganodermic acid, and ganodermic alcohols, are derived from lanosterol and have antitumor, antimetastatic, cytotoxic, and enzyme-inhibitory effects (Venturella et al. 2021). Some of the predominant bioactive compound and their chemical structure are mentioned in Fig. 1.

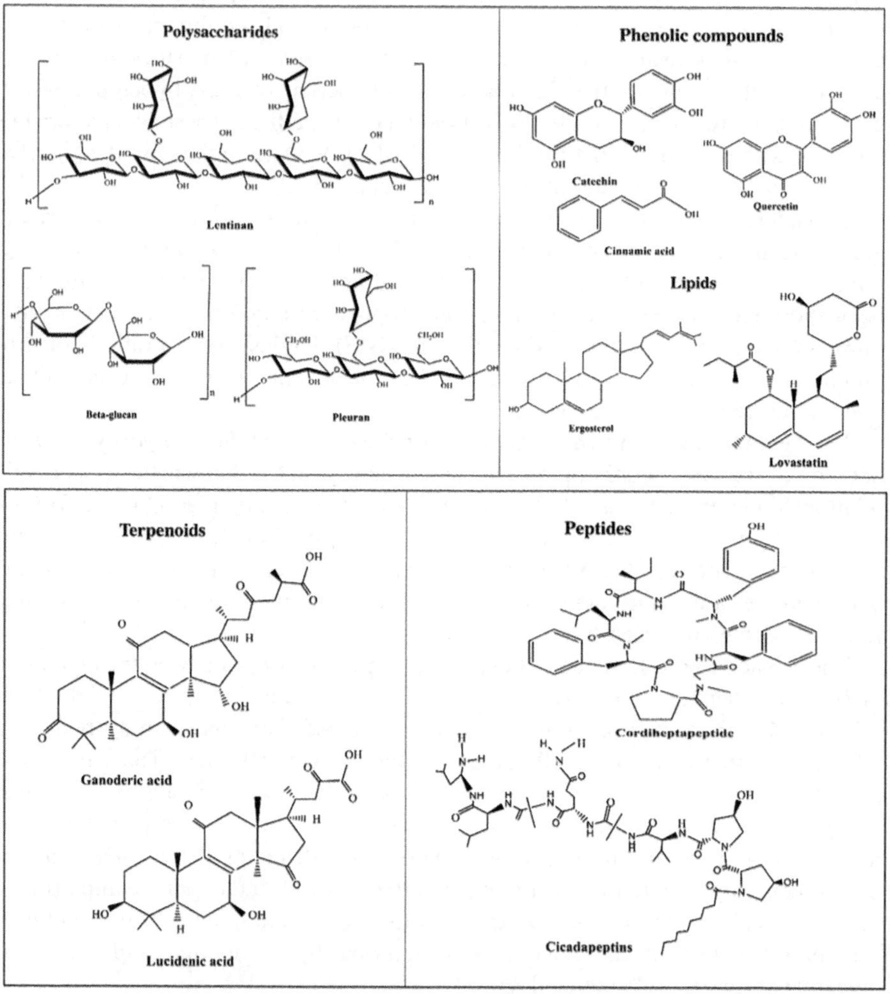

Fig. 1 Chemical structure of mushroom bioactive compounds

One of the biggest genera of macrofungi, *Agaricus*, contains several edible species with high nutritional and medicinal value (Usman et al. 2021). *A. bisporus*, also known as button mushroom, is a useful food source containing several significant bioactive substances (Khan et al. 2015). Its food and medicinal qualities make it one of the most significant mushrooms (Usman et al. 2021).

3.1 Role of Mushroom Bioactives in Human Health

For a very long time, people have been using edible mushrooms for food, and they are known for their taste, texture, and therapeutic qualities. Except for *Ganoderma lucidum*, which has a hard, bare-fruiting body, most mushroom-fruiting bodies contain about 90% moisture. Their dry matter has low levels of fats (2 to 6%) and high levels of proteins (19 to 35%) and carbohydrates (50–65%) (Rathore et al. 2017). These classes of complex carbohydrates may be found in fruiting bodies, sclerotium, cultured mycelium, and even culture broth in mushrooms at different stages of development (Yin et al. 2020). Compared to saturated fatty acids, polyunsaturated fatty acids make up over 75% of all fatty acids found in mushrooms, with oleic and linoleic acids being the most prevalent. Mushroom protein contains every essential amino acid for human development (Kayode et al. 2015).

Furthermore, mushrooms are a good source of calcium, which is necessary for overall health, and a sufficient amount of other minerals like potassium, phosphorus, magnesium, iron, zinc, and copper (Rathore et al. 2017). These macromolecular carbohydrates are found in many different types of animals, microbes, and plants and are also the main constituents of medicinal mushrooms (Phan et al. 2019). In addition to these dietary components, mushrooms also include a variety of bioactive substances, including phenolic compounds, proteins, peptides, polysaccharides, proteoglycans, terpenes, lectins, etc. As a result, there is now a significantly wider variety of edible mushroom cultivation (Yadav and Negi, 2021). An overview of the different activities of mushroom bioactives in different fields is projected in Fig. 2.

3.1.1 Mushroom Polysaccharides

Most bioactive compounds with immunomodulatory activity isolated from mushrooms in recent decades have been based on polysaccharides with or without side chain modifications (polysaccharide proteins and polysaccharopeptides). The leading chains of polymer structures, degrees of branching, conformations, molecular weights, and other physical characteristics of the immunomodulating polysaccharides are highly diverse, and these factors collectively have a substantial impact on each polysaccharide's bioactivity and mode of action (Chakraborty et al. 2019). While the chemical structure of the core polysaccharide in other polysaccharide-based compounds exhibiting immunomodulatory characteristics is similar, the branches may have distinct conjugates or different branching linkages. These

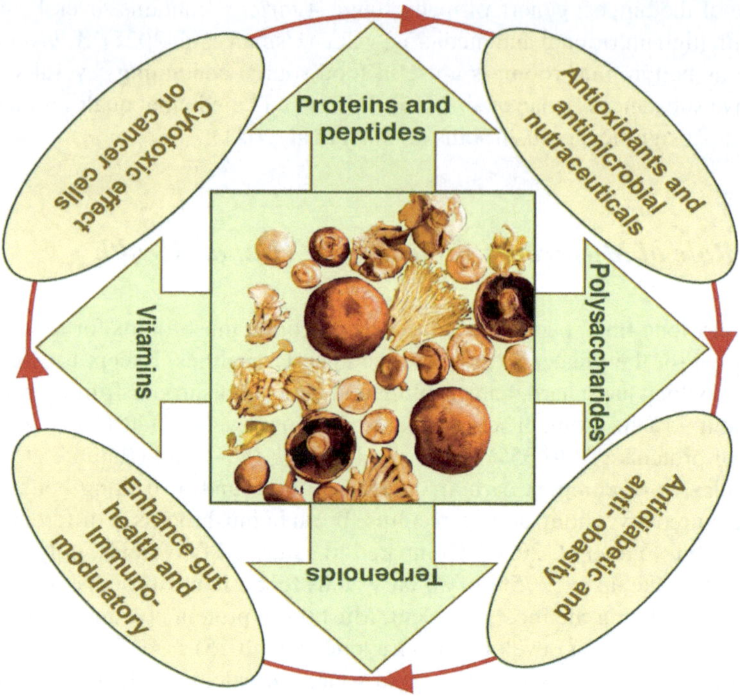

Fig. 2 Mushrooms and their therapeutic properties

polysaccharide-conjugate complexes, which have α(1–4)- and β(1–3)-glycosidic bonds to protein constituents, are known as heteroglucans. The well-known polysaccharide-protein complexes generated by *Agaricus brazie* and *Macrocybe gigantea* (syn. *Tricholoma giganteum*) have also demonstrated immunomodulatory properties (Zhao et al. 2020). Either homopolysaccharides or homoglycans or heteropolysaccharides or heteroglycans are classified as polysaccharides. In terms of configuration and/or linkage position, homo- and hetero-PSs can have homolinkages or heterolinkages (Kour et al. 2022).

3.1.2 Mushroom Proteins and Peptides and Their Properties

A crucial component of the bioactive elements found in mushrooms, proteins, and peptides have a variety of pharmacological effects, including immunomodulatory, anti-inflammatory, antitumor, and antimicrobial ones. As such, they provide essential health benefits. The species, makeup of the substrate, pileus size, and harvest time all affect the amount of protein in the mushroom (Samsudin and Abdullah 2019). On a dry weight basis, mushrooms contain approximately 19–35 g of highly digestible protein per 100 g (Rathore et al. 2017). Extracted from fungi (*F. velutipes* and a few *Ganoderma* species), fungal immunomodulatory proteins (FIPs) are a novel class of bioactive proteins with excellent curative potential (Bao et al. 2018). Furthermore, mushrooms have a good concentration of each of the nine amino acids that humans need, including phenylalanine, lysine, isoleucine, leucine, valine, histidine, threonine, and methionine (Rathore et al. 2017). Though they often come in lower on the protein scale than animal products, mushrooms have a relatively high protein content (Shbeeb et al. 2019).

3.1.3 Mushrooms and Lipids

The fat content of mushrooms is extremely low, ranging from 2 to 6 g/100 g^{-1} on a dry weight basis. They are mainly composed of unsaturated fatty acids rather than saturated ones, mainly oleic acid (C18:1) and linoleic acid (C18:2), with minimal linolenic acid. The essential aromatic compound 1-octen-3-ol, which gives mushrooms their distinct flavor, is derived from linoleic acid (Rathore et al. 2017).

3.1.4 Terpenes and Terpenoids

Terpenoids, which are made up of connected isoprene units, are some of the strong biologically active substances found in mushrooms. They fall into one of four categories: sesquiterpenoids, diterpenoids, triterpenoids, and monoterpenoids (Anke 2020). Terpenes are categorized as volatile and unsaturated hydrocarbons widely isolated from mushrooms and are primarily responsible for their anti-inflammatory properties. These terpenoids also have pharmacological properties that include anti-inflammatory, antiviral, antimalarial, anticancer, and anticholinesterase properties (Üstün et al. 2018). Terpenoids derived from mushrooms have recently attracted scientific interest as promising therapeutic products with a variety of pharmacological activities, including anti-inflammatory, antiviral, anti-neurodegenerative, anticancer, liver-protective, lipid-lowering, and antioxidant (Kour et al. 2022). Although terpenoids belong to a broad and diverse class of organic compounds, the fundamental building block of all of them is isoprene, which has five carbon atoms and the molecular formula (C5H6)n (Enshasy et al. 2013). The triterpenoids derived from *G. lucidum* and *G. lingzhi* are arguably the most well-known of these compounds. These triterpenoids have anti-inflammatory and antidrug nephrotoxic properties (Zhao et al. 2020).

3.2 Nutraceutical and Cosmetic Products by Using Mushroom Bioactives

In recent years, mushrooms have emerged as one of the primary sources of functional foods and medications (Grimm and Wösten 2018). A food or component of a food that offers health or medical benefits, such as illness prevention and treatment, is referred to as a nutraceutical. With their high nutritional and functional value, mushrooms are considered a delicacy and widely used in functional and nutraceutical food ingredients (Üstün et al. 2018). Apart from their involvement in manufacturing functional foods, the bioactive nutraceuticals derived from mushrooms were primarily investigated and created as dietary supplements. These supplements can potentially prevent the onset of severe diseases and shield the body from harmful environmental conditions (Ma et al. 2018). Pharmaceutical companies like Zhejiang Fangge Pharmaceutical Co., Ltd. in China; FineCo Ltd. in South Korea; ecoNugenics, Mushroom Wisdom, MycoFormulas, Gourmet Mushrooms, Healing Edge Sciences, Inc., and Aloha Medicinal Inc. in the USA; linkNUTRITION in the UK; Myko San in Crotia; Pleuran, GlycaNova AS in Norway; and Concord Mushroom Supplements in Australia are among those companies currently developing and marketing mushroom polysaccharide-based extracts, medications, and supplements for anticancer purposes (Rai et al. 2021). In Germany, medicinal mushrooms are governed by food law rather than pharmaceutical law. Before authorization, pharmaceutical companies are required to furnish the authorities with information regarding the product's safety, efficacy, and pharmaceutical quality (Üstün et al. 2018). The use of mushrooms as nutraceuticals has increased during the past 20 years, and numerous edible species have been thoroughly studied and verified for use in medicine (Rathore et al. 2017). There are several dietary supplements based on mushrooms on the market these days. These consist of (1) powders made from artificial cultivation of fruiting bodies, hot water, or hydroalcoholic extracts of these fruiting bodies; (2) dried and powder form preparations of the combination of substrate, mycelium, and primordial mushroom; (3) biomass or mycelium extracted from liquid cultures grown in fermentation tanks or bioreactors; (4) capsules or tablets from naturally dried mushrooms fruiting bodies; and (5) spores and their extracts (Üstün et al. 2018). Various mushroom nutraceuticals are available in the market (Rathore et al. 2017).

3.3 Mushroom with Cosmetic Values

Studies have yet to be conducted to investigate the biotechnological potential of mushroom extract for creating numerous cosmetic products (Taofiq et al. 2016b). The cosmetics industry continues to work nonstop to separate specific key components from mushrooms, validate their aesthetic qualities, and then use those ingredients in lotions and creams (Taofiq et al. 2016a). It has been discovered that tyrosinase enzyme activity inhibition can result in skin whitening (Usman et al. 2021). Cosmetic formulas containing extracts from mushrooms inhibit the development of bacterial strains that harm the skin. A few of the base cream's bioactive ingredients, which are derived from mushrooms, improve the appearance and are crucial in the fight against hyperpigmentation, inflammation, and aging skin (Taofiq et al. 2016a). Because of its potent antioxidant qualities, ethanolic extract from *A. bisporus* effectively treats skin aging. Antibacterial activity of *A. bisporus* extracts is demonstrated against pathogenic microorganisms, such as methicillin-resistant and methicillin-susceptible *Staphylococcus aureus*. It is well known that these bacteria invade the skin after an injury or inflammation (Usman et al. 2021). When applied to specific bacterial strains, the base cream made from the *A. bisporus* extract demonstrated strong antioxidant potential and antibacterial activity (Taofiq et al. 2016b). Phenolic compounds have antibacterial properties that typically cause DNA damage and rupture bacterial membranes (Usman et al. 2021).

4 Signaling Pathways Affected by Mushroom Bioactives

Mushrooms exhibit anticancer properties by stimulating immune cells called lymphocytes, which fight cancer. The active compounds, particularly polysaccharides like lentinan and schizophyllan, are valued for their potential anticancer effects (Hamza et al. 2022). Cancer claims over six million lives annually globally, with 9.9 million deaths reported in 2020 (Yue et al. 2021). Modifiable risk factors contribute to various cancers affecting the breast, cervix, colon, kidney, lung, esophagus, pancreas, skin, and stomach (Kour et al. 2022).

Chronic diseases are primarily responsible for the majority of deaths and disabilities. Recently, mushroom-derived polysaccharides, glycopeptides, and other proteins have displayed promising potential in combating these chronic conditions. The distinct characteristics of these polysaccharides and proteins, including their structural attributes, branching patterns, bond configurations, and conformation types, play a significant role in influencing cellular responses. These biomolecules exhibit specific interactions with cellular proteins, thereby regulating various signaling pathways. This innovative approach involving mushroom-derived biomolecules presents a notable strategy for chronic diseases such as cardiovascular diseases, cancer, diabetes mellitus, and regulation of gut microbiota (Table 2) (Zhang et al. 2022).

4.1 Anticarcinogenic Activity

Cancer is characterized by the abnormal and uncontrolled proliferation of cells, which can be induced by various factors, including both biotic and abiotic stressors. Several polysaccharides and proteins have been demonstrated to effectively

Table 2 Isolated mushroom bioactives and respective function

Mushroom bioactives	Species	Function	References
Ergosterol peroxide	*Grifola frondosa*	Antidiabetic	Chen et al. (2019)
Lyophilized powder	*L. edodes*	Antidiabetic	Xia et al. (2023)
Polysaccharides	*Ganoderma lucidum*	Anti-inflammatory, hepatoprotective	Ahmad (2018)
Lentinan	*Lentinus edodes*	Antiproliferative, promotes apoptosis	Yuan et al. (2023)
Terpenoid (Hericenones and Erinacines)	*Hericium erinaceus*	Nerve growth factor (NGF)	Thongbai et al. (2015)
Polyphenols	*Inonotus obliquus*	The antioxidant effect, glycation inhibitors	Doi et al. (2023)
Polysaccharides, sterol, protein, polypeptide	*Cordyceps sinensis*	Immunomodulatory effect	Liu et al. (2015)
Polysaccharides	*Agaricus bisporus*	Efficient result for hepatic injury	Huang et al. (2016)
Polysaccharides	*Pleurotus ostreatus*	Antidiabetic	Zhang et al. (2016)
1,6-β-D-glucans	*Agaricus bisporus* and *Agaricus brasiliensis*	Immunostimulatory	Smiderle et al. (2013)
Polysaccharides	*Pleurotus ostreatus*	Regulate dyslipidemia	Zhang et al. (2017)
Polysaccharides	*Pleurotus eryngii*	Anticarcinogenic	Ren et al. (2016)
Polysaccharides	*G. lucidum*	Inhibiting prostate cancer	Liang et al. (2015)
Peptides latcripin-7A	*L. edodes*	Inhibiting cell cycle at the G1 phase in gastric cancer cells	Din et al. (2021)
Polysaccharide extract	*L. edodes*	Regulate cholesterol metabolism	Gil-Ramírez et al. (2016)
Polysaccharides	*L. edodes*	Breast cancer	Larypoor (2022)
Polysaccharides	*Grifola frondosa*	Antidiabetic	Guo et al. (2020)
Ethanolic extract	*Ganoderma neojaponicum*	Colon cancer	Lau et al. (2021)

suppress cell proliferation and promote apoptosis, a programmed cell death process. The cell cycle encompasses distinct phases and is a crucial component of cell development and division. These phases include the (G1) phase, where the cell increases in size; the (S) phase, during which DNA synthesis takes place; the (G2) phase, where the cell readies itself for division; and the (M) phase, which is the mitotic phase leading to cell division. After these phases, cells can enter a resting phase (G0). Recent research has revealed that polysaccharides extracted from mushrooms can arrest the cell cycle at the intermediary phase and inhibit the growth of cancer cells, showing promise as a potential treatment approach (Liu et al. 2020; Vetriventhan et al. 2020; Zhang et al. 2022). Apoptosis, an inherent biological process, has been demonstrated to be modulated by diverse mushroom polysaccharides and glycopeptides (Fig. 2). Notably, polysaccharides derived from *Flammulina velutipes* exhibit noteworthy anticarcinogenic properties. The aqueous extract from *Ganoderma lucidum*, characterized by a substantial polysaccharide concentration, exerts inhibitory effects on prostate cancer cells. This inhibition is manifested through the downregulation of androgen receptor and PCa-specific antigen mRNA expression, coupled with the induction of caspase-3/7 activation (Liang et al. 2015).

Trametes robiniophila, a distinct mushroom variety, has been described with noteworthy anticancer properties. This has been determined by comprehensive in vitro and in vivo studies, wherein the mushroom demonstrated the capacity to suppress cancer-inducing cells directly. Moreover, it enhances the expression of CD4+ T cells. Noteworthy outcomes were observed, including the enhancement of cytokines and macrophages. Clinical investigations have further elucidated the potential of collaborative treatments involving *Trametes robiniophila* and other established anticancer therapies. The amalgamation of these modalities demonstrated a synergistic effect, resulting in superior anticarcinogenic efficacy (Pan et al. 2019). In conclusion, the studies strongly suggest that mushrooms and their extracts constitute promising reservoirs for developing novel pharmaceutical agents, offering hope for use in medical treatments.

Lentinan (extracted from *L. edodes*), a bioactive compound derived from mushrooms, has been prominently known for its anticancer properties in various studies. Research indicates that lentinan induces the production of tumor necrosis factor-alpha (TNF-α), a key cytokine that influences tumor cells. Clinical studies suggest that lentinan can potentially enhance the quality of life for individuals with cancer (Antonelli et al. 2021). However, there is limited available data regarding the clinical evidence and optimal dosage of lentinan. Further research is needed to better understand its effectiveness and appropriate use in a clinical setting.

4.2 *Antidiabetic Activity*

Diabetes mellitus represents a metabolic condition characterized by an elevation in circulating glucose levels, arising either from unregulated insulin secretion or reduced responsiveness of cellular receptors to the signals triggered by this hormone. These metabolic changes affect how the body deals with carbohydrates, fats, and proteins. In type 2 diabetes, most complications arise from high blood sugar levels, which is the leading cause of the health issues associated with the condition (Stojkovic et al. 2019). An effective approach to managing type 2 diabetes involves targeting the breakdown of complex carbohydrates by pancreatic α-amylase and regulating glucose absorption by inhibiting the intestinal α-glucosidase enzyme. Commercial drugs are available to inhibit α-glucosidase, helping to regulate blood sugar levels in individuals with diabetes. Still, they have some side effects, so recent trends show that people are more inclined to buy natural products. A study on various medicinal mushroom methanol and hexane extracts shows inhibition activities on amylase and glucosidase enzymes in type 2 diabetes. Among all studied mushroom extracts, *C. rutilus* hexane extract (IC50: 0.05 ± 0.01 mg/mL) demonstrated the highest inhibitory activity on a-amylase, while *P. ostreatus* hexane (IC50: 0.10 ± 0.01 mg/mL) showed the highest inhibitory activity on α-glucosidase (Deveci et al. 2021).

4.3 Antiviral Effect

The global health crisis was characterized by the swift spread of the coronavirus, which first emerged in Wuhan, China, in December 2019. As the virus swiftly traversed the globe, scientists explored unconventional paths for potential treatments. Natural compounds from medicinal mushrooms showed promise for creating new antiviral drugs. Different dietary treatments for various COVID-19 symptoms were studied and compared to the potential therapeutic benefits of mushroom products. A key research finding suggests that altered forms of β-glucan exhibit a significant impact on suppressing viral replication and hold promise for future investigations. Additionally, terpenoids, lectins, glycoproteins, lentinan, galactomannan, and polysaccharides derived from mushrooms emerged as potentially effective prophylactic or therapeutic agents against COVID-19. These results contribute to the growing body of knowledge seeking alternative approaches to combat the global health challenge posed by the COVID-19 pandemic (Arunachalam et al. 2022). Lectins are proteins that bind to carbohydrates. They are categorized based on their structure, origin, and the sugars by which they bind. The specific sugars they attach to determine their function and potential medical uses. These proteins, which bind to sugars, have shown various medicinal activities like fighting tumors, viruses, and bacteria; modulating the immune system; and managing diabetes. Interestingly, despite their potential, lectins from mushrooms haven't been thoroughly explored as antiviral agents. Discovering new antiviral substances is crucial for public health emergencies like the pandemic and future viral outbreaks. Lectins work against viruses by affecting different stages in the virus's life cycle (Fig. 3), including attachment, entry, and replication (El-Maradny et al. 2021).

5 Future Implications

Various components found in mushrooms, such as polysaccharides, proteins, and glycopeptides, promote health and are used in products like nutraceuticals, pharmaceuticals, and cosmeceuticals. In this chapter, we thoroughly discussed the medicinal properties (pharmacology) of these mushroom components, how they work to

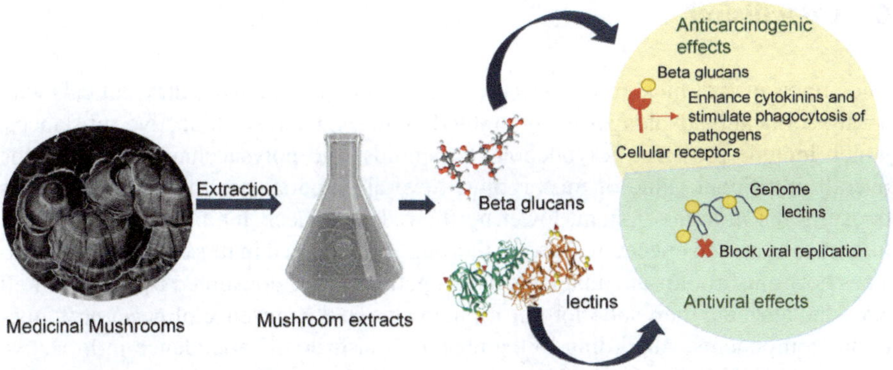

Fig. 3 Medicinal compounds extracted from mushroom extract contributing anticarcinogenic and antiviral mechanism

promote health, and the latest developments and trends in their use. In recent years, different techniques for extracting these components have been discovered to achieve higher yields. However, using these components can be challenging because they are unstable. To overcome this challenge and achieve better results, it is crucial to have an effective drug delivery system in place. This system helps ensure that the mushroom components are delivered in a way that maximizes their stability and, consequently, their health-promoting benefits.

Innovative methods have been used to deliver drugs effectively, mainly through techniques like encapsulation with polymers, nano-formulations, and liposomes. All these methods enhance the stability and are effective for the targeted delivery of bioactives. Microencapsulation is another prominent approach for mushroom bioactives. Nano-formulation, as the name suggests, has been developed by immensely small particles that enhance the effectiveness of the developed product; some examples are a combination of polysaccharides extracted from various mushrooms and cellulose nanofiber film which shows increased strength, antioxidant activity, and decreased water vapor permeability (Fang et al. 2018; Mingyi et al. 2019). Another case in which polysaccharides extracted from *Ganoderma lucidum* nanoprecipitation were employed to convert them into nanoparticles. This process enhanced the free radicals scavenging (such as DPPH, hydroxyl radical, and superoxide anion radical) and improves their stability even under harsh temperature, ionic strength, or pH (Qin et al. 2018). In summary, advanced technology has played an essential role in developing future healthcare products and drugs by using mushrooms. Understanding the pharmacology of mushroom components and addressing stability challenges through effective drug delivery systems are crucial for unlocking the full potential of these natural compounds.

6 Conclusion

The importance of mushrooms as a plentiful supply of bioactive nutraceuticals with a range of health advantages is highlighted in this chapter. All bioactive substances, which include proteins, polyphenols, terpenoids, and polysaccharides, add to the overall nutritional value of mushrooms. Several potential health benefits, such as boosting the immune system, lowering the risk of cancer, inhibiting of growth of tumor cells, and cosmeceuticals were thoroughly discussed in this chapter. It emphasizes how mushrooms are adaptable as a functional food consumed by people of all ages and attribute their superior nutritional value to the presence of necessary functional components. According to the chapter, mushrooms' abundance in bioactive compounds makes them valuable and comprehensive health food items.

Furthermore, it is suggested that mushrooms be incorporated into nutraceuticals and functional foods by utilizing their nutritional benefits to create products that promote health and well-being. However, it highlights the need for more investigation into the physiological and molecular characteristics of mushrooms' bioactive components to understand the development of functional foods. With continued research and developments, mushrooms could become a commonly recognized nutraceutical food for daily supplementation.

Acknowledgments Authors thank NIFTEM authorities for providing a Ph.D. fellowship to Mrs. Charu Agnihotri.

References

Abreu H, Simas FF, Smiderle FR, Sovrani V, Dallazen JL, Maria-Ferreira D et al (2019) Gelling functional property, anti-inflammatory, and antinociceptive bioactivities of β-D-glucan from the edible mushroom Pholiota nameko. Int J Biol Macromol 122:1128–1135. https://doi.org/10.1016/j.ijbiomac.2018.09.062

Acharya K, Ghosh S, Dutta AK (2016) Pharmacognostic standardization based on physicochemical and molecular parameters of a medicinal mushroom Schizophyllum commune. Orient Pharm Exp Med 16(4):259–266. https://doi.org/10.1007/s13596-016-0241-y

Ahmad MF (2018) Ganoderma lucidum: persuasive biologically active constituents and their health endorsement. Biomed Pharmacother 107:507–519. https://doi.org/10.1016/j.biopha.2018.08.036

Ahmad I, Arif M, Mimi X, Zhang J, Ding Y, Lyu F (2023) Therapeutic values and nutraceutical properties of shiitake mushroom (Lentinula edodes): a review. Trends Food Sci Technol 134:123

Anke T (2020) Secondary metabolites from mushrooms. J Antibiot 73(10):655–656. https://doi.org/10.1038/s41429-020-0358-6

Antonelli M, Donelli D, Firenzuoli F (2021) Lentinan for integrative cancer treatment: an umbrella review. Proceedings 79(1):1. https://doi.org/10.3390/IECBM2020-08733

Anusiya G, Gowthama Prabu U, Yamini NV, Sivarajasekar N, Rambabu K, Bharath G, Banat F (2021) A review of the therapeutic and biological effects of edible and wild mushrooms. Bioengineered 12(2):11239–11268. https://doi.org/10.1080/21655979.2021.2001183

Arunachalam K, Sasidharan SP, Yang X (2022) A concise review of mushrooms antiviral and immunomodulatory properties that may combat against COVID-19. Food Chem Adv 1:100023. https://doi.org/10.1016/j.focha.2022.100023

Assemie A, Abaya G (2022) The effect of edible mushroom on health and their biochemistry. Int J Microbiol 2022:8744788. https://doi.org/10.1155/2022/8744788

Bach F, Helm CV, Bellettini MB, Maciel GM, Haminiuk CW, Isidoro. (2017) Edible mushrooms: a potential source of essential amino acids, glucans and minerals. Int J Food Sci Technol 52(11):2382–2392. https://doi.org/10.1111/ijfs.13522

Badalyan SM, Barkhudaryan A, Rapior S (2019) Recent progress in research on the pharmacological potential of mushrooms and prospects for their clinical application. In: Medicinal mushrooms: recent progress in research and development, pp 1–70

Bao D-P, Bai R, Gao Y-N, Wu Y, Wang Y (2018) Computational insights into the molecular mechanism of the high immunomodulatory activity of LZ-8 protein isolated from the Lingzhi or Reishi medicinal mushroom Ganoderma lucidum (Agaricomycetes). Int J Med Mushrooms 20(6):537

Baosong C, Sixian W, Gaoqiang L, Li B, Ying H, Ruilin Z, Hongwei L (2020) Anti-inflammatory diterpenes and steroids from peels of the cultivated edible mushroom Wolfiporia Cocos. Phytochem Lett 36:11–16. https://doi.org/10.1016/j.phytol.2020.01.005

Chakraborty I, Sen IK, Mondal S, Rout D, Bhanja SK, Maity GN, Maity P (2019) Bioactive polysaccharides from natural sources: a review on the antitumor and immunomodulating activities. Biocatal Agric Biotechnol 22:101425. https://doi.org/10.1016/j.bcab.2019.101425

Chaturvedi VK, Agarwal S, Gupta KK, Ramteke PW, Singh MP (2018) Medicinal mushroom: boon for therapeutic applications. 3 Biotech 8(8):334. https://doi.org/10.1007/s13205-018-1358-0

Chen, Yuqing, Liu, Dan, Wang, Dingyi, Lai, Shanshan, Zhong, Ruting, Liu, Yuanyuan, . . . Zhao, Chao. (2019). Hypoglycemic activity and gut microbiota regulation of a novel polysaccharide from Grifola frondosa in type 2 diabetic mice. Food Chem Toxicol, 126, 295–302. doi: https://doi.org/10.1016/j.fct.2019.02.034

Chong PS, Fung M-L, Wong KH, Lim LW (2020) Therapeutic potential of Hericium erinaceus for depressive disorder. Int J Mol Sci 21(1):163

Chugh, Rishi Man, Mittal, Pooja, Mp, Namratha, Arora, Tanu, Bhattacharya, Tanima, Chopra, Hitesh, . . . Gautam, Rupesh K. (2022). Fungal mushrooms: a natural compound with therapeutic applications. Front Pharmacol, 13, 925387

Davis R, Taylor A, Nally R, Benson KF, Stamets P, Jensen GS (2020) Differential immune activating, anti-inflammatory, and regenerative properties of the aqueous, ethanol, and solid fractions of a medicinal mushroom blend. J Inflamm Res Volume 13:117–131

Deveci E, Çayan F, Tel-Çayan G, Duru ME (2021) Inhibitory activities of medicinal mushrooms on α-amylase and α-glucosidase-enzymes related to type 2 diabetes. S Afr J Bot 137:19–23. https://doi.org/10.1016/j.sajb.2020.09.039

Din SR, Ud N, Azhar M, Ramzan MN, Saleem MZ, Ghayas H, Ahmad B et al (2021) Latcripin-7A from Lentinula edodes C91-3 induces apoptosis, autophagy, and cell cycle arrest at G1 phase in human gastric cancer cells via inhibiting PI3K/Akt/mTOR signaling. Eur J Pharmacol 907:174305. https://doi.org/10.1016/j.ejphar.2021.174305

Doi N, Araki K, Fukuta Y, Kuwagaito Y, Yamauchi Y, Sasai Y et al (2023) Anti-glycation and antioxidant effects of Chaga mushroom decoction extracted with a fermentation medium. Food Sci Technol Res 29(2):155–161. https://doi.org/10.3136/fstr.FSTR-D-22-00120

Du B, Zhu F, Xu B (2018) An insight into the anti-inflammatory properties of edible and medicinal mushrooms. J Funct Foods 47:334–342. https://doi.org/10.1016/j.jff.2018.06.003

El-Maradny YA, El-Fakharany EM, Abu-Serie MM, Hashish MH, Selim HS (2021) Lectins purified from medicinal and edible mushrooms: insights into their antiviral activity against pathogenic viruses. Int J Biol Macromol 179:239–258. https://doi.org/10.1016/j.ijbiomac.2021.03.015

Elsayed A, Enshasy E, Hesham W, Mohammad AM, Aziz R (2014) Mushrooms: a potential natural source of anti-inflammatory compounds for medical applications. Mediators Inflamm 2014:805841. https://doi.org/10.1155/2014/805841

Enshasy E, Hesham A, Hatti-Kaul R (2013) Mushroom immunomodulators: unique molecules with unlimited applications. Trends Biotechnol 31(12):668–677. https://doi.org/10.1016/j.tibtech.2013.09.003

Fang D, Deng Z, Jung J, Hu Q, Zhao Y (2018) Mushroom polysaccharides-incorporated cellulose nanofiber films with improved mechanical, moisture barrier, and antioxidant properties. J Appl Polym Sci 135(15):46166. https://doi.org/10.1002/app.46166

Ferreira ICFR, Morales P, Barros L (2016) Wild plants, mushrooms and nuts: functional food properties and applications. Wiley, Hoboken, NJ

Fukushima-Sakuno E (2020) Bioactive small secondary metabolites from the mushrooms Lentinula edodes and Flammulina velutipes. J Antibiot 73(10):687–696

Gil-Ramírez A, Caz V, Smiderle FR, Martin-Hernandez R, Largo C, Tabernero M et al (2016) Water-soluble compounds from Lentinula edodes influencing the HMG-CoA reductase activity and the expression of genes involved in the cholesterol metabolism. J Agric Food Chem 64(9):1910–1920. https://doi.org/10.1021/acs.jafc.5b05571

Grimm D, Wösten HAB (2018) Mushroom cultivation in the circular economy. Appl Microbiol Biotechnol 102:7795–7803

Guo W-L, Deng J-C, Pan Y-Y, Xu J-X, Hong J-L, Shi F-F et al (2020) Hypoglycemic and hypolipidemic activities of Grifola frondosa polysaccharides and their relationships with the modulation of intestinal microflora in diabetic mice induced by high-fat diet and streptozotocin. Int J Biol Macromol 153:1231–1240. https://doi.org/10.1016/j.ijbiomac.2019.10.253

Gupta S, Kaplan A (2018) Solving the mystery of the yellow zone of the asthma action plan. NPJ Primary Care Respir Med 28(1):1. https://doi.org/10.1038/s41533-017-0067-1

Hamza A, Ghanekar S, Kumar DS (2022) Current trends in health-promoting potential and biomaterial applications of edible mushrooms for human wellness. Food Biosci 51:102290

Hamza A, Ghanekar S, Kumar DS (2023) Current trends in health-promoting potential and biomaterial applications of edible mushrooms for human wellness. Food Biosci 51:102290. https://doi.org/10.1016/j.fbio.2022.102290

Hao R, Zhou X, Zhao X, Lv X, Zhu X, Gao N et al (2023) Flammulina velutipes polysaccharide counteracts cadmium-induced gut injury in mice via modulating gut inflammation, gut microbiota and intestinal barrier. Sci Total Environ 877:162910. https://doi.org/10.1016/j.scitotenv.2023.162910

Huang J, Ou Y, Yew TW, David L, Jingna L, Bo L et al (2016) Hepatoprotective effects of polysaccharide isolated from Agaricus bisporus industrial wastewater against CCl4-induced hepatic injury in mice. Int J Biol Macromol 82:678–686. https://doi.org/10.1016/j.ijbiomac.2015.10.014

Hyder MS, Dutta SD (2021) Mushroom-derived polysaccharides as antitumor and anticancer agent: a concise review. Biocatal Agric Biotechnol 35:102085

Iman V, Mohan S, Abdelwahab SI, Karimian H, Nordin N, Fadaeinasab M et al (2016) Anticancer and anti-inflammatory activities of girinimbine isolated from Murraya koenigii. Drug Des Devel Ther 11:103–121. https://doi.org/10.2147/DDDT.S115135

Kaur M, Sharma S, Kaur R, Kaur J, Singh A (2023) Effect of UV-B irradiation on bioconversion of ergosterol to vitamin D2 and its impact on nutritional properties of oyster mushroom (Pleurotus Florida). Int J Food Sci Technol 58(10):5114–5120. https://doi.org/10.1111/ijfs.16610

Kawagishi H, Zhuang C, Yunoki R (2008) Compounds for dementia from Hericium erinaceum. Drugs Future 33(2):149

Kayode RMO, Olakulehin TF, Adedeji BS, Ahmed O, Aliyu TH, Badmos AHA (2015) Evaluation of amino acid and fatty acid profiles of commercially cultivated oyster mushroom (Pleurotus sajor-caju) grown on gmelina wood waste. Niger Food J 33(1):18–21. https://doi.org/10.1016/j.nifoj.2015.04.001

Khan AA, Gani A, Shah A, Masoodi FA, Hussain PR, Wani IA, Khanday FA (2015) Effect of γ-irradiation on structural, functional and antioxidant properties of β-glucan extracted from button mushroom (Agaricus bisporus). Innov Food Sci Emerg Technol 31:123–130

Khursheed R, Singh SK, Wadhwa S, Gulati M, Awasthi A (2020) Therapeutic potential of mushrooms in diabetes mellitus: role of polysaccharides. Int J Biol Macromol 164:1194–1205. https://doi.org/10.1016/j.ijbiomac.2020.07.145

Kou R-W, Xia B, Wang Z-J, Li J-N, Yang J-R, Gao Y-Q et al (2022) Triterpenoids and meroterpenoids from the edible Ganoderma resinaceum and their potential anti-inflammatory, antioxidant and anti-apoptosis activities. Bioorg Chem 121:105689. https://doi.org/10.1016/j.bioorg.2022.105689

Kour H, Kour D, Kour S, Singh S, Hashmi J, Azhar S, Yadav AN et al (2022) Bioactive compounds from mushrooms: emerging bioresources of food and nutraceuticals. Food Biosci 50:102124. https://doi.org/10.1016/j.fbio.2022.102124

Larypoor M (2022) Investigation of HER-3 gene expression under the influence of carbohydrate biopolymers extract of shiitake and reishi in MCF-7 cell line. Mol Biol Rep 49(7):6563–6572. https://doi.org/10.1007/s11033-022-07496-w

Lau M-F, Chua K-H, Sabaratnam V, Kuppusamy UR (2021) In vitro and in silico anticancer evaluation of a medicinal mushroom, Ganoderma neo-japonicum Imazeki, against human colonic carcinoma cells. Biotechnol Appl Biochem 68(4):902–917. https://doi.org/10.1002/bab.2013

Li H, Tian Y, Menolli N Jr, Ye L, Karunarathna SC, Perez-Moreno J et al (2021) Reviewing the world's edible mushroom species: a new evidence-based classification system. Compr Rev Food Sci Food Saf 20(2):1982–2014. https://doi.org/10.1111/1541-4337.12708

Liang Z-E, Yi Y-J, Guo Y-T, Wang R-C, Hu Q-L, Xiong X-Y (2015) Inhibition of migration and induction of apoptosis in LoVo human colon cancer cells by polysaccharides from Ganoderma lucidum. Mol Med Rep 12(5):7629–7636. https://doi.org/10.3892/mmr.2015.4345

Liu Y, Wang J, Wang W, Zhang H, Zhang X, Han C (2015) The chemical constituents and pharmacological actions of <i>Cordyceps sinensis</i>. Evid Based Complement Alternat Med 2015:575063. https://doi.org/10.1155/2015/575063

Liu J-Y, Hou X-X, Li Z-Y, Shan S-H, Chang MC, Feng C-P, Wei Y (2020) Isolation and structural characterization of a novel polysaccharide from Hericium erinaceus fruiting bodies and its arrest of cell cycle at S-phage in colon cancer cells. Int J Biol Macromol 157:288–295. https://doi.org/10.1016/j.ijbiomac.2020.04.162

Ma G, Yang W, Zhao L, Pei F, Fang D, Hu Q (2018) A critical review on the health promoting effects of mushrooms nutraceuticals. Food Sci Human Wellness 7(2):125–133

Martins A (2016) The numbers behind mushroom biodiversity. In: Wild plants, mushrooms and nuts, pp 15–63

Mingyi Y, Belwal T, Devkota HP, Li L, Luo Z (2019) Trends of utilizing mushroom polysaccharides (MPs) as potent nutraceutical components in food and medicine: a comprehensive review. Trends Food Sci Technol 92:94–110. https://doi.org/10.1016/j.tifs.2019.08.009

Moumita S, Das B (2022) Assessment of the prebiotic potential and bioactive components of common edible mushrooms in India and formulation of synbiotic microcapsules. LWT 156:113050

Muszyńska B, Grzywacz-Kisielewska A, Kała K, Gdula-Argasińska J (2018) Anti-inflammatory properties of edible mushrooms: a review. Food Chem 243:373–381. https://doi.org/10.1016/j.foodchem.2017.09.149

Mwangi RW, Macharia JM, Wagara IN, Bence RL (2022) The antioxidant potential of different edible and medicinal mushrooms. Biomed Pharmacother 147:112621. https://doi.org/10.1016/j.biopha.2022.112621

Pan J, Yang C, Jiang Z, Huang J (2019) Trametes robiniophila Murr: a traditional Chinese medicine with potent anti-tumor effects. Cancer Manag Res 11:1541–1549. https://doi.org/10.2147/CMAR.S193174

Phan C-W, Tan EY-Y, Sabaratnam V (2019) Bioactive molecules in edible and medicinal mushrooms for human wellness. In: Bioactive molecules in food. Springer, Berlin, pp 1597–1620

Qin Y, Xiong L, Li M, Liu J, Wu H, Qiu H et al (2018) Preparation of bioactive polysaccharide nanoparticles with enhanced radical scavenging activity and antimicrobial activity. J Agric Food Chem 66(17):4373–4383. https://doi.org/10.1021/acs.jafc.8b00388

Rai SN, Mishra D, Singh P, Vamanu E, Singh MP (2021) Therapeutic applications of mushrooms and their biomolecules along with a glimpse of in silico approach in neurodegenerative diseases. Biomed Pharmacother 137:111377

Rathore H, Prasad S, Sharma S (2017) Mushroom nutraceuticals for improved nutrition and better human health: a review. PharmaNutrition 5(2):35–46. https://doi.org/10.1016/j.phanu.2017.02.001

Reis D, Echer E, Schenkel PC, Camassola M (2022) Effects of bioactive compounds from Pleurotus mushrooms on COVID-19 risk factors associated with the cardiovascular system. J Integr Med 20:385

Reka B, Sindhu D, Asha S (2021) Bio-active compounds and its immunomodulatory properties, vol 11, p 35

Ren D, Wang N, Guo J, Yuan L, Yang X (2016) Chemical characterization of Pleurotus eryngii polysaccharide and its tumor-inhibitory effects against human hepatoblastoma HepG-2 cells. Carbohydr Polym 138:123–133. https://doi.org/10.1016/j.carbpol.2015.11.051

Samsudin NIP, Abdullah N (2019) Edible mushrooms from Malaysia; a literature review on their nutritional and medicinal properties. Int Food Res J 26(1):11–31

Sande D, Oliveira GP, MAFE M, Martins BA, MTNS L, Takahashi JA (2019) Edible mushrooms as a ubiquitous source of essential fatty acids. Food Res Int 125:108524. https://doi.org/10.1016/j.foodres.2019.108524

Sen S, Karati D, Priyadarshini R, Dua TK, Paul P, Sahu R, Nandi G (2023) Cordyceps sinensis (yarsagumba): pharmacological properties of a mushroom. Pharmacol Res 8:100294. https://doi.org/10.1016/j.prmcm.2023.100294

Shbeeb DA, Farahat MF, Ismail HM (2019) Macronutrients analysis of fresh and canned Agaricus bisporus and Pleurotus ostreatus mushroom species sold in Alexandria markets, Egypt. Prog Nutr 21:203–209

Singdevsachan SK, Auroshree P, Mishra J, Baliyarsingh B, Tayung K, Thatoi H (2016) Mushroom polysaccharides as potential prebiotics with their antitumor and immunomodulating properties: a review. Bioact Carbohydr Diet Fibre 7(1):1–14. https://doi.org/10.1016/j.bcdf.2015.11.001

Smiderle FR, Alquini G, Tadra-Sfeir MZ, Iacomini M, Wichers HJ, Griensven V, Leo JLD (2013) Agaricus bisporus and Agaricus brasiliensis (1→6)-β-d-glucans show immunostimulatory activity on human THP-1 derived macrophages. Carbohydr Polym 94(1):91–99. https://doi.org/10.1016/j.carbpol.2012.12.073

Sousa P, Tavares-Valente D, Amorim M, Azevedo-Silva J, Pintado M, Fernandes J (2023) β-Glucan extracts as high-value multifunctional ingredients for skin health: a review. Carbohydr Polym 322:121329. https://doi.org/10.1016/j.carbpol.2023.121329

Stastny J, Marsik P, Tauchen J, Bozik M, Mascellani A, Havlik J et al (2022) Antioxidant and anti-inflammatory activity of five medicinal mushrooms of the genus Pleurotus. Antioxidants 11(8):1569

Stojkovic D, Smiljkovic M, Ciric A, Glamoclija J, Van Griensven L, Ferreira ICFR, Sokovic M (2019) An insight into antidiabetic properties of six medicinal and edible mushrooms: inhibition of α-amylase and α-glucosidase linked to type-2 diabetes. S Afr J Bot 120:100–103. https://doi.org/10.1016/j.sajb.2018.01.007

Taofiq O, González-Paramás AM, Martins A, Barreiro MF, Ferreira ICFR (2016a) Mushrooms extracts and compounds in cosmetics, cosmeceuticals and nutricosmetics—a review. Ind Crop Prod 90:38–48

Taofiq O, Heleno SA, Calhelha RC, Alves MJ, Barros L, Barreiro MF et al (2016b) Development of mushroom-based cosmeceutical formulations with anti-inflammatory, anti-tyrosinase, antioxidant, and antibacterial properties. Molecules 21(10):1372

Taofiq O, Martins A, Barreiro MF, Ferreira ICFR (2016c) Anti-inflammatory potential of mushroom extracts and isolated metabolites. Trends Food Sci Technol 50:193–210. https://doi.org/10.1016/j.tifs.2016.02.005

Thongbai B, Rapior S, Hyde KD, Wittstein K, Stadler M (2015) Hericium erinaceus, an amazing medicinal mushroom. Mycol Prog 14(10):91. https://doi.org/10.1007/s11557-015-1105-4

Usman M, Murtaza G, Ditta A (2021) Nutritional, medicinal, and cosmetic value of bioactive compounds in button mushroom (Agaricus bisporus): a review. Appl Sci 11(13):5943

Üstün NŞ, Bulam S, Pekşen A (2018) The use of mushrooms and their extracts and compounds in functional foods and nutraceuticals. In: Türkmen A (ed) International Technology Sciences and Design Symposium, vol 1, pp 1205–1222

Ványolós A, Muszyńska B, Chuluunbaatar B, Gdula-Argasińska J, Kała K, Hohmann J (2020) Extracts and steroids from the edible mushroom Hypholoma lateritium exhibit anti-inflammatory properties by inhibition of COX-2 and activation of Nrf2. Chem Biodivers 17(9):e2000391. https://doi.org/10.1002/cbdv.202000391

Venturella G, Ferraro V, Cirlincione F, Gargano ML (2021) Medicinal mushrooms: bioactive compounds, use, and clinical trials. Int J Mol Sci 22(2):634

Vetriventhan M, Azevedo VCR, Upadhyaya HD, Nirmalakumari A, Kane-Potaka J, Anitha S et al (2020) Genetic and genomic resources, and breeding for accelerating improvement of small millets: current status and future interventions. Nucleus 63(3):217–239. https://doi.org/10.1007/s13237-020-00322-3

Xia F, Cao SL, Wang MW, Sun YJ (2023) Optimizing extraction, structural characterization, and in vitro hypoglycemic activity of a novel polysaccharide component from Lentinus edodes. Food Biosci 55:103007. https://doi.org/10.1016/j.fbio.2023.103007

Yadav D, Negi PS (2021) Bioactive components of mushrooms: processing effects and health benefits. Food Res Int 148:110599. https://doi.org/10.1016/j.foodres.2021.110599

Yin C, Noratto GD, Fan X, Chen Z, Yao F, Shi D, Gao H (2020) The impact of mushroom polysaccharides on gut microbiota and its beneficial effects to host: a review. Carbohydr Polym 250:116942. https://doi.org/10.1016/j.carbpol.2020.116942

Yuan M, Li C, Xiao X, Wan D, Xi B, Jiang X, Zhang J (2023) Effect of lentinan on proliferation and apoptosis of human astrocytoma U251 cells. Pol J Pathol 74(2):136–140. https://doi.org/10.5114/pjp.2022.119264

Yue GG-L, Lau CB-S, Leung P-C (2021) Medicinal plants and mushrooms with immunomodulatory and anticancer properties—a review on Hong Kong's experience. Molecules 26(8):2173

Zhang Y, Hu T, Zhou H, Zhang Y, Jin G, Yang Y (2016) Antidiabetic effect of polysaccharides from Pleurotus ostreatus in streptozotocin-induced diabetic rats. Int J Biol Macromol 83:126–132. https://doi.org/10.1016/j.ijbiomac.2015.11.045

Zhang Y, Wang Z, Jin G, Yang X, Zhou H (2017) Regulating dyslipidemia effect of polysaccharides from Pleurotus ostreatus on fat-emulsion-induced hyperlipidemia rats. Int J Biol Macromol 101:107–116. https://doi.org/10.1016/j.ijbiomac.2017.03.084

Zhang S, Lei L, Zhou Y, Ye F-y, Zhao G-h (2022) Roles of mushroom polysaccharides in chronic disease management. J Integr Agric 21(7):1839–1866. https://doi.org/10.1016/S2095-3119(21)63871-6

Zhao S, Gao Q, Rong C, Wang S, Zhao Z, Liu Y, Xu J (2020) Immunomodulatory effects of edible and medicinal mushrooms and their bioactive immunoregulatory products. J Fungi 6(4):269

Economic Benefits of Natural Products in Modern Drug Discovery in the Twenty-First Century

Shradha S. Tiwari, Sandip M. Honmane, Rohit R. Sarda, Surendra G. Gattani, Paresh R. Mahaparale, and Md Enamul Hoque

Contents

S. S. Tiwari (✉)
School of Pharmacy, MIT Vishwaprayag University, Solapur, Maharashtra, India
e-mail: shraddha.life10@gmail.com

S. M. Honmane
Annasaheb Dange College of B Pharmacy, Sangli, Maharashtra, India

R. R. Sarda
Dayanand College of Pharmacy, Latur, Maharashtra, India

S. G. Gattani
School of Pharmacy, SRTMU, Nanded, Maharashtra, India

P. R. Mahaparale
Government College of Pharmacy, Aurangabad, Maharashtra, India

M. E. Hoque (✉)
Department of Biomedical Engineering, Military Institute of Science and Technology, Dhaka, Bangladesh
e-mail: enamul1973@gmail.com; enamul1973@bme.mist.ac.bd

© The Author(s), under exclusive license to Springer Nature Singapore Pte Ltd. 2024

A. Das Talukdar et al. (eds.), *Traditional Resources and Tools for Modern Drug Discovery*, Interdisciplinary Biotechnological Advances,
https://doi.org/10.1007/978-981-97-4600-2_25

Abstract Natural products can influence the economic diversity and sustainable livelihood of a country. The natural products obtained from plants, animals, marine sources, and microorganisms possess potential therapeutic properties. Plant- derived medicines and extracts can be used to treat various pathological conditions. Composite plant extracts without isolation of the active ingredient can be used as pharmaceuticals. Natural products obtained from microorganisms are the vital source for antibiotics, immune modulators, anticancer agents, anthelminthics, etc. Herbal extracts can be used for wellness and are very effective as cosmeceuticals. Modern drug discovery involves the use of one or two active compounds to design pharmaceuticals. Plant- and animal- based natural products can be used in next-generation modern drug discovery. Natural products obtained from various plants like *Santalum album* (sandalwood), *Crocus sativus* (saffron), *Juglans regia* (walnut), *Curcuma longa* (turmeric), few species of *Aspergillus* produces products like kojic acid which serve as fungal metabolic shows abundant economic benefits. Currently, there are various challenging diseases like cancer, HIV, diabetes mellitus, etc. affecting significantly the global health. For these diseases, potent natural pharmaceuticals can be designed as novel herbal formulations. Modern nanotechnology can be used to design herbal drug delivery. Paclitaxel and morphine are very- well-known drugs obtained from extracts of plants that exhibit higher economic value. The use of natural products in modern drug discovery can boost up the economy. This chapter discusses various natural products in medicines, their efficacy, advantages, and economic benefits in modern drug discovery. Opportunities, risks, challenges, and future prospective of natural products in modern drug discovery are also highlighted.

Keywords Natural products · Drug discovery · Paclitaxel · Morphine · Modern technology

1 Introduction

Natural products offer a vast array of chemical structures, many of which are not easily synthesized in the laboratory. This diversity provides a prosperous resource of compounds for drug discovery efforts, potentially leading to the development of novel therapeutics with unique mechanisms of action. Natural products, derived from plants, fungi, microorganisms, and animals, represent an incredibly diverse reservoir of chemical compounds. These compounds have evolved over millions of years in response to various ecological pressures, resulting in a vast array of molecular structures with diverse biological activities (Dias et al. 2012).

Approximately 35% of the worldwide medicine market each year consists of products derived from natural sources or their derivatives. This primarily comprises plant-derived medications approximately 25%, with contributions also coming from microorganisms around 13% and animal sources around 3% (Calixto 2019).

The chemical structures of natural products encompass a wide range of scaffolds, including complex polycyclic molecules, macrocycles, alkaloids, terpenoids, polyketides, and peptides, among others. These structures often possess unique stereochemistry, functional groups, and three-dimensional arrangements, which contribute to their biological activity and pharmacological properties. Importantly, many of these chemical features are difficult to mimic through purely synthetic approaches, making natural products invaluable sources of novel chemical entities for drug discovery efforts (Dzobo 2022).

The diversity of natural products provides researchers with a rich source of compounds to explore for potential therapeutic applications. Screening natural product libraries against biological targets can yield hits with a diverse array of pharmacological effects, including antibacterial, antiviral, antifungal, anti-inflammatory, anticancer, and neuroprotective effects, among others. Importantly, the unique chemical structures of natural products often translate into distinct mechanisms of action, enabling the development of therapeutics that target specific disease pathways or molecular targets (Atanasov et al. 2021; Chopra and Dhingra 2021).

In addition to their potential as lead compounds, natural products also serve as inspiration for the design of synthetic analogs and derivatives. Medicinal chemists often draw inspiration from the molecular architectures and biosynthetic pathways of natural products to design novel synthetic compounds with improved drug-like properties and therapeutic potential. This combinatorial approach, combining the structural diversity of natural products with the synthetic versatility of organic chemistry, has led to the development of numerous clinically important drugs, including semisynthetic derivatives of natural products like the antibiotics penicillin and cephalosporin (Lin and Kück 2022).

By harnessing the unique structural features and biological activities of natural products, scientists can uncover new leads with diverse mechanisms of action; this opens avenues for exploring novel medications to address an extensive spectrum of illnesses and medical issues (Silber et al. 2016).

Many of the drugs in clinical use today have their origins in natural products. Penicillin is obtained from *Penicillium* molds, aspirin is obtained from bark of the willow tree, morphine is isolated from opium poppy (*Papaver somniferum*), and paclitaxel is isolated from pacific yew (*Taxus brevifolia*). This track record of success underscores the potential of natural compounds serve as reservoirs for novel pharmaceutical agents and advocate for ongoing investigation in this domain (Stage et al. 2018).

Natural products are often found in abundance in nature, making them relatively inexpensive to acquire compared to synthetic compounds. Additionally, the chemical structures of natural products may require less modification to achieve pharmacological activity, potentially reducing the cost and time required for drug development.

The sustainable harvesting of natural products can provide economic opportunities for local communities while promoting conservation efforts. By supporting environmentally friendly practices, the pharmaceutical industry can contribute to both economic development and biodiversity conservation. Natural products often represent novel chemical structures that can be patented, providing pharmaceutical companies with intellectual property protection and potential commercialization opportunities. This can incentivize investment in natural product research and development, driving innovation in drug discovery.

Natural products have the potential to be repurposed for new therapeutic indications, leveraging their diverse biological activities. This strategy can notably diminish both the time and expenses involved in drug development, given that the safety profiles of numerous natural products are already extensively documented. Natural products can serve as starting points for combinatorial chemistry approaches, allowing researchers to generate libraries of structurally diverse compounds for screening against disease targets. This strategy enhances the efficiency of drug discovery efforts by rapidly exploring a large chemical space (Wainwright et al. 2022).

With growing interest in natural and botanical remedies, products derived from natural sources may enjoy a competitive advantage in the marketplace. This consumer preference can drive demand for natural product-based drugs, creating economic opportunities for pharmaceutical companies. Natural products have the potential to address unmet medical needs, particularly in resource-limited settings where access to healthcare is limited. By developing drugs derived from natural sources, pharmaceutical companies can make significant contributions to global health and economic development. The economic advantages underscore the significance of natural products in contemporary endeavors to discover new drugs and highlights the potential of this approach to deliver innovative therapeutics with significant societal and economic impact (Aware et al. 2022).

Examples of globally top-selling medicines derived from natural products encompass antibiotics and antifungal agents like erythromycin, clarithromycin, amoxicillin, and amphotericin B. Additionally, there are anticancer agents such as paclitaxel, docetaxel, and camptothecin, as well as cholesterol-lowering drugs like atorvastatin, simvastatin, and lovastatin. Furthermore, immunosuppressants such as tacrolimus and cyclosporin A, along with antihypertensive agents like captopril and enalapril, are notable examples (Harvey 2000; Strohl 2000).

2 Natural Products in Medicines

Natural products play an increasingly pivotal role in drug discovery, attributed to their extensive molecular diversity and unique biofunctionality. Naturally obtained products from plants, animals, microorganism, insects, and marine source have been extensively used in pharmaceuticals. Paclitaxel is the chemotherapeutic drug used in cancer commercially known as Taxol®, initially extracted from *Taxus brevifolia* Nutt (Pacific yew), and represents a significant milestone in cancer treatment. Another anticancer medication, Docetaxel (Taxotere®), is derived from *Taxus baccata* (European yew), marking another breakthrough in pharmaceutical innovation. Vincristine and vinblastine vinca alkaloids obtained from the plant *Catharanthus roseus* (Madagascar periwinkle) are antitumor drugs (Naeem et al. 2022).

Quinine (*Cinchona* bark) and artemisinin (*Artemisia annua*) are the antimalarial drugs obtained from plants, and digoxin is obtained from the flower of *Digitalis lanata*. Penicillin is a potent antibiotic obtained from fungus, tetracycline obtained from bacterial species *Streptomyces aureofaciens*, doxorubicin from bacteria *Streptomyces peucetius*, and cyclosporine obtained from aerobic fungi *Tolypocladium inflatum*. Roughly 33% of drugs approved by the FDA in the last two decades have origins in natural substances or their modified forms (Thomford et al. 2018).

Honey, propolis, royal jelly, bee pollen, beeswax, and bee venom obtained from honey bees can be used in medicine and cosmeceuticals. The products obtained from natural source can be used for skin care and treatment (Kurek-Górecka et al. 2020).

Commonly used analgesic drug, aspirin, derived from the bark of the willow tree species *Salix alba*, stands as a natural medicinal compound that has profoundly impacted the pharmaceutical industry. It serves as a primary example of the transition from traditional herbal remedies to synthetically produced medicinal compounds.

Cannabidiol is anti-inflammatory, analgesic drug produced from the cannabis plant is currently undergoing clinical trials that has been studied for its probable use in the treatment of epilepsy, anxiety, chronic pain, etc. (Chaachouay and Zidane 2024).

Resveratrol is a natural polyphenol, anti-inflammatory, antidiabetic drug, used in the treatment of various neurodegenerative disease obtained from grape, apple, blueberries, and plum (Koushki et al. 2018).

The plant *Cannabis sativa* L. contains various metabolites like cannabinoids; this plant is being used in the treatment of various diseases like asthma, epilepsy, fatigue, glaucoma, etc. (Radwan et al. 2021).

Benign prostatic hyperplasia (BPH) is a urinary condition. Various natural remedies, including pumpkin seed, willow herb, tomato, maritime pine bark, *Pygeum africanum* bark, rye pollen, saw palmetto fruit, and nettle root, are utilized in its treatment and management (Csikós et al. 2021).

Mangiferin is a compound obtained from *Mangifera indica* (mango) and honeybush; *Cyclopia* sp. is a strong antioxidant, chemopreventive, antimicrobial, antipyretic, anti-inflammatory, and antiviral agent (Ochocka et al. 2017; Navarro et al. 2019).

Turmeric or *Curcuma longa* contains curcumin which has antimicrobial, antioxidant and preservative effects. It is used in the treatment and management of various diseases like osteoarthritis, dyslipidemia, diabetes, metabolic syndrome, skin diseases, inflammation, endothelial dysfunction, autoimmune disease, respiratory disease, nonalcoholic fatty liver disease, cancer, depression, premenstrual syndrome, and hyperuricemia (Panahi et al. 2019; Patel et al. 2020).

Quercetin is a flavonoid obtained from brassica, berries, spring onions, ginkgo biloba, grapes, apple, onion, tomatoes, and tea. It shows anti obesity, anti tumor, anti-inflammatory, antidiabetes, anti- allergic, anti- aging, and antioxidant effects (Shin et al. 2019).

Lycopene is carotenoid obtained from tomatoes, grapes, cranberry, papaya, etc. It is good antioxidant and rich source of vitamin A. It is employed in the management of diverse cardiovascular conditions, asthma, and cancer (Story et al. 2010; Przybylska 2020).

Gingerol is a polyphenol found in ginger (*Zingiber officinale*) used for the treatment of arthritis, hypertension, fever, and cardiopathy. It exhibits anti cancer and anti- inflammatory activity.

Red ginger exhibits antihypertensive, antimicrobial, immunomodulatory, antihyperlipidemic, antihyperuricemic, and cytotoxic activities (Sugimoto et al. 2018; Zhang et al. 2022a).

Apigenin is flavone glycoside mainly found in *Artemisia*, *Matricaria*, *Achillea*, and *Tanacetum* genera of *Asteraceae* species. Thyme, basil, celery, parsley, chamomile, onion, and tea are the source of apigenin. It possesses various biological activity including antioxidant, cardioprotective, antitumor, neuroprotective, anti-allergic, antimicrobial, anti-inflammatory, and antigenotoxic activity (Ali et al. 2017).

Embelin obtained from *Embelia ribes* Burm. possesses anti-inflammatory, analgesic, antioxidant, antidiabetic, antibacterial, anticonvulsant, hepatoprotective, anxiolytic, and antifertility effects (Li et al. 2019).

Naringenin is a flavonoid obtained from citrus fruits like grapefruits and oranges and in vegetables like tomatoes and figs. Naringenin is pharmacologically anti-inflammatory, hepatoprotective, anticancer, antimicrobial, and anti-atherogenic (Kumar and Bhan 2020; Salehi et al. 2019).

In the recent drug discovery, instead of entire plant extracts, only single compound is used. In traditional medicine, it is considered that plant metabolite induce synergistic therapeutic effect; thus, entire plant extracts were used to get enhanced therapeutic effects. Herbal drug extract of the plants *Ganoderma lucidum*, *Sophora flavescens*, and *Glycyrrhiza uralensis* has been found effective in the treatment of bronchoconstriction and in asthma (Wen et al. 2005).

3 Types of Natural Products for Drug Formulations

Throughout history, natural products have served as crucial reservoirs of therapeutic agents, remaining an invaluable asset in the ongoing process of drug discovery and development.

3.1 Plant-Derived Compounds

3.1.1 Polymers

Naturally derived products can be used as an therapeutic agent, cosmeceuticals, excipients, polymers, solubilizer, emulsifying agent, suspending agent, formulation aid, etc. for the formulation of dosage form. Researchers have extensively explored biodegradable natural polymers as the optimal option for encapsulating and delivering drugs (Idrees et al. 2020; Zhu et al. 2021; Sun 2019).

Alginate is a biopolymer, polysaccharide extracted from brown seaweed algae and bacteria. It is used in wound healing and tissue engineering. It has drug delivery applications like gelling agent and biomedical applications (Adamiak and Sionkowska 2023; Abourehab et al. 2022). Alginates have huge pharmaceutical and biomedical applications; it can be used in the formulation of microparticles, nanoparticles, tablets, wound healing, etc. (Kruk and Winnicka 2022).

Acacia is the dried extrudate obtained from *Acacia senegal* and other related species. Tragacanth is obtained from *Astragalus gummifer*. Acacia and tragacanth are widely used in pharmaceuticals as extrusion spheronization aids in producing pellet. Gum ghatti is swellable polysaccharide obtained from *Anogeissus latifolia*, composed of calcium and magnesium salt of ghattic acid; it is widely used in pharmaceuticals (Singh et al. 2024; Mughal et al. 2011).

Gum tragacanth, acacia, gum ghatti, xanthan gum, and guar gum can be used as emulsifying and suspending agents. Natural gums like guar gum and xanthan gum can also be used to formulate controlled release and sustained release formulations to retard the drug release (Mughal et al. 2011).

Pectin is the natural polysaccharide obtained from citrus fruit peels; it has gelling, thickening, and emulsification properties. Pectin can be used as antibacterial and antimicrobial agent (Li et al. 2021; Ciriminna et al. 2020).

3.2 Animal -Derived Compounds

Animal-derived polymers play crucial roles in pharmaceutical formulations, contributing to drug stability, delivery, and efficacy. However, with the increasing demand for alternative sources and concerns regarding animal welfare and allergenic potential, there is ongoing research into developing synthetic or plant-derived alternatives to these animal-derived polymers.

Chitin and chitosan are naturally occurring polymers derived from marine organisms like lobsters and mollusks, as well as from sources such as insects, fungi, crab shells, and yeast. They consist of lengthy polymeric chains formed by glycosidic linkages and can be applied in innovative drug delivery methods and adapted for novel delivery systems. Chitosan, derived from the deacetylation of chitin present in the exoskeletons of crustaceans, is incorporated into pharmaceuticals due to its mucoadhesive characteristics. It is used in drug delivery systems and wound dressings and as a coating for tablets (Ciriminna et al. 2020).

Collagen and gelatin serve as extensively utilized polymers in pharmaceutical applications. Collagen and its derivatives exhibit considerable potential for various biological activities. As the primary structural protein in both vertebrates and invertebrates, collagen constitutes approximately 30% of the total protein content in animal organisms. Collagen- based biomaterials are used in skin care, wound healing, tissue augmentation, skin repair, etc. Collagen-based scaffolds are used in tissue engineering and regenerative medicine. Gelatin, on the other hand, is a diverse blend of peptides originating from collagen. This derivative serves various purposes, ranging from culinary uses to pharmaceutical applications, owing to its unique properties such as gel formation and film-building capabilities. Additionally, collagen and gelatin play crucial roles in the maintenance of skin elasticity, joint health, and tissue regeneration, making them indispensable components in numerous industries including cosmetics, food, and medicine. It is used in manufacturing of capsule, tablets, ointment, and creams (Younes and Rinaudo 2015; Liu et al. 2015; Ruszczak and Schwartz 1999).

Hyaluronic acid (HA) is a glycosaminoglycan found in human skin extracellular matrix and is a biodegradable polymer used in tissue regeneration, cosmeceuticals having anti-aging and anti-inflammatory agent, intra-articular injections for joint disorders, used in wound healing, etc. It has huge biomedical applications (Graça et al. 2020; Abatangelo et al. 2020).

Albumin is the protein, derived from blood plasma (serum albumin) or egg whites (egg albumin), and is used in pharmaceuticals as a stabilizer for vaccines, as a carrier for drug delivery systems and in cell culture media (Chen et al. 2021).

Glycosaminoglycans (GAGs) are the polysaccharides, such as heparin and chondroitin sulfate; are found in connective tissues; and are used in pharmaceuticals for their anticoagulant properties (heparin) and as supplements for joint health (chondroitin sulfate) (Lima et al. 2017).

4 Efficacy and Advantages of Natural Products in Pharmaceuticals

Natural compounds, obtained from animals, fungi, plants, and microorganisms, have been utilized in pharmaceuticals for centuries due to their efficacy and various advantages. Some of the key benefits and advantages of natural products in pharmaceuticals include the following:

4.1 Therapeutic Efficacy

Many natural products have demonstrated therapeutic efficacy through centuries of traditional use and scientific research. Metabolites derived from plants, such as alkaloids, flavonoids, and terpenoids, often possess potent pharmacological activities, including antimicrobial, anti-inflammatory, anti depressant, anticancer, and analgesic properties (Roy et al. 2022).

4.2 Chemical Diversity

Natural products offer a vast array of chemical structures and biological activities, providing an abundant reservoir of lead compounds for drug discovery and development. The diverse chemical scaffolds found in natural products offer potential for the development of novel pharmaceuticals with unique mechanisms of action and therapeutic targets (Prasathkumar et al. 2021).

4.3 Lower Toxicity and Side Effects

Natural products tend to have lower toxicity and fewer adverse effects compared to many synthetic drugs. This is often attributed to their evolutionary adaptation for interaction with biological systems, resulting in better tolerability and reduced risk of toxicity (Karimi et al. 2015).

4.4 Bioavailability and Pharmacokinetics

Some natural products exhibit favorable pharmacokinetic properties, including enhanced bioavailability and tissue distribution, which can improve their efficacy and therapeutic outcomes. Bioenhancers, which are natural substances that enhance the absorption of drugs, have been discovered. They work by modifying either the permeability of membranes or the metabolism that occurs before the drug enters the systemic circulation, ultimately increasing the amount of unchanged drug in the bloodstream. Aloe vera gel, chitosan, cod-liver oil, menthol, sodium deoxycholate, curcumin (turmeric extract), grapefruit extract, lycopene (carotenoid), lysergol (alkaloids), peppermint oil, piperine (alkaloid), and quercetin (flavonoid) are examples of bioenhancers; these improve solubility and permeability characteristics of drug, leading to improved absorption and systemic availability (Peterson et al. 2019; Kesarwani et al. 2013).

4.5 Combination Therapy and Synergistic Effects

Natural products offer the potential for combination therapy and synergistic effects when used along other natural products or traditional medications. Various natural metabolites and bioactive compounds combined with synthetic anticancer drugs have enhanced anticancer efficacy. This can enhance therapeutic efficacy, overcome drug resistance, and reduce the dosage required, thereby minimizing adverse effects and improving patient compliance. Retigeric acid B is a triterpene acid obtained from the *Lobaria* and is an anticancer drug in combination with cisplatin which is also anticancer which has found enhanced efficacy against prostate cancer (Cheon and Ko 2022; Akter et al. 2021).

4.6 Sustainability and Environmental Impact

With growing concerns about environmental sustainability and ecological impact, the use of natural products in pharmaceuticals offers a more environmentally friendly alternative to some synthetic drugs. The conservation and sustainable utilization of medicinal plant resources serve as a trustworthy guide for preserving and responsibly utilizing medicinal plants. Environmentally friendly and sustainable harvesting practices, cultivation of medicinal plants, and conservation efforts can help preserve biodiversity and support local communities. Medicinal plants offer substantial socioeconomic and health advantages, particularly in developing nations characterized by elevated poverty levels and restricted access to healthcare services (Chen et al. 2016).

4.7 Cultural Acceptance and Traditional Knowledge

Natural products often have cultural significance and are deeply rooted in traditional medicine systems worldwide including Ayurveda, Chinese medicine, Kampo, Unani, traditional Korean medicine, etc. Integrating traditional knowledge with modern scientific research can lead to the discovery of valuable medicinal compounds and therapeutic strategies, enhancing cultural acceptance and community engagement in healthcare.

In general, natural products remain pivotal in pharmaceutical research and development, presenting hopeful pathways for exploration and discovery of novel drugs and the improvement of existing therapeutic options. Their proven efficacy, chemical diversity, and various advantages make them valuable resources in the quest for safer and more effective medicines (Alves and Rosa 2007; Yuan et al. 2016; Gao et al. 2007; Fabricant and Farnsworth 2001).

5 Natural Products in Modern Drug Discovery and Research

Ayurvedic medicines primarily sourced their raw materials from plants, typically in the form of crude drugs like dried herbal powders, extracts, or mixtures. Additionally, various vibrant tribal communities across the country have a longstanding tradition of utilizing herbs in ethnobotanical practices (Dzobo 2022).

Natural products and their metabolites have been used for drug targeting in precision medicine. Precision medicine through targeted therapy aims to address specific disease characteristics for more effective treatment. The Human Genome and The Cancer Genome Atlas (TCGA) strive to elucidate the role of genetics and gene expression in disease development and response to therapy. While there was optimism that genetic insights would revolutionize drug discovery and lead to highly effective treatments, this expectation has not been met. Drug discovery often employs the "gene- to- screen" approach, which hinges on genes as the primary drivers of cellular characteristics, making them viable targets for drug screening (Chaachouay and Zidane 2024). Natural substances and their structurally similar counterparts have made significant contributions to pharmacotherapy, mainly in the treatment of infectious diseases, especially antibiotics (Atanasov et al. 2021).

G- protein-coupled receptors (GPCRs) are significant targets for drug development due to their role in regulating essential physiological functions and their suitability for drug targeting. Natural products have remained invaluable resources in the search for new drugs. Peptides originating from various natural sources such as plants, animals, fungi, and bacteria have an impact on GPCR signaling. Among these naturally occurring compounds, peptides derived from plants, cone snails, snakes, spiders, scorpions, fungi, and bacteria are of significant interest. These peptides hold promise as starting points for developing ligands that modulate GPCR activity. Atropine, caffeine, cannabidiol, and codeine are the natural products acting on GPCR. Peptides sourced from nature, such as those rich in cysteine from plants or derived from venomous creatures like cone snails, snakes, spiders, and scorpions, along with peptides found in marine fungi and bacteria, serve as ligands for G-protein-coupled receptors (GPCRs) (Gruber et al. 2010; Muratspahić et al. 2019).

To address present and emerging global health challenges, it is imperative to employ a blend of inventive drug design and cutting-edge technologies, such as artificial intelligence, in the development of new medications. AI-driven molecular design and lead discovery, inspired by natural products, are revolutionizing drug discovery, offering an innovative approach to finding new medications (Mullowney et al. 2023).

Natural products (NPs) are highly valued for their diverse structures and biological activities, making them important for developing both small molecules and macrocyclic drugs. They have historically been a significant source of novel therapeutics, with a large portion of drugs in the pharmaceutical pipeline originating from NPs or their derivatives. However, despite their proven track record, natural product research has waned in many major pharmaceutical companies (Dzobo 2022; Saldívar-González et al. 2021).

6 Natural Products in Cosmeceuticals

6.1 Reasons for Choosing Natural and Organic Cosmetics (Natrue n.d.)

Natural and organic cosmetics are profoundly impacting the cosmetics industry and redefining the beauty landscape. While "beauty" has traditionally been the prime term associated with cosmetic products, an increasing number of consumers are looking for "sustainable," "environmentally friendly," and "ethical" qualities in the products they purchase. Natural and organic cosmetics genuinely embody these principles, not only prioritizing human well-being but also demonstrating a commitment to environmental and biodiversity conservation. If you contemplate adopting more natural and organic cosmetics for self-care, NATRUE provides three persuasive reasons to switch from conventional to natural options.

6.1.1 Provide Natural and Organic Care for Your Body

Numerous mainstream cosmetic products incorporate significant quantities of petroleum-derived compounds, such as phthalates, oxybenzones, or parabens. While these substances are subject to regulation to ensure their safe application in cosmetics, they do not align well with the principles of genuinely natural and organic cosmetics. Such cosmetics should predominantly feature minerals, organic plant extracts, products from microorganisms, or their derivatives. Embracing sustainable, organic, and biodegradable materials is key in the formulation of natural and organic cosmetics assures customers of the authenticity, origin, and quality of the products. These complex mixes, for instance, ingredients sourced from plant extracts and oils, offer benefits like protecting, nourishing, and hydrating the skin.

Natural and organic cosmetics not only protect and care for our skin, but they also help us maintain our whole health. Their portrayal of nature through natural fragrances and subtle textures draws us closer to it and awakens our senses. Indeed, numerous natural essential oils derived from plants, such as lavender, rose, and mint, are utilized in aromatherapy to energize, calm, and soothe our bodies. Natural or organic cosmetics are more than just beauty items; they are a reproduction of nature in perfect harmony with our bodies and our surroundings.

6.1.2 A More Sustainable and Eco-Friendly Approach

Many conventional cosmetics contain petroleum-based components, such as aluminum and lead, which require extensive mining that poses a threat to wildlife habitats and soil. In contrast, natural and organic cosmetics primarily use plant and flower extracts, which are cultivated and sustainably produced. The use of various natural and organic raw materials requires the preservation of biodiversity and sustainable production practices. Many NATRUE member companies support organic farming and biodynamic agriculture, which provide sustainable sourcing and contribute to biodiversity preservation.

Natural and organic cosmetics prioritize harmony with nature, including the use of eco-friendly packaging and materials. To reduce packaging waste, companies are using recycled materials and creative packaging designs, such as cream deodorants, solid shampoo bars, and toothpaste tablets. Refilling alternatives are becoming more popular in response to consumer demand for sustainable products that adhere to circular economy principles.

6.1.3 The Ethical and Social Dimensions Are Far More Than Merely Cosmetics

The use of local, regional, or national production in natural and organic beauty brands is a common characteristic of small- and medium-sized enterprises (SMEs). This approach has the potential to establishing shorter and more sustainable supply chains which can enhance brand identity by fostering a more personal and authentic connection with consumers. These brands often collaborate with local farming communities in developing countries to source an ingredient, which not only supports the social and economic development of these communities but also guarantees the sustainable utilization of plant-derived natural resources sourced from plants and flowers that are sustainable. Some leading international natural and organic cosmetic brands are Weleda and PRIMA.

Natural cosmeceuticals, derived from sources such as animals, plants, and microorganisms, are gaining popularity, with surveys indicating a significant projected increase in demand for these products (Shah et al. 2024). Throughout history, people across diverse cultures have utilized plant-based substances, or natural products, to enhance their appearance. Dating back to ancient Egypt around 3000 B.C., cosmetics have evolved to become integral to daily beauty regimens worldwide. Initially, natural ingredients were utilized in beauty cosmetics, with paints and dyes added on occasion. Natural compounds have returned with current trends in cosmetic products derived primarily from plant sources. Plant products (natural products) have long been used as a resource of medicine and food. A diverse array of natural products find application in cosmetic formulations, serving purposes ranging from acne treatment and antioxidant properties to anti-inflammatory and anti-aging effects. Additionally, they are employed in hair care, including stimulating hair growth, providing hair color, and addressing scalp issues like dandruff, as well as in skin protection and toiletry products. Essential oils, derived from plants, are particularly noteworthy in this regard, serving as key ingredients in fragrances, hair care items, and skin moisturizers. Furthermore, natural substances like *Emblica officinalis* (amla), *Acacia concinna* (shikakai), and *Callicarpa macrophylla* (priyangu) have found widespread use in skin and hair care products due to their ability to mitigate adverse effects.

Furthermore, Indian women continue to use natural items such as *Pterocarpus santalinus* L. and *Curcuma longa* for skincare, *Lawsonia inermis* L. for hair coloring, and a variety of natural oils such as coconut, olive, shea butter, jojoba, and essential oils in perfumes. The significance of natural products in cosmetics is underscored in the latest book chapter (Desam and Al-Rajab 2021).

7 Economic Benefits and Revenues of Natural Products and Constituents in Modern Drug Discovery

The economic potential of natural products in contemporary drug discovery in the twenty-first century shows promise (Coy-Barrera et al. 2023; Akram et al. 2023; Rosita and Bazán 2023; Singh et al. 2023). Derived from plants, fungi, animals, and microorganisms, these substances have historically been pivotal in drug discovery, providing a rich source of bioactive compounds with therapeutic properties. The utilization of computational and cheminformatics approaches has revolutionized the exploration of natural products, enhancing their potential in drug discovery pipelines. Scientists are increasingly focusing on the therapeutic potentials of natural compounds, mimicking their structures to develop novel drugs for complex disorders like neurodegenerative diseases. Despite ethical concerns, the success of natural products in clinical trials against neurodegenerative disorders underscores their significant economic potential in modern drug development.

Natural products offer opportunity to use locally available components and formulations to achieve significant societal advantages. They frequently provide the sole economical medical treatment in many impoverished places of the world and are excellent local sources of nourishment. They can be employed in products that provide beauty and well-being benefits. The effective use of local natural sources as ingredients has the potential to contribute to solutions to societal challenges such as climate change, environmental costs of packaging and transportation, poor dietary choices, and a general lack of sustainability in many of our current economic models. Natural items may also provide alternate remedies to the COVID-19 pandemic that are less expensive and more accessible than medicines in many parts of the world (OpenTox 2020).

In addition to its immediate benefits to local communities, traditional medicine has become a valuable resource for discovering new drugs through modern scientific methodologies and approaches. This not only holds the potential to benefit people globally but also presents low- and middle-income countries (LMICs) with an opportunity for economic growth and a demonstration of strength and advancement through industrialization.

Traditional medicines inspire drug discoveries in LMICs, whereas a comprehensive and effective R&D structure can speed up the process and save money. The discovery of artemisinin in China provided a solid pattern for promoting the sustainable development of traditional medicines, as well as a good example of how to carry out traditional medicine R&D in low-resource situations (Liu et al. 2022).

The worldwide medication market is worth approximately 1.1 trillion US dollars. Natural products account for around 35% of all medicines. Brazil has the world's biggest biodiversity, with over 50,000 species of higher plants (Calixto 2019).

8 Risks and Challenges of Using Natural Products in Drug Delivery

Primary hurdles in advancing natural products within drug discovery pipelines encompass technical obstacles in screening, isolating, characterizing, and optimizing compounds. These challenges have resulted in dwindling interest from the pharmaceutical sector (Gaudry et al. 2023; Mudunuru et al. 2023). Extraction of plant components can reduce curative capacity and therapeutic efficacy. The effects of these formulations on biological systems can vary depending on the context, underscoring the importance of comprehending their molecular mechanisms. Investigating natural chemicals and plant-based substances poses challenges due to their complex mixtures of metabolites, which can alter their activity outside their native environment. Therefore, adopting a balanced strategy is crucial in pharmaceutical research involving natural products to maintain their efficacy.

Additionally, the need for feature-alignment steps based on retention time in current processing workflows hinders comparison of samples from different batches or instrumental setups (Anwar et al. 2022). Drug resistance remains a significant hurdle in successful cancer treatment, emphasizing the importance of identifying and developing appropriate therapeutics (Tresina et al. 2021). Furthermore, the difficulties in collecting marine organisms for natural product extraction due to their habitats and uncultivable nature pose biological challenges in studying marine natural products (Llorach-Pares et al. 2022). These challenges highlight the need for innovative technologies and computational approaches to streamline natural product-based drug discovery processes.

Natural excipients offer advantages like biodegradability, low toxicity, and compatibility with active pharmaceutical ingredients, enhancing safety and effectiveness (Chintamaneni et al. 2023). However, issues such as poor solubility, bioavailability, stability, and limited research on pharmacokinetics and toxicology hinder the optimal application of natural products like dihydromyricetin (DHM) in pharmaceutical formulations (Zhang et al. 2022b). To overcome these challenges, polymeric nanoparticles have emerged as efficient carriers for natural bioactive agents, enhancing stability, controlled release, and therapeutic efficacy while mitigating toxicity concerns (Elmowafy et al. 2023). Additionally, combining natural compounds with chemotherapeutics in lipid-based nanocarriers shows promise in synergistic anticancer outcomes, addressing limitations of conventional therapies like multidrug resistance and nonspecific distribution (Teixeira et al. 2023). Selecting appropriate drug delivery systems for herbal cosmetics is crucial to ensure efficacy, stability, and safety of the final product (Gadhire et al. 2023).

Extracting plant components may diminish their curative abilities and medicinal efficacy. These preparations exhibit diverse effects on biological systems in various contexts, underscoring the necessity to investigate their molecular mechanisms. Natural compounds and plant-based substances pose challenges in study due to their composition of metabolites and alteration of activity when removed from their natural environment. To effectively harness natural products for pharmaceutical development, a balanced approach is indispensable in preserving their activity.

9 Opportunities for Natural Products in Drug Discovery

Natural products continue to hold significant promise in drug discovery, with recent advancements revitalizing interest in their potential applications (Mudunuru et al. 2023; Kam et al. 2023; Kĕmuling et al. 2023). Despite challenges like technical barriers and declining industry pursuit in the past, new technologies such as genome mining, improved analytical tools, and innovative screening methods are overcoming these obstacles (Philipp 2022). Natural products offer vast opportunities in drug discovery due to their historical significance, aided by modern computational methods, overcoming previous challenges and showing promise for future pharmaceutical development. The discovery of bioactive natural products remains crucial in combating diseases, and novel workflows like deuterium adduct bioactivity screening (DABS) are accelerating this process, leading to the identification of potent compounds like zillamycin (Rosita and Bazán 2023). Recent drug launches based on natural products underscore their impact across various therapeutic areas, highlighting nature as a rich source of pharmacophores. Technological advancements, including computational approaches, are further enhancing the exploration of natural products and constituents for drug discovery, promising a vast future scope in the pipeline.

Throughout history, natural products and their derivatives have been pivotal in pharmacotherapy, particularly in treating cancer and infectious diseases. Recent technological progressions, including enhanced analytical instruments, genome exploration, and microbial cultivation methods, are reigniting enthusiasm in natural products as potential drug candidates. Total synthesis of bioactive natural products using carbohydrates as chiral sources has been achieved, leading to the development of analogs and the establishment of large-scale preparations of medicinally useful compounds. Research collaboration and the development of new strategies for total synthesis are enhancing the synthesis of biologically active natural products. Natural products have historically been crucial in drug discovery, particularly for treating cancer and infectious diseases.

Technological advancements like improved analytical tools and genome mining are revitalizing significance in natural constituents and products as drug leads. The total synthesis of bioactive natural products using carbohydrates as chiral sources has been achieved, leading to the development of analogs and large-scale preparations of medicinally useful compounds.

10 Conclusion and Future Prospects

Natural products have gained significant attention in modern drug discovery due to their economic benefits (Coy-Barrera et al. 2023; Akram et al. 2023; Rosita and Bazán 2023). They offer a wealth of active principles that have historically played crucial roles in drug development. Scientists are increasingly exploring the therapeutic potentials of natural compounds, mimicking their structures to develop novel drugs. These products have shown promising success in addressing neurodegenerative conditions like Alzheimer's and Parkinson's diseases, presenting a cost-effective and potentially efficient alternative to synthetic medications. However, the integration of natural compounds in drug discovery is not without challenges, including ethical considerations that need to be addressed in the future. Overall, the economic values of natural medicinal products are significant, offering potential benefits for both drug development and society at large.

References

Abatangelo G, Vindigni V, Avruscio G, Pandis L, Brun P (2020) Hyaluronic acid: redefining its role. Cells 9(7):1743. PMID: 32708202; PMCID: PMC7409253. https://doi.org/10.3390/cells9071743

Abourehab MAS, Rajendran RR, Singh A, Pramanik S, Shrivastav P, Ansari MJ, Manne R, Amaral LS, Deepak A (2022) Alginate as a promising biopolymer in drug delivery and wound healing: a review of the state-of-the-art. Int J Mol Sci 23(16):9035. PMID: 36012297; PMCID: PMC9409034. https://doi.org/10.3390/ijms23169035

Adamiak K, Sionkowska A (2023) State of innovation in alginate-based materials. Mar Drugs 21(6):353. PMID: 37367678; PMCID: PMC10302983. https://doi.org/10.3390/md21060353

Akram M, Rashid A, Zainab R, Laila U, Khalil MT, Anwar H, Thotakura N, Riaz M (2023) Application and research of natural products in modern medical treatment. J Mod Pharmacol Pathol 1:7. https://doi.org/10.53964/jmpp.2023007

Akter R, Najda A, Rahman MH, Shah M, Wesołowska S, Hassan SSU, Mubin S, Bibi P, Saeeda S (2021) Potential role of natural products to combat radiotherapy and their future perspectives. Molecules 26(19):5997. Retraction in: Molecules 2023 Dec 14;28(24): PMID: 34641542; PMCID: PMC8512367. https://doi.org/10.3390/molecules26195997

Ali F, Rahul F, Naz F, Jyoti S, Siddique YH (2017) Health functionality of apigenin: a review. Int J Food Prop 20:1197–1238. https://doi.org/10.1080/10942912.2016.1207188

Alves RR, Rosa IM (2007) Biodiversity, traditional medicine and public health: where do they meet? J Ethnobiol Ethnomed 3:14. https://doi.org/10.1186/1746-4269-3-14

Anwar S, Malik JA, Ahmed S, Kameshwar VA, Alanazi J, Alamri A, Ahemad N (2022) Can natural products targeting EMT serve as the future anticancer therapeutics? Molecules 27:7668. https://doi.org/10.3390/molecules27227668

Atanasov AG, Zotchev SB, Dirsch VM et al (2021) Natural products in drug discovery: advances and opportunities. Nat Rev Drug Discov 20:200–216. https://doi.org/10.1038/s41573-020-00114-z

Aware CB, Patil DN, Suryawanshi SS, Mali PR, Rane MR, Gurav RG, Jadhav JP (2022) Natural bioactive products as promising therapeutics: a review of natural product-based drug development. S Afr J Bot 151(Part B):512–528.,ISSN 0254-6299. https://doi.org/10.1016/j.sajb.2022.05.028

Calixto JB (2019) The role of natural products in modern drug discovery. An Da Acad Bras De Ciências 91:e20190105. https://doi.org/10.1590/0001-3765201920190105

Chaachouay N, Zidane L (2024) Plant-derived natural products: a source for drug discovery and development. Drugs Drug Candidates 3:184–207. https://doi.org/10.3390/ddc3010011

Chen SL, Yu H, Luo HM, Wu Q, Li CF, Steinmetz A (2016) Conservation and sustainable use of medicinal plants: problems, progress, and prospects. Chinas Med 11:37. PMID: 27478496; PMCID: PMC4967523. https://doi.org/10.1186/s13020-016-0108-7

Chen CB, Hammo B, Barry J, Radhakrishnan K (2021) Overview of albumin physiology and its role in pediatric diseases. Curr Gastroenterol Rep 23(8):11. https://doi.org/10.1007/s11894-021-00813-6

Cheon C, Ko SG (2022) Synergistic effects of natural products in combination with anticancer agents in prostate cancer: a scoping review. Front Pharmacol 13:963317. PMID: 36172195; PMCID: PMC9510769. https://doi.org/10.3389/fphar.2022.963317

Chintamaneni PV, Thangavelu P, Chaitanya MV, Ali HS, Usamo FB (2023) Natural products as pharmaceutical additives in drug delivery systems. In: Plant polysaccharides as pharmaceutical excipients, pp 45–81. https://doi.org/10.1016/b978-0-323-90780-4.00024-3

Chopra B, Dhingra AK (2021) Natural products: a lead for drug discovery and development. Phytother Res 35(9):4660–4702. Epub 2021 Apr 13. https://doi.org/10.1002/ptr.7099

Ciriminna R, Fidalgo A, Meneguzzo F, Presentato A, Scurria A, Nuzzo D, Alduina R, Ilharco LM, Pagliaro M (2020) Pectin: a long-neglected broad-Spectrum antibacterial. ChemMedChem 15(23):2228–2235. Epub 2020 Sep 28. https://doi.org/10.1002/cmdc.202000518

Coy-Barrera E, Ogungbe IV, Schmidt TJ (2023) Natural products for drug discovery in the 21st century: innovations for novel therapeutics. Molecules 28:3690. https://doi.org/10.3390/molecules28093690

Csikós E, Horváth A, Ács K, Papp N, Balázs VL, Dolenc MS, Kenda M, Kočevar Glavač N, Nagy M, Protti M, Mercolini L, Horváth G, Farkas Á, On Behalf Of The Oemonom (2021) Treatment of benign prostatic hyperplasia by natural drugs. Molecules 26(23):7141. PMID: 34885733; PMCID: PMC8659259. https://doi.org/10.3390/molecules26237141

Desam NR, Al-Rajab AJ (2021) The importance of natural products in cosmetics. In: Pal D, Nayak AK (eds) Bioactive natural products for pharmaceutical applications. Advanced structured materials, vol 140. Springer, Cham. https://doi.org/10.1007/978-3-030-54027-2_19

Dias DA, Urban S, Roessner U (2012) A historical overview of natural products in drug discovery. Metabolites 2(2):303–336. PMID: 24957513; PMCID: PMC3901206. https://doi.org/10.3390/metabo2020303

Dzobo K (2022) The role of natural products as sources of therapeutic agents for innovative drug discovery. In: Comprehensive pharmacology, pp 408–422. Epub 2022 Jun 9. PMCID: PMC8016209. https://doi.org/10.1016/B978-0-12-820472-6.00041-4

Elmowafy M, Shalaby K, Elkomy MH, Alsaidan OA, Gomaa HAM, Abdelgawad MA, Mostafa EM (2023) Polymeric nanoparticles for delivery of natural bioactive agents: recent advances and challenges. Polymers 15:1123. https://doi.org/10.3390/polym15051123

Fabricant DS, Farnsworth NR (2001) The value of plants used in traditional medicine for drug discovery. Environ Health Perspect 109:69–75. https://doi.org/10.1289/ehp.01109s169

Gadhire P, Bais SK, Kolekar S (2023) Novel herbal drug delivery in cosmetics. Int J Adv Res Sci Commun Technol 3:307. https://doi.org/10.48175/ijarsct-7982

Gao XM, Zhang TM, Zhang JR, Guo JS, Zhong GS (2007) Chinese Materia Medica. China Press of Traditional Chinese Medicine, Beijing

Gaudry A, Pagni M, Mehl F, Moretti S, Quiros-Guerrero L-M, Rutz A et al (2023) A sample-centric and knowledge-driven computational framework for natural products drug discovery. ChemRxiv. https://doi.org/10.26434/chemrxiv-2023-sljbt

Graça MFP, Miguel SP, Cabral CSD, Correia IJ (2020) Hyaluronic acid-based wound dressings: a review. Carbohydr Polym 241:116364. Epub 2020 Apr 27. https://doi.org/10.1016/j.carbpol.2020.116364

Gruber CW, Muttenthaler M, Freissmuth M (2010) Ligand-based peptide design and combinatorial peptide libraries to target G protein-coupled receptors. Curr Pharm Des 16(28):3071–88. https://doi.org/10.2174/138161210793292474. PMID: 20687879; PMCID: PMC4939874

Harvey A (2000) Strategies for discovering drugs from previously unexplored natural products. Drug Discov Today 5:294–299

Idrees H, Zaidi SZJ, Sabir A, Khan RU, Zhang X, Hassan SU (2020) A review of biodegradable natural polymer-based nanoparticles for drug delivery applications. Nanomaterials (Basel) 10(10):1970. PMID: 33027891; PMCID: PMC7600772. https://doi.org/10.3390/nano10101970

Kam A, Loo S, Lee SMY (2023) Editorial on Special Issue "Natural products for drug discovery and development". Processes 11:1784. https://doi.org/10.3390/pr11061784

Karimi A, Majlesi M, Rafieian-Kopaei M (2015) Herbal versus synthetic drugs; beliefs and facts. J Nephropharmacol 4(1):27–30. PMID: 28197471; PMCID: PMC5297475

Kĕmuling R, Dyah A, Raje. (2023) Bioactive natural product discovery via deuterium adduct bioactivity screening. ACS Chem Biol 18:1192. https://doi.org/10.1021/acschembio.3c00083

Kesarwani K, Gupta R, Mukerjee A (2013) Bioavailability enhancers of herbal origin: an overview. Asian Pac J Trop Biomed 3(4):253–266. PMID: 23620848; PMCID: PMC3634921. https://doi.org/10.1016/S2221-1691(13)60060-X

Koushki M, Amiri-Dashatan N, Ahmadi N, Abbaszadeh HA, Rezaei-Tavirani M (2018) Resveratrol: a miraculous natural compound for diseases treatment. Food Sci Nutr 6(8):2473–2490. PMID: 30510749; PMCID: PMC6261232. https://doi.org/10.1002/fsn3.855

Kruk K, Winnicka K (2022) Alginates combined with natural polymers as valuable drug delivery platforms. Mar Drugs 21(1):11. PMID: 36662184; PMCID: PMC9861938. https://doi.org/10.3390/md21010011

Kumar R, Bhan AT (2020) Naringenin suppresses chemically induced skin cancer in two-stage skin carcinogenesis mouse model. Nutr Cancer 72(6):976–983. https://doi.org/10.1080/01635581.2019.1656756

Kurek-Górecka A, Górecki M, Rzepecka-Stojko A, Balwierz R, Stojko J (2020) Bee products in dermatology and skin care. Molecules 25(3):556. PMID: 32012913; PMCID: PMC7036894. https://doi.org/10.3390/molecules25030556

Li Z, Chen S-J, Yu X-A et al (2019) Pharmacokinetic and bioavailability studies of embelin after intravenous and oral administration to rats. Evid Based Complement Alternat Med 2019:9682495. https://doi.org/10.1155/2019/9682495

Li DQ, Li J, Dong HL, Li X, Zhang JQ, Ramaswamy S, Xu F (2021) Pectin in biomedical and drug delivery applications: a review. Int J Biol Macromol 185:49–65. Epub 2021 Jun 17. https://doi.org/10.1016/j.ijbiomac.2021.06.088

Lima M, Rudd T, Yates E (2017) New applications of heparin and other Glycosaminoglycans. Molecules 22(5):749. PMID: 28481236; PMCID: PMC6154012. https://doi.org/10.3390/molecules22050749

Lin X, Kück U (2022) Cephalosporins as key lead generation beta-lactam antibiotics. Appl Microbiol Biotechnol 106(24):8007–8020. PMID: 36401643; PMCID: PMC9712332. https://doi.org/10.1007/s00253-022-12272-8

Liu D, Nikoo M, Boran G, Zhou P, Regenstein JM (2015) Collagen and gelatin. Annu Rev Food Sci Technol 6:527–557. Epub 2015 Mar 23. https://doi.org/10.1146/annurev-food-031414-111800

Liu G, Xie Y, Sun Y et al (2022) Drug research and development opportunities in low- and middle-income countries: accelerating traditional medicine through systematic utilization and comprehensive synergy. Infect Dis Poverty 11:27. https://doi.org/10.1186/s40249-022-00954-4

Llorach-Pares L, Nonell-Canals A, Avila C, Sanchez-Martinez M (2022) Computer-aided drug design (CADD) to De-Orphanize marine molecules: finding potential therapeutic agents for neurodegenerative and cardiovascular diseases. Mar Drugs 20:53. https://doi.org/10.3390/md20010053

Mudunuru S, Gurubilli CSR, Sharma VR, Srinivasa Rao G (2023) Synthesis of biologically active compounds derived from natural products. Int J Pharmacogn Chem 4(2):1–6. https://doi.org/10.46796/ijpc.v4i2.444

Mughal MA, Iqbal Z, Neau SH (2011) Guar gum, xanthan gum, and HPMC can define release mechanisms and sustain release of propranolol hydrochloride. AAPS PharmSciTech 12(1):77–87. Epub 2010 Dec 21. PMID: 21174179; PMCID: PMC3066369. https://doi.org/10.1208/s12249-010-9570-1

Mullowney MW, Duncan KR, Elsayed SS et al (2023) Artificial intelligence for natural product drug discovery. Nat Rev Drug Discov 22:895–916. https://doi.org/10.1038/s41573-023-00774-7

Muratspahić E, Freissmuth M, Gruber CW (2019) Nature-derived peptides: a growing niche for GPCR ligand discovery. Trends Pharmacol Sci 40(5):309–326. Epub 2019 Apr 5. PMID: 30955896; PMCID: PMC6624132. https://doi.org/10.1016/j.tips.2019.03.004

Naeem A, Hu P, Yang M, Zhang J, Liu Y, Zhu W, Zheng Q (2022) Natural products as anticancer agents: current status and future perspectives. Molecules 27(23):8367. PMID: 36500466; PMCID: PMC9737905. https://doi.org/10.3390/molecules27238367

Natrue (n.d.). https://natrue.org/3-reasons-to-switch-to-natural-and-organic-cosmetics/

Navarro M, Arnaez E, Moreira I et al (2019) Polyphenolic characterization, antioxidant, and cytotoxic activities of Mangifera indica cultivars from Costa Rica. Food Secur 8(9):384. https://doi.org/10.3390/foods8090384

Ochocka R, Hering A, Stefanowicz-Hajduk J, Cal K, Barańska H (2017) The effect of mangiferin on skin: penetration, permeation and inhibition of ECM enzymes. PLoS One 12(7):e0181542. https://doi.org/10.1371/journal.pone.0181542

OpenTox (2020). https://www.opentox.net/events/virtual-conference-2020/program

Panahi Y, Fazlolahzadeh O, Atkin SL et al (2019) Evidence of curcumin and curcumin analogue effects in skin diseases: a narrative review. J Cell Physiol 234(2):1165–1178. https://doi.org/10.1002/jcp.27096

Patel SS, Acharya A, Ray R, Agrawal R, Raghuwanshi R, Jain P (2020) Cellular and molecular mechanisms of curcumin in prevention and treatment of disease. Crit Rev Food Sci Nutr 60(6):887–939. https://doi.org/10.1080/10408398.2018.1552244

Peterson B, Weyers M, Steenekamp JH, Steyn JD, Gouws C, Hamman JH (2019) Drug bioavailability enhancing agents of natural origin (bioenhancers) that modulate drug membrane permeation and pre-systemic metabolism. Pharmaceutics 11(1):33. PMID: 30654429; PMCID: PMC6359194. https://doi.org/10.3390/pharmaceutics11010033

Philipp K (2022) Keynote lecture 2. "Opportunities for natural product sciences in modern drug discovery". Planta Med. https://doi.org/10.1055/s-0042-1758904

Prasathkumar M, Anisha S, Dhrisya C, Becky R, Sadhasivam S (2021) Therapeutic and pharmacological efficacy of selective Indian medicinal plants—a review. Phytomed Plus 1(2):100029.,ISSN 2667-0313,. https://doi.org/10.1016/j.phyplu.2021.100029

Przybylska S (2020) Lycopene–a bioactive carotenoid offering multiple health benefits: a review. Int J Food Sci Technol 55(1):11–32. https://doi.org/10.1111/ijfs.14260

Radwan MM, Chandra S, Gul S, ElSohly MA (2021) Cannabinoids, phenolics, terpenes and alkaloids of *cannabis*. Molecules 26(9):2774. PMID: 34066753; PMCID: PMC8125862. https://doi.org/10.3390/molecules26092774

Rosita B, Bazán A (2023) Computational methods in natural products-based drug discovery. https://doi.org/10.1007/978-981-99-1316-9_5

Roy A, Khan A, Ahmad I, Alghamdi S, Rajab BS, Babalghith AO, Alshahrani MY, Islam S, Islam MR (2022) Flavonoids a bioactive compound from medicinal plants and its therapeutic applications. Biomed Res Int 2022:5445291. PMID: 35707379; PMCID: PMC9192232. https://doi.org/10.1155/2022/5445291

Ruszczak Z, Schwartz RA (1999) Collagen uses in dermatology—an update. Dermatology 199(4):285–289. https://doi.org/10.1159/000018305

Saldívar-González FI, Aldas-Bulos VD, Medina-Franco JL, Plisson F (2021) Natural product drug discovery in the artificial intelligence era. Chem Sci 13(6):1526–1546. PMID: 35282622; PMCID: PMC8827052. https://doi.org/10.1039/d1sc04471k

Salehi B, Fokou PVT, Sharifi-Rad M et al (2019) The therapeutic potential of naringenin: a review of clinical trials. Pharmaceuticals 12(1):11. https://doi.org/10.3390/ph12010011

Shah KC, Verma K, Trivedi JD, Mehta SH, Kumar S, Dhameliya TM, Kathuria D (2024) In: Kathuria D, Sharma A, Verma M, Nayik GA (eds) Bioprospecting of natural sources for cosmeceuticals. Royal Society of Chemistry, London, pp 1–17

Shin EJ, Lee JS, Hong S, Lim T-G, Byun S (2019) Quercetin directly targets JAK2 and PKCδ and prevents UV-induced photoaging in human skin. Int J Mol Sci 20(21):5262. https://doi.org/10.3390/ijms20215262

Silber J, Kramer A, Labes A, Tasdemir D (2016) From discovery to production: biotechnology of marine fungi for the production of new antibiotics. Mar Drugs 14(7):137. PMID: 27455283; PMCID: PMC4962027. https://doi.org/10.3390/md14070137

Singh S, Chib S, Akhtar MJ, Kumar B, Chawla PA, Bhatia R (2023) Paradigms and success stories of natural products in drug discovery against neurodegenerative disorders (NDDs). Curr Neuropharmacol 22:992. https://doi.org/10.2174/1570159X21666230105110834

Singh R, Priya H, Kumar SR, Trivedi D, Prasad N, Ahmad F, Chengaiyan JG, Haque S, Rana SS (2024) Gum Ghatti: a comprehensive review on production, processing, remarkable properties, and diverse applications. ACS Omega 9(9):9974–9990. PMID: 38463282; PMCID: PMC10918680. https://doi.org/10.1021/acsomega.3c08198

Stage TB, Bergmann TK, Kroetz DL (2018) Clinical pharmacokinetics of paclitaxel monotherapy: an updated literature review. Clin Pharmacokinet 57(1):7–19. PMID: 28612269; PMCID: PMC8572663. https://doi.org/10.1007/s40262-017-0563-z

Story EN, Kopec RE, Schwartz SJ, Harris GK (2010) An update on the health effects of tomato lycopene. Annu Rev Food Sci Technol 1:189–210. https://doi.org/10.1146/annurev.food.102308.124120

Strohl WR (2000) The role of natural products in a modern drug discovery program. Drug Discov Today 5:39–41

Sugimoto K, Takeuchi H, Nakagawa K, Matsuoka Y (2018) Hyperthermic effect of ginger (Zingiber officinale) extract-containing beverage on peripheral skin surface temperature in women. Evid Based Complement Alternat Med 2018:1–8. https://doi.org/10.1155/2018/3207623

Sun Z (2019) Hyperbranched polymers in modifying natural plant fibers and their applications in polymer matrix composites—A review. J Agric Food Chem 67(32):8715–8724. Epub 2019 Aug 2. https://doi.org/10.1021/acs.jafc.9b03436

Teixeira PV, Fernandes E, Soares TB, Adega F, Lopes CM, Lúcio M (2023) Natural compounds: co-delivery strategies with chemotherapeutic agents or nucleic acids using lipid-based nanocarriers. Pharmaceutics 15:1317. https://doi.org/10.3390/pharmaceutics15041317

Thomford NE, Senthebane DA, Rowe A, Munro D, Seele P, Maroyi A, Dzobo K (2018) Natural products for drug discovery in the 21st century: innovations for novel drug discovery. Int J Mol Sci 19(6):1578. PMID: 29799486; PMCID: PMC6032166. https://doi.org/10.3390/ijms19061578

Tresina PS, Selvam MS, Rajesh A, Doss A, Mohan VR (2021) Natural products in drug discovery: approaches and development. J Pharm Res 33:93. https://doi.org/10.9734/JPRI/2021/V33I35A31879

Wainwright CL, Teixeira MM, Adelson DL, Braga FC, Buenz EJ, Campana PRV, David B, Glaser KB, Harata-Lee Y, Howes M-JR, Izzo AA, Maffia P, Mayer AMS, Mazars C, Newman DJ, Lughadha EN, Pádua RM, Pimenta AMC, Parra JAA, Qu Z, Shen H, Spedding M, Wolfender J-L (2022) Future directions for the discovery of natural product-derived immunomodulating drugs: an IUPHAR positional review. Pharmacol Res 177:106076. https://doi.org/10.1016/j.phrs.2022.106076

Wen MC, Wei CH, Hu ZQ, Srivastava K, Ko J, Xi ST, Mu DZ, Du JB, Li GH, Wallenstein S et al (2005) Efficacy and tolerability of anti-asthma herbal medicine intervention in adult patients with moderate-severe allergic asthma. J Allergy Clin Immunol 116:517–524

Younes I, Rinaudo M (2015) Chitin and chitosan preparation from marine sources. Structure, properties and applications. Mar Drugs 13(3):1133–1174. PMID: 25738328; PMCID: PMC4377977. https://doi.org/10.3390/md13031133

Yuan H, Ma Q, Ye L, Piao G (2016) The traditional medicine and modern medicine from natural products. Molecules 21(5):559. PMID: 27136524; PMCID: PMC6273146. https://doi.org/10.3390/molecules21050559

Zhang S, Kou X, Zhao H, Mak KK, Balijepalli MK, Pichika MR (2022a) *Zingiber officinale* var. *rubrum*: red Ginger's medicinal uses. Molecules 27(3):775. PMID: 35164040; PMCID: PMC8840670. https://doi.org/10.3390/molecules27030775

Zhang R, Zhang H, Shi H, Zhang D, Zhang Z, Liu H (2022b) Strategic developments in the drug delivery of natural product dihydromyricetin: applications, prospects, and challenges. Drug Deliv 29:3052. https://doi.org/10.1080/10717544.2022.2125601

Zhu S, Tang Y, Lin C, Liu XY, Lin Y (2021) Recent advances in patterning natural polymers: from nanofabrication techniques to applications. Small Methods 5(3):e2001060. Epub 2021 Feb 19. https://doi.org/10.1002/smtd.202001060

GPSR Compliance

The European Union's (EU) General Product Safety Regulation (GPSR) is a set of rules that requires consumer products to be safe and our obligations to ensure this.

If you have any concerns about our products, you can contact us on ProductSafety@springernature.com

In case Publisher is established outside the EU, the EU authorized representative is:

Springer Nature Customer Service Center GmbH
Europaplatz 3
69115 Heidelberg, Germany

Batch number: 10712230

Printed by Printforce, the Netherlands